Grundlehren der mathematischen Wissenschaften 351

A Series of Comprehensive Studies in Mathematics

For further volumes:
www.springer.com/series/138

Anton Bovier • Frank den Hollander

Metastability

A Potential-Theoretic Approach

Anton Bovier
Institut für Angewandte Mathematik
Rheinische Friedrich-Wilhelms-Universität
Bonn, Germany

Frank den Hollander
Mathematisch Instituut
Universiteit Leiden
Leiden, The Netherlands

ISSN 0072-7830 ISSN 2196-9701 (electronic)
Grundlehren der mathematischen Wissenschaften
ISBN 978-3-319-24775-5 ISBN 978-3-319-24777-9 (eBook)
DOI 10.1007/978-3-319-24777-9

Library of Congress Control Number: 2015959720

Mathematics Subject Classification (2010): 60K35, 60J45, 82C26

Springer Cham Heidelberg New York Dordrecht London

Printed on acid-free paper

Springer is part of Springer Science+Business Media (www.springer.com)

Preface

C'est une grande folie de vouloir être sage tout seul.
(François de La Rochefoucauld, Réflexions)

Metastability is a wide-spread phenomenon in the dynamics of non-linear systems—physical, chemical, biological or economic—subject to the action of temporal random forces typically referred to as *noise*. In the narrower perspective of statistical physics, metastable behaviour can be seen as the dynamical manifestation of a first-order *phase transition*, i.e., a crossover that involves a jump in some intrinsic physical parameter such as the energy density or the magnetisation. Attempts to understand and model metastable systems mathematically go back to the early 20th century, notably through the work of H. Eyring and H.A. Kramers, who were concerned with metastable phenomena occurring in chemical reactions.

The modern mathematical approach to metastability was pioneered by M.I. Freidlin and A.D. Wentzell in the late 1960's and early 1970's. They introduced the theory of *large deviations on path-space* in order to analyse the long-term behaviour of dynamical systems under the influence of weak random perturbations. Their realisation that metastable behaviour is controlled by large deviations of the random processes driving the dynamics has permeated most of the mathematical literature on the subject since. A comprehensive account of this development, referred to as the *pathwise approach to metastability*, is given in their 1984 monograph *Random Perturbations of Dynamical Systems* [115]. At around the same time the application of these ideas in a statistical physical context was initiated in a paper by M. Cassandro, A. Galves, E. Olivieri and M.E. Vares [51], which in turn triggered a whole series of papers on metastability of Markovian lattice models. This further development is treated at length in the 2005 monograph *Large Deviations and Metastability* by E. Olivieri and M.E. Vares [198], which provides the key elements of the symbiosis between statistical physics, large deviation theory and metastability.

The present book is concerned with an alternative way to tackle metastability, initiated around 2000 by A. Bovier, M. Eckhoff, V. Gayrard and M. Klein [33], referred to now as the *potential-theoretic approach to metastability*. Here, the pathwise view taken in the Freidlin-Wentzell theory is largely discarded. Instead of aiming at identifying the most likely paths and estimating their probabilities, it interprets the metastability phenomenon as a sequence of visits of the path to different *metastable sets*, and focuses on the precise analysis of the respective *hitting probabilities and hitting times* of these sets with the help of potential theory. The fact that this requires the solution of *Dirichlet problems* in typically high-dimensional spaces has probably acted as a deterrent for a long time, and has prevented an efficient use of the ensuing methods at a much earlier stage. The key point in the potential-theoretic approach is the realisation that, in the specific setting related to metastability, most questions of interest can be reduced to the computation of *capacities*, and that these capacities in turn can be estimated by exploiting powerful *variational principles*. In this way, the metastable *dynamics* of the system can essentially be understood via an analysis of its *statics*. This constitutes a major simplification, and acts as a guiding principle. In addition, potential theory also allows to deduce detailed information on the spectral characteristics of the generator of the dynamics, which are typically assumed in the so-called *spectral approach to metastability* initiated by Davies [73, 74] in the 1980's.

The setting of this book is the theory of Markov processes, for the most part, *reversible* Markov processes. Within this limitation, however, there is a wide range of models that are adequate to describe a variety of different real-world systems. The models we aim at range from finite-state Markov chains, finite-dimensional diffusions and stochastic partial differential equations, via mean-field dynamics with and without disorder, to stochastic spin-flip and particle-hopping dynamics and probabilistic cellular automata. Our main aim is to unveil the *common universal features* of these systems with respect to their metastable behaviour.

The book is divided into nine parts:

- *Part I* presents the metastability phenomenon in its various manifestations, with emphasis on its universal aspects. A brief overview of the history of the subject is given, including a comparison of the pathwise, the spectral, the potential-theoretic and the computational approach. Two paradigmatic models are presented: the Kramers model of Brownian motion in a double-well potential and the two-state Markov chain. These models serve as a *red thread* through the book, in the sense that the much more complex and real-world models treated later still exhibit a metastable behaviour that is in many respects similar. An outline of which models will be treated in the book and which main techniques will be used to analyse them is provided, as well as a brief perspective on metastability in areas other than interacting particle systems.
- *Part II* provides the necessary background on Markov processes (and can be skipped by readers with a basic knowledge of probability theory). Here, the central theme is the relation between Markov processes, martingales, and Dirichlet problems. A brief outline of large deviation theory is provided, as well as a description of three *variational principles* for capacities that play a key role in the study of metastability: the Dirichlet principle, the Thomson principle and the Berman-Konsowa principle.
- *Part III* contains the core of the theory. Here, we give the definition of metastable systems and metastable sets in terms of properties of capacities, and we describe the consequences of these definitions for the distribution of metastable hitting

times and for the spectral properties of the associated Markov generators. We also introduce and discuss the basic techniques that can be used to compute capacities and equilibrium potentials, and to estimate harmonic functions.

Parts IV–VIII highlight the key models that can be treated with the help of these techniques. It is here that the potential-theoretic approach to metastability fully comes to life.

- *Part IV* studies *diffusions with small noise*, both finite-dimensional (random walks and stochastic differential equations) and infinite-dimensional (stochastic partial differential equations).
- *Part V* describes *coarse-graining techniques* applied to the Curie-Weiss model *in large volumes at positive temperatures*, both for a non-random and a random magnetic field.
- *Part VI* focusses on *lattice systems in small volumes at low temperatures*. In this setting, energy dominates entropy. An abstract set-up is put forward, and universal metastability theorems are derived under general hypotheses. These hypotheses are subsequently proved for Ising spins subject to Glauber dynamics and lattice gases subject to Kawasaki dynamics.
- *Part VII* extends the results in Part VI to *lattice systems in large volumes at low temperatures*. In large volumes, spatial entropy comes into play, which complicates the analysis. Both for Glauber dynamics and Kawasaki dynamics the key quantities controlling metastable behaviour can be identified, but at the cost of a severe restriction on the starting measure of the dynamics.
- *Part VIII* looks at metastable behaviour of *lattice systems in small volumes at high densities*, in particular the zero-range process.
- *Part IX* lists a number of challenges for future research, both within metastability and beyond. It describes systems that are presently too hard to deal with in detail, but are expected to come within reach in the next few years. In particular, we look at post-nuclear growth for Ising spins subject to Glauber dynamics (limiting shape of large droplets) and at continuum particle systems with pair interactions (crystallisation), for which a number of results are already available.

Along the way we will encounter a variety of ideas and techniques from probability theory, analysis and combinatorics, including martingale theory, variational calculus and isoperimetric inequalities. It is the combination of physical insight and mathematical tools that allows for making progress, in the best of the tradition of mathematical physics.

Throughout the book we only consider classical stochastic dynamics. It would be interesting to consider quantum stochastic dynamics as well, but this is beyond the scope of the book. We also do not address issues related to numerical simulation, which is rather delicate due to the extremely long time scales involved.

It is a pleasure to thank the colleagues with whom we have worked on metastability over the past 15 years: Florent Barret, Alessandra Bianchi, Michael Eckhoff, Alessandra Faggionato, Alexandre Gaudillière, Véronique Gayrard, Dmitry Ioffe, Sabine Jansen, Oliver Jovanovski, Markus Klein, Roman Kotecký, Francesco

Manzo, Sylvie Méléard, Patrick Müller, Francesca Nardi, Rebecca Neukirch, Enzo Olivieri, Elena Pulvirenti, Elisabetta Scoppola, Martin Slowik, Cristian Spitoni, Siamak Taati and Alessio Troiani. Special thanks are due to Aernout van Enter for reading the entire text and providing a host of valuable comments.

Bonn, Germany
Leiden, The Netherlands
June 4, 2015

Anton Bovier
Frank den Hollander

Logical structure of the monograph

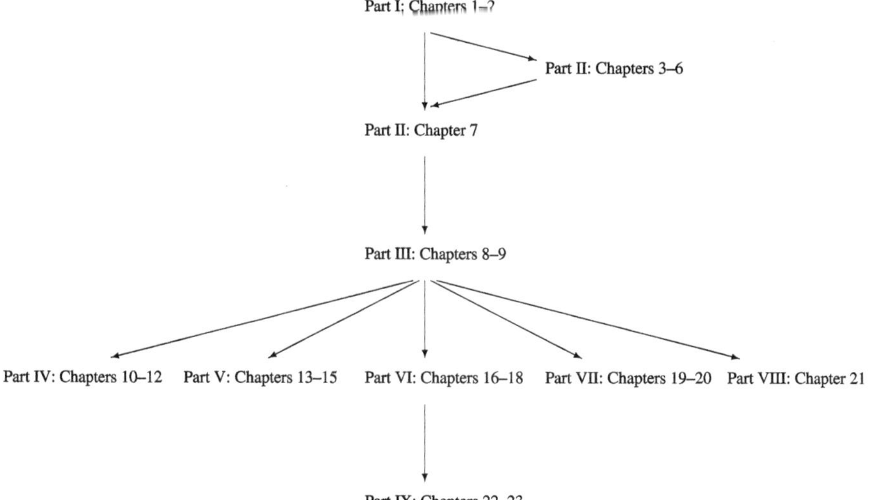

Acknowledgements

Anton Bovier was supported by the German Research Foundation (DFG) through the Collaborative Research Centers 611 *Singular Phenomena and Scaling in Mathematical Models* and 1060 *The Mathematics of Emergent Effects*, by the Hausdorff Center for Mathematics (HCM) in Bonn, by the German-Israeli Foundation (GIF), and by the Lady Davis Fellowship Trust (Haifa and Jerusalem). Frank den Hollander was supported by the Netherlands Organisation for Scientific Research (NWO) through Gravitation Grant 024.002.003-NETWORKS, and by the European Research Council (ERC) through Advanced Grant 267356-VARIS *Variational Approach to Random Interacting Systems*.

The writing of this book started in 2011 while Frank den Hollander held a Bonn Research Chair at the HCM, and continued in 2012 while Anton Bovier held a Kloosterman Chair at the Mathematical Institute of Leiden University. Regular visits back and forth took place in 2013–2015. The authors thank their home institutions for hospitality.

Contents

Part I
Introduction

Part I contains an introduction to the concept of metastability. Chapter 1 provides background and motivation, including a brief historical account and a brief description of the main ideas driving the pathwise, the spectral and the potential-theoretic approach to metastability. Chapter 2 sketches the aims and the scopes of the book, including a description of two paradigmatic models and an outline of the various models to be considered later in the book, organised according to the conceptual and technical challenges they involve.

Chapter 1
Background and Motivation

In Science—in fact, in most things—it is usually best to begin at the beginning. In some things, of course, it's better to begin at the other end. For instance, if you wanted to paint a dog green, it might be best to begin with the tail, as it doesn't bite at that end.
(Lewis Carroll, Sylvie and Bruno Concluded)

We begin with a brief description of the *phenomenon* of metastability (Sects. 1.1–1.2) and a brief historical perspective of the *mathematical theories* that were developed to obtain a quantitative understanding of this phenomenon (Sect. 1.3).

1.1 Phenomenology

Metastability is a widespread phenomenon that arises in a large variety of systems—physical, chemical, biological or economic. A simple experiment anyone can do at home goes as follows. Fill a plastic bottle with distilled water and put it into the freezer. After an hour or so, carefully take it out of the freezer. If you are lucky, then the water still is liquid, but the temperature is down to somewhere between minus 5 and minus 10 degrees centigrade. Now slowly pour the water out of the bottle and into a bowl. When done carefully, you should see a very fast freezing of the water as it hits the bowl. What happens is that the water is *undercooled*, i.e., the *stable state* of the water would have been ice, but you found it in the freezer in a *metastable state*. This state is very sensitive to perturbations, and the shaking you subject it to when pouring it into the bowl triggers an immediate transition to the stable state, which is ice. Should you have left the bottle in the freezer unperturbed, such a transition would eventually have happened *spontaneously*. In fact, if you were to watch the bottle in the freezer over a long time, then you would eventually see this quick freezing happen (you may need a lot of patience). Moreover, if you repeat this experiment many times, then you will observe that the time until freezing is rather variable, and is much longer than the time of the actual freezing itself. It is reported that similar phenomena occur in very still and clean mountain lakes in winter. The water cools well below the freezing point until suddenly the lake freezes over.

A. Bovier, F. den Hollander, *Metastability*,
Grundlehren der mathematischen Wissenschaften 351,
DOI 10.1007/978-3-319-24777-9_1

Similarly, in chemistry the mixing of two reactive compounds (like oxygen and hydrogen) may lead to a metastable state that can persist for a very long time, but when triggered (by a spark) transits very rapidly to the stable state (water). In economics, stock prizes may persist for a long time on high levels, in spite of economists warnings that the market is "overheated", until a "crash" occurs and prices drop within days or hours to much lower levels. In economics jargon, there was a "bubble" that has collapsed. Such phenomena are ubiquitous. The common features are: a large variability in the moment of the onset of some dramatic change in the properties of the system, a much shorter time for the actual transition (i.e., between the onset of a noticeable change and the moment a new state is reached), and *unpredictability* of the time of the onset of the transition.

What is behind all of this? A simple thought experiment reveals a possible mechanism. Suppose that in a mountain range there are two valleys, A and B. Reaching valley B from valley A requires climbing up 1 km on a steep slope. An experienced mountaineer will easily make this journey in a fairly predictable time, say, 4 hours. Now suppose that there is a tourist visiting valley A, a drunken tourist who wanders around in the valley without any particular purpose and occasionally climbs up the slope towards valley B. However, as he does so, he will encounter certain obstacles or will get tired, and just slurps back to the base of the valley (where, for the sake of the argument, he will have some drinks to retain his confused state of mind). As our tourist is not terribly interested in getting to the second valley anyway, we may assume that he does not learn anything from his excursions up the slope and after each visit to the local pub finds himself in just the same condition as before. Let us now assume that, after many days, we find our tourist in valley B. What has happened? Well, after many failed excursions uphill, and equally many returns to the pub, on a lucky day he just happened to climb straight up the slope and then tumbled down into valley B. Should anyone have observed this final successful climb, he might not have been able to distinguish the tourist from the experienced mountaineer, who would have taken the same path on purpose in the first place. A rough estimate reveals how long it took the tourist to get to valley B: the number of attempts (returns to the pub) will be on average $1/p$ when p is the probability to get over the edge before returning to the pub. If the average time Δ of such an unsuccessful excursion is not too tiny (say, 30 minutes), then the average time Δ/p until the final crossing can run into years when p is small (as is to be expected). Moreover, given the fact that our tourist is not learning anything (and given that no other conditions are changing over time, such as the weather), the fact that at each time back to the pub the tourist is back to where he started implies that the number of failed attempts, and hence the total time until the final crossing, are essentially unpredictable. Thus, in this simple example we recognise and understand all the features of the metastability phenomenon mentioned above. As we will see later, this little thought experiment indeed captures all the crucial features behind metastability.

The first main challenge of *mathematics* is of a *qualitative* nature, namely, to explain why in a large variety of systems the same type of metastable behaviour is observed. Many such systems can be described from first principles as many-body systems subject to classical or quantum dynamics. While the corresponding equations of motion are known, they are typically very hard to analyse, in particular, over the extremely long time intervals in which metastable behaviour occurs. Also, metastability manifestly exhibits *randomness* (the unpredictable time of the occurrence of the transition), the source of which may be difficult to extract from the underlying deterministic dynamics. It may be due to quantum effects, or external perturbations of a (non-closed) system, or the effect of unresolved high-frequency degrees of freedom. A first simplification is to pass to a description of the system as a *stochastic dynamics*. The justification of such a description is an interesting topic in itself, which will not be addressed in the present book. Rather, a stochastic model of the dynamics of the systems we are interested in will be the starting point of our analysis of metastability. Even more restrictively, we will limit our analysis to *Markov processes*. Still, even within this restricted setting, there is a wide variety of different models where metastability emerges and where the explanation of the underlying universality is possible.

The second main challenge is of a *quantitative* nature. Given the parameters of some underlying model, we would like to be able to compute *as precisely as possible* the quantities controling the metastable phenomena, in particular, the distribution of the times of the transitions between metastable and stable states. Again, this is hard because most metastable systems of practical relevance are many-body systems whose dynamics is not easy to capture, neither analytically nor numerically, and because extremely long time scales may be involved. (See Newman and Barkema [194] for an overview on Monte Carlo methods in statistical physics.) Understanding metastability on the quantitative level is of considerable practical interest, as it affects the behaviour and functioning of many systems in nature.

1.2 Condensation and magnetisation: from gases to ferromagnets

From the point of view of *statistical mechanics*, metastability is the dynamical signature of a *first-order phase transition*. In *equilibrium* statistical physics, a first-order phase transition is said to occur if a system is sensitive to the change of a parameter (or a boundary condition), in the sense that certain *extensive variables* (such as density or magnetisation) show a discontinuity as functions of certain *intensive variables* (such as pressure or magnetic field). Dynamically, this sensitivity manifests itself in the fact that, as the parameter is varied across the phase transition curve, the system remains for a considerable amount of time (typically random) in the "old phase" before it suddenly changes to the "new phase", the true equilibrium phase. In other words, the extensive variables change their value as a function of time with a random delay. Thus, the study of metastability can be seen as part of *non-equilibrium* statistical physics. Let us discuss this in a bit more detail in an example.

The most commonly observed occurrence of metastability is the phenomenon of condensation of over-saturated water vapour (rainfall). The common explanation of what is going on can be found in elementary physics textbooks. If water vapour is

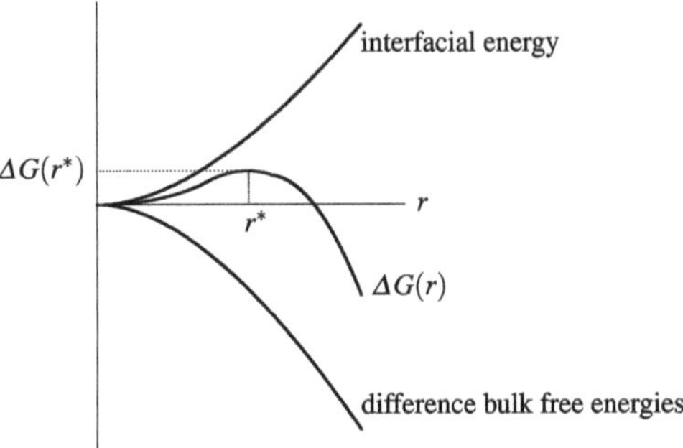

Fig. 1.1 Effective free energy $\Delta G(r)$ of a droplet as a function of its radius r (*middle curve*). The threshold for condensation is the critical radius r^*

cooled below the critical temperature, then the free energy of the gas-phase is larger than that of the liquid-phase. Therefore, thermodynamics predicts a transition from the gas-phase to the liquid-phase. However, this transition can only be achieved by an aggregation of water molecules. This aggregation has to start somewhere in the system with the formation of small droplets of liquid. The key point is that the effective free energy of such a droplet is made up of two terms: (1) the difference between the bulk free energies of the two phases; (2) the interfacial energy between the two phases. This leads to a formula of the type (see Fig. 1.1)

$$\Delta G(r) = \text{difference bulk free energies} + \text{interfacial energy} = -\Delta r^d + \sigma r^{d-1}, \tag{1.2.1}$$

where $\Delta, \sigma > 0$ represent the effect of (1) and (2) per unit volume, respectively, per unit surface, and r is the radius of the droplet.

The function in (1.2.1) is increasing and positive up to a value r^*, the *critical radius*, and is decreasing afterwards. This means that it is unfavourable for the system to have small droplets and favourable to have large droplets: indeed droplets with a radius smaller than r^* tend to evaporate while droplets with a radius larger than r^* tend to grow. But how is it possible to create a large enough droplet of liquid-phase within the gas-phase? The answer is: by *thermal fluctuations*, i.e., the system on some small scale temporarily violates the laws of thermodynamics and evolves in directions that locally increase the free energy. In this way it can form critical droplets of liquid, and once this is done these droplets can continue to grow in full agreement with the laws of thermodynamics. If the parameters of the system are such that r^* is large, then the fluctuations that produce such supercritical droplets are very rare, which leads to a long lifetime of the *metastable* gas-phase. Thus, the crossover is triggered by the appearance of a *critical droplet* of the new phase inside the old phase, which subsequently grows and invades the system. Just as in the example of our drunken tourist, *the transition from the metastable state is characterised by many unsuccessful attempts of the system to create a critical droplet.*

The adhoc notion of “thermal fluctuations” may appear rather mysterious, and does require explanation and theory. We will see in Chaps. 18 and 20, in the context of *stochastic* models for the dynamics of lattice gases, that the above picture arises

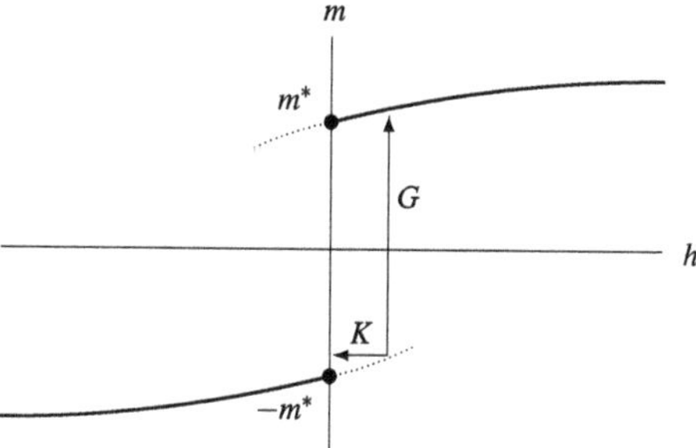

Fig. 1.2 Hysteresis in ferromagnets: plot of the magnetisation m versus the magnetic field h. The *dotted pieces* refer to the magnetisation of the metastable states. The *arrows* refer to the metastable crossovers. The *symbols* G and K stand for Glauber and Kawasaki dynamics (to be treated in Part VI), for which the magnetisation is not preserved, respectively, is preserved

very naturally and can be fully quantified. In this context, the excess free energy of a critical droplet is called the *free energy barrier* for the onset of the phase transition. The presence of such a barrier is the reason for the metastable behaviour, and thermal fluctuations are the driving force for transitions out of the metastable state. The formation of a critical droplet (i.e., a droplet of critical radius) is the minimal effort these fluctuations have to make to initiate the phase transition dynamically. Of course, the same explanation applies when a liquid freezes and—reversed in time—when a liquid evaporates or a solid melts. The fine details are different, but the overall picture is the same.

Another situation where the same principles are at work is magnetic hysteresis, which is treated in Chaps. 13, 17 and 19 (see Fig. 1.2). When a ferromagnetic material is placed in a magnetic field h it *magnetises*, i.e., the atomic magnetic moments ("spins") tend to align with the field. At temperatures below the so-called Curie temperature, this magnetised state persists (forever) even when the field is turned off: the spontaneous magnetisation is $m^\star$. This persistence is the sign of a first-order phase transition. Moreover, even when afterwards the direction of the field is inverted, the magnetisation will remain in the old direction and will only align with the new direction after some time. The reason is the same as for the supersaturation of a gas: the ferromagnetic material has to create local droplets with the opposite magnetisation, and these droplets become energetically favourable and hence start to grow only after they have acquired some minimal size. The creation of such critical droplets is again the work of thermal fluctuations.

Ferromagnets are particularly easy to manipulate and very precise measurements are possible. Figure 1.2 is the paradigmatic figure for metastable behaviour, and is held to be ubiquitous in all situations of metastability.

1.3 Historical perspective

The study of metastability has a long and rich history. In this section we give a brief summary of the most important developments.

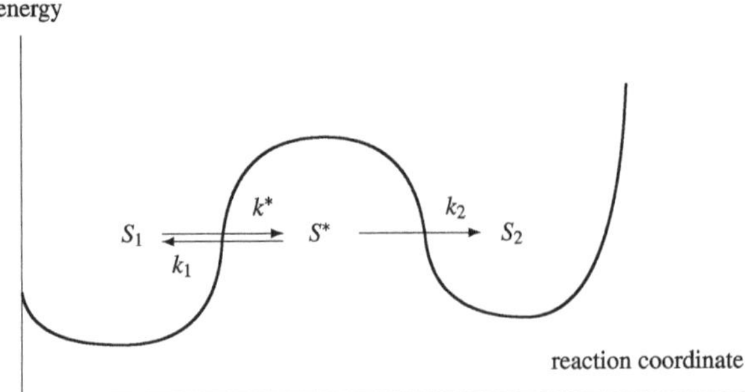

Fig. 1.3 Chemical reaction from state S_1 to state S_2 via transition state S^* with reaction rates k^*, k_1 and k_2

1.3.1 Early achievements

The earliest attempt at a *quantitative* description of metastability dates back to the work of van 't Hoff [229], within the context of chemical reaction-rate theory (see Fig. 1.3). In 1884 he proposed a formula for the temperature dependence of the rate constant R associated with a chemical reaction, of the form

$$R = \exp[-E/kT]. \tag{1.3.1}$$

Here, E is the *activation energy* associated with the reaction (in joules per molecule), T is the absolute temperature (in degrees Kelvin), and k is the Boltzmann constant. (If the activation energy is measured in joules per mole, then k is to be replaced by what is called the gas constant.) In 1889 Arrhenius [8] proposed a refinement of (1.3.1), namely,

$$R = A\,\exp[-E/kT], \tag{1.3.2}$$

where the prefactor A is called the *amplitude*. He also provided the following physical interpretation of (1.3.2). For molecules to react they must first acquire a minimum amount of energy, say E. At absolute temperature T, the fraction of molecules that have a kinetic energy larger than E is proportional to $\exp[-E/kT]$, according to the Maxwell-Boltzmann distribution of statistical mechanics. Hence, $\exp[-E/kT]$ is the probability that a single collision causes a reaction. If A is interpreted as the average number of collisions per time unit, then R is the average number of collisions per time unit that cause a reaction, and the inverse $1/R$ is the *average reaction time*.

Equation (1.3.2) goes under the name of *Arrhenius equation* or *Arrhenius law*. The same equation applies to other situations where an energy barrier is involved, such as the phenomena of condensation and magnetisation mentioned in Sect. 1.2. Still other examples are the motion of dislocations in crystals, the ageing of spin glasses and the folding of proteins, which underlines the *universal character* of the Arrhenius formula. In Part VI we will see that (1.3.2) provides an excellent approximation of the average metastable crossover time for a large class of models with a stochastic dynamics in small volumes at low temperatures.

Several modifications of the Arrhenius equation have been proposed over the years. One modification is a temperature-dependence of the prefactor of the form $A(T/T_0)^\alpha$, with T_0 a reference temperature and $\alpha \in \mathbb{R}$ a dimensionless exponent. In Part VII we will encounter a model with a stochastic dynamics in large volume at low temperature where this form of the prefactor is needed, with A proportional to the volume and $\alpha = 1$. In general, however, this form of the prefactor is neither easy to explain theoretically nor easy to verify experimentally. Another modification is a stretched exponential of the form

$$R = A \exp\bigl[-(E/kT)^{\bar{\alpha}}\bigr], \tag{1.3.3}$$

where $\bar{\alpha} \in (0, 1)$ is a dimensionless exponent. Such an equation appears when the reaction is controlled by a range of activation energies (occurring e.g. in disordered systems) or a range of space-time scales (occurring e.g. in Mott multi-range random hopping). The system is said to be "ageing" as it explores larger and larger activation energies and space-time scales. In this book we will not deal with models that require this modification.

In 1940 Kramers proposed a toy model of a chemical reaction based on Brownian motion in a double-well potential [157]. Using this model, he was able to derive explicit expressions for E and A in (1.3.2) in terms of the shape of the potential (see Sect. 2.1.1 for details). This work was the first to provide a *mathematical* verification of the Arrhenius equation based on a *mesoscopic model* that replaces the microscopic collisions of the molecules involved in the chemical reaction by a Brownian motion, in the spirit of Einstein's explanation of Brownian motion. Various refinements of the Kramers formula, e.g. to higher dimensions and to different choices of the noise, were obtained in the 1960's and 1970's. See the 1981 monograph *Stochastic Processes in Physics and Chemistry* by van Kampen [228] for an overview. These refinements in turn led to the theory of random perturbations of dynamical systems developed by Freidlin and Wentzell (see Chap. 2 for details), in which explicit expressions for E and A were derived in much greater generality. This line of research eventually led to the so-called *pathwise approach* to metastability, which will be discussed in more detail in Sect. 1.3.2 below.

In 1966, Lebowitz and Penrose [162] provided a mathematical explanation of the gas-liquid phase transition within the context of the so-called *van der Waals-Maxwell theory*. They proposed a spin-model with a long-range interaction called "Kac-potential" and showed that the free energy of this model correctly predicts the pressure-versus-volume phase diagram, including the line of coexistence constructed via the "Maxwell's equal area rule". In 1971, Penrose and Lebowitz [199] proposed a framework for a rigorous theory of metastability for particle systems. They characterised metastable states via three conditions: (1) the system has only one stable state (the thermodynamic phase); (2) the lifetime of the metastable state is very long; (3) the crossover from the metastable state to the stable state is an "irreversible" process, in the sense that the return time is much longer than the decay time. The main tool in [199] is the *restricted ensemble*, which is defined to be the Gibbs measure *conditioned* on the particle configuration lying in a suitable subset R of the configuration space, representing the metastable state, e.g. corresponding to a supersaturated vapour whose density is conditioned to lie below the density of the liquid. The rate at which the stochastic dynamics brings the system outside R is maximal at time zero. This incipient rate plays the role of an escape rate λ. The lifetime of the metastable state is identified with $1/\lambda$, and is an inherently *dynamical* quantity. The choice of R must be such that: (1) the Gibbs measure conditioned on R describes a pure phase; (2) λ is very small; (3) R has a very small weight under the unconditional Gibbs measure. For the spin-model with Kac-potential, Penrose and Lebowitz were able to compute λ explicitly (on a rough scale) and show that $1/\lambda$ coincides with the *activation free energy* needed to move out of R. Based on these results, an early attempt to axiomatise metastability was made by Sewell [218, Chap. 6]. For further details we refer the reader to Penrose and Lebowitz [200] and to Olivieri and Vares [198, Sect. 4.1].

1.3.2 The pathwise approach

The pathwise approach to metastability was initiated in the late 1960's and early 1970's by Freidlin and Wentzell. They introduced the theory of *large deviations on path space* in order to analyse the long-term behaviour of *dynamical systems* under the influence of weak random perturbations. Their realisation that metastable behaviour is controlled by large deviations of the random processes driving the dynamics has permeated most of the mathematical literature on the subject since. A comprehensive account of this development is given in their 1984 monograph *Random Perturbations of Dynamical Systems* [115]. The application of these ideas in a statistical physics context was pioneered in 1984 by Cassandro, Galves, Olivieri and Vares [51]. They realised that the theory put forward by Freidlin and Wentzell could be applied to study metastable behaviour of *interacting particle systems*. This paper led to a flurry of results for a variety of Markovian lattice models, which are described at length in the 2005 monograph *Large Deviations and Metastability* [198] by Olivieri and Vares. This work provides the key elements of the symbiosis between statistical physics, large deviation theory and metastability.

The advantage of the pathwise approach is that it gives very detailed information on the metastable behaviour of the system. By identifying the most likely path between metastable states (typically, the global minimiser of some "action integral" that constitutes the large deviation rate function in path space), the time of the crossover can be determined and information can be obtained on what the system does before and after the crossover ("tube of typical trajectories"). The drawback of the pathwise approach is that it is generally hard to identify and control the rate function, especially for systems with a spatial interaction, for which the dynamics is non-local. Consequently, the pathwise approach typically leads to relatively crude results on the crossover time.

1.3.3 The spectral approach

In the 1980's, Davies [70–74] proposed an axiomatic approach to metastability based on *spectral properties* of generators of reversible Markov processes (in some L^2-space). He showed that metastable behaviour arises when the spectrum of the generator consists of a cluster of very small real eigenvalues, separated by a comparatively wide gap from the rest of the spectrum. Under additional assumptions on boundedness of the corresponding eigenfunctions, he showed that the eigenfunctions allow for a decomposition of the state space into "metastable" sets, and that the motion of the Markov process between these sets is slow, with time-scales that are given by the inverses of the corresponding eigenvalues. In the 1990's, these results were developed further by Gaveau and Schulman [124], and Gaveau and Moreau [123]. While the spectral approach to metastability is conceptually nice and natural, it is typically very difficult to verify the assumptions made on the spectrum.

1.3.4 The potential-theoretic approach

The potential-theoretic approach to metastability was initiated in 2001 in a paper by Bovier, Eckhoff, Gayrard and Klein [33]. Here, the pathwise view is largely discarded. Instead of aiming at identifying the most likely paths realising a metastable crossover and estimating their probabilities, it interprets the metastability phenomenon as a sequence of visits of the path to different metastable sets, and focuses on a precise analysis of the respective hitting probabilities and hitting times of these sets with the help of *potential theory*. Phrased differently, it translates the problem of understanding the metastable behaviour of Markov processes to the study of equilibrium potentials and capacities of *electric networks*.

More precisely, the configurations of the system are viewed as the vertices of the network and the transitions between pairs of configurations as the edges of the network. The transition probabilities are represented by the conductances of the associated edges. In this language, the hitting probability of a set of configurations as a function of the starting configuration of the Markov process can be expressed in terms of the equilibrium potential on the network when the potential is set to 1 on the vertices of the target set and to 0 on the starting vertex. The average hitting time of the set can then be expressed in terms of the equilibrium potential and the capacity associated with the target set and the starting vertex. For metastable sets it turns out that *the average hitting time is essentially the inverse of the capacity.*

A key observation in the potential-theoretic approach is the fact that capacities can be estimated by exploiting powerful *variational principles*. In fact, dual variational principles are available that express the capacity both as a supremum (over potentials) and as an infimum (over flows). This opens up the possibility to derive sharp lower bounds and upper bounds on the capacity via a judicious choice of test functions. In fact, with the proper physical insight, test functions can be found for which the lower bounds and the upper bounds are asymptotically equivalent (in an appropriate limit corresponding to a metastable regime). A second key observation is that the relevant equilibrium potentials can, to the extent necessary, be in turn bounded from above and below by capacities with the help of *renewal* equations. This is absolutely crucial, as it avoids the formidable problem of solving the boundary value problems through which the equilibrium potentials are defined. Effectively, it means that estimates of the average crossover time can be derived that are much sharper than those obtained via the pathwise approach.

Capacities are expressed with the help of *Dirichlet forms*, which are functionals of the space of potentials, respectively, flows, and correspond to the energy associated with the network. These Dirichlet forms have the dimension of the configuration space, and thus are typically very high-dimensional. However, it turns out that the ensuing high-dimensional variational principles for the capacity often can be *reduced* to low-dimensional variational principles when the system is metastable. This comes from the fact that metastable crossovers occur near saddles connecting metastable sets of configurations and, consequently, the equilibrium potential is very close to 1 or to 0 away from these saddles. As a result, the full variational principle *reduces* to a simpler variational principle, which only lives on the configurations close to the saddle and captures the fine details of the dynamics when it makes the crossover. In Parts IV–VIII we will see plenty of examples of this reduction. In some cases the simpler variational problem is so low-dimensional that it can be solved explicitly.

The quantitative success of the potential-theoretic approach, relying on tractable variational principles for capacities, also entails its effective limitation to the case of *reversible* Markov processes. While variational characterisations of capacities are known also for *non-reversible* Markov processes (see Sect. 7.3), they are far more complicated and difficult to use than their reversible counterparts. Some attempts in this direction have been made by Eckhoff [101, 102], and more recently by Gaudillière and Landim [121] and Slowik [220]. This area is wide open for future research.

Historically, the potential-theoretic approach has its roots in the early work by Kramers [157], who performed precise computations of metastable crossover times in the context of a Brownian motion in a double-well potential. Such explicit solutions of the Dirichlet problems involved are, however, possible only in the one-dimensional setting. There have been numerous computations in higher-dimensional settings, based on formal perturbation theory, which can be seen as precursors of the potential-theoretic approach (see e.g. Matkowsky and Schuss [181], Matkowsky, Schuss and Tier [182], Knessl, Matkowsky, Schuss and Tier [153], and the discussion in Maier and Stein [169]).

The potential-theoretic approach also connects nicely to the spectral approach. As we will see in Chaps. 8 and 11, in many cases the spectral assumptions of Davies are a consequence of metastability as characterised by capacities.

1.3.5 The computational approach

As mentioned above, there is great interest in quantitative numerical computations for specific systems that exhibit metastable phenomena. Since metastability is driven by rare events and involves excessively long time-scales, doing a simulation is extremely challenging and requires highly sophisticated techniques. Some of the methods developed so far have relations to or are motivated by theoretical work, in particular, the spectral approach (see e.g. Schütte, Huisinga and Meyn [216]). The so-called *transition path theory*, developed in the 2000's, uses ideas similar to those appearing in the potential-theoretic approach, but relies on numerical methods to compute harmonic functions (see e.g. E and Vanden-Eijnden [100], Ren and Vanden-Eijnden [99], Metzner, Schütte and Vanden-Eijnden [184]). Covering this huge field is beyond the scope of the present book.

Chapter 2
Aims and Scopes

What a convenient thing it would be if all thieves had the same shape! It's so confusing to have some of them quadrupeds and others bipeds! (Lewis Carroll, Sylvie and Bruno)

While *classical mechanics* is concerned with deterministic equations of motion, *statistical mechanics* adopts the view that many-particle systems can best be described with the help of probabilistic techniques that do justice to the intrinsic complexity of these systems and to our incomplete knowledge of the precise microscopic state they are in. The aim of *equilibrium statistical mechanics* is to describe many-particle systems through Gibbs distributions, i.e., probability distributions on configuration spaces given by Boltzmann weight factors based on interaction Hamiltonians.

A key target is the computation of the *free energy* of the system as a function of macroscopic or mesoscopic parameters. (Gibbs distributions minimise the free energy according to the Gibbs variational principle.) The idea is that the free energy, which captures the *equilibrium* (= static) properties of the system, implicitly contains information that is pertinent to the *non-equilibrium* (= dynamic) properties of the system as well, since both depend on the energy landscape encoded in the interaction Hamiltonian, with the Gibbs distribution being invariant under the dynamics. A guiding principle of the present book is to make this idea as precise as possible within the context of metastability. On the intuitive level, we deal with Gibbs distributions that put *relatively* large weight on disjoint sets with different macroscopic properties ("metastable states"), separated by regions of small weight through which the system cannot move easily ("saddles"). To make this intuition precise we must make certain assumptions on the dynamics. Typically we must assume that the dynamics is either *local* or *diffusive*, meaning that long-range jumps are either rare or are excluded.

In Sect. 2.1 we describe two paradigmatic models for metastability that serve as a *red thread* through much of the book. In Sect. 2.2 we explain the importance of model reduction, linking complex realistic models to simple toy models that capture the metastable behaviour on an aggregate level. This reduction is necessary to understand the universality behind metastable phenomena. In Sect. 2.3, we give a brief outline of the *variational point of view* that is central to the potential-theoretic approach to metastability—the subject of the present monograph—thereby expanding further on what was already written in Sect. 1.3.4. In Sect. 2.4 we provide a list

A. Bovier, F. den Hollander, *Metastability*,
Grundlehren der mathematischen Wissenschaften 351,
DOI 10.1007/978-3-319-24777-9_2

of the models to be considered in Parts IV–VIII, with a brief indication of what we prove about them and how we organise them. In Sect. 2.5 we mention a number of related topics that are not treated in this book but are slowly coming within reach of mathematical theory.

2.1 Two paradigmatic models

In Sects. 2.1.1–2.1.2 we describe two paradigmatic models for metastability: the Kramers model for Brownian motion in a double-well potential and finite-state Markov chains with exponentially small transition probabilities.

2.1.1 Kramers model: Brownian motion in a double-well

One of the first *mathematical models* for metastability was proposed in 1940 by Kramers [157]. It consists of the one-dimensional *diffusion equation* (or Langevin equation in physics terminology)

$$dX_t = b(X_t)\,dt + \sqrt{2\varepsilon}\,dB_t, \tag{2.1.1}$$

where X_t denotes the position at time t of a "particle" diffusing in a drift field $b = -W'$, with $W\colon \mathbb{R} \to \mathbb{R}$ a *double-well potential*, i.e., a function with two local minima and two steep walls (see Fig. 2.1), and B_t denotes the position at time t of a standard Brownian motion.[1]

Equation (2.1.1) has become the *paradigm of metastability*. Kramers was able to settle essentially all the interesting questions related to this model. In particular, he derived the so-called *Kramers formula* for the average transition time from a local minimum at u to a global minimum at v via a saddle point at z^* (see Fig. 2.1):

$$\mathbb{E}_u[\tau_v] = \big[1 + o(1)\big] \frac{2\pi}{\sqrt{[-W''(z^*)]W''(u)}} \exp\big[\big(W(z^*) - W(u)\big)/\varepsilon\big]. \tag{2.1.2}$$

This formula fits the classical Arrhenius law with *activation energy* $E = W(z^*) - W(u)$, *amplitude* $A = 2\pi/\sqrt{[-W''(z^*)]W''(u)}$ and inverse temperature $\beta = 1/kT = 1/\varepsilon$. Note that the flatter W is near z^* and u, the larger is the amplitude: flatness slows down the crossover at z^* and increases the number of returns to u.

[1]In fact, (2.1.1) emerges as a special case of the more general equation considered by Kramers, namely, the Ornstein-Uhlenbeck equation $dX_t = V_t dt$, $\mu^{-1} dV_t = -dX_t + b(X_t)dt + \sqrt{2\varepsilon} dB_t$, where V_t denotes the velocity at time t and μ is a friction parameter. This equation gives rise to (2.1.1) in the limit as $\mu \to \infty$. Thus, (2.1.1) can be seen as the equation of motion of a particle moving under the influence of a friction force, a gradient force and a random force in the limit where the friction becomes infinitely strong.

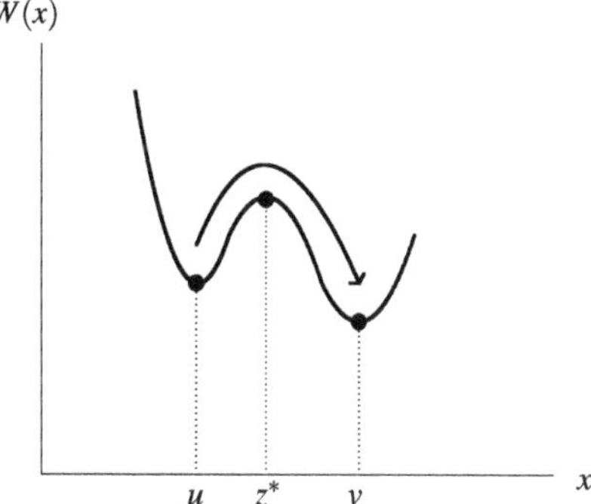

Fig. 2.1 A double-well potential with a local minimum at u, a global minimum at v and a saddle point at z^*

Formula (2.1.2) exhibits a structure that is typical for metastable systems. There is an *exponential term*, here given by $\exp[(W(z^*) - W(u))/\varepsilon]$, which provides the leading asymptotic behaviour. The pathwise approach to metastability, which is based on large deviation theory, typically is capable to identify this term, sometimes referred to as the *exponential asymptotics*, by showing that

$$\varepsilon \ln \mathbb{E}_u[\tau_v] = \big[1 + o(1)\big]\big(W\big(z^*\big) - W(u)\big), \quad \varepsilon \downarrow 0. \tag{2.1.3}$$

However, identifying the *prefactor*, here given by $2\pi/\sqrt{[-W''(z^*)]W''(u)}$, is in general a far more subtle problem. It is the ambition of the potential-theoretic approach exhibited in this book to provide a *unified framework* that allows to obtain rigorous asymptotic formulas as in (2.1.2) with an explicit *prefactor* for a wide class of metastable systems. We will see plenty of examples in Parts VI–VII.

The multi-dimensional generalisation of (2.1.2) is attributed to Eyring and is called the *Eyring-Kramers formula* (see Glasstone, Laidler and Eyring [127], Weidenmüller and Zhang [236], Maier and Stein [170]). Actually, Eyring's so-called transition-state theory [106] is based on quantum-mechanical considerations and is different from the classical theory of Kramers. It interprets the potential as a restricted quantum-mechanical free energy. For a historical discussion, see Pollak and Talkner [201].

2.1.2 Finite-state Markov processes

The model of Kramers describes the evolution of an *effective order parameter* of a metastable system driven by *diffusive noise*. It is clear that in this model the "particle" spends most of its time close to the two local minima of the double-well potential. This suggests a further simplification of the picture, namely, a reduction to a *two-state system*, with the two states u, v representing the two wells of the potential. The time at which this system jumps from state u to state v and backwards

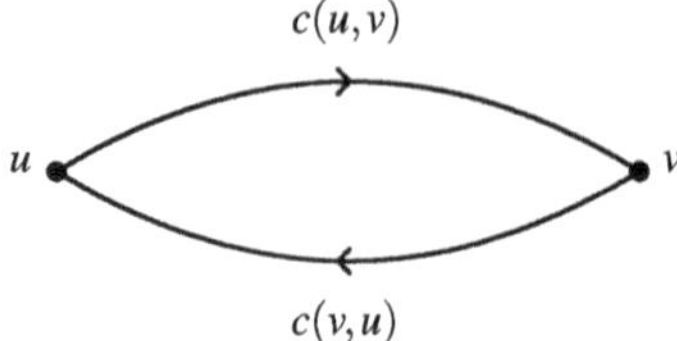

Fig. 2.2 Transition rates for the two-state Markov chain

can be reasonably approximated by the first hitting time τ_v of the local minimum v starting from the local minimum u, and vice versa for τ_u. As we will see in Parts IV–V, in the limit as $\varepsilon \downarrow 0$ the times τ_v and τ_u normalised by their expectations tend to exponentially distributed random variables. This means that a rough approximation of the long-term behaviour of the Kramers model is given by a *continuous-time Markov chain* with state space $\{u, v\}$ and transition rates (see Fig. 2.2)

$$\begin{aligned} c(u, v) &= \mathrm{e}^{-r(u,v)/\varepsilon}, \qquad & r(u, v) &= W\big(z^*\big) - W(u), \\ c(v, u) &= \mathrm{e}^{-r(v,u)/\varepsilon}, \qquad & r(v, u) &= W\big(z^*\big) - W(v). \end{aligned} \tag{2.1.4}$$

The average crossover times $\mathbb{E}_u[\tau_v] = 1/c(u, v)$ and $\mathbb{E}_v[\tau_u] = 1/c(v, u)$ capture the leading order asymptotics of (2.1.1), as expressed in (2.1.3).

The above setting can be easily generalised to systems with multiple metastable states. An effective model for such systems would be a continuous-time Markov chain with a *finite* state space $\mathscr{M} = \{m_1, \dots, m_n\}$ and transition rates $c(m_i, m_j) = \exp[-r(m_i, m_j)/\varepsilon]$, $i, j = 1, \dots, n$. The basic task of a theory of metastability is to determine these transition rates from first principles. This idea was properly formalised by Freidlin and Wentzell [115] in the context of small random perturbations of dynamical systems. In their theory the coefficients $r(m_i, m_j)$ are computed with the help of the theory of *large deviations on path space* (see Chap. 6).

Finite-state Markov chains with exponentially small transition rates have become a subject of interest by themselves. By allowing the transition rates to be either exponentially small *or* equal to one, the above picture is capable of describing models from statistical physics, in particular, *spin-flip systems* and *lattice gases* in finite volumes at low temperatures, with ε playing the role of temperature (see Part VI).

The analysis of the metastability properties of finite-state Markov chains is a non-trivial problem in itself. In the early 1990's an intense activity in this direction started with the work of Catoni and Cerf [53] and Olivieri and Scoppola [196, 197]. The methods used were, once again, large deviations on the path space of these Markov chains. The difficulties that arise in the analysis of specific models are essentially of a combinatorial nature: the optimal paths for transitions between metastable states need to be identified and to be counted. This leads to interesting problems, such as the *discrete isoperimetric inequalities* studied in Alonso and Cerf [4], which we will encounter in Part VI. Only later, in the 2000's, was it noted that potential theory is very well suited to simplify the analysis and to draw sharper results from the same input, as first pointed out by Bovier and Manzo [39] and later amplified in Bovier, den Hollander and Nardi [31].

2.2 Model reduction

Kramers model and finite-state Markov chains can both be seen as simple *toy models* that ought to be *derivable* from more complex *realistic models* of interest. Ideally, we would like to start with many-body systems of interacting quantum particles. This, however, is beyond present-day technology. The most complex models we will consider in this book are classical interacting particle systems, in particular spin-flip systems and lattice gases. These are Markov processes with a high-dimensional (sometimes even infinite-dimensional) state space. Typically, the noise on the level of the microscopic dynamics is not small, and the large-scale dynamics of the system depends on the interplay between energetic and entropic effects.

It is generally accepted in the physics and chemistry literature that reduced models, describing the time evolution of the system on an *intermediate aggregate level* of mesoscopic variables, provides a good description of metastable behaviour. Examples of such models are stochastic differential or partial differential equations with small noise. Ideally, such effective dynamics should be derived with the help of *coarse-graining techniques*, in the spirit of the renormalisation group theory in equilibrium statistical mechanics (see the monograph by Presutti [202]). However, this derivation is quite problematic, partly because renormalisation maps typically do not preserve the Markovian nature of the dynamics. An even more serious issue is that, while at least formally *deterministic* evolution equations (like the Allen-Cahn equation [3] treated in Chap. 12) can be derived as scaling limits (i.e., laws of large numbers in probabilistic language), a proper understanding of metastability requires that we move beyond the deterministic limit and retain at least part of the *random* perturbations of the dynamics. In the literature this goes by the name of *diffusion limits*. However, there are subtle and poorly understood issues regarding the proper choice of the noise term. In this book we will treat diffusion processes with small noise as interesting models in their own right in Part IV. The issue of the derivation of mesoscopic dynamics from microscopic dynamics in the mean-field setting will be touched upon in Part V.

2.3 Variational point of view

The focus of this book is on the potential-theoretic approach to metastability. The basic ideas are classical: many probabilistic quantities can be represented as solutions of Dirichlet problems. The usefulness of this observation may appear to be limited, as it amounts to having to solve partial differential equations or discrete analogues thereof. In general, no explicit analytic solutions of such problems are available. Two notable exceptions are one-dimensional diffusions (which is the reason for the solvability of the Kramers model) and one-dimensional nearest-neighbour random walks.

The power of the potential-theoretic approach arises from the fact that it avoids to solve the Dirichlet problem. Instead, it makes use of a representation formula for the Green function in terms of *capacities*, the invariant measure and harmonic functions. Since renewal arguments can be used to control harmonic functions by capacities, the key objects of the theory are capacities and the invariant measure. The great advantage of this approach materialises in the context of *reversible* Markov processes, i.e., Markov processes whose semi-groups are self-adjoint operators in an L^2-space with respect to an invariant measure. This provides the main weapon of the method: the *Dirichlet principle* expresses capacities as infima of the *Dirichlet form* over classes of functions that are constrained by boundary conditions. The usefulness of this variational principle has long been recognised, e.g. in the analysis of finite-state Markov chains. The book by Doyle and Snell [96] is an excellent source for this material. For a more recent exposition, see Levin, Peres and Wilmer [163]. Part II of the book provides the background on potential theory of reversible Markov processes that is necessary to deal with problems of metastability.

As a variational problem, the Dirichlet principle is a simple instrument to turn physical intuition into upper bounds, and the sharpness of these upper bounds is limited by diligence and imagination only. A particularly nice aspect of the Dirichlet problem is that it satisfies certain monotonicity properties with respect to underlying parameters. In fact, on this basis Berman and Konsowa [23] derived a *dual variational principle* that expresses capacities (in the case of a discrete state space) in terms of suprema over flows (similar to, but different from the better known *Thomson principle*), which we call the *Berman-Konsowa principle*. As an upshot, the latter allows for the derivation of lower bounds that complement the upper bounds obtained via the Dirichlet principle. It is a rather remarkable fact that in many examples upper and lower bounds can be obtained for the metastable crossover time that differ by a multiplicative factor of the form $1+o(1)$ only, where $o(1)$ tends to zero as the time scale of the metastable system tends to infinity. We will see these ideas at work in a variety of examples throughout the book. Part III outlines the basic techniques that are needed to implement these ideas.

A *key observation* is that the analysis of the Dirichlet principle and the Berman-Konsowa principle in essence is part of *equilibrium* statistical physics, since it deals with acquiring the relevant knowledge of the free energy landscape of the system. Potential theory links this knowledge to the metastable dynamics of the system, which is part of *non-equilibrium* statistical physics.

2.4 Specific models

The following models will be considered in Parts IV–VIII.

- In Part IV we study *diffusions with small noise*. Chapter 10 deals with diffusions on lattices with small spacings, the simplest setting in which the potential-theoretic approach to metastability can be applied. Under certain regularity assumptions on the transition probabilities (for discrete time) or transition rates (for continuous time), we carry out a detailed calculation of metastable crossover times. Chapter 11 considers finite-dimensional diffusions on subsets of $\mathbb{R}^d$ and sharpens the classical results of Freidlin-Wentzell theory by using the potential-theoretic approach. The Kramers formula is generalised to a d-dimensional diffusion in a general potential satisfying minimal structural assumptions, a link is made with the principal eigenvalue of the generator of the diffusion, and the exponential distribution of the crossover time is established. Chapter 12 looks at stochastic partial differential equations, which are the infinite-dimensional analogues of the diffusions dealt with in Chap. 11, and shows that similar results apply for a particular example called the Allen-Cahn equation. The theory is complete in one dimension, but suffers from difficulties in higher dimensions, where the noise has to be "truncated properly".

- In Part V we deal with models that allow for *coarse-graining*, i.e., a lumping of states that leads to a simpler Markov process on a reduced state space. Chapter 13 analyses the Curie-Weiss model (the archetype model for ferromagnetism) subject to Glauber spin-flip dynamics. The metastable behaviour of the magnetisation can be fully computed in the limit as the volume tends to infinity, at any subcritical temperature, and turns out to be similar to that of the Kramers model. Chapters 14–15 extend the analysis to the random-field Curie-Weiss model. If the support of the distribution of the magnetic field is finite (Chap. 14), then this model behaves similarly as the Curie-Weiss model, with a large-volume metastable behaviour that is like the Kramers model when the dimension is equal to the size of the support. The computations become much more complicated when the support is infinite (Chap. 15), in which case delicate coupling techniques are required.
- Part VI looks at lattice models subject to a Metropolis dynamics in a *finite volume in the limit as the temperature tends to zero*. Chapter 16 explains how the potential-theoretic approach can be used to prove that these models have the same metastable behaviour as the two-state Markov chain, provided a number of minimal hypotheses are satisfied. Two other dynamics are briefly discussed as well, namely, heat-bath dynamics and probabilistic cellular automata, for which the same universal metastable behaviour can be derived under similar hypotheses. Chapter 17 settles the hypotheses in the case of Glauber dynamics for Ising spins, Chapter 18 in the case of Kawasaki dynamics for lattice gas particles.
- Part VII looks at nucleation in lattice systems that *grow to infinity as the temperature tends to zero*. Spatial entropy comes into play: in large volumes, even at low temperatures, entropy is competing with energy because the metastable state and the states that evolve from it under the dynamics have a non-trivial spatial structure. The main idea is that the system exhibits "homogeneous nucleation", i.e., after the large volume is divided up into smaller (but still large) subvolumes, the system is found to behave more or less independently in different subvolumes. Chapter 19 looks at Glauber dynamics, Chap. 20 at Kawasaki dynamics. For the latter, the computations are delicate and require a proof of "equivalence of ensembles" in a dynamical setting.
- Part VIII describes lattice systems at *high densities*. The focus in Chap. 21 is on the zero-range process, which consists of a collection of particles performing continuous-time simple random walks with on-site attraction and no on-site repulsion. We consider the limit where the particle density is high, show that the

process spends most of its time in a "condensed state", i.e., a configuration where most of the particles pile up on a single site, and prove that the process evolves via a "metastable hopping" of this pile from one site to another. Both the hopping time and the hopping distribution are computed.

The different parts on applications can essentially be read independently and have the following substructure:

- Part IV: (diffusions with small noise)

  (10) Discrete diffusions
  (11) Continuous diffusions $*$
  (12) Stochastic partial differential equations $*$

- Part V: (coarse-graining in large volumes at positive temperatures)

  (13) Curie-Weiss mean-field model
  (14) Curie-Weiss in discrete random magnetic field
  (15) Curie-Weiss in continuous random magnetic field $*$

- Part VI: (lattice systems in small volumes at low temperatures)

  (16) General theory
  (17) Glauber dynamics
  (18) Kawasaki dynamics

- Part VII: (lattice systems in large volumes at low temperatures)

  (19) Glauber dynamics $*$
  (20) Kawasaki dynamics $*$

- Part VIII: (lattice systems in small volumes at high densities)

  (21) Zero-range dynamics $*$

The chapters without $*$ concern models where the state space is *simple* (e.g. discrete and finite) and a complete description of the metastable behaviour is achieved. The chapters with $*$ concern models where the state space is *not simple* (e.g. continuous and infinite) and only partial results are obtained.

2.5 Related topics

Apart from being manifest in interacting particle systems, metastability is an important feature of complex systems in general. Topics within reach that will not be considered in this monograph include:

- *Ageing*: A random dynamics goes through a cascade of metastable states in which it gets trapped on increasingly larger space-time scales. As a consequence, the decay of space-time correlation functions depends on the age of the system. Examples are random walks in random environments, used to describe spin glass dynamics (Bouchaud, Cugliandolo, Kurchan and Mézard [30], Ben Arous, Bovier and Gayrard [18]).

- *Conformational dynamics*: Large (bio)-molecules undergo transitions between metastable states (= conformations) under the influence of thermal noise. There is a strong application-driven interest in the numerical identification of these states and of their lifetimes (Grassberger, Barkema, Nadler [129], E, Ren and Vanden-Eijnden [99], Schütte and Sarich [217]). It is a challenge to develop a rigorous mathematical framework to analyse such systems (Caputo, Lacoin, Martinelli, Simenhaus and Toninelli [50]).
- *Population dynamics*: Selective sweeps in genetic populations, triggered by mutations that drive the population from one dominant trait to another, can be viewed as transitions between metastable states (Dawson and Greven [75]). Viruses moving through a complex network may cause an epidemic. The epidemic may be interpreted as a metastable state of the network, which lasts until the virus disappears. This metastable state depends sensitively on the size and the architecture of the network (Chatterjee and Durrett [56], Mourrat and Valesin [187]).
- *Gene regulatory networks*: The genetic information encoded in DNA fixes the topology of the network. Transitions between the various phenotypic states can be understood as crossovers between metastable states of the corresponding dynamical system subject to noise (Kauffmann [149], Huang [141]).

Part II
Markov Processes

The playground for our expedition into metastability is the theory of Markov processes. Part II presents a summary introduction to this subject, with special emphasis on what will be needed in Part III to describe metastability. The simplest examples are Markov processes in discrete time and discrete space. The general theory is developed from there.

Chapter 3 recalls some basic notions from probability theory. Chapters 4–5 look at Markov processes in discrete, respectively, continuous time, with the focus on generators and semigroups, martingales, and Itō calculus. Chapter 6 gives a brief introduction to large deviations, and looks at path large deviations for finite- and infinite-dimensional diffusion professes via action integrals. Chapter 7 collects the main ingredients from potential theory that are needed in the rest of the book: with capacity playing a central role in the study of metastable transition times, and variational principles for capacities being the main vehicles to estimate capacities.

Readers with a background in probability theory can skip Chaps. 3–6.

Chapter 3
Some Basic Notions from Probability Theory

On peut même dire, à parler en rigueur, que presque toutes nos connaissances ne sont que probables; et dans le petit nombre de choses que nous pouvons savoir avec certitude, dans les sciènces mathématiques elles-mêmes, les principaux moyens de parvenir à la vérité, l'induction et l'analogie, se fondent sur les probabilités, en sorte que le système entier des connaissances humaines se rattache à la théorie exposé dans cet essai.
(Pierre Simon de Laplace, Théorie Analytique des Probabilités)

In this chapter we recall some basic notions from probability theory in order to set notation and to have easy references for later use. Proofs are mostly omitted. Readers who are unfamiliar with the concepts appearing below should consult basic textbooks on probability theory and stochastic processes (see Sect. 3.6 for a list of possible references). Readers who are familiar may skip to Sect. 4.

Section 3.1 defines key ingredients such as probability spaces, random variables, integrals and Radon-Nikodým derivative. Section 3.2 defines stochastic processes and states the Daniell-Kolmogorov extension theorem. Section 3.3 defines conditional expectation, conditional probability and conditional probability measure. Sections 3.4–3.5 list the main properties of martingales in discrete time, respectively continuous time.

3.1 Probability and measures

3.1.1 Probability spaces

A *space* Ω is an arbitrary non-empty set. Elements of Ω are denoted by ω. If $A \subset \Omega$ is a subset of Ω, then we denote by $\mathbb{1}_A$ the indicator function of the set A, i.e.,

$$\mathbb{1}_A(\omega) = \begin{cases} 1, & \text{if } \omega \in A, \\ 0, & \text{if } \omega \in A^c = \Omega \backslash A. \end{cases} \tag{3.1.1}$$

A. Bovier, F. den Hollander, *Metastability*,
Grundlehren der mathematischen Wissenschaften 351,
DOI 10.1007/978-3-319-24777-9_3

Definition 3.1 Let Ω be a space. A family $\mathscr{A} = \{A_\lambda\}_{\lambda \in I}$, with $A_\lambda \subset \Omega$ for all $\lambda \in I$ with I an arbitrary set, is called a *class* of Ω. A non-empty class of Ω is called an *algebra* if:

(i) $\Omega \in \mathscr{A}$.
(ii) For all $A \in \mathscr{A}$, $A^c \in \mathscr{A}$.
(iii) For all $A, B \in \mathscr{A}$, $A \cup B \in \mathscr{A}$.

If $\mathscr{A}$ is an algebra and

(iv) $\bigcup_{n\in\mathbb{N}} A_n \in \mathscr{A}$ whenever $A_n \in \mathscr{A}$ for all $n \in \mathbb{N}$,

then $\mathscr{A}$ is called a σ*-algebra*.

Definition 3.2 A space Ω, together with a σ-algebra $\mathscr{F}$ of subsets of Ω, is called a *measurable space* $(\Omega, \mathscr{F})$.

Definition 3.3 Let $(\Omega, \mathscr{F})$ be a measurable space. A map $\mu\colon \mathscr{F} \to [0, \infty]$ is called a (positive) *measure* if

(i) $\mu(\emptyset) = 0$.
(ii) For any countable family $\{A_n\}_{n\in\mathbb{N}}$ of mutually disjoint elements of $\mathscr{F}$,

$$\mu\left(\bigcup_{n\in\mathbb{N}} A_n\right) = \sum_{n\in\mathbb{N}} \mu(A_n). \tag{3.1.2}$$

A measure μ is called finite if $\mu(\Omega) < \infty$. A measure is called σ-finite if there exists a sequence $(\Omega_n)_{n\in\mathbb{N}}$ of subsets of Ω such that $\Omega = \bigcup_{n\in\mathbb{N}} \Omega_n$ and $\mu(\Omega_n) < \infty$ for all $n \in \mathbb{N}$.

A triple $(\Omega, \mathscr{F}, \mu)$ is called a *measure space*.

Definition 3.4 Let $(\Omega, \mathscr{F})$ be a measurable space. A positive measure $\mathbb{P}$ on $(\Omega, \mathscr{F})$ that satisfies $\mathbb{P}(\Omega) = 1$ is called a *probability measure*. A triple $(\Omega, \mathscr{F}, \mathbb{P})$, with Ω a set, $\mathscr{F}$ a σ-algebra of subsets of Ω and $\mathbb{P}$ a probability measure on $(\Omega, \mathscr{F})$, is called a *probability space*.

In most instances one is concerned with the canonical setting where Ω is a topological space and $\mathscr{F} = \mathscr{B}(\Omega)$ is the Borel-σ-algebra of Ω.

Definition 3.5 Let E be a topological space. The *Borel-σ-algebra* $\mathscr{B}(E)$ of E is the smallest σ-algebra that contains all the open sets of E.

One says that the Borel-σ-algebra is *generated* by the open sets of E.

A topological space endowed with a metric and its metric topology is called a *metric space*. A metric space E is called complete if any Cauchy sequence in E converges in E. E is called *separable* if it contains a countable subset that is dense in E. The standard setting of probability theory is a complete, separable and topological space whose topology is equivalent to some metric topology. Such a space is called a *Polish space*.

The crucial theorem permitting the construction of measures is *Carathéodory's theorem*:

Theorem 3.6 (Carathéodory's theorem) *Let Ω be a set and let $\mathscr{A}$ be an algebra on Ω. Let $\mu_0 : \mathscr{A} \to [0,\infty]$ be a countably additive map. Then there exists a measure μ on $(\Omega, \sigma(\mathscr{A}))$ such that $\mu = \mu_0$ on $\mathscr{A}$. If μ_0 is σ-finite, then μ is unique.*

A measure defined on a Borel-σ-algebra is sometimes called a *Borel measure.*

The most important issue that arises in applications is to characterise a measure with the minimal amount of information possible. The basic tool here is *Dynkin's theorem.*

Definition 3.7 Let Ω be a space. A class $\mathscr{T}$ of Ω is called a Π-system if it is closed under finite intersections. A class $\mathscr{G}$ of Ω is called a λ-system if

(i) $\Omega \in \mathscr{G}$.
(ii) If $A, B \in \mathscr{G}$, and $A \supset B$, then $A \setminus B \in \mathscr{G}$.
(iii) If $A_n \in \mathscr{G}$ and $A_n \subset A_{n+1}$, then $\lim_{n\to\infty} A_n \in \mathscr{G}$.

Theorem 3.8 (Dynkin's theorem) *If $\mathscr{T}$ is a Π-system and $\mathscr{G}$ is a λ-system, then $\mathscr{G} \supset \mathscr{T}$ implies that $\mathscr{G}$ contains the smallest σ-algebra containing $\mathscr{T}$.*

The most useful application of Dynkin's theorem is the observation that if two probability measures are equal on a Π-system that generates the σ-algebra, then they are equal on the σ-algebra (since the set on which the two measures coincide forms a λ-system containing $\mathscr{T}$).

Dynkin's lemma has a sometimes useful analogue for so-called monotone classes of functions.

Theorem 3.9 (Monotone class theorem) *Let $\mathscr{H}$ be a class of bounded, measurable functions from Ω to $\mathbb{R}$. Assume that*

(i) *$\mathscr{H}$ is a vector space over $\mathbb{R}$.*
(ii) *$1 \in \mathscr{H}$.*
(iii) *If $f_n \geq 0$ are in $\mathscr{H}$ and $f_n \uparrow f$, where f is bounded, then $f \in \mathscr{H}$.*

If $\mathscr{H}$ contains the indicator functions of every element of a Π-system $\mathscr{S}$, then $\mathscr{H}$ contains any bounded $\sigma(\mathscr{S})$-measurable function.

3.1.2 Random variables

Definition 3.10 Let $(\Omega, \mathscr{F})$ and $(E, \mathscr{G})$ be two measurable spaces. A map $X\colon \Omega \to E$ is called measurable from $(\Omega, \mathscr{F})$ to $(E, \mathscr{G})$ if $X^{-1}(A) = \{\omega \in \Omega\colon X(\omega) \in A\} \in \mathscr{F}$ for all $A \in \mathscr{G}$.

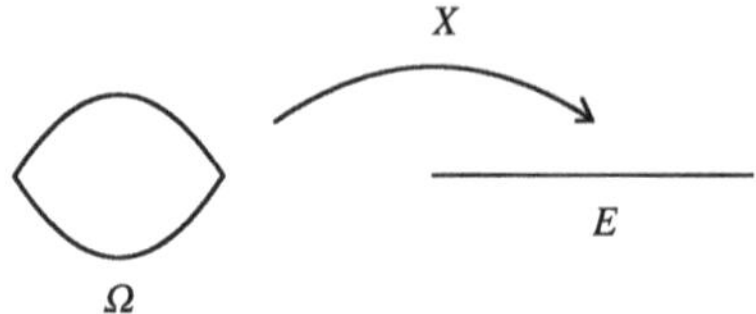

Fig. 3.1 A random variable X is a measurable map from Ω to E

The notion of measurability implies that a measurable map is capable of transporting a measure from one space to another. Namely, if $(\Omega, \mathscr{F}, \mathbb{P})$ is a probability space and f is a measurable map from $(\Omega, \mathscr{F})$ to $(E, \mathscr{G})$, then

$$\mathbb{P}_X = \mathbb{P} \circ X^{-1} \tag{3.1.3}$$

defines a probability measure on $(E, \mathscr{G})$, called the *induced measure*. Indeed, for any $B \in \mathscr{G}$, by definition

$$\mathbb{P}_X(B) = \mathbb{P}\big(X^{-1}(B)\big) \tag{3.1.4}$$

is well defined because $X^{-1}(B) \in \mathscr{F}$.

The standard notion of a *random variable* refers to a measurable function from some measurable space to the Borel space $(\mathbb{R}, \mathscr{B}(\mathbb{R}))$. One generally extends this notion by calling any measurable map from a measurable space $(\Omega, \mathscr{F})$ to a measurable space $(E, \mathscr{B}(E))$, with E a topological space or a metric space, an *E-valued random variable* or an E-valued *Borel function* (see Fig. 3.1). Thus, one has an *abstract probability space* $(\Omega, \mathscr{F}, \mathbb{P})$ on which all kinds of random variables—be it real numbers, infinite sequences, functions or measures—are defined simultaneously.

An important notion is that of the σ-algebra generated by random variables.

Definition 3.11 Let $(\Omega, \mathscr{F})$ be a measurable space, and let $(E, \mathscr{B}(E))$ be a topological space equipped with its Borel-σ-algebra. Let X be an E-valued random variable. Then $\sigma(X)$ is the smallest σ-algebra such that X is measurable from $(\Omega, \sigma(X))$ to $(E, \mathscr{B}(E))$.

3.1.3 Integrals

We next recall the notion of the integral of a measurable function (respectively, the expectation value of a random variable). To do so we first introduce the notion of *simple functions*:

Definition 3.12 A measurable function $g\colon \Omega \to \mathbb{R}$ is called simple if it takes only finitely many values, i.e., if there are numbers $k \in \mathbb{N}$ and $w_1, \ldots, w_k \in \mathbb{R}$, and a partition $A_1, \ldots, A_k \in \mathscr{F}$ of Ω (i.e., $\bigcup_{i=1}^k A_i = \Omega$ and $A_i \cap A_j = \emptyset$ for all $1 \le i < j \le k$) such that $A_i = \{\omega \in \Omega : g(\omega) = w_i\}$ for $1 \le i \le k$. In that case we can write

$$g(\omega) = \sum_{i=1}^{k} w_i \, \mathbb{1}_{A_i}(\omega). \tag{3.1.5}$$

The space of simple measurable functions is denoted by $\mathscr{E}_+$.

It is obvious what the integral of a simple function should be.

Definition 3.13 Let $(\Omega, \mathscr{F}, \mu)$ be a measure space and $g = \sum_{i=1}^k w_i \mathbb{1}_{A_i}$ a simple function. Then

$$\int_\Omega g \, d\mu = \sum_{i=1}^{k} w_i \, \mu(A_i). \tag{3.1.6}$$

The integral of a general measurable function is defined via approximation with simple functions.

Definition 3.14

(i) Let f be non-negative and measurable. Then

$$\int_\Omega f \, d\mu = \sup_{\substack{g \in \mathscr{E}_+ \\ g \le f}} \int_\Omega g \, d\mu \in \mathbb{R} \cup \{\infty\}. \tag{3.1.7}$$

(ii) For f measurable, put

$$f(\omega) = f(\omega) \, \mathbb{1}_{f(\omega) \ge 0} + f(\omega) \, \mathbb{1}_{f(\omega) < 0} = f_+(\omega) - f_-(\omega). \tag{3.1.8}$$

If either $\int_\Omega f_+ \delta\mu < \infty$ or $\int_\Omega f_- d\mu < \infty$, then define

$$\int_\Omega f \, d\mu = \int_\Omega f_+ \, d\mu - \int_\Omega f_- \, d\mu. \tag{3.1.9}$$

(iii) A function f is called integrable, or *absolutely integrable*, if

$$\int_\Omega |f| \, d\mu < \infty. \tag{3.1.10}$$

We next state the key properties of the integral. The most fundamental property is the monotone convergence theorem, which justifies the definition above.

Theorem 3.15 (Monotone convergence) *Let $(\Omega, \mathscr{F}, \mu)$ be a measure space and f a real-valued non-negative measurable function. Let $(f_n)_{n \in \mathbb{N}}$ be a non-decreasing* *sequence of non-negative measurable functions that converge pointwise to f. Then*

$$\int_\Omega f\,d\mu = \lim_{n\to\infty} \int_\Omega f_n\,d\mu. \tag{3.1.11}$$

The monotone convergence theorem allows us to provide an "explicit construction" of the integral originally used by Lebesgue as a definition.

Lemma 3.16 *Let f be a non-negative measurable function. Then*

$$\int_\Omega f\,d\mu = \lim_{n\to\infty} \left[\sum_{k=0}^{n2^n-1} 2^{-n}k\,\mu\big(\omega \in \Omega:\ 2^{-n}k \le f(\omega) < 2^{-n}(k+1)\big) + n\,\mu\big(\omega \in \Omega:\ f(\omega) \ge n\big) \right]. \tag{3.1.12}$$

The following lemma is known as *Fatou's lemma*:

Lemma 3.17 (Fatou's lemma) *Let $(f_n)_{n\in\mathbb{N}}$ be a sequence of measurable non-negative functions. Then*

$$\int_\Omega \liminf_{n\to\infty} f_n\,d\mu \le \liminf_{n\to\infty} \int_\Omega f_n\,d\mu. \tag{3.1.13}$$

Equally important is *Lebesgue's dominated convergence theorem*:

Theorem 3.18 (Lebesgue's dominated convergence) *Let $(f_n)_{n\in\mathbb{N}}$ be a sequence of absolutely integrable functions, and let f be a measurable function such that*

$$\lim_{n\to\infty} f_n(\omega) = f(\omega) \quad \text{for } \mu\text{-almost all } \omega. \tag{3.1.14}$$

Let $g \ge 0$ be a measurable function such that $\int_\Omega g\,d\mu < \infty$ and

$$\big|f_n(\omega)\big| \le g(\omega) \quad \text{for } \mu\text{-almost all } \omega. \tag{3.1.15}$$

Then f is absolutely integrable with respect to μ and

$$\lim_{n\to\infty} \int_\Omega f_n\,d\mu = \int_\Omega f\,d\mu. \tag{3.1.16}$$

3.1.4 Spaces of integrable functions

We briefly summarise some frequently used notions concerning spaces of integrable functions. Given a measure space $(\Omega, \mathscr{F}, \mu)$ and a $p \in [1, \infty]$, we define for a

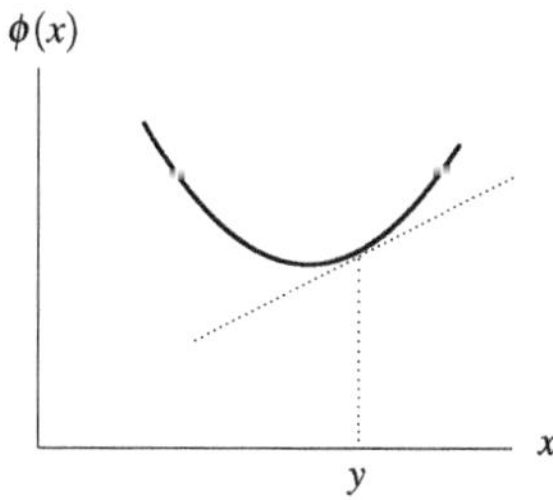

Fig. 3.2 A convex function

measurable function $X\colon \Omega \to \mathbb{R}$,

$$\|X\|_{p,\mu} = \|X\|_p = \big(\mathbb{E}\big[|X|^p\big]\big)^{1/p} = \left(\int_\Omega |X|^p d\mu\right)^{1/p}. \tag{3.1.17}$$

The set of functions X such that $\|X\|_{p,\mu} < \infty$ is denoted by $\mathscr{L}^p(\Omega, \mathscr{F}, \mu) = \mathscr{L}^p$.

Theorem 3.19 (Minkowski's inequality) *For all $X, Y \in \mathscr{L}^p$ and $p \in [1, \infty]$,*

$$\|X + Y\|_p \le \|X\|_p + \|Y\|_p. \tag{3.1.18}$$

Theorem 3.20 (Hölder's inequality) *For all measurable functions X, Y and $p, q \in [1, \infty]$ such that $\frac{1}{p} + \frac{1}{q} = 1$,*

$$\big|\mathbb{E}[XY]\big| \le \|X\|_p \, \|Y\|_q. \tag{3.1.19}$$

Both inequalities follow from one of the most important inequalities in integration theory: Jensen's inequality (see Fig. 3.2).

Theorem 3.21 (Jensen's inequality) *Let $(\Omega, \mathscr{F}, \mu)$ be a probability space, X an absolutely integrable random variable, and $\varphi\colon \mathbb{R} \to \mathbb{R}$ a convex function. Then, for any $c \in \mathbb{R}$,*

$$\mathbb{E}\big[\varphi\big(X - \mathbb{E}[X] + c\big)\big] \ge \varphi(c). \tag{3.1.20}$$

In particular,

$$\mathbb{E}\big[\varphi(X)\big] \ge \varphi\big(\mathbb{E}[X]\big). \tag{3.1.21}$$

Proof If φ is convex then, for every y, there is a straight line below φ that touches φ at $(y, \varphi(y))$, i.e., there exists an $m \in \mathbb{R}$ such that $\varphi(x) \ge \varphi(y) + (x - y)m$. Choosing $x = X - \mathbb{E}[X] + c$ and $y = c$, and taking expectations on both sides, we get (3.1.20). □

We leave it as an exercise to deduce Minkowski's inequality and Hölder's inequality from Jensen's inequality.

Since Minkowski's inequality is really a triangle inequality and linearity is trivial, we would be inclined to think that $\|\cdot\|_p$ is a norm and $\mathscr{L}^p$ is a normed space. In fact, the only problem is that $\|X\|_p = 0$ does not imply $X = 0$, since X may be non-zero on sets of μ-measure zero. Therefore, to define a normed space, we consider equivalence classes of functions in $\mathscr{L}^p$ by calling two functions X, X' *equivalent* when $X - X'$ is non-zero only on a set of measure zero. The space of these equivalence classes is called $L^p = L^p(\Omega, \mathscr{F}, \mu)$.

The following fact about L^p-spaces will be useful.

Lemma 3.22 *$L^p(\Omega, \mathscr{F}, \mu)$ is a Banach space (i.e., a complete normed vector space).*

The case $p = 2$ is particularly nice, in that L^2 is not only a Banach space but also a Hilbert space. The point is that Hölder's inequality for $p = 2$ yields

$$\mathbb{E}[XY] \leq \sqrt{\mathbb{E}[X^2]\,\mathbb{E}[Y^2]} = \|X\|_2 \|Y\|_2. \tag{3.1.22}$$

This means that on L^2 there exists a quadratic form $(\cdot, \cdot)_\mu$,

$$(X, Y)_\mu = \int_\Omega XY d\mu = \mathbb{E}[XY], \tag{3.1.23}$$

that has the properties of a *scalar product*. The L^2-norm being the derived norm, we have $\|X\|_2 = \sqrt{(X, X)_\mu}$.

3.1.5 Convergence

A central issue in probability theory is the notion of *convergence* of probability measures and random variables. The most commonly used concept of convergence of probability measures is that of *weak convergence*.

Definition 3.23 Let $(\mu_n)_{n\in\mathbb{N}}$ be a sequence of probability measures defined on some topological space S, and let μ be a probability measure defined on the same space. The sequence $(\mu_n)_{n\in\mathbb{N}}$ is said to *converge weakly* to μ if and only if, for all continuous functions $f\colon S \to \mathbb{R}$,

$$\lim_{n\to\infty} \mu_n(f) = \mu(f). \tag{3.1.24}$$

If $S = \mathbb{R}$, then this is equivalent to saying that the sequence of probability distribution functions $(F_n)_{n\in\mathbb{N}}$ defined by $F_n(x) = \mu_n((-\infty, x])$ converges to the probability distribution function F defined by $F(x) = \mu((-\infty, x])$, at every point $x \in \mathbb{R}$ of continuity of F.

As to convergence of random variables, the standard concepts of pointwise or even uniform convergence are often too rigid and need to be replaced by weaker notions.

The following notion of *convergence in law* is natural when we do not think of random variables as functions on sample spaces, but rather only care about the probability distributions they induce.

Definition 3.24 Let $(X_n)_{n\in\mathbb{N}}$ be a sequence of random variables with values in some topological space, and let X be a random variable on the same space. The sequence $(X_n)_{n\in\mathbb{N}}$ is said to *converge in law* to X,

$$X_n \xrightarrow{\mathcal{D}} X, \tag{3.1.25}$$

if and only if the induced probability measures $(\mathbb{P}_{X_n})_{n\in\mathbb{N}}$ converge weakly to $\mathbb{P}_X$.

The following notion of *convergence in probability* takes the functional aspect of random variables more seriously.

Definition 3.25 Let $(X_n)_{n\in\mathbb{N}}$ be a sequence of random variables with values in some topological space, and let X be a random variable on the same space. The sequence $(X_n)_{n\in\mathbb{N}}$ is said to *converge in probability* to X if and only if, for any $\varepsilon > 0$,

$$\lim_{n\to\infty} \mathbb{P}\big(|X_n - X| \geq \varepsilon\big) = 0. \tag{3.1.26}$$

A stronger notion, which comes closer to pointwise convergence, is that of *almost sure convergence.*

Definition 3.26 Let $(X_n)_{n\in\mathbb{N}}$ be a sequence of random variables with values in some topological space, and let X be a random variable on the same space. The sequence $(X_n)_{n\in\mathbb{N}}$ is said to *converge almost surely* to X if and only if

$$\mathbb{P}\Big(\lim_{n\to\infty} X_n = X\Big) = 1. \tag{3.1.27}$$

Clearly, almost sure convergence implies convergence in probability, and convergence in probability implies convergence in law.

Finally, there is the notion of *convergence in L^p*.

Definition 3.27 Let $(X_n)_{n\in\mathbb{N}}$ be a sequence of random variables with values in some normed space, and let X be a random variable on the same space. Let $p \in (0,\infty)$. The sequence $(X_n)_{n\in\mathbb{N}}$ is said to *converges to X in L^p* if and only if

$$\lim_{n\to\infty} \mathbb{E}\big[|X_n - X|^p\big] = 0. \tag{3.1.28}$$

Even almost sure convergence does not imply convergence of the integral of the random variable without extra conditions. Lebesgue's dominated convergence theorem provides a sufficient condition. There exists a useful improvement of the dominated convergence theorem that leads us to the important notion of *uniform integrability*.

Definition 3.28 Let $(\Omega, \mathcal{F}, \mathbb{P})$ be a probability space. A class $\mathcal{C}$ of real-valued random variables X is called *uniformly integrable* when for every $\varepsilon > 0$ there exists a $K = K(\varepsilon) < \infty$ such that

$$\mathbb{E}\left[|X|\mathbb{1}_{|X|>K}\right] < \varepsilon \quad \forall X \in \mathcal{C}. \tag{3.1.29}$$

Note, in particular, that if $\mathcal{C}$ is uniformly integrable, then there exists a constant $C < \infty$ such that $\mathbb{E}[|X|] \leq C$ for all $X \in \mathcal{C}$.

Theorem 3.29 *Let X_n, $n \in \mathbb{N}$, and X be integrable random variables on a probability space $(\Omega, \mathcal{F}, \mathbb{P})$. Then $\lim_{n\to\infty} \mathbb{E}[|X_n - X|] = 0$ if and only if*

(i) *$X_n \to X$ in probability as $n \to \infty$.*
(ii) *The family $X_n, n \in \mathbb{N}$, is uniformly integrable.*

A simple criterion for uniform integrability is L^p-boundedness for $p > 1$.

Lemma 3.30 *Let $\mathcal{C}$ be a class of random variables. Assume that, for some $p > 1$,*

$$\sup_{X\in\mathcal{C}} \mathbb{E}\left[|X|^p\right] = c < \infty. \tag{3.1.30}$$

Then $\mathcal{C}$ is uniformly integrable.

Proof By Hölder's inequality and Chebychev's inequality, for all $X \in \mathcal{C}$ and $p, q > 1$ such that $1/p + 1/q = 1$,

$$\mathbb{E}\left[|X|\mathbb{1}_{|X|>K}\right] \leq \mathbb{E}\left[|X|^p\right]^{1/p} \mathbb{P}\left(|X| > K\right)^{1/q} \leq c^{1/p} c^{p/q} K^{-p/q}, \tag{3.1.31}$$

which tends to zero uniformly in $\mathcal{C}$ as $K \to \infty$. □

3.1.6 Radon-Nikodým derivative

It is possible to modify a measure μ on a measurable space $(\Omega, \mathcal{F})$ with the help of a measurable function X. Indeed, set

$$\mu^X(A) = \int_A X\, d\mu, \quad A \in \mathcal{F}. \tag{3.1.32}$$

If μ is the Lebesgue measure, then μ^X is the absolutely continuous measure with density X. In general, if $\mathcal{O}$ is a measurable set with $\mu(\mathcal{O}) = 0$, then also $\mu^X(\mathcal{O}) = 0$. In words, a μ-null set is also a μ^X-null set. The latter property leads us to the notion of absolute continuity between general measures.

Definition 3.31 Let μ, ν be two measures on a measurable space $(\Omega, \mathcal{F})$.

(i) ν is *absolutely continuous* with respect to μ, written $\nu \ll \mu$, if and only if all μ-null sets are ν-null sets.
(ii) Two measures μ, ν are *equivalent* if $\mu \ll \nu$ and $\nu \ll \mu$.
(iii) A measure ν is *singular* with respect to μ if there exists a set $\mathcal{O} \in \mathcal{F}$ such that $\mu(\mathcal{O}) = 0$ and $\nu(\mathcal{O}^c) = 0$.

The following important theorem, called the *Radon-Nikodým theorem*, asserts that relative absolute continuity is equivalent to the existence of a density.

Theorem 3.32 (Radon-Nikodým theorem) *Let μ, ν be two σ-finite measures on a measurable space $(\Omega, \mathcal{F})$. Then the following two statements are equivalent:*

(i) $\nu \ll \mu$.
(ii) *There exists a non-negative measurable function X such that $\nu = \mu^X$.*

Moreover, X is unique up to null sets.

Definition 3.33 If $\nu \ll \mu$, then a positive measurable function X such that $\nu = \mu^X$ is called the *Radon-Nikodým derivative* of ν with respect to μ, denoted by

$$X = \frac{d\nu}{d\mu}. \tag{3.1.33}$$

The following property of the Radon-Nikodým derivative is very important.

Lemma 3.34 *Let μ, ν be two σ-finite measures on $(\Omega, \mathcal{F})$ such that $\nu \ll \mu$. If X is $\mathcal{F}$-measurable and ν-integrable, then*

$$\int_A X \, d\nu = \int_A X \frac{d\nu}{d\mu} \, d\mu, \quad A \in \mathcal{F}. \tag{3.1.34}$$

3.2 Stochastic processes

Stochastic processes are the main models for systems that exhibit metastability. In this section we recall some basis facts.

3.2.1 Definition of stochastic processes

There are various equivalent ways in which stochastic processes can be defined, and it is useful to keep these in mind. The standard way is as follows. We begin with an abstract probability space $(\Omega, \mathcal{F}, \mathbb{P})$. Next, we need a measurable space $(S, \mathcal{B}(S))$ (typically a Polish space together with its Borel σ-algebra), where S is called the *state space*. Finally, we need a set I, called the *index set*. A stochastic process with

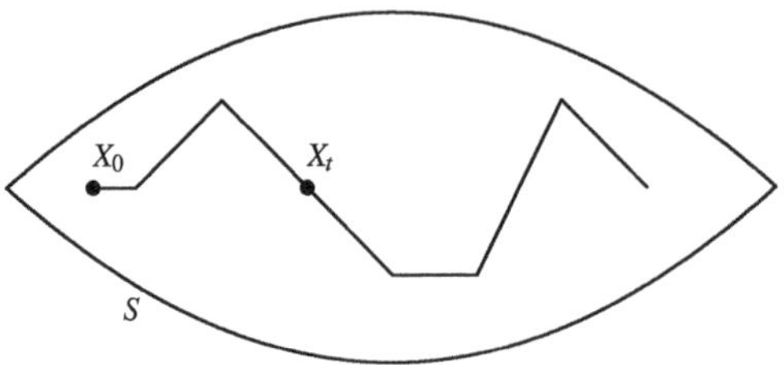

Fig. 3.3 The path of a stochastic process X taking values in S

state space S and index set I is a collection of $(S, \mathscr{B}(S))$-valued *random variables* $(X_t)_{t\in I}$ defined on $(\Omega, \mathscr{F}, \mathbb{P})$. If I is either $\mathbb{N}_0 = \mathbb{N} \cup \{0\}$, $\mathbb{Z}$, $\mathbb{R}_+ = [0,\infty)$ or $\mathbb{R}$, then we may think of it as *time*. Depending on whether I is discrete or continuous, we refer to $(X_t)_{t\in I}$ as a stochastic process with *discrete* or *continuous* time (see Fig. 3.3).

Given a stochastic process as defined above, we can take a different perspective and, for each $\omega \in \Omega$, view $X(\omega)$ as a map from I to S,

$$X(\omega)\colon\ I \to S, \quad t \mapsto X_t(\omega). \tag{3.2.1}$$

We call such a function a *sample path* of X, or a *realisation* of X. Here we want to view the stochastic process as a random variable taking values in the space of functions,

$$X\colon\ \Omega \to S^I, \quad \omega \mapsto X(\omega), \tag{3.2.2}$$

where we view S^I as the space of functions from I to S. To complete this image, we need to endow S^I with a σ-algebra, called $\mathscr{B}^I$. How should we choose this σ-algebra? Our picture will be that X maps $(\Omega, \mathscr{F})$ to $(S^I, \mathscr{B}^I)$.

Lemma 3.35 *Let $\mathscr{B}^I$ be the smallest σ-algebra that contains all subsets of S^I of the form*

$$C(A,t) = \left\{x \in S^I\colon\ x_t \in A\right\} \tag{3.2.3}$$

with $A \in \mathscr{B} = \mathscr{B}(S)$, $t \in I$. Then $\mathscr{B}^I$ is the smallest σ-algebra such that all functions $X_t\colon\ \Omega \to S$, $t \in I$, are measurable, i.e., $\mathscr{B}^I = \sigma(X_t, t \in I)$.

Definition 3.36 If $J \subset I$ is finite and $B \in \mathscr{B}^J$, then we call

$$C(B,J) = \left\{x \in S^I\colon\ x_J = \{x_t\}_{t\in J} \in B\right\} \tag{3.2.4}$$

a *cylinder set* or, more precisely, a finite-dimensional cylinder set. If B is of the form $B = \times_{t\in J} A_t$, $A_t \in \mathscr{B}$, then we call it a *special cylinder.*

It is clear that $\mathscr{B}^I$ contains all finite-dimensional cylinder sets. But, of course, it contains much more. We call $\mathscr{B}^I$ the product σ-algebra, or the algebra generated by the cylinder sets.

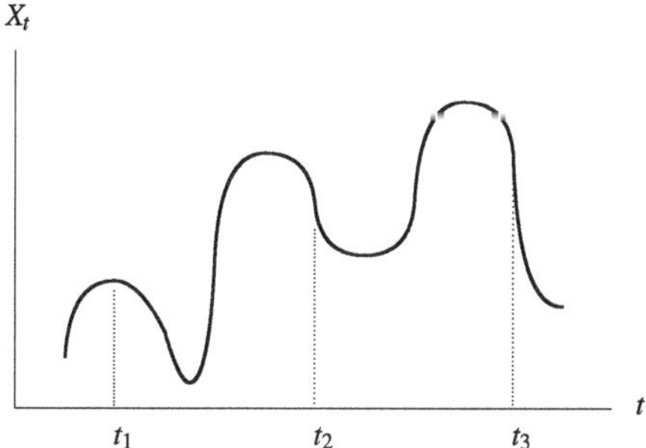

Fig. 3.4 A stochastics process is determined by its finite-dimensional distributions

If we view X as a map from Ω to the set of all S-valued functions on I, then we can define the probability distribution induced by $\mathbb{P}$ on the space $(S^I, \mathscr{B}^I)$,

$$\mathbb{P}_X = \mathbb{P} \circ X^{-1}, \tag{3.2.5}$$

as the distribution of the random variable X.

3.2.2 The Daniell-Kolmogorov extension theorem

The most fundamental observation is that stochastic processes are determined by their observation on finitely many points in time. For $J \subset I$ we will denote by π_J the canonical projection from S^I to S^J, i.e., $\pi_J X \in S^J$ such that $(\pi_J X)_t = X_t$ for all $t \in J$. Naturally, on S^J we can define the distributions

$$\mathbb{P}_X^J = \mathbb{P} \circ (\pi_J X)^{-1}. \tag{3.2.6}$$

Definition 3.37 Let $F(I)$ denote the set of finite non-empty subsets of I. Then the collection of probability measures

$$\left\{\mathbb{P}_X^J : J \in F(I)\right\} \tag{3.2.7}$$

is called the collection of finite-dimensional distributions of X (or finite-dimensional marginal distributions, or finite-dimensional marginals).

Note that the finite-dimensional distributions determine $\mathbb{P}_X$ on the algebra of finite-dimensional cylinder sets. Hence, by Dynkin's theorem, they determine the distribution on the σ-algebra $\mathscr{B}^I$, which is nice. What is even nicer is that we can also go the other way and *construct* the law of a stochastic process from specified finite-dimensional distributions (see Fig. 3.4). This is the content of the following fundamental theorem due to Kolmogorov and Daniell.

Theorem 3.38 (Daniell-Kolmogorov extension theorem) *Let S be a Polish space, and let $\mathscr{B} = \mathscr{B}(S)$ be its Borel-σ-algebra. Let I be a set. Suppose that, for each*

$J \in F(I)$, there exists a probability measure $\mathbb{P}^J$ on $(S^J, \mathscr{B}^J)$ such that, for any $J_1 \subset J_2 \in F(I)$,

$$\mathbb{P}^{J_1} = \mathbb{P}^{J_2} \circ \pi_{J_1}^{-1}, \tag{3.2.8}$$

where π_{J_1} denotes the canonical projection from S^{J_2} to S^{J_1}. Then there exists a unique measure $\mathbb{P}$ on $(S^I, \mathscr{B}^I)$ such that, for all $J \in F(I)$,

$$\mathbb{P} \circ \pi_J^{-1} = \mathbb{P}^J. \tag{3.2.9}$$

Note that we need not distinguish cases according to the nature of the set I.

3.3 Conditional expectations

The notions of conditional expectations and conditional probabilities are central to the theory of Markov processes and we collect them here.

3.3.1 Definition of conditional expectations

Definition 3.39 Consider a probability space $(\Omega, \mathscr{F}, \mathbb{P})$. Let $\mathscr{G} \subset \mathscr{F}$ be a sub-σ-algebra of $\mathscr{F}$. Let X be a random variable, i.e., a $\mathscr{F}$-measurable (real-valued) function on Ω such that $\mathbb{E}[|X|] < \infty$. We say that a function Y is a *conditional expectation of X given $\mathscr{G}$*, written $Y = \mathbb{E}(X|\mathscr{G})$, if

(i) Y is $\mathscr{G}$-measurable.
(ii) For all $A \in \mathscr{G}$,

$$\mathbb{E}[\mathbb{1}_A Y] = \mathbb{E}[\mathbb{1}_A X]. \tag{3.3.1}$$

If two functions Y, Y' both satisfy the conditions of a conditional expectation, then they can differ only on sets of probability zero, i.e., $\mathbb{P}(Y = Y') = 1$. Such different realisations of a conditional expectation are called "versions". The following theorem guarantees the existence of conditional expectations.

Theorem 3.40 *Let $(\Omega, \mathscr{F}, \mathbb{P})$ be a probability space, let X be a random variable such that $\mathbb{E}[|X|] < \infty$, and let $\mathscr{G} \subset \mathscr{F}$ be a sub-σ-algebra of $\mathscr{F}$. Then*

(i) *There exists a $\mathscr{G}$-measurable function $\mathbb{E}[X|\mathscr{G}]$, unique up to sets of measure zero, called the conditional expectation of X given $\mathscr{G}$, such that for all $A \in \mathscr{G}$,*

$$\int_A \mathbb{E}[X|\mathscr{G}]\, d\mathbb{P} = \int_A X\, d\mathbb{P}. \tag{3.3.2}$$

(ii) *If X is absolutely integrable, and Z is an absolutely integrable $\mathcal{G}$-measurable random variable such that, for some Π-System $\mathcal{D}$ with $\sigma(\mathcal{D}) = \mathcal{G}$,*

$$\mathbb{E}[Z] = \mathbb{E}[X], \quad \text{and} \quad \int_A Z\, d\mathbb{P} = \int_A X\, d\mathbb{P} \quad \forall A \in \mathcal{D}, \tag{3.3.3}$$

then $Z = \mathbb{E}[X|\mathcal{G}]$ a.s.

(Recall from Definition 3.7 that a Π-system is a set that is closed under finite intersections.)

In many cases the σ-algebra $\mathcal{G}$ with respect to which we are conditioning is the σ-algebra $\sigma(Y)$ generated by some other random variable Y. In those cases we will write

$$\mathbb{E}\big[X|\sigma(Y)\big] = \mathbb{E}[X|Y] \tag{3.3.4}$$

and call this the conditional expectation of X given Y.

3.3.2 Elementary properties of conditional expectations

Conditional expectations share most of the properties of ordinary expectations. The following is a list of elementary properties:

Lemma 3.41 *Let $(\Omega, \mathcal{F}, \mathbb{P})$ be a probability space and let $\mathcal{G} \subset \mathcal{F}$ be a sub-σ-algebra. Then:*

(i) *If X is $\mathcal{G}$-measurable, then $\mathbb{E}[X|\mathcal{G}] = X$ a.s.*
(ii) *The map $X \mapsto \mathbb{E}[X|\mathcal{G}]$ is linear.*
(iii) $\mathbb{E}[\mathbb{E}[X|\mathcal{G}]] = \mathbb{E}[X]$.
(iv) *If $\mathcal{B} \subset \mathcal{G}$ is a σ-algebra, then $\mathbb{E}[\mathbb{E}[X|\mathcal{G}]|\mathcal{B}] = \mathbb{E}[X|\mathcal{B}]$ a.s.*
(v) $|\mathbb{E}[X|\mathcal{G}]| \le \mathbb{E}[|X|\,|\mathcal{G}]$ *a.s.*
(vi) *If $X \le Y$, then $\mathbb{E}[X|\mathcal{G}] \le \mathbb{E}[Y|\mathcal{G}]$ a.s.*

The following theorem summarises the most important properties of conditional expectations with regard to taking limits.

Theorem 3.42 *Let X_n, $n \in \mathbb{N}$, and Y be absolutely integrable random variables on a probability space $(\Omega, \mathcal{F}, \mathbb{P})$, and let $\mathcal{G} \subset \mathcal{F}$ be a sub-σ-algebra. Then:*

(i) *If $0 \le X_n \uparrow X$ a.s. as $n \to \infty$, then $\mathbb{E}[X_n|\mathcal{G}] \uparrow \mathbb{E}[X|\mathcal{G}]$ a.s. as $n \to \infty$.*
(ii) *If $X_n \ge 0$ a.s. for all n, then*

$$\mathbb{E}\Big[\liminf_{n\to\infty} X_n|\mathcal{G}\Big] \le \liminf_{n\to\infty} \mathbb{E}[X_n|\mathcal{G}]. \tag{3.3.5}$$

(iii) *If $X_n \to X$ a.s. as $n \to \infty$ and $|X_n| \le |Y|$ for all n, then $\mathbb{E}[X_n|\mathcal{G}] \to \mathbb{E}[X|\mathcal{G}]$ a.s. as $n \to \infty$.*

Of course, these are just the analogues of the three basic convergence theorems for ordinary expectations. A useful further property is the following lemma.

Lemma 3.43 *Let X be integrable and let Y be bounded and $\mathscr{G}$-measurable. Then*

$$\mathbb{E}[XY|\mathscr{G}] = Y\mathbb{E}[X|\mathscr{G}] \quad a.s. \tag{3.3.6}$$

There is a natural connection between independence and conditional expectation.

Lemma 3.44 *Two σ-algebras $\mathscr{G}_1, \mathscr{G}_2$ are independent if and only if, for all $\mathscr{G}_2$-measurable integrable random variables X,*

$$\mathbb{E}[X|\mathscr{G}_1] = \mathbb{E}[X]. \tag{3.3.7}$$

By choosing $X = \mathbb{1}_A$, $A \in \mathscr{G}_2$, we see that (3.3.7) reduces to the independence of events.

3.3.3 Conditional probability measures

From conditional expectations we want to construct conditional probability measures. As before, we consider a probability space $(\Omega, \mathscr{F}, \mathbb{P})$ and a sub-σ-algebra $\mathscr{G}$. For any $A \in \mathscr{F}$, we can define

$$\mathbb{P}(A|\mathscr{G}) = \mathbb{E}[\mathbb{1}_A|\mathscr{G}], \tag{3.3.8}$$

and call it the *conditional probability of A given $\mathscr{G}$*. This is a $\mathscr{G}$-measurable function that satisfies (see Fig. 3.5)

$$\int_G \mathbb{P}(A|\mathscr{G})\, d\mathbb{P} = \int_G \mathbb{1}_A\, d\mathbb{P} = \mathbb{P}(A \cap G), \quad G \in \mathscr{F}. \tag{3.3.9}$$

It clearly inherits from the conditional expectation the following properties:

(i) $0 \leq \mathbb{P}(A|\mathscr{G}) \leq 1$ a.s.

(ii) $\mathbb{P}(A|\mathscr{G}) = 0$ a.s. if and only if $\mathbb{P}(A) = 0$. Moreover, $\mathbb{P}(A|\mathscr{G}) = 1$ a.s. if and only if $\mathbb{P}(A) = 1$.

(iii) If $A_n \in \mathscr{F}$, $n \in \mathbb{N}$, are disjoint sets then

$$\mathbb{P}\left(\bigcup_{n\in\mathbb{N}} A_n|\mathscr{G}\right) = \sum_{n\in\mathbb{N}} \mathbb{P}(A_n|\mathscr{G}) \quad \text{a.s.} \tag{3.3.10}$$

(iv) If $A_n \in \mathscr{F}$, $n \in \mathbb{N}$, such that $\lim_{n\to\infty} A_n = A$ then

$$\lim_{n\to\infty} \mathbb{P}(A_n|\mathscr{G}) = \mathbb{P}(A|\mathscr{G}) \quad \text{a.s.} \tag{3.3.11}$$

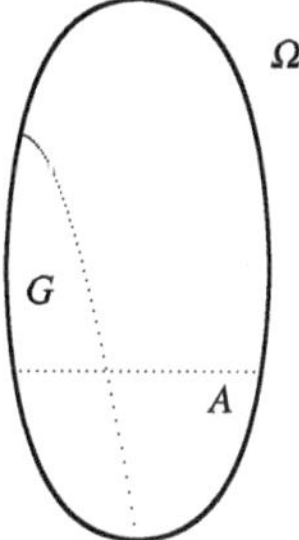

Fig. 3.5 Conditional probability of A given $\mathscr{G}$, as defined in (3.3.9)

These observations bring us close to viewing conditional probabilities as $\mathscr{G}$-measurable functions taking values in the set of probability measures, at least for almost all ω. The problem, however, is that the requirement of σ-additivity, which seems to be satisfied because of (iii), is in fact problematic: (iii) says that for any sequence $(A_n)_{n\in\mathbb{N}}$ there exists a set Ω' of full measure such that

$$\mathbb{P}\Big(\bigcup_{n\in\mathbb{N}} A_n|\mathscr{G}\Big)(\omega) = \sum_{n\in\mathbb{N}} \mathbb{P}(A_n|\mathscr{G})(\omega) \quad \forall\, \omega \in \Omega'. \tag{3.3.12}$$

However, Ω' may depend on $(A_n)_{n\in\mathbb{N}}$ and, since the space is not countable, it is unclear whether there exists a set of full measure on which (3.3.12) holds for *all* sequences. These considerations lead us to the definition of so-called *regular conditional probabilities*.

Definition 3.45 Let $(\Omega, \mathscr{F}, \mathbb{P})$ be a probability space and let $\mathscr{G}$ be a sub-σ-algebra. A *regular conditional probability measure* or *regular conditional probability* on $\mathscr{F}$ given $\mathscr{G}$ is a function $P(\omega, A)$, defined for all $A \in \mathscr{F}$ and all $\omega \in \Omega$, such that:

(i) For each $\omega \in \Omega$, $P(\omega, \cdot)$ is a probability measure on $(\Omega, \mathscr{F})$.
(ii) For each $A \in \mathscr{F}$, $P(\cdot, A)$ is a $\mathscr{G}$-measurable function coinciding with the conditional probability $\mathbb{P}(A|\mathscr{G})$ almost everywhere.

The point is that if we have a regular conditional probability, then we can express conditional expectations as expectations with respect normal probability measures.

Theorem 3.46 *With the notation above, if $P_\omega[A] = P(\omega, A)$ is a regular conditional probability on $\mathscr{F}$ given $\mathscr{G}$, then for an $\mathscr{F}$-measurable integrable random variable X,*

$$\mathbb{E}[X|\mathscr{G}](\omega) = \int_\Omega X\, dP_\omega \quad a.s. \tag{3.3.13}$$

The question remains when regular conditional probabilities exist. A central result is the existence when Ω is a Polish space.

Theorem 3.47 *Let $(\Omega, \mathcal{B}(\Omega), \mathbb{P})$ be a probability space with Ω a Polish space and $\mathcal{B}(\Omega)$ its Borel-σ-algebra. Let $\mathcal{G} \subset \mathcal{B}(\Omega)$ be a sub-σ-algebra. Then there exists a regular conditional probability $P(A, \omega)$ given $\mathcal{G}$.*

3.4 Martingales in discrete time

One of the most fundamental and useful concepts in the theory of stochastic processes is that of a *martingale*. Since this will play a major rôle in the remainder of the book, we will spend some time to expose its main properties. As some subtleties arise in continuous time, we begin with the simpler case of discrete time.

3.4.1 Definitions

Definition 3.48 Let $(\Omega, \mathcal{F})$ be a measurable space. A family of sub-σ-algebras $(\mathcal{F}_n)_{n \in \mathbb{N}_0}$ of $\mathcal{F}$ that satisfies

$$\mathcal{F}_0 \subset \mathcal{F}_1 \subset \mathcal{F}_2 \subset \cdots \subset \mathcal{F}_\infty = \sigma\left(\bigcup_{n \in \mathbb{N}_0} \mathcal{F}_n\right) \subset \mathcal{F}, \tag{3.4.1}$$

is called a *filtration* of the σ-algebra $\mathcal{F}$. A quadruple $(\Omega, \mathcal{F}, \mathbb{P}, (\mathcal{F}_n)_{n \in \mathbb{N}_0})$ is called a filtered (probability) space.

Filtrations and stochastic processes are closely linked together, in two ways.

Definition 3.49 A stochastic process $X = (X_n)_{n \in \mathbb{N}_0}$ is called *adapted* to the filtration $(\mathcal{F}_n)_{n \in \mathbb{N}_0}$ if X_n is $\mathcal{F}_n$-measurable for every n.

Definition 3.50 Let $X = (X_n)_{n \in \mathbb{N}_0}$ be a stochastic process on $(\Omega, \mathcal{F}, \mathbb{P})$. The *natural filtration* $(\mathcal{W}_n)_{n \in \mathbb{N}_0}$ with respect to X is the smallest filtration such that X is adapted to it, i.e.,

$$\mathcal{W}_n = \sigma(X_0, \ldots, X_n). \tag{3.4.2}$$

Definition 3.51 A stochastic process X on a filtered space is called a *martingale* if and only if the following hold:

(i) The process X is adapted to the filtration $(\mathcal{F}_n)_{n \in \mathbb{N}_0}$.
(ii) For all $n \in \mathbb{N}_0$, $\mathbb{E}[|X_n|] < \infty$.
(iii) For all $n \in \mathbb{N}$,

$$\mathbb{E}[X_n | \mathcal{F}_{n-1}] = X_{n-1} \quad \text{a.s.} \tag{3.4.3}$$

If (i) and (ii) hold but, instead of (iii), it is only true that $\mathbb{E}[X_n|\mathscr{F}_{n-1}] \geq X_{n-1}$, respectively, $\mathbb{E}[X_n|\mathscr{F}_{n-1}] \leq X_{n-1}$, then the process X is called a *submartingale*, respectively, a *supermartingale*.

We will next head for the fundamental theorem stating the impossibility of "winning games" built on martingales.

Definition 3.52 A stochastic process $C = (C_n)_{n\in\mathbb{N}_0}$ is called *previsible* if C_n is $\mathscr{F}_{n-1}$-measurable for all $n \in \mathbb{N}$.

Given an adapted stochastic process X and a previsible process C, we can define the *discrete stochastic integral*

$$W_n = \sum_{k=1}^{n} C_k(X_k - X_{k-1}) = (C \bullet X)_n. \tag{3.4.4}$$

The most pertinent fact about martingales is their stability under the martingale transform:

Theorem 3.53 *The martingale transform has the following properties:*

(i) *Let C be a uniformly bounded non-negative previsible process and let X be a supermartingale. Then $C \bullet X$ is a supermartingale that vanishes at zero.*
(ii) *Let C be a uniformly bounded previsible process and let X be a martingale. Then $C \bullet X$ is a martingale that vanishes at zero.*
(iii) *Both in* (i) *and* (ii) *the condition of uniform boundedness of C can be replaced by boundedness in $\mathscr{L}^2$ of C and X.*

3.4.2 Upcrossings and convergence

It is essentially a consequence of Theorem 3.53 that uniformly integrable martingales converge almost surely. This is the content of Doob's supermartingale convergence theorem:

Theorem 3.54 (Supermartingale convergence) *Let X be an $\mathscr{L}^1$-bounded supermartingale (i.e., $\sup_{n\in\mathbb{N}_0} \mathbb{E}[|X_n|] < \infty$). Then a.s. $X_\infty = \lim_{n\to\infty} X_n$ exists and is a finite random variable.*

The Doob convergence theorem implies that non-negative supermartingales converge a.s. This is because the supermartingale property ensures that $\mathbb{E}[|X_n|] = \mathbb{E}[X_n] \leq \mathbb{E}[X_0]$, so the uniform boundedness in $\mathscr{L}^1$ is always guaranteed.

The reason behind the (super)martingale convergence theorem is the Doob's upcrossing lemma, which states that in (super)martingales oscillations are necessarily linked to growth. Let $a < b$, and let $U_N(X, [a,b])$ be the number of times X crosses the interval $[a,b]$ from below up to time N (see Fig. 3.6).

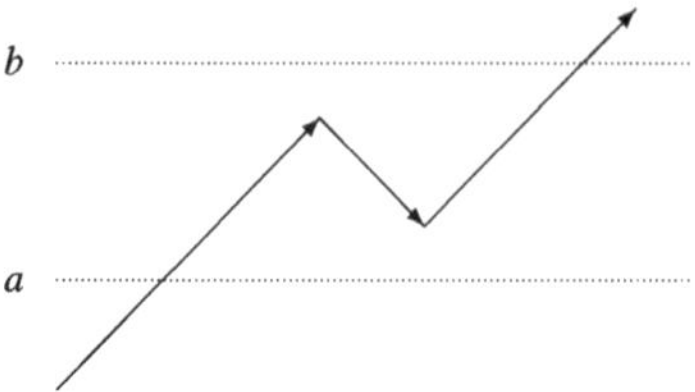

Fig. 3.6 An upcrossing

Lemma 3.55 (Doob's upcrossing lemma) *Let X be a supermartingale. Then*

$$(b-a)\,\mathbb{E}\big[U_N\big(X,[a,b]\big)\big] \le \mathbb{E}\big[|X_N-a|\mathbb{1}_{X_N<a}\big]. \tag{3.4.5}$$

The following is an immediate consequence of Lemma 3.55.

Corollary 3.56 *Let X be an $\mathscr{L}^1$-bounded supermartingale. For $[a,b]$ an interval, define $U_\infty(X,[a,b]) = \lim_{n\to\infty} U_n(X,[a,b])$. Then*

$$(b-a)\,\mathbb{E}\big[U_\infty\big(X,[a,b]\big)\big] \le a + \sup_{n\in\mathbb{N}_0} \mathbb{E}\big[|X_n|\big] < \infty. \tag{3.4.6}$$

In particular, $\mathbb{P}(U_\infty(X,[a,b])=\infty)=0$.

For proofs of the above results and further details, see Rogers and Williams [208].

3.4.3 Maximum inequalities

In this section we derive some fundamental inequalities for martingales. One of the most useful ones is the following *maximum inequality*.

Theorem 3.57 (Doob's maximum inequality) *Let Z be a non-negative submartingale. Then, for $c>0$ and $n\in\mathbb{N}_0$,*

$$c\,\mathbb{P}\Big(\max_{0\le k\le n} Z_k \ge c\Big) \le \mathbb{E}[Z_n\mathbb{1}_{\max_{0\le k\le n} Z_k\ge c}] \le \mathbb{E}[Z_n]. \tag{3.4.7}$$

Proof Since this inequality is fundamental for many applications, we will include its proof. The second inequality in (3.4.7) is trivial. To prove the first inequality, define the sequence of disjoint events given by $F_0=\{Z_0\ge c\}$ and

$$F_k = \bigcap_{0\le \ell<k} \{Z_\ell<c\}\cap\{Z_k\ge c\} = \big\{\omega\in\Omega:\ \min(0\le \ell\le n:\ X_\ell\ge c)=k\big\}. \tag{3.4.8}$$

Then

$$F = \Big\{\sup_{0\le k\le n} Z_k \ge c\Big\} = \bigcup_{k=0}^{n} F_k. \tag{3.4.9}$$

Clearly, $F_k \in \mathscr{F}_k$. Moreover, on F_k we know that $Z_k \geq c$. Thus

$$\mathbb{E}[Z_n \mathbb{1}_{F_k}] \geq \mathbb{E}[Z_k \mathbb{1}_{F_k}] \geq c\,\mathbb{P}(F_k), \tag{3.4.10}$$

where the first inequality uses the submartingale property of Z. Hence

$$\mathbb{E}[Z_n \mathbb{1}_F] = \sum_{k=0}^{n} \mathbb{E}[Z_n \mathbb{1}_{F_k}] \geq c \sum_{k=0}^{n} \mathbb{P}(F_k) = c\,\mathbb{P}(F), \tag{3.4.11}$$

which implies the claim. □

If $(M_n)_{n\in\mathbb{N}_0}$ is a martingale and f is a convex function, then $(f(M_n))_{n\in\mathbb{N}_0}$ is a submartingale. This observation allows us to obtain useful inequalities from the one in Theorem 3.57. In particular, the so-called Kolmogorov inequality follows by choosing $f(X) = X^2$. Another useful choice is the exponential function $f(X) = e^X$.

A useful trick when dealing with stochastic processes is to "extract the martingale part". There are several such decompositions. The following, called the *Doob decomposition*, is very important and its continuous-time analogue is fundamental for the theory of stochastic integration.

Theorem 3.58 (Doob decomposition)

(i) *Let $X = (X_n)_{n\in\mathbb{N}_0}$ be an adapted process on a filtered space $(\Omega, \mathscr{F}, \mathbb{P}, (\mathscr{F}_n)_{n\in\mathbb{N}_0})$ with $X_n \in \mathscr{L}^1$ for all $n \in \mathbb{N}_0$. Then X can be written in the form*

$$X = X_0 + M + A, \tag{3.4.12}$$

where M is a martingale with $M_0 = 0$ and A is a previsible process with $A_0 = 0$. (Here, (3.4.12) is to be understood as $X_0 = X_0$ and $X_n = X_0 + M_n + A_n$, $n \in \mathbb{N}$.) This decomposition is unique modulo indistinguishability, i.e., if $X = X_0 + M' + A'$ for some other M', A', then

$$\mathbb{P}\big(M_n = M_n' \text{ and } A_n = A_n'\ \forall n \in \mathbb{N}_0\big) = 1. \tag{3.4.13}$$

(ii) *The process X is a submartingale if and only if A is an increasing process, in the sense that*

$$\mathbb{P}(A_n \leq A_{n+1}\ \forall n \in \mathbb{N}_0) = 1. \tag{3.4.14}$$

Proof The proof is easy. All we need to do is to derive explicit formulas for M and A. Assume that a decomposition of the claimed form exists. Then

$$\begin{aligned}\mathbb{E}[X_n - X_{n-1}|\mathscr{F}_{n-1}] &= \mathbb{E}[M_n - M_{n-1}|\mathscr{F}_{n-1}] + \mathbb{E}[A_n - A_{n-1}|\mathscr{F}_{n-1}]\\ &= 0 + A_n - A_{n-1}\end{aligned} \tag{3.4.15}$$

by the martingale and predictability properties. Therefore

$$A_n = \sum_{k=1}^{n} \mathbb{E}[X_k - X_{k-1} | \mathcal{F}_{k-1}] \quad \text{a.s.} \tag{3.4.16}$$

Now simply *define* A_n by (3.4.16) and M_n by $M_n = X_n - X_0 - A_n$. Then, clearly, M is a martingale, and A is by construction predictable. This completes the proof of assertion (i). Assertion of (ii) is obvious from (3.4.15). □

An immediate application of the decomposition theorem is a maximum inequality without positivity assumption.

Lemma 3.59 *If X is either a submartingale or a supermartingale, then*

$$c\,\mathbb{P}\Big(\sup_{0\le k\le n} |X_k| \ge 3c\Big) \le 4\mathbb{E}\big[|X_0|\big] + 3\mathbb{E}\big[|X_n|\big], \quad n \in \mathbb{N}_0, c > 0. \tag{3.4.17}$$

Proof We consider the case where X is a submartingale (the case of a supermartingale is identical by passing to $-X$). Then there is a Doob decomposition

$$X = X_0 + M + A \tag{3.4.18}$$

with A an increasing process. Hence

$$\sup_{0\le k\le n} |X_k| \le |X_0| + \sup_{0\le k\le n} |M_k| + \sup_{0\le k\le n} |A_k| = |X_0| + \sup_{0\le k\le n} |M_k| + A_n. \tag{3.4.19}$$

Note that $|M|$ is a non-negative submartingale, so for the supremum of $|M_k|$ we can use Theorem 3.57. We use the simple observation that if $x + y + z > 3c$, then at least one of the x, y, z must exceed c. Therefore

$$\begin{aligned} c\,\mathbb{P}\Big(\sup_{0\le k\le n} |X_k| \ge 3c\Big) &\le c\,\mathbb{P}\big(|X_0| \ge c\big) + c\,\mathbb{P}\Big(\sup_{0\le k\le n} |M_k| \ge c\Big) + c\,\mathbb{P}(A_n \ge c) \\ &\le \mathbb{E}\big[|X_0|\big] + \mathbb{E}\big[|M_n|\big] + \mathbb{E}[A_n]. \end{aligned} \tag{3.4.20}$$

We have

$$\mathbb{E}\big[|M_n|\big] = \mathbb{E}\big[|X_n - X_0 - A_n|\big] \le \mathbb{E}\big[|X_n|\big] + \mathbb{E}\big[|X_0|\big] + \mathbb{E}[A_n] \tag{3.4.21}$$

and

$$\mathbb{E}[A_n] = \mathbb{E}[X_n - X_0 - M_n] = \mathbb{E}[X_n - X_0] \le \mathbb{E}\big[|X_n|\big] + \mathbb{E}\big[|X_0|\big]. \tag{3.4.22}$$

Inserting these two bounds into (3.4.20), we get the claim. □

The Doob decomposition gives rise to two important processes associated with a martingale M, namely, the *bracket process* $\langle M\rangle$ and the *quadratic variation process* $[M]$.

Definition 3.60 Let M be a martingale in $\mathscr{L}^2$ with $M_0 = 0$. Then M^2 is a submartingale with Doob decomposition

$$M^2 = N + \langle M \rangle, \tag{3.4.23}$$

where N is a martingale that vanishes at zero and $\langle M \rangle$ is a previsible process that vanishes at zero.

Note that boundedness in $\mathscr{L}^1$ of $\langle M \rangle$ is equivalent to boundedness in $\mathscr{L}^2$ of M. From the formulas associated with the Doob decomposition we deduce that

$$\langle M \rangle_n - \langle M \rangle_{n-1} = \mathbb{E}\big[M_n^2 - M_{n-1}^2 | \mathscr{F}_{n-1}\big] = \mathbb{E}\big[(M_n - M_{n-1})^2 | \mathscr{F}_{n-1}\big]. \tag{3.4.24}$$

Definition 3.61 For M as before, define

$$[M]_n = \sum_{k=1}^{n} (M_k - M_{k-1})^2. \tag{3.4.25}$$

Lemma 3.62 *If M is as before, then*

$$M^2 - [M] = V = (C \bullet M), \tag{3.4.26}$$

where V is a martingale, and $C_n = 2M_{n-1}$. If M is bounded in $\mathscr{L}^2$, then V is bounded in $\mathscr{L}^1$.

3.4.4 Stopping times and stopped martingales

Our analysis of metastability in Part III relies crucially on the analysis of the first times when a Markov process hits certain sets. These *first hitting times* are special cases of so-called *stopping times*: a time whose occurrence can be determined based on the outcome of the process until that time alone.

Definition 3.63 A map $\tau\colon \Omega \to \mathbb{N}_0 \cup \{\infty\}$ is called a stopping time (with respect to a filtration $(\mathscr{F}_n)_{n \in \mathbb{N}_0}$) if

$$\{T = n\} \in \mathscr{F}_n \quad \forall n \in \mathbb{N}_0 \cup \{\infty\}. \tag{3.4.27}$$

The most important examples of stopping times are hitting times. Let X be an adapted process, and let $B \in \mathscr{B}$. Define

$$\tau_B = \inf\{n \in \mathbb{N}\colon\ X_n \in B\}, \tag{3.4.28}$$

i.e., the *first hitting time* of B when the location of X_0 is not counted (see Fig. 3.7). Then τ_B is a stopping time. To see this, note that if $n \in \mathbb{N}$, then

$$\{\tau_B = n\} = \big\{\omega \in \Omega\colon\ X_n(\omega) \in B,\ X_k(\omega) \neq B\ \forall\, 1 \le k < n\big\}. \tag{3.4.29}$$

Fig. 3.7 First hitting locations in B (indicated by $*$) when X_0 (indicated by $\bullet$) is not in B, respectively, is in B

This event is manifestly in $\mathcal{F}_n$. The event $\{\tau_B = \infty\}$ occurs if and only if $\{X_n \notin B \ \forall n \in \mathbb{N}\} \subset \mathcal{F}_\infty$.

In principle all stopping times can be realised as first hitting times of some process. To do so, simply define

$$I_{[T,\infty)}(n,\omega) = \begin{cases} 1, & \text{if } n \geq T(\omega), \\ 0, & \text{otherwise.} \end{cases} \tag{3.4.30}$$

This process is adapted, and $T = \tau_1$.

It is sometimes convenient to have the notion of a σ-algebra of events that take place before a stopping time.

Definition 3.64 The pre-T-σ-algebra $\mathcal{F}_T$ is the set of events $F \subset \Omega$ such that

$$F \cap \{T \leq n\} \in \mathcal{F}_n \quad \forall n \in \mathbb{N}_0 \cup \{\infty\}. \tag{3.4.31}$$

Pre-T-σ-algebras will play an important rôle in the formulation of the strong Markov property. There are some useful elementary facts associated with this concept.

Lemma 3.65 *Let S, T be stopping times.*

(i) *If X is an adapted process, then X_T is $\mathcal{F}_T$-measurable.*
(ii) *If $S < T$, then $\mathcal{F}_S \subset \mathcal{F}_T$.*
(iii) $\mathcal{F}_{T \wedge S} = \mathcal{F}_T \cap \mathcal{F}_S$.
(iv) *If $F \in \mathcal{F}_{S \vee T}$, then $F \cap \{S \leq T\} \in \mathcal{F}_T$.*
(v) $\mathcal{F}_{S \vee T} = \sigma(\mathcal{F}_T, \mathcal{F}_S)$.

The interplay of stopping times and martingale properties is of fundamental importance in potential theory, to be described in Chap. 7. We next discuss this in some detail via an example taken from finance. Consider a supermartingale X. We want to play a strategy that depends on a stopping time T, say, we keep one "unit of stock" until the random time T:

$$C_n = C_n^T = \mathbb{1}_{n \leq T}. \tag{3.4.32}$$

Note that $C^T = (C_n^T)_{n\in\mathbb{N}_0}$ is a previsible process, namely,

$$\{C_n^T = 0\} = \{T \le n-1\} \in \mathscr{F}_{n-1}, \tag{3.4.33}$$

and, since C_n^T only takes the values 0 and 1, this inclusion suffices to show that $C_n^T \in \mathscr{F}_{n-1}$. The "wealth process" associated with this strategy is $C^T \bullet X = ((C^T \bullet X)_n)_{n\in\mathbb{N}_0}$ with

$$\left(C^T \bullet X\right)_n = X_{T\wedge n} - X_0. \tag{3.4.34}$$

If we define the stopped process $X^T = (X_n^T)_{n\in\mathbb{N}_0}$ via

$$X_n^T(\omega) = X_{T(\omega)\wedge n}(\omega), \tag{3.4.35}$$

then we have alternatively

$$C^T \bullet X = X^T - X_0. \tag{3.4.36}$$

Since C^T is positive and bounded, Theorem 3.53 leads us to the following statement.

Theorem 3.66

(i) *If X is a supermartingale and T is a stopping time, then the stopped process $X^T = (X_{T\wedge n})_{n\in\mathbb{N}_0}$, is a supermartingale. In particular,*

$$\mathbb{E}[X_{T\wedge n}] \le \mathbb{E}[X_0] \quad \forall n \in \mathbb{N}_0. \tag{3.4.37}$$

(ii) *If X is a martingale and T is a stopping time, then X^T is a martingale. In particular,*

$$\mathbb{E}[X_{T\wedge n}] = \mathbb{E}[X_0] \quad \forall n \in \mathbb{N}_0. \tag{3.4.38}$$

Note that Theorem 3.66 does not assert that $\mathbb{E}[X_T] \le \mathbb{E}[X_0]$. The following theorem gives conditions under which this inequality holds.

Theorem 3.67 (Doob's optional stopping theorem)

(i) *Let T be a stopping time, and let X be a supermartingale. Then X_T is integrable and*

$$\mathbb{E}[X_T] \le \mathbb{E}[X_0], \tag{3.4.39}$$

provided one of the following conditions holds:

(a) *T is bounded (i.e., there exists an $N \in \mathbb{N}$ such that $T(\omega) \le N$ for all $\omega \in \Omega$).*
(b) *X is bounded and T is a.s. finite.*
(c) *$\mathbb{E}[T] < \infty$ and, for some $K < \infty$,*

$$\left|X_n(\omega) - X_{n-1}(\omega)\right| \le K \quad \forall n \in \mathbb{N},\ \omega \in \Omega. \tag{3.4.40}$$

(ii) *If X is a martingale, then $\mathbb{E}[X_T] = \mathbb{E}[X_0]$ in any of the situations above.*

Proof We already know that $\mathbb{E}[X_{n\wedge T}] - \mathbb{E}[X_0] \le 0$ for all $n \in \mathbb{N}$. In case (a), we know that $T \wedge N = T$, and so $\mathbb{E}[X_T] = \mathbb{E}[X_{T\wedge N}] \le \mathbb{E}[X_0]$, as claimed. In case (b), we start from $\mathbb{E}[X_{n\wedge T}] - \mathbb{E}[X_0] \le 0$ and let $n \to \infty$. Since T is almost surely finite, we have $\lim_{n\to\infty} X_{n\wedge T} = X_T$ a.s., and since X_n is uniformly bounded, we get

$$\lim_{n\to\infty} \mathbb{E}[X_{T\wedge n}] = \mathbb{E}\Big[\lim_{n\to\infty} X_{T\wedge n}\Big] = \mathbb{E}[X_T], \tag{3.4.41}$$

which implies the result. In case (c), we observe that

$$|X_{T\wedge n} - X_0| = \left|\sum_{k=1}^{T\wedge n} (X_k - X_{k-1})\right| \le KT, \tag{3.4.42}$$

and $\mathbb{E}[KT] < \infty$ by assumption. Thus, we can again take the limit $n \to \infty$ and use Lebesgue's dominated convergence theorem to justify that the inequality survives.

Finally, to justify (ii), use that if X is a martingale, then both X and $-X$ are supermartingales. The ensuing two inequalities imply the desired equality. □

Theorem 3.67 may look strange, since it seems to contradict the "no winning strategy". Indeed, take the simple random walk $(S_n)_{n\in\mathbb{N}_0}$ starting from $S_0 = 0$ and define the stopping time $T = \inf\{n\colon\ S_n = 10\}$. Then, clearly, $X_T = 10 \neq \mathbb{E}[X_0] = 0$. So, using (c) we must conclude that $\mathbb{E}[T] = \infty$. In fact, the "sure" gain when we achieve our goal is offset by the fact that on average it takes infinitely long to reach this goal (of course, most games will end quickly, but chances are that some may take very long).

Case (c) in Theorem 3.67 is the situation we hope to have the most often. The following lemma states that $\mathbb{E}[T] < \infty$ whenever the probability of the event leading to T is eventually sufficiently large.

Lemma 3.68 *Suppose that T is a stopping time and that there exist $N \in \mathbb{N}$ and $\varepsilon > 0$ such that*

$$\mathbb{P}(T \le n + N \,|\, \mathscr{F}_n) > \varepsilon \quad a.s. \quad \forall n \in \mathbb{N}_0. \tag{3.4.43}$$

Then $\mathbb{E}[T] < \infty$.

Proof For $k \in \mathbb{N}$ we can write, by iteration,

$$\begin{aligned}
\mathbb{P}(T > kN) &= \mathbb{E}[\mathbb{1}_{T>(k-1)N}\mathbb{1}_{T>kN}] \\
&= \mathbb{E}\big[\mathbb{E}[\mathbb{1}_{T>(k-1)N}\mathbb{1}_{T>kN} | \mathscr{F}_{(k-1)N}]\big] \\
&= \mathbb{E}\big[\mathbb{1}_{T>(k-1)N}\,\mathbb{E}[\mathbb{1}_{T>kN} | \mathscr{F}_{(k-1)N}]\big] \\
&\le (1-\varepsilon)\,\mathbb{E}[\mathbb{1}_{T>(k-1)N}] \\
&\le (1-\varepsilon)^k.
\end{aligned} \tag{3.4.44}$$

The exponential decay of this probability implies the claim. □

Finally, we state *Doob's supermartingale inequality* for non-negative supermartingales.

Theorem 3.69 (Doob's supermartingale inequality) *Let X be a non-negative supermartingale and T a stopping time. Then*

$$\mathbb{E}[X_T] \leq \mathbb{E}[X_0]. \tag{3.4.45}$$

Moreover, for any $c > 0$,

$$c\,\mathbb{P}\Big(\sup_{k\in\mathbb{N}_0} X_k > c\Big) \leq \mathbb{E}[X_0]. \tag{3.4.46}$$

Proof We know that $\mathbb{E}[X_{T\wedge n}] \leq \mathbb{E}[X_0]$. Using Fatou's lemma, we may pass to the limit $n \to \infty$. For (3.4.45), set $T = \inf\{n \in \mathbb{N}_0 \colon X_n > c\}$. Clearly, $X_T \geq c$ if $\sup_{k\in\mathbb{N}_0} X_k > c$, and zero otherwise. Thus, $\mathbb{E}[X_T] \geq c\,\mathbb{P}(\sup_{k\in\mathbb{N}_0} X_k > c)$, and (3.4.46) follows from (3.4.45). □

3.5 Martingales in continuous time

In principle, most of the results that hold for martingales in discrete time carry over to continuous time. There are, however, a number of subtleties that need to be taken care of.

3.5.1 Càdlàg functions

The first question we need to settle is the choice of function space where the processes live in. Often this is the set of continuous functions, but in general this set is too restrictive. It turns out that a good choice is the set of so-called càdlàg functions.

Definition 3.70 A function $f\colon \mathbb{R}_+ \to \mathbb{R}$ is called a càdlàg function ("continue à droite, limites à gauche") if

(i) For every $t \in \mathbb{R}_+$, $f(t) = \lim_{s\downarrow t} f(s)$.
(ii) For every $t > 0$, $f(t-) = \lim_{s\uparrow t} f(s)$ exists.

It will be important to be able to extend functions defined on countable sets to càdlàg functions. Abbreviate $\mathbb{Q}_+ = \mathbb{Q} \cap \mathbb{R}_+$.

Definition 3.71 A function $y\colon \mathbb{Q}_+ \to \mathbb{R}$ is called *regularisable* if

(i) For every $t \in \mathbb{R}_+$, $\lim_{q\downarrow t} y(q)$ exists and is finite.
(ii) For every $t > 0$, $y(t-) = \lim_{q\uparrow t} y(s)$ exists and is finite.

Without going into further details, we state the fact that regularisability is a measurable property.

Lemma 3.72 *Let $(Y_q)_{q\in\mathbb{Q}_+}$ be a stochastic process defined on $(\Omega, \mathscr{F}, \mathbb{P})$, and let*

$$G = \{\omega \in \Omega \colon\ q \mapsto Y_q(\omega) \text{ is regularisable}\}. \tag{3.5.1}$$

Then $G \in \mathscr{F}$.

Next, we observe that from a regularisable function we can readily obtain a càdlàg function by taking limits from the right.

Theorem 3.73 *Let $y\colon\ \mathbb{Q}_+ \to \mathbb{R}$ be a regularisable function. Define, for $t \in \mathbb{R}_+$,*

$$f(t) = \lim_{q \downarrow t} y(q). \tag{3.5.2}$$

Then f is càdlàg.

3.5.2 Filtrations, supermartingales and càdlàg processes

We begin with a probability space $(\Omega, \mathscr{G}, \mathbb{P})$. We define a continuous-time filtration $(\mathscr{G}_t)_{t\in\mathbb{R}_+}$, similarly as in the discrete-time setting.

Definition 3.74 A filtration $(\mathscr{G}_t)_{t\in\mathbb{R}_+}$ of $(\Omega, \mathscr{G}, \mathbb{P})$ is an increasing family of sub-σ-algebras such that, for $0 \le s < t$,

$$\mathscr{G}_s \subset \mathscr{G}_t \subset \mathscr{G}_\infty = \sigma\left(\bigcup_{r\in\mathbb{R}_+} \mathscr{G}_r\right) \subset \mathscr{G}. \tag{3.5.3}$$

A quadruple $(\Omega, \mathscr{G}, \mathbb{P}, (\mathscr{G}_t)_{t\in\mathbb{R}_+})$ a called a filtered space.

Definition 3.75 A stochastic process $(X_t)_{t\in\mathbb{R}_+}$ is called *adapted* to the filtration $(\mathscr{G}_t)_{t\in\mathbb{R}_+}$ if X_t is $\mathscr{G}_t$-measurable for every $t \in \mathbb{R}_+$.

Definition 3.76 A stochastic process X on a filtered space is called a *martingale* if the following hold:

(i) The process X is adapted to the filtration $(\mathscr{G}_t)_{t\in\mathbb{R}_+}$.
(ii) For all $t \in \mathbb{R}_+$, $\mathbb{E}[|X_t|] < \infty$.
(iii) For all $0 \le s \le t$,

$$\mathbb{E}[X_t|\mathscr{G}_s] = X_s \quad \text{a.s.} \tag{3.5.4}$$

Sub- and supermartingales are defined in the same way, with $=$ in (3.5.4) replaced by $\ge$, respectively, $\le$.

So far almost nothing has changed with respect to the discrete-time setting. Note, in particular, that if we take a monotone sequence of times $(t_n)_{n\in\mathbb{N}_0}$, then $(Y_n)_{n\in\mathbb{N}_0} = (X_{t_n})_{n\in\mathbb{N}_0}$ is a discrete-time (sub/super)martingale whenever $(X_t)_{t\in\mathbb{R}_+}$ is a continuous-time (sub/super)martingale.

The next lemma is important because it connects martingale properties to càdlàg properties.

Lemma 3.77 *Let Y be a supermartingale on a filtered space $(\Omega, \mathscr{G}, \mathbb{P}, (\mathscr{G}_t)_{t\in\mathbb{R}_+})$. Let $t \in \mathbb{R}_+$, and let $q(-n)$, $n \in \mathbb{N}$, be such that $q(-n) \downarrow t$ as $n \to \infty$. Then*

$$\lim_{q(-n)\downarrow t} Y_{q(-n)} \tag{3.5.5}$$

exists a.s. and in $\mathscr{L}^1$.

Proof This is an application of the Lévy-Doob downward theorem (see Rogers and Williams [208, Chaps. II.51 and II.63]). □

Spaces of càdlàg functions are the natural setting for stochastic processes.

Definition 3.78 A stochastic process is called a càdlàg process if all its sample paths are càdlàg functions; càdlàg processes that are (sub/super)martingales are called càdlàg (sub/super)martingales.

Note that we do not just require that almost all sample paths are càdlàg.

3.5.3 The Doob regularity theorem

We will now show that the setting of càdlàg functions is suitable for the theory of martingales.

Theorem 3.79 (Doob's regularity theorem) *Let $Y = (Y_t)_{t\in\mathbb{R}_+}$ be a supermartingale defined on a filtered space $(\Omega, \mathscr{G}, \mathbb{P}, (\mathscr{G}_t)_{t\in\mathbb{R}_+})$. Define the set*

$$G = \big\{\omega \in \Omega : \text{ the map } \mathbb{Q}_+ \ni q \mapsto Y_q(\omega) \in \mathbb{R} \text{ is regularisable}\big\}. \tag{3.5.6}$$

Then $G \in \mathscr{G}$ and $\mathbb{P}(G) = 1$. The process X defined by

$$X_t(\omega) = \begin{cases} \lim_{q\downarrow t} Y_q(\omega), & \text{if } \omega \in G, \\ 0, & \text{otherwise}, \end{cases} \tag{3.5.7}$$

is a càdlàg process.

One might hope that Theorem 3.79 settles all problems related to continuous-time martingales. Simply start with any supermartingale and pass to the càdlàg regularization. However, a problem of measurability arises. This can be seen in the trivial example of a process with a single jump. Let Y_t be defined for $\omega \in \Omega$ as

$$Y_t(\omega) = \begin{cases} 0, & \text{if } t \leq 1, \\ q(\omega), & \text{if } t > 1, \end{cases} \tag{3.5.8}$$

where $\mathbb{E}[q] = 0$. Let $(\mathscr{G}_t)_{t\in\mathbb{R}_+}$ be the natural filtration associated with this process. Clearly, $\mathscr{G}_t = \{\emptyset, \Omega\}$ for $t \leq 1$ and Y is a martingale with respect to this filtration. The càdlàg version of this process is

$$X_t(\omega) = \begin{cases} 0, & \text{if } t < 1, \\ q(\omega), & \text{if } t \geq 1. \end{cases} \tag{3.5.9}$$

Now, $X = (X_t)_{t\in\mathbb{R}_+}$ is not adapted to the filtration $(\mathscr{G}_t)_{t\in\mathbb{R}_+}$, since X_1 is not measurable with respect to $\mathscr{G}_1$. This problem cannot be remedied by a simple modification on sets of measure zero because $\mathbb{P}(X_1 = Y_1) < 1$. In particular, X is not a martingale with respect to the filtration $(\mathscr{G}_t)_{t\in\mathbb{R}_+}$, because

$$\mathbb{E}[X_{1+\varepsilon}|\mathscr{G}_1] = 0 \neq X_1 \quad \forall \varepsilon > 0. \tag{3.5.10}$$

We thus see that the right-continuous regularisation of Y at the point of the jump anticipates information from the future. If we want to develop a theory on càdlàg processes, then we must take this into account and introduce a richer filtration that contains this information.

Definition 3.80 Let $(\Omega, \mathscr{G}, \mathbb{P}, (\mathscr{G}_t)_{t\in\mathbb{R}_+})$ be a filtered space. Define, for $t \in \mathbb{R}_+$,

$$\mathscr{G}_{t+} = \bigcap_{s>t} \mathscr{G}_s = \bigcap_{\mathbb{Q}\ni q>t} \mathscr{G}_q, \tag{3.5.11}$$

and let

$$\mathscr{N}(\mathscr{G}_\infty) = \left\{G \in \mathscr{G}_\infty : \mathbb{P}(G) \in \{0, 1\}\right\}. \tag{3.5.12}$$

Then the *partial augmentation* $(\mathscr{H}_t)_{t\in\mathbb{R}_+}$ of the filtration $(\mathscr{G}_t)_{t\in\mathbb{R}_+}$ is defined as

$$\mathscr{H}_t = \sigma\left(\mathscr{G}_{t+}, \mathscr{N}(\mathscr{G}_\infty)\right). \tag{3.5.13}$$

The following lemma, which is obvious from the construction of càdlàg versions, justifies this definition.

Lemma 3.81 *If Y is a supermartingale with respect to the filtration $(\mathscr{G}_t)_{t\in\mathbb{R}_+}$ and X is its càdlàg version defined in Theorem* 3.79, *then X is adapted to the partially augmented filtration $(\mathscr{H}_t)_{t\in\mathbb{R}_+}$.*

A natural question is whether in this setting X is a supermartingale. The next theorem answers this question in the affirmative and is to be seen as the completion of Theorem 3.79.

Theorem 3.82 *With the assumptions and notations of Lemma* 3.81, X *is a supermartingale with respect to the filtration* $(\mathscr{H}_t)_{t\in\mathbb{R}_+}$. *Moreover, X is a modification of Y if and only if Y is right-continuous, in the sense that*

$$\lim_{s\downarrow t}\mathbb{E}\big[|Y_t - Y_s|\big] = 0 \quad \forall t \in \mathbb{R}_+. \tag{3.5.14}$$

Henceforth we will work on filtered spaces that are already partially augmented, i.e., our standard setting (called "the usual setting" in Rogers and Williams [208]) is as follows.

Definition 3.83 A filtered càdlàg space is a quadruple $(\Omega, \mathscr{F}, \mathbb{P}, (\mathscr{F}_t)_{t\in\mathbb{R}_+})$, where $(\Omega, \mathscr{F}, \mathbb{P})$ is a probability space and $(\mathscr{F}_t)_{t\in\mathbb{R}_+}$ is a filtration that satisfies the following properties:

(i) $\mathscr{F}$ is $\mathbb{P}$-complete (i.e., contains all sets of $\mathbb{P}$-measure zero).
(ii) $\mathscr{F}_0$ contains all sets of $\mathbb{P}$-measure 0.
(iii) $\mathscr{F}_t = \mathscr{F}_{t+}$, i.e., $t \mapsto \mathscr{F}_t$ is right-continuous.

If $(\Omega, \mathscr{G}, \mathbb{P}, (\mathscr{G}_t)_{t\in\mathbb{R}_+})$ is a filtered space, then the minimal enlargement of this space satisfying conditions (i), (ii) and (iii) is called the right-continuous regularisation of this space.

Theorem 3.84 *The process X constructed in Theorem* 3.79 *is a supermartingale with respect to the filtration* $(\mathscr{F}_t)_{t\in\mathbb{R}_+}$.

We finally give a version of Doob's regularity theorem for processes defined on càdlàg spaces.

Theorem 3.85 *Let $(\Omega, \mathscr{F}, \mathbb{P}, (\mathscr{F}_t)_{t\in\mathbb{R}_+})$ be a filtered càdlàg space. Let Y be an adapted supermartingale. Then Y has a càdlàg modification Z if and only if the map $t \mapsto \mathbb{E}[Y_t]$ is right-continuous, in which case Z is a càdlàg supermartingale.*

3.5.4 Convergence theorems and martingale inequalities

Key results on discrete-time martingale theory were Doob's forward and backward convergence theorems and the maximum inequalities. We will now consider the corresponding results in continuous time.

Theorem 3.86 (Supermartingale convergence) *Let X be a càdlàg supermartingale with respect to a filtered space $(\Omega, \mathscr{G}, \mathbb{P}, (\mathscr{G}_t)_{t\in\mathbb{R}_+})$. Assume that* $\sup_{t\in\mathbb{R}_+}\mathbb{E}[|X_t|] < \infty$. *Then*

$$\lim_{t\to\infty} X_t = X_\infty \tag{3.5.15}$$

exists almost surely in $\mathbb{R}$.

In a similar way the maximum inequalities for càdlàg submartingales can be inferred from their discrete-time counterparts.

Theorem 3.87 (Doob's maximum inequality) *Let Z be a non-negative càdlàg submartingale on a filtered space. Then, for any $c > 0$ and $t \in \mathbb{R}_+$,*

$$\mathbb{P}\Big(\sup_{0\le s\le t} Z_s \ge c\Big) \le c^{-1}\mathbb{E}[Z_t \mathbb{1}_{\sup_{0\le s\le t} Z_s \ge c}] \le c^{-1}\mathbb{E}[Z_t]. \tag{3.5.16}$$

3.5.5 Stopping times

The notions around stopping times introduced in this section are important in the theory of Markov processes. We need to be careful in the continuous-time setting, even though we closely follow the discrete-time setting.

We consider a filtered space $(\Omega, \mathcal{G}, \mathbb{P}, (\mathcal{G}_t)_{t\in\mathbb{R}_+})$.

Definition 3.88 A map $T\colon\ \Omega \to [0,\infty]$ is called a $(\mathcal{G}_t)_{t\in\mathbb{R}_+}$-stopping time if

$$\{T \le t\} = \big\{\omega \in \Omega\colon\ T(\omega) \le t\big\} \in \mathcal{G}_t \quad \forall 0 \le t \le \infty. \tag{3.5.17}$$

If T is a stopping time, then the *pre-T-σ-algebra* $\mathcal{G}_T$ is the set of all $\Lambda \in \mathcal{G}$ such that

$$\Lambda \cap \{T \le t\} \in \mathcal{G}_t \quad \forall 0 \le t \le \infty. \tag{3.5.18}$$

With this definition we have the usual properties of pre-T-σ-algebras:

Lemma 3.89 *Let S, T be stopping times.*

(i) *If $S \le T$, then $\mathcal{G}_S \subset \mathcal{G}_T$.*
(ii) *$\mathcal{G}_{T\wedge S} = \mathcal{G}_T \cap \mathcal{G}_S$.*
(iii) *If $F \in \mathcal{G}_{S\vee T}$, then $F \cap \{S \le T\} \in \mathcal{G}_T$.*
(iv) *$\mathcal{G}_{S\vee T} = \sigma(\mathcal{G}_T, \mathcal{G}_S)$.*

It will be useful to also talk about stopping times with respect to the filtration $(\mathcal{G}_{t+})_{t\in\mathbb{R}_+}$.

Definition 3.90 A map $T\colon\ \Omega \to [0,\infty]$ is called a $(\mathcal{G}_{t+})_{t\in\mathbb{R}_+}$-stopping time if

$$\{T < t\} = \big\{\omega \in \Omega\colon\ T(\omega) < t\big\} \in \mathcal{G}_t \quad \forall 0 \le t \le \infty. \tag{3.5.19}$$

If T is a $(\mathcal{G}_{t+})_{t\in\mathbb{R}_+}$-stopping time, then the *pre-T-σ-algebra* $\mathcal{G}_{T+}$ is the set of all $\Lambda \in \mathcal{G}$ such that

$$\Lambda \cap \{T < t\} \in \mathcal{G}_t \quad \forall 0 \le t \le \infty. \tag{3.5.20}$$

Lemma 3.91 *Let $(S_n)_{n\in\mathbb{N}}$ be a sequence of $(\mathcal{G}_t)_{t\in\mathbb{R}_+}$-stopping times.*

(i) *If $S_n \uparrow S$, then S is a $(\mathcal{G}_t)_{t\in\mathbb{R}_+}$-stopping time.*
(ii) *If $S_n \downarrow S$, then S is a $(\mathcal{G}_{t+})_{t\in\mathbb{R}_+}$-stopping time and $\mathcal{G}_{S+} = \bigcap_{n\in\mathbb{N}} \mathcal{G}_{S_n+}$.*

Definition 3.92 A process $(X_t)_{t\in\mathbb{R}_+}$ is called $(\mathcal{G}_t)_{t\in\mathbb{R}_+}$-*progressive* if for every $t \in \mathbb{R}_+$ the restriction of the map $(s,\omega) \mapsto X_s(\omega)$ to $[0,t] \times \Omega$ is $\mathcal{B}([0,t]) \times \mathcal{G}_t$-measurable.

The notion of a progressive process is stronger than that of an adapted process. Its importance arises from the fact that T-stopped progressive processes are measurable with respect to their respective pre-T-σ-algebra. The nice fact is that in the càdlàg setting all works well.

Lemma 3.93 *An adapted càdlàg process in a metrisable space $(S, \mathcal{B}(S))$ is progressive.*

Lemma 3.94 *If X is progressive with respect to the filtration $(\mathcal{G}_t)_{t\in\mathbb{R}_+}$ and T is a $(\mathcal{G}_t)_{t\in\mathbb{R}_+}$-stopping time, then X_T is $\mathcal{G}_T$-measurable.*

3.5.6 First hitting time and first entrance time

In the case of discrete-time Markov processes we have seen that hitting times of certain sets provide particularly important examples of stopping times. We will now extend this discussion to the continuous-time setting. It is important to distinguish between the notions of hitting time and entrance time. These differ in the way the position of the process at time 0 is treated.

Definition 3.95 Let X be a stochastic process with values in a measurable space $(E, \mathcal{E})$. Let $\Gamma \in \mathcal{E}$. We call

$$\Delta_\Gamma(\omega) = \inf\{t \in \mathbb{R}_+ : X_t(\omega) \in \Gamma\} \tag{3.5.21}$$

the *first entrance time* of the set Γ, and

$$\tau_\Gamma(\omega) = \inf\{t \in \mathbb{R}_+ \backslash \{0\} : X_t(\omega) \in \Gamma\} \tag{3.5.22}$$

the *first hitting time* of the set Γ. In both cases the infimum is understood to be ∞ if the process never enters Γ.

Recall that in the discrete-time setting we only worked with τ_Γ, which is in fact the more important notion. Here is an example of a stopping time.

Lemma 3.96 *Let E be a metric space and let F be a closed set. Let X be a continuous adapted process. Then Δ_F is a $(\mathcal{G}_t)_{t\in\mathbb{R}_+}$-stopping time and τ_F is a $(\mathcal{G}_{t+})_{t\in\mathbb{R}_+}$-stopping time.*

Proof Let ρ denote the metric on E. Then the map $x \mapsto \rho(x, F)$ is continuous, and hence the map $\omega \mapsto \rho(X_q(\omega), x)$ is $\mathcal{G}_q$-measurable for $q \in \mathbb{Q}_+$. Since the paths $t \mapsto X_t(\omega)$ are continuous, we have $\Delta_F(\omega) \leq t$ if and only if

$$\inf_{q \in \mathbb{Q} \cap [0,t]} \rho\big(X_q(\omega), F\big) = 0 \tag{3.5.23}$$

and so Δ_F is measurable with respect to $(\mathcal{G}_t)_{t \in \mathbb{R}_+}$. For τ_F the situation is slightly different at time zero. Indeed, let $\Delta_F^r = \inf\{t \geq r : X_t \in F\}$, $r > 0$. Obviously, from the previous result we have that D_F^r is a $(\mathcal{G}_t)_{t \in \mathbb{R}_+}$-stopping time. On the other hand, $\{\tau_F > 0\}$ if and only if there exists a $\delta > 0$ such that $\Delta_F^r > \delta$ for all $\mathbb{Q} \ni r > 0$. But, clearly, the event

$$A_\delta = \bigcap_{\mathbb{Q} \ni r > 0} \{\Delta_F^r > \delta\} \tag{3.5.24}$$

is $\mathcal{G}_\delta$-measurable, and so the event

$$\{\tau_F = 0\} = \{\tau_F > 0\}^c = \bigcap_{\delta > 0} A_\delta^c \tag{3.5.25}$$

is $\mathcal{G}_{0+}$-measurable, and so τ_F is a $(\mathcal{G}_{t+})_{t \in \mathbb{R}_+}$-stopping time. □

To see where the difference between Δ_F and τ_F comes from, consider the process starting at the boundary of F. Then $\Delta_F = 0$, while τ_F may or may not be zero: it could be that the process immediately leaves F and only returns after some positive time t, in which case $\tau_F > 0$, or it may stay for awhile in F, in which case $\tau_F = 0$. To distinguish between the two cases, we must look a little bit into the future (recall Fig. 3.7).

3.5.7 *Optional stopping and optional sampling*

We have seen in the theory of discrete-time Markov processes that martingale properties of processes stopped at stopping times are important. We need similar results for càdlàg processes. We again work on a filtered càdlàg space $(\Omega, \mathcal{F}, \mathbb{P}, (\mathcal{F}_t)_{t \in \mathbb{R}_+})$ on which all the processes will be defined and adapted. The key result is the following *optional sampling theorem.*

Theorem 3.97 (Optional sampling theorem) *Let X be a càdlàg submartingale and let T, S be $(\mathcal{F}_t)_{t \in \mathbb{R}_+}$-stopping times. Then, for each $M < \infty$,*

$$\mathbb{E}\big[X(T \wedge M) | \mathcal{F}_S\big] \geq X(S \wedge T \wedge M) \quad a.s. \tag{3.5.26}$$

If, in addition,

(i) *T is finite a.s.,*

(ii) $\mathbb{E}[|X(T)|] < \infty$,
(iii) $\lim_{M\to\infty} \mathbb{E}[X(M)\mathbb{1}_{T>M}] = 0$,

then

$$\mathbb{E}\big[X(T)|\mathscr{F}_S\big] \geq X(S \wedge T) \quad a.s. \tag{3.5.27}$$

Equality holds for martingales.

A special case of Theorem 3.97 implies the following corollary.

Corollary 3.98 *Let X be a càdlàg (sub/super)martingale, and let T be a stopping time. Then $X^T = (X_{T\wedge t})_{t\in\mathbb{R}_+}$ is a (sub/super)martingale.*

In the case of uniformly integrable supermartingales, we get the *Doob optional sampling theorem*:

Theorem 3.99 *Let X be a uniformly integrable or a non-negative càdlàg supermartingale. Let S and T be stopping times with $S \leq T$. Then $X_T \in \mathscr{L}^1$ and*

$$\mathbb{E}[X_\infty|\mathscr{F}_T] \leq X_T \quad a.s. \tag{3.5.28}$$

and

$$\mathbb{E}[X_T|\mathscr{F}_S] \leq X_S \quad a.s., \tag{3.5.29}$$

with equality when X is a uniformly integrable martingale.

An important concept connecting martingales and stopping times is that of a *local martingale*.

Definition 3.100 A stochastic process M is called a *local martingale* if there exists a sequence of stopping times $(\tau_n)_{n\in\mathbb{N}}$, with $\tau_n \leq \tau_{n+1}$ and $\lim_{n\to\infty} \tau_n = \infty$, such that the processes $M^{\tau_n} = (M_{t\wedge\tau_n})_{t\in\mathbb{R}_+}$ are martingales. The same terminology applies to sub- and super-martingales, as well as to various integrability properties.

3.6 Bibliographical notes

1. Standard textbooks on probability theory are Feller [108, 109], Billingsley [28], Chow and Teicher [58], Bauer [13], Kallenberg [146]. In these books the proofs can be found that were omitted in this chapter.

2. The presentation in Sects. 3.4–3.5 largely follows the book of Rogers and Williams [208].

Chapter 4
Markov Processes in Discrete Time

I think that method of government ought to answer well. You see, the Kings would be sure to make Laws contradicting each other: so the Subject could never be punished, because, whatever he did he'd be obeying some Law.

(Lewis Carroll, Sylvie and Bruno Concluded)

Markov processes are the basic class of stochastic processes that we will use to model metastable systems. Similarly to what we saw in Chap. 3, there is a substantial difference in the mathematical difficulties involved in dealing with *discrete time* and *continuous time*. In this chapter we give an outline of the theory of discrete-time Markov processes (also called Markov chains). In Chap. 5 we will deal with continuous-time Markov processes.

Section 4.1 gives the main definitions and lists some key facts. Section 4.2 looks at the link between Markov processes and martingales. Section 4.3 lists a few properties that are specific to the setting where the state space is countable.

4.1 Markov processes: main definitions and key facts

4.1.1 Definition and elementary properties

Markov processes $X = (X_t)_{t \in I}$ with $I = \mathbb{N}_0$ or $I = \mathbb{R}_+$ are stochastic analogues of dynamical systems. As such they must satisfy two basic properties: (1) they must be *causal*, i.e., we want to be able to write down an expression for the law of X_t given the σ-algebra $\mathscr{F}_{t-} = \sigma(X_s, 0 \leq s < t)$; (2) they must be *forgetful of the past*, i.e., given the value of X_s at some time $0 \leq s < t$, the law of X_t is independent of the values of X_u at all times $0 \leq u < s$ (see Fig. 4.1).

The basic definition is in fact independent of the nature of the time parameter.

Definition 4.1 A stochastic process with state space S and index set $I = \mathbb{N}_0$ or $I = \mathbb{R}_+$ is called a *Markov process* if the following holds. For any $t \geq s \geq 0$, there exists a probability kernel $P_{s,t}\colon S \times \mathscr{B} \to [0, 1]$, satisfying:

A. Bovier, F. den Hollander, *Metastability*,
Grundlehren der mathematischen Wissenschaften 351,
DOI 10.1007/978-3-319-24777-9_4

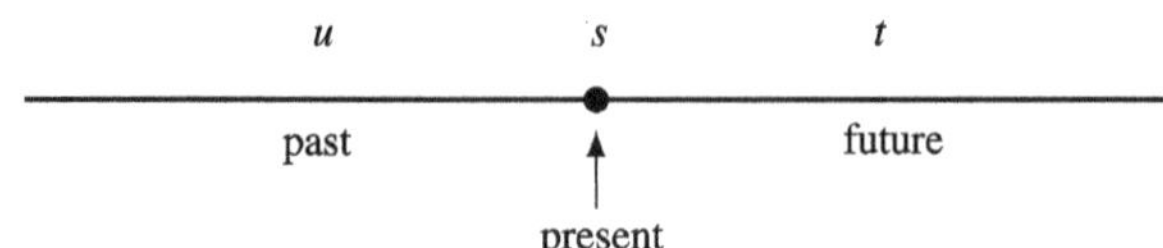

Fig. 4.1 Illustration of the Markov property: the future depends on the present not the past

(i) For any $x \in S$, $P_{s,t}(x, \cdot)$ is a probability measure on $(S, \mathscr{B})$.
(ii) For any $A \in \mathscr{B}$, $P_{s,t}(\cdot, A)$ is a measurable function on S, such that, for any $t \geq s \geq 0$,

$$\mathbb{P}(X_t \in A|\mathscr{F}_s)(\omega) = P_{s,t}\big(X_s(\omega), A\big). \tag{4.1.1}$$

In the case of discrete time, i.e., for index set $I = \mathbb{N}_0$, the compatibility conditions impose severe restrictions on the kernels $P_{s,t}$ that allow us to consider only the *one-step transition kernel* $\mathscr{P}_{t-1} = P_{t-1,t}$. Indeed, a stochastic process X with state space S and index set $\mathbb{N}_0$ is a discrete-time Markov process with one-step transition kernels $(\mathscr{P}_t)_{t\in\mathbb{N}}$ if, for all $A \in \mathscr{B}$ and $t \in \mathbb{N}$,

$$\mathbb{P}(X_t \in A|\mathscr{F}_{t-1})(\omega) = \mathscr{P}_{t-1}\big(X_{t-1}(\omega), A\big), \quad \mathbb{P}\text{-a.s.} \tag{4.1.2}$$

This requirement fixes the law $\mathbb{P}$ up to one more probability measure on $(S, \mathscr{B})$, namely, the *initial distribution* p_0.

Theorem 4.2 *Let $(S, \mathscr{B})$ be a Polish space, let P be a transition kernel on $\mathbb{N}_0 \times \mathbb{N}_0 \times S \times \mathscr{B}$, and let p_0 be a probability measure on $(S, \mathscr{B})$. Then there exists a unique stochastic process satisfying* (4.1.2) *such that $\mathbb{P}(X_0 \in A) = p_0(A)$ for all A.*

Proof In view of the Daniell-Kolmogorov extension theorem (Theorem 3.38), we have to show that our requirements fix all finite-dimensional distributions, and that these satisfy the compatibility conditions. This is essentially a problem of notation. We will need to be able to derive formulas for

$$\mathbb{P}(X_{t_n} \in A_n, \ldots, X_{t_1} \in A_1) \tag{4.1.3}$$

for $0 \leq t_1 < t_2 < \cdots < t_n$, $A_1, \ldots, A_n \in \mathscr{B}$ and $n \in \mathbb{N}$. To get started, we consider

$$\mathbb{P}(X_t \in A|\mathscr{F}_s), \quad 0 \leq s < t, \tag{4.1.4}$$

and use that, by the elementary properties of conditional expectations,

$$\begin{aligned}
\mathbb{P}(X_t \in A|\mathscr{F}_s) &= \mathbb{E}\big[\mathbb{P}(X_t \in A|\mathscr{F}_{t-1})|\mathscr{F}_s\big] \\
&= \mathbb{E}\big[\mathscr{P}_{t-1}\big(X_{t-1}(\omega), A\big)|\mathscr{F}_s\big] \\
&= \mathbb{E}\left[\mathbb{E}\left[\int_S \mathscr{P}_{t-1}(x_{t-1}, A)\mathscr{P}_{t-2}\big(X_{t-2}(\omega), dx_{t-1}\big)\mathscr{F}_{t-2}\right]\bigg|\mathscr{F}_s\right]
\end{aligned}$$

$$= \int_S \mathscr{P}_{t-1}(x_{t-1}, A) \int_S \mathscr{P}_{t-2}(x_{t-2}, dx_{t-1}) \dots$$

$$\dots \int_S \mathscr{P}_{s+1}(x_{s+1}, dx_{s+2}) \mathscr{P}_s\big(X_s(\omega), dx_{s+1}\big), \tag{4.1.5}$$

where we refrain from writing "a.s.", which applies to all equations relating to conditional expectations. We thus have

$$P_{s,t}(x, A) = \int_S \mathscr{P}_{t-1}(x_{t-1}, A) \int_S \mathscr{P}_{t-2}(x_{t-2}, dx_{t-1}) \dots$$

$$\dots \int_S \mathscr{P}_{s+1}(x_{s+1}, dx_{s+2}) \mathscr{P}_s(x, dx_{s+1}). \tag{4.1.6}$$

With this object, we can proceed to more complicated expressions:

$$\begin{aligned}
&\mathbb{P}(X_{t_n} \in A_n, \dots, X_{t_1} \in A_1) \\
&= \mathbb{E}\big[\mathbb{P}(X_{t_n} \in A_n | \mathscr{F}_{t_{n-1}}) \mathbb{1}_{A_{n-1}}(X_{t_{n-1}}) \dots \mathbb{1}_{A_1}(X_{t_1})\big] \\
&= \mathbb{E}\big[\mathbb{E}\big[P_{t_{n-1},t_n}(X_{t_{n-1}}, A_n) | \mathscr{F}_{t_{n-1}}\big] \mathbb{1}_{A_{n-1}}(X_{t_{n-1}}) \dots \mathbb{1}_{A_1}(X_{t_1})\big] \\
&= \mathbb{E}\Bigg[\mathbb{E}\bigg[\int_{A_{n-1}} P_{t_{n-1},t_n}(x_{n-1}, A_n) P_{t_{n-2},t_{n-1}}\big(X_{t_{n-2}}(\omega), dx_{n-1}\big) \big| \mathscr{F}_{t_{n-2}}\bigg] \\
&\quad \times \mathbb{1}_{A_{n-2}}(X_{t_{n-2}}) \dots \dots \mathbb{1}_{A_1}(X_{t_1})\Bigg] \\
&= \int_{A_{n-1}} P_{t_{n-1},t_n}(x_{n-1}, A_n) \int_{A_{n-2}} P_{t_{n-2},t_{n-1}}(x_{n-2}, dx_{n-1}) \dots \\
&\quad \dots \int_{A_1} P_{t_1,t_2}(x_1, dx_2) \int_S P_{0,t_1}(x_0, dx_1) P_0(dx_0).
\end{aligned} \tag{4.1.7}$$

Thus, we get the desired expression for the marginal distributions in terms of the transition kernel P and the initial distribution p_0. The compatibility relations follow from the following obvious, but important, property of the transition kernels. □

Lemma 4.3 *The transition kernels $P_{s,t}$ satisfy the Chapman-Kolmogorov equations*:

$$P_{s,t}(x, A) = \int_S P_{r,t}(y, A) P_{s,r}(x, dy), \quad t > r > s. \tag{4.1.8}$$

Proof This is obvious from the definition. □

The proof of the compatibility relations is now also obvious; if some of the A_i, $1 \le i \le n$, are equal to S, then we can use (4.1.8) and recover the expressions for the lower-dimensional marginals. □

4.1.2 Markov processes with stationary transition probabilities

In general, we call a stochastic process whose index set supports the action of a group (or semigroup) *stationary* with respect to the action of this group (or semigroup) if all finite-dimensional distributions are invariant under the simultaneous shift of all time indices. Specifically, if our index set I is $\mathbb{R}_+$ or $\mathbb{Z}$ or $\mathbb{N}_0$, then a stochastic process is stationary if, for all $\ell \in \mathbb{N}$, $s_1, \dots, s_\ell \in I$, $A_1 \dots, A_\ell \in \mathscr{B}$ and $t \in I$,

$$\mathbb{P}(X_{s_1} \in A_1, \dots, X_{s_\ell} \in A_\ell) = \mathbb{P}(X_{s_1+t} \in A_1, \dots, X_{s_\ell+t} \in A_\ell). \tag{4.1.9}$$

We can express this property also as follows. For $t \in I$, define the process $X \circ \theta_t$ by $(X \circ \theta_t)_s = X_{t+s}$. Then X is stationary if and only if, for all $t \in I$, the processes X and $X \circ \theta_t$ have the same finite-dimensional distributions.

In the case of Markov processes, a necessary (but not sufficient) condition for stationarity is the stationarity of the transition kernels.

Definition 4.4 A Markov process with discrete time $I = \mathbb{N}_0$ and state space S is said to have *stationary transition probabilities* if its one-step transition kernel $\mathscr{P}_t$ is independent of t, i.e., there exists a probability kernel $P(x, A)$ such that

$$\mathscr{P}_t(x, A) = P(x, A) \tag{4.1.10}$$

for all $t \in \mathbb{N}_0$, $x \in S$ and $A \in \mathscr{B}$.

With the notation $P_{s,t}$ for the transition kernel from time s to time t, we can alternatively state that a Markov process has *stationary transition probabilities* if there exists a family of transition kernels $P_t(x, A)$ such that

$$P_{s,t}(x, A) = P_{t-s}(x, A) \tag{4.1.11}$$

for all $s, t \in \mathbb{N}_0$ with $0 \le s < t$, $x \in S$ and $A \in \mathscr{B}$. Note that $\mathscr{P}_t$ and P_t are different objects and should not be confused.

A key concept for Markov processes with stationary transition probabilities is that of an *invariant* distribution and *invariant* measure.

Definition 4.5 Let P be the transition kernel of a Markov process with stationary transition probabilities. Then a probability measure π on $(S, \mathscr{B})$ is called an *invariant distribution* if

$$\int_S \pi(dx) P(x, A) = \pi(A) \tag{4.1.12}$$

for all $A \in \mathscr{B}$. More generally, a positive and σ-finite measure π satisfying (4.1.12) is called an *invariant measure*.

Lemma 4.6 *A Markov process with stationary transition probabilities and initial distribution $p_0 = \pi$ is a stationary stochastic process if and only if π is an invariant distribution.*

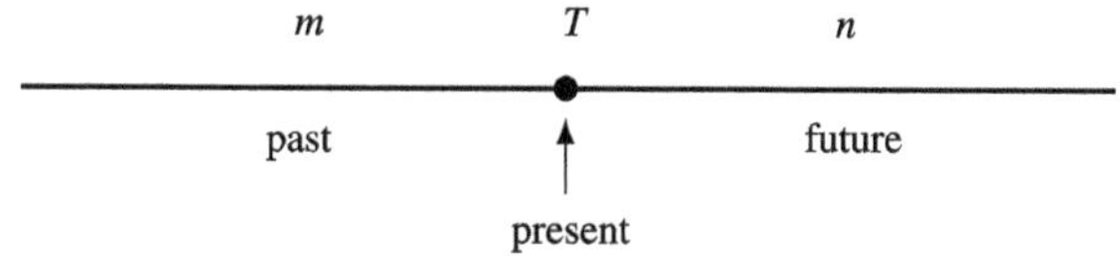

Fig. 4.2 Illustration of the strong Markov property: the future depends on the present not the past, even when the present occurs at a random stopping time T. Recall Fig. 4.1

There always is at least one invariant measure. When S is finite, this invariant measure can be chosen to be a probability measure. However, when S is infinite, it may not be possible to do so.

4.1.3 The strong Markov property

The setting of Markov processes is highly suitable for the application of the notion of stopping times introduced in Chap. 3. In fact, one of the important properties of Markov processes is that we can split the past and the future also at random times (see Fig. 4.2).

Theorem 4.7 *Let X be a Markov process with stationary transition probabilities. The X satisfies the* strong Markov property: *Let $(\mathscr{F}_n)_{n\in\mathbb{N}_0}$ be a filtration to which X is adapted, and let T be a stopping time. Let F and G be $\mathscr{F}$-measurable functions, and in addition let F be measurable with respect to the pre-T-σ-algebra $\mathscr{F}_T$. Then*

$$\mathbb{E}[\mathbb{1}_{T<\infty} F G\circ\theta_T|\mathscr{F}_0] = \mathbb{E}\big[\mathbb{1}_{T<\infty} F \mathbb{E}'\big[G|\mathscr{F}'_0\big](X_T)|\mathscr{F}_0\big], \tag{4.1.13}$$

where $\mathbb{E}'$ refers to an independent copy X' of the Markov chain X.

Proof We have

$$\begin{aligned}\mathbb{E}[\mathbb{1}_{T<\infty} F G\circ\theta_T|\mathscr{F}_0] &= \mathbb{E}\big[\mathbb{E}[\mathbb{1}_{T<\infty} F G\circ\theta_T|\mathscr{F}_T]|\mathscr{F}_0\big]\\ &= \mathbb{E}\big[\mathbb{1}_{T<\infty} F \mathbb{E}[G\circ\theta_T|\mathscr{F}_T]|\mathscr{F}_0\big].\end{aligned} \tag{4.1.14}$$

But $\mathbb{E}[G\circ\theta_T|\mathscr{F}_T]$ depends only on X_T. Moreover, by stationarity $\mathbb{E}[G\circ\theta_T|\mathscr{F}_T] = \mathbb{E}'[G|\mathscr{F}'_0](X_T)$, and so the claim of the theorem follows. □

4.2 Markov processes and martingales

In this section we show how Markov transition kernels can be seen as operators acting on spaces of measures, respectively, spaces of functions.

4.2.1 Semigroups

For μ a σ-finite measure on S and P a Markov transition kernel, we define the measure μP as

$$(\mu P)(A) = \int_S P(x, A) d\mu(x), \tag{4.2.1}$$

and similarly for the t-step transition kernel

$$(\mu P_t)(A) = \int_S P_t(x, A) d\mu(x). \tag{4.2.2}$$

By the Markov property, we have

$$(\mu P_t)(A) = (\mu P^t)(A). \tag{4.2.3}$$

The action on measures has the following natural interpretation in terms of the process: if $\mathbb{P}(X_0 \in A) = \mu(A)$, then

$$\mathbb{P}(X_t \in A) = (\mu P_t)(A). \tag{4.2.4}$$

Alternatively, for f a bounded and measurable function on S, we define

$$(Pf)(x) = \int_S f(y) P(x, dy) \tag{4.2.5}$$

and

$$(P_t f)(x) = \int_S f(y) P_t(x, dy), \tag{4.2.6}$$

where again

$$P_t f = P^t f. \tag{4.2.7}$$

We say that $(P_t)_{t \in \mathbb{N}_0}$ is a semigroup acting on the space of measures, respectively, on the space of bounded measurable functions. The interpretation of the action on functions is as follows.

Lemma 4.8 *Let $(P_t)_{t \in \mathbb{N}_0}$ be a Markov semigroup acting on bounded measurable functions f. Then*

$$(P_t f)(x) = \mathbb{E}\big[f(X_t) | \mathscr{F}_0\big](x) = \mathbb{E}_x\big[f(X_t)\big]. \tag{4.2.8}$$

Proof We need to prove (4.2.8) only for $t = 1$. But, by definition,

$$\mathbb{E}_x\big[f(X_1)\big] = \int_S f(y) \mathbb{P}(X_1 \in dy | \mathscr{F}_0)(x) = \int_S f(y) P(x, dy), \tag{4.2.9}$$

which proves the claim. □

4.2.2 The martingale problem

Note that, by telescopic expansion, we have the elementary formula

$$P_t f - f = \sum_{s=0}^{t-1} P_s(P - \mathbb{1})f = \sum_{s=0}^{t-1} P_s(Lf), \quad t \in \mathbb{N}_0, \tag{4.2.10}$$

where we call $L = P - \mathbb{1}$ the (discrete-time) generator of our Markov process (this formula will turn out to have a complete analogue in the continuous-time setting). An interesting consequence is the following observation, referred to as the *martingale problem* in discrete time.

Lemma 4.9 *Let L be the generator of a Markov process X, and let f be a bounded measurable function. Then*

$$M_t = f(X_t) - f(X_0) - \sum_{s=0}^{t-1} (Lf)(X_s), \quad t \in \mathbb{N}_0, \tag{4.2.11}$$

is a martingale.

Proof Let $t, r \in \mathbb{N}_0$. Then

$$\begin{aligned}
\mathbb{E}[M_{t+r}|\mathscr{F}_t] &= \mathbb{E}\big[f(X_{t+r})|\mathscr{F}_t\big] - \mathbb{E}\big[f(X_0)|\mathscr{F}_t\big] - \sum_{s=0}^{t+r-1} \mathbb{E}\big[(Lf)(X_s)|\mathscr{F}_t\big] \\
&= (P_r f)(X_t) - f(X_t) + f(X_t) - f(X_0) \\
&\quad - \sum_{s=t}^{t+r-1} \mathbb{E}\big[(Lf)(X_s)|\mathscr{F}_t\big] - \sum_{s=0}^{t-1} \mathbb{E}\big[(Lf)(X_s)|\mathscr{F}_t\big] \\
&= f(X_t) - f(X_0) - \sum_{s=0}^{t-1} (Lf)(X_s) \\
&\quad + (P_r f)(X_t) - f(X_t) - \sum_{s=0}^{r-1} (P_r Lf)(X_t) \\
&= M_t + 0,
\end{aligned} \tag{4.2.12}$$

by (4.2.10), which proves the lemma. □

Note that (4.2.11) is the Doob decomposition of the process $(f(X_t))_{t\in\mathbb{N}_0}$ because $(\sum_{s=0}^{t-1}(Lf)(X_s))_{t\in\mathbb{N}_0}$ is a previsible process. Check that this fact follows directly from (3.4.16).

What is important about the latter observation is that it gives rise to a characterisation of the generator that will turn out to be extremely useful in the continuous-time setting. Namely, we can ask whether the requirement that $(M_t)_{t\in\mathbb{N}_0}$ be a martingale given a family of pairs (f, Lf) fully characterises a Markov process.

Theorem 4.10 (Martingale problem) *Let X be a discrete-time stochastic process on a filtered space such that X is adapted. Then X is a Markov process with transition kernel $P = \mathbb{1} + L$ if and only if for all bounded measurable functions f the expression on the right-hand side of* (4.2.11) *is a martingale.*

Proof Lemma 4.9 already provides the "only if" part, so it remains to show the "if" part. First, if we assume that X is a Markov process, then setting $r = 1$ and $t = 0$ in (4.2.12) and taking conditional expectations given $\mathcal{F}_0$, we see from Lemma 4.8 that $\mathbb{E}(f(X_1)) = f(X_0) + (Lf)(X_0)$, which implies that the transition kernel must be $\mathbb{1} + L$.

It remains to show that X is indeed a Markov process. For this we want to show that

$$\mathbb{E}\big[f(X_{t+s})|\mathcal{F}_t\big] = \big((\mathbb{1} + L)^s f\big)(X_t) = \big(P^s f\big)(X_t), \tag{4.2.13}$$

from the martingale problem formulation. To see this, we just use the calculation in (4.2.12) to see that

$$\begin{aligned}
\mathbb{E}\big[f(X_{t+r})|\mathcal{F}_t\big] &= \mathbb{E}[M_{t+r}|\mathcal{F}_t] + f(X_0) \\
&\quad + \sum_{s=0}^{t-1}(Lf)(X_s) + \sum_{s=t}^{t+r-1} \mathbb{E}\big[(Lf)(X_s)|\mathcal{F}_t\big] \\
&= M_t + f(X_0) + \sum_{s=0}^{t-1}(Lf)(X_s) + \sum_{s=t}^{t+r-1} \mathbb{E}\big[(Lf)(X_s)|\mathcal{F}_t\big] \\
&= f(X_t) + \sum_{s=0}^{r-1} \mathbb{E}\big[(Lf)(X_{t+s})|\mathcal{F}_t\big].
\end{aligned} \tag{4.2.14}$$

For $r = 1$,

$$\mathbb{E}\big[f(X_{t+1})|\mathcal{F}_t\big] = f(X_t) + (Lf)(X_t) = \big((\mathbb{1} + L)f\big)(X_t) = (Pf)(X_t), \tag{4.2.15}$$

which is (4.2.13) for $s = 1$. Now proceed by induction: assume that (4.2.13) holds for all bounded measurable functions for $s \le r - 1$. We must show that it then also holds for $s = r$. To do this, we use (4.2.14) for the last sum in (4.2.14),

$$\sum_{s=0}^{r-1} \mathbb{E}\big[(Lf)(X_{t+s})|\mathcal{F}_t\big] = \sum_{s=0}^{r-1}\big(P_s(Lf)\big)(X_t) = \big(P^r f\big)(X_t) - f(X_t), \tag{4.2.16}$$

where we undid the telescoping sum in (4.1.10). Insertion into (4.2.14) yields (4.2.13) for $s = r$. Hence (4.2.13) holds for all r, by induction. □

The full strength of Theorem 4.10 will become apparent in the continuous-time setting, for which it remains valid. A crucial point is that it will not even be necessary to consider all bounded measurable functions: a sufficiently rich class will do. This allows us to formulate martingale problems even when we cannot write down the generator in an explicit form.

4.2.3 Harmonic functions and martingales

We have seen that measures μ satisfying $\mu L = 0$ are of special importance in the theory of Markov processes. Also of central importance are functions f that satisfy $Lf = 0$. In this section we will assume that the transition kernels of our Markov processes have bounded support, i.e., there is a $K < \infty$ such that $|X_{t+1} - X_t| \leq K < \infty$ for all $t \in \mathbb{N}_0$, a.s.

Definition 4.11 Let L be the generator of a Markov process. A measurable function satisfying

$$(Lf)(x) = 0 \quad \forall x \in S \tag{4.2.17}$$

is called a *harmonic function*. A function is called *subharmonic* or *superharmonic* if $Lf \geq 0$, respectively, $Lf \leq 0$.

Theorem 4.12 *Let X be a Markov process with generator L. Then a non-negative function f is:*

(i) *harmonic when $(f(X_t))_{t\in\mathbb{N}_0}$ is a martingale;*
(ii) *subharmonic when $(f(X_t))_{t\in\mathbb{N}_0}$ is a submartingale;*
(iii) *superharmonic when $(f(X_t))_{t\in\mathbb{N}_0}$ is a supermartingale.*

Proof Simply use Lemma 4.9. □

Theorem 4.12 establishes a profound relationship between potential theory and martingales. This link will be developed further in Chap. 5.

A nice application of Theorem 4.12 is the *maximum principle*.

Theorem 4.13 (Maximum principle) *Let X be a Markov process and let D be a bounded open domain such that $\mathbb{E}[\tau_{D^c}] < \infty$. Assume that f is a non-negative subharmonic function on D. Then*

$$\sup_{x\in D} f(x) \leq \sup_{x\in D^c} f(x). \tag{4.2.18}$$

Proof Define $T = \tau_{D^c}$. Then

$$\mathbb{E}\big[f(X_T)|\mathcal{F}_0\big](x) \geq f(x). \tag{4.2.19}$$

Since $X_T \in D^c$, it must be true that

$$\sup_{y \in D_c} f(y) \geq \mathbb{E}\big[f(X_T)|\mathcal{F}_0\big](x) \geq f(x) \quad \forall x \in D, \tag{4.2.20}$$

which proves the claim. In (4.2.19) we again used the Doob optional stopping theorem, (Theorem 3.67(i, b)). □

Theorem 4.13 can be phrased as saying that (sub)harmonic functions take their maximum on the boundary (since the set D^c in (4.2.18) can be replaced by the subset $\partial D \subset D^c$ such that $\mathbb{P}_x(X_T \in \partial D) = 1$ for all $x \in D$). The above proof is an example of how intrinsically analytic results can be proven with the help of probabilistic arguments. The next section will further develop this theme.

4.2.4 The Doob transform

Let us consider a discrete-time Markov process X with generator $P - \mathbb{1}$. We may want to consider modifications of the process obtained by conditioning on certain events to occur. One type of conditioning is to reach some specific set in a particular location. For instance, consider a random walk on a finite interval conditioned to exit the interval on a specific side. When and how can we do this, and what is the nature of the resulting process? In particular, is the resulting process again a Markov process and, if so, what is its generator?

As an example, let us condition a Markov process to hit a domain B for the first time in a subset $A \subset B$. We may assume that $\mathbb{E}[\tau_B] < \infty$. Define $h(x) = \mathbb{P}_x[\tau_A = \tau_B]$, $x \notin B$, and note that h is harmonic. Let $\mathbb{P}$ be the law of X. Let us define a new law $\mathbb{P}^h$ on the space of paths as follows: if Y is a $\mathcal{F}_t$-measurable random variable, then

$$\mathbb{E}^h[Y|\mathcal{F}_0] = \frac{1}{h(X_0)}\mathbb{E}\big[h(X_t)Y|\mathcal{F}_0\big]. \tag{4.2.21}$$

Lemma 4.14 *With the notation above, if Y is a $\mathcal{F}_{\tau_B - 1}$-measurable function, then*

$$\mathbb{E}^h_x[Y] = \mathbb{E}_x[Y|\tau_A = \tau_B]. \tag{4.2.22}$$

Proof This is an application of the strong Markov property. We have

$$\begin{aligned}
\mathbb{E}^h_x[Y] &= \frac{1}{h(x)}\mathbb{E}_x\big[Yh(X_{\tau_B - 1})\big] = \frac{1}{h(x)}\mathbb{E}_x\big[Y\mathbb{E}'\big[\mathbb{1}_{\tau_A = \tau_B}|\mathcal{F}'_0\big](X_{\tau_B - 1})\big] \\
&= \frac{1}{h(x)}\mathbb{E}_x\big[Y\mathbb{E}[\mathbb{1}_{\tau_A = \tau_B}|\mathcal{F}_{\tau_B - 1}]\big] = \frac{1}{\mathbb{P}_x(\tau_A = \tau_B)}\mathbb{E}_x[Y\mathbb{1}_{\tau_A = \tau_B}] \\
&= \mathbb{E}_x[Y|\tau_A = \tau_B].
\end{aligned} \tag{4.2.23}$$

Here, the first equality is just the definition of h and reproduces the form of the right-hand side of the strong Markov property, the second equality is the strong Markov property (recall Theorem 4.7), while the fourth equality uses the fact that the event $\{\tau_A = \tau_B\}$ depends only on what happens after $\tau_B - 1$, and so $\mathbb{1}_{\tau_A=\tau_B}\theta_{\tau_B-1} = \mathbb{1}_{\tau_A=\tau_B}$. □

Let us next look at the transformed law $\mathbb{P}^h$ in the general case. The first property to check is whether $\mathbb{P}^h$ is defined in a consistent way. Some thought shows that it suffices to prove the following lemma.

Lemma 4.15 *Let Y be $\mathscr{F}_s$-measurable. Then, for any $t \geq s \geq 0$,*

$$\mathbb{E}^h[Y|\mathscr{F}_0] = \frac{1}{h(X_0)}\mathbb{E}\big[h(X_s)Y|\mathscr{F}_0\big] = \frac{1}{h(X_0)}\mathbb{E}\big[h(X_t)Y|\mathscr{F}_0\big]. \tag{4.2.24}$$

In particular, $\mathbb{P}^h[\Omega|\mathscr{F}_0] = 1$.

Proof Just introduce the conditional expectation

$$\mathbb{E}\big[h(X_t)Y|\mathscr{F}_0\big] = \mathbb{E}\big[\mathbb{E}\big[h(X_t)Y|\mathscr{F}_s\big]|\mathscr{F}_0\big] = \mathbb{E}\big[Y\mathbb{E}\big[h(X_t)|\mathscr{F}_s\big]|\mathscr{F}_0\big], \tag{4.2.25}$$

and use that $(h(X_t))_{t\in\mathbb{N}_0}$ is a martingale by Theorem 4.12, to get

$$\mathbb{E}\big[h(X_t)Y|\mathscr{F}_0\big] = \mathbb{E}\big[h(X_s)Y|\mathscr{F}_0\big], \tag{4.2.26}$$

from which the claim follows. □

Lemma 4.15 shows, in particular, why it is important that h be a harmonic function.

Next, we ask the question whether the law $\mathbb{P}^h$ is a Markov chain. To this end we turn to the martingale problem. We will show that there exists a generator L^h such that

$$M_t^h = f(X_t) - f(X_0) - \sum_{s=0}^{t-1}(L^h f)(X_s), \quad t \in \mathbb{N}_0, \tag{4.2.27}$$

is a martingale under the law $\mathbb{E}^h$, i.e., we show that

$$\mathbb{E}^h\big[M_t^h|\mathscr{F}_r\big] = M_r^h, \quad t > r \geq 0. \tag{4.2.28}$$

First note that, by (4.2.21),

$$\begin{aligned}\mathbb{E}^h\big[M_t^h|\mathscr{F}_r\big] = {} & \frac{1}{h(X_r)}\mathbb{E}\big[h(X_t)f(X_t)|\mathscr{F}_r\big] - f(X_0) - \sum_{s=0}^{r-1}(L^h f)(X_s) \\ & - \sum_{s=r}^{t-1}\frac{1}{h(X_r)}\mathbb{E}\big[h(X_s)(L^h f)(X_s)|\mathscr{F}_r\big]. \end{aligned} \tag{4.2.29}$$

The two middle terms are part of M_r^h and so we must compute $\mathbb{E}[f(X_t)h(X_t)|\mathscr{F}_r]$. This is done by applying Lemma 4.9 for the law $\mathbb{P}$ and the function fh, which yields

$$\mathbb{E}\big[f(X_t)h(X_t)|\mathscr{F}_r\big] = f(X_r)h(X_r) + \sum_{s=r}^{t-1} \mathbb{E}\big[\big(L(fh)\big)(X_s)|\mathscr{F}_r\big]. \tag{4.2.30}$$

Inserting this into (4.2.29), we obtain

$$\begin{aligned} \mathbb{E}^h\big[M_t^h|\mathscr{F}_r\big] &= f(X_r) - f(X_0) - \sum_{s=0}^{r-1}\big(L^h f\big)(X_s) \\ &\quad + \frac{1}{h(X_r)} \sum_{s=r}^{t-1} \big[\mathbb{E}\big[\big(L(fh)\big)(X_s)|\mathscr{F}_r\big] - \mathbb{E}\big[h(X_s)\big(L^h f\big)(X_s)|\mathscr{F}_r\big]\big] \\ &= M_r^h \\ &\quad + \frac{1}{h(X_r)} \sum_{s=r}^{t-1} \big[\mathbb{E}\big[\big(L(fh)\big)(X_s)|\mathscr{F}_r\big] - \mathbb{E}\big[h(X_s)\big(L^h f\big)(X_s)|\mathscr{F}_r\big]\big]. \end{aligned} \tag{4.2.31}$$

The second term will vanish if we choose L^h such that

$$(Lf)(x) = h(x)^{-1}\big(L(hf)\big)(x),$$

i.e.,

$$L^h f(x) = \frac{1}{h(x)} \int_S P(x, dy)h(y)f(y) - f(x). \tag{4.2.32}$$

Thus we see that, under $\mathbb{P}^h$, X solves the martingale problem corresponding to the generator L^h, and hence is a Markov chain with transition kernel $P^h = L^h + \mathbb{1}$. The process X under $\mathbb{P}^h$ is called the (Doob) h-transform of the original Markov process.

4.3 Markov processes with countable state space

In this section we provide some results on Markov processes with countable state space, in particular, we introduce the notions of recurrence and transience, and discuss the existence and uniqueness of invariant distributions.

It will be useful to adopt a notation that is close to the matrix notation of finite Markov processes, namely,

$$P\big(i, \{j\}\big) = p(i, j), \tag{4.3.1}$$

which is called the *transition matrix*. We start with the following elementary concepts.

Definition 4.16 Two states $i, j \in S$ the state space of a Markov process *communicate* if there exist $k, k' \geq 0$ such that $(p^k)_{ij} > 0$ and $(p^{k'})_{ji} > 0$.

It is easy to see that *communication* is an equivalence relation. We call the equivalence classes *communicating* classes.

Definition 4.17 A Markov process with countable state space is called *irreducible* if and only if its state space forms a single communicating class.

Another concept of importance is *periodicity*.

Definition 4.18 A state $i \in S$ of a Markov chain has period $d(i)$ if and only if $d(i)$ is the largest common divisor of all numbers $n \in \mathbb{N}$ such that $(P^n)_{i,i} > 0$. A state of period 1 is called *aperiodic*.

It is easy to see that the function d is constant on communicating classes. We will henceforth place ourselves in the setting of irreducible Markov processes. In a countably infinite state space the following notions are fundamental.

Definition 4.19 Let X be an irreducible Markov process with countable state space S.

(i) X is called *transient* if

$$\mathbb{P}_i(\tau_i < \infty) < 1 \quad \forall i \in S. \tag{4.3.2}$$

(ii) X is called *recurrent* if it is not transient.
(iii) X is called *positive recurrent* if

$$\mathbb{E}_i[T_i] < \infty \quad \forall i \in S. \tag{4.3.3}$$

(iv) X is called *ergodic* if it is irreducible, positive recurrent and aperiodic.

The notions of recurrence and transience can be defined for single states rather than for the entire process. In the case of irreducible Markov processes, all states have the same characteristics.

Some simple consequences of the above definition are the following.

Lemma 4.20 *Let X be an irreducible Markov process with a countable state space S. Then X is transient if and only if*

$$\mathbb{P}_\ell(X_t = \ell \text{ infinitely often}) = 0 \quad \forall \ell \in S. \tag{4.3.4}$$

Positive recurrent Markov processes possess a unique invariant probability distribution.

Lemma 4.21 *Let X be an irreducible positive recurrent Markov process with a countable state space S. Then*

$$\mu(j) = \frac{\mathbb{E}_\ell[\sum_{t=1}^{\tau_\ell} \mathbb{1}_{X_t=j}]}{\mathbb{E}_\ell[\tau_\ell]} \quad \forall j, \ell \in S, \tag{4.3.5}$$

is the unique invariant probability distribution of X.

Proof Define $\nu_\ell(j) = \mathbb{E}_\ell[\sum_{t=1}^{\tau_\ell} \mathbb{1}_{X_t=j}]$. We first show that ν is an invariant measure. Indeed, $1 = \sum_{m\in S} \mathbb{1}_{X_{t-1}=m}$, and hence, by the strong Markov property,

$$\begin{aligned}
\nu_\ell(j) = \mathbb{E}_\ell\left[\sum_{t=1}^{\tau_\ell} \mathbb{1}_{X_t=j}\right] &= \mathbb{E}_\ell\left[\sum_{m\in S}\sum_{t=1}^{\tau_\ell} \mathbb{1}_{X_t=j}\mathbb{1}_{X_{t-1}=m}\right] \\
&= \sum_{m\in S} \mathbb{E}_\ell\left[\sum_{t=1}^{\tau_\ell} \mathbb{1}_{X_{t-1}=m}\mathbb{P}(X_t = j|\mathcal{F}_{t-1})(m)\right] \\
&= \sum_{m\in S} \mathbb{E}_\ell\left[\sum_{t=1}^{\tau_\ell} \mathbb{1}_{X_{t-1}=m}\right] p(m, j) \\
&= \sum_{m\in S} \mathbb{E}_\ell\left[\sum_{t=1}^{\tau_\ell} \mathbb{1}_{X_t=m}\right] p(m, j) \\
&= \sum_{m\in S} \nu_\ell(m) p(m, j).
\end{aligned} \tag{4.3.6}$$

Thus, ν_ℓ solves the equation for the invariant measure. It remains to show that ν_ℓ is normalisable. However,

$$\sum_{j\in S} \nu_\ell(j) = \mathbb{E}_\ell[\tau_\ell] < \infty, \tag{4.3.7}$$

by assumption. Hence $\nu_\ell(j)/\sum_{i\in S}\nu_\ell(i) = \mu(j)$ is an invariant probability distribution.

We next show uniqueness. First, note that for any irreducible Markov process with countable state space, if ν is an invariant measure and $\nu(i) = 0$ for some $i \in S$, then $\nu = 0$. Indeed, if $\nu(j) > 0$ for some j, then there exists a finite t such that $P^t_{ji} > 0$, and so $\nu(i) \geq \nu(j)P^t_{ji} > 0$, which is a contradiction. Next, note that $\nu_\ell(\ell) = 1$. We can actually show that ν_ℓ is the only invariant measure such that $\nu_\ell(\ell) = 1$, which implies the desired uniqueness result as follows. Below we show that $\nu(j) \geq \nu_\ell(j)$ for all $j \in S$ for any other invariant measure ν such that $\nu(\ell) = 1$. Since $\nu - \nu_\ell$ is a positive invariant measure as well and is zero at ℓ, it must vanish identically, which implies that $\nu = \nu_\ell$.

We have

$$\nu(i) = \sum_{j_1\neq\ell} p(j_1, i)\nu(j) + p(\ell, i), \tag{4.3.8}$$

since $\nu(\ell) = 1$ by hypothesis. Write

$$p(\ell, i) = \mathbb{E}_\ell[\mathbb{1}_{\tau_\ell \geq 1} \mathbb{1}_{X_1 = i}]. \tag{4.3.9}$$

Iterate (4.3.8) to get

$$\begin{aligned}
\nu(i) &= \sum_{j_1, j_2 \neq \ell} p(j_2, j_1) p(j_1, i) \nu(j_2) + \sum_{j_1 \neq \ell} p(\ell, j_1) p(j_1, i) + \mathbb{E}_\ell[\mathbb{1}_{\tau_\ell \geq 1} \mathbb{1}_{X_1 = i}] \\
&= \sum_{j_1, j_2 \neq \ell} p(j_2, j_1) p(j_1, i) \nu(j_2) + \mathbb{E}_\ell\left[\sum_{t=1}^{2 \wedge \tau_\ell} \mathbb{1}_{X_t = i}\right].
\end{aligned} \tag{4.3.10}$$

Further iteration yields, for any $n \in \mathbb{N}$,

$$\begin{aligned}
\nu(i) &= \sum_{j_1, j_2, \dots j_n \neq \ell} p(j_n, j_{n-1}) \dots p(j_2, j_1) p(j_1, i) \nu(j_n) + \mathbb{E}_\ell\left[\sum_{t=1}^{n \wedge \tau_\ell} \mathbb{1}_{X_t = i}\right] \\
&\geq \mathbb{E}_\ell\left[\sum_{t=1}^{n \wedge \tau_\ell} \mathbb{1}_{X_t = i}\right].
\end{aligned} \tag{4.3.11}$$

Let $n \to \infty$ and use (4.3.7) to get $\nu(i) \geq \nu_\ell(i)$. □

Corollary 4.22 *For an irreducible positive recurrent Markov process with countable state space S,*

$$\mu(j) = \frac{1}{\mathbb{E}_j[\tau_j]}, \quad j \in S. \tag{4.3.12}$$

Proof Just set $\ell = j$ in the definition of $\mu(j)$, and note that $\nu_j(j) = \mathbb{E}_j[\sum_{t=1}^{\tau_j} \mathbb{1}_{X_t = x}] = 1$. □

The invariant distribution determines the long-time behaviour of the Markov process. We state the following two *ergodic theorems* without proof.

Theorem 4.23 (Ergodic theorem; strong form) *Let X be an irreducible positive recurrent Markov process with invariant probability distribution μ. Then, for any bounded measurable function $f : S \to \mathbb{R}$,*

$$\lim_{n \to \infty} \frac{1}{n} \sum_{k=1}^{n} f(X_k) = \int_S f \, d\mu \quad a.s. \tag{4.3.13}$$

Theorem 4.24 (Ergodic theorem; weak form) *Let X be an irreducible aperiodic and positive recurrent Markov process with transition matrix P and invariant probability distribution μ. Then, for any initial law π_0,*

$$\lim_{n \to \infty} \left(\pi_0 P^n\right)_i = \mu(i), \quad i \in S. \tag{4.3.14}$$

4.4 Bibliographical notes

1. There are many good textbooks on Markov processes. Nice modern treatments can be found in Norris [195], Stroock [222], or Levin, Peres and Wilmer [163]. Two classic texts are those by Kemeny and Snell [151], or Kemeny, Snell and Knapp [150].

2. The idea of characterising Markov processes by an associated martingale problem goes back to Stroock and Varadhan [223].

Chapter 5
Markov Processes in Continuous Time

> *"Why cannot you explain the process?" he inquired. Mein Herr was ready with a quite unanswerable reason. "Because you have no words, in your language, to convey the ideas that are needed. I could explain it in – in – but you would not understand it!"* (Lewis Carroll, Sylvie and Bruno Concluded)

In this chapter we review some important aspects of Markov processes in continuous time. Their study involves much more analytic work than in the discrete-time setting. However, there is also a lot of common structure. The basic definition of a Markov process was already given in Chap. 4 (see Definition 4.1). Clearly, one-step transition kernels no longer make sense and we need to look for the appropriate analogue of the *generator* of the process. In this chapter we will only consider the case of Markov processes with stationary transition kernels.

Section 5.1 takes a brief look at Markov jump processes. Section 5.2 lists a few basic properties of Brownian motion. Section 5.3 gives the definition of general Markov processes via generators and semigroups, and focusses on a special class called Feller-Dynkin processes, emphasising the central rôle of the strong Markov property. Section 5.4 introduces and studies the so-called martingale problem, which is a powerful way to construct general Markov processes, and addresses the issues of existence and uniqueness. Section 5.5 gives a brief summary on Itō calculus, preparing for Sect. 5.6 that introduces and studies stochastic differential equations, states existence and uniqueness criteria for strong solutions, and presents the Doob h-transform in this setting. Section 5.7 looks at stochastic partial differential equations.

5.1 Markov jump processes

The simplest class of continuous-time Markov processes are *Markov jump processes*. They are constructed "explicitly" from Markov processes in discrete time. The idea is simple: take a discrete-time Markov process and randomise the waiting times between the successive moves in such a way as to obtain a continuous-time Markov process.

A. Bovier, F. den Hollander, *Metastability*,
Grundlehren der mathematischen Wissenschaften 351,
DOI 10.1007/978-3-319-24777-9_5

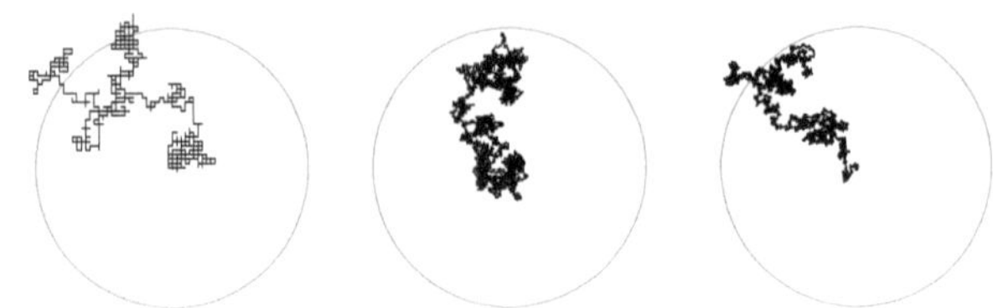

Fig. 5.1 Simulation of simple random walk on $\mathbb{Z}^2$ with $n = 10^3$, 10^4 and 10^5 steps. The circles have radius $n^{1/2}$ in units of the step size. Brownian motion on $\mathbb{R}^2$ is the continuum limit of simple random walk on $\mathbb{Z}^2$. (Courtesy Bill Casselman and Gordon Slade)

To be more precise, let $(Y_n)_{n\in\mathbb{N}_0}$, be a discrete-time Markov process with state space S, transition kernel P (also called jump distribution) and initial distribution μ. Let $m\colon\ S \to \mathbb{R}_+$ be a uniformly bounded and measurable function. Let $e_{i,x}$, $i \in \mathbb{N}_0$, $x \in S$, be a family of independent exponential random variables with mean $m(x)$, defined on the same probability space $(\Omega, \mathscr{F}, \mathbb{P})$ as $(Y_n)_{n\in\mathbb{N}_0}$, and assume that the Y_n and the $e_{i,x}$ are mutually independent. Define

$$S(n) = \sum_{i=0}^{n-1} e_{i,Y_i}, \qquad n \in \mathbb{N}_0, \tag{5.1.1}$$

which is called the *clock process*: $S(n)$ represents the time at which the n-th jump takes place. Define the inverse function

$$S^{-1}(t) = \sup\big\{n \in \mathbb{N}_0\colon\ S(n) \le t\big\}, \qquad t \in \mathbb{R}_+, \tag{5.1.2}$$

and set

$$X(t) = Y_{S^{-1}(t)}. \tag{5.1.3}$$

Theorem 5.1 *The process $(X(t))_{t\in\mathbb{R}_+}$ defined through* (5.1.3) *is a continuous-time Markov process with càdlàg paths.*

5.2 Brownian motion

Markov jump processes are not the most general continuous-time Markov processes. For instance, stochastic processes with continuous paths, such as diffusions, are excluded. The simplest and most important example is *Brownian motion*. In this section we give a brief recapitulation of its basic properties.

The main reason why Brownian motion is so important is that it arises as the universal limit of a large class of *discrete-time* processes, in particular, *random walks*, (see Fig. 5.1)

$$S_n = \sum_{k=1}^{n} X_k, \qquad n \in \mathbb{N}_0, \tag{5.2.1}$$

with X_k, $k \in \mathbb{N}$, i.i.d. random variables. Let us focus on the centred case: $\mathbb{E}(X_1) = 0$. The central limit theorem states that $Z_n = n^{-1/2} S_n$ converges in distribution to a Gaussian random variable, provided $\mathbb{E}[X_1^2] = \sigma^2 < \infty$. A natural question that goes beyond this observation is to ask whether the entire path $\{S_n, n \in \mathbb{N}\}$ converges to a limiting object. It is clear that if we rescale like

$$Z_n(t) = (n\sigma^2)^{-1/2} \sum_{k=1}^{\lfloor tn \rfloor} X_k, \quad t \in (0, 1], \tag{5.2.2}$$

then

$$Z_n(t) \xrightarrow{\mathscr{D}} B_t, \quad t \in (0, 1], \tag{5.2.3}$$

where B_t is a centred Gaussian random variable with variance t. Moreover, for $\ell \in \mathbb{N}$ and a finite collection of indices $0 = t_0 < t_1 < \cdots < t_\ell$, define $Y_n(i) = Z_n(t_i) - Z_n(t_{i-1})$. Then the random variables $Y_n(i)$ are independent, and it is easy to see that they jointly converge, as $n \to \infty$, to a family of independent, centred Gaussian random variables with variances $t_i - t_{i-1}$. This implies that the finite-dimensional distributions of the processes $(Z_n(t))_{t \in [0,1]}$ converge to the finite-dimensional distributions of the Gaussian process, $(B_t)_{t \in [0,1]}$, with covariance $\mathbb{E}[B_s B_t] = s \wedge t$. The latter is called Brownian motion and has very interesting properties.

5.2.1 Definition of Brownian motion

Definition 5.2 An $\mathbb{R}$-valued stochastic process $(B_t)_{t \in \mathbb{R}_+}$ defined on a probability space $(\Omega, \mathscr{F}, \mathbb{P})$ is called a 1-dimensional Brownian motion starting in 0 if and only if

(o) $B_0 = 0$ a.s.
(i) For any $p \in \mathbb{N}$ and any $0 = t_0 < t_1 < \cdots < t_p < \infty$, the random variables $B_{t_1}, B_{t_2} - B_{t_1}, \ldots, B_{t_p} - B_{t_{p-1}}$ are independent and each $B_{t_i} - B_{t_{i-1}}$ is a centred Gaussian random variable with variance $t_i - t_{i-1}$.
(ii) For any $\omega \in \Omega$, the map $t \mapsto B_t(\omega)$ is continuous (i.e., $B(\omega)\colon \mathbb{R}_+ \to \mathbb{R}$ is a continuous function).

Alternatively, we can describe Brownian motion as follows.

Lemma 5.3 *Brownian motion in* 1 *dimension is the Gaussian process* $(B_t)_{t \in \mathbb{R}_+}$ *with values in* $\mathbb{R}$ *such that*

(o) $B_0 = 0$.
(i) *For any* $t \in \mathbb{R}_+$, $\mathbb{E}[B_t] = 0$.
(ii) *For any* $t, s \geq 0$, $\mathbb{E}[B_t B_s] = t \wedge s$.
(iii) *For any* $\omega \in \Omega$, *the map* $t \mapsto B_t(\omega)$ *is continuous.*

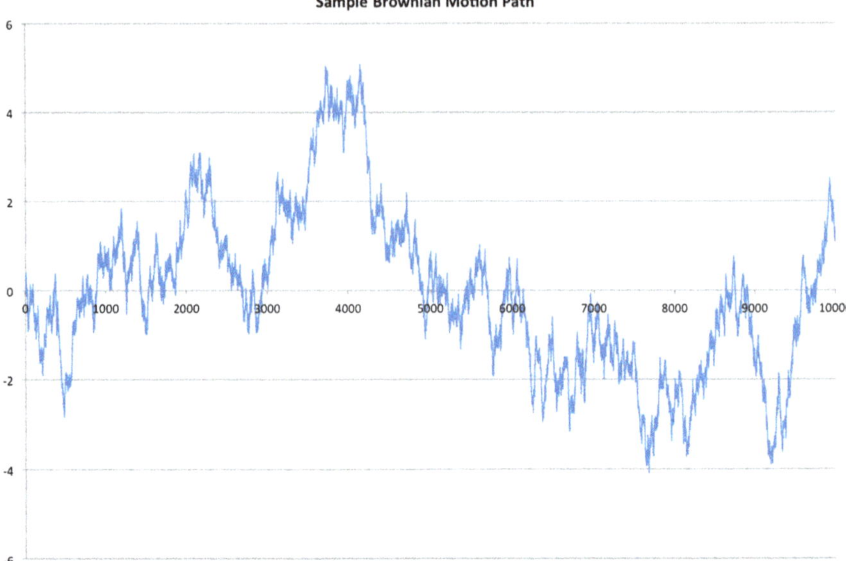

Fig. 5.2 A sample of a Brownian motion path in 1 dimension

Proof Let B be Brownian motion as defined in Definition 5.2. Then properties (o), (i) and (iii) are obviously satisfied. To show that (ii) holds, assume without loss of generality that $t > s$. Then

$$\mathbb{E}[B_t B_s] = \mathbb{E}\big[(B_t - B_s)B_s + B_s^2\big] = 0 + s = t \wedge s, \tag{5.2.4}$$

where we use that $B_t - B_s$ and B_s are independent and centred, and B_s has variance s.

To prove the converse, i.e., to prove that any stochastic process with the above properties is Brownian motion, we can simply use the fact that the law of a Gaussian process is uniquely determined by its mean and its covariance. Therefore the stochastic process has the same law as Brownian motion (see Fig. 5.2), and has continuous paths by property (iii), so it is Brownian motion. □

Once we have Brownian motion in dimension 1, we can trivially define Brownian motion in dimension d.

Definition 5.4 Brownian motion in d dimensions, written $B = (B^{(1)}, \ldots, B^{(d)})$, is a stochastic process indexed by $\mathbb{R}_+$ with values in $\mathbb{R}^d$ such that the components $B^{(i)}$ are mutually independent Brownian motions in $\mathbb{R}$.

If B is Brownian motion in $\mathbb{R}^d$ and $a \in \mathbb{R}^d$, then the process $a + B$ is called *Brownian motion started in a*.

Theorem 5.5 *Brownian motion exists.*

There are a number of ways to prove the existence of Brownian motion, and we refer the reader to the literature for proofs. In a way, the most appealing proof is via Donsker's theorem, which constructs Brownian motion as the limit of sums of i.i.d. random variables via an interpolated version of (5.2.2).

Having constructed the random variable $(B_t)_{t\in\mathbb{R}_+}$ in $C(\mathbb{R}_+,\mathbb{R}^d)$, we can define its distribution, the so-called *Wiener measure*. We want to construct this as a measure on the space of continuous functions equipped with its Borel σ-algebra. For this it is useful to observe the following.

Lemma 5.6 *The smallest σ-algebra $\mathscr{C}$ on $C(\mathbb{R}_+,\mathbb{R}^d)$ that makes all the coordinate functions $t\mapsto w(t)$ measurable coincides with the Borel-σ-algebra $\mathscr{B}=\mathscr{B}(C(\mathbb{R}_+,\mathbb{R}^d))$ of the metrisable space $C(\mathbb{R}_+,\mathbb{R}^d)$ equipped with the topology of uniform convergence on compact sets.*

Proof First, $\mathscr{C}\subset\mathscr{B}$ because all functions $t\mapsto w(t)$ are continuous and hence measurable with respect to the Borel-σ-algebra $\mathscr{B}$. To prove that $\mathscr{B}\subset\mathscr{C}$, note that the topology of uniform convergence is equivalent to the metric topology relative to the metric

$$d\big(w,w'\big)=\sum_{n\in\mathbb{N}}2^{-n}\sup_{0\leq t\leq n}\big(\big|w(t)-w'(t)\big|\wedge 1\big),\qquad w,w'\in C\big(\mathbb{R}_+,\mathbb{R}^d\big). \tag{5.2.5}$$

We thus have to show that any ball with respect to this distance is measurable with respect to $\mathscr{C}$. But since w,w' are continuous functions, we have

$$\sup_{t\in[0,n]}\big(\big|w(t)-w'(t)\big|\wedge 1\big)=\sup_{t\in[0,n]\cap\mathbb{Q}}\big(\big|w(t)-w'(t)\big|\wedge 1\big), \tag{5.2.6}$$

and so we see that balls are indeed in $\mathscr{C}$. □

Note that, by construction, the map $\omega\mapsto B(\omega)$ is measurable because the maps $\omega\mapsto B_t(\omega)$ are measurable for all t, and by the definition of $\mathscr{C}$ the coordinate maps $B\mapsto B_t$ are measurable for all t. Therefore the following definition makes sense.

Definition 5.7 Let $(B_t)_{t\in\mathbb{R}_+}$ be a Brownian motion in $\mathbb{R}^d$ defined on a probability space $(\Omega,\mathscr{F},\mathbb{P})$. The probability measure on $(C(\mathbb{R}_+,\mathbb{R}^d),\mathscr{B}(C(\mathbb{R}_+,\mathbb{R}^d)))$ given as the image of $\mathbb{P}$ under the map $\omega\mapsto(B_t(\omega))_{t\in\mathbb{R}_+}$ is called the *d-dimensional Wiener measure*.

Note that uniqueness of the Wiener measure is an immediate consequence of the Daniell-Kolmogorov theorem (Theorem 3.38), since we already know that the finite-dimensional distributions are fixed by the prescription of the covariances.

An important property of Brownian motion is the following scale invariance.

Lemma 5.8 *For any $a\in\mathbb{R}_+$, the processes $B=(B_t)_{t\in\mathbb{R}_+}$ and $A=(A_t)_{t\in\mathbb{R}_+}$ with $A_t=a^{-1}B_{ta^2}$ have the same distribution.*

Proof Obviously, A is a Gaussian process. It is also obvious that the time change and the multiplication by a constant preserve the continuity of the paths, the starting position 0 and the fact that the process has mean zero. Thus, it suffices to show that B and A have the same covariance. But

$$\mathbb{E}[A_t A_s] = a^{-2}\mathbb{E}[B_{a^2 t} B_{a^2 s}] = a^{-2}\big[(a^2 t) \wedge (a^2 s)\big] = s \wedge t, \tag{5.2.7}$$

which is the covariance of B. □

5.2.2 Martingale and Markov properties

Brownian motion is a martingale. The proof of this fact is elementary.

Theorem 5.9 *Brownian motion defined on a probability space $(\Omega, \mathscr{F}, \mathbb{P})$ is a continuous-time martingale, in the sense that, if $(\mathscr{F}_t)_{t\in\mathbb{R}_+}$ is a filtration of $\mathscr{F}$ such that B is adapted, then*

$$\mathbb{E}[B_t | \mathscr{F}_s] = B_s, \quad 0 \le s < t. \tag{5.2.8}$$

Next, we show that Brownian motion is also a Markov process. For the definition of a continuous-time Markov process, we use the obvious generalisation of a discrete-time Markov processes.

Definition 5.10 A stochastic process X with state space S and index set $\mathbb{R}_+$ is called a *continuous-time Markov process* if there exists a two-parameter family of probability kernels $P_{s,t}$ satisfying the Chapman-Kolmogorov equations

$$P_{s,t}(x, A) = \int_S P_{r,t}(y, A) P_{s,r}(x, dy), \quad \forall r \in (s,t),\ A \in \mathscr{B}, \tag{5.2.9}$$

such that, for all $A \in \mathscr{B}$ and $0 \le s < t$,

$$\mathbb{P}(X_t \in A \mid \mathscr{F}_s)(\omega) = P_{s,t}\big(X_s(\omega), A\big) \quad \text{a.s.} \tag{5.2.10}$$

Theorem 5.11 *Brownian motion in dimension d is a continuous-time Markov process with transition kernel*

$$P_{s,t}(x, A) = \frac{1}{(2\pi(t-s))^{d/2}} \int_A \exp\left(-\frac{\|y-x\|^2}{2(t-s)}\right) dy. \tag{5.2.11}$$

The proof is left as an exercise.

We now come, again somewhat informally, to the martingale problem associated with Brownian motion.

Theorem 5.12 (Martingale problem) *Let $f \in C^2(\mathbb{R}_+, \mathbb{R}^d)$ with bounded second derivatives. Let B be Brownian motion and Δ the Laplacian. Then*

$$M_t = f(B_t) - f(B_0) - \frac{1}{2}\int_0^t (\Delta f)(B_s)ds \tag{5.2.12}$$

is a martingale.

Proof For simplicity we only consider the case $d = 1$ (the general case works the same way). We proceed as in the discrete-time case:

$$\begin{aligned}
\mathbb{E}[M_{t+r} \mid \mathscr{F}_t] &= f(B_t) - f(B_0) - \frac{1}{2}\int_0^t f''(B_s)ds \\
&\quad + \mathbb{E}\big[f(B_{t+r}) - f(B_t) \mid \mathscr{F}_t\big] - \frac{1}{2}\int_0^r \mathbb{E}\big[f''(B_{t+s}) \mid \mathscr{F}_t\big]ds \\
&= M_t + \frac{1}{\sqrt{2\pi r}}\int_{\mathbb{R}} f(y)\exp\left(-\frac{(y-B_t)^2}{2r}\right)dy - f(B_t) \\
&\quad - \frac{1}{2}\int_0^r \frac{1}{\sqrt{2\pi s}}\left[\int_{\mathbb{R}} f''(y)\exp\left(-\frac{(y-B_t)^2}{2s}\right)dy\right]ds.
\end{aligned} \tag{5.2.13}$$

The last inequality holds since, via integration by parts,

$$\begin{aligned}
&\frac{1}{\sqrt{2\pi s}}\int_{\mathbb{R}} f''(y)\exp\left(-\frac{(y-x)^2}{2s}\right)dy \\
&= \int_{\mathbb{R}} f(y)\frac{d^2}{dy^2}\left[\frac{1}{\sqrt{2\pi s}}\exp\left(-\frac{(y-x)^2}{2s}\right)\right]dy \\
&= \frac{1}{\sqrt{2\pi}}\int_{\mathbb{R}} f(y)[-s^{-3/2} + (y-x)^2 s^{-5/2}]\exp\left(-\frac{(y-x)^2}{2s}\right)dy \\
&= 2\int_{\mathbb{R}} f(y)\frac{d}{ds}\left[\frac{1}{\sqrt{2\pi s}}\exp\left(-\frac{(y-x)^2}{2s}\right)\right]dy.
\end{aligned} \tag{5.2.14}$$

Integrating the last expression in (5.2.14) over s, from 0 to r, we get

$$\frac{2}{\sqrt{2\pi r}}\int_{\mathbb{R}} f(y)\exp\left(-\frac{(x-y)^2}{2r}\right)dy - 2f(x), \tag{5.2.15}$$

where we use that

$$\lim_{h\downarrow 0}\frac{1}{\sqrt{2\pi h}}\int_{\mathbb{R}} f(y)\exp\left(-\frac{(x-y)^2}{2h}\right)dy = f(x). \tag{5.2.16}$$

Inserting (5.2.14)–(5.2.15) into (5.2.13), we get $\mathbb{E}[M_{t+r} \mid \mathscr{F}_t] = M_t$, which concludes the proof. □

Note that we really only used that the function

$$e(x,t) = \frac{1}{\sqrt{2\pi t}} \exp\left(-\frac{\|x\|^2}{2t}\right) \tag{5.2.17}$$

satisfies the (parabolic) partial differential equation

$$\frac{\partial}{\partial t} e(x,t) = \frac{1}{2} \Delta e(x,t), \tag{5.2.18}$$

with the (singular) initial condition

$$e(x,0) = \delta(x), \tag{5.2.19}$$

where δ denotes the Dirac-delta function, defined by $\int_{\mathbb{R}^d} \delta(x) f(x) dx = f(0)$ for any bounded measurable function f. The function $e(x,t)$ is called the *heat kernel* associated with Brownian motion.

Theorem 5.12 suggests to call $\frac{1}{2}\Delta$ the *generator* of Brownian motion, and to think of (5.2.13) as the associated martingale problem. We will put this on firm ground in the next sections.

5.3 General Markov processes

We briefly review the most important aspects of the general theory. For S a metric space, let $B(S,\mathbb{R}) = B(S)$ denote the space of real-valued, bounded and measurable functions on S, $C(S,\mathbb{R}) = C(S)$ the space of continuous functions on S, $C_b(S,\mathbb{R}) = C_b(S)$ the space of bounded continuous functions on S, and $C_0(S,\mathbb{R}) = C_0(S)$ the space of bounded continuous functions on S that vanish at infinity. Clearly, $C_0(S) \subset C_b(S) \subset C(S) \subset B(S)$.

5.3.1 Semigroups and generators

The main building block for a time-homogeneous Markov process is the so-called transition kernel $P\colon \mathbb{R}_+ \times S \times \mathscr{B} \to [0,1]$. As in discrete time, the compatibility conditions impose the Chapman-Kolmogorov equations for transition kernels.

Definition 5.13 A *Markov transition function* $(P_t)_{t\in\mathbb{R}_+}$ is a family of kernels $P_t\colon S \times \mathscr{B} \to [0,1]$ with the following properties:

(i) For each $t \in \mathbb{R}_+$ and $x \in S$, $P_t(x,\cdot)$ is a measure on $(S,\mathscr{B})$ with $\mathbb{P}_t(x,S) \le 1$.
(ii) For each $t \in \mathbb{R}_+$ and $A \in \mathscr{B}$, $P_t(\cdot,A)$ is a $\mathscr{B}$-measurable function on S.
(iii) For each $s,t \in \mathbb{R}_+$,

$$P_{s+t}(x,A) = \int P_t(y,A) P_s(x,dy). \tag{5.3.1}$$

It is useful to shift from measures of sets to integrals of functions. This motivates the following equivalent definition of a time-homogeneous Markov process.

Definition 5.14 A stochastic process X with state space S and index set $\mathbb{R}_+$ is a continuous-time homogeneous Markov process on a filtered space $(\Omega, \mathscr{F}, \mathbb{P}, (\mathscr{F}_t)_{t\in\mathbb{R}_+})$ with transition function $(P_t)_{t\in\mathbb{R}_+}$ if it is adapted to $(\mathscr{F}_t)_{t\in\mathbb{R}_+}$ and, for all bounded $\mathscr{B}$-measurable functions f and all $s, t \in \mathbb{R}_+$,

$$\mathbb{E}\big[f(X_{t+s})|\mathscr{F}_s\big](\omega) = (P_t f)\big(X_s(\omega)\big) \quad \text{a.s.} \tag{5.3.2}$$

It is convenient to think of the transition kernels as bounded linear operators on $B(S, \mathbb{R})$, acting as

$$(P_t f)(x) = \int_S P_t(x, dy) f(y). \tag{5.3.3}$$

The Chapman-Kolmogorov equations in Definition 5.13(iii) then takes the simple form $P_s P_t = P_{t+s}$, and $(P_t)_{t\in\mathbb{R}_+}$ can be seen as a *semigroup* of bounded linear operators. Note that we also have the dual action of P_t on the space of probability measures via

$$(\mu P_t)(A) = \int_S \mu(dx) P_t(x, A). \tag{5.3.4}$$

This gives the duality relation

$$(\mu P_t)(f) = \int_S \mu(dx)(P_t f)(x) = \mu(P_t f), \quad f \in B(S, \mathbb{R}). \tag{5.3.5}$$

The condition $\mathbb{P}_t(x, S) \le 1$ may look surprising, since we would expect $\mathbb{P}_t(x, S) = 1$. However, it is sometimes convenient to consider the more general situation where the process may leave the state space, i.e., may "die".

Equation (5.3.1) allows us to think of Markov transition functions as operators on the Banach space of bounded measurable functions.

Definition 5.15 A family $(P_t)_{t\in\mathbb{R}_+}$ of bounded linear operators on $B(S, \mathbb{R})$ is called a *sub-Markov semigroup* if, for all $t \in \mathbb{R}_+$,

(i) $P_t\colon\ B(S, \mathbb{R}) \to B(S, \mathbb{R})$.
(ii) If $0 \le f \le 1$, then $0 \le P_t f \le 1$.
(iii) For all $s \in \mathbb{R}_+$, $P_{t+s} = P_t P_s$.
(iv) If $f_n \downarrow 0$, then $P_t f_n \downarrow 0$.

A sub-Markov semigroup is called *normal* if $P_0 = 1$, and is called *honest* if $P_t 1 = 1$ for all $t \in \mathbb{R}_+$.

Our aim will be to construct the generator of the semigroup. We are looking for an operator $\mathscr{L}$ such that $P_t = \exp(t\mathscr{L})$, where "exp" is the exponential map, defined through its Taylor expansion. This is a good enough way to construct a semigroup from a bounded generator $\mathscr{L}$. We will see shortly that this works well for Markov jump processes with bounded jump rates $m(x)$, $x \in S$. The general case, however, is a bit more involved. The proper setting in which the relation between semigroup and generator can be generalised is that of a so-called *strongly continuous contraction semigroup*.

Definition 5.16 Let B_0 be a Banach space. A family $(P_t)_{t\in\mathbb{R}_+}$ of bounded linear operators from B_0 to B_0 is called a *strongly continuous contraction semigroup* if the following conditions are verified:

(i) For all $f \in B_0$, $\lim_{t\downarrow 0} \|P_t f - f\| = 0$.
(ii) $\|P_t\| \leq 1$ for all $t \in \mathbb{R}_+$.
(iii) $P_t P_s = P_{t+s}$, for all $s, t \in \mathbb{R}_+$.

Here $\|\cdot\|$ denotes the operator norm corresponding to the norm on B_0.

We can now define the notion of infinitesimal generator.

Definition 5.17 Let B_0 be a Banach space and let $(P_t)_{t\in\mathbb{R}_+}$ be a strongly continuous contraction semigroup. We say that f is in the domain $\mathscr{D}(\mathscr{L})$ of $\mathscr{L}$, if there exists a function $g \in B_0$, such that

$$\lim_{t\downarrow 0}\left\|t^{-1}(P_t f - f) - g\right\| = 0. \tag{5.3.6}$$

For such f we set $\mathscr{L} f = g$, where g is the function that satisfies (5.3.6).

Note that we define $\mathscr{D}(\mathscr{L})$ at the same time as $\mathscr{L}$. In general, $\mathscr{L}$ will be an unbounded operator (e.g. a differential operator) whose domain is strictly smaller than B_0. Some authors describe the generator of a Markov process as a collection of pairs of functions (f, g) satisfying (5.3.6).

In the case of Markov jump processes, we can identify the generator easily.

Lemma 5.18 *Let X be a Markov jump process with jump distribution P and jump rates $m(x)$, $x \in S$. Then X has a generator with domain $B(0)$ given by*

$$(\mathscr{L} f)(x) = m(x)\int_S \big(f(y) - f(x)\big)P(x, dy). \tag{5.3.7}$$

Conversely, the semigroup $(P_t)_{t\in\mathbb{R}_+}$ is given by

$$(P_t f)(x) = \big(\exp(t\mathscr{L})f\big)(x) = \sum_{n\in\mathbb{N}_0} \frac{t^n}{n!}(\mathscr{L}^n f)(x). \tag{5.3.8}$$

For a proof see Ethier and Kurtz [104].

The classical link between generators and strongly continuous contraction semigroups is the *Hille-Yosida theorem*.

Theorem 5.19 (Hille-Yosida theorem) *A linear operator $\mathscr{L}$ on a Banach space B_0 is the generator of a strongly continuous contraction semigroup if and only if the following hold:*

(i) *The domain $\mathscr{D}(\mathscr{L})$ of $\mathscr{L}$ is dense in B_0.*
(ii) *$\mathscr{L}$ is* dissipative, *i.e., for all $\lambda > 0$ and all $f \in \mathscr{D}(\mathscr{L})$,*

$$\left\|(\lambda - \mathscr{L})f\right\| \geq \lambda \|f\|. \tag{5.3.9}$$

(iii) *There exists a $\lambda > 0$ such that* range$(\lambda - \mathscr{L}) = B_0$.

The proof of the Hille-Yosida theorem is quite involved and functional analytic in nature. It makes use of the concept of *resolvent*, which provides the constructive link between generator and semigroup. The proof of the Hille-Yosida theorem also provides a construction of the semigroup from the generator, but we will not need this here.

5.3.2 Feller-Dynkin processes

We next turn to a class of Markov semigroups that will be seen to have nice properties. Our state space is a locally compact Hausdorff space S with a countable basis (e.g. $S = \mathbb{R}^d$). We do not need to assume compactness, but we will need to consider the one-point compactification of S obtained by adding a "coffin state" ∂, making $S^\partial = S \cup \partial$ into a compact metrisable space.

We will place ourselves in the setting where the Hille-Yosida theorem works, and make a specific choice for the underlying Banach space, namely, we will work on the space $C_0(S)$ of continuous functions vanishing at infinity. This requires that we put a restriction on the semigroups so that they preserve $C_0(S)$. The latter is known as the *Feller property*.

Definition 5.20 A *Feller-Dynkin semigroup* is a strongly continuous sub-Markov semigroup $(P_t)_{t\in\mathbb{R}_+}$ acting on the space $C_0(S)$, in particular, for all $t \in \mathbb{R}_+$,

$$P_t\colon\ C_0(S) \to C_0(S). \tag{5.3.10}$$

It is an analytic fact (coming from the Riesz representation theorem) that to any strongly continuous contraction semigroup there corresponds a sub-Markov kernel $P_t(x,dy)$ such that $(P_t f)(x) = \int_S P_t(x,dy) f(y)$ for all $f \in C_0(S)$. The key result is the existence theorem for Feller-Dynkin processes.

Theorem 5.21 *Let $(P_t)_{t\in\mathbb{R}_+}$ be a Feller-Dynkin semigroup on $C_0(S)$. Then there exists a strong Markov process with values in S^∂, càdlàg paths and transition kernels $(P_t)_{t\in\mathbb{R}_+}$. (The unique existence of the Markov process on the level of finite-dimensional distributions does not require the Feller property.)*

Proof The Daniell-Kolmogorov extension theorem guarantees the existence of a unique process on the product space $(S^{\partial})^{\mathbb{R}_+}$, provided the finite-dimensional marginals satisfy the compatibility conditions. This is easily verified, just as in the discrete-time setting, by using the Chapman-Kolmogorov equations.

We want to show that the paths of the process are regularisable. For this we need to bring martingales into the game, and also need the notion of the *resolvent* R_λ, associated with a strongly continuous contraction semigroup, defined through

$$R_\lambda = \int_0^\infty \mathrm{e}^{-\lambda t} P_t dt, \quad \lambda \in \mathbb{R}_+. \tag{5.3.11}$$

Lemma 5.22 *Let $g \in C_0(S)$ and $g \geq 0$. Set $h = R_1 g$. Then*

$$0 \leq \mathrm{e}^{-t} P_t h \leq h. \tag{5.3.12}$$

If $Y = (Y_t)_{t\in\mathbb{R}_+}$ is the corresponding Markov process, then $(\mathrm{e}^{-t}h(Y_t))_{t\in\mathbb{R}_+}$ is a supermartingale.

Proof The lower bound in (5.3.12) is clear, since P_t maps non-negative functions to non-negative functions. To get the upper bound, write

$$\mathrm{e}^{-s} P_s h = \mathrm{e}^{-s} P_s R_1 g = \mathrm{e}^{-s} P_s \int_0^\infty \mathrm{e}^{-u} P_u g du = \int_s^\infty \mathrm{e}^{-u} P_u g du \leq R_1 g = h. \tag{5.3.13}$$

Now, $(\mathrm{e}^{-t}h(Y_t))_{t\in\mathbb{R}_+}$ is a supermartingale, since

$$\mathbb{E}\big[\mathrm{e}^{-s-t} h(Y_{t+s}) | \mathcal{F}_t\big] = \mathrm{e}^{-s-t} (P_s h)(Y_t) \leq \mathrm{e}^{-t} h(Y_t), \tag{5.3.14}$$

where in the last step we use (5.3.12). □

As a consequence of Lemma 5.22, the functions $\mathrm{e}^{-q}h(Y_q)$ are regularisable, i.e., $\lim_{q\downarrow t} \mathrm{e}^{-q}h(Y_q)$ exists for all t almost surely. We can therefore take a countable dense subset $\{g_i\}_{i\in\mathbb{N}}$ of elements of $C_0(S)$, set $h_i = R_1 g_i$, and observe that the set $\mathcal{H} = \{h_i\}_{i\in\mathbb{N}}$ separates points in S^{∂}, while almost surely $\mathrm{e}^{-q}h_i(Y_q)$ is regularisable for all $i \in \mathbb{N}$. But then $X_t = \lim_{q\downarrow t} Y_q$ exists for all t almost surely and is a càdlàg process.

Finally, we establish that X is a modification of Y. To do this, let $f, g \in C_0(S)$. Then

$$\begin{aligned}\mathbb{E}\big[f(Y_t)g(X_t)\big] &= \lim_{q\downarrow t} \mathbb{E}\big[f(Y_t)g(Y_q)\big] = \lim_{q\downarrow t} \mathbb{E}\big[f(Y_t)(P_{q-t}g)(Y_t)\big] \\ &= \mathbb{E}\big[f(Y_t)g(Y_t)\big],\end{aligned} \tag{5.3.15}$$

where the first equality uses the definition of X_t and the third equality uses the strong continuity of P_t. By an application of the monotone class theorem (Theorem 3.9) this implies that $\mathbb{E}[f(Y_t, X_t)] = \mathbb{E}[f(Y_t, Y_t)]$ for any bounded measurable function on $S^{\partial} \times S^{\partial}$, and hence that $\mathbb{P}(X_t = Y_t) = 1$. □

Theorem 5.21 allows us to consider Feller-Dynkin Markov processes defined on the space of càdlàg functions with values in S^∂ (with the additional property that, if $X_t = \partial$ or $X_{t-} = \partial$, then $X_s = \partial$, for all $s \geq t$). We will henceforth do so (with the usual right-continuous filtration).

5.3.3 *The strong Markov property*

Our Feller-Dynkin processes have the Markov property. In particular, if ζ is an $\mathcal{F}_t$-measurable function and $f \in C_0(S)$, then

$$\mathbb{E}\big[\zeta f(X_{t+s})\big] = \mathbb{E}\big[\zeta (P_s f)(X_t)\big]. \tag{5.3.16}$$

However, we want more, namely, like in the case of discrete-time Markov processes, we want to be able to split past and future at stopping times. Let θ_t be the shift acting on Ω via

$$X(\theta_t \omega)_s = (\theta_t X)(\omega)_s = X(\omega)_{s+t}. \tag{5.3.17}$$

Then we have the following *strong Markov property*:

Theorem 5.23 *Let T be an $\mathcal{F}_{t+}$-stopping time, and let $\mathbb{P}$ be the law of a Feller-Dynkin Markov process X. Then, for all bounded random variables η, if T is a stopping time, then*

$$\mathbb{E}[\theta_T \eta | \mathcal{F}_{T+}] = \mathbb{E}_{X_T}[\eta], \tag{5.3.18}$$

or equivalently, for all $\mathcal{F}_{T_+}$-measurable bounded random variables ξ,

$$\mathbb{E}[\xi \theta_T \eta] = \mathbb{E}\big[\xi \mathbb{E}_{X_T}[\eta]\big]. \tag{5.3.19}$$

Proof Consider the dyadic approximation of the stopping time T defined as

$$T^{(n)}(\omega) = \begin{cases} k2^{-n}, & \text{if } (k-1)2^{-n} \leq T(\omega) < k2^{-n},\ k \in \mathbb{N}, \\ \infty, & \text{if } T(\omega) = \infty. \end{cases} \tag{5.3.20}$$

For $\Lambda \in \mathcal{F}_{T+}$, set

$$\Lambda_{n,k} = \big\{\omega \in \Omega:\ T^{(n)}(\omega) = 2^{-n}k\big\} \cap \Lambda \in \mathcal{F}_{k2^{-n}}. \tag{5.3.21}$$

Let f be a continuous function on S. Then

$$\begin{aligned} \mathbb{E}\big[f(X_{T^{(n)}+s})\mathbb{1}_\Lambda\big] &= \sum_{k \in \mathbb{N}\cup\{\infty\}} \mathbb{E}\big[f(X_{k2^{-n}+s})\mathbb{1}_{\Lambda_{n,k}}\big] \\ &= \sum_{k \in \mathbb{N}\cup\{\infty\}} \mathbb{E}\big[(P_s f)(X_{k2^{-n}})\mathbb{1}_{\Lambda_{n,k}}\big] \\ &= \mathbb{E}\big[(P_s f)(X_{T^{(n)}})\mathbb{1}_\Lambda\big]. \end{aligned} \tag{5.3.22}$$

Finally let $n \to \infty$. By the right-continuity of paths, we have

$$\lim_{n\to\infty} X_{T^{(n)}+s} = X_{T+s}, \quad s \in \mathbb{R}_+. \tag{5.3.23}$$

Since f is continuous, it also follows that

$$\lim_{n\to\infty} f(X_{T^{(n)}+s}) = f(X_{T+s}), \quad s \in \mathbb{R}_+. \tag{5.3.24}$$

Since, by the Feller property, $P_s f$ is also continuous, it further follows that

$$\lim_{n\to\infty} (P_s f)(X_{T^{(n)}}) = (P_s f)(X_T), \quad s \in \mathbb{R}_+, \tag{5.3.25}$$

and so from (5.3.22), by dominated convergence,

$$\mathbb{E}\big[f(X_{T+s})\mathbb{1}_A\big] = \mathbb{E}\big[(P_s f)(X_T \mathbb{1}_A\big], \quad s \in \mathbb{R}_+. \tag{5.3.26}$$

To conclude the proof we need only generalise (5.3.26) to more general functions. But this can be done in the usual manner via the monotone class theorem, and presents no particular difficulties. Indeed, we first check that $\mathbb{1}_A$ can be replaced by any bounded $\mathscr{F}_{T+}$-measurable function. Next, we show through explicit computation that instead of $f(X_{T+s})$ we can put $\prod_{i=1}^n f_i(X_{T+s_i})$, and finally we can again use the monotone class theorem to conclude the proof for the general case. □

Note that working with Feller semigroups has payed off!

5.4 The martingale problem

For discrete-time Markov processes we have encountered a characterisation in terms of the *martingale problem*. While this proved to be quite handy, there was nothing profoundly important about its use. This very much changes in the continuous-time setting. In fact, the martingale problem characterisation of Markov processes, originally proposed by Stroock and Varadhan, turns out to be the "proper" way to deal with the theory in many respects.

5.4.1 Generators and cores

In principle, the Hille-Yosida theorem gives us precise criteria for recognising when a given linear operator generates a strongly continuous contraction semigroup and hence a Markov process. However, if we look at the conditions more carefully, then we realise that in many situations it will be impractical to verify them. The domain of a generator is usually far too large to allow for a description of the action of the generator on all of its elements. For instance, for Brownian motion we want to think of the generator as $\frac{1}{2}$ times the Laplacian on the space $C^2(\mathbb{R}_+, \mathbb{R}^d)$. But this operator is closed only in $d = 1$, but not in $d \geq 2$, so already in this case we enter into subtle issues we would rather like to avoid.

Let us first discuss this issue from a functional analytic point of view. To that end we need to recall a few notions from operator theory.

Definition 5.24 Let G, C be two linear operators with domains $\mathscr{D}(G), \mathscr{D}(C)$, respectively. We say that C is an *extension* of G if

(i) $\mathscr{D}(G) \subset \mathscr{D}(C)$.
(ii) $Gf = Cf$ for all f in $\mathscr{D}(G)$.

Definition 5.25 A linear operator G on a Banach space B_0 is called *closed* if its *graph*, which is the set

$$\Gamma(G) = \{(f, Gf) \colon \ f \in \mathscr{D}(G)\} \subset B_0 \times B_0, \tag{5.4.1}$$

is closed in $B_0 \times B_0$. Equivalently, G is closed, if, for any sequence $(f_n)_{n\in\mathbb{N}}$ in $\mathscr{D}(G)$ such that $\lim_{n\to\infty} f_n = f$ and $\lim n \to \infty G f_n = g$, it is true that $f \in \mathscr{D}(G)$ and $g = Gf$.

We call the *closure* $\overline{\mathscr{L}}$ of a linear operator $\mathscr{L}$ the minimal extension of $\mathscr{L}$ that is closed. An operator that has a closed linear extension is called *closable*.

Lemma 5.26 *A dissipative linear operator $\mathscr{L}$ on B_0 whose domain $\mathscr{D}(\mathscr{L})$ is dense in B_0 is closable and the closure of* range$(\lambda - \mathscr{L})$ *is equal to* range$(\lambda - \overline{\mathscr{L}})$ *for all* $\lambda > 0$.

Proof Let $(f_n)_{n\in\mathbb{N}}$ be a sequence in $\mathscr{D}(\mathscr{L})$ such that $f_n \to f$ and $\mathscr{L} f_n \to g$ as $n \to \infty$. We would like to associate with any such f the value g and then define $\mathscr{L} f = g$ for all achievable f, which would then be the desired closed extension of $\mathscr{L}$. So, all we need to show is that, if $f_n' \to f$ and $\mathscr{L} f_n' \to g'$, then $g' = g$. In fact, it suffices to show that, if $f_n \to 0$ and $\mathscr{L} f_n \to g$, then $g = 0$. To do this, consider a sequence of functions $(g_n)_{n\in\mathbb{N}}$ in $\mathscr{D}(\mathscr{L})$ such that $g_n \to g$. Such a sequence exists because $\mathscr{D}(\mathscr{L})$ is dense in B_0. Using the dissipativity of $\mathscr{L}$, we get

$$\left\|(\lambda - \mathscr{L}) g_n - \lambda g\right\| = \lim_{k\to\infty} \left\|(\lambda - \mathscr{L})(g_n + \lambda f_k)\right\| \geq \lim_{k\to\infty} \lambda \|g_n + \lambda f_k\| = \lambda \|g_n\|, \tag{5.4.2}$$

where in the first inequality we use that $0 = \lim_{k\to\infty} f_k$ and $g = \lim_{k\to\infty} \mathscr{L} f_k$. Dividing by λ and taking the limit $\lambda \to \infty$, we obtain

$$\|g_n\| \leq \|g_n - g\|. \tag{5.4.3}$$

Since $g_n - g \to 0$, this implies $g_n \to 0$.

The identification of the closure of the range with the range of the closure follows from the observation made in Definition 5.25 that the range of a dissipative operator is closed if and only if the operator is closed. □

As a consequence of Lemma 5.26, if a dissipative linear operator $\mathscr{L}$ on B_0 is closable and the range of $\lambda - \mathscr{L}$ is dense in B_0, then its closure is the generator of a strongly continuous contraction semigroup on B_0. These observations motivate the definition of a *core* of a linear operator.

Definition 5.27 Let $\mathscr{L}$ be a linear operator on a Banach space B_0. A subspace $D \subset \mathscr{D}(\mathscr{L})$ is called a *core* for $\mathscr{L}$ if the closure of the restriction of $\mathscr{L}$ to D, written $\mathscr{L}_D$, is equal to $\mathscr{L}$.

Lemma 5.28 *Let $\mathscr{L}$ be the generator of a strongly continuous contraction semigroup on B_0. Then a subspace $D \subset \mathscr{D}(\mathscr{L})$ is a core for $\mathscr{L}$ if and only if D is dense in B_0 and, for some $\lambda > 0$, range$(\lambda - \mathscr{L}_D)$ is dense in B_0.*

Proof The claim follows from the preceding observations. □

The following is a very useful characterisation of a core in our context.

Lemma 5.29 *Let $\mathscr{L}$ be the generator of a strongly continuous contraction semigroup $(P_t)_{t\in\mathbb{R}_+}$ on B_0. Let D be a dense subset of $\mathscr{D}(\mathscr{L})$. If, for all $t \in \mathbb{R}_+$, $P_t \colon D \to D$, then D is a core. In fact, it suffices that there is a dense subset $D_0 \subset D$ such that P_t maps D_0 into D.*

Proof Let $f \in D_0$ and set

$$f_n = \frac{1}{n}\sum_{k=0}^{n^2} \mathrm{e}^{-\lambda k/n} P_{k/n} f. \tag{5.4.4}$$

We have $f_n \in D$. By strong continuity,

$$\begin{aligned}\lim_{n\to\infty} (\lambda - \mathscr{L}) f_n &= \lim_{n\to\infty} \frac{1}{n}\sum_{k=0}^{n^2} \mathrm{e}^{-\lambda k/n} P_{k/n}(\lambda - \mathscr{L}) f \\ &= \int_0^\infty dt\, \mathrm{e}^{-\lambda t} P_t(\lambda - \mathscr{L}) f = R_\lambda(\lambda - \mathscr{L}) f = f. \end{aligned} \tag{5.4.5}$$

Thus, for any $f \in D_0$, there exists a sequence of functions $((\lambda - \mathscr{L}) f_n)_{n\in\mathbb{N}}$ in range$(\lambda - \mathscr{L}_D)$ that converges to f. Hence the closure of the range of $(\lambda - \mathscr{L}_D)$ contains D_0. Since D_0 is dense in B_0, the assertion follows from Lemma 5.28. □

Example Let $\mathscr{L}$ be the generator of Brownian motion. We claim that $C^\infty = C^\infty(\mathbb{R}^d)$ is a core for $\mathscr{L}$ and $\mathscr{L}$ is the closure of $\frac{1}{2}\Delta$ on this core. Indeed, C^∞ is dense in the space of continuous functions. To show that C^∞ is a core, by Lemma 5.29 we need only show that P_t maps C^∞ to C^∞. But this is obvious from the explicit formula for the transition kernel of Brownian motion in Theorem 5.11.

To check that the restriction of $\mathscr{L}$ to C^∞ is $\frac{1}{2}\Delta$ is a simple calculation. Hence $\mathscr{L}$ is the closure of $\frac{1}{2}\Delta$.

The above results are nice when we already know the semigroup. In more complicated situations we may only be able to write down the action of what we want to be the generator of the Markov process on some small subspace of functions. The question is: How can we find out whether this specifies a (unique) strongly continuous contraction semigroup on the full space of functions, e.g. $C_0(S)$? We may be able to show that it is dissipative, but is range$(\lambda - \mathscr{L})$ dense in $C_0(S)$? The martingale problem formulation is a powerful tool to address this question.

5.4.2 The martingale problem

We begin with a relatively simple observation.

Lemma 5.30 *Let X be a Feller-Dynkin process with transition functions $(P_t)_{t\in\mathbb{R}_+}$ and generator $\mathscr{L}$. Define, for $f, g \in B(S)$,*

$$M_t = f(X_t) - \int_0^t g(X_s)ds. \tag{5.4.6}$$

If $f \in \mathscr{D}(\mathscr{L})$ and $g = \mathscr{L}f$, then $(M_t)_{t\in\mathbb{R}_+}$ is an $(\mathscr{F}_t)_{t\in\mathbb{R}_+}$-martingale.

Proof The proof runs as in the discrete-time setting. Write

$$\begin{aligned}
\mathbb{E}[M_{t+u}|\mathscr{F}_t] &= \mathbb{E}\big[f(X_{t+u})|\mathscr{F}_t\big] - \int_0^t (\mathscr{L}f)(X_s)ds - \int_t^{t+u} \mathbb{E}\big[(\mathscr{L}f)(X_s)|\mathscr{F}_t\big]ds \\
&= \int_S P_u(X_t,dy)f(y) - \int_0^t (\mathscr{L}f)(X_s)ds \\
&\quad - \int_0^u \left[\int_S P_s(X_t,dy)(\mathscr{L}f)(y)\right] ds \\
&= f(X_t) - \int_0^t (\mathscr{L}f)(X_s)ds \\
&\quad + \int_S P_u(X_t,dy)f(y) - f(X_t) - \int_0^u \left[\int_S P_s(X_t,dy)(\mathscr{L}f)(y)\right] ds \\
&= M_t + \int_S P_u(X_t,dy)f(y) - f(X_t) - \int_0^u (P_s\mathscr{L}f)(X_t)ds. \qquad (5.4.7)
\end{aligned}$$

But

$$(P_s\mathscr{L}f)(x) = \frac{d}{ds}(P_sf)(x), \tag{5.4.8}$$

and so

$$\int_S P_u(X_t, dy) f(y) - f(X_t) - \int_0^u (P_s \mathscr{L} f)(X_t) ds = 0, \tag{5.4.9}$$

from which the claim follows. □

By "the martingale problem" we will mean the inverse problem associated with the above observation.

Definition 5.31 Given a linear operator $\mathscr{L}$ with domain $\mathscr{D}(\mathscr{L})$ and range$(\mathscr{L}) \subset C_b(S)$, an S-valued càdlàg process on a filtered càdlàg space $(\Omega, \mathscr{F}, \mathbb{P}, (\mathscr{F}_t)_{t\in\mathbb{R}_+})$ is called *a solution of the martingale problem associated with the operator* $\mathscr{L}$ if, for any $f \in \mathscr{D}(\mathscr{L})$, $(M_t)_{t\in\mathbb{R}_+}$ defined by (5.4.6) is an $(\mathscr{F}_t)_{t\in\mathbb{R}_+}$-martingale.

Before we continue, we need some additional notions for convergence in Banach spaces.

Definition 5.32 A sequence $(f_n)_{n\in\mathbb{N}}$ in $B(S)$ is said to converge *bounded pointwise* (bp) to a function $f \in B(S)$ if and only if

(i) $\sup_{n\in\mathbb{N}} \|f_n\|_\infty < \infty$.
(ii) For every $x \in S$, $\lim_{n\to\infty} f_n(x) = f(x)$.

A set $M \in B(S)$ is called bp-closed, if, for any sequence $(f_n)_{n\in\mathbb{N}}$ in M such that $\text{bp} - \lim_{n\to\infty} f_n = f \in B(S)$, it is true that $f \in M$. The bp-closure of a set $D \subset B(S)$ is the smallest bp-closed set in $B(S)$ that contains D. A set M is called bp-dense if its closure is $B(S)$.

Lemma 5.33 *Let* $(f_n)_{n\in\mathbb{N}}$ *be such that* $\text{bp}-\lim_{n\to\infty} f_n = f$ *and* $\text{bp}-\lim_{n\to\infty} \mathscr{L} f_n = \mathscr{L} f$. *If*

$$\left(f_n(X_t) - \int_0^t (\mathscr{L} f_n)(X_s) ds \right)_{t\in\mathbb{R}_+} \tag{5.4.10}$$

is a martingale for all $n \in \mathbb{N}$, *then*

$$\left(f(X_t) - \int_0^t (\mathscr{L} f)(X_s) ds \right)_{t\in\mathbb{R}_+} \tag{5.4.11}$$

is a martingale.

Proof Straightforward. □

The implication of Lemma 5.33 is that in order to find a unique solution of the martingale problem it suffices to know the generator on a core.

Theorem 5.34 *Let* $\mathscr{L}_1$ *be an operator with domain* $\mathscr{D}(\mathscr{L}_1)$ *and range* range$(\mathscr{L}_1)$, *and let* $\mathscr{L}$ *be an extension of* $\mathscr{L}_1$. *Suppose that the bp-closures of the graphs of* $\mathscr{L}_1$

and $\mathscr{L}$ are the same. Then a stochastic process X is a solution of the martingale problem for $\mathscr{L}$ if and only if it is a solution of the martingale problem for $\mathscr{L}_1$.

Proof This follows from Lemma 5.33. □

The strategy will be to understand when the martingale problem has a unique solution, and to show that this solution is a Markov process. It will be comforting to see that only dissipative operators can give rise to the solution of a martingale problem.

We first prove a result that gives an equivalent characterisation of the martingale problem.

Lemma 5.35 *Let $\mathscr{L}$ be the generator of a continuous-time Markov process X. Suppose that $f \in \mathscr{D}(\mathscr{L})$ and $k\colon S \to \mathbb{R}$ is continuous and bounded from below. Then*

$$\left(f(X_t) - \int_0^t (\mathscr{L}f)(X_s)\,ds\right)_{t\in\mathbb{R}_+} \tag{5.4.12}$$

is a martingale if and only if

$$\left(\mathrm{e}^{-\int_0^t k(X_s)ds} f(X_t) + \int_0^t \mathrm{e}^{-\int_0^s k(X_r)dr}\big[k(X_s)f(X_s) - (\mathscr{L}f)(X_s)\big]ds\right)_{t\in\mathbb{R}_+} \tag{5.4.13}$$

is a martingale.

Proof To prove this lemma we need the following theorem.

Theorem 5.36 *Let M be a càdlàg local martingale (recall Definition* 3.100), *and let V be a continuous and adapted process that is locally of bounded variation. Then $W = (W_t)_{t\in\mathbb{R}_+}$ with*

$$W_t = \int_0^t V_s dM_s = V_t M_t - V_0 M_0 - \int_0^t M_s dV_s \tag{5.4.14}$$

is a càdlàg local martingale as well.

Proof By the definition of local martingales, a.s. we can find an increasing sequence of stopping times $(\tau_n)_{n\in\mathbb{N}}$ with $\lim_{n\to\infty}\tau_n = \infty$ such that M^{τ_n} are martingales for each $n \in \mathbb{N}$, where M^{τ_n} is the martingale M stopped at time τ_n. We may, moreover, assume that $|M^{\tau_n}| \le n$ and

$$R_V(t) = \sup_{m\in\mathbb{N}} \sup_{0\le u_0\le\cdots\le u_m\le t} \sum_{k=0}^{m-1} |V_{u_{k+1}} - V_{u_k}| \le n. \tag{5.4.15}$$

We have

$$\int_0^t V_s dM_s^{\tau_n} = \lim_{n\to\infty} \sum_{k=0}^{m-1} V_{u_k^n}\left[M_{u_{k+1}^n}^{\tau_n} - M_{u_k^n}^{\tau_n}\right], \tag{5.4.16}$$

where $(u_k^n)_{n\in\mathbb{N}}$ is any sequence of partitions of $[0,t]$ such that

$$\lim_{n\to\infty} \max_{0\le k\le n} \left|u_{k+1}^n - u_k^n\right| = 0. \tag{5.4.17}$$

This limit exists since, by elementary reshuffling,

$$\sum_{k=0}^{m-1} V_{u_k^n}\left[M_{u_{k+1}^n}^{\tau_n} - nM_{u_k^n}^{\tau_n}\right] = V_t^{\tau_n} M_t^{\tau_n} - V_0 M_0^{\tau_n} - \sum_{k=0}^{m-1} M_{u_{k+1}^n}^{\tau_n}\left[V_{u_{k+1}^n}^{\tau_n} - V_{u_k^n}^{\tau_n}\right]. \tag{5.4.18}$$

Since V is of bounded variation and M^{τ_n} is bounded, the latter sum converges to $\int_0^t M_s dV_s$, both a.s. and in $\mathscr{L}^1$, as $n\to\infty$. As a consequence, the same is true for the left-hand side. Since, for any $n\in\mathbb{N}$, the left-hand side is a martingale, this property remains true in the limit as $n\to\infty$. The limit as $n\to\infty$ exists because $\lim_{n\to\infty}\tau_n = \infty$ a.s. □

The proof of Lemma 5.35 follows from Theorem 5.36. Indeed, choose $M_t = f(X_t) - \int_0^t (\mathscr{L}f)(X_s)ds$ and $V_t = \exp(-\int_0^t k(X_s)ds)$. A tedious but straightforward computation (which uses Fubini's theorem) shows that the expression in (5.4.13) is of the form $V_t X_t - \int_0^t X_s dV_s$ and hence defines a martingale. □

Corollary 5.37 *Let $(\mathscr{F}_t)_{t\in\mathbb{R}_+}$ be a filtration and X an adapted process. Let $f,g\in B(S)$. Then, for $\lambda>0$, (5.4.6) is a martingale if and only if*

$$\left(e^{-\lambda t} f(X_t) + \int_0^t e^{-\lambda s}\left[\lambda f(X_s) - g(X_s)\right]ds\right)_{t\in\mathbb{R}_+} \tag{5.4.19}$$

is a martingale.

We use this corollary to establish the following.

Lemma 5.38 *Let $\mathscr{L}$ be a linear operator with domain and range in $B(S)$. If a solution of the martingale problem for $\mathscr{L}$ exists for any initial condition $X_0 = x\in S$, then $\mathscr{L}$ is dissipative.*

Proof Let $f\in\mathscr{D}(\mathscr{L})$ and $g=\mathscr{L}f$. Use that (5.4.19) is a martingale with $\lambda>0$. Taking expectations and letting $t\to\infty$, we get

$$f(X_0) = f(x) = \mathbb{E}\left[\int_0^\infty e^{-\lambda s}\left[\lambda f(X_s) - g(X_s)\right]ds\right] \tag{5.4.20}$$

and hence

$$|f(x)| \le \int_0^\infty e^{-\lambda s}\mathbb{E}|\lambda f(X_s) - g(X_s)|ds \le \int_0^\infty e^{-\lambda s}\|\lambda f - g\|ds = \lambda^{-1}\|\lambda f - g\|, \tag{5.4.21}$$

which shows that $\lambda\|f\| \le \|(\lambda - \mathscr{L})f\|$ and proves that $\mathscr{L}$ is dissipative. □

We already know that martingales typically have a càdlàg modification. Provided the set of functions on which we have defined our martingale problem is sufficiently rich, this property ought to carry over to the solution of the martingale problem as well. The following theorem shows when this is true.

Theorem 5.39 *Suppose that S is separable, $\mathscr{D}(\mathscr{L}) \subset C_b(S)$, $\mathscr{D}(\mathscr{L})$ is separating and contains a countable subset that separates points. If X is a solution of the associated martingale problem, and if for any $\varepsilon > 0$ and $T < \infty$ there exists a compact set $K_{\varepsilon,T} \subset S$ such that*

$$\mathbb{P}\big(\forall t \in [0,T] \cap \mathbb{Q}:\ X_t \in K_{\varepsilon,T}\big) > 1 - \varepsilon, \tag{5.4.22}$$

then X has a càdlàg modification.

Proof See Ethier and Kurtz [104, Chap. 4, Theorem 3.6]. □

5.4.3 Uniqueness

We have seen that solutions of the martingale problem provide candidates for nice Markov processes. The two main issues to understand are when a martingale problem has a *unique* solution and whether this solution represents a Markov process. When talking about uniqueness we will always assume that an initial distribution μ is given. Thus, the data for the martingale problem is a pair $(\mathscr{L}, \mu)$, where $\mathscr{L}$ is a linear operator with its domain $\mathscr{D}(\mathscr{L})$ and μ is a probability measure on S.

The following result is hardly surprising.

Theorem 5.40 *Let S be separable and let $\mathscr{L}$ be a linear dissipative operator on $B(S)$ with domain $\mathscr{D}(\mathscr{L}) \subset B(S)$. Suppose there exists a $\mathscr{L}'$ with domain $\mathscr{D}(\mathscr{L}') \subset \mathscr{D}(\mathscr{L})$ such that $\mathscr{L}$ is an extension of $\mathscr{L}'$. Let $\overline{\mathscr{D}(\mathscr{L}')} = \overline{\mathrm{range}(\lambda - \mathscr{L}')} = D$, and let D be separating. Let X be a solution of the martingale problem for $(\mathscr{L}, \mu)$. Then X is a Markov process whose semigroup on D is generated by the closure of $\mathscr{L}'$, and the martingale problem for $(\mathscr{L}, \mu)$ has a unique solution.*

Proof See Ethier and Kurtz [104, Sect. 4.4]. □

Finally we can establish a uniqueness criterion and the strong Markov property for solutions of martingale problems.

Theorem 5.41 *Let S be a separable space and let $\mathscr{L}$ be a linear operator on $B(S)$. Suppose that for any initial distribution μ, any two solutions X, Y of the martingale problem for $(\mathscr{L}, \mu)$ have the same one-dimensional distributions, i.e., $\mathbb{P}(X_t \in A) = \mathbb{P}(Y_t \in A)$ for any $t \in \mathbb{R}_+$ and any Borel set A. Then the following hold:*

(i) *Any solution of the martingale problem for $\mathscr{L}$ is a Markov process and any two solutions of the martingale problem with the same initial distribution have the same finite-dimensional distributions (i.e., uniqueness holds).*
(ii) *If $\mathscr{D}(\mathscr{L}) \subset C_b(S)$ and X is a solution of the martingale problem with càdlàg sample paths, then for any a.s. finite stopping time τ,*

$$\mathbb{E}\big[f(X_{t+\tau})|\mathscr{F}_\tau\big] = \mathbb{E}\big[f(X_{t+\tau})|X_\tau\big] \quad \forall f \in B(S). \tag{5.4.23}$$

(iii) *If, in addition to the assumptions in (ii), there exists a càdlàg solution of the martingale problem for any initial measure $\mu = \delta_x$, $x \in S$, then the strong Markov property holds, i.e.,*

$$\mathbb{E}\big[f(X_{t+\tau})|\mathscr{F}_\tau\big] = (P_t f)(X_\tau). \tag{5.4.24}$$

Proof See Ethier and Kurtz [104, Sect. 4.4.]. □

Note that in Theorem 5.41 we made no assumptions on the choice of $\mathscr{D}(\mathscr{L})$. In particular, it need not separate points, as in Theorem 5.40. The latter is in fact implicit in the requirement that uniqueness of the one-dimensional marginals holds. This leads us to the following observation: for a martingale problem uniqueness of the one-dimensional marginals implies uniqueness of the finite-dimensional marginals.

5.4.4 Existence

We have seen that a uniquely solvable martingale problem provides a way to construct a Markov process. We therefore need to find ways to produce solutions of martingale problems. The best way to do this is through approximations and weak convergence.

Lemma 5.42 *Let $\mathscr{L}$ be a linear operator with domain and range in $C_b(S)$. Let $(\mathscr{L}_n)_{n\in\mathbb{N}}$ be a sequence of linear operators with domain and range in $B(S)$. Assume that for any $f \in \mathscr{D}(A)$ there exists a sequence $(f_n)_{n\in\mathbb{N}}$ with $f_n \in \mathscr{D}(\mathscr{L}_n)$ such that*

$$\lim_{n\to\infty} \|f_n - f\| = 0 \quad \text{and} \quad \lim_{n\to\infty} \|\mathscr{L}_n f_n - \mathscr{L} f\| = 0. \tag{5.4.25}$$

If, for each $n \in \mathbb{N}$, X^n is a solution of the martingale problem for $\mathscr{L}_n$ with càdlàg sample paths and X^n converges to X weakly, then X is a càdlàg solution of the martingale problem for $\mathscr{L}$.

Proof Let $k \in \mathbb{N}$, and let $0 \le t_1 < \cdots < t_k \le t < s$ be elements of the set $\mathscr{C}(X) = \{u \in \mathbb{R}_+ : \mathbb{P}(X_u = X_{u-}) = 1\}$. Let $h_1, \ldots, h_k \in C_b(S)$, and let f, f_n be as in the hypothesis of the lemma. Then

$$\begin{aligned} &\mathbb{E}\left[\left(f(X_s) - f(X_t) - \int_t^s (\mathscr{L}f)(X_u)du\right)\prod_{i=1}^k h_i(X_{t_i})\right] \\ &= \lim_{n\to\infty} \mathbb{E}\left[\left(f_n(X_s^n) - f_n(X_t^n) - \int_t^s (\mathscr{L}f_n)(X_u^n)du\right)\prod_{i=1}^k h_i(X_{t_i}^n)\right] = 0. \end{aligned} \tag{5.4.26}$$

The complement of the set $\mathscr{C}(X)$ is at most countable, and hence (5.4.26) carries over to all points $0 \le t_1 < \cdots < t_k \le t < s$. But this implies that X solves the martingale problem for $\mathscr{L}$. □

The usefulness of Lemma 5.42 is based on the following lemma, which implies that we can use Markov jump processes as approximations.

Lemma 5.43 *Let S be compact and let $\mathscr{L}$ be a dissipative operator on $C(S)$ with dense domain and $\mathscr{L}1 = 0$. Then there exists a sequence of positive contraction operators $(T_n)_{n\in\mathbb{N}}$ on $B(S)$ given by transition kernels such that, for $f \in \mathscr{D}(\mathscr{L})$,*

$$\lim_{n\to\infty} n(T_n - 1)f = \mathscr{L}f. \tag{5.4.27}$$

Proof Here is a rough sketch of the proof, which is closely related to the Hille-Yosida theorem. From $\mathscr{L}$ we construct the resolvent $(n - \mathscr{L})^{-1}$ on the range of $(n - \mathscr{L})$. For a dissipative $\mathscr{L}$, the operators $n(n - \mathscr{L})^{-1}$ are bounded (by 1) on range$(n - \mathscr{L})$. Hence, by the Hahn-Banach theorem, they can be extended to $C(S)$ as bounded operators. Using the Riesz representation theorem, we can then associate with $n(n - \mathscr{L})^{-1}$ a probability measure μ_n via

$$\left(n(n - \mathscr{L})^{-1} f\right)(x) = \int_S f(y)\mu_n(x, dy), \tag{5.4.28}$$

and so $n(n - \mathscr{L})^{-1} = T_n$ defines a Markov transition kernel. Finally, it remains to show that $n(T_n - 1)f = n\mathscr{L}(n - \mathscr{L})^{-1}f = T_n\mathscr{L}f$ converges to $\mathscr{L}f$ for $f \in \mathscr{D}(\mathscr{L})$, which is straightforward. □

The point of Lemma 5.43 is that it shows that the martingale problem for $\mathscr{L}$ can be approximated by martingale problems with *bounded* generators of the form

$$(\mathscr{L}_n f)(x) = n\int_S \left[f(y) - f(x)\right]\mu_n(x, dy), \tag{5.4.29}$$

where $\mathscr{L}_n$ is the generator of a *Markov jump process*. For such a generator, the construction of a solution can be done explicitly in various ways, e.g. by letting the transition kernel be the convergent series for $\exp(t\mathscr{L}_n)$.

5.5 Itō calculus

An important class of stochastic processes that exhibit metastability are solutions of stochastic differential equations, to be treated in Sect. 5.6. This requires the notion of stochastic integrals. In this section we give a brief outline of the main concepts. For further reading, see e.g. the monographs by Karatzas and Shreve [148] and by Itō and McKean [142].

In this section we will work on a filtered space $(\Omega, \mathscr{F}, \mathbb{P}, (\mathscr{F}_t)_{t\in\mathbb{R}_+})$ that satisfies the conditions of the "usual setting" of Definition 3.83. We will be interested to define *stochastic integrals* of the form

$$\int_0^t X_s dM_s, \tag{5.5.1}$$

where M is a martingale and X is a progressive process. In fact, the full ambition of stochastic analysis is to find the largest class of pairs of processes M and X for which such an integral can be reasonably defined, which leads to the notion of *semi-martingale*, but here we will limit our ambition to the considerably simpler case when M is a *continuous* square-integrable martingale, i.e., when M has a.s. continuous paths and $\mathbb{E}[M_t^2] < \infty$ for all $t \in \mathbb{R}_+$. This includes the important case where M is a Brownian motion.

In the sequel we will sometimes state results only for martingales. But these can all be extended to local martingales.

5.5.1 *Square-integrable continuous martingales*

The definition of the stochastic integral was already provided in Theorem 5.36, where the integral $\int_0^t V_s dM_s$ was defined through a Stieltjes-integral in the case where V_t has (locally) bounded variation. We thus see that the challenge is to define stochastic integrals when also the integrand is not of bounded variation. Before doing so we need to return briefly to the theory of martingales.

Let M be a càdlàg martingale. We want to define its quadratic variation process $[M]$ in analogy with the discrete-time setting. This will be contained in the following fundamental result.

Theorem 5.44 *Let M be a continuous square-integrable martingale. Then there exists a unique increasing process $[M]$ such that the process $M^2 - [M]$ is a uniformly integrable continuous martingale.*

Proof We will only consider the case where M is continuous. We may also assume that M is bounded: otherwise we consider the martingale stopped when it exceeds a finite value N. Define stopping times

$$T_0^n = 0, \qquad T_{k+1}^n = \inf\{t > T_k^n : |M_t - M_{T_k^n}| \geq 2^{-n}\}. \tag{5.5.2}$$

Set $t_k^n = t \wedge T_k^n$. Assuming that $M_0 = 0$, we can write (by telescopic expansions)

$$M_t^2 = 2\sum_{k\in\mathbb{N}} M_{t_{k-1}^n}(M_{t_k^n} - M_{t_{k-1}^n}) + \sum_{k\in\mathbb{N}}(M_{t_k^n} - M_{t_{k-1}^n})^2. \tag{5.5.3}$$

Let

$$H_t^n = \sum_{k\in\mathbb{N}} M_{T_{k-1}^n} \mathbb{1}_{T_{k-1}^n < t \le T_k^n}. \tag{5.5.4}$$

Note that $H^n = (H_t^n)_{t\in\mathbb{R}_+}$ is *left-continuous*, which makes it previsible. The first term in the right-hand side of (5.5.3) is $(H^n \bullet M)_t$, the so-called *discrete stochastic integral* (see (3.4.4)), and we know from Theorem 3.53 that this is an L^2-bounded martingale. We define

$$A_t^n = \sum_{k\in\mathbb{N}}(M_{t_k^n} - M_{t_{k-1}^n})^2. \tag{5.5.5}$$

Then

$$M_t^2 = 2\big(H^n \bullet M\big)_t + A_t^n. \tag{5.5.6}$$

By construction, H^n approximates M well:

$$\sup_{t\in\mathbb{R}_+}\left|H_t^n - H_t^{n+1}\right| \le 2^{-n-1}, \qquad \sup_{t\in\mathbb{R}_+}\left|H_t^n - M_t\right| \le 2^{-n}. \tag{5.5.7}$$

The sets $J_n(\omega) = \{T_k^n(\omega), k \in \mathbb{N}\}$ refine each other, i.e., $J_n(\omega) \subset J_{n+1}(\omega)$, and

$$A_{T_k^n}^n \le A_{T_{k+1}^n}^n. \tag{5.5.8}$$

Now it is elementary to see that

$$\begin{aligned}
\mathbb{E}\big[\big(\big((H^n - H^{n+1}) \bullet M\big)_\infty\big)^2\big] &= \mathbb{E}\Big[\sum_{k\in\mathbb{N}}\big(H_{t_{k-1}}^n - H_{t_{k-1}}^{n+1}\big)^2(M_{t_k} - M_{t_{k-1}})^2\Big] \\
&\le 2^{-2n-2}\,\mathbb{E}\Big[\sum_{k\in\mathbb{N}}(M_{t_k} - M_{t_{k-1}})^2\Big] \\
&= 2^{-2n-2}\mathbb{E}\big[M_\infty^2\big].
\end{aligned} \tag{5.5.9}$$

Thus, the continuous martingales $(H^n \bullet M)$ converge uniformly as $n \to \infty$ to a continuous martingale, say N. This implies that the processes A^n converge as $n \to \infty$ to some continuous process, say A, and

$$M_t^2 = 2N_t + A_t. \tag{5.5.10}$$

Due to the fact that the sets $J_n(\omega)$ form refinements and that A_n increases on the stopping times T_k^n, it follows that

$$A_{T_k^n} \le A_{T_{k+1}^n} \tag{5.5.11}$$

for all k, n. So A is increasing on the closure of $J(\omega) = \bigcup_{n \in \mathbb{N}} J_n(\omega)$. Thus, if $J(\omega)$ is dense, then A is increasing. The remaining option is that the complement of $J(\omega)$ contains some open interval I. But in that case, since no T_k^n is in I, M must be constant on I, and so must A. Thus, A is a continuous increasing process such that $M^2 - A$ is a continuous martingale, and hence $A = [M]$.

It remains to show the uniqueness of the process $[M]$. For this we use the following lemma.

Lemma 5.45 *Suppose that M is a continuous local martingale that has paths of finite variation. If $M_0 = 0$, then $M_t = 0$ for all t.*

Proof Again, by stopping M at $\tau_n = \inf\{t \in \mathbb{R}_+ : V_M(t) > n\}$, where

$$V_M(t) = \lim_{n \to \infty} \sum_{k \in \mathbb{N}} |M_{u_k^n} - M_{u_{k-1}^n}| \tag{5.5.12}$$

is the total variation process, we may assume that M has bounded total variation. Then, obviously,

$$A_t^n = \sum_{k \in \mathbb{N}} (M_{t_k^n} - M_{t_{k-1}^n})^2 \leq 2^{-n} \sum_{k \in \mathbb{N}} |M_{t_k^n} - M_{t_{k-1}^n}| \leq 2^{-n} V_M(t), \tag{5.5.13}$$

which tends to zero as $n \to \infty$. Thus, M^2 is a martingale. So $\mathbb{E}[M_t^2] = 0$ for all t, and a positive random variable of zero mean is zero a.s. □

Uniqueness of $[M]$ follows from the above observations. Indeed, assume that there are two processes A, A' with the desired properties. Then $A - A'$ is the difference of two uniformly integrable martingales, and hence is itself a uniformly integrable martingale. On the other hand, since A and A' are increasing and hence are of finite variation, their difference is of finite variation, and thus is identically zero by Lemma 5.45. □

It will be convenient to note the following fact.

Theorem 5.46 *Let M be a càdlàg martingale. Then, for any $t \in \mathbb{R}_+$ and any sequence of partitions $\{u_k^n\}$ of the interval $[0, t]$ such that $\lim_{n \to \infty} \max_{k \in \mathbb{N}} |u_k^n - u_{k-1}^n| = 0$,*

$$\sum_{k \in \mathbb{N}} (M_{u_{k+1}^n} - M_{u_k^n})^2 \stackrel{\mathscr{D}}{\to} [M]_t. \tag{5.5.14}$$

Moreover, if M is square integrable, then the convergence also holds in $\mathscr{L}^1$.

The proof of this theorem is somewhat technical and will not be included. See e.g. Ethier and Kurtz [104].

For the case where M is Brownian motion, we have the following fact.

Lemma 5.47 *If B is standard Brownian motion, then $[B]_t = t$.*

Recall from the discrete-time theory that there were two brackets associated with a martingale: $\langle M\rangle$ and $[M]$. The first corresponds to the process given by Theorem 5.44, the second is the quadratic variation process. In the case of continuous martingales, they are the same.

5.5.2 Stochastic integrals for simple processes

We have already seen that the stochastic integral can be defined as a Stieltjes integral for integrators of bounded variation. We will now show the crucial connection between the quadratic variation process of the stochastic integral and the process $[M]$. We begin with the case where the integrand, X is a step function.

Definition 5.48 A stochastic process is called *simple process*, if it has sample paths that are step functions paths of the form

$$X_t(\omega) = \sum_{i=1}^{\infty} x_i(\omega)\mathbb{1}_{t_{i-1}<t\leq t_i}, \tag{5.5.15}$$

for some increasing sequence $t_i \in \mathbb{R}_+$ and random variables $x_i \in \mathbb{R}$ that are $\mathscr{F}_{t_{i-1}}$-measurable. The space of simple processes is denoted by $\mathscr{E}_b$.

Clearly, the stochastic integral for such a function is defined and equals

$$\int_0^t X_s dM_s = \sum_{t_i\leq t} x_i\big(M(t_i)-M(t_{i-1})\big) + x_{m(t)+1}\big(M(t)-M(t_{m(t)})\big),$$

where $m(t) = \max\{m \in \mathbb{N} : t_m \leq t\}$.

The following lemma states the crucial properties of stochastic integrals.

Lemma 5.49 *On some filtered probability space, let M be a continuous square-integrable martingale and $X \in \mathscr{E}_b$ a simple process. Then $\int_0^t X_s dM_s$ as defined above is a continuous square-integrable martingale and*

$$\left[\int_0^{\cdot} X dM\right]_t = \int_0^t X^2 d[M]. \tag{5.5.16}$$

Proof The proof is straightforward and will be skipped. See Ethier and Kurtz [104, Sect. 5.2.]. □

The stochastic integral for more general integrands should share the properties stated in Lemma 5.49. The natural goal is to extend the integral to integrands X for which the objects characterising it make sense. Note, in particular, that it follows from (5.5.16) that

$$\mathbb{E}\left[\left(\int_0^t X_s dM_s\right)^2\right] = \mathbb{E}\left[\int_0^t X_s^2 d[M]_s\right]. \tag{5.5.17}$$

This means that the map $X \mapsto \int_0^t X_s dM_s$, from the space of left-continuous step-functions equipped with the norm

$$\|X\|_{2,d[M]} = \left(\mathbb{E}\left[\int_0^t X_s^2 d[M]_s\right]\right)^{1/2} \tag{5.5.18}$$

to the space of local square-integrable martingales with the norm $\mathscr{L}^2(\mathbb{P})$, is an *isometry*, called the *Itō isometry*. We will extend this isometry to all of $\mathscr{L}^2(d[M])$ to define the *Itō integral*.

Theorem 5.50 *Let M be a continuous square-integrable local martingale, and let $X \in \mathscr{L}^2(d[M])$. Then there exists a unique continuous square-integrable local martingale $(\int_0^t X_s dM_s)_{t \in \mathbb{R}_+}$ such that for any sequence of left-continuous step-functions $(X_n)_{n \in \mathbb{N}}$ that satisfies*

$$\sum_{n \in \mathbb{N}} \left\{\mathbb{E}\left[\int_0^n (X_n - X)_s^2 d[M]_s\right]\right\}^{1/2} < \infty, \tag{5.5.19}$$

it is true that

$$\lim_{n \to \infty} \sup_{0 \le t \le T} \left|\int_0^t (X_n - X)_s dM_s\right| = 0, \tag{5.5.20}$$

a.s. and in $\mathscr{L}^2$. Moreover,

$$\left[\int_0^\cdot X_s dM_s\right]_t = \int_0^t X_s^2 d[M]_s. \tag{5.5.21}$$

Proof See Ethier and Kurtz [104, Sect. 5.2, Theorem 2.3]. □

Remark 5.51 Theorem 5.50 extends the isometry $X \mapsto \int X dM$ from the dense set of left-continuous bounded step-functions to the full space $\mathscr{L}^2(d[M])$.

Remark 5.52 Theorem 5.50 is not the end of the possible extensions of the definition of stochastic integrals. Using localisation arguments as indicated in the definition of the bracket $[M]$, we can extend the space of integrators to continuous local martingales without the assumption of square integrability.

5.5.3 *Itō formula*

We now come to a most useful formula involving the notion of stochastic integrals, the celebrated *Itō formula*. This formula is the analogue for functions of stochastic processes with unbounded variation of the fundamental theorem of calculus.

We consider a stochastic process X of the form

$$X = X_0 + V + M, \tag{5.5.22}$$

where V is a continuous adapted process of bounded variation, M is a local martingale (which we may assume to be in $\mathscr{L}^2$ by Remark 5.51), and $V_0 = M_0 = 0$. Let $f : \mathbb{R}_+ \times \mathbb{R} \to \mathbb{R}$ be continuously differentiable in the first argument and twice continuously differentiable in the second argument.

Theorem 5.53 (Itō formula) *With the assumptions stated above, the following holds*:

$$\begin{aligned} f(t, X_t) - f(0, X_0) = &\int_0^t \frac{\partial}{\partial s} f(s, X_s) ds + \int_0^t \frac{\partial}{\partial x} f(s, X_s) dV_s \\ &+ \int_0^t \frac{\partial}{\partial x} f(s, X_s) dM_s + \frac{1}{2} \int_0^t \frac{\partial^2}{\partial x^2} f(s, X_s) d[M]_s. \end{aligned} \tag{5.5.23}$$

Remark 5.54 The Itō formula can be stated more conveniently in differential form as

$$df(t, X_t) = \frac{\partial}{\partial t} f(t, X_t) dt + \frac{\partial}{\partial x} f(t, X_t) dX_t + \frac{1}{2} \frac{\partial^2}{\partial x^2} f(t, X_t) d[X]_t, \tag{5.5.24}$$

with the understanding that $d[X] = d[M]$, since the quadratic variation of the finite variation process V is zero.

Proof See Ethier and Kurtz [104, Sect. 5.2]. □

5.6 Stochastic differential equations

In this section we construct stochastic processes $X = (X_t)_{t \in \mathbb{R}_+}$ in $\mathbb{R}^d$ that solve differential equations of the form

$$dX_t = b(t, X_t) dt + \sigma(t, X_t) dB_t \tag{5.6.1}$$

with a prescribed initial condition $X_0 = x_0$, where B is Brownian motion. The interpretation of (5.6.1) is not entirely straightforward due to the term $\sigma(t, X_t) dB_t$.

We will therefore interpret it as the integral equation

$$X_t = x_0 + \int_0^t b(s, X_s)ds + \int_0^t \sigma(s, X_s)dB_s, \tag{5.6.2}$$

where the integral with respect to B is understood as the Itō stochastic integral. In the most general setting the *drift functions* b and the *diffusion functions* σ are assumed to be locally bounded and measurable.

The questions we are interested in are existence and uniqueness of solutions of (5.6.2), as well as properties of these solutions. In Sects. 5.6.1–5.6.2 below we discuss the notion of *strong* solution. For stochastic differential equations there is also the notion of *weak* solution, but we will skip that here.

5.6.1 Strong solutions

Abbreviate $W = C(\mathbb{R}_+, \mathbb{R}^n)$, and let $\mathscr{H} = \mathscr{B}(W)$ be its corresponding Borel-σ-algebra. Consider the filtration $(\mathscr{H}_t)_{t\in\mathbb{R}_+}$ with $\mathscr{H}_t$ the σ-algebra generated by the paths up to time t. We denote by $\bar{\mathscr{H}_t}$ the augmentation of $\mathscr{H}_t$ with respect to the Wiener measure. The formal set-up for a stochastic differential equation involves an initial condition and a Brownian motion, all of which require a probability space denoted by

$$\big(\Omega, \mathscr{F}, \mathbb{P}, (\mathscr{F}_t)_{t\in\mathbb{R}_+}, \xi, B\big), \tag{5.6.3}$$

where

(i) $(\Omega, \mathscr{F}, \mathbb{P}, (\mathscr{F}_t)_{t\in\mathbb{R}_+})$ is a filtered space satisfying the usual conditions.
(ii) B is a Brownian motion on $\mathbb{R}^d$ adapted to $(\mathscr{F}_t)_{t\in\mathbb{R}_+}$.
(iii) ξ is an $\mathscr{F}_0$-measurable random variable.

The *canonical* set-up has $\Omega = \mathbb{R}^n \times W$ and $\mathbb{P} = \mu \times \mathbb{Q}$, where μ is the law of ξ, $\mathbb{Q}$ is the Wiener measure, and $\mathscr{F}_t$ is the usual augmentation of $\mathscr{F}_t^0 = \sigma\{\xi, B_s, 0 \le s \le t\}$.

The precise definition of *pathwise uniqueness* of an SDE is as follows.

Definition 5.55 For an SDE pathwise uniqueness holds if the following is true. For any set-up $(\Omega, \mathscr{F}, \mathbb{P}, (\mathscr{F}_t)_{t\in\mathbb{R}_+}, \xi, B)$ and any two continuous semi-martingales X and X' such that

$$\int_0^t \big(|b(s, X_s)| + |\sigma(s, X_s)|^2\big)ds < \infty \tag{5.6.4}$$

solving the SDE with initial condition ξ and Brownian motion B,

$$\mathbb{P}\big(X_t = X_t' \ \forall t \in \mathbb{R}_+\big) = 1. \tag{5.6.5}$$

If an SDE admits for any set-up $(\Omega, \mathscr{F}, \mathbb{P}, (\mathscr{F}_t)_{t\in\mathbb{R}_+}, \xi, B)$ exactly one continuous semi-martingale as solution, then we say that it is *exact*.

The notion of *strong solution* is naturally associated with the setting of exact SDEs.

Definition 5.56 A strong solution of an SDE is a function

$$F\colon \mathbb{R}^n \times W \to W \tag{5.6.6}$$

such that

$$F^{-1}(\mathscr{H}_t) \subset \mathscr{B}(\mathbb{R}^n) \times \bar{\mathscr{H}_t} \quad \forall t \in \mathbb{R}_+ \tag{5.6.7}$$

and on any set-up $(\Omega, \mathscr{F}, \mathbb{P}, (\mathscr{F}_t)_{t\in\mathbb{R}_+}, \xi, B)$ the process $X = F(\xi, B)$ solves the SDE.

5.6.2 Existence and uniqueness of strong solutions

Existence and uniqueness results in the strong sense can be proven in a similar way as for the case of ordinary differential equations, with the help of Gronwall's inequality and Picard's iteration scheme. We recall Gronwall's lemma, whose proof is elementary.

Lemma 5.57 *Let $f, g\colon [0, b) \to \mathbb{R}_+$ be continuous nonnegative maps. Suppose that, for some $A \geq 0$, f satisfies*

$$f(t) \leq A + \int_0^t f(s)g(s)ds, \quad t \in [0, b). \tag{5.6.8}$$

Then

$$f(t) \leq A \exp\left(\int_0^t g(s)ds\right), \quad t \in [0, b). \tag{5.6.9}$$

In particular, if (5.6.8) *holds with $A = 0$, then $f = 0$.*

Lemma 5.57 is at the heart of all uniqueness proofs for differential equations. The idea is always the same: given a differential equation $x' = F(x, s)$, try to find a norm $\|x\|$ and a Lipschitz bound $\|F(s, x) - F(s, y)\| \leq g(s)\|x - y\|$, at least for small times s and small $\|x - y\|$. For two solutions x, y with initial values x_0, y_0, set $f(t) = \|x_t - y_t\|$. Then

$$f(t) \leq \|x_0 - y_0\| + \int_0^t g(s)f(s)ds, \tag{5.6.10}$$

so that we can apply Gronwall's lemma to get (at least locally)

$$f(t) \leq \|x_0 - y_0\| \exp\left(\int_0^t g(s)ds\right). \tag{5.6.11}$$

This implies uniqueness of solutions with the same initial conditions.

The general approach is to assume local Lipschitz conditions to prove existence of solutions for finite times, and then glue solutions together until a possible explosion. Let us give the basic uniqueness and existence results due to Itō.

Theorem 5.58 *Assume that b and σ are bounded and measurable, and that there exists an open set $U \subset \mathbb{R}$, $T > 0$ and $K < \infty$ such that*

$$\left|b(t,x) - b(t,y)\right| + \left|\sigma(t,x) - \sigma(t,y)\right| \leq K|x-y|, \quad x, y \in U, \ 0 \leq t \leq T. \tag{5.6.12}$$

Let X, Y be two solutions of (5.6.2) (*with the same Brownian motion B*), *and set*

$$\tau = \inf\{t \in \mathbb{R}_+ : \ X_t \notin U \text{ or } Y_t \notin U\}. \tag{5.6.13}$$

If $\mathbb{E}[(X_0 - Y_0)^2] = 0$, then

$$\mathbb{P}(X_{t\wedge\tau} = Y_{t\wedge\tau} \ \forall 0 \leq t \leq T) = 1. \tag{5.6.14}$$

Proof The proof is based on Gronwall's lemma and runs very much like its deterministic analogue. As norm we choose a uniform $\mathscr{L}^2$-bound:

$$\begin{aligned}
\mathbb{E}\Big[\max_{0\leq s\leq t}(X_{s\wedge\tau} - Y_{s\wedge\tau})^2\Big] &\leq 2\,\mathbb{E}\big[(X_0 - Y_0)^2\big] \\
&\quad + 4\,\mathbb{E}\Bigg[\max_{0\leq s\leq t}\bigg(\int_0^{s\wedge\tau}\big(\sigma(u,X_u) - \sigma(u,Y_u)\big)dB_u\bigg)^2\Bigg] \\
&\quad + 4\,\mathbb{E}\Bigg[\max_{0\leq s\leq t}\bigg(\int_0^{s\wedge\tau}\big(b(u,X_u) - b(u,Y_u)\big)du\bigg)^2\Bigg] \\
&\leq 16\,\mathbb{E}\bigg[\int_0^{t\wedge\tau}\big(\sigma(u,X_u) - \sigma(u,Y_u)\big)^2 du\bigg] \\
&\quad + 4t\,\mathbb{E}\bigg[\int_0^{t\wedge\tau}\big(b(u,X_u) - b(u,Y_u)\big)^2 du\bigg] \\
&\leq 4K^2(t+4)\mathbb{E}\bigg[\int_0^{t\wedge\tau}(X_u - Y_u)^2 du\bigg] \\
&\leq 4K^2(t+4)\int_0^t \mathbb{E}\Big[\max_{0\leq u\leq s}(X_{u\wedge\tau} - Y_{u\wedge\tau})^2\Big]ds.
\end{aligned} \tag{5.6.15}$$

The first inequality uses that $(a+b)^2 \leq 2a^2 + 2b^2$, the second inequality uses the Cauchy-Schwarz inequality for the drift term and the Doob L^2-maximum inequality for the diffusion term, the third inequality uses the Lipschitz condition, while the fourth inequality uses Fubini's theorem.

We see that $f(s) = \mathbb{E}[\max_{0\leq s\leq t}(X_{s\wedge\tau} - Y_{s\wedge\tau})^2]$ satisfies the hypothesis of Gronwall's lemma with $A = 0$, so that

$$\mathbb{E}\Big[\max_{0\leq t\leq T}(X_{t\wedge\tau} - Y_{t\wedge\tau})^2\Big] = 0. \tag{5.6.16}$$

By Chebyshev's inequality this implies that $\mathbb{P}(\max_{0\leq t\leq T}|X_t - Y_t| = 0) = 1$, as claimed. □

Finally, existence of solutions (for finite times) can be proved via the usual Picard iteration scheme under Lipschitz and growth conditions.

Theorem 5.59 *Let b, σ satisfy the Lipschitz conditions in* (5.6.12) *and assume that*

$$\max_{0\leq t\leq T}\left[\left|b(t,x)\right|^2 + \left|\sigma(t,x)\right|^2\right] \leq K^2\left(1+|x|^2\right). \tag{5.6.17}$$

Let ξ be a random vector with finite second moment, independent of B, and let $(\mathcal{F}_t)_{t\in\mathbb{R}_+}$ be the usual augmentation of the filtration associated with B and ξ. Then there exists a continuous $(\mathcal{F}_t)_{t\in\mathbb{R}_+}$-adapted process X that is a strong solution of the SDE with initial condition ξ. Moreover, X is square-integrable, i.e., for any $T>0$ there exists a $C(T,K)$ such that, for all $0\leq t\leq T$,

$$\mathbb{E}\left[\|X_t\|^2\right] \leq C(K,T)\left(1+\mathbb{E}\left[\|\xi\|^2\right]\right) e^{C(K,T)t}. \tag{5.6.18}$$

Proof We define a map F from the space of continuous adapted processes X that are uniformly square-integrable on $[0,T]$ to itself via

$$F(X)_t = \xi + \int_0^t b(s,X_s)ds + \int_0^t \sigma(s,X_s)dB_s. \tag{5.6.19}$$

Note that the square integrability of $F(X)$ needs the growth conditions in (5.6.17). As in (5.6.15),

$$\begin{aligned}\mathbb{E}\Big[\sup_{0\leq t\leq T}\|F(X)_t - F(Y)_t\|^2\Big] &\leq 2\mathbb{E}\left[\sup_{0\leq t\leq T}\left\|\int_0^t \big(\sigma(X_s)-\sigma(Y_s)\big)dB_s\right\|^2\right]\\ &\quad +2\mathbb{E}\left[\sup_{0\leq t\leq T}\left\|\int_0^t \big(b(X_s)-b(Y_s)\big)ds\right\|^2\right]\\ &\leq 2K^2(1+T)\int_0^T \mathbb{E}\Big[\sup_{0\leq s\leq t}\|X_s - Y_s\|^2\Big]dt,\end{aligned} \tag{5.6.20}$$

and, by iteration of this inequality, there exists a C (depending on K,T), such that

$$\mathbb{E}\Big[\sup_{0\leq t\leq T}\|F^k(X)_t - F^k(Y)_t\|^2\Big] \leq \frac{C^kT^{2k}}{k!}\mathbb{E}\Big[\sup_{0\leq t\leq T}\|X_t - Y_t\|^2\Big], \tag{5.6.21}$$

where F^k is the k-th iterate of F. Thus, for k sufficiently large, F^k is a contraction, and hence has a unique fixed point that solves the SDE. We can construct this solution as follows. Choose $X_t^{(0)} = \xi$, $X^{(k)} = F(X^{(k-1)})$, $k\in\mathbb{N}$. From the preceding inequality

$$\mathbb{E}\Big[\sup_{0\le t\le T}\|X_t^{(k+1)}-X_t^{(k)}\|^2\Big]\le\frac{C^kT^{2k}}{k!}\mathbb{E}\big[1+\xi^2\big].\tag{5.6.22}$$

Apply the same arguments to estimate

$$\mathbb{E}\big[\|X_t^{(k+1)}\|^2\big]\le K\,\mathbb{E}\big[\|\xi\|^2\big]+KT\int_0^t\big(1+\mathbb{E}\big[\|X_s^{(k)}\|^2\big]\big)ds.\tag{5.6.23}$$

Iterate this inequality,

$$\mathbb{E}\big[\|X_t^{(k+1)}\|^2\big]\le\mathbb{E}\big[\|\xi\|^2\big]+\big(1+\mathbb{E}\big[\|\xi\|^2\big]\big)\sum_{i=1}^k\frac{KT^{i+1}}{i!}\le KT\big(1+\mathbb{E}\big[\|\xi\|^2\big]\big)\,\mathrm{e}^{KTt},\tag{5.6.24}$$

to get the growth bound in (5.6.18) with $C(K,T)=KT$. □

5.6.3 The Doob transform

An important way in which a drift can be produced is via conditioning. We have already seen this in the case of discrete-time Markov processes. We will again see that the martingale formulation plays a useful rôle. As in the discrete-time setting, the key result is the following.

Theorem 5.60 *Let X be a Markov process, i.e., a solution of the martingale problem for an operator $\mathscr{L}$, and let h be a strictly positive harmonic function. Define the measure $\mathbb{P}^h$ such that, for any $\mathscr{F}_t$-measurable random variable Y,*

$$\mathbb{E}_x^h[Y]=\frac{1}{h(x)}\mathbb{E}_x\big[h(X_t)Y\big].\tag{5.6.25}$$

Then $\mathbb{P}^h$ is the law of a solution of the martingale problem for the operator $\mathscr{L}^h$ defined by

$$\big(\mathscr{L}^hf\big)(x)=\frac{1}{h(x)}(\mathscr{L}hf)(x).\tag{5.6.26}$$

As an important example, let us consider the case of Brownian motion in a regular domain $D\subset\mathbb{R}^d$ killed at the boundary ∂D. We assume that h is a harmonic function on D, and we let τ_D be the first exit time of D. Then

$$\mathscr{L}^h=\frac{1}{2}\Delta+\frac{\nabla h}{h}\cdot\nabla,\tag{5.6.27}$$

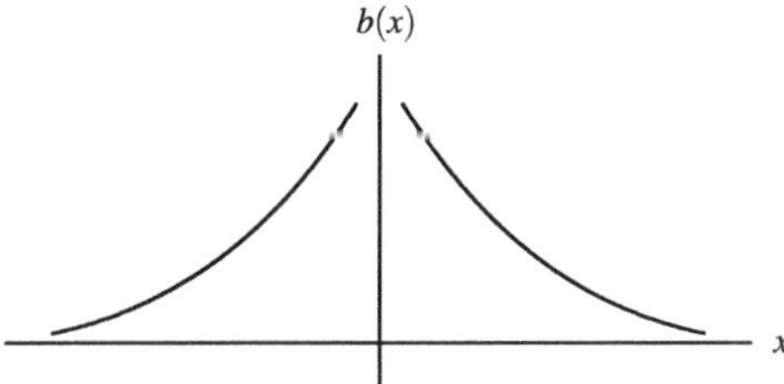

Fig. 5.3 Drift function $b\colon \mathbb{R} \to \mathbb{R}$ for Brownian motion conditioned to never hit the origin: $b(x) = 1/x$

and hence, under the law $\mathbb{P}^h$, the Brownian motion becomes the solution of the SDE

$$dX_t = \frac{\nabla h(X_t)}{h(X_t)}\,dt + dB_t. \tag{5.6.28}$$

On the other hand, we have seen that if $h(x)$ is the probability of some event, e.g. $h(x) = \mathbb{P}_x(X_{\tau_D} \in A)$ for some $A \in \partial D$, then

$$\mathbb{P}^h(\cdot) = \mathbb{P}(\cdot\,|X_{\tau_D} \in A). \tag{5.6.29}$$

This means that the Brownian motion conditioned to exit D at a given location can be represented as the solution of an SDE with a specific drift. For instance, let $d = 1$ and $D = (0, R)$. Consider the Brownian motion conditioned to leave D at R. It is well known that

$$\mathbb{P}_x(X_{\tau_D} = R) = x/R, \quad x \in D. \tag{5.6.30}$$

Thus, the conditioned Brownian motion solves

$$dX_t = \frac{1}{X_t}\,dt + dB_t. \tag{5.6.31}$$

We can let $R \to \infty$ without changing the SDE. Hence, the solution of (5.6.31) is Brownian motion conditioned to never return to the origin (see Fig. 5.3). This is reasonable, because the strength of the drift away from zero goes to infinity near 0. Still, it is remarkable that the conditioning can be reproduced by the application of a proper drift.

5.6.4 The Girsanov theorem

The Girsanov theorem is a particularly useful tool to study properties of stochastic processes that can be seen as modifications of Brownian motions. For simplicity we consider only the one-dimensional setting, but the obvious extension to the multi-dimensional setting holds as well.

Suppose that we are given a filtered space $(\Omega, \mathscr{F}, \mathbb{P}, (\mathscr{F}_t)_{t\in\mathbb{R}_+})$ satisfying the usual assumptions, a Brownian motion B and an adapted process X that is square-integrable with respect to dt, i.e., X is an integrand for B. Suppose that we want to study the process

$$W_t = B_t - \int_0^t X_s \, ds. \tag{5.6.32}$$

We may think of $X_s = b(s, B_s)$ for some bounded measurable function b, the simplest example being $b(s, X_s) = b$, in which case

$$W_t = B_t - bt,$$

which is Brownian motion with a constant drift b.

How can we compute properties of W? In particular, can we find a new probability measure, written $\widetilde{\mathbb{P}}$, such that under this new measure W becomes simple? The Girsanov theorem is a striking affirmative answer to this question.

Theorem 5.61 (Girsanov theorem) *Let B, X, W be as above. Define*

$$Z_t = Z_t(X) = \exp\left(\int_0^t X_s \, dB_s - \tfrac{1}{2} \int_0^t X_s^2 \, ds\right) \tag{5.6.33}$$

and let $\widetilde{\mathbb{P}}$ be defined by

$$\widetilde{\mathbb{P}}_T(A) = \mathbb{E}\big[Z_T(X) \mathbb{1}_A\big]. \tag{5.6.34}$$

If $(Z_t)_{0 \le t \le T}$ is a martingale, then the process $(W_t)_{0 \le t \le T}$ is a Brownian motion under $\widetilde{\mathbb{P}}_T$.

Remark 5.62 We may check, using the Itō formula, that Z_t solves

$$dZ_t = Z_t \, X_t \, dB_t. \tag{5.6.35}$$

To see why, let $f(t, x) = \mathrm{e}^{x - \int_0^t X_s^2 \, ds}$ and $Y_t = \int_0^t X_s \, dB_s$. Then Y is a martingale with bracket $[Y]_t = \int_0^t X_s^2 \, ds$, and $Z_t = f(t, Y_t)$. By the Itō formula,

$$df(t, Y_t) = f(t, Y_t)\big(-\tfrac{1}{2} X_t^2 \, dt + dY_t + \tfrac{1}{2} d[Y]_t\big) = Z_t \, X_t \, dB_t. \tag{5.6.36}$$

Hence $(Z_t)_{0 \le t \le T}$ is a positive local martingale and so, by Fatou's lemma, a supermartingale. It is a martingale whenever $\mathbb{E}[Z_t] = 1$ for all t.

Proofs of the Girsanov theorem can be found in most standard textbooks on stochastic analysis, e.g. in Karatzas and Shreve [148].

5.7 Stochastic partial differential equations

The theory of stochastic partial differential equations (SPDEs) is substantially more involved than that of stochastic differential equations (SDEs), and in many respects

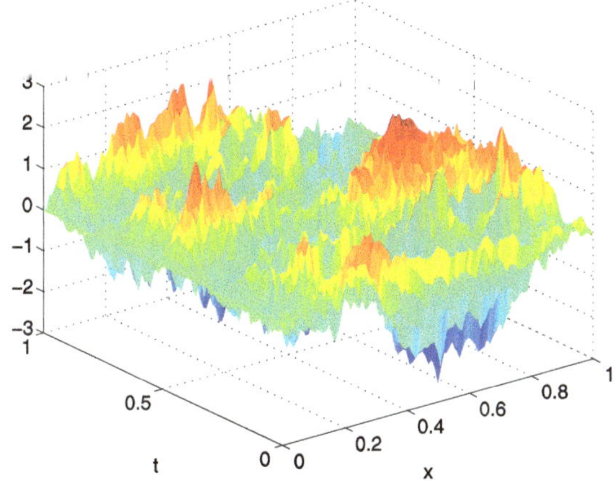

Fig. 5.4 A space-time plot $(x,t) \mapsto u(x,t)$ of the stochastic Allen-Cahn equation

is not yet finalised. For recent developments, see Hairer [135]. In Chap. 12 we will discuss metastability for one example system, namely, the *stochastic Allen-Cahn equation* (see Fig. 5.4 for a visualisation). In this section we present the relevant background.

5.7.1 The stochastic Allen-Cahn equation

Formally, the stochastic Allen-Cahn equation is a partial differential equation of the form

$$\frac{\partial}{\partial t} u(x,t) = \frac{1}{2} D \frac{\partial^2}{\partial x^2} u(x,t) - V'\big(u(x,t)\big) + \sqrt{2\varepsilon}\, \frac{\partial^2}{\partial x \partial t} W(x,t). \tag{5.7.1}$$

Here, $x \in [0,1]$ is the space-coordinate, $t \in \mathbb{R}_+$ is the time-coordinate, $D > 0$ is the coupling constant, $V\colon \mathbb{R} \to \mathbb{R}$ is the potential, $\varepsilon > 0$ is the noise-strength, and W is the space-time *Brownian sheet*, i.e., the centred Gaussian process indexed by $[0,1] \times \mathbb{R}_+$ with covariance

$$\mathbb{E}\big[W(x,t)W(y,s)\big] = (x \wedge y)(t \wedge s). \tag{5.7.2}$$

The initial condition is given by $u(x,0) = u_0(x)$, $x \in [0,1]$, with u_0 a continuous function. We also need to choose boundary conditions, e.g. periodic boundary conditions $u(0,t) = u(1,t)$, $t \in \mathbb{R}_+$, or von Neumann boundary conditions $\partial_x u(0,t) = \partial_x u(1,t) = 0$, $t \in \mathbb{R}_+$.

We need some further assumptions on the potential V.

Assumption 5.63

- V is C^3 on $\mathbb{R}$.
- V is convex at infinity, i.e., there exist $R, c > 0$ such that

$$V''(u) > c > 0, \quad |u| > R. \tag{5.7.3}$$

- V is polynomial of finite degree.

The SPDE in (5.7.1) can be seen as the stochastic perturbation of an infinite-dimensional gradient system,

$$\frac{\partial}{\partial t}u = -D_\phi F, \tag{5.7.4}$$

where, for ϕ a differentiable function,

$$F(\phi) = \int_0^1 \left[\tfrac{1}{2}D\left|\phi'(x)\right|^2 + V\big(\phi(x)\big)\right] dx, \tag{5.7.5}$$

and $D_\phi F$ is the Fréchet derivative of F. Of course, (5.7.1) is an informal expression because the derivatives of the Brownian sheet do not exist. Formally, we think of it as a Gaussian process such that

$$\mathbb{E}\left[\frac{\partial^2 W(x,t)}{\partial x \partial t}\frac{\partial^2 W(y,s)}{\partial y \partial s}\right] = \delta(x-y)\delta(t-s), \tag{5.7.6}$$

where $\delta(\cdot)$ is the Dirac function, but it is clear that (5.7.1) requires even more interpretation than an SDE.

To get an idea of what is at stake, consider first the linear equation

$$\frac{\partial v(x,t)}{\partial t} = \tfrac{1}{2}D\frac{\partial^2 v(x,t)}{\partial^2 x} + \sqrt{2\varepsilon}\,\frac{\partial^2 W(x,t)}{\partial x \partial t}, \tag{5.7.7}$$

with initial condition $v(x,0) = 0$. Naturally, we expect to be able to solve this equation with the help of the Fourier transform. Indeed, space-time white noise can be constructed as a Fourier series as follows. Let $B_n(t)$, $n \in \mathbb{Z}$, be i.i.d. Brownian motions. Set

$$\frac{\partial W(t,x)}{\partial x} = \sum_{n\in\mathbb{Z}} e^{(2\pi i)nx} B_n(t). \tag{5.7.8}$$

On the level of formal computations, this process has the desired correlation structure. Therefore, denoting by

$$\hat{v}(n,t) = \frac{1}{2\pi}\int_0^1 e^{-(2\pi i)nx} v(x,t)dx \quad n \in \mathbb{Z}, \tag{5.7.9}$$

the spatial Fourier coefficients of v, we find that these satisfy the stochastic ordinary differential equations

$$d\hat{v}(n,t) = -\tfrac{1}{2}D(2\pi n)^2\hat{v}(n,t)\,dt + \sqrt{2\varepsilon}\,dB_n(t), \tag{5.7.10}$$

with initial condition $\hat{v}(n,0) = 0$. Note that these equations are uncoupled for different n. The equations in (5.7.10) are interpreted as Itō-SDEs, so we are on firm ground. The solution of (5.7.10) is readily found to be

$$\hat{v}(n,t) = \sqrt{2\varepsilon}\,\mathrm{e}^{-\frac{1}{2}D(2\pi n)^2 t}\int_0^t \mathrm{e}^{\frac{1}{2}D(2\pi n)^2 s}\,dB_n(s). \tag{5.7.11}$$

This suggests the representation of the solution v in the form

$$v(x,t) = \sum_{n\in\mathbb{Z}} \mathrm{e}^{(2\pi i)nx}\,\hat{v}(n,t). \tag{5.7.12}$$

A quick check whether this series represents a bona fide stochastic process is the computation of the variance of the spatial L^2-norm:

$$\begin{aligned}\mathbb{E}\big[\|v(x,t)\|_2^2\big] &= 2\varepsilon\sum_{n\in\mathbb{Z}} \mathrm{e}^{-D(2\pi n)^2 t}\,\mathbb{E}\bigg[\int_0^t \mathrm{e}^{\frac{1}{2}D(2\pi n)^2 s}\,dB_n(s)\bigg]^2 \qquad (5.7.13)\\ &= 2\varepsilon\sum_{n\in\mathbb{Z}} \mathrm{e}^{-D(2\pi n)^2 t}\int_0^t \mathrm{e}^{D(2\pi n)^2 s}\,ds\\ &= 2\varepsilon\sum_{n\in\mathbb{Z}} \frac{1}{D(2\pi n)^2}\big(1-\mathrm{e}^{-D(2\pi n)^2 t}\big).\end{aligned}$$

Clearly, the sum converges.

Remark 5.64 If we do the same analysis in dimension $d > 1$, then we obtain

$$\sum_{n\in\mathbb{Z}^d} \frac{1}{D(2\pi n)^2}\big(1-\mathrm{e}^{-D(2\pi n)^2 t}\big), \tag{5.7.14}$$

which diverges.

So, at least in dimension $d = 1$, we can construct a proper stochastic process taking values in L^2 that can be reasonably considered a solution of the linear SPDE in (5.7.7). This process has much nicer properties than the space-time white noise itself, which suggests that it is better to *define* solutions of the non-linear equation through solutions of the linear equation. This goes as follows. Denote by $p_t(x,y)$ the density of the semi-group generated by $D\partial^2/\partial^2 x$ on $[0,1]$ (the heat kernel with suitable boundary conditions). Then a solution of the inhomogeneous linear equation

$$du(x,t) - \tfrac{1}{2}D\,\frac{\partial^2 u(x,t)}{\partial^2 x}\,dt = r(x,t)\,dt \tag{5.7.15}$$

with initial condition $u_0(x)$ can be written as

$$u(x,t) = \int_0^1 dy\; g_t(x,y)u_0(y) + \int_0^t ds \int_0^1 dy\; g_{t-s}(x,y)r(y,s). \tag{5.7.16}$$

Write the non-linear equation in (5.7.1) as

$$\frac{\partial u(x,t)}{\partial t} - \tfrac{1}{2}D\,\frac{\partial^2 u(x,t)}{\partial^2 x} = -V'\big(u(x,t)\big) + \sqrt{2\varepsilon}\,\frac{\partial^2 W(x,t)}{\partial x\,\partial t}. \tag{5.7.17}$$

Next, think of the entire right-hand side as an inhomogeneous term (i.e., ignore the fact that the right-hand side involves the solution itself). Then we can represent the solution of (5.7.17) as

$$\begin{aligned} u(x,t) &= \int_0^1 dy\; g_t(x,y)u_0(y) - \int_0^t ds \int_0^1 dy\; g_{t-s}(x,y)V'\big(u(y,s)\big) \\ &\quad + \sqrt{2\varepsilon}\int_0^t ds \int_0^1 g_{t-s}(x,y)\,dW(y,s) \\ &= \int_0^1 dy\; g_t(x,y)u_0(y) - \int_0^t ds \int_0^1 dy\; g_{t-s}(x,y)V'\big(u(y,s)\big) \\ &\quad + v(x,t). \end{aligned} \tag{5.7.18}$$

Here, the last term is taken as the solution of the linear equation in (5.7.7) we just constructed. The other terms in the right-hand side are ordinary integrals, so all the expressions make sense. A so-called *mild solution* is a process that satisfies this equation, i.e., instead of the ill-defined SPDE driven by space-time white noise, we now have a non-linear integral equation driven by the more regular noise $v(x,t)$.

Definition 5.65 A random field u is a *mild solution* of (5.7.1) if:

(i) Almost surely u is continuous on $[0,1]\times\mathbb{R}_+$ and predictable.
(ii) For all $(x,t)\in[0,1]\times\mathbb{R}_+$,

$$u(x,t) = \int_0^1 dy\; g_t(x,y)u_0(y) - \int_0^t ds \int_0^1 dy\; g_{t-s}(x,y)V'\big(u(y,s)\big) + v(x,t), \tag{5.7.19}$$

with $v(x,t)$ given by (5.7.12).

Existence, uniqueness and regularity of the solution of (5.7.1) are contained in the following theorem, where $C_{bc}([0,1])$ denotes the set of continuous functions on $[0,1]$ respecting the chosen boundary conditions.

Theorem 5.66 *For every initial condition $u_0 \in C_{bc}([0,1])$, the SPDE in (5.7.1) has a unique mild solution. Moreover, for all $T > 0$ and $p \geq 1$,*

$$\mathbb{E}\Big[\sup_{[0,T]\times[0,1]} \big|u(x,t)\big|^p\Big] \leq C(T,p). \tag{5.7.20}$$

The random field u is 2α-Hölder in space and α-Hölder in time for every $\alpha \in (0, \frac{1}{4})$.

The only complication that arises comes from the fact that V' is not globally Lipschitz. However, due to Assumption 5.63, we have

$$-xV'(x) < C, \tag{5.7.21}$$

which allows us to prove global existence via a localisation argument, as in the analogous SDE cases.

Remark 5.67 As noted earlier, $v(x,t)$ is regular only in dimension $d = 1$. In dimensions $d > 1$, the construction breaks down. The source of the problem is the strong spatial irregularity of the noise. Therefore in applications often SPDEs with stronger spatial correlations are considered. The task is to choose the right noise for the application at hand.

5.7.2 Discretisation

From our perspective, the stochastic Allen-Cahn equation should arise as the limit of a spatially discrete system. This works out well in dimension $d = 1$. The discrete system consists of $N \in \mathbb{N}$ coupled stochastic differential equations of the form

$$dX_j(t) = -\frac{\partial}{\partial X_j} F_{D,N}\big(X(t)\big)dt + \sqrt{2\varepsilon}\, dB_j(t), \quad j \in \Lambda, \tag{5.7.22}$$

where $\Lambda = \mathbb{Z}/N\mathbb{Z} = \{1, \dots, N\}$, $X_j(t)$ are the components of $X(t) \in \mathbb{R}^N$, and

$$F_{D,N}(x) = \sum_{j\in\Lambda} V(x_j) + \tfrac{1}{4} D \sum_{j\in\Lambda} (x_j - x_{j+1})^2, \tag{5.7.23}$$

with B_j, $j \in \Lambda$, independent Brownian motions.

In order to obtain a limit as $N \to \infty$, we need to rescale the potential and the coupling constant. To that end we replace $F_{D,N}(x)$ by

$$F_{D,N}(x) = N^{-1} F_{N^2 D,N}(x). \tag{5.7.24}$$

If we replace the unit lattice by a lattice of spacing $1/N$, i.e., $(x_j)_{j\in\Lambda}$ is the discretisation of a real-valued function x on $[0,1]$ given by $x_j = x(j/N)$, then the resulting potential converges formally to

$$\lim_{N\to\infty} F_{D,N}(x) = \int_0^1 V\big(x(s)\big)\,ds + \tfrac{1}{2}D\int_0^1 \big[x'(s)\big]^2\,ds \tag{5.7.25}$$

with $x(0) = x(1)$. We need to rescale the Brownian noise by a factor $1/\sqrt{N}$. We may relate this to a space-time white noise by setting formally

$$N^{-1/2}B_j(t) = \int_{(j-1)/N}^{j/N} W(x,t)dx, \quad j \in \Lambda. \tag{5.7.26}$$

Finally, we must accelerate time by a factor N, i.e., we set $X^N(t) = X(tN)$. The resulting discrete equations take the form

$$dX_j^N(t) = -\tfrac{1}{2}DN^2\big[X_{j+1}^N(t)+X_{j-1}^N-2X_j^N(t)\big]dt - V'\big(X_j^N(t)\big)dt+\sqrt{2\varepsilon N}d\widetilde{B}_j(t), \tag{5.7.27}$$

with $\tilde{B}_j$, $j \in \Lambda$, independent Brownian motions.

We now define the function $u^N(x,t)$: $[0,1]\times\mathbb{R}_+ \to \mathbb{R}$ such that, for any given $t \in \mathbb{R}_+$, $u^N(\cdot,t)$ is the linear interpolation between the points $(j/N, X_j^N(t))$, $j \in \Lambda$. Then u^N can also be represented as a mild solution of the discrete system, which allows us to prove convergence to a mild solution of the SPDE. To do this, we proceed again by solving the linear equations

$$dY_j^N(t) = -\tfrac{1}{2}D\,N^2\big[Y_{j+1}^N(t) + Y_{j-1}^N - 2Y_j^N(t)\big]\,dt + \sqrt{2\varepsilon N}\,d\widetilde{B}_j(t) \tag{5.7.28}$$

with the help of Fourier series. Set $\widehat{Y}_n^N = N^{-1}\sum_{j\in\Lambda} \mathrm{e}^{-(2\pi i)nj/N} Y_j^N$ and, consequently, $Y_j^N = \sum_{n\in\Lambda} \mathrm{e}^{(2\pi i)nj/N}\widehat{Y}_n^N$. Note that also

$$\sqrt{N}\,\widetilde{B}_j(t) = \sum_{n\in\Lambda} \mathrm{e}^{(2\pi i)nj/N} B_n(t), \tag{5.7.29}$$

where the B_n, $n \in \Lambda$, are the independent Brownian motions from (5.7.8). A simple computation yields that the Fourier modes satisfy the SDEs (with zero initial condition)

$$d\widehat{Y}_n^N(t) = -\tfrac{1}{2}DN^2 2\big[\cos(2\pi n/N) - 2\big]\widehat{Y}_n^N(t)\,dt + \sqrt{2\varepsilon}\,dB_n(t). \tag{5.7.30}$$

Abbreviate $\widehat{\Delta}_n^N = N^2 2[\cos(2\pi n/N) - 2]$. Then these equations can be solved as

$$\widehat{Y}_n^N(t) = \sqrt{2\varepsilon}\,\mathrm{e}^{-\frac{1}{2}D\widehat{\Delta}_n^N t}\int_0^t \mathrm{e}^{\frac{1}{2}D\widehat{\Delta}_n^N s}\,dB_n(s). \tag{5.7.31}$$

Define

$$v_j^N(t) = \sum_{n\in\Lambda} \mathrm{e}^{(2\pi i)nj/N}\widehat{Y}_n^N(t). \tag{5.7.32}$$

As in the continuous case, we get the ℓ^2-bound

$$\mathbb{E}\Big[N^{-1}\sum_{j\in\Lambda}|v_j^N(t)|^2\Big]=\sum_{n\in\Lambda}\frac{1-e^{-D\widehat{\Delta}_n^N t}}{D\widehat{\Delta}_n^N}. \tag{5.7.33}$$

It is easy to see that this expression converges to the right-hand side of (5.7.13) as $N\to\infty$. In fact, denoting by $v^N(x,t)$ the linear interpolation of the points $(j/N, v_j^N(t))$, $j\in\Lambda$, we can show that this process converges to $v(x,t)$ in L^2.

Finally we can write the discrete equations in their mild form as

$$X_j^N(t)=\sum_{k\in\Lambda}p_t^N(j,k)X_k(0)-\sum_{k\in\Lambda}\int_0^t p_{t-s}^N(j,k)V'\big(X_k^N(s)\big)\,ds+v_j^N(t), \tag{5.7.34}$$

where p^N is the semi-group associated with the discrete Laplacian. Note that the noise is coupled to the mild form of the SPDE in that both are driven by the same Brownian motions.

The above formulation can be further embellished by writing it for the interpolations u^N defined by putting $u^N(j/N,t)=X_j^N(t)$ for $j\in\Lambda$ and using linear interpolation. Define (for von Neumann boundary conditions)

$$\kappa_N(x)=\frac{\lfloor Nx\rfloor}{N}+\frac{1}{2N}. \tag{5.7.35}$$

Let p^N be the linear interpolation of p^N on $[0,1]\times[0,1]$ along the discretisation points.

Lemma 5.68 *For every $u_0\in C_{bc}([0,1])$ and $N\in\mathbb{N}$ the function u^N defined on $[0,1]\times\mathbb{R}^+$ satisfies the equation*

$$\begin{aligned}u^N(x,t)&=\int_0^1 dy\ g_t^N\big(x,\kappa_N(y)\big)u_0\big(\kappa_N(y)\big)\\&\quad-\int_0^t ds\int_0^1 dy\ g_{t-s}^N\big(x,\kappa_N(y)\big)V'\big(u^N\big(\kappa_N(y),s\big)\big)+v^N(x,t).\end{aligned} \tag{5.7.36}$$

For all $T>0$ and $p\geq 1$,

$$\sup_{N\in\mathbb{N}}\mathbb{E}\Big[\sup_{(x,t)\in[0,1]\times[0,T]}\big|u^N(x,t)\big|^p\Big]\leq C(T,p). \tag{5.7.37}$$

The following theorem asserts the convergence of the solution of (5.7.36) to the solution of (5.7.1).

Theorem 5.69 *For all $u_0\in C_{bc}^3([0,1])$, $T>0$ and $p\geq 1$,*

$$\lim_{N\to\infty}u^N=u, \tag{5.7.38}$$

where convergence holds in the following senses:

- *In* L^p, *i.e.*, $\lim_{N\to\infty}\mathbb{E}[\|u^N - u\|^p_{\infty,T}]^{\frac{1}{p}} = 0$.
- *Almost surely in* $C([0,1]\times[0,T])$, *i.e., for every* $\eta\in(0,\frac{1}{2})$ *there exists an almost surely finite random variable* Ξ *such that*

$$\left\|u^N - u\right\|_{\infty,T} \le \frac{\Xi}{N^\eta}, \tag{5.7.39}$$

where $\|w\|_{\infty,T} = \sup_{t\in[0,T]}\sup_{x\in[0,1]}|w(x,t)|$.

The convergence of the solutions also implies the convergence of the hitting times of the discrete approximations to those of the SPDE. A precise statement is as follows. Let u_0 be the initial condition of the solution of (5.7.1) and ϕ a continuous function. For $\rho > 0$, define the hitting times

$$\begin{aligned} \tau(\rho) &= \inf\bigl\{t>0\colon \left\|u(t)-\phi\right\|_\infty < \rho\bigr\}, \\ \tau^N(\rho) &= \inf\bigl\{t>0\colon \left\|u^N(t)-\phi^N\right\|_\infty < \rho\bigr\}, \end{aligned} \tag{5.7.40}$$

where ϕ^N is the linear approximation of ϕ.

Theorem 5.70 *Suppose that* $\lim_{N\to\infty}\|\phi^N - \phi\|_\infty = 0$ *and that there exists a* ρ_0 *such that, for every* $0<\rho<\rho_0$,

$$\mathbb{E}_{u_0}\bigl[\tau(\rho)\bigr] < \infty. \tag{5.7.41}$$

Then, for every $0<\rho<\rho_0$,

$$\lim_{N\to\infty}\tau^N(\rho) = \tau(\rho) \quad a.s., \qquad \lim_{N\to\infty}\mathbb{E}_{u_0^N}\bigl[\tau^N(\rho)\bigr] = \mathbb{E}_{u_0}\bigl[\tau(\rho)\bigr]. \tag{5.7.42}$$

The proof of this theorem is straightforward and can be found in Barret [11].

5.8 Bibliographical notes

1. Much of the exposition in this chapter is taken from Ethier and Kurtz [104], Roger and Williams [207, 208] and Karatzas and Shreve [148]. The martingale problem formulation is due to Stroock and Varadhan [223].

2. The conditions for existence stated in Theorems 5.59 are not necessary. In particular, growth conditions are important only when the solutions can reach regions where the coefficients become too large. Formulations of weaker hypotheses for existence and uniqueness can be found in Jacod and Shiryaev [143, Chap. 14]. Their verification in concrete cases can be tricky.

3. Examples of introductory textbooks on SPDEs are Da Prato [68], Holden [139] and Röckner [203]. A classical treatise is the St. Flour lecture notes by Walsh [235].

4. The Allen-Cahn (or Ginzburg-Landau) equation models the behaviour of an elastic string in a potential with viscous stochastic forcing (see e.g. Funaki [117]). It also has interpretations in quantum field theory (see Fajona [107], Cassandro, Olivieri and Picco [52]), and in statistical physics as a reaction-diffusion equation modelling phase transitions and evolution of interfaces (see Brassesco [41], Brassesco and Buttà [42]).

5. The existence, uniqueness and regularity of the solution of the stochastic Allen-Cahn equation stated in Proposition 5.66 was proved in Gyöngy and Pardoux [134]. The convergence of the finite discretisation in Theorem 5.69 was proved in Funaki [117] and Gyöngy [133] for V with V' globally Lipschitz. Barret [11] extended this result to the setting of Assumption 5.63.

6. Existence of strong solutions via renormalisation for a class of SPDEs containing the Allen-Cahn equation on the two-dimensional torus was shown by Da Prato and Debussche [67]. There are interesting recent developments. Hairer [135] proposes a renormalisation strategy that allows to give sense to the white noise case in dimensions $d = 2, 3$.

Chapter 6
Large Deviations

> *"A large cage!" the Professor promptly replied. "Bring a large cage", he said to the people generally, "with strong bars of steel, and a portcullis made to go up and down like a mouse-trap! Does anyone happen to have such a thing about him?"* (Lewis Carroll, Sylvie and Bruno Concluded)

This chapter gives a summary introduction to large deviations. Although large deviation theory is not our main interest in this monograph, it is an essential element in our conceptual understanding of metastability. Moreover, it provides tools to obtain estimates, which often serve as preliminary steps towards more refined estimates.

Section 6.1 recalls the main ingredients of large deviation theory in a general setting (without proofs). Section 6.2 gives a full derivation of path large deviations for diffusion processes (under strong regularity assumptions). Section 6.3 takes a brief look at path large deviations for stochastic partial differential equations. Section 6.4 formulates the extension to path large deviations for Markov processes (without proofs). Section 6.5 gives a brief outline of the Freidlin-Wentzell theory of metastability, collects some properties of associated action integrals, and looks at crossing and exit problems that are crucial for a proper understanding of metastability.

6.1 Large deviation principles

Definition 6.1 A family of probability measures $(\mu_\varepsilon)_{\varepsilon>0}$ on a Polish space $\mathscr{X}$ is said to satisfy the large deviation principle (LDP) with rate function $I\colon \mathscr{X} \to [0,\infty]$ if

(i) I has compact level sets and is not identically infinite,
(ii) $\liminf_{\varepsilon\downarrow 0} \varepsilon \ln \mu_\varepsilon(O) \geq -I(O)$ for all $O \subseteq \mathscr{X}$ open,
(iii) $\limsup_{\varepsilon\downarrow 0} \varepsilon \ln \mu_\varepsilon(C) \leq -I(C)$ for all $C \subseteq \mathscr{X}$ closed,

where $I(S) = \inf_{x\in S} I(x)$, $S \subseteq \mathscr{X}$.

Informally, the LDP says that if $B_\delta(x)$ is a ball of radius $\delta > 0$ centred at $x \in \mathscr{X}$, then

$$\mu_\varepsilon\big(B_\delta(x)\big) = \mathrm{e}^{-[1+o(1)]I(x)/\varepsilon} \tag{6.1.1}$$

A. Bovier, F. den Hollander, *Metastability*,
Grundlehren der mathematischen Wissenschaften 351,
DOI 10.1007/978-3-319-24777-9_6

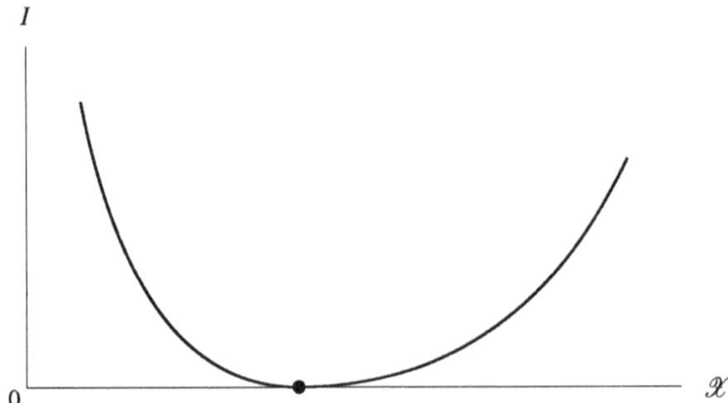

Fig. 6.1 Paradigmatic picture of a rate function with a unique zero

when $\varepsilon \downarrow 0$ followed by $\delta \downarrow 0$ (see Fig. 6.1).

If in (i) the level sets of I are assumed to be closed only and in (iii) the inequality is assumed to hold for compact sets only, then it is said that the *weak LDP* holds. Strengthening a weak LDP to an LDP amounts to establishing *exponential tightness*, i.e., to proving that for every $N < \infty$ there exists a compact set $K_N \subseteq \mathscr{X}$ such that $\limsup_{\varepsilon \downarrow 0} \varepsilon \ln \mu_\varepsilon([K_N]^c) \leq -N$.

The LDP is the workhorse for the computation of averages of exponential functionals, as contained in the following lemma.

Lemma 6.2 (Varadhan's lemma) *If* $(\mu_\varepsilon)_{\varepsilon>0}$ *satisfies the LDP on* $\mathscr{X}$ *with rate function* I*, then*

$$\lim_{\varepsilon \downarrow 0} \varepsilon \ln \int_{\mathscr{X}} \mathrm{e}^{F(x)/\varepsilon}\, \mu_\varepsilon(dx) = \Lambda_F, \quad \forall F \in C_b(\mathscr{X}), \tag{6.1.2}$$

where $C_b(\mathscr{X})$ *is the space of bounded continuous functions on* $\mathscr{X}$*, and*

$$\Lambda_F = \sup_{x \in \mathscr{X}} \big[F(x) - I(x)\big]. \tag{6.1.3}$$

The result in Lemma 6.2 can be extended to include F that are unbounded and/or discontinuous, provided certain tail estimates on μ_ε are available. Varadhan's lemma has the following inverse.

Lemma 6.3 (Bryc's lemma) *Suppose that* $(\mu_\varepsilon)_{\varepsilon>0}$ *is exponentially tight and the limit in* (6.1.2) *exists for all* $F \in C_b(\mathscr{X})$*. Then* $(\mu_\varepsilon)_{\varepsilon\geq 0}$ *satisfies the LDP with rate function* I *given by*

$$I(x) = \sup_{F \in C_b(\mathscr{X})} \big[F(x) - \Lambda_F\big], \quad x \in \mathscr{X}. \tag{6.1.4}$$

There are several "forward principles" that allow LDP's to be generated from one another. A key example is the *contraction principle* (see Fig. 6.2).

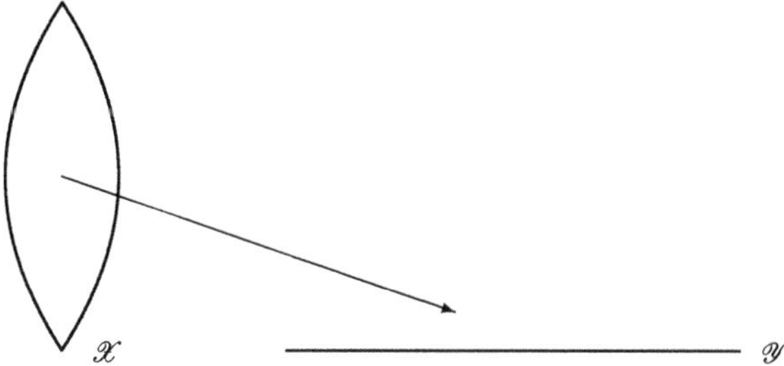

Fig. 6.2 Illustration of the contraction principle

Lemma 6.4 (Contraction principle) *Let $(\mu_\varepsilon)_{\varepsilon>0}$ satisfy the LDP on $\mathscr{X}$ with rate function I. Let $\mathscr{Y}$ be a second Polish space, and let $T\colon\ \mathscr{X} \to \mathscr{Y}$ be a continuous map from $\mathscr{X}$ to $\mathscr{Y}$. Then the family of probability measures $(\nu_\varepsilon)_{\varepsilon>0}$ on $\mathscr{Y}$ defined by $\nu = \mu \circ T^{-1}$ satisfies the LDP on $\mathscr{Y}$ with rate function J given by*

$$J(y) = \inf_{\substack{x\in\mathscr{X}\\ T(x)=y}} I(x), \quad y \in \mathscr{Y}. \tag{6.1.5}$$

Another example is via exponential tilting:

Lemma 6.5 *Let $(\mu_\varepsilon)_{\varepsilon>0}$ satisfy the LDP on $\mathscr{X}$ with rate function I, and let $F \in C_b(\mathscr{X})$. Then the family of probability measures $(\nu_\varepsilon)_{\varepsilon>0}$ on $\mathscr{X}$ defined by*

$$\nu_\varepsilon(dx) = \frac{1}{N_\varepsilon}\,\mathrm{e}^{F(x)/\varepsilon}\,\mu_\varepsilon(dx), \qquad N_\varepsilon = \int_X \mathrm{e}^{F(x)/\varepsilon}\,\mu_\varepsilon(dx), \tag{6.1.6}$$

satisfies the LDP on $\mathscr{X}$ with rate function J given by

$$J(x) = \Lambda_F - \big[F(x) - I(x)\big], \quad x \subset \mathscr{X}. \tag{6.1.7}$$

A final example is the *Dawson-Gärtner projective limit LDP*:

Theorem 6.6 (Dawson-Gärtner projective limit LDP) *Let $(\mu_\varepsilon)_{e>0}$ be a family of probability measures on $\mathscr{X}$. Let $(\pi^N)_{N\in\mathbb{N}}$ be a nested family of projections acting on $\mathscr{X}$, such that $\bigcup_{n\in\mathbb{N}} \pi^N\mathscr{X} = \mathscr{X}$, and let*

$$\mathscr{X}^N = \pi^N\mathscr{X}, \quad \mu_\varepsilon^N = \mu_\varepsilon \circ \big(\pi^N\big)^{-1}, \quad N \in \mathbb{N}. \tag{6.1.8}$$

If, for each $N \in \mathbb{N}$, the family $(\mu_\varepsilon^N)_{\varepsilon>0}$ satisfies the LDP on $\mathscr{X}^N$ with rate function I^N, then $(\mu_\varepsilon)_{\varepsilon>0}$ satisfies the LDP on $\mathscr{X}$ with rate function I given by

$$I(x) = \sup_{N\in\mathbb{N}} I^N\big(\pi^N x\big), \quad x \in \mathscr{X}. \tag{6.1.9}$$

Since

$$I^N(y) = \inf_{\{x\in\mathscr{X}:\ \pi^N(x)=y\}} I(x), \quad y \in \mathscr{X}^N, \tag{6.1.10}$$

the supremum in (6.1.9) is monotone in N by the nestedness property. The projective limit LDP can, for instance, be used to extend a suitably nested sequence of LDP's on finite-dimensional spaces to an LDP on an infinite-dimensional space.

LDPs can be formulated on general topological spaces $\mathscr{X}$, although typically this comes at the cost of more technicalities. Conversely, more can be said when $\mathscr{X}$ has more structure. For instance, if $\mathscr{X}$ is vector space, then the rate function can be identified as the *Legendre transform* of a (generalised) cumulant generating function. When $\mathscr{X} = \mathbb{R}^d$, this is known as the *Gärtner-Ellis theorem*:

Theorem 6.7 (Gärtner-Ellis theorem) *Let $(\mu_\varepsilon)_{\varepsilon>0}$ be a family of probability measures on $\mathbb{R}^d$, $d \geq 1$, with the following properties*:

(i) $\phi(u) = \lim_{\varepsilon\downarrow 0} \varepsilon \ln \int_{\mathbb{R}^d} \mathrm{e}^{\langle u,x\rangle/\varepsilon}\, \mu_\varepsilon(dx)$ *exists in* $\mathbb{R}$ *for all* $u \in \mathbb{R}^d$, *where* $\langle\cdot,\cdot\rangle$ *denotes the standard inner product on* $\mathbb{R}^d$.
(ii) $u \mapsto \phi(u)$ *is differentiable on* $\mathbb{R}^d$.

Then $(\mu_\varepsilon)_{\varepsilon>0}$ satisfies the LDP on $\mathbb{R}^d$ with a convex rate function ϕ^* given by

$$\phi^*(x) = \sup_{u\in\mathbb{R}^d} \big[\langle u,x\rangle - \phi(u)\big], \quad x \in \mathbb{R}^d. \tag{6.1.11}$$

There is a version of Theorem 6.7 where the domain of ϕ is not all of $\mathbb{R}^d$, in which case additional assumptions must be made.

Two special cases of Theorem 6.7 deserve to be mentioned. Let $(X_i)_{i\in\mathbb{N}}$, be i.i.d. $\mathbb{R}$-valued random variables with common law ρ. Let $\mathscr{M}_1(\mathbb{R})$ denote the space of probability measures on $\mathbb{R}$ (which is a subset of the vector space of signed measures on $\mathbb{R}$).

- *Cramér's Theorem*: Let μ_n denote the law of the *empirical average* $n^{-1}\sum_{i=1}^n X_i$. If $M(\lambda) = \int \mathrm{e}^{\lambda x}\rho(dx) < \infty$ for all $\lambda \in \mathbb{R}$, then $(\mu_n)_{n\in\mathbb{N}}$ satisfies the LDP on $\mathbb{R}$ with rate $\varepsilon = n^{-1}$ and rate function

$$I(x) = \sup_{\lambda\in\mathbb{R}}\big[\lambda x - \ln M(\lambda)\big], \quad x \in \mathbb{R}. \tag{6.1.12}$$

- *Sanov's Theorem*: Let μ_n denote the law of the *empirical distribution* $n^{-1}\sum_{i=1}^n \delta_{X_i}$. Then $(\mu_n)_{n\in\mathbb{N}}$ satisfies the LDP on $\mathscr{M}_1(\mathbb{R})$ with rate $\varepsilon = n^{-1}$ and rate function

$$I(\nu) = \int_{\mathbb{R}} \nu(dx) \ln\left[\frac{d\nu}{d\rho}(x)\right], \quad \nu \in \mathscr{M}_1(\mathbb{R}), \tag{6.1.13}$$

with the right-hand side infinite when ν is not absolutely continuous with respect to ρ.

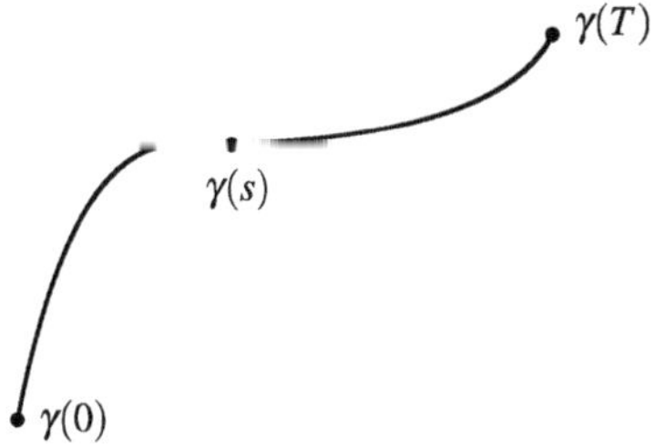

Fig. 6.3 A path γ over the time interval $[0, T]$

6.2 Path large deviations for diffusion processes

The general theory in Sect. 6.1 serves as the framework for many concrete examples. In this section we take a look at large deviations on path space for diffusion processes. In Sect. 6.2.1 we derive *Schilder's theorem* for Brownian motion. The proof is written out in detail in order to convey the background of this theorem. In Sects. 6.2.2–6.2.3 we show how to extend Schilder's theorem to diffusions.

6.2.1 Brownian motion

Brownian motion $B = (B_t)_{t\in\mathbb{R}_+}$ on $\mathbb{R}^d$ starting at the origin will typically be at a distance of order $\sqrt{t}$ from the origin at time t, in particular, B_t/t converges to zero a.s. as $t \to \infty$. We are interested in computing the probability that B follows an exceptional path for which B_t lives on space-scale t. To formalise this idea, we fix a time-horizon $T > 0$ and a smooth path $\gamma\colon\ [0, T] \to \mathbb{R}^d$ starting at the origin (see Fig. 6.3), and we estimate the probability

$$\mathbb{P}\Big(\sup_{s\in[0,T]} \big\|\varepsilon B_{s/\varepsilon} - \gamma(s)\big\| \leq \delta\Big), \quad \varepsilon \downarrow 0, \tag{6.2.1}$$

where $\|\cdot\|$ denotes the Euclidean norm on $\mathbb{R}^d$. The following result is known as *Schilder's theorem.* Let $C_0([0, T])$ be the space of continuous functions $f\colon\ [0, T] \to \mathbb{R}^d$ such that $f(0) = 0$ equipped with the supremum norm $\|f\|_\infty = \sup_{s\in[0,T]} \|f(s)\|$.

Theorem 6.8 (Schilder's theorem) *Set $B^\varepsilon = (B^\varepsilon_s)_{s\in[0,T]}$ with $B^\varepsilon_s = \varepsilon B_{s/\varepsilon}$. Then $(B^\varepsilon)_{\varepsilon>0}$ satisfies the LDP on $C_0([0, T])$ with rate function I given by*

$$I(\gamma) = \begin{cases} \frac{1}{2}\int_0^T \|\dot\gamma(s)\|^2 ds, & \text{if } \gamma \in H_1, \\ \infty, & \text{otherwise}, \end{cases} \tag{6.2.2}$$

where H_1 is the space of absolutely continuous functions with square-integrable derivative equipped with the norm $\|f\|_{H_1} = [\int_0^T \|\dot f(s)\|^2 ds]^{1/2}$.

Proof First, we prove a lower bound for the probability in (6.2.1).

Lemma 6.9 *For every $\gamma \in H_1$,*

$$\liminf_{\varepsilon \downarrow 0} \varepsilon \ln \mathbb{P}\big(\big\| B^\varepsilon - \gamma \big\|_\infty < \delta\big) \geq -I(\gamma) \quad \forall \delta > 0. \tag{6.2.3}$$

Proof Fix $\delta > 0$. Note that $(\varepsilon B_{s/\varepsilon})_{s\in[0,T]}$ has the same distribution as $(\varepsilon^{1/2} B_s)_{s\in[0,T]}$. Hence

$$\mathbb{P}\big(\big\| B^\varepsilon - \gamma \big\|_\infty < \delta\big) = \mathbb{P}\big(\big\| B - \varepsilon^{-1/2} \gamma \big\|_\infty < \varepsilon^{-1/2} \delta\big). \tag{6.2.4}$$

To estimate the probability in the right-hand side, we observe that, by the Girsanov theorem (Theorem 5.61), the process $\widetilde{B} = (\widetilde{B}_s)_{s\in[0,T]}$ defined by

$$\widetilde{B}_s = B_s - \varepsilon^{-1/2} \gamma(s) \tag{6.2.5}$$

is a Brownian motion under the measure $\mathbb{Q}$ defined through the Radon-Nikodým derivative

$$\frac{d\mathbb{Q}}{d\mathbb{P}} = \exp\left[\varepsilon^{-1/2} \int_0^T \dot{\gamma}(s) dB_s - \varepsilon^{-1} \tfrac{1}{2} \int_0^T \|\dot{\gamma}(s)\|^2 ds\right]. \tag{6.2.6}$$

Hence, abbreviating

$$Z(B, \gamma) = \int_0^T \dot{\gamma}(s) dB_s, \tag{6.2.7}$$

we get

$$\begin{aligned}
\mathbb{P}\big(\big\| B - \varepsilon^{-1/2} \gamma \big\|_\infty < \varepsilon^{-1/2} \delta\big) &= \mathbb{P}\big(\|\widetilde{B}\|_\infty < \varepsilon^{-1/2} \delta\big) \\
&= \mathbb{E}_{\mathbb{Q}}\big[\exp\big(-\varepsilon^{-1/2} Z(\widetilde{B}, \gamma) - \varepsilon^{-1} I(\gamma)\big) \mathbb{1}_{\|\widetilde{B}\|_\infty < \varepsilon^{-1/2}\delta}\big] \\
&= \exp\big(-\varepsilon^{-1} I(\gamma)\big)\, \mathbb{Q}\big(\|\widetilde{B}\|_\infty < \varepsilon^{-1/2} \delta\big) \\
&\quad \times \mathbb{E}_{\mathbb{Q}}\big[\exp\big(-\varepsilon^{-1/2} Z(\widetilde{B}, \gamma)\big) \mid \|\widetilde{B}\|_\infty < \varepsilon^{-1/2} \delta\big] \\
&= \exp\big(-\varepsilon^{-1} I(\gamma)\big)\, \mathbb{P}\big(\|B\|_\infty < \varepsilon^{-1/2} \delta\big) \\
&\quad \times \mathbb{E}_{\mathbb{P}}\big[\exp\big(-\varepsilon^{-1/2} Z(B, \gamma)\big) \mid \|B\|_\infty < \varepsilon^{-1/2} \delta\big].
\end{aligned} \tag{6.2.8}$$

From Jensen's inequality we have that

$$\begin{aligned}
&\mathbb{E}_{\mathbb{P}}\big[\exp\big(-\varepsilon^{-1/2} Z(B, \gamma)\big) \mid \|B\|_\infty < \varepsilon^{-1/2} \delta\big] \\
&\quad \geq \exp\big(-\varepsilon^{-1/2}\, \mathbb{E}_{\mathbb{P}}\big[Z(B, \gamma) \mid \|B\|_\infty < \varepsilon^{-1/2} \delta\big]\big) = 1.
\end{aligned} \tag{6.2.9}$$

On the other hand, it is easy to see (applying Doob's maximum in equality for submartingales in Theorem 3.57) that

$$\lim_{\varepsilon \downarrow 0} \mathbb{P}\big(\|B\|_\infty < \varepsilon^{-1/2} \delta\big) = 1 \tag{6.2.10}$$

and hence

$$\liminf_{\varepsilon\downarrow 0} \varepsilon \ln \mathbb{P}\Big(\big\| B - \varepsilon^{-1/2}\gamma \big\|_\infty < \varepsilon^{-1/2}\delta\Big) \geq -I(\gamma). \tag{6.2.11}$$

Together with (6.2.4) this gives the desired lower bound. □

Next, we prove an upper bound for the probability in (6.2.1), which is stated in a somewhat particular form.

Lemma 6.10 *Let* $K_\lambda = \{\gamma \in H_1 \colon I(\gamma) \leq \lambda\}$, $\delta, \lambda \in [0,\infty)$. *Then*

$$\limsup_{\varepsilon\downarrow 0} \varepsilon \ln \mathbb{P}\Big(\inf_{\gamma\in K_\lambda} \big\| B^\varepsilon - \gamma \big\|_\infty > \delta\Big) \leq -\lambda. \tag{6.2.12}$$

Proof Fix $\lambda \in [0,\infty)$. For $n \in \mathbb{N}$, set $t_k = (k/n)T$, $k = 0,\dots,n$. Let $L^\varepsilon = (L^\varepsilon_s)_{s\in[0,T]}$ be the linear interpolation of $B^\varepsilon = (B^\varepsilon_s)_{s\in[0,T]}$ such that $B^T_{t_k} = L^T_{t_k}$ for $k = 0,\dots,n$. Then

$$\begin{aligned}
\mathbb{P}\big(\big\| B^\varepsilon - L^\varepsilon \big\|_\infty > \delta\big) &\leq \sum_{k=1}^n \mathbb{P}\Big(\max_{s\in[t_{k-1},t_k]} \big\| B^\varepsilon_s - L^\varepsilon_s \big\| > \delta\Big)\\
&= n\,\mathbb{P}\Big(\max_{s\in[0,T/n]} \big\| B^\varepsilon_s - (sn/T)B^\varepsilon_{T/n} \big\| > \delta\Big)\\
&= n\,\mathbb{P}\Big(\max_{s\in[0,1]} \| B_s - sB_1 \| > \delta(n/T\varepsilon)^{1/2}\Big)\\
&\leq n\,\mathbb{P}\Big(\max_{s\in[0,1]} \| B_s \| > \tfrac12\delta(n/T\varepsilon)^{1/2}\Big),
\end{aligned} \tag{6.2.13}$$

where we use that $\max_{s\in[0,1]} \|B_s\| \leq \frac12 x$ implies $\max_{s\in[0,1]} \|B_s - sB_1\| \leq x$. The last probability can be estimated by using the following exponential inequality for one-dimensional Brownian motion:

$$\mathbb{P}_{d=1}\Big(\max_{s\in[0,t]} |B_s| > xt\Big) \leq 2\exp\big(-\tfrac12 x^2 t\big), \quad t \in \mathbb{R}_+. \tag{6.2.14}$$

This is easily obtained by using that $Z = (Z_t)_{t\in\mathbb{R}_+}$ with $Z_t = \exp(B_t - \frac12 t)$ is a martingale and by applying the Doob maximum inequality for submartingales in (3.57). Inserting (6.2.14) into (6.2.13), we get

$$\begin{aligned}
\mathbb{P}\Big(\max_{s\in[0,1]} \|B_s\| > \tfrac12\delta(n/\varepsilon T)^{1/2}\Big) &\leq d\,\mathbb{P}_{d=1}\Big(\max_{s\in[0,1]} |B_s| > \tfrac12\delta(n/Td\varepsilon)^{1/2}\Big)\\
&\leq 2d\,\mathrm{e}^{-\frac{\delta^2 n}{8Td\varepsilon}},
\end{aligned} \tag{6.2.15}$$

and so

$$\limsup_{\varepsilon\downarrow 0} \varepsilon \ln \mathbb{P}\big(\big\| B^\varepsilon - L^\varepsilon \big\|_\infty > \delta\big) \leq -\frac{\delta^2 n}{8Td}. \tag{6.2.16}$$

On the other hand,

$$\limsup_{\varepsilon\downarrow 0} \varepsilon \ln \mathbb{P}\big(I(L^\varepsilon) > \lambda\big) \le -\lambda. \tag{6.2.17}$$

Indeed, we have

$$I(L^\varepsilon) = \frac{n}{2T} \sum_{k=1}^{n} \big\| B^\varepsilon_{t_k} - B^\varepsilon_{t_{k-1}} \big\|^2 = \tfrac{1}{2}\varepsilon \sum_{i=1}^{dn} \eta_i^2, \tag{6.2.18}$$

where η_i, $i \in \mathbb{N}$, are i.i.d. standard normal random variables. Since

$$\mathbb{E}\big[\mathrm{e}^{\rho \frac{1}{2}\eta_i^2}\big] \le C_\rho < \infty \quad \forall\, 0 < \rho < 1, \tag{6.2.19}$$

it follows that

$$\mathbb{P}\left(\tfrac{1}{2}\varepsilon \sum_{i=1}^{dn} \eta_i^2 > \lambda \right) \le \mathrm{e}^{-\rho\lambda/\varepsilon}\, \mathbb{E}\big[\mathrm{e}^{\rho \sum_{i=1}^{dn} \frac{1}{2}\eta_i^2}\big] \le \mathrm{e}^{-\rho\lambda/\varepsilon} (C_\rho)^{dn}, \tag{6.2.20}$$

which yields (6.2.17) after letting $\varepsilon \downarrow 0$ followed by $\rho \uparrow 1$. Combining (6.2.16)–(6.2.17), and using that

$$\mathbb{P}\Big(\inf_{\gamma\in K_\lambda} \big\| B^\varepsilon - \gamma \big\|_\infty > \delta \Big) \le \mathbb{P}\big(\big\| B^\varepsilon - L^\varepsilon \big\|_\infty > \delta \big) + \mathbb{P}\big(I(L^\varepsilon) > \lambda\big), \tag{6.2.21}$$

we get

$$\limsup_{\varepsilon\downarrow 0} \varepsilon \ln \mathbb{P}\Big(\inf_{\gamma\in K_\lambda} \big\| B^\varepsilon - \gamma \big\|_\infty > \delta \Big) \le \left(-\frac{\delta^2 n}{8Td} \right) \vee (-\lambda), \tag{6.2.22}$$

which yields the claim after we let $n \to \infty$. □

Finally, we show that I has compact level sets.

Lemma 6.11 *K_λ, $\lambda \in [0,\infty)$, are compact.*

Proof We have

$$\gamma \in H_1 \iff \sup_{N\in\mathbb{N}} \sup_{0\le t_1<\cdots<t_N\le T} \frac{1}{2} \sum_{i=1}^{N-1} \frac{|\gamma(t_{i+1}) - \gamma(t_i)|^2}{|t_{i+1} - t_i|^2} < \infty, \tag{6.2.23}$$

with the supremum being equal to $I(\gamma)$. Let $(\gamma_n)_{n\in\mathbb{N}}$ be a convergent sequence in $C_0([0,T])$ with limit γ. Then

$$\begin{aligned}
I(\gamma) &= \sup_{N\in\mathbb{N}} \sup_{0\le t_1<\cdots<t_N\le T} \frac{1}{2}\sum_{i=1}^{N-1} \frac{|\gamma(t_{i+1})-\gamma(t_i)|^2}{|t_{i+1}-t_i|^2} \\
&= \sup_{N\in\mathbb{N}} \sup_{0\le t_1<\cdots<t_N\le T} \frac{1}{2}\sum_{i=1}^{N-1} \lim_{n\to\infty} \frac{|\gamma_n(t_{i+1})-\gamma_n(t_i)|^2}{|t_{i+1}-t_i|^2} \\
&\le \lim_{n\to\infty} \sup_{N\in\mathbb{N}} \sup_{0\le t_1<\cdots<t_N\le T} \frac{1}{2}\sum_{i=1}^{N-1} \frac{|\gamma_n(t_{i+1})-\gamma_n(t_i)|^2}{|t_{i+1}-t_i|^2} \\
&= \lim_{n\to\infty} I(\gamma_n),
\end{aligned} \tag{6.2.24}$$

i.e., $\gamma \mapsto I(\gamma)$ is lower semi-continuous, which implies that K_λ, $\lambda \in [0,\infty)$, are closed. Next, for any $\lambda\in[0,\infty)$, $\gamma\in K_\lambda$, and $0\le u<v\le T$,

$$\|\gamma(v)-\gamma(u)\| = \left\| \int_u^v \dot\gamma(s)ds \right\| \le \left[(v-u)\int_u^v \|\dot\gamma(s)\|^2 ds \right]^{1/2} \le [2(v-u)\lambda]^{1/2}. \tag{6.2.25}$$

Hence K_λ is uniformly bounded and uniformly equi-continuous. Therefore, by the Arzelà-Ascoli theorem, K_λ is compact. □

We are now ready to prove the LDP in Theorem 6.8. Since I is not identically infinite, Part (i) in Definition 6.1 follows from Lemma 6.11. To get Part (ii), note that for every open set $O\subset C_0([0,1])$ and every $\gamma\in O$, there exists a $\delta=\delta(O,\gamma)>0$ such that $\{\gamma'\colon\ \|\gamma'-\gamma\|_\infty<\delta\}\subseteq O$. Hence, Lemma 6.9 implies that

$$\liminf_{\varepsilon\downarrow 0} \varepsilon \ln \mathbb{P}(O) \ge \liminf_{\varepsilon\downarrow 0} \varepsilon \ln \mathbb{P}\big(\|B^\varepsilon-\gamma\|_\infty<\delta\big) \ge -I(\gamma), \tag{6.2.26}$$

which yields Part (ii) after we take the supremum over $\gamma\in O$. Part (iii) is derived as follows. Since K_λ, $\lambda\in[0,\infty)$, are compact, we know that

$$\forall\lambda,\delta'>0\ \exists\delta=\delta(\lambda,\delta')>0\colon\ K^\delta_{\lambda-2\delta'} \subseteq K_{\lambda-\delta'}, \tag{6.2.27}$$

where $K^\delta_\lambda = \{\gamma'\colon\ \inf_{\gamma\in K_\lambda} \le \|\gamma'-\gamma\|_\infty \le \delta\}$ is the δ-blow-up of K_λ. Pick $C\subseteq C_0([0,T])$ closed, and let $\lambda_C = \inf_{\gamma\in C} I(\gamma)$. Then $C\subseteq [K_{\lambda_C-\delta'}]^c$ for all $\delta'>0$. Hence (6.2.27) gives

$$\mathbb{P}(C) \le \mathbb{P}([K_{\lambda_C-\delta'}]^c) \le \mathbb{P}\big([K^\delta_{\lambda-2\delta'}]^c\big), \quad \delta=\delta(\lambda_C,\delta'). \tag{6.2.28}$$

By applying Lemma 6.10, we get

$$\limsup_{\varepsilon\downarrow 0} \varepsilon \ln \mathbb{P}(C) \le \limsup_{\varepsilon\downarrow 0} \varepsilon \ln \mathbb{P}\Big(\inf_{\gamma\in K_{\lambda_C - 2\delta'}} \| B^\varepsilon - \gamma \|_\infty > \delta \Big)$$
$$\le -\big(\lambda_C - 2\delta'\big). \tag{6.2.29}$$

Letting $\delta' \downarrow 0$, we get Part (iii). □

6.2.2 Brownian motion with drift

We next show how to pass to the analogous result for a Brownian motion with a drift, namely, we consider the SDE

$$X_t^\varepsilon = B_t^\varepsilon + \int_0^t b\big(X_s^\varepsilon\big) ds, \quad t \in \mathbb{R}_+, \tag{6.2.30}$$

where $b\colon \mathbb{R}^d \to \mathbb{R}^d$ is assumed to be globally Lipschitz.

Theorem 6.12 *Set $X^\varepsilon = (X_s^\varepsilon)_{s\in[0,T]}$. Then $(X^\varepsilon)_{\varepsilon>0}$ satisfies the LDP on $C_0([0,T])$ with rate function $\widetilde{I}$ given by*

$$\widetilde{I}(\gamma) = \tfrac{1}{2} \int_0^T \big\| \dot{\gamma}(s) - b\big(\gamma(s)\big) \big\|^2 ds. \tag{6.2.31}$$

Proof The easiest way to set up the proof is to consider the map $F\colon C_0([0,T]) \to C_0([0,T])$ given by

$$F(\gamma) = f, \tag{6.2.32}$$

where f is the solution of the integral equation

$$f(t) = \gamma(t) + \int_0^t b\big(f(s)\big) ds, \quad t \in [0,T]. \tag{6.2.33}$$

We may use Gronwall's lemma to show that F is continuous. Therefore $X^\varepsilon = F(B^\varepsilon)$ in distribution, and hence

$$\mathbb{P}\big(X^\varepsilon \in A\big) = \mathbb{P}\big(B^\varepsilon \in F^{-1}(A)\big), \quad A \subseteq C_0\big([0,T]\big). \tag{6.2.34}$$

Since under a continuous map the inverse image of an open (closed) set is again an open (closed) set, we can use Schilder's theorem (Theorem 6.8) and the contraction principle (Lemma 6.4) to obtain that $(X^\varepsilon)_{\varepsilon>0}$ satisfies the LDP with rate function $\widetilde{I} = I \circ F^{-1}$, i.e.,

$$\widetilde{I}(\gamma) = \inf\big\{ I(f)\colon\ F(\gamma) = f \big\}. \tag{6.2.35}$$

Since

$$F^{-1}(f)(t) = \gamma(t) = f(t) - \int_0^t b(f(s))ds, \quad t \in [0, T], \tag{6.2.36}$$

we get the claim. □

6.2.3 Diffusion processes

It is possible to push the argument in Sect. 6.2.2 and consider the SDE

$$X_t^\varepsilon = \int_0^t b(X_s^\varepsilon)ds + \int_0^t \sigma(X_s^\varepsilon)dB_s^\varepsilon, \quad t \in \mathbb{R}_+, \tag{6.2.37}$$

where $b\colon \mathbb{R}^d \to \mathbb{R}^d$ is assumed to be bounded and globally Lipschitz, and $\sigma\colon \mathbb{R}^d \to \mathbb{M}_d$ (with $\mathbb{M}_d$ the space of $d \times d$ $\mathbb{R}$-valued matrices) is assumed to have entries $\sigma_{ij}\colon \mathbb{R}^d \to \mathbb{R}$, $1 \le i, j \le d$, that are bounded and globally Lipschitz as well. Moreover, the *diffusion matrix* $a = \sigma\sigma^\dagger$ (the product of σ and its transpose $\sigma^\dagger$) is assumed to be *uniformly elliptic*, i.e., there exists a $\delta > 0$ such that $\langle a(x)y, y\rangle \ge \delta\|y\|^2$ for all $x, y \in \mathbb{R}^d$.

Theorem 6.13 *Set $X^\varepsilon = (X_s^\varepsilon)_{s\in[0,T]}$. Then $(X^\varepsilon)_{\varepsilon>0}$ satisfies the LDP on $C_0([0,1])$ with rate function $\widehat{I}$ given by*

$$\widehat{I}(\gamma) = \begin{cases} \frac{1}{2}\int_0^T \langle[\dot\gamma - (b\circ\gamma)],\, a^{-1}(\gamma)[\dot\gamma - (b\circ\gamma)]\rangle(s)ds, & \text{if } \gamma \in H_1, \\ \infty, & \text{otherwise,} \end{cases} \tag{6.2.38}$$

where a^{-1} is the inverse of a, and $\langle\cdot,\cdot\rangle$ is the standard inner product on $\mathbb{R}^d$.

Theorem 6.13 can be deduced from Theorem 6.12 with the help of a *time-change argument*. To see how, first suppose that $b = 0$ and for simplicity take $d = 1$. Then $[B^\varepsilon]_{[X^\varepsilon]_t} = [X^\varepsilon]_t$, where $[\cdot]$ is the quadratic variation (recall Theorem 5.44). Let $i(t)$ be such that $(X^\varepsilon_{i(t)})_{t\in\mathbb{R}_+}$ is Brownian motion. Then $[X^\varepsilon]_{i(t)} = t$. On the other hand,

$$[X^\varepsilon]_{i(t)} = \int_0^{i(t)} \sigma(X_s^\varepsilon)^2 ds. \tag{6.2.39}$$

Hence differentiation gives $\sigma(X^\varepsilon_{i(t)})^2 = 1$, which together with $X^\varepsilon_{i(t)} = B_t^\varepsilon$ shows that

$$i(t) = \int_0^t \sigma(B_s^\varepsilon)^{-2} ds. \tag{6.2.40}$$

Since $i(t)$ is measurable with respect to the filtration generated by $(B_s^\varepsilon)_{s\in[0,t]}$, it follows that $B^\varepsilon_{i^{-1}(t)}$ has the same distribution as X_t^ε and hence is a weak solution of (6.2.37) with $b = 0$. Schilder's theorem for this time-changed Brownian motion yields the claim in (6.2.38) with $b = 0$ (and $a = \sigma^2$). After adding the drift, we get (6.2.38).

Finally, it is possible to allow b, σ to be *time-dependent*. For the rate functions this simply amounts to writing $b(\gamma(s), s)$ and $\sigma(\gamma(s), s)$ in the formulas. We refer to the literature for assumptions and proofs.

6.3 Path large deviations for stochastic partial differential equations

The LDPs in Sect. 6.2 can be extended from SDEs to SPDEs. We focus on the class of SPDEs that was described in Sect. 5.7.

Return to (5.7.1). For $(x, t) \in [0, 1] \times \mathbb{R}_+$, consider the family of SPDEs

$$\frac{\partial}{\partial t} u^\varepsilon(x,t) = \frac{1}{2} D \frac{\partial^2}{\partial x^2} u^\varepsilon(x,t) - V'\big(u^\varepsilon(x,t)\big) + \sqrt{2\varepsilon}\, \frac{\partial^2}{\partial x \partial t} W(x,t), \tag{6.3.1}$$

where $\varepsilon > 0$ is a parameter that scales the strength of the noise. We want to think of a mild solution as a random variable taking values in a Banach space. To do so, define, for $\alpha \in (0, \infty)$, the Banach space

$$\mathscr{B}_\alpha = \big\{ f \in C\big([0,1]\big) \colon \|f\|_\alpha < \infty \big\} \tag{6.3.2}$$

equipped with the norm

$$\|f\|_\alpha = \sup_{x \in [0,1]} \big|f(x)\big| + \sup_{x,y \in [0,1]} \frac{|f(x) - f(y)|}{|x - y|^\alpha}. \tag{6.3.3}$$

As initial condition for (6.3.1) we take $u(\cdot, 0) = \xi(\cdot)$ with $\xi \in \mathscr{B}_\alpha$. Fix $T > 0$. Define the space $W_2^{1,2}$ as

$$W_2^{1,2} = \left\{ \gamma \colon [0,1] \times [0,T] \to \mathbb{R} \colon \int_0^T dt \int_0^1 dx \left| \frac{\partial \gamma(x,t)}{\partial t} \right|^2 < \infty \right\}. \tag{6.3.4}$$

Recall that, by Definition 5.65 and Theorem 5.66, a mild solution of our SPDE is a continuous map with values in $\mathscr{B}_\alpha$, for any $\alpha \in (0, \frac{1}{4})$. Thus, a family of mild solutions $(u^\varepsilon)_{\varepsilon > 0}$ is a family of random variables with values in the Banach space $C([0,T], \mathscr{B}_\alpha)$. The following theorem asserts that these satisfy an LDP.

Theorem 6.14 *The family $(u^\varepsilon)_{\varepsilon>0}$ satisfies the LDP on $C([0,T], \mathscr{B}_\alpha)$ with rate function I given by*

$$I(\gamma) = \begin{cases} \frac{1}{2} \int_0^T dt \int_0^1 dx \, |\frac{\partial}{\partial t}\gamma(x,t) - \frac{1}{2} D \frac{\partial^2}{\partial x^2} \gamma(x,t) + V'(x, \gamma(x,t))|^2, \\ \qquad \textit{if } \gamma \in W_2^{1,2},\ \gamma(\cdot, 0) = \xi(\cdot), \\ \infty, \quad \textit{otherwise}. \end{cases} \tag{6.3.5}$$

Here is a sketch of how Theorem 6.14 comes about. The starting point is the LDP for the Brownian sheet, which is the analogue of Schilder's theorem for Brownian motion (Theorem 6.8). To state this LDP, let $\mathscr{H}$ be the space of all $h \in C([0,1] \times [0,T])$ such that there exists an $\dot{h} \in L^2([0,1] \times [0,T])$ with

$$h(x,t) = \int_0^x dy \int_0^t ds\, \dot{h}(y,s), \quad x \in [0,1],\, t \in [0,T]. \tag{6.3.6}$$

Theorem 6.15 *$(\sqrt{2\varepsilon}\, W)_{\varepsilon>0}$ satisfies the LDP on $C([0,1] \times [0,T])$ with rate function I_0 given by*

$$I_0(h) = \begin{cases} \frac{1}{2}\int_0^T dt \int_0^1 dx\, |\dot{h}(x,t)|^2, & \text{if } h \in \mathscr{H}, \\ \infty, & \text{otherwise.} \end{cases} \tag{6.3.7}$$

The LDP in Theorem 6.14 follows from Theorem 6.15 via the *contraction principle* (Lemma 6.4), and identifies the rate function as

$$I(\gamma) = \inf\{I_0(h)\colon\ h \in \mathscr{H},\ \mathscr{T}(h) = \gamma\}, \tag{6.3.8}$$

where $\mathscr{T}$ is the map from $\mathscr{H}$ into $C([0,T], \mathscr{B}_\alpha)$ such that $\mathscr{T}(h) = \gamma$ is the solution of

$$\gamma(x,t) = \int_0^1 dy\, g_t(x,y)\xi(y) + \int_0^t ds \int_0^1 dy\, g_{t-s}(x,y)\big[-V'\big(\gamma(y,s)\big) + \dot{h}(y,s)\big] \tag{6.3.9}$$

with $g_t(x,y)$ the density of the semi-group generated by $\frac{1}{2}D\partial^2/\partial^2 x$ on $[0,1]$ (the heat kernel). Here, (6.3.8) and (6.3.9) are the infinite-dimensional analogues of (6.2.33) and (6.2.35). The fact that (6.3.8) is the same as (6.3.5) follows from the same type of inversion argument as in (6.2.36).

6.4 Path large deviations for Markov processes

In this section we state a path LDP for a general class of discrete-time Markov processes subject to certain regularity conditions. This LDP will turn out to be useful in Chap. 10.

Fix $d \geq 1$. For $\varepsilon > 0$, let $Z^\varepsilon = (Z_n^\varepsilon)_{n\in\mathbb{N}_0}$ be the *time-inhomogeneous* Markov process on $\varepsilon\mathbb{Z}^d$, starting at the origin, with transition kernel

$$\begin{aligned} p^\varepsilon(x,y;n) &= \mathbb{P}\big(Z_{n+1}^\varepsilon = y \mid Z_n^\varepsilon = x\big) \\ &= \begin{cases} \exp[-q^\varepsilon(x, \varepsilon^{-1}(y-x); \varepsilon n)], & \text{if } \varepsilon^{-1}(y-x) \in \Delta, \\ 0, & \text{otherwise,} \end{cases} \qquad x,y \in \varepsilon\mathbb{Z}^d,\ n \in \mathbb{N}_0, \end{aligned} \tag{6.4.1}$$

where Δ is a finite subset of $\mathbb{Z}^d$, and $q^\varepsilon\colon\ \mathbb{R}^d \times \Delta \times [0,\infty) \to (0,\infty)$, $\varepsilon > 0$, is a family of functions that are assumed to be bounded away from 0 and ∞, to be globally Lipschitz in the first and in the third coordinate, uniformly in the second coordinate and in $\varepsilon > 0$, and to be such that

$$\lim_{\varepsilon \downarrow 0} q^{\varepsilon} = q \quad \text{for some } q\colon \mathbb{R}^d \times \Delta \times [0,\infty) \to (0,\infty), \tag{6.4.2}$$

with the convergence uniform in all three coordinates.

For $u, v, v^* \in \mathbb{R}^d$ and $t \in \mathbb{R}_+$, define

$$\begin{aligned} \mathscr{L}(u,v;t) &= \ln \sum_{w \in \Delta} \mathrm{e}^{-q(u,w;t)+\langle v,w\rangle}, \\ \mathscr{L}^*\big(u,v^*;t\big) &= \sup_{v \in \mathbb{R}^d} \big[\big\langle v,v^*\big\rangle - \mathscr{L}(u,v;t)\big]. \end{aligned} \tag{6.4.3}$$

Fix $T > 0$. Let $\bar{Z}^{\varepsilon} = (\bar{Z}^{\varepsilon}(s))_{s\in[0,T]}$ denote the linear interpolation of

$$\big(Z^{\varepsilon}\big(\big\lceil \varepsilon^{-1} s\big\rceil\big)\big)_{s\in[0,T]}. \tag{6.4.4}$$

Theorem 6.16 *For every $T > 0$, $(\bar{Z}^{\varepsilon})_{\varepsilon>0}$ satisfies the LDP on $C_0([0,T])$ with rate function I given by*

$$I(\gamma) = \begin{cases} \int_0^T \mathscr{L}^*(\gamma(s),\dot{\gamma}(s);s)ds, & \text{if } \gamma \in D_0([0,T]), \\ \infty, & \text{otherwise}, \end{cases} \tag{6.4.5}$$

where $D_0([0,T]) \subseteq C_0([0,T])$ is the space of absolutely continuous functions with integrable derivative equipped with the norm $\|f\|_{D_0([0,T])} = \int_0^T \|\dot{f}(s)\|ds$.

A simple example to which the above setting applies is simple random walk on $\varepsilon\mathbb{Z}^d$, for which we choose $\Delta = \{x \in \mathbb{Z}^d\colon \|x\| = 1\}$ and $q^{\varepsilon} = \ln(2d)$. For this case Theorem 6.16 reduces to Mogul'skiĭ's theorem [186] for simple random walk, the analogue of Schilder's theorem for Brownian motion (see Dembo and Zeitouni [79]).

6.5 Freidlin-Wentzell theory

In this section we give a brief indication of how large deviations on path space are used in the pathwise approach to metastability of Freidlin and Wentzell (recall Sect. 1.3.2).

6.5.1 Properties of action functionals

The rate functions in Sects. 6.2–6.4 have the form of a classical action functional in Newtonian mechanics, i.e., they are of the form

$$I(\gamma) = \int_0^T \mathscr{L}\big(\gamma(s), \dot{\gamma}(s), s\big) ds, \tag{6.5.1}$$

for some *Lagrangian* $\mathscr{L}$. In Theorem 6.12, for instance, $\mathscr{L}$ takes on the special form

$$\mathscr{L}\big(\gamma(s), \dot{\gamma}(s), s\big) = \tfrac{1}{2}\big\|\dot{\gamma}(s) - b\big(\gamma(s)\big)\big\|_2^2. \tag{6.5.2}$$

The principle of least action in classical mechanics states that the system follows the trajectory of minimal action subject to boundary conditions. This leads to the *Euler-Lagrange equations*

$$\frac{d^2}{ds^2}\gamma(s) - 2\dot{\gamma}(s)\frac{db(\gamma(s))}{d\gamma(s)} = -b\big(\gamma(s)\big)\frac{db(\gamma(s))}{d\gamma(s)}. \tag{6.5.3}$$

which in the case of (6.5.2) take the form

$$\frac{d^2}{ds^2}\gamma(s) = 2b\big(\gamma(s)\big)\frac{d}{d\gamma(s)}b\big(\gamma(s)\big). \tag{6.5.4}$$

We can readily identify a special class of solutions of this second-order differential equation, namely, solutions of the first order differential equation

$$\dot{\gamma}(s) = b\big(\gamma(s)\big). \tag{6.5.5}$$

These solutions have the property that they yield absolute minima of the action functional, since they satisfy

$$\mathscr{L}\big(\gamma(s), \dot{\gamma}(s)\big) = 0. \tag{6.5.6}$$

Of course, being first-order, this equation admits only one boundary (or initial) condition.

6.5.2 Crossing and exit problems

A typical question we may ask is the following: What is the probability of a solution connecting two points $u, v \in \mathbb{R}^d$ in time T? The LDP in Theorem 6.12 provides the answer, namely,

$$\lim_{\delta\downarrow 0}\lim_{\varepsilon\downarrow 0} \varepsilon \ln \mathbb{P}\big(X_T^\varepsilon \in B_\delta(v) \mid X_0^\varepsilon \in B_\delta(u)\big) = -\inf_{\gamma:\ \gamma(0)=u,\, \gamma(T)=v} I(\gamma), \tag{6.5.7}$$

where $B_\delta(x)$ is the ball of radius $\delta > 0$ around $x \in \mathbb{R}^d$. This leads us to solve (6.5.4) subject to the boundary conditions $\gamma(0) = u$ and $\gamma(T) = v$. Unfortunately, not all solutions of (6.5.4) also solve (6.5.5) as they can have positive action, meaning that the event under consideration has an exponentially small probability. However, under certain conditions we may find a zero-action solution, for instance, when we do not fix the time of arrival at v:

$$\lim_{\delta\downarrow 0}\lim_{\varepsilon\downarrow 0}\varepsilon\ln\mathbb{P}\big(X_s^\varepsilon\in B_\delta(v)\text{ for some }s\in[0,T]\mid X_0^\varepsilon\in B_\delta(u)\big)$$
$$=-\inf_{\gamma:\,\gamma(0)=u,\,\gamma(s)=v\text{ for some }s\in[0,T]} I(\gamma). \tag{6.5.8}$$

Clearly, the infimum will be zero if the solution of (6.5.5) with $\gamma(0)=u$ has the property that $\gamma(s)=v$ for some $s\in[0,T]$.

Suppose that we consider an event as in (6.5.8) that admits a zero-action path γ with $\gamma(0)=u$ and $\gamma(T)=v$. Define the time-reversed path $\bar\gamma(s)=\gamma(T-s)$, $s\in[0,T]$. Clearly, $\dot{\bar\gamma}(s)=-\dot\gamma(T-s)$. Hence a simple calculation, via (6.5.1)–(6.5.2), shows that

$$I(\gamma)-I(\bar\gamma)=2\int_0^T b\big(\gamma(s)\big)\dot\gamma(s)ds=2\int_\gamma b(x)dx. \tag{6.5.9}$$

Let us now specialise to the case where b is the gradient of a potential F, i.e., $b(x)=\nabla F(x)$, $x\in\mathbb{R}^d$. Then

$$\int_\gamma b(x)dx=F\big(\gamma(T)\big)-F\big(\gamma(0)\big)=F(v)-F(u). \tag{6.5.10}$$

Hence

$$I(\gamma)-I(\bar\gamma)=2\big[F(v)-F(u)\big]. \tag{6.5.11}$$

If $I(\gamma)=0$, then $I(\bar\gamma)=2[F(u)-F(v)]$, and this is the minimal possible value for any path going from v to u. Thus, there is the remarkable fact that *the most likely path going uphill in a potential is the time-reversal of the solution of the gradient flow*.

So far we have considered paths of a fixed time length T. Freidlin and Wentzell allowed paths of arbitrary length and introduced the notion of *quasi-potential*:

$$V(u,v)=\inf_{T<\infty}\inf_{\gamma:\,\gamma(0)=u,\,\gamma(T)=v} I(\gamma),\qquad u,v\in\mathbb{R}^d. \tag{6.5.12}$$

They showed that, uniformly in $w_\delta\in B_{\delta/2}(u)$,

$$-V(u,v)=\lim_{\delta\downarrow 0}\lim_{\varepsilon\downarrow 0}\varepsilon\ln\mathbb{P}\big(\inf\big\{t>\sigma_{B_\delta(u)}:\ X_t^\varepsilon\in B_\delta(v)\big\} \tag{6.5.13}$$
$$\le\inf\big\{t>\sigma_{B_\delta(u)}:\ X_t^\varepsilon\in B_{\delta/2}(u)\big\}\mid X_0^\varepsilon=w_\delta\big),$$

where $\sigma_{B_\delta(u)}=\inf\{t\in\mathbb{R}_+:\ X_t^\varepsilon\notin B_\delta(u)\}$ is the first exit time of the ball $B_\delta(u)$. The probability in (6.5.13) is the proper version of the escape probability from u to v.

In the setting of Fig. 6.4, we have

$$V(u,v)=V\big(u,z^*\big)+V\big(z^*,v\big), \tag{6.5.14}$$

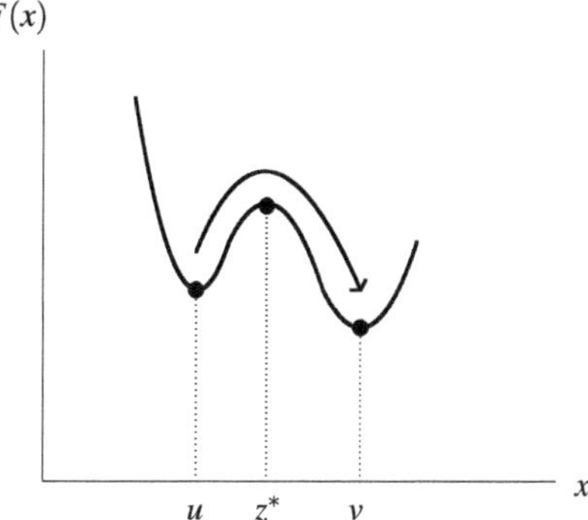

Fig. 6.4 A one-dimensional example of a potential F (recall Fig. 2.1)

where, by (6.5.11),

$$V\big(u,z^*\big) = V\big(z^*,u\big) + 2\big[F\big(z^*\big) - F(u)\big], \tag{6.5.15}$$

while

$$V\big(z^*,u\big) = V\big(z^*,v\big) = 0. \tag{6.5.16}$$

Hence $V(u,v) = 2[F(z^*) - F(u)]$, i.e., the exponential asymptotics of the escape probability from u to v is given by twice the height of the potential barrier from u to v.

Let $\tau_{B_\delta(v)}$ be the first hitting time of the ball $B_\delta(v)$. With the help of a simple renewal argument, (6.5.13) can be shown to imply that, for every $\rho > 0$ and uniformly in $w_\delta \in B_{\delta/2}(u)$,

$$\lim_{\delta\downarrow 0}\lim_{\varepsilon\downarrow 0} \mathbb{P}\big(\mathrm{e}^{2[F(z^*)-F(u)-\rho]/\varepsilon} \le \tau_{B_\delta(v)} \le \mathrm{e}^{2[F(z^*)-F(u)+\rho]/\varepsilon} \mid X_0^\varepsilon = w_\delta\big) = 1. \tag{6.5.17}$$

6.5.3 Metastability

The discussion in Sects. 6.5.1–6.5.2 forms the basis of the treatment of metastability in *Freidlin-Wentzell theory*. In this theory, any constant or periodic solution of (6.5.6) is a candidate for a metastable state. If γ is such a solution, then it is called *unstable* when there exists another solution $\tilde\gamma$ and a family of functions $(\gamma_n)_{n\in\mathbb{N}}$ such that

$$\gamma_n(t) = \begin{cases} \gamma(t), & t \le -n, \\ \tilde\gamma(t), & t \ge n, \end{cases} \tag{6.5.18}$$

while $\inf_{n\in\mathbb{N}} I(\gamma_n) = 0$. In other words, a solution is unstable when it can be deformed into another solution at an arbitrarily small cost. Otherwise, the solution is called stable. In the context of the Markov process, a stable solution is interpreted as a *metastable state*, also called a *cycle*. For us the most interesting situations correspond to fixed points, i.e., solutions of (6.5.6) that are constant in time. In the case of a reversible Markov process these are the only possible solutions.

A system is called metastable when it has at least two metastable states. In the presence of noise there exist (exponentially unlikely) trajectories that constitute transitions between these states. The variational problem in (6.5.12) with u, v metastable states (respectively, its obvious extension when u, v are not fixed points), provides the asymptotics of the transition probabilities between them, while (6.5.17) provides control over the transition times between them. This, in a nutshell, is the basis of the Freidlin-Wentzell theory of metastability. The strong point of this theory is its great versatility. In particular, no assumption of reversibility needs to be made. The weak point is the poor level of precision, i.e., *only the exponential asymptotics* of characteristic quantities such as hitting times is obtained.

Freidlin-Wentzell theory does not offer the tools to go beyond the exponential asymptotics. The goal of the present monograph is to position potential theory as the key mathematical framework for obtaining sharper asymptotics, and to outline the main ideas and techniques that are available to tackle concrete models.

6.6 Bibliographical notes

1. Section 6.1 is a crash course on large deviation theory. Definition 6.1 is due to Varadhan. Lemmas 6.2–6.5 and Theorems 6.7–6.6 are key instruments, and are easy to prove. For further reading we refer to the monographs by Varadhan [232], Ellis [103], Deuschel and Stroock [91], Dembo and Zeitouni [79], and den Hollander [80].

2. Theorem 6.13 in Sect. 6.2 lies at the heart of Freidlin-Wentzell theory. For further reading we refer to the monographs by Freidlin and Wentzell [115], Dupuis and Ellis [98], and Feng and Kurtz [110].

3. The LDP in Theorem 6.14 is derived in Sowers [221]. Extensions to larger classes of SPDEs were obtained by Kallianpur and Xiong [147] and by Chenal and Millet [57].

4. Theorem 6.16 in Sect. 6.4 is taken from Bovier and Gayrard [37] and will be needed in Chap. 10. For extensions to general dynamical systems, see the monograph by Kifer [152].

5. In Theorem 6.16 it is possible to restrict the Markov process to $\varepsilon\mathbb{Z}^d \cap D$ with D a convex subset of $\mathbb{R}^d$ and allow for singular behaviour near the boundary of D. This will be a natural setting for the application to mean-field models, which will be treated in Part IV. It is further straightforward to extend Theorem 6.16 to continuous space and/or time under certain additional regularity conditions.

6. The application of large deviation theory to the problem of metastability in the work of Freidlin and Wentzell [115] initiated the rigorous mathematical treatment of metastability. This development was picked up by many authors. A pivotal paper is Cassandro, Galves, Olivieri and Vares [51], which established the link to interacting particle systems. A comprehensive account of metastability from this point of view is given in the monograph by Olivieri and Vares [198].

Chapter 7
Potential Theory

Mais la méthode la plus générale et la plus directe pour résoudre les questions de probabilité consiste à les faire dépendre d'equations aux différences.
(Pierre Simon de Laplace, Théorie Analytique des Probabilités)

The martingale problem and the stopping times that were described in Chaps. 4–5 provide the key link between Markov processes and Dirichlet problems. This chapter gives a detailed account of this connection. Although, once again, the basic principles are the same in discrete and in continuous time, we split the presentation: discrete time and countable state space (Sect. 7.1), continuous time and general state space (Sect. 7.2). The mixed cases are similar and are left to the reader. Once we have built up the necessary tools, we provide three variational formulas for the capacity referred to as the *Dirichlet principle*, the *Thomson principle* and the *Berman-Konsowa principle* (Sect. 7.3). These will be crucial for the metastable analyses carried out in Parts IV–VIII. The variational principles can be extended to the non-reversible setting, but become harder to work with (Sect. 7.4).

7.1 The Dirichlet problem: discrete time

7.1.1 Definition

In this section we place ourselves in the setting of a discrete-time Markov process $X = (X_n)_{n\in\mathbb{N}_0}$ on a countable state space S with transition kernel P and generator $L = P - \mathbb{1}$. To avoid complications we will always assume that X is irreducible. We will use the notation of Chap. 4. Let $D \subset S$, $g\colon D \to \mathbb{R}$, $\bar{g}\colon D^c \to \mathbb{R}$ and $k\colon D \to [-K, \infty)$ with $-\infty < K < 1$, where $D^c = S\backslash D$. We call the following pair of equations for an unknown function f a *Dirichlet problem* (see Fig. 7.1):

$$\begin{aligned}(-Lf)(x) + k(x)f(x) &= g(x), \quad \forall x \in D,\\ f(x) &= \bar{g}(x), \quad \forall x \in D^c.\end{aligned} \tag{7.1.1}$$

The following theorem provides a stochastic representation for the solution of such Dirichlet problems.

A. Bovier, F. den Hollander, *Metastability*,
Grundlehren der mathematischen Wissenschaften 351,
DOI 10.1007/978-3-319-24777-9_7

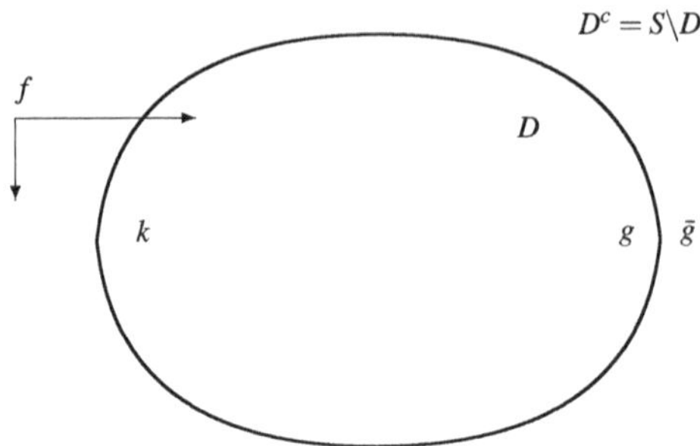

Fig. 7.1 Dirichlet problem for $f\colon S \to \mathbb{R}$ with source $k\colon D \to [-K,\infty)$ and boundary conditions $g\colon D \to \mathbb{R}$ and $\bar{g}\colon D^c \to \mathbb{R}$

Theorem 7.1 *Let X be a discrete-time Markov process with generator L. Assume that D is such that*

$$\mathbb{E}_x\left[\tau_{D^c}(1-K)^{-\tau_{D^c}}\right] < \infty \quad \forall x \in D, \tag{7.1.2}$$

where $\tau_{D^c} = \inf\{t \in \mathbb{N}\colon X(t) \in D^c\}$. Then the Dirichlet problem (7.1.1) *has a unique solution given by*

$$f(x) = \mathbb{E}_x\left[\left(\prod_{u=0}^{\tau_{D^c}-1} \frac{1}{1+k(X_u)}\right)\bar{g}(X_{\tau_{D^c}}) + \sum_{s=0}^{\tau_{D^c}-1}\left(\prod_{u=0}^{s}\frac{1}{1+k(X_u)}\right)g(X_s)\right],$$

$$x \in D, \tag{7.1.3}$$

with the convention that the empty product equals 1.

Proof The most convenient way to prove Theorem 7.1 is via the martingale problem characterisation of Markov processes. Indeed, as in Lemma 4.9, we check that, for any $k\colon S \to \mathbb{R}$ bounded from below,

$$\begin{aligned} M_t = & \left(\prod_{u=0}^{t-1}\frac{1}{1+k(X_u)}\right)f(X_t) - f(X_0) \\ & + \sum_{s=0}^{t-1}\left(\prod_{u=0}^{s}\frac{1}{1+k(X_u)}\right)\left[(-Lf)(X_s) + k(X_s)f(X_s)\right] \end{aligned} \tag{7.1.4}$$

is a martingale. Moreover, Doob's optional stopping theorem applies to $M_{\tau_{D^c}}$ under condition (7.1.2) (recall Theorem 3.67(ii)), and so $\mathbb{E}_x[M_{\tau_{D^c}}] = M_0 = 0$. □

Note that the solution of the Dirichlet problem is unique, unless the homogeneous problem

$$\begin{aligned} (-Lf)(x) + k(x)f(x) &= 0, \quad \forall x \in D, \\ f(x) &= 0, \quad \forall x \in D^c, \end{aligned} \tag{7.1.5}$$

admits a non-zero solution. The most interesting case for us is when $k = \lambda$ is constant. In that case, if (7.1.5) admits a non-zero solution, then λ is called an *eigenvalue* and the corresponding solution an *eigenfunction* of the Dirichlet problem. A solution of the homogeneous Dirichlet boundary value problem with $k = 0$,

$$\begin{aligned} (-Lf)(x) &= 0, \quad && \forall x \in D, \\ f(x) &= \bar{g}(x), \quad && \forall x \in D^c, \end{aligned} \tag{7.1.6}$$

is called a *harmonic function* (see Sect. 4.2.3).

One of the most important applications of Theorem 7.1 concerns the case $k = 0$, $g = 1$ and $\bar{g} = 0$. This yields the following characterisation of the mean exit time.

Corollary 7.2 *Let $D \subset S$ and set*

$$w(x) = \begin{cases} \mathbb{E}_x[\tau_{D^c}], & x \in D, \\ 0, & x \in D^c. \end{cases} \tag{7.1.7}$$

Then w is the unique solution of the Dirichlet problem

$$\begin{aligned} (-Lw)(x) &= 1, \quad && x \in D, \\ w(x) &= 0, \quad && x \in D^c. \end{aligned} \tag{7.1.8}$$

7.1.2 Green function, equilibrium potential and measure

The objects we introduce now will turn out to be fundamental in the study of metastability. We consider the case where the solution of the Dirichlet problem in (7.1.1) is unique. For simplicity, we restrict ourselves to the case where $k = \lambda$ is constant. Then the solution to (7.1.1) can be written in the form

$$f(x) = \sum_{z \in D} G^{\lambda}_{D^c}(x,z) g(z) + \sum_{z \in D^c} H^{\lambda}_{D^c}(x,z) \bar{g}(z), \quad x \in D, \tag{7.1.9}$$

where

$$G^{\lambda}_{D^c}(x,z) = \mathbb{E}_x \left[\sum_{s=0}^{\tau_{D^c}-1} (1+\lambda)^{-s-1} \mathbb{1}_{X_s = z} \right], \quad x, z \in D, \tag{7.1.10}$$

is called the *Green function*, and

$$\begin{aligned} H^{\lambda}_{D^c}(x,z) &= \mathbb{E}_x\left[(1+\lambda)^{-\tau_{D^c}} \mathbb{1}_{X_{\tau_{D^c}} = z}\right] \\ &= \sum_{s \in \mathbb{N}_0} (1+\lambda)^{-s}\, \mathbb{P}_x(\tau_{D^c} = s,\, X_s = z), \quad x \in D,\, z \in D^c, \end{aligned} \tag{7.1.11}$$

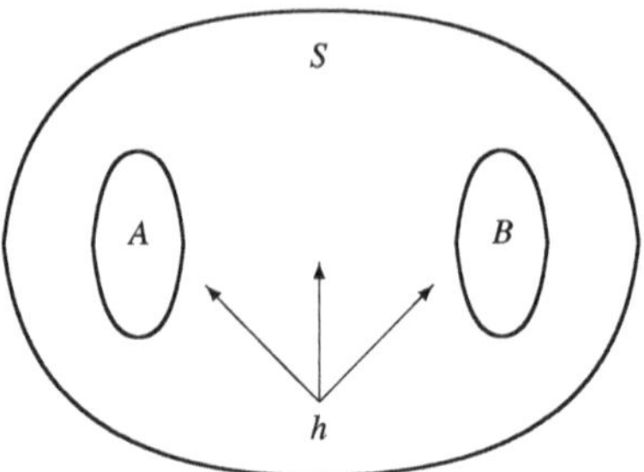

Fig. 7.2 Dirichlet problem for $h\colon S \to [0,1]$ with boundary conditions $h = 1$ on A and $h = 0$ on B

is called the *Poisson kernel*. Clearly, the Green function can also be characterised as the solution of the problem

$$\begin{aligned} (-LG^{\lambda}_{D^c})(x,z) + \lambda G^{\lambda}_{D^c}(x,z) &= \mathbb{1}_z(x), \quad &\forall x \in D, \\ G^{\lambda}_{D^c}(x,z) &= 0, &\forall x \in D^c. \end{aligned} \tag{7.1.12}$$

In the special case $\lambda = 0$ we have the following appealing representation of the Green function $G^0_{D^c} = G_{D^c}$ and the Poisson kernel $H^0_{D^c} = H_{D^c}$:

$$G_{D^c}(x,z) = \mathbb{E}_x\left[\sum_{s=0}^{\tau_{D^c}-1} \mathbb{1}_{X_s=z}\right], \quad x,z \in D, \tag{7.1.13}$$

$$H_{D^c}(x,z) = \mathbb{P}_x(X_{\tau_{D^c}} = z), \quad x \in D,\ z \in D^c, \tag{7.1.14}$$

i.e., for the Markov process starting at $x \in D$, $G_{D^c}(x,z)$ represents the average number of visits to $z \in D$ before it exits from D, while $H_{D^c}(x,z)$ represents the probability that it enters D^c at z.

The following object will be absolutely central in our study of metastability. Let $A, B \subset S$ be two non-empty disjoint subsets. Consider the Dirichlet problem (see Fig. 7.2)

$$\begin{aligned} (-Lh)(x) &= 0, \quad &\forall x \in S\backslash(A \cup B), \\ h(x) &= 1, &\forall x \in A, \\ h(x) &= 0, &\forall x \in B. \end{aligned} \tag{7.1.15}$$

Suppose that (7.1.15) has a unique solution, e.g. because $\mathbb{E}_x[\tau_{A\cup B}] < \infty$ for all $x \in S$. The harmonic function that solves (7.1.15) is denoted by $h_{A,B}(x)$ and is called the *equilibrium potential*. The representation of the solution given in (7.1.9) and (7.1.13)–(7.1.14), with $D = S\backslash(A \cup B)$, $D^c = A \cup B$, $g = 0$, $\bar{g}(x) = 1$, $x \in A$, and $\bar{g}(x) = 0$, $x \in B$, implies that

$$h_{A,B}(x) = \mathbb{E}_x\big[\mathbb{1}_A(X_{\tau_{A\cup B}})\big] = \mathbb{P}_x(\tau_A < \tau_B), \quad x \in S\backslash(A \cup B). \tag{7.1.16}$$

This equation gives an analytic representation for the probability in the right-hand side when $x \in S\backslash(A \cup B)$. Using the Markov property, we can get a similar expression when $x \in A \cup B$. Namely, for $x \in A \cup B$,

$$\begin{aligned}\mathbb{P}_x(\tau_A < \tau_B) &= \sum_{y \in S\backslash(A\cup B)} p(x,y)\mathbb{P}_y(\tau_A < \tau_B) + \sum_{y\in A} p(x,y)\\ &= \sum_{y\in S} p(x,y)h_{A,B}(y) = (Ph_{A,B})(x)\\ &= (Lh_{A,B})(x) + h_{A,B}(x).\end{aligned} \tag{7.1.17}$$

The latter can be written for $x \in B$ as

$$(Lh_{A,B})(x) = \mathbb{P}_x(\tau_A < \tau_B) - 0 = \mathbb{P}_x(\tau_A < \tau_B), \tag{7.1.18}$$

and for $x \in A$ as

$$(-Lh_{A,B})(x) = 1 - \mathbb{P}_x(\tau_A < \tau_B) = \mathbb{P}_x(\tau_B < \tau_A). \tag{7.1.19}$$

The quantity

$$e_{A,B}(x) = (-Lh_{A,B})(x), \quad x \in A, \tag{7.1.20}$$

is called the *equilibrium measure* on A, and is the second central object in our study of metastability.

The following simple observation provides a fundamental connection between the objects we have introduced so far, and leads to a different representation of the equilibrium potential. Pretend that the equilibrium measure $e_{A,B}$ is already known. Then the equilibrium potential solves the *inhomogeneous* Dirichlet problem

$$\begin{aligned}(-Lh)(x) &= e_{A,B}(x), \quad \forall x \in S\backslash B,\\ h(x) &= 0, \quad \forall x \in B.\end{aligned} \tag{7.1.21}$$

The solution of (7.1.21) can be written in terms of the Green function.

Theorem 7.3 *With the notation introduced above,*

$$h_{A,B}(x) = \sum_{y\in A} G_B(x,y)e_{A,B}(y), \quad x \in S. \tag{7.1.22}$$

Relation (7.1.22) can be used to express the Green function in terms of the equilibrium measure and the equilibrium potential: simply choose $A = \{a\}$, to get

$$G_B(x,a) = \frac{h_{a,B}(x)}{e_{a,B}(a)}, \quad x \in S. \tag{7.1.23}$$

Note that $e_{a,B}(a) = \mathbb{P}_a(\tau_B < \tau_a)$ has the meaning of an *escape probability* from a to B. The full power of Theorem 7.3 will come out in the *reversible* case, which we discuss next.

7.1.3 Reversibility

Considerable simplifications occur when we assume a certain symmetry property of the transition kernels known as *reversibility* or, in physics terminology, *detailed balance*.

Definition 7.4 A Markov process with countable state space S and transition kernel $P = \{p(x, y), x, y \in S\}$, is called *reversible* if there exists a non-zero $\mu\colon S \to \mathbb{R}_+$ such that

$$\mu(x)p(x, y) = \mu(y)p(y, x) \quad \forall x, y \in S. \tag{7.1.24}$$

The function μ is called the reversible measure of the Markov process.

The function space $L^2(S, \mu)$ is a natural space to work on when the Markov process is reversible with respect to μ.

Lemma 7.5 *Let $f \in L^2(S, \mu)$, where μ is invariant with respect to P. Then $Pf \in L^2(S, \mu)$.*

Proof The claim follows from the fact that P is a contraction in the L^2-norm:

$$\begin{aligned}\sum_{x\in S}\mu(x)[(Pf)(x)]^2 &= \sum_{x\in S}\mu(x)\Big[\sum_{y\in S}p(x, y)f(y)\Big]^2 \\ &\le \sum_{x\in S}\mu(x)\sum_{y'\in S}p(x, y')f(y')^2\sum_{y''\in S}p(x, y'') \\ &\le \sum_{x\in S}\mu(x)\sum_{y'\in S}p(x, y')f(y')^2 = \sum_{y'\in S}\mu(y')f(y')^2,\end{aligned} \tag{7.1.25}$$

where we use the Cauchy-Schwarz inequality and the invariance of μ, i.e., $\mu P = \mu$. □

Reversibility can be expressed by saying that the transition kernel P acts as a self-adjoint operator on the Hilbert space $L^2(S, \mu)$.

Lemma 7.6 *If μ is a reversible probability measure for P, then μ is an invariant probability measure for P.*

Proof Clearly, $f = 1$ is in $L^2(S, \mu)$. Hence, for all bounded measurable functions g,

$$\sum_{x,y\in S}\mu(x)p(x, y)g(y) = \sum_{x,y\in S}p(y, x)\mu(y)g(y) = \sum_{y\in S}\mu(y)g(y), \tag{7.1.26}$$

and so μ is invariant. □

We next come to the definition of the Dirichlet form, which plays a central rôle in the potential-theoretic approach to metastability.

Lemma 7.7 *Let L be the generator of a Markov process with reversible measure μ. Then L defines a non-negative-definite quadratic form*

$$\mathscr{E}(f,g)=\sum_{x\in S}\mu(x)f(x)(-Lg)(x),\quad f,g\in L^2(S,\mu), \tag{7.1.27}$$

called the Dirichlet form.

Proof In the discrete case it suffices to write out $\mathscr{E}(f,g)$ explicitly. Namely, by reversibility,

$$\begin{aligned}\mathscr{E}(f,g)&=\sum_{x,y\in S}\mu(x)p(x,y)f(x)\big[g(x)-g(y)\big]\\&=\sum_{x,y\in S}\mu(x)p(x,y)f(y)\big[g(y)-g(x)\big].\end{aligned} \tag{7.1.28}$$

Symmetrising between the first and the second expression, we get

$$\begin{aligned}\mathscr{E}(f,g)&=\tfrac{1}{2}\sum_{x,y\in S}\mu(x)p(x,y)\big\{f(x)\big[g(x)-g(y)\big]+f(y)\big[g(y)-g(x)\big]\big\}\\&=\tfrac{1}{2}\sum_{x,y\in S}\mu(x)p(x,y)\big[f(x)-f(y)\big]\big[g(x)-g(y)\big].\end{aligned} \tag{7.1.29}$$

This expression manifestly is a non-negative-definite quadratic form. □

An important rôle will be played by the analogue of the two Green identities for sums.

Lemma 7.8 *Let $f,g\in L^2(S,\mu)$ and $D\subset S$. Assume that P is reversible with respect to μ. Then*

(i) (*first Green identity*)

$$\begin{aligned}&\tfrac{1}{2}\sum_{x,y\in D}\mu(x)p(x,y)\big[f(x)-f(y)\big]\big[g(x)-g(y)\big]\\&\quad=\sum_{x\in D}\mu(x)f(x)(-Lg)(x)-\sum_{x\in D,y\in D^c}\mu(x)p(x,y)f(x)\big[g(x)-g(y)\big]\end{aligned} \tag{7.1.30}$$

(ii) (*second Green identity*)

$$\sum_{x\in D}\mu(x)\big[f(x)(Lg)(x)-g(x)(Lf)(x)\big]$$

$$= \sum_{x\in D, y\in D^c} \mu(x)p(x,y)\big[f(x)g(y) - g(x)f(y)\big]$$

$$= \sum_{y\in D^c} \mu(y)\big[g(y)(Lf)(y) - f(y)(Lg)(y)\big]. \tag{7.1.31}$$

Proof To prove the first Green identity, we proceed as in the proof of Lemma 7.7. If $D = S$, then the proof gives (7.1.30) without the last term. If $D \subsetneq S$, then in order to produce the full action of L we must add the terms that involve $y \in D^c$.

The first equality in the second Green identity is a trivial consequence of the first Green identity. To get the second equality, use reversibility, add terms that involve $x \in D^c$ to produce the full action of L, and use that these terms add up to zero. Note that the equality between the first and the last line is just the statement that L is symmetric in $L^2(S, \mu)$. □

An illustration of what can be done with the Green identities is the following formula for the Poisson kernel in terms of the Green function.

Lemma 7.9 (Poisson kernel and Green function) *If P is reversible with respect to μ and $D \subset S$, then the Poisson kernel defined in* (7.1.11) *satisfies*

$$H_{D^c}(z,y) = \sum_{x\in D} \frac{\mu(x)}{\mu(z)} p(x,y)\big[G_{D^c}(x,z) - G_{D^c}(y,z)\big], \quad z \in D, y \in D^c. \tag{7.1.32}$$

Proof Fix $z \in D$. In (7.1.31), choose for f the solution of the Dirichlet problem in (7.1.6), and choose $g(x) = G_{D^c}(x,z)$, $x \in S$. With this choice, by (7.1.12) with $\lambda = 0$, the first line in (7.1.31) simply becomes $-\mu(z)f(z)$. The second line reads

$$\sum_{x\in D, y\in D^c} \mu(x)p(x,y)\big[f(x)G_{D^c}(y,z) - G_{D^c}(x,z)f(y)\big]$$

$$= - \sum_{x\in D, y\in D^c} \mu(x)p(x,y)G_{D^c}(x,z)\bar{g}(y), \tag{7.1.33}$$

where we use that $G_{D^c}(y,z) = 0$ and $f(y) = \bar{g}(y)$ for $y \in D^c$, again by (7.1.6) and (7.1.12) with $\lambda = 0$. Hence

$$f(z) = \sum_{y\in D^c} \left(\sum_{x\in D} \frac{\mu(x)}{\mu(z)} p(x,y)G_{D^c}(x,z) \right) \bar{g}(y). \tag{7.1.34}$$

From this expression the Poisson kernel $H_{D^c}(z,y)$ can be read off as the sum between the brackets, where we recall (7.1.9) and use that $\bar{g}$ is arbitrary. Since $G_{D^c}(y,z) = 0$ for $y \in D^c$, we thus obtain (7.1.32). □

A nice aspect of (7.1.32) is that by reversibility it correctly extends to D, namely, $H_{D^c}(z,y) = 0$ for $z, y \in D$.

As a second application of the Green identities, we obtain the following alternative to Theorem 7.3.

Theorem 7.10 *If P is reversible with respect to μ, then for all non-empty disjoint sets $A, B \subset S$,*

$$h_{A,B}(x) = \sum_{y \in A} \frac{\mu(y)}{\mu(x)} G_B(y,x) e_{A,B}(y), \quad x \in S. \tag{7.1.35}$$

In particular, if f is a solution of the Dirichlet problem

$$\begin{aligned} (-Lf)(x) &= g(x), \quad &\forall x \in B^c, \\ f(x) &= 0, \quad &\forall x \in B, \end{aligned} \tag{7.1.36}$$

then

$$\sum_{y \in A} \nu_{A,B}(y) f(y) = \frac{1}{\mathrm{cap}(A,B)} \sum_{x \in S} \mu(x) h_{A,B}(x) g(x), \tag{7.1.37}$$

where $\nu_{A,B}$ is the probability measure on A given by

$$\nu_{A,B}(y) = \frac{\mu(y) e_{A,B}(y)}{\mathrm{cap}(A,B)}, \quad y \in A, \tag{7.1.38}$$

with normalisation factor

$$\mathrm{cap}(A,B) = \sum_{x \in A} \mu(x) e_{A,B}(x). \tag{7.1.39}$$

Proof The key observation is that not only L but also its inverse L^{-1} is symmetric in $L^2(S,\mu)$. This implies that

$$\mu(x) G_B(x,y) = \mu(y) G_B(y,x), \quad x, y \in S, \tag{7.1.40}$$

and yields (7.1.35) via (7.1.22). Multiplying both sides of (7.1.35) by $\mu(x)g(x)$, summing over $x \in S$, and noting that $\sum_{x \in S} G_B(y,x) g(x) = f(y)$, we get (7.1.37) apart from the normalisation factor $\mathrm{cap}(A,B)$. Dividing by this quantity, we obtain (7.1.37). □

The measure $\nu_{A,B}$ is called the *last-exit biased distribution* on A for the transition from A to B. The number $\mathrm{cap}(A,B)$ is called the *capacity of the pair* (A,B).

The following corollary of Theorem 7.10 provides a formula for mean hitting times, which plays a *crucial rôle* in our study of metastability.

Corollary 7.11 *Let $A, B \subset S$ be non-empty and disjoint. Then, for reversible Markov processes,*

$$\sum_{x \in A} \nu_{A,B}(x)\, \mathbb{E}_x[\tau_B] = \frac{1}{\mathrm{cap}(A,B)} \sum_{y \in S} \mu(y) h_{A,B}(y). \tag{7.1.41}$$

In particular, for $A=\{x\}$,

$$\mathbb{E}_x[\tau_B]=\frac{1}{\mathrm{cap}(x,B)}\sum_{y\in S}\mu(y)h_{x,B}(y). \tag{7.1.42}$$

Proof Note that the representation in (7.1.3) shows that the solution f of (7.1.1) with $k=0$, $\bar{g}=0$ and $g=1$ is

$$f(x)=\mathbb{E}_x[\tau_B], \quad x\in S. \tag{7.1.43}$$

Inserting this into (7.1.37), we get (7.1.41). □

In Theorem 7.10 *capacity* made its first appearance. The first Green identity provides an important *alternative representation of capacity in terms of the Dirichlet form.*

Lemma 7.12 *Let* $A,B\subset S$ *be non-empty and disjoint. Then* $\mathrm{cap}(A,B)$ *defined in* (7.1.39) *can be expressed as*

$$\mathrm{cap}(A,B)=\mathscr{E}(h_{A,B},h_{A,B}). \tag{7.1.44}$$

Proof This is obvious from the definition of the Dirichlet form in Lemma 7.7, the definition of the equilibrium measure in (7.1.20), the definition of the capacity in (7.1.39) and the definition of the equilibrium potential in (7.1.15) defining the equilibrium potential $h_{A,B}$. □

Note that Lemma 7.12 becomes useful through the alternative representation of the Dirichlet form given in (7.1.29).

We close by listing a few relations, linking hitting probabilities and capacities, that will be needed in Chap. 8.

Lemma 7.13

(i) $\mu(x)\mathbb{P}_x(\tau_B<\tau_x)=\mathrm{cap}(x,B)$ *for* $x\in S,\ B\subset S\backslash\{x\}$.
(ii) $\mathbb{P}_y(\tau_x<\tau_B)/\mathbb{P}_x(\tau_y<\tau_B)=\mathrm{cap}(x,B)/\mathrm{cap}(y,B)$ *for* $x,y\in S$, $B\subset S\backslash\{x,y\}$.
(iii) $\mathbb{P}_y(\tau_B<\tau_x)\le\mathrm{cap}(x,B)/\mathrm{cap}(x,y)$ *for* $x,y\in S,\ B\subset S\backslash\{x,y\}$.

Proof (i) It follows from (7.1.19)–(7.1.20) with $A=\{x\}$ that $e_{x,B}(x)=(-Lh_{x,B})(x)$ $=\mathbb{P}_x(\tau_B<\tau_x)$. It follows from (7.1.39) with $A=\{x\}$ that $\mathrm{cap}(x,B)=\mu(x)e_{x,B}(x)$.

(ii) Use the second Green identity in (7.1.31), with $D=\{x\}$, $g=h_{x,B}$, $f=h_{y,B}$ and $x,y\in S\backslash B$, to get

$$\mu(x)(Lh_{x,B})(x)h_{y,B}(x)=\mu(y)(Lh_{y,B})(y)h_{x,B}(y), \quad x,y\in S\backslash B. \tag{7.1.45}$$

Since $h_{y,B}(x)=\mathbb{P}_x(\tau_y<\tau_B)$ by (7.1.16) with $A=\{y\}$, we get the claim.

(iii) Again use the second Green identity, this time with $D = \{x\}$, $g = h_{x,y}$, $f = h_{B,x}$ and $x, y \in S\backslash B$, to get

$$\mu(y)(Lh_{y,x})(y)h_{B,x}(y) = \sum_{z\in B} \mu(z)(Lh_{B,x})(z)h_{y,x}(z), \quad x, y \in S\backslash B. \tag{7.1.46}$$

The left-hand side equals $\mathrm{cap}(y, x)\mathbb{P}_y(\tau_B < \tau_x)$, the right-hand side is bounded from above by $\sum_{z\in B} \mu(z)(Lh_{B,x})(z)$, which equals $\mathrm{cap}(B, x)$ by (7.1.39). □

7.1.4 One-dimensional nearest-neighbour random walks

An important example where explicit computations are possible is that of a Markov process with state space $S \subseteq \mathbb{Z}$ for which transitions are allowed between nearest-neighbour sites only. Such Markov processes are referred to as *birth-death processes*. We denote the transition rates by $p(x, y)$, $y = x \pm 1$. In this case, there is a strictly positive invariant measure μ such that $\mu(x)p(x, y) = \mu(y)p(y, x)$ for all $x, y \in S$.

Equilibrium potential

Due to the one-dimensional nature of our Markov process, the only equilibrium potentials we have to compute are of the form

$$h_{b,a}(x) = \mathbb{P}_x(\tau_b < \tau_a), \quad a < x < b. \tag{7.1.47}$$

This satisfies the one-dimensional discrete boundary-value problem

$$\begin{aligned} p(x, x+1)\big[h(x+1) - h(x)\big] + p(x, x-1)\big[h(x-1) - h(x)\big] &= 0, \quad a < x < b,\\ h(a) &= 0,\\ h(b) &= 1. \end{aligned} \tag{7.1.48}$$

Note that the first equation can be conveniently rewritten as

$$p(x, x+1)\big[h(x+1) - h(x)\big] = p(x, x-1)\big[h(x) - h(x-1)\big]. \tag{7.1.49}$$

Setting $d(x) = h(x) - h(x-1)$, we get

$$d(x+1) = \frac{p(x, x-1)}{p(x, x+1)} d(x), \tag{7.1.50}$$

so that

$$d(x) = \prod_{z=a+1}^{x-1} \frac{p(z, z-1)}{p(z, z+1)} d(a+1). \tag{7.1.51}$$

Using reversibility, we can write the product as

$$\prod_{z=a+1}^{x-1} \frac{p(z,z-1)}{p(z,z+1)} = \prod_{z=a+1}^{x-1} \frac{\mu(z)}{\mu(z+1)} \frac{p(z,z-1)}{p(z+1,z)} = \frac{\mu(a+1)}{\mu(x)} \frac{p(a+1,a)}{p(x,x-1)}. \tag{7.1.52}$$

But

$$h(x) = d(x) + d(x-1) + d(x-2) + \cdots + d(a+1) + h(a), \tag{7.1.53}$$

so that

$$h(x) = R(a,x)\,\mu(a+1)\,p(a+1,a)\,d(a+1) + h(a), \tag{7.1.54}$$

where we abbreviate

$$R(u,v) = \sum_{y=u+1}^{v} \frac{1}{\mu(y)} \frac{1}{p(y,y-1)}, \qquad u < v. \tag{7.1.55}$$

Now $h(a) = 0$, and so it remains to determine $d(a+1)$ from the condition $h(b) = 1$, i.e.,

$$1 = R(a,b)\,\mu(a+1)\,p(a+1,a)\,d(a+1). \tag{7.1.56}$$

Combining this with (7.1.54), we get

$$h_{b,a}(x) = \frac{R(a,x)}{R(a,b)}, \qquad a < x < b. \tag{7.1.57}$$

Capacity

We continue by computing capacities. The equilibrium measure is given by the formula

$$\begin{aligned} &e_{b,a}(a) \\ &\quad = p(a,a+1)h_{b,a}(a+1) + p(a,a-1)h_{b,a}(a-1) = p(a,a+1)h_{b,a}(a+1), \end{aligned} \tag{7.1.58}$$

since $h_{b,a}(a-1) = 0$. Inserting (7.1.57), we get

$$e_{b,a}(a) = p(a,a+1)\frac{R(a,a+1)}{R(a,b)} = \frac{\frac{p(a,a+1)}{\mu(a+1)p(a+1,a)}}{R(a,b)} = \frac{1}{\mu(a)R(a,b)}. \tag{7.1.59}$$

Consequently, for the capacity we get

$$\mathrm{cap}(a,b) = \frac{1}{R(a,b)}. \tag{7.1.60}$$

Remark 7.14 Formula (7.1.60) suggests another common *electrostatic* interpretation of capacities, namely, as *conductances*. In fact, if we interpret $\mu(x)p(x,x-1)=\mu(x-1)p(x-1,x)$ as the conductance of the resistor $(x-1,x)$, then, by Ohm's law, (7.1.60) represents the conductance of the chain of resistors from a to b.

Mean hitting time

Inserting (7.1.57) and (7.1.60) into (7.1.42) (with $A=\{x\}$ and $B=\{a\}$), we get

$$\mathbb{E}_x[\tau_a] = R(a,x)\left(\sum_{y=a+1}^{x-1} \mu(y)\frac{R(a,y)}{R(a,x)} + \sum_{y=x}^{\infty}\mu(y)\right), \quad a<x. \tag{7.1.61}$$

This formula will be used in Chap. 13 to compute the metastable crossover time for the Curie-Weiss model. The latter will be shown to link up nicely with Kramers formula for Brownian motion in a double-well potential, as discussed in Sect. 2.1.1. See also Sect. 7.2.5.

7.2 The Dirichlet problem: continuous time

In the case of continuous time a number of technical problems arise that make the theory a bit more delicate. Structurally, however, all remains the same.

7.2.1 Definition

Much of what we discussed in Sect. 7.1 carries over to continuous-time Markov processes. The basic representation theorem for solutions of Dirichlet problems is provided through the martingale problem characterisation of general Markov processes, as discussed in Sect. 5.4.

We consider a continuous-time Markov process $X=(X_t)_{t\in\mathbb{R}_+}$ with state space S and generator $\mathscr{L}$. Let $D\subset S$ be an open set and specify continuous functions $g,k\colon D\to\mathbb{R}$ and $\bar{g}\colon D^c\to\mathbb{R}$. The question is whether we can find a continuous function $f\colon S\to\mathbb{R}$ such that

$$\begin{aligned}(-\mathscr{L}f)(x)+k(x)f(x)&=g(x), &&\forall x\in D,\\ f(x)&=\bar{g}(x), &&\forall x\in D^c.\end{aligned} \tag{7.2.1}$$

The Dirichlet problem in (7.2.1) can also be posed when $\bar{g}$ is not a continuous function. In that case the continuity requirement must be replaced by the condition that, for all $x\in\partial D$, if $\lim_{n\to\infty}x_n=x$ in D, then $\lim_{n\to\infty}f(x_n)=\bar{g}(x)$.

The analogue of Theorem 7.1 is the following basic representation theorem.

Theorem 7.15 *Let $\mathscr{L}$ be the generator of a continuous-time Markov process X, and assume that the associated martingale problem has a unique solution. Assume that $f \in \mathscr{D}(\mathscr{L})$, the domain of $\mathscr{L}$, solves the Dirichlet problem in (7.2.1), and let X be a solution of the martingale problem associated with $\mathscr{L}$. Let $\tau_{D^c} = \inf\{t > 0\colon X_t \notin D\}$. If*

$$\mathbb{E}_x\left[\tau_{D^c} \exp\left(\inf_{x\in D} k(x)\tau_{D^c}\right)\right] < \infty, \quad x \in D, \tag{7.2.2}$$

then

$$\begin{aligned} f(x) = \mathbb{E}_x\bigg[\bar{g}(X_{\tau_{D^c}})\exp\left(-\int_0^{\tau_{D^c}} k(X_s)ds\right) \\ + \int_0^{\tau_{D^c}} g(X_t)\exp\left(-\int_0^t k(X_s)ds\right)dt\bigg], \quad x \in D. \end{aligned} \tag{7.2.3}$$

Proof We use Lemma 5.35 with $g = \mathscr{L}f$, where f solves the Dirichlet problem. Condition (7.2.2) is, like its analogue (7.1.2) in the discrete-time case, sufficient to imply that the optional sampling theorem holds. □

As in (7.1.9) for the discrete case, we rewrite (7.2.3) for the case $k(x) = \lambda$ as

$$f(x) = \int_D G^\lambda_{D^c}(x, dz)\, g(z) + \int_{D^c} H^\lambda_{D^c}(x, dz)\, \bar{g}(z), \quad x \in D, \tag{7.2.4}$$

where

$$G^\lambda_{D^c}(x, dz) = \mathbb{E}_x\left[\int_0^{\tau_{D^c}} \mathrm{e}^{-\lambda t} \mathbb{1}_{X_t\in dz}\, dt\right], \quad x, z \in D, \tag{7.2.5}$$

is called the *Green function* and

$$\begin{aligned} H^\lambda_{D^c}(x, dz) &= \mathbb{E}_x\left[\mathrm{e}^{-\lambda\tau_{D^c}} \mathbb{1}_{X_{\tau_{D^c}}\in dz}\right] \\ &= \int_0^\infty \mathrm{e}^{-\lambda t}\mathbb{P}_x(\tau_{D^c} \in dt,\, X_t \in dz), \quad x \in D,\, z \in D^c, \end{aligned} \tag{7.2.6}$$

is called the *Poisson kernel*.

For the sequel it is useful to separate the discussion into two parts, dealing with the two main classes of Markov processes we need later on: Markov processes with countable state space (Sect. 7.2.2), and diffusion processes (Sect. 7.2.3).

7.2.2 Countable state space

In a countable state space S, all continuous-time Markov processes are essentially time changes of discrete-time Markov processes (see Sect. 5.1). The generator $\mathscr{L}$

can be written in terms of jump rates $c(x,y)$, $x,y \in S$, with $c(x,x)=0$, $x \in S$, in the form

$$(\mathscr{L}f)(x) = \sum_{y\in S} c(x,y)\big[f(y)-f(x)\big], \quad x \in S. \tag{7.2.7}$$

As long as the total jump rates

$$c(x) = \sum_{y\in S} c(x,y), \quad x \in S, \tag{7.2.8}$$

are bounded from above, very little changes from the discrete-time setting, and most formulas remain unaltered. We need the hitting times

$$\tau_A = \inf\{t>0\colon\ X_t \in A,\ \exists\, 0<s<t\colon\ X_s \neq X_0\}, \quad A \subset S. \tag{7.2.9}$$

Note that this definition makes sure that if $X_0 \in A$, then τ_A is not identically zero.

The Green function and the Poisson kernel take the form

$$G^{\lambda}_{D^c}(x,y) = \mathbb{E}_x\left[\int_0^{\tau_{D^c}} \mathrm{e}^{-\lambda t}\,\mathbb{1}_{X_t=y}\,dt\right], \quad x,y \in D, \tag{7.2.10}$$

$$H^{\lambda}_{D^c}(x,y) = \mathbb{E}_x\big[\mathrm{e}^{-\lambda\tau_{D^c}}\,\mathbb{1}_{X(\tau_{D^c})=y}\big], \quad x \in D,\ y \in D^c. \tag{7.2.11}$$

The solution $h_{A,B}$ of the Dirichlet problem in (7.1.15) with L replaced by $\mathscr{L}$ is again the equilibrium potential with the probabilistic interpretation in (7.1.16). We define the *equilibrium measure* on A as

$$e_{A,B}(x) = (-\mathscr{L}h_{A,B})(x), \quad x \in A, \tag{7.2.12}$$

but its probabilistic interpretation is slightly altered: a moment of reflection shows that

$$\begin{aligned}
\mathbb{P}_x(\tau_A < \tau_B) &= \frac{1}{c(x)}(\mathscr{L}h_{A,B})(x), && x \in B,\\
\mathbb{P}_x(\tau_B < \tau_A) &= \frac{1}{c(x)}(-\mathscr{L}h_{A,B})(x), && x \in A.
\end{aligned} \tag{7.2.13}$$

Indeed, $\frac{1}{c(x)}\mathscr{L}$ is the generator of the underlying discrete-time Markov process. Apart from the factor $\frac{1}{c(x)}$, all formulas derived for the reversible discrete-time case remain unaltered.

7.2.3 *Diffusion processes*

Matters become more involved for an uncountable state space. We will restrict our discussion to the case of elliptic diffusion processes in $\mathbb{R}^d$,

$$dX_t = b(X_t)dt + \sigma(X_t)dB_t, \tag{7.2.14}$$

where b is a time-independent drift vector and σ is a time-independent dispersion matrix. We have seen in Chap. 5 that solutions of this equation are strong Markov processes with a generator whose restriction to $C^2(\mathbb{R}^d)$ is given by

$$(\mathscr{L}f)(x) = \frac{1}{2}\sum_{i,j=1}^{d} a_{ij}(x)\frac{\partial^2 f(x)}{\partial x_i \partial x_j} + \sum_{i=1}^{d} b_i(x)\frac{\partial f(x)}{\partial x_i}, \tag{7.2.15}$$

where the *diffusion matrix* a is given by

$$a_{ij}(x) = \sum_{k=1}^{d} \sigma_{ik}(x)\sigma_{kj}(x). \tag{7.2.16}$$

In the sequel we will always assume that the dispersion matrix σ is non-degenerate and hence the diffusion matrix a is strictly positive, i.e., for all $x \in \mathbb{R}^d$, $a(x)$ defines a strictly positive quadratic form. For this case the operator $\mathscr{L}$ is called *elliptic*. If, for some open domain $D \subset \mathbb{R}^d$,

$$\sum_{i,j} a_{ij}(x)\xi_i\xi_j \geq \delta\|\xi\|_2^2, \quad x \in D, \tag{7.2.17}$$

then we call $\mathscr{L}$ *uniformly elliptic* in D.

The classical *Dirichlet problem* associated with an elliptic operator $\mathscr{L}$ and an open domain D is described as follows (where we assume that D is bounded). Let $g, k\colon \bar{D} \to \mathbb{R}$ and $\bar{g}\colon \partial D \to \mathbb{R}$ be continuous functions. We want to find a continuous function $f\colon \bar{D} \to \mathbb{R}$ such that

$$\begin{aligned} (-\mathscr{L}f)(x) + k(x)f(x) &= g(x), \quad \forall x \in D, \\ f(x) &= \bar{g}(x), \quad \forall x \in \partial D. \end{aligned} \tag{7.2.18}$$

Theorem 7.15 applies to this situation, but it is somewhat delicate to check when the assumptions are satisfied.

For the case $k(x) \leq 0$, condition (7.2.2) is ensured (for bounded domains) by a rather weak ellipticity condition.

Lemma 7.16 *Let $D \subset \mathbb{R}^d$ be open and bounded. Assume that, for some $1 \leq \ell \leq d$,*

$$\min_{x \in \bar{D}} a_{\ell\ell}(x) > 0. \tag{7.2.19}$$

Then $\mathbb{E}_x[\tau_{D^c}] < \infty$ for all $x \in D$.

Proof Set $a = \min_{x\in\bar{D}} a_{\ell\ell}(x)$, $b = \max_{x\in\bar{D}} \|b(x)\|$ and $q = \min_{x\in\bar{D}} x_\ell$. Let $\nu > 2b/a$. Consider the smooth function $h(x) = -\mu e^{\nu x_\ell}$, $x \in \mathbb{R}^d$, with $\mu > 0$. Clearly,

$$(-\mathscr{L}h)(x) = \mu e^{\nu x_\ell}\left(\tfrac{1}{2}\nu^2 a_{\ell\ell}(x) + \nu b_\ell(x)\right) \geq \tfrac{1}{2}\mu\nu a e^{\nu q}(\nu - 2b/a). \tag{7.2.20}$$

Choose μ such that the right-hand side is larger than 1, so that $(-\mathscr{L}h)(x) \geq 1$ for all $x \in D$. Since

$$h(X_{t\wedge\tau_{D^c}}) + \int_0^{t\wedge\tau_{D^c}} (-\mathscr{L}h)(X_s)ds \tag{7.2.21}$$

is a martingale, it follows that

$$\mathbb{E}_x\left[\int_0^{t\wedge\tau_{D^c}} (-\mathscr{L}h)(X_s)ds\right] = h(x) - \mathbb{E}_x\big[h(X_{t\wedge\tau_{D^c}})\big], \tag{7.2.22}$$

hence

$$\mathbb{E}_x[t\wedge\tau_{D^c}] \leq h(x) - \mathbb{E}_x\big[h(X_{t\wedge\tau_{D^c}})\big], \tag{7.2.23}$$

and so

$$\mathbb{E}_x[t\wedge\tau_{D^c}] \leq \max_{y\in\bar{D}} |h(y)| < \infty. \tag{7.2.24}$$

Passing to the limit $t \to \infty$, we get $\mathbb{E}_x[\tau_{D^c}] < \infty$. □

Theorem 7.15 gives us a stochastic representation formula for solutions of the Dirichlet problem, under the assumption that such solutions exist and also a weak solution of the SDE in (7.2.14) exists. We may ask whether we can actually use this formula to prove the existence of solutions of the Dirichlet problem. This is indeed the case under certain regularity conditions on the boundary of D. A sufficient criterion is given in the following proposition.

Theorem 7.17 *A point $z \in \partial D$ is regular if there exists a cone A with tip z such that $A \cap B_r(z) \subset D^c$ for some $r > 0$, where $B_r(z)$ is the ball of radius r centred at z. If all points of ∂D are regular, then existence and uniqueness of the solution of the Dirichlet problem holds.*

7.2.4 Reversible Markov processes

We now return to reversible Markov processes in the general continuous-time setting. Matters are similar as in discrete time, but formulations are slightly different.

Reversibility

Let $(P_t)_{t\in\mathbb{R}_+}$ be a strongly continuous contraction semigroup acting on the space $B(S)$ of bounded measurable functions on S. Assume that a measure μ on S is invariant with respect to $(P_t)_{t\in\mathbb{R}_+}$. Then the action of $(P_t)_{t\in\mathbb{R}_+}$ can be extended to $L^2(S,\mu)$. The following lemmas are the analogues of Lemmas 7.5–7.7 for discrete time, and their proofs can be copied.

Lemma 7.18 *Let $f \in L^2(S, \mu)$, where μ is invariant with respect to $(P_t)_{t\in\mathbb{R}_+}$. Then $P_t f \in L^2(S, \mu)$ for all $t \in \mathbb{R}_+$.*

Having an L^2-action of P_t, we can define its adjoint P_t^* via

$$\int_S \mu(dx) f(x)(P_t g)(x) = \int_S \mu(dx)\big(P_t^* f\big)(x) g(x), \quad f, g \in L^2(S, \mu). \tag{7.2.25}$$

We may check that $(P_t^*)_{t\in\mathbb{R}_+}$ is itself a Markov semigroup that generates the time-reversal of X, in the sense that $(P_t^* f)(X_t) = f(X_0)$.

Definition 7.19 A measure μ on S is called reversible with respect to $(P_t)_{t\in\mathbb{R}_+}$ if $P_t^* = P_t$ for some $t > 0$ (and hence for all $t > 0$).

Lemma 7.20 *If μ is a reversible probability measure for $(P_t)_{t\in\mathbb{R}_+}$, then μ is an invariant probability measure for $(P_t)_{t\in\mathbb{R}_+}$.*

The notions that were introduced above extend from the semigroup to the generator. Thus, for an invariant measure μ, we can define the adjoint $\mathscr{L}^*$ of a generator $\mathscr{L}$ via

$$\begin{aligned} &\int_S \mu(dx)\big(\mathscr{L}^* g\big)(x) f(x) = \int_S \mu(dx)(\mathscr{L} f)(x) g(x), \\ &\forall\, f, g \in \mathscr{D}(\mathscr{L})\colon\ \mathscr{L} f, \mathscr{L} g \in L^2(S, \mu). \end{aligned} \tag{7.2.26}$$

If μ is a probability measure, then the second condition is automatically verified. A reversible Markov process is therefore characterised by the fact that its generator is self-adjoint in $L^2(S, \mu)$ for some invariant measure μ.

Lemma 7.21 *Let μ be a reversible measure. Then the generator $\mathscr{L}$ defines a non-negative-definite quadratic form,*

$$\mathscr{E}(f, g) = \int_S \mu(dx) g(x)(-\mathscr{L} f)(x), \tag{7.2.27}$$

called the Dirichlet form.

Proof By the fact that $\mathscr{L}$ is self-adjoint, $\mathscr{E}(f, f)$ is real for all $f \in \mathscr{D}(\mathscr{L})$. Moreover, if $\mathscr{E}(f, f) < \infty$, then

$$\mathscr{E}(f, f) = \lim_{t\downarrow 0} t^{-1} \int_S \mu(dx) f(x)\big[f(x) - (P_t f)(x)\big]. \tag{7.2.28}$$

But

$$\begin{aligned} \int_S \mu(dx) f(x)\big[f(x) - (P_t f)(x)\big] &= \|f\|_{2,\mu}^2 - \int \mu(dx) f(x)(P_t f)(x) \\ &\geq \|f\|_{2,\mu}^2 - \|f\|_{2,\mu}\|P_t f\|_{2,\mu} \geq \|f\|_{2,\mu}^2 - \|f\|_{2,\mu}\|f\|_{2,\mu} = 0, \end{aligned} \tag{7.2.29}$$

where we use Cauchy-Schwarz and Lemma 7.18. Combining (7.2.28)–(7.2.29), we get $\mathscr{E}(f,f) \geq 0$. Since $\mathscr{L}$ is positive and self-adjoint, it can be written in the form $\mathscr{L} = A^t A$ with A positive. Hence the Dirichlet form has the form

$$\mathscr{E}(f,g) = \int_S \mu(dx)(Af)(x)(Ag)(x), \tag{7.2.30}$$

and is manifestly non-negative definite. □

Reversible diffusions

First, we note that the formal adjoint in $L^2(dx)$ of the operator $\mathscr{L}$ given in (7.2.15) is

$$\begin{aligned}(\mathscr{L}^* g)(x) &= \frac{1}{2}\sum_{i,j}\frac{\partial^2}{\partial x_i \partial x_j}\big[a_{ij}(x)g(x)\big] - \sum_i \frac{\partial}{\partial x_i}\big[b_i(x)g(x)\big] \\ &= \frac{1}{2}\sum_{i,j} a_{ij}(x)\frac{\partial^2 g(x)}{\partial x_i \partial x_j} \\ &\quad + \sum_i \left(\sum_j \frac{\partial a_{ij}(x)}{\partial x_j} - b_i(x)\right)\frac{\partial g(x)}{\partial x_i} \\ &\quad + \left(\frac{1}{2}\sum_{ij}\frac{\partial^2 a_{ij}(x)}{\partial x_i \partial x_j} - \sum_i \frac{\partial b_i(x)}{\partial x_i}\right) g(x). \end{aligned} \tag{7.2.31}$$

Hence $\mathscr{L}^* = \mathscr{L}$ if and only if

$$\sum_j \frac{\partial a_{ij}(x)}{\partial x_j} = 2b_i(x), \quad i = 1, \dots, d, \tag{7.2.32}$$

which thus is the condition for the diffusion to be reversible with respect to Lebesgue measure.

Next, we look for a reversible measure of the form $\mu(dx) = \mathrm{e}^{-F(x)}dx$. Then μ is reversible if and only if, for all $g \in \mathscr{D}(\mathscr{L})$,

$$\big(\mathscr{L}^*\big(g\mathrm{e}^{-F}\big)\big)(x) = \mathrm{e}^{-F(x)}(\mathscr{L}g)(x). \tag{7.2.33}$$

A simple computation via integration by parts shows that

$$\begin{aligned}\big(\mathscr{L}^*\big(g\mathrm{e}^{-F}\big)\big)(x) &= \mathrm{e}^{-F(x)}\frac{1}{2}\sum_{i,j} a_{ij}(x)\frac{\partial^2 g(x)}{\partial x_i \partial x_j} \\ &\quad - \mathrm{e}^{-F(x)}\sum_{i,j} a_{ij}(x)\frac{\partial F(x)}{\partial x_i}\frac{\partial g(x)}{\partial x_j}\end{aligned}$$

$$+\,\mathrm{e}^{-F(x)}\frac{1}{2}\sum_{i,j}a_{ij}(x)\left[\frac{\partial^2 F(x)}{\partial x_i \partial x_j}+\frac{\partial F(x)}{\partial x_i}\frac{\partial F(x)}{\partial x_j}\right]g(x)$$

$$+\,\mathrm{e}^{-F(x)}\sum_i\left(\sum_j\frac{\partial a_{ij}(x)}{\partial x_j}-b_i(x)\right)\left(-\frac{\partial F(x)}{\partial x_i}g(x)+\frac{\partial g(x)}{\partial x_i}\right)$$

$$+\,\mathrm{e}^{-F(x)}\left(\frac{1}{2}\sum_{ij}\frac{\partial^2 a_{ij}(x)}{\partial x_i\partial x_j}-\sum_i\frac{\partial b_i(x)}{\partial x_i}\right)g(x). \tag{7.2.34}$$

The condition for reversibility is therefore

$$\sum_j\left(-a_{ij}(x)\frac{\partial F(x)}{\partial x_j}+\frac{\partial a_{ij}}{\partial x_j}\right)=2b_i(x),\quad i=1,\dots,d, \tag{7.2.35}$$

or

$$b_i(x)=\frac{1}{2}\mathrm{e}^{F(x)}\sum_j\frac{\partial}{\partial x_j}\left(a_{ij}(x)\mathrm{e}^{-F(x)}\right),\quad i=1,\dots,d. \tag{7.2.36}$$

Inserting this relation into (7.2.15), we see that the operator $\mathcal{L}$ can be written in the form

$$(\mathcal{L}g)(x)=\frac{1}{2}\mathrm{e}^{F(x)}\sum_{i,j}\left(\frac{\partial}{\partial x_i}a_{ij}(x)\mathrm{e}^{-F(x)}\frac{\partial}{\partial x_j}\right)g(x). \tag{7.2.37}$$

In the simplest case where $a_{ij}=\delta_{ij}$, (7.2.36) reads

$$b_i(x)=-\frac{1}{2}\frac{\partial}{\partial x_i}F(x),\quad i=1,\dots,d, \tag{7.2.38}$$

i.e., the drift b is the gradient of the potential $-F$ (up to the factor $\frac{1}{2}$). In that case the generator $\mathcal{L}$ takes the suggestive form

$$(\mathcal{L}g)(x)=\frac{1}{2}\mathrm{e}^{F(x)}\left(\nabla\mathrm{e}^{-F(x)}\nabla\right)g(x). \tag{7.2.39}$$

The corresponding Dirichlet form can be written as

$$\mathcal{E}(f,g)=-\int_S\mu(dx)f(x)(\mathcal{L}g)(x)=\frac{1}{2}\int_S\mu(dx)\big\langle\nabla f(x),\nabla g(x)\big\rangle, \tag{7.2.40}$$

where $\langle\cdot,\cdot\rangle$ denotes the standard inner product in $\mathbb{R}^d$. In the case of general a we just need to use the inner product relative to a, i.e.,

$$\mathcal{E}(f,g)=-\int_S\mu(dx)f(x)(\mathcal{L}g)(x)=\frac{1}{2}\int_S\mu(dx)\sum_{i,j}a_{ij}(x)\frac{\partial f(x)}{\partial x_i}\frac{\partial g(x)}{\partial x_j}. \tag{7.2.41}$$

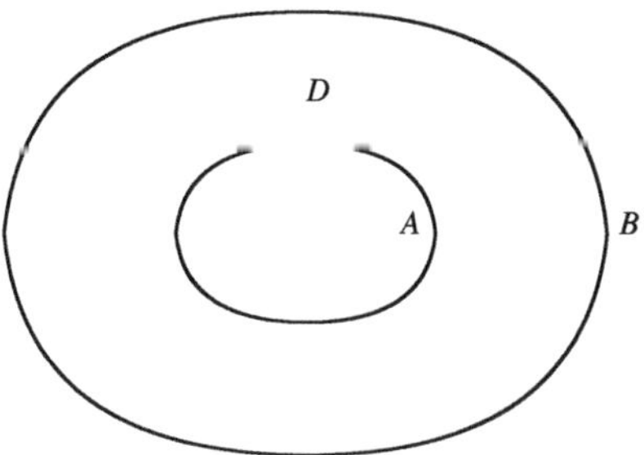

Fig. 7.3 Capacitor D between A and B

Equilibrium measure, equilibrium potential and capacity

In the following we return to the general case of an SDE corresponding to a generator that is a uniformly elliptic differential operator $\mathscr{L}$ with coefficients satisfying Lipschitz conditions (so that unique strong solutions of the SDE exist).

Let D be an open regular domain in $\mathbb{R}^d$ with $\partial D = A \cup B$, where A, B are non-empty and disjoint (see Fig. 7.3). Then the solution of the Dirichlet problem

$$\begin{aligned} (-\mathscr{L}h)(x) &= 0, \quad x \in D, \\ h(x) &= 1, \quad x \in A, \\ h(x) &= 0, \quad x \in B, \end{aligned} \tag{7.2.42}$$

is denoted by $h_{A,B}$ and is called the *equilibrium potential* of the *capacitor* (A, B). As in the discrete-time case,

$$h_{A,B}(x) = \mathbb{P}_x(\tau_A < \tau_B), \quad x \in D. \tag{7.2.43}$$

Remark 7.22 The above names come from the classical case where $\mathscr{L} = \frac{1}{2}\Delta$, for which the Dirichlet problem is a problem of electrostatics. The sets A and B correspond to two metal plates attached to a battery that imposes a constant voltage (potential difference) between the plates. The solution of this problem describes the electrostatic potential, whose gradient is the electrostatic field.

Next, we consider the inhomogeneous Dirichlet problem,

$$\begin{aligned} (-\mathscr{L}f)(x) &= g(x), \quad x \in D, \\ f(x) &= 0, \qquad x \in \partial D. \end{aligned} \tag{7.2.44}$$

We have seen in Theorem 7.15 that if (7.2.44) has a unique solution, then this solution has the probabilistic representation

$$f(x) = \mathbb{E}_x\left[\int_0^{\tau_{D^c}} g(X_t)dt\right], \quad x \in D. \tag{7.2.45}$$

The Green kernel will often have a density with respect to Lebesgue measure, i.e.,

$$G_{D^c}(x,dy) = G_{D^c}(x,y)dy, \quad x,y \in D. \tag{7.2.46}$$

In that case $G_{D^c}(x,y)$ is called the *Green function*. For the special case of (7.2.4) with $\lambda = 0$, $g = 1$ and $\bar{g} = 0$, (7.2.45) yields the relation

$$\mathbb{E}_x[\tau_{D^c}] = \int_D G_{D^c}(x,y)\,dy. \tag{7.2.47}$$

Let us next look at the relation between the equilibrium potential and the Dirichlet form in the case of a reversible diffusion. We want to compute $\mathscr{E}(h_{A,B},h_{A,B})$. We might be tempted to think that $\mathscr{E}(h_{A,B},h_{A,B}) = 0$, because $(\mathscr{L}h_{A,B})(x) = 0$ except on the sets ∂A and ∂B. But on these sets $\mathscr{L}h_{A,B}$ is singular because $h_{A,B}$ is not differentiable. Therefore we may interpret $\mathscr{L}h_{A,B}$ as a measure that is concentrated on A and B. Since $h_{A,B}$ vanishes on ∂B, we get

$$\mathscr{E}(h_{A,B},h_{A,B}) = \int_{\partial A} \mu(x)(-\mathscr{L}h_{A,B})(dx). \tag{7.2.48}$$

The measure $e_{A,B}(dx) = (-\mathscr{L}h_{A,B})(dx)$ is called the *equilibrium measure* associated with the *capacitor* (A,B).

To understand the above observation better, let us return to the case $a_{ij} = \delta_{ij}$. We then have the following integral formulas known as the *Green identities*, which constitute the analogue of Lemma 7.8.

Lemma 7.23 *Let D be a regular domain, let $f,g \in C^2(D)$, and let $\mathscr{L}$ be the reversible operator given by* (7.2.37). *Then*

(i) (*first Green identity*)

$$\int_D dx\, \mathrm{e}^{-F(x)}\big[\langle \nabla f(x), \nabla g(x)\rangle - g(x)(2\mathscr{L}f)(x)\big] = \int_{\partial D} \mathrm{e}^{-F(x)} g(x)\partial_{n(x)} f(x)\,d\sigma_D(x) \tag{7.2.49}$$

(ii) (*second Green identity*)

$$\int_D dx\, \mathrm{e}^{-F(x)}\big[f(x)(2\mathscr{L}g)(x) - g(x)(2\mathscr{L}f)(x)\big] = \int_{\partial D} \mathrm{e}^{-F(x)}\big[g(x)\partial_{n(x)} f(x) - f(x)\partial_{n(x)} g(x)\big]\,d\sigma_D(x) \tag{7.2.50}$$

hold with

$$\partial_{n(x)} = \sum_{i,j} n_i(x) a_{ij}(x) \frac{\partial}{\partial x_j}, \tag{7.2.51}$$

where $n(x)$ denotes the inner normal unit vector at $x \in \partial D$. In the case $a_{ij} = \delta_{ij}$, $\partial_{n(x)}$ is the usual normal derivative at x.

Proof For the case $F = 0$ and $a_{ij} = \delta_{ij}$, both formulas are classical and can be found in any standard textbook on potential theory. The extension to the general case is by straightforward computation. □

As in the discrete case, the Green identities give rise to a representation of the Poisson kernel in terms of the Green function.

Lemma 7.24 *If $\mathscr{L}$ is reversible with respect to μ and $D \subset \mathbb{R}^d$ is open and regular, then the solution of the Dirichlet boundary value problem*

$$\begin{aligned} -(\mathscr{L}f)(x) &= 0, \qquad x \in D, \\ f(x) &= \bar{g}(x), \quad x \in \partial D, \end{aligned} \tag{7.2.52}$$

is given by

$$f(x) = \int_{\partial D} \bar{g}(y)\, \mathrm{e}^{F(x)-F(y)} \partial_{n(y)} G_{D^c}(y,x) d\sigma_D(y), \quad x \in D, \tag{7.2.53}$$

i.e.,

$$H_{D^c}(x,dy) = \mathrm{e}^{F(x)-F(y)} \partial_{n(y)} G_{D^c}(y,x) d\sigma_D(y), \quad x, y \in D. \tag{7.2.54}$$

Using the first Green identity, we can state a precise relation between the equilibrium potential and the capacity. Namely, setting $f = g = h_{A,B}$ in (7.2.49), we see that

$$\begin{aligned} \mathscr{E}(h_{A,B}, h_{A,B}) &= \int_{\partial A} dx\, \mathrm{e}^{-F(x)} h_{A,B}(x)(-\mathscr{L}h_{A,B})(x) \\ &= \int_{\partial A} \mathrm{e}^{-F(x)} \partial_{n(x)} h_{A,B}(x) d\sigma_A(x), \end{aligned} \tag{7.2.55}$$

i.e., on A the equilibrium measure $e_{A,B}$ is given by

$$e_{A,B}(dx) = \partial_{n(x)} h_{A,B}(x) d\sigma_A(x). \tag{7.2.56}$$

Remark 7.25 The quantity

$$\mathrm{cap}(A,B) = \int_A \mathrm{e}^{-F(x)} \partial_{n(x)} h_{A,B}(x) d\sigma_A(x) = \int_A \mathrm{e}^{-F(x)} e_{A,B}(dx) \tag{7.2.57}$$

is called the *capacity* of the capacitor (A, B), which in electrical language is the total charge on the plate A. Using (7.2.55), we see that, alternatively, the capacity is the total *energy* of the potential $h_{A,B}$.

We have the analogue of Lemma 7.12.

Lemma 7.26 *Let $A, B \subset S$ be non-empty and disjoint. Then*

$$\mathrm{cap}(A, B) = \mathcal{E}(h_{A,B}, h_{A,B}). \tag{7.2.58}$$

Last-exit distribution and equilibrium measure

It will be nice to have a probabilistic interpretation of the equilibrium measure that explains why $-\mathcal{L}h_{A,B}$ really is a surface measure.

By the definition of the generator $\mathcal{L}$, we formally have

$$\begin{aligned}(-\mathcal{L}h_{A,B})(x) &= \lim_{t\downarrow 0} t^{-1}\big((1-P_t)h_{A,B}\big)(x)\\ &= \lim_{t\downarrow 0} t^{-1}\mathbb{E}_x\big[1-\mathbb{P}_{X_t}(\tau_A<\tau_B)\big]\\ &= \lim_{t\downarrow 0} t^{-1}\mathbb{E}_x\big[\mathbb{P}_{X_t}(\tau_B<\tau_A)\big].\end{aligned} \tag{7.2.59}$$

For $x \in D$ the limit exists and equals zero. For $x \in A$, however, the limit does not exist, but we will make sense of it in a weak sense. To that end, let us define the *last exit time* T_A from A prior to arrival in B as

$$T_A = \sup\{0 < t < \tau_B \colon X_t \in A\}, \tag{7.2.60}$$

with the convention that $\sup\emptyset = 0$. This is not a stopping time, and

$$\mathbb{P}_x(T_A > 0) = \mathbb{P}_x(\tau_A < \tau_B) = h_{A,B}(x), \quad x \in D. \tag{7.2.61}$$

Note that we can write the expectation in the last line of (7.2.59) as

$$\mathbb{E}_x\big[\mathbb{P}_{X_t}(\tau_B<\tau_A)\big] = \mathbb{P}_x(0 < T_A < t), \quad x \in D \cup A. \tag{7.2.62}$$

Set

$$\psi_t(x) = t^{-1}\mathbb{P}_x(0 < T_A < t), \quad x \in D \cup A. \tag{7.2.63}$$

Define the *last exit distribution* $\ell(x, dy)$ on A by

$$\ell(x, dy) = \mathbb{P}_x(X_{T_A -} \in dy,\, T_A > 0), \quad x \in D \cup A,\ y \in A. \tag{7.2.64}$$

Lemma 7.27 *Let f be continuous on $\bar{D} = D \cup \partial D$. Then*

$$\lim_{t\downarrow 0}\int_{D\cup A} G_B(x,y)\psi_t(y)f(y)dy = \int_A \ell(x,dy)f(y), \quad x \in D\cup A. \tag{7.2.65}$$

Proof Without loss of generality, let $f \geq 0$. Fix $x \in D \cup A$. Using the integral representation of the Green function in (7.2.5), we get

$$\int_{D\cup A} G_B(x,y)\psi_t(y)f(y)dy = \mathbb{E}_x\left[\int_0^{\tau_B}\psi_t(X_s)f(X_s)ds\right] \tag{7.2.66}$$

$$= t^{-1} \int_0^\infty \mathbb{E}_x\big[f(X_s)\mathbb{P}_{X_s}(0 < T_A < t)\big]\,ds = t^{-1} \int_0^\infty \mathbb{E}_x\big[f(X_s)\mathbb{1}_{s<T_A<s+t}\big]ds$$

$$= \mathbb{E}_x\left[\mathbb{1}_{0<T_A\le t}\, t^{-1} \int_0^{T_A} f(X_s)ds\right] + \mathbb{E}_x\left[\mathbb{1}_{T_A>t}\, t^{-1} \int_{T_A-t}^{T_A} f(X_s)ds\right].$$

Both terms in the last line are obviously uniformly bounded as $t \downarrow 0$. Moreover,

$$\mathbb{E}_x\left[\mathbb{1}_{0<T_A\le t}\, t^{-1} \int_0^{T_A} f(X_s)ds\right] \le C\,\mathbb{P}_x[0 < T_A \le t] \downarrow 0, \quad t \downarrow 0, \tag{7.2.67}$$

and, by the continuity of f,

$$\lim_{t\downarrow 0} \mathbb{E}_x\left[\mathbb{1}_{T_A>t}\, t^{-1} \int_{T_A-t}^{T_A} f(X_s)ds\right] = \mathbb{E}_x\big[\mathbb{1}_{T_A>0}\, f(X_{T_A-})\big]. \tag{7.2.68}$$

The right-hand side equals $\int_A \ell(x,dy) f(y)$. □

From Lemma 7.27 we deduce that the family of measures $\psi_t(y)dy$, $t > 0$, converges as $t \downarrow 0$ to a measure $e(dy)$ on A, which satisfies

$$G_B(x,y)e(dy) = \ell(x,dy), \quad x \in D \cup A,\ y \in A. \tag{7.2.69}$$

Integrating this formula over A, we arrive at the expression

$$\int_A G_B(x,y)e(dy) = \int_A \ell(x,dy) = h_{A,B}(x), \quad x \in D \cup A. \tag{7.2.70}$$

Hence $e(dy) = e_{A,B}(dy)$, the equilibrium measure that was introduced in (7.2.56).

In conclusion, we have proven the following analogue of Theorem 7.3.

Theorem 7.28 *For D and A, B as before,*

$$h_{A,B}(x) = \int_A G_B(x,y)e_{A,B}(dy), \quad x \in D \cup A \cup B. \tag{7.2.71}$$

Note that (7.2.71) holds on A because $\int_A \ell(x,dy) = 1$, $x \in A$, and on B because $G_B(x,y) = 0$, $x \in B$.

It is instructive to view Theorem 7.28 in the following way. We have already seen that we may think of $-\mathscr{L}h_{A,B}$ as a measure. The solution of the formal Dirichlet problem

$$\begin{aligned} (-\mathscr{L}h)(dx) &= e_{A,B}(dx), & x &\in D \cup A, \\ h(x) &= 0, & x &\in B \end{aligned} \tag{7.2.72}$$

in terms of the Green function is precisely the expression in (7.2.71).

Using reversibility, we obtain from Theorem 7.28 the following analogue of Theorem 7.10.

Theorem 7.29 *For D and A, B as before,*

$$h_{A,B}(x) = \int_A \frac{\mu(y)}{\mu(x)} G_B(y,x) e_{A,B}(dy), \quad x \in D \cup A \cup B. \tag{7.2.73}$$

The formula for the Green function gives corresponding formulas for solutions of Dirichlet problems. For instance, if for some function g we consider the Dirichlet problem

$$\begin{aligned} (-\mathscr{L} f)(x) &= g(x), & x \in D \cup A, \\ f(x) &= 0, & x \in B, \end{aligned} \tag{7.2.74}$$

then $f(x) = \int_{D\cup A} dy\, G_B(x,y) g(y)$. By reversibility,

$$G_B(x,y) = \mathrm{e}^{F(x)-F(y)} G_B(y,x),$$

and so

$$\begin{aligned} &\int_{D\cup A} \mathrm{e}^{-F(x)} h_{A,B}(x) g(x)\, dx \\ &= \int_{D\cup A} dx\, \mathrm{e}^{-F(x)} g(x) \int_A G_B(y,x)\, \mathrm{e}^{F(x)-F(y)}\, e_{A,B}(dy) \\ &= \int_A \mathrm{e}^{-F(y)}\, e_{A,B}(dy) \int_{D\cup A} G_B(y,x) g(x)\, dx \\ &= \int_A \mathrm{e}^{-F(y)}\, e_{A,B}(dy) f(y). \end{aligned} \tag{7.2.75}$$

Introducing the probability measure

$$\nu_{A,B}(dy) = \frac{\mathrm{e}^{-F(y)}\, e_{A,B}(dy)}{\operatorname{cap}(A,B)}, \quad y \in A, \tag{7.2.76}$$

we get

$$\int_A \nu_{A,B}(dy) f(y) = \frac{1}{\operatorname{cap}(A,B)} \int_{D\cup A} \mathrm{e}^{-F(x)}\, h_{A,B}(x) g(x)\, dx\,. \tag{7.2.77}$$

By picking $g = 1$, we get the following analogue of Corollary 7.11 linking crossover times to capacity.

Corollary 7.30 *For D and A, B as before,*

$$\int_A \nu_{A,B}(dy) \mathbb{E}_y[\tau_B] = \frac{1}{\operatorname{cap}(A,B)} \int_{D\cup A} dx\, \mathrm{e}^{-F(x)} h_{A,B}(x). \tag{7.2.78}$$

Proof Setting $w(y) = \mathbb{E}_y[\tau_B]$, for $y \notin B$, w solves the Dirichlet problem (7.2.74) with $g = 1$. Thus (7.2.78) is immediate from (7.2.77). □

7.2.5 *One-dimensional diffusions*

As in the case of nearest-neighbour random walks on $\mathbb{Z}$, diffusions on $\mathbb{R}$ allow for explicit solutions. In fact, the continuous case is even easier than the discrete case, which was explained in Sect. 7.1.4.

All homogeneous boundary value problems can, by linearity, be reduced to a computation of the equilibrium potential $h_{c,\{a,b\}}$ for $c \in (a,b)$, which is the solution of the Dirichlet problem

$$\begin{aligned} (-\mathscr{L}h)(x) &= 0, \quad x \in (a,b)\backslash c, \\ h(x) &= 0, \quad x \in \{a,b\}, \\ h(c) &= 1, \end{aligned} \tag{7.2.79}$$

where for later reference we choose the generator to be of the form

$$(-\mathscr{L}h)(x) = -\varepsilon a(x) h''(x) - b(x) h'(x) \tag{7.2.80}$$

for $\varepsilon > 0$, with $a(x) > 0$ and $b(x) \in \mathbb{R}$. The case $a(x) = 1$ corresponds to the classical Kramers equation (2.1.1). It follows from the general formula in (7.2.35) that the invariant measure for this diffusion is given by

$$\mu(dx) = \frac{1}{a(x)} \exp\left[\int_0^x \frac{b(z)}{a(z)} dz/\varepsilon \right], \tag{7.2.81}$$

up to normalisation. Set

$$\int_0^x \frac{b(z)}{a(z)} dz = -\widetilde{F}(x). \tag{7.2.82}$$

Then it is easy to verify that the Dirichlet form is given by

$$\mathscr{E}(f,g) = \tfrac{1}{2}\varepsilon \int_{\mathbb{R}} \mathrm{e}^{-\widetilde{F}(x)/\varepsilon} f'(x) g'(x)\, dx. \tag{7.2.83}$$

To compute the equilibrium potential $h_{c,\{a,b\}}$, we must solve the second-order differential equation

$$\varepsilon a(x) h''(x) + b(x) h'(x) = 0, \tag{7.2.84}$$

which reduces to the first-order differential equation

$$\varepsilon a(x) u'(x) + b(x) u(x) = 0 \tag{7.2.85}$$

after we set $u = h'$. Clearly, (7.2.85) has the general solution

$$u(x) = C_1 \mathrm{e}^{\widetilde{F}(x)/\varepsilon}, \tag{7.2.86}$$

and so the general solution of (7.2.84) is

$$h(x) = C_1 \int_0^x e^{\widetilde{F}(r)/\varepsilon} dr + C_2 \tag{7.2.87}$$

with C_1 and C_2 integration constants to be determined from the boundary conditions. In particular, for the equilibrium potential we have

$$h_{c,\{a,b\}}(x) = \begin{cases} \frac{\int_a^x e^{\widetilde{F}(r)/\varepsilon} dr}{\int_a^c e^{\widetilde{F}(r)/\varepsilon} dr}, & a < x < c, \\ \frac{\int_x^b e^{\widetilde{F}(r)/\varepsilon} dr}{\int_c^b e^{\widetilde{F}(r)/\varepsilon} dr}, & c < x < b. \end{cases} \tag{7.2.88}$$

Hence the capacity $\mathrm{cap}(c, \{a, b\})$ is readily computed as

$$\mathrm{cap}\big(c, \{a, b\}\big) = \mathscr{E}(h_{c,\{a,b\}}, h_{c,\{a,b\}}) = \frac{\varepsilon}{2 \int_a^c e^{\widetilde{F}(r)/\varepsilon} dr} + \frac{\varepsilon}{2 \int_c^b e^{\widetilde{F}(r)/\varepsilon} dr}. \tag{7.2.89}$$

From Lemma 7.28 we get the following formula for the Green function on (a, b):

$$G_{\{a,b\}}(y, x) = \frac{e^{-\widetilde{F}(x)/\varepsilon}}{a(x)} \frac{h_{y,\{a,b\}}(x)}{\mathrm{cap}(y, \{a, b\})}, \tag{7.2.90}$$

where the second equality uses (7.2.57). Note that this computation is a nice alternative to the usual method of variation of constants used to obtain the Green function.

Now, if $\lim_{x\downarrow-\infty} \widetilde{F}(x) = \infty$, then $\lim_{a\downarrow-\infty} \int_a^x e^{\widetilde{F}(r)/\varepsilon} dr = \infty$, and we get

$$\lim_{a\downarrow-\infty} h_{c,\{a,b\}}(x) = \begin{cases} 1, & -\infty < x < c, \\ \frac{\int_x^b e^{\widetilde{F}(r)/\varepsilon} dr}{\int_c^b e^{\widetilde{F}(r)/\varepsilon} dr}, & c < x < b, \end{cases} \tag{7.2.91}$$

and

$$\lim_{a\downarrow-\infty} \mathrm{cap}\big(c, \{a, b\}\big) = \frac{\varepsilon}{2 \int_c^b e^{\widetilde{F}(r)/\varepsilon} dr}. \tag{7.2.92}$$

Hence

$$\lim_{a\downarrow-\infty} G_{\{a,b\}}(y, x) = \begin{cases} 2(\varepsilon a(x))^{-1} e^{-\widetilde{F}(x)/\varepsilon} \int_x^b e^{\widetilde{F}(r)/\varepsilon} dr, & y < x < b, \\ 2(\varepsilon a(x))^{-1} e^{-\widetilde{F}(x)/\varepsilon} \int_y^b e^{\widetilde{F}(r)/\varepsilon} dr, & x < y < b. \end{cases} \tag{7.2.93}$$

Integrating over $x \in (-\infty, b)$, we get

$$\mathbb{E}_y[\tau_b] = 2 \int_{-\infty}^b e^{-\widetilde{F}(x)/\varepsilon} \frac{1}{\varepsilon a(x)} \bigg(\int_{x\vee y}^b e^{\widetilde{F}(r)/\varepsilon} \, dr \bigg) dx, \quad y \in (-\infty, b). \tag{7.2.94}$$

Note that, for $a(x) = 1$ as in (2.1.1), the definition of $\widetilde{F}$ in (7.2.82) reduces to $b = -\widetilde{F}'$, i.e., $\widetilde{F} = F$ with F the potential. If F is chosen to be a double-well

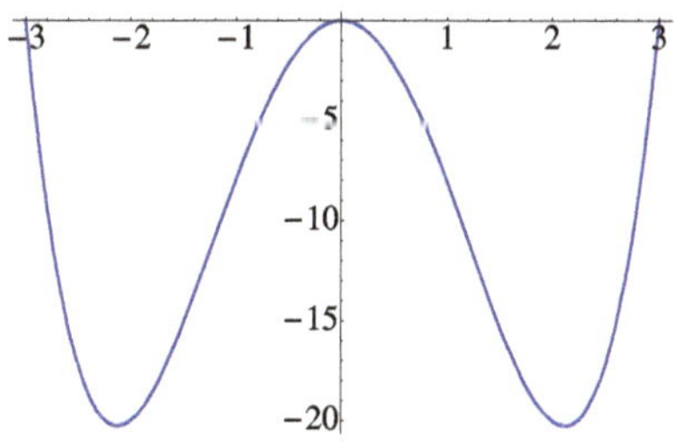

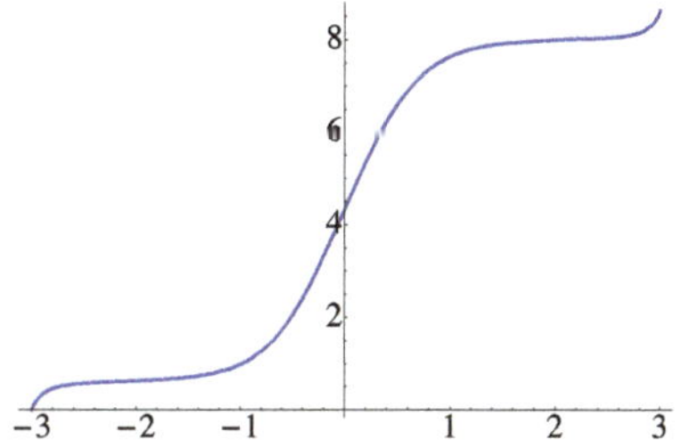

Fig. 7.4 Example of the setting in Remark 7.32 with $a = -3$ and $b = 3$: a potential $x \mapsto F(x)$ on $[-3, 3]$ and its associated equilibrium potential $x \mapsto h_{3,-3}(x) = \mathbb{P}_x(\tau_3 < \tau_{-3})$

potential, then (7.2.94) yields, in the limit as $\varepsilon \downarrow 0$ and with the help of elementary Laplace asymptotics, the *Kramers formula* in (2.1.2) advertised in Sect. 2.1.

Remark 7.31 Note that we chose an arbitrary normalisation for the invariant measure, which influences the value of the capacity. It does not, however, affect the value of physical quantities, in particular, the Green function and the mean hitting time.

Remark 7.32 If instead of (7.2.79) we take the Dirichlet problem

$$\begin{aligned} (-\mathscr{L}h)(x) &= 0, \quad x \in (a,b), \\ h(a) &= 0, \\ h(b) &= 1, \end{aligned} \tag{7.2.95}$$

then (7.2.94) becomes

$$\mathbb{E}_y[\tau_b] = 2\int_y^b \mathrm{e}^{-\widetilde{F}(x)/\varepsilon} \frac{1}{\varepsilon a(x)} \left(\int_y^x \mathrm{e}^{\widetilde{F}(r)/\varepsilon}\, dr \right) dx, \quad y \in (a,b). \tag{7.2.96}$$

See Fig. 7.4 for an example.

7.3 Variational principles

As was pointed out earlier, variational principles are at the heart of our endeavor to obtain sharp estimates on key quantities in metastable systems. We have already seen that such quantities can be expressed as solutions to PDE's (or discrete analogues of PDE's), but finding these is hard. Variational principles provide tools to get good estimates without an explicit solution. In this section we discuss three variational principles: the *Dirichlet principle*, the *Thomson principle* and the *Berman-Konsowa principle*.

7.3.1 The Dirichlet principle

In Sects. 7.1–7.2 we have seen that the Dirichlet form computed on the equilibrium potential gives the capacity. We will now show that the equilibrium potential is the solution of a variational problem.

Theorem 7.33 (Dirichlet principle) *Let D and A, B be as in the definition of the Dirichlet problem in* (7.1.15) *and* (7.2.42) (*see Fig.* 7.3). *Let $\mathscr{H}_{A,B}$ be the space of continuous functions f on $\bar{D}$ such that*

(i) $\mathscr{E}(f, f) < \infty$.
(ii) $f \geq 1$ *on A and* $f \leq 0$ *on B.*

Assume that the corresponding Dirichlet problem has a unique solution, the equilibrium potential $h_{A,B}$. Then

$$\mathrm{cap}(A, B) = \inf_{f \in \mathscr{H}_{A,B}} \mathscr{E}(f, f). \tag{7.3.1}$$

Moreover, if $\mathscr{H}_{A,B} \neq \emptyset$, then the infimum in (7.3.1) *is attained uniquely at the equilibrium potential, i.e.,* $\mathrm{cap}(A, B) = \mathscr{E}(h_{A,B}, h_{A,B})$.

Proof We write the proof in the diffusion setting, but the same arguments work in general. Suppose that $\mathscr{H}_{A,B} \neq \emptyset$. Let g be a function with $\mathscr{E}(g, g) < \infty$ such that $g \geq 0$ on A and $g \leq 0$ on B. Then, for $h \in \mathscr{H}_{A,B}$ and $\varepsilon > 0$ (recall (7.2.55)), using the second Green identity (7.2.50),

$$\begin{aligned}
&\mathscr{E}(h + \varepsilon g, h + \varepsilon g) - \mathscr{E}(h, h) = \varepsilon\big[\mathscr{E}(h, g) + \mathscr{E}(g, h)\big] + \varepsilon^2 \mathscr{E}(g, g) \\
&= \varepsilon \int_{\partial A} \mathrm{e}^{-F(x)} g(x) \partial_{n(x)} h(x)\, d\sigma_A(x) + \varepsilon \int_{\partial B} \mathrm{e}^{-F(x)} g(x) \partial_{n(x)} h(x)\, d\sigma_B(x) \\
&\quad + 2\varepsilon \int_D \mu(dx) g(x) (\mathscr{L} h)(x) + \varepsilon^2 \mathscr{E}(g, g).
\end{aligned} \tag{7.3.2}$$

If $h = h_{A,B}$ is the equilibrium potential, then the boundary integrals are non-negative and the first term in the last line vanishes. Since the second term in the last line is non-negative, it follows that h is a global minimum of $\mathscr{E}$ in $\mathscr{H}_{A,B}$. Finally, suppose that there is another function f such that $\mathscr{E}(f, f) = \mathscr{E}(h, h)$. Then the identity

$$\mathscr{E}\big(\tfrac{f+h}{2}, \tfrac{f+h}{2}\big) + \mathscr{E}\big(\tfrac{f-h}{2}, \tfrac{f-h}{2}\big) = \tfrac{1}{2}\mathscr{E}(f, f) + \tfrac{1}{2}\mathscr{E}(h, h) \tag{7.3.3}$$

implies that

$$\mathscr{E}\big(\tfrac{f+h}{2}, \tfrac{f+h}{2}\big) = \mathscr{E}(h, h) - \mathscr{E}\big(\tfrac{f-h}{2}, \tfrac{f-h}{2}\big). \tag{7.3.4}$$

Since h is a global minimum, this equality can only hold if

$$\mathscr{E}(f - h, f - h) = 0. \tag{7.3.5}$$

But, by (7.2.40) (recall that we are in the diffusion setting), the latter means that $\|\nabla(f-h)\|_2 = 0$ μ-a.s., i.e., $f-h$ is constant μ-a.s. Because of condition (ii), it follows that $f = h$ μ-a.s. □

The Dirichlet principle is a powerful tool for asymptotic computations of capacities via upper and lower bounds. An elementary *upper bound* is the following.

Corollary 7.34 *For any* $f \in \mathscr{H}_{A,B}$,

$$\mathrm{cap}(A,B) \le \mathscr{E}(f,f). \tag{7.3.6}$$

Since $\mathscr{E}(f,f)$ is a sum (or an integral) of non-negative terms (recall (7.1.29), (7.2.30) and (7.2.40)), a *lower bound* can be obtained by dropping some of these terms. Upper and lower estimates of this type, which are flexible, will turn out to be very important in Parts IV–VIII.

7.3.2 The Thomson principle

A classical reverse variational principle is due to Thomson.

Theorem 7.35 (Thomson principle, Version 1) *Assume that* A, B *are such that the corresponding Dirichlet problem has a unique solution* $h_{A,B}$. *Let* $\mathscr{T}_{A,B}$ *denote the space of super-harmonic functions on* D^c *that take values in* $[0,1]$, *i.e.,*

$$\mathscr{T}_{A,B} = \left\{h\colon\ S \to [0,1],\ h \in L^2(S,\mu)\colon\ (\mathscr{L}h)(x) \le 0\ \forall x \in S\backslash D\right\}. \tag{7.3.7}$$

Then

$$\mathrm{cap}(A,B) = \sup_{h \in \mathscr{T}_{A,B}} \frac{\mathscr{E}(1_A,h)^2}{\mathscr{E}(h,h)}, \tag{7.3.8}$$

and the supremum is attained at $h = h_{A,B}$.

Proof The proof is simple. By (7.2.26–7.2.27), for all $h \in \mathscr{T}_{A,B}$,

$$\mathscr{E}(h_{A,B},h) = \int_S \mu(dx) h_{A,B}(x)(-\mathscr{L}h)(x) \ge \int_A \mu(dx)(-\mathscr{L}h)(x) = \mathscr{E}(1_A,h). \tag{7.3.9}$$

On the other hand, by the Cauchy-Schwarz inequality,

$$\mathscr{E}(h_{A,B},h)^2 \le \mathscr{E}(h,h)\mathscr{E}(h_{A,B},h_{A,B}), \tag{7.3.10}$$

and hence, for all $h \in \mathscr{T}_{A,B}$,

$$\mathrm{cap}(A,B) = \mathscr{E}(h_{A,B},h_{A,B}) \ge \frac{\mathscr{E}(1_A,h)^2}{\mathscr{E}(h,h)}. \tag{7.3.11}$$

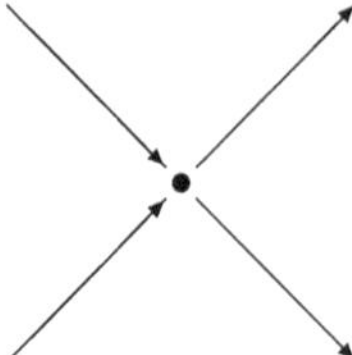

Fig. 7.5 Kirchhoff's law says that the in-flow and the out-flow are the same for all vertices that are not wired to the outside

Thus, the right-hand side of (7.3.8) is a lower bound for $\mathrm{cap}(A, B)$. Since, by definition (see (7.2.55–7.2.57)), $\mathrm{cap}(A, B) = \mathscr{E}(1_A, h_{A,B})$, the lower bound in (7.3.11) is attained for $h = h_{A,B}$. □

The Thomson principle is much more difficult to exploit than the Dirichlet principle, since it imposes the constraint of super-harmonicity on the test functions. Guessing good super-harmonic functions is not easy.

In the setting of Markov processes with countable state space, there is an alternative (and better known) formulation of the Thomson principle in terms of *flows* (see Fig. 7.5).

Definition 7.36 Let $\Gamma = (S, E)$ be a graph with edge set E and vertex set S. Let $A, B \subset S$ be non-empty and disjoint. A map $f\colon\ E \to \mathbb{R}$ is called a *unit flow* from A to B when

(i) *Kirchhoff's law* holds: the flows into and out of vertices in $S\backslash(A \cup B)$ are the same, i.e.,

$$\sum_{\substack{y\in S:\\(y,x)\in E}} f\big((y,x)\big) = \sum_{\substack{z\in S:\\(x,z)\in E}} f\big((x,z)\big) \quad \forall x \in S\backslash(A \cup B). \tag{7.3.12}$$

(ii) The total flow out of A and into B is one, i.e.,

$$\sum_{x\in A}\sum_{\substack{z\in S:\\(x,z)\in E}} f\big((x,z)\big) = 1 = \sum_{x\in B}\sum_{\substack{y\in S:\\(y,x)\in E}} f\big((y,x)\big). \tag{7.3.13}$$

Note that in the discrete case,

$$(\mathscr{L}h)(x) = \sum_{y\in S} p(x,y)\big[h(y) - h(x)\big], \quad x \in S, \tag{7.3.14}$$

is a sum over edge-functions on the graph of the Markov process. The Dirichlet form can therefore be written as

$$\mathscr{E}(h,g) = \frac{1}{2}\sum_{(x,y)\in E} \mu(x)p(x,y)\big[h(y)-h(x)\big]\big[g(y)-g(x)\big] \quad (7.3.15)$$
$$= \frac{1}{2}\sum_{(x,y)\in E} \frac{\{\mu(x)p(x,y)[h(y)-h(x)]\}\{\mu(x)p(x,y)[g(y)-g(x)]\}}{\mu(x)p(x,y)}.$$

Defining the functional $\mathscr{D}$ on pairs of edge functions u, v by

$$\mathscr{D}(u,v) = \frac{1}{2}\sum_{(x,y)\in E} \frac{1}{\mu(x)p(x,y)} u\big((x,y)\big)v\big((x,y)\big), \quad (7.3.16)$$

we have

$$\mathscr{E}(h,g) = \mathscr{D}(\mu p\nabla h, \mu p\nabla g), \quad (7.3.17)$$

with the obvious definition of $\mu p\nabla$. In particular,

$$\mathrm{cap}(A,B) = \mathscr{D}(\mu p\nabla h_{A,B}, \mu p\nabla h_{A,B}). \quad (7.3.18)$$

On the other hand, for any unit flow f we have

$$\mathscr{D}(\mu p\nabla h_{A,B}, f) = \sum_{(x,y)\in E} \big[h_{A,B}(x) - h_{A,B}(y)\big] f\big((x,y)\big)$$
$$= \sum_{x\in A}\sum_{\substack{y\in S:\\(x,y)\in E}} f\big((x,y)\big) = 1. \quad (7.3.19)$$

Applying the Cauchy-Schwarz inequality as in (7.3.10) again, we get

$$\mathrm{cap}(A,B) \geq \frac{\mathscr{D}(\mu p\nabla h_{A,B}, f)^2}{\mathscr{D}(f,f)} = \frac{1}{\mathscr{D}(f,f)} \quad (7.3.20)$$

for any unit flow f.

Theorem 7.37 (Thomson principle, Version 2) *For Markov processes with countable state space, with the notation above,*

$$\mathrm{cap}(A,B) = \sup_{f\in\mathscr{U}_{A,B}} \frac{1}{\mathscr{D}(f,f)}, \quad (7.3.21)$$

where $\mathscr{U}_{A,B}$ is the space of all unit flows from A to B. The supremum is attained for the harmonic unit flow

$$f_{h_{A,B}}\big((x,y)\big) = \frac{\mu(x)p(x,y)[h_{A,B}(y) - h_{A,B}(x)]_+}{\mathrm{cap}(A,B)}. \quad (7.3.22)$$

Proof In view of (7.3.20), we only need to verify that equality holds for the particular choice of the harmonic unit flow. To check that $\mathscr{D}(f_{h_{A,B}}, f_{h_{A,B}}) = 1/\mathrm{cap}(A,B)$ is immediate. We only need to verify that $f_{h_{A,B}}$ is a unit flow from A to B.

Lemma 7.38 *Let h be a harmonic function with respect to $\mathscr{L}$. Then ϕ_h defined by*

$$\phi_h\big((x,y)\big) = \mu(x)p(x,y)\big[h(y)-h(x)\big]_+ \tag{7.3.23}$$

is a flow.

Proof For $y \in S$, compute

$$\begin{aligned}
&\sum_{x\in S}\big[\phi_h\big((x,y)\big)-\phi_h\big((y,x)\big)\big]\\
&\quad=\sum_{x\in S}\big\{\mu(x)p(x,y)\big[h(y)-h(x)\big]_+ - \mu(y)p(y,x)\big[h(x)-h(y)\big]_+\big\}\\
&\quad=\sum_{x\in S}\mu(y)p(y,x)\big\{\big[h(y)-h(x)\big]_+ - \big[h(x)-h(y)\big]_+\big\}\\
&\quad=-\sum_{x\in S}\mu(y)p(y,x)\big[h(x)-h(y)\big] = \mu(y)(-\mathscr{L}h)(y)=0,
\end{aligned} \tag{7.3.24}$$

which says that ϕ_h is a flow. □

By Lemma 7.38 and since $\sum_{x\in A}\sum_{y\in S}\phi h_{A,B}(x,y)=\mathrm{cap}(A,B)$, $f_{h_{A,B}}$ is a unit flow from A to B. This proves the theorem. □

Remark 7.39 Note that the proof of Lemma 7.38 implies that for any function g the edge-function ϕ_g defined in (7.3.23) satisfies

$$\sum_{x\in S}\big[\phi_g\big((x,y)\big)-\phi_g\big((y,x)\big)\big]=\mu(y)(-\mathscr{L}g)(y), \quad y\in S. \tag{7.3.25}$$

7.3.3 The Berman-Konsowa principle

Berman and Konsowa [23] obtained another variational principle, for the case of discrete-time Markov processes, which generates lower bounds that improve on those obtained from the Thomson principle. Its derivation is quite different, and actually starts from the Dirichlet principle.

We work in the same setting as in Sect. 7.3.2.

Definition 7.40 Given a graph $\Gamma=(S,E)$ and non-empty disjoint subsets $A, B \subset S$, an edge function $f\colon E\to[0,\infty)$ is called a loop-free unit flow from A to B when:

(i) f is a unit flow from A to B.
(ii) Any path γ of edges from A to B such that $f(e)>0$ for all $e\in\gamma$ is self-avoiding. In particular, if $f((x,y))>0$, then $f((y,x))=0$.

First we observe that a loop-free unit flow gives rise to a directed Markov process. For $x \in S$, let $\mathscr{F}(x) = \sum_{y\in S:\,(x,y)\in E} f((x,y))$ be the total flow out of x, and assume that $\mathscr{F}(x) > 0$ for all $x \in S$. For $(x,y) \in E$ and $x \in S\backslash B$, let

$$q^f\big((x,y)\big) = \frac{f((x,y))}{\mathscr{F}(x)}, \tag{7.3.26}$$

and put $q^f((x,y)) = 0$ for $x \in B$. We construct a Markov process with law $\mathbb{P}^f$, initial distribution $\mathbb{P}^f(X_0 = x) = \mathscr{F}(x)\mathbb{1}_{x\in A}$ and transition matrix q^f that is killed in B. $\mathbb{P}^f$ can also be seen as a probability distribution on self-avoiding paths from A to B, with

$$\mathbb{P}^f(\gamma) = \mathscr{F}(\gamma_0) \prod_{i=0}^{|\gamma|-1} q^f\big((\gamma_i, \gamma_{i+1})\big). \tag{7.3.27}$$

Lemma 7.41 *Let $e \in E$. Then*

$$\mathbb{P}^f(e \in \gamma) = f(e). \tag{7.3.28}$$

Proof Let $e = (x,y)$. Then, by the Markov property and the fact that the paths are self-avoiding, the probability in question equals the probability that a path hits x and immediately moves to y:

$$\mathbb{P}^f(e \in \gamma) = \mathbb{P}^f(\tau_x < \tau_B) q^f\big((x,y)\big). \tag{7.3.29}$$

Use (7.3.27) to write

$$\mathbb{P}^f(\tau_x < \tau_B) = \sum_{\gamma:\, A\to x} \mathscr{F}(\gamma_0) \prod_{i=0}^{|\gamma|-1} q^f\big((\gamma_i, \gamma_{i+1})\big). \tag{7.3.30}$$

The summation over paths has to be carried out carefully. To that end, recursively define the sets

$$\begin{aligned} &A_0 = A, \\ &A_n = \big\{z \in S\backslash A\colon \exists_{y\in A_{n-1}} f\big((y,z)\big) > 0,\ \forall_{y\notin A_0\cup\cdots\cup A_{n-1}} f\big((y,z)\big) = 0\big\}. \end{aligned} \tag{7.3.31}$$

Note that, due to the loop-freeness of the flow, for any $z \in S$ there exists a unique $n^*(z)$ such that $z \in A_{n^*(z)}$. Set

$$G(z) = \mathbb{P}^f(\tau_z < \tau_B). \tag{7.3.32}$$

Then the Markov property implies the recursive identity

$$G(z) = \sum_{y\in A_0\cup\cdots\cup A_{n^*(z)-1}} G(y) q^f\big((y,z)\big). \tag{7.3.33}$$

We prove by induction that $G(z) = \mathscr{F}(z)$ for all $z \in S$. Indeed, for $z \in A$ we have $G(z) = \mathscr{F}(z)$ by our choice of the initial condition. It therefore suffices to show that if $G(z) = \mathscr{F}(z)$ holds for all $z \in A_k$, $0 \le k \le n$, then it also holds for $z \in A_{n+1}$. Now, by (7.3.31–7.3.33), for $z \in A_{n+1}$ we have

$$G(z) = \sum_{y \in A_0 \cup \cdots \cup A_n} G(y) q^f((y,z)) = \sum_{y \in A_0 \cup \cdots \cup A_n} f((y,z)), \tag{7.3.34}$$

where we use the induction hypothesis. However, for $z \in A_{n+1}$ we also have

$$\sum_{y \in A_0 \cup \cdots \cup A_n} f(y,z) = \sum_{y \in S} f((y,z)) = \sum_{w \in S} f(z,w) = \mathscr{F}(z), \tag{7.3.35}$$

where we use that the flow satisfies Kirchhoff's law. Thus, we have completed the induction step and have proven that

$$\mathbb{P}^f(\tau_x < \tau_B) = \mathscr{F}(x), \quad x \in S. \tag{7.3.36}$$

Combine (7.3.26), (7.3.29) and (7.3.36) to get the claim. □

Remark 7.42 Lemma 7.41 is the only place where the flow property of f is used. In terms of flows the observation in (7.3.36) is the probabilistic interpretation of the fact that $\mathscr{F}(x)$ is the total flow into x.

Provided $f(e) > 0$, we can divide (7.3.28) by $f(e)$, to obtain

$$\mathbb{1}_{f(e)>0} = \sum_{\gamma} \mathbb{P}^f(\gamma) \frac{\mathbb{1}_{e \in \gamma}}{f(e)}. \tag{7.3.37}$$

Now pick any function $h \in \mathscr{H}_{A,B}$. Then

$$\begin{aligned} \mathscr{E}(h,h) &\ge \sum_{(x,y) \in E} \mu(x) p(x,y) \left[h(x) - h(y)\right]^2 \mathbb{1}_{f((x,y))>0} \\ &= \sum_{(x,y) \in E} \mu(x) p(x,y) \left[h(x) - h(y)\right]^2 \sum_{\gamma} \mathbb{P}^f(\gamma) \frac{\mathbb{1}_{(x,y) \in \gamma}}{f((x,y))} \\ &= \sum_{\gamma} \mathbb{P}^f(\gamma) \sum_{(x,y) \in \gamma} \frac{\mu(x) p(x,y)}{f((x,y))} \left[h(x) - h(y)\right]^2. \end{aligned} \tag{7.3.38}$$

From this we can derive a lower bound on the capacity, namely, we take the infimum over h and interchange the sum over γ with the infimum over h:

$$\mathrm{cap}(A,B) \ge \inf_{h \in \mathscr{H}_{A,B}} \sum_{\gamma} \mathbb{P}^f(\gamma) \sum_{(x,y) \in \gamma} \frac{\mu(x) p(x,y)}{f((x,y))} \left[h(x) - h(y)\right]^2$$

$$\geq \sum_\gamma \mathbb{P}^f(\gamma) \inf_{h\in\mathscr{H}_{A,B}} \sum_{(x,y)\in\gamma} \frac{\mu(x)p(x,y)}{f((x,y))}\left[h(x)-h(y)\right]^2$$

$$= \sum_\gamma \mathbb{P}^f(\gamma)\Bigg(\sum_{(x,y)\in\gamma} \frac{f((x,y))}{\mu(x)p(x,y)}\Bigg)^{-1}. \tag{7.3.39}$$

In the last step we use the explicit solution of the Dirichlet problem on the one-dimensional path γ. We readily see that equality holds when we insert the harmonic unit flow (see (7.3.22)). Thus, we have proved the following theorem.

Theorem 7.43 (Berman-Konsowa principle) *Let $\mathbb{U}_{A,B}$ denote the set of loop-free unit flows from A to B. Then*

$$\mathrm{cap}(A,B) = \sup_{f\in\mathbb{U}_{A,B}} \mathbb{E}^f\Bigg[\Bigg(\sum_{(x,y)\in\gamma} \frac{f((x,y))}{\mu(x)p(x,y)}\Bigg)^{-1}\Bigg]. \tag{7.3.40}$$

Remark 7.44 The Berman-Konsowa principle improves the Thomson principle. Namely, by Jensen's inequality and Lemma 7.41,

$$\begin{aligned}&\mathbb{E}^f\Bigg[\Bigg(\sum_{(x,y)\in\gamma} \frac{f((x,y))}{\mu(x)p(x,y)}\Bigg)^{-1}\Bigg]\\ &\quad\geq \Bigg(\mathbb{E}^f\Bigg[\sum_{(x,y)\in\gamma} \frac{f((x,y))}{\mu(x)p(x,y)}\Bigg]\Bigg)^{-1} = \Bigg(\frac{1}{2}\sum_{(x,y)} \frac{f((x,y))^2}{\mu(x)p(x,y)}\Bigg)^{-1} = \frac{1}{\mathscr{D}(f,f)}.\end{aligned} \tag{7.3.41}$$

Hence, every choice of f yields a better lower bound via the Berman-Konsowa principle than via the Thomson principle.

The more serious advantage of the Berman-Konsowa principle is the fact that the bounds can often be evaluated explicitly. The sums appearing in the right-hand side of (7.3.40) are straightforward, live on the flow realising the supremum, and are independent of the realisation of the Markov chain, so that the expectation over $\mathbb{E}^f$ becomes trivial. This will be explained in the examples that are treated in Parts IV–VIII.

7.4 Variational principles in the non-reversible setting

Variational representations for capacities are known also in the non-reversible setting, but they are much more involved and therefore far less useful. Here is a brief account.

We assume that there exists a unique ergodic invariant measure μ. We denote by

$$p^*(x,y) = \frac{\mu(y)}{\mu(x)} p(y,x), \quad x,y \in S, \tag{7.4.1}$$

the transition probabilities of the *time-reversed* Markov process, and by

$$p^s(x,x) = \tfrac{1}{2}\big[p(x,y) + p^*(x,y)\big] \tag{7.4.2}$$

the transition probabilities of the *symmetrised* Markov process. Analogously, we write $\mathscr{L}^*$ and $\mathscr{L}^s$ for the generators of these processes. Note that from the definition of capacity (see (7.1.39)) we get that, for any $f \in \mathscr{H}_{A,B}$,

$$\mathrm{cap}(A,B) = (f, -\mathscr{L} h_{A,B})_\mu. \tag{7.4.3}$$

In particular,

$$\mathrm{cap}(A,B) = \big(h^*_{A,B}, -\mathscr{L} h_{A,B}\big)_\mu = \big(-\mathscr{L}^* h^*_{A,B}, h_{A,B}\big)_\mu = \mathrm{cap}^*(A,B). \tag{7.4.4}$$

Define the norms

$$\|f\|^2_{H^1} = \mathscr{E}(f,f), \tag{7.4.5}$$

and

$$\|f\|^2_{H^{-1}} = \sup_{g \in H^1} \big[2(f,g)_\mu - \mathscr{E}(g,g)\big]. \tag{7.4.6}$$

Note that on the space of functions with zero mean we have $\|f\|^2_{H^{-1}} = (f, \mathscr{L}^{-1} f)$, while otherwise the H^{-1}-norm is infinite. An application of the Cauchy-Schwarz inequality yields the bound

$$|(f,g)_\mu| \le \|f\|_{H^1} \|g\|_{H^{-1}}. \tag{7.4.7}$$

Using this bound with f replaced by $-\mathscr{L}^* f$ and g by $h_{A,B}$, we can show that

$$\big(-\mathscr{L}^* f, h_{A,B}\big)^2_\mu \le \mathrm{cap}(A,B) \sup_{h \in \mathscr{G}_{A,B}} \big[2\big(-\mathscr{L}^* f, h\big)_\mu - \|h\|^2_{H^1}\big], \tag{7.4.8}$$

where $\mathscr{G}_{A,B}$ denotes the space of functions that are constant on the sets A and B. Furthermore, for any $f \in \mathscr{H}_{A,B}$ we have, via (7.4.3),

$$\big(-\mathscr{L}^* f, h_{A,B}\big)_\mu = (f, -\mathscr{L} h_{A,B})_\mu = \mathrm{cap}(A,B), \tag{7.4.9}$$

so that we obtain the bound

$$\mathrm{cap}(A,B) \le \inf_{f \in \mathscr{H}_{A,B}} \sup_{h \in \mathscr{G}_{A,B}} \big[2\big(-\mathscr{L}^* f, h\big)_\mu - \|h\|^2_{H^1}\big]. \tag{7.4.10}$$

It now suffices to choose $f = \frac{1}{2}(h_{A,B} + h^*_{A,B})$, where $h^*_{A,B}$ is the equilibrium potential for the adjoint generator $\mathscr{L}^*$, to verify that the infimum is attained at f. This yields the following *Dirichlet principle* for the non-reversible case.

Theorem 7.45 (Dirichlet principle: non-reversible case) *For non-empty disjoint sets $A, B \subset S$,*

$$\operatorname{cap}(A,B) = \inf_{f \in \mathscr{H}_{A,B}} \sup_{h \in \mathscr{G}_{A,B}} \left[2\left(-\mathscr{L}^* f, h\right)_\mu - \|h\|_{H^1}^2\right]. \tag{7.4.11}$$

The Thomson principle in the form of Theorem 7.35 carries over to the non-reversible case.

Another version of both the Dirichlet principle and the Thomson principle is the following theorem, whose proof can be found in Slowik [220]. We need the following notations: for $g\colon S \to \mathbb{R}$, set

$$\Psi_g\big((x,y)\big) = \mu(x) p^s(x,y)\big[g(x) - g(y)\big] \tag{7.4.12}$$

and

$$\Phi_g\big((x,y)\big) = \mu(x) p(x,y) g(x) - \mu(y) p(y,x) g(y). \tag{7.4.13}$$

Theorem 7.46 (Dirichlet and Thomson principles: non-reversible case) *Consider a Markov process with a countable state space S. Let $A, B \subset S$ be non-empty and disjoint. Then:*

(i) *The Dirichlet principle holds, in the sense that*

$$\operatorname{cap}(A,B) = \inf_{f \in \mathscr{H}_{A,B}} \inf_{\psi \in \mathscr{U}^0_{A,B}} \mathscr{D}(\Phi_f - \psi, \Phi_f - \psi), \tag{7.4.14}$$

where $\mathscr{H}_{A,B}$ is the space of functions defined in Theorem 7.33 and $\mathscr{U}^0_{A,B}$ is the space of zero-flows. The infima are attained at $f = \frac{1}{2}(h_{A,B} + h^*_{A,B})$ and $\psi = \Phi_f - \Psi_{h_{A,B}}$.

(ii) *The Thomson principle holds, in the sense that*

$$\operatorname{cap}(A,B) = \sup_{g \in G^0_{A,B}} \sup_{\phi \in \mathscr{U}^1_{A,B}} \frac{1}{\mathscr{D}(\phi - \Phi_g, \phi - \Phi_g)}, \tag{7.4.15}$$

where $G^0_{A,B}$ is the space of functions that vanish on $A \cup B$ and $\mathscr{U}^1_{A,B}$ is the space of unit flows. The suprema are attained for $\phi = \phi_{A,B} + \Phi_g$ and $g = \frac{1}{2}(h^*_{A,B} - h_{A,B})/\operatorname{cap}(A,B)$, where $\phi_{A,B}$ is the harmonic flow.

Whether or not these variational principles are useful in connection with metastability remains to be seen. They are substantially more involved than their analogues in the reversible case, where the minimiser has a transparent probabilistic interpretation that makes it easy to come up with good guesses for test functions.

7.5 Bibliographical notes

1. The connection between Markov processes and potential theory goes back to Kakutani [144, 145]. A fundamental treatment is given in the monograph by Doob [95]. For a presentation of the theory in the context of discrete Markov processes, see e.g. the book by Doyle and Snell [96].

2. There is a formula for the mean hitting time that does not require reversibility, as was noted by Gaveau and Moreau [123]. It suffices to recall that (7.1.23) holds in general, to get

$$\mathbb{E}_a[\tau_B] = \sum_{x \in S} \frac{\mu(x) h_{x,B}(a)}{\mathrm{cap}(x, B)}. \tag{7.5.1}$$

Note that this formula is not quite as nice as (7.1.41), but in principle it constitutes an alternative. For more details, see the PhD thesis of Eckhoff [101, 102]. Fernández, Manzo, Nardi, Scoppola and Sohier [111] and Fernández, Manzo, Nardi and Scoppola [112] develop a theory of metastability without reversibility, based on certain assumptions involving slow escape, fast thermalisation and fast recurrence, and provide examples of dynamics for which these assumptions can be verified.

3. The Dirichlet form $\mathscr{E}$ can be extended to the set $\{f : \mathscr{E}(f, f) < \infty\}$, which typically is larger than the domain of $\mathscr{L}$. An entire theory is available that allows us to use this fact to construct a Markov process from a Dirichlet form. For a detailed treatment, see e.g. the monograph by Fukushima, Oshida and Takeda [116].

4. In textbooks, Green identities are given for the case $\mathscr{L} = \frac{1}{2}\Delta$ only. We have not been able to find a reference where they are stated in general in explicit form.

5. The derivation in Sect. 7.2.4 is taken from the monograph by Sznitman [226].

6. For irregular domains, existence and uniqueness issues are more delicate. For further reading, we refer the reader to the monograph by Karatzas and Shreve [148].

7. The approach in Sect. 7.3.3 was developed by Bianchi, Bovier and Ioffe [24] following the original paper by Berman and Konsowa [23]. In den Hollander and Jansen [82] the Berman-Konsowa principle is extended to arbitrary reversible Markov jump processes on Polish spaces. The latter paper contains an appendix in which the physical interpretations of the Dirichlet, Thomson and Berman-Konsowa variational principles are elaborated.

8. The connection between the Berman-Konsowa principle and the Thomson principle has been worked out by Slowik [219]. Remark 7.44 comes from that paper.

9. The history of variational principles in the non-reversible case appears to be a little obscure. The Dirichlet principle in the form of (7.4.11) was obtained by

Doyle [97]. Our presentation follows the exposition given by Slowik [220]. The Dirichlet principle in the form of (7.4.14) is given by Landim [159] and Gaudillière and Landim [121], while the Thomson principle in (7.4.15) apparently appears for the first time in Slowik [220].

10. For a detailed discussion of Theorem 7.17, see Karatzas and Shreve [148].

11. A host of material on reversible Markov processes with countable state space is presented in the online-book by Aldous and Fill [2].

Part III
Metastability

Part III develops the theory of metastability for Markov processes.

Chapter 8 provides key definitions and basic properties. The starting point is a *definition* of a metastable Markov process based on *capacities*. After that, *renewal estimates* are used to derive bounds on harmonic functions in terms of capacities, and it is shown that capacities are *approximately ultrametric*. This in turn leads to estimates on *mean hitting times*. Finally, metastability is linked to the *spectrum* of the generator of the Markov process, which is shown to decompose into a cluster of small real eigenvalues that are separated from the rest of the spectrum by a gap. This in turn leads to the exponential law for the metastable crossover time.

Chapter 9 collects basic techniques. Upper and lower bounds on capacity are derived with the help of the Dirichlet principle. Coarse-graining techniques are used to describe metastability of Markov chains with a high degree of symmetry, like the Curie-Weiss model. Regularity estimates on harmonic functions are derived for Markov processes with uncountable state spaces, like elliptic diffusions. These in turn are linked to coupling methods.

Chapter 8
Key Definitions and Basic Properties

La véritable éloquence consiste à dire tout ce qu'il faut, et à ne dire que ce qu'il faut.
(François de La Rochefoucauld, Réflexions)

In this chapter we introduce the basic setup for our approach to metastability. The guiding principle is to provide a definition of metastable sets, representing metastable states in model systems, that is verifiable in concrete models and implies the type of behaviour that is associated with metastability. The intuitive picture we have in mind comes from the paradigmatic Brownian motion in a double-well (or a multi-well) potential in one dimension. Here, the metastable states correspond to "valleys" of the potential, labeled by the local minima of the potential. Our aim is to give a definition that applies in far more general situations.

Section 8.1 defines metastable sets and provides the characterisation of metastability in terms of capacities. Section 8.2 shows how renewal estimates can be used to obtain upper and lower bounds on the equilibrium potential in terms of capacity and establishes the approximate ultrametricity of capacity. Section 8.3 uses these results to obtain sharp bounds on mean hitting times. Section 8.4 makes the link with spectral theory. Section 8.5, finally, mentions some problems that come up for uncountable state spaces.

8.1 Characterisation of metastability

Consider a Markov process X with state space S and discrete or continuous time. Let $\mathbb{P}$ denote the law of X and $\mathbb{P}_x$ the law of X conditioned on $X_0 = x$. We will typically assume that X is uniquely ergodic with invariant measure μ. For $D \subset S$, let τ_D denote the *first hitting time* of X in D, i.e.,

$$\tau_D = \inf\{t > 0\colon\ X(t) \in D\}. \tag{8.1.1}$$

The fundamental feature we would like to associate with metastability is the existence of two well-separated time scales and the partition of the state space into disjoint sets S_i, $i \in I$, such that, when X starts in S_i, on a *short time scale* it reaches

A. Bovier, F. den Hollander, *Metastability*,
Grundlehren der mathematischen Wissenschaften 351,
DOI 10.1007/978-3-319-24777-9_8

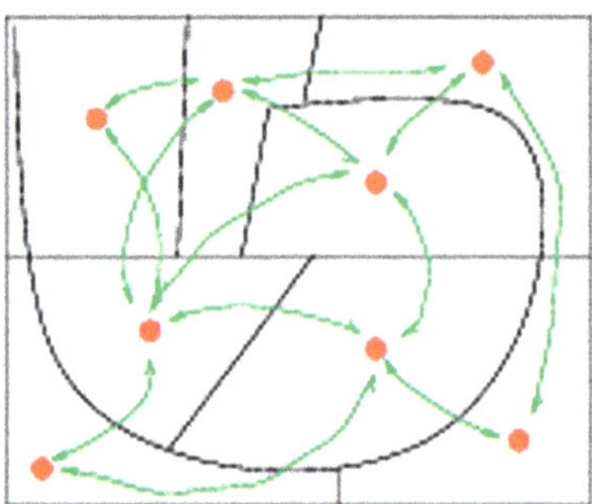

Fig. 8.1 Picture of a metastable set (*dots*) labelling the metastable valleys, and transitions between these valleys (*arrows*)

some sort of local equilibrium concentrated on S_i, while on a *long time scale* it exits S_i and moves to some S_j with $j \neq i$, where it again reaches local equilibrium, etc. We may think of the dynamics as "hopping" between *quasi-invariant sets* (see Fig. 8.1). To capture this picture, an appealing way is to characterise the rapid approach to local equilibrium by saying that, in a suitable sense, X is locally recurrent or Harris recurrent: each S_i contains a small set $B_i \subset S_i$ that is revisited by X very frequently before it moves out of S_i.

On this basis, an intuitively appealing definition of metastability could be the following:

- *A family of Markov processes is called metastable if there exists a collection of disjoint sets $B_i \subset S$, $i \in I$, such that*

$$\frac{\sup_{x \notin \bigcup_{i\in I} B_i} \mathbb{E}_x[\tau_{\bigcup_{i\in I} B_i}]}{\inf_{i\in I} \inf_{x\in B_i} \mathbb{E}_x[\tau_{\bigcup_{j\in I\setminus i} B_j}]} = o(1). \tag{8.1.2}$$

Here, $o(1)$ should be thought of as a small intrinsic parameter that characterises the "degree" of metastability, since typically we deal with a family of Markov processes indexed by a parameter (like temperature, system size, etc.) that allows us to make (8.1.2) as small as we like.

The definition in (8.1.2) characterises metastability in terms of a physical property, namely, hitting times of the system. Certainly we would want such a property to hold for a system to be called metastable. However, the problem is that (8.1.2) is not immediately verifiable, since mean hitting times are generally difficult to compute. Indeed, one of our goals is to compute mean hitting times, and so (8.1.2) would put us in a circular set-up. It is thus desirable to have an equivalent definition involving more manageable quantities.

The relations in Corollaries 7.11 and 7.30 between mean hitting times and capacities suggest an *alternative characterisation of metastability through capacities*. We will see that this characterisation entails many advantages. We first give a tentative definition of metastable sets.

- A family of Markov processes is called metastable if there exists a collection of disjoint sets $B_i \subset S$, $i \in I$, such that

$$\frac{\sup_{i\in I}\inf_{x\in B_i}\mathbb{P}_x(\tau_{\bigcup_{j\in I\setminus i}B_j} < \hat{\tau}_{B_i})}{\inf_{x\notin\bigcup_{i\in I}B_i}\mathbb{P}_x(\tau_{\bigcup_{i\in I}B_i} < \hat{\tau}_{B(x)})} = o(1), \tag{8.1.3}$$

where $B(x)$ is a sufficiently large neighbourhood of x, and $\hat{\tau}_{B(x)} = \inf\{t > \tau_{B(x)^c} : X_t \in B(x)\}$ is the first time X returns to $B(x)$ after having left it.

Remark 8.1 Note that here and in the sequel we always include the *stable* set in the collection of metastable sets, contrary to what is common practice. In particular, if there is only one metastable set, then it is stable, and the system exhibits no metastable behaviour on this level of resolution.

This definition leaves some questions open. Should we take the supremum over $x \in B_i$ in the numerator rather than the infimum? What should be the choice for $B(x)$? How can we relate the probabilities appearing in the definition to capacities, as advertised?

It will emerge that the usefulness of the definition depends crucially on further properties of the sets B_i, $i \in I$, and on local mixing properties of the process. Before we continue this discussion, we turn to the simplest case from which we can derive much of our intuition: Markov processes in discrete time with countable state spaces.

An important goal will be to derive general properties of metastable systems. Since (8.1.3) implies frequent returns to the small starting set B_i before the transition to a set B_j, $j \neq i$, we expect an exponential law for the transition times. We further expect that the process of successive visits to the sets B_i, $i \in I$, asymptotically is a Markov process on I.

Everything becomes easy and transparent when the state space S is finite, and we can replace the sets B_i, $i \in I$, and $B(x)$, $x \in S$, in (8.1.3) by single points. It will be useful to understand this simple setting first.

The following definition of a set of metastable points applies (see Fig. 8.1).

Definition 8.2 (Metastable points) Suppose that $|S| < \infty$. A Markov processes X is said to be ρ-metastable with respect to a set of points $\mathscr{M} \subset S$ if

$$|S|\,\frac{\sup_{x\in\mathscr{M}}[\mathrm{cap}(x,\mathscr{M}\backslash x)/\mu(x)]}{\inf_{y\notin\mathscr{M}}[\mathrm{cap}(y,\mathscr{M})/\mu(y)]} \le \rho \ll 1. \tag{8.1.4}$$

Remark 8.3 Definition 8.2 is useful because, as we will see later, it involves quantities that are either known or are controllable. It becomes intuitively even more appealing after we note that (8.1.4) can be written alternatively as

$$|S|\,\frac{\sup_{x\in\mathscr{M}}\mathbb{P}_x(\tau_{\mathscr{M}\backslash x} < \tau_x)}{\inf_{y\notin\mathscr{M}}\mathbb{P}_y(\tau_{\mathscr{M}} < \tau_y)} \le \rho \ll 1, \tag{8.1.5}$$

where to go from (8.1.4) to (8.1.5) we use Lemma 7.13(i). Note the appearance of the cardinality of the state space in (8.1.4)–(8.1.5). The definition makes sense when we have a sequence of processes where the cardinality of the state space is either fixed or increases slowly. If $|S| = \infty$, but there exists a subset $S_0 \subset S$ with $\mathrm{cap}(\mathcal{M}, S_0^c) \ll \max_{x \in \mathcal{M}} \mathrm{cap}(x, \mathcal{M} \backslash x)$, then $|S|$ can be replaced by $|S_0|$ in Definition 8.2. The reader may verify this fact in the proofs below. The intuitive reason is that, under this assumption, the process will have visited all metastable points long before it leaves the set S_0.

We want to show that if a process is metastable in the sense of Definition 8.2, then we can express mean hitting times of subsets of metastable points *in terms of capacities and the invariant measure alone*. This is based on the key formula in (7.1.41), which here reads, for $J \subset \mathcal{M}, n \in \mathcal{M}$,

$$\mathbb{E}_n[\tau_J] = \frac{1}{\mathrm{cap}(n, J)} \sum_{y \in S} \mu(y) h_{n,J}(y). \tag{8.1.6}$$

The main work is to control the sum over the equilibrium potential in (8.1.6). To do this, we show in Sect. 8.2 how to control the equilibrium potential in terms of capacities. In Sect. 8.3 we use these estimates to derive bounds on mean hitting times.

8.2 Renewal estimates and ultrametricity

The estimation of the equilibrium potential through capacities is based on a renewal argument that is simple in the case of a discrete state space.

Lemma 8.4 (Renewal estimate on equilibrium potential) *Let $A, B \subset S$ be non-empty disjoint sets, and let $x \notin A \cup B$. Then*

$$\max\left(1 - \frac{\mathrm{cap}(x, B)}{\mathrm{cap}(x, A)}, 0\right) \leq h_{A,B}(x) \leq \min\left(\frac{\mathrm{cap}(x, A)}{\mathrm{cap}(x, B)}, 1\right). \tag{8.2.1}$$

Proof The upper bound follows from the estimate

$$\begin{aligned} h_{A,B}(x) = \mathbb{P}_x(\tau_A < \tau_B) &= \frac{\mathbb{P}_x(\tau_A < \tau_{B \cup x})}{1 - \mathbb{P}_x(\tau_x < \tau_{A \cup B})} \\ &= \frac{\mathbb{P}_x(\tau_A < \tau_{B \cup x})}{\mathbb{P}_x(\tau_{A \cup B} < \tau_x)} \leq \frac{\mathbb{P}_x(\tau_A < \tau_x)}{\mathbb{P}_x(\tau_B < \tau_x)} = \frac{\mathrm{cap}(x, A)}{\mathrm{cap}(x, B)}, \end{aligned} \tag{8.2.2}$$

where the second equality comes from counting the returns to x without a hit of A or B. The lower bound follows from the upper bound via the symmetry relation $h_{A,B}(x) = 1 - h_{B,A}(x)$. □

An important consequence of the renewal estimate in Lemma 8.4 is the *approximate ultrametricity of capacities.*

Lemma 8.5 *Let $D \subset S$ and $x, y \in S \backslash D$. If* $\mathrm{cap}(x, D) \leq \delta\, \mathrm{cap}(x, y)$ *for* $0 < \delta < 1$, *then*

$$1 - \delta \leq \frac{\mathrm{cap}(x, D)}{\mathrm{cap}(y, D)} \leq \frac{1}{1-\delta}. \tag{8.2.3}$$

Proof By Lemma 7.13(ii), we have

$$\frac{\mathrm{cap}(x, D)}{\mathrm{cap}(y, D)} = \frac{\mathbb{P}_y(\tau_x < \tau_D)}{\mathbb{P}_x(\tau_y < \tau_D)}. \tag{8.2.4}$$

Trivially, the right-hand side can be sandwiched as

$$1 - \mathbb{P}_y(\tau_D < \tau_x) \leq \frac{\mathbb{P}_y(\tau_x < \tau_D)}{\mathbb{P}_x(\tau_y < \tau_D)} \leq \frac{1}{1 - \mathbb{P}_x(\tau_D < \tau_y)}. \tag{8.2.5}$$

But from the renewal bound in Lemma 8.4 we have

$$\mathbb{P}_x(\tau_D < \tau_y) = h_{D,y}(x) \leq \frac{\mathrm{cap}(x, D)}{\mathrm{cap}(x, y)} \leq \delta. \tag{8.2.6}$$

Substitution into the right-hand side of (8.2.5) yields the upper bound in (8.2.3). On the other hand, by Lemma 7.13(iii) we also have

$$\mathbb{P}_y(\tau_D < \tau_x) \leq \frac{\mathrm{cap}(x, D)}{\mathrm{cap}(x, y)} \leq \delta. \tag{8.2.7}$$

Substitution into the left-hand side of (8.2.5) yields the lower bound in (8.2.3). □

Lemma 8.5 has the following corollary, which is the version of the approximate ultrametric triangle inequality we are looking for.

Corollary 8.6 (Approximate ultrametricity of capacities) *For all distinct* $x, y, z \in S$,

$$\mathrm{cap}(x, y) \geq \tfrac{1}{2} \min\big(\mathrm{cap}(x, z), \mathrm{cap}(y, z)\big). \tag{8.2.8}$$

Proof Suppose that the claim is false. Then there exist distinct $x, y, z \in S$ with $\mathrm{cap}(x, y) < \frac{1}{2}\mathrm{cap}(x, z)$ and $\mathrm{cap}(x, y) < \frac{1}{2}\mathrm{cap}(y, z)$. Lemma 8.5 with $\delta = \frac{1}{2}$ therefore implies that

$$\tfrac{1}{2} \leq \frac{\mathrm{cap}(x, y)}{\mathrm{cap}(y, z)} \leq 2, \qquad \tfrac{1}{2} \leq \frac{\mathrm{cap}(x, y)}{\mathrm{cap}(x, z)} \leq 2, \tag{8.2.9}$$

which yields a contradiction. □

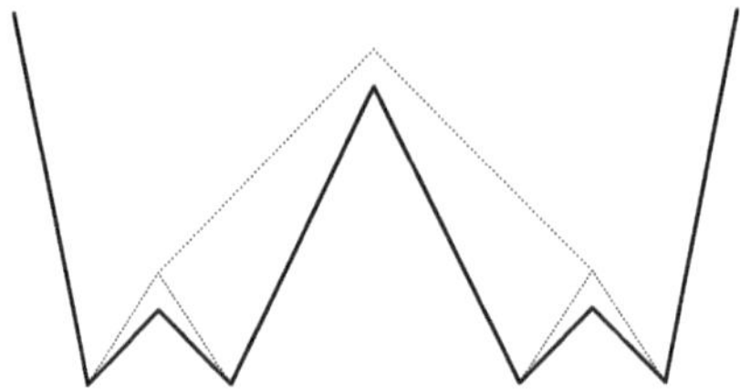

Fig. 8.2 Ultrametricity of valleys

It is useful to have the notion of a *valley* around a point in $\mathscr{M}$, which will serve as an attractor for the dynamics (see Fig. 8.2). For $m \in \mathscr{M}$, let

$$A(m) = \Big\{ z \in S : \mathbb{P}_z[\tau_m = \tau_{\mathscr{M}}] = \sup_{n \in \mathscr{M}} \mathbb{P}_z[\tau_n = \tau_{\mathscr{M}}] \Big\}. \tag{8.2.10}$$

Note that valleys may overlap, but from Lemma 8.5 it follows that their intersection has a negligible mass under the invariant distribution. The following estimate holds.

Lemma 8.7 *Let $m, n \in \mathscr{M}$ and $x \notin \{m, n\}$. If*

$$\mathbb{P}_x(\tau_m < \tau_x) \geq \varepsilon \quad \textit{and} \quad \mathbb{P}_x(\tau_n < \tau_x) \geq \varepsilon, \tag{8.2.11}$$

then

$$\mu(x) \leq 2\varepsilon^{-1} \operatorname{cap}(m, n). \tag{8.2.12}$$

Proof It follows from (7.1.18)–(7.1.20) and (7.1.39) that (8.2.11) implies $\operatorname{cap}(n, x) \geq \varepsilon \mu(x)$ and $\operatorname{cap}(m, x) \geq \varepsilon \mu(x)$. Hence

$$\operatorname{cap}(m, n) \geq \tfrac{1}{2} \varepsilon \mu(x) \tag{8.2.13}$$

by Corollary 8.6. □

Corollary 8.8 *Assume that $x \in S \backslash \mathscr{M}$ has the property $\mathbb{P}_x(\tau_m = \tau_{\mathscr{M}}) = \mathbb{P}_x(\tau_n = \tau_{\mathscr{M}}) = \max_{\ell \in \mathscr{M}} \mathbb{P}_x(\tau_\ell = \tau_{\mathscr{M}})$. Then*

$$\mu(x) \leq 2\rho \min\{\mu(m), \mu(n)\}, \tag{8.2.14}$$

with ρ from Definition 8.2.

Proof Since $\sum_{\ell \in \mathscr{M}} \mathbb{P}_x(\tau_\ell = \tau_{\mathscr{M}}) = 1$, we have

$$\mathbb{P}_x(\tau_m = \tau_{\mathscr{M}}) \geq 1/|\mathscr{M}| \quad \text{and} \quad \mathbb{P}_x(\tau_n = \tau_{\mathscr{M}}) \geq 1/|\mathscr{M}|. \tag{8.2.15}$$

Moreover, by the renewal estimate,

$$\mathbb{P}_x(\tau_m = \tau_{\mathscr{M}}) \leq \frac{\mathbb{P}_x(\tau_m < \tau_x)}{\mathbb{P}_x(\tau_{\mathscr{M}} < \tau_x)}, \tag{8.2.16}$$

and similarly with m replaced by n. Hence the hypotheses of Lemma 8.7 are satisfied with $\varepsilon = \mathbb{P}_x(\tau_{\mathscr{M}} < \tau_x)/|\mathscr{M}|$, and so

$$\mu(x) \leq \frac{2|\mathscr{M}|\,\mathrm{cap}(m,n)}{\mathbb{P}_x(\tau_{\mathscr{M}} < \tau_x)} \leq 2\rho \min\big(\mu(m), \mu(n)\big), \tag{8.2.17}$$

where the last inequality follows from Lemma 7.13(i) and Definition 8.2. □

In view of Corollary 8.8 we may modify the definition of the valleys $A(m)$, $m \in \mathscr{M}$, by reassigning their overlaps in an arbitrary fashion so that they become disjoint.

We will make frequent use of the following corollary as well.

Corollary 8.9 *Let $J \subset \mathscr{M}$, $m \in \mathscr{M}\backslash J$ and $y \in A(m)\backslash m$. Then either*

$$\tfrac{1}{2} \leq \frac{\mathrm{cap}(m,J)}{\mathrm{cap}(y,J)} \leq 2, \tag{8.2.18}$$

or

$$1 \leq 2|\mathscr{M}|\,\frac{\mathrm{cap}(m,J)}{\mathrm{cap}(y,\mathscr{M})}. \tag{8.2.19}$$

Proof By Lemma 8.5 with $D = J$ and $\delta = \frac{1}{2}$, if $\mathrm{cap}(m,J) \leq \frac{1}{2}\mathrm{cap}(m,y)$, then (8.2.18) holds. To get (8.2.19), use that

$$\mathrm{cap}(y,\mathscr{M}) \leq \sum_{n\in\mathscr{M}} \mathrm{cap}(y,n) \leq |\mathscr{M}| \max_{n\in\mathscr{M}} \mathrm{cap}(y,n). \tag{8.2.20}$$

Since $y \in A(m)$, the maximum must be achieved for $n = m$, which gives $\mathrm{cap}(y,\mathscr{M}) \leq |\mathscr{M}|\,\mathrm{cap}(y,m)$. Combining this with $\mathrm{cap}(m,J) > \frac{1}{2}\mathrm{cap}(m,y)$, we get the claim. □

8.3 Estimates on mean hitting times

A pleasant feature of the definition of metastability in terms of capacities is that it allows us to exploit Corollary 7.11 and link mean hitting times to capacities in a simple way. We derive rough bounds in Sect. 8.3.1 and subsequently sharpen them in Sect. 8.3.2.

8.3.1 Rough bounds

We begin with an *a priori bound* on the mean hitting time of $\mathscr{M}$.

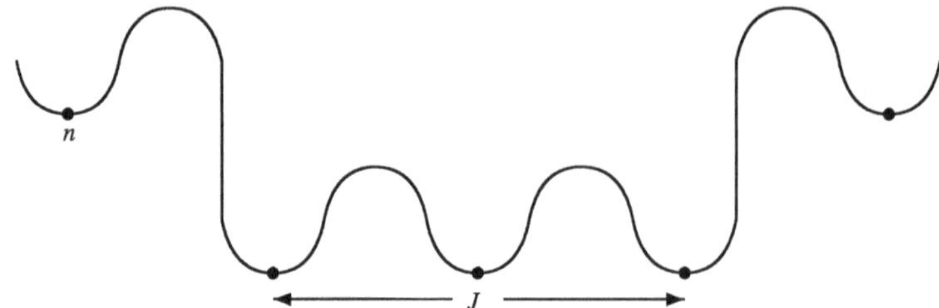

Fig. 8.3 Choice of $J \subset \mathscr{M}$ and $n \in \mathscr{M} \backslash J$

Lemma 8.10 *For reversible Markov processes with* $|S| < \infty$,

$$\sup_{z \notin \mathscr{M}} \mathbb{E}_z[\tau_{\mathscr{M}}] \leq |S| \sup_{y \notin \mathscr{M}} \frac{\mu(y)}{\operatorname{cap}(y, \mathscr{M})}. \tag{8.3.1}$$

Proof Recall from (7.1.13) that $\mathbb{E}_z[\tau_{\mathscr{M}}] = \sum_{y \in S \backslash \mathscr{M}} G_{\mathscr{M}}(z, y)$. Using the representation in (7.1.23) for the Green function $G_{\mathscr{M}}(z, y)$, we get, for $z \notin \mathscr{M}$,

$$\mathbb{E}_z[\tau_{\mathscr{M}}] = \sum_{y \in S \backslash \mathscr{M}} \frac{h_{y,\mathscr{M}}(z)}{e_{y,\mathscr{M}}(y)} \leq |S| \sup_{y \in S \backslash \mathscr{M}} \frac{1}{e_{y,\mathscr{M}}(y)}, \tag{8.3.2}$$

which yields (8.3.1) after we recall from (7.1.39) that $e_{y,\mathscr{M}}(y) = \operatorname{cap}(y, \mathscr{M})/\mu(y)$. □

In the proof we used the trivial bound $h_{y,\mathscr{M}}(z) \leq 1$. This explains the remark made after Definition 8.2: with additional work the term $|S|$ can be replaced by the cardinality of a smaller set of y's where $h_{y,\mathscr{M}}(z)$ is close to 1.

8.3.2 Sharp bounds

We next turn to the computation of mean hitting times from a point $n \in \mathscr{M}$ to some subset $J \subset \mathscr{M}$ (see Fig. 8.3). Return to (8.1.6). Decompose the sum in the right-hand side as

$$\sum_{y \in S} \mu(y) h_{n,J}(y) = \sum_{m \in \mathscr{M}} \mu(m) W_{n,J}(m), \quad W_{n,J}(m) = \sum_{y \in A(m)} \frac{\mu(y)}{\mu(m)} h_{n,J}(y), \tag{8.3.3}$$

where $A(m)$ is the set defined in (8.2.10), modified so that $\bigcup_{m \in \mathscr{M}} A(m)$ becomes a disjoint union (recall the remarks made below (8.2.10) and (8.2.17)). Lemmas 8.11 and 8.13 below provide technical estimates of the quantities in (8.3.3). After the statement and the proof of these lemmas we will explain how these estimates must be read, and in what regimes they reduce to simpler estimates.

The first technical lemma gives bounds on $h_{n,J}(y)$ and $\mu(y)/\mu(m)$ in the different sets $A(m)$, $m \in \mathcal{M}$. Abbreviate

$$a = a(m) = \inf_{y \in A(m)} \left[\mathrm{cap}(y, \mathcal{M}) / \mu(y) \right]. \tag{8.3.4}$$

Lemma 8.11 *Let $J \subset \mathcal{M}$ and $n \in \mathcal{M} \backslash J$.*

(i) *If $m = n$, then $h_{m,J}(m) = 1$, and for $y \in A(m) \backslash m$ either*

$$h_{n,J}(y) \geq 1 - 2|\mathcal{M}| \frac{\mathrm{cap}(m, J)}{\mathrm{cap}(y, \mathcal{M})} \tag{8.3.5}$$

or

$$\mu(y) \leq 2|\mathcal{M}| \frac{1}{a} \mathrm{cap}(m, J). \tag{8.3.6}$$

(ii) *If $m \in J$, then $h_{n,J}(m) = 0$, and for $y \in A(m) \backslash m$ either*

$$\mu(y) h_{n,J}(y) \leq 2|\mathcal{M}| \frac{1}{a} \mathrm{cap}(m, n) \tag{8.3.7}$$

or

$$\mu(y) \leq 2|\mathcal{M}| \frac{1}{a} \mathrm{cap}(m, n). \tag{8.3.8}$$

(iii) *If $m \notin J \cup n$, then for $y \in A(m)$ either*

$$1 - 4 \frac{\mathrm{cap}(m, J)}{\mathrm{cap}(m, n)} \leq h_{n,J}(y) \leq 4 \frac{\mathrm{cap}(m, n)}{\mathrm{cap}(m, J)} \tag{8.3.9}$$

or

$$\mu(y) \leq 2\,|\mathcal{M}|\,\frac{1}{a} \max\big(\mathrm{cap}(m, J), \mathrm{cap}(m, n)\big). \tag{8.3.10}$$

Proof The values of $h_{n,J}(y)$ for $y \in J \cup n$ are trivial from the definition of the equilibrium potential. By Lemma 8.4, for $J \subset \mathcal{M}$, $n \in \mathcal{M}$ and $y \notin \mathcal{M}$,

$$1 - \frac{\mathrm{cap}(y, J)}{\mathrm{cap}(y, n)} \leq h_{n,J}(y) \leq \frac{\mathrm{cap}(y, n)}{\mathrm{cap}(y, J)}. \tag{8.3.11}$$

(i) To get the first assertion, use Corollary 8.9. In the first case, this yields

$$h_{m,J}(y) \geq 1 - 2 \frac{\mathrm{cap}(m, J)}{\mathrm{cap}(y, m)} \geq 1 - 2\,|\mathcal{M}|\frac{\mathrm{cap}(m, J)}{\mathrm{cap}(y, \mathcal{M})}, \tag{8.3.12}$$

where we use (8.2.20). In the second case, we get (8.3.6) via the definition of a.

(ii) Use the upper bound in (8.3.11) to get

$$h_{n,J}(y) \leq \frac{\mathrm{cap}(y, n)}{\mathrm{cap}(y, J)}, \tag{8.3.13}$$

and use Corollary 8.9 with $J = n$. In the first case, $\text{cap}(n, y) \leq 2\,\text{cap}(m, n)$, and hence

$$h_{n,J}(y) \leq 2\,\frac{\text{cap}(n,m)}{\text{cap}(y,J)}. \tag{8.3.14}$$

From here (8.3.7) follows as in (i). In the second case, (8.3.8) is again straightforward.

(iii) Write the two renewal bounds from Lemma 8.4,

$$1 - \frac{\text{cap}(y,J)}{\text{cap}(y,n)} \leq h_{n,J}(y) \leq \frac{\text{cap}(y,n)}{\text{cap}(y,J)}, \tag{8.3.15}$$

and again use Corollary 8.9. If (8.2.18) holds both for $J = J$ and $J = n$, then we can replace y by m in the numerators and denominators of (8.3.15) at the cost of a factor 4 to get (8.3.9). If, on the other hand, (8.2.19) holds, then we get (8.3.10) just as in the previous cases. □

Remark 8.12 Case (iii) is special in as much as it does not give sharp estimates when $\text{cap}(m,J) \approx \text{cap}(m,n)$. If this situation occurs and the corresponding terms contribute to leading order, then we cannot get sharp estimates with the tools exploited above, and better estimates on the equilibrium potential are needed.

The second technical lemma uses the estimates in Lemma 8.11 to obtain estimates of $W_{n,J}(m)$ in (8.3.3).

Lemma 8.13 *Let $J \subset \mathscr{M}$ and $n \in \mathscr{M}\backslash J$.*

(i) *If $m = n$, then*

$$W_{n,J}(n) \leq \frac{\mu(A(n))}{\mu(n)} \tag{8.3.16}$$

and

$$W_{n,J}(n) \geq \frac{\mu(A(n))}{\mu(n)}\left(1 - \frac{\text{cap}(n,J)}{\mu(A(n))} 4|\mathscr{M}|\,\frac{1}{a}\,\big|A(n)\big|\right). \tag{8.3.17}$$

(ii) *If $m \in J$, then*

$$W_{n,J}(m) \leq C|\mathscr{M}|\,\frac{1}{a}\,\frac{\text{cap}(m,n)}{\mu(m)}\left|\left\{y \in A(m)\colon\ \mu(y) \geq |\mathscr{M}|\,\frac{1}{a}\,\frac{\text{cap}(m,n)}{\mu(n)}\right\}\right| \tag{8.3.18}$$

for some $C \in (0,\infty)$ independent of ρ in Definition 8.2.

(iii) *If $m \notin J \cup n$, then*

$$W_{n,J}(m) \leq \frac{\mu(A(m))}{\mu(m)}. \tag{8.3.19}$$

Moreover:

(iii1) *If* $\text{cap}(m, J) \leq \frac{1}{2}\text{cap}(m, n)$, *then*

$$W_{n,J}(m) \geq \left(1 - 4\frac{\text{cap}(m, J)}{\text{cap}(m, n)}\right)\left(\frac{\mu(A(m))}{\mu(m)} - C|\mathcal{M}|\frac{1}{a}\frac{\text{cap}(m, J)}{\mu(m)}\right). \tag{8.3.20}$$

(iii2) *If* $\text{cap}(m, J) \geq \frac{1}{2}\text{cap}(m, n)$, *then*

$$W_{n,J}(m) \leq \frac{\mu(A(m))}{\mu(m)}\frac{\text{cap}(m, n)}{\text{cap}(m, J} + 2|\mathcal{M}||A(m)|\frac{1}{a}\frac{\text{cap}(m, J)}{\mu(m)}. \tag{8.3.21}$$

Proof The proof consists of just inserting the bounds from Lemma 8.11. □

Lemma 8.13 looks complicated. Ignoring small terms, we see that the statement boils down to the following:

(i) The starting valley always contributes

$$W_{n,J}(n) \approx \frac{\mu(A(n))}{\mu(n)}, \tag{8.3.22}$$

which gives a contribution of $\frac{\mu(A(n))}{\text{cap}(n,J)}$ to the mean hitting time.

(ii) For $m \in J$,

$$W_{n,J}(m) \lesssim \frac{\text{cap}(m, n)}{\mu(m)}\frac{1}{a}|S|. \tag{8.3.23}$$

This gives a contribution to the mean hitting time of order at most $|S|/a$, which by assumption is small compared to that coming from (i).

(iii) For $m \notin J \cup n$,

$$W_{n,J}(m) \lesssim \frac{\mu(A(m))}{\mu(m)}. \tag{8.3.24}$$

(iii1) This bound is achieved when $\text{cap}(m, J) \ll \text{cap}(m, n)$. In this case the contribution to the hitting time is $\frac{\mu(A(m))}{\text{cap}(n,J)}$, which is small compared to the one from (i) only if $\mu(m) \ll \mu(n)$.

(iii2) If $\text{cap}(m, J) \gg \text{cap}(m, n)$, then the bound can be improved to

$$W_{n,J}(m) \leq \frac{\mu(A(m))}{\mu(m)}\frac{\text{cap}(m, n)}{\text{cap}(m, J)} + \frac{2|\mathcal{M}||A(m)|}{a}\frac{\text{cap}(m, J)}{\mu(m)}. \tag{8.3.25}$$

The second term is always harmless, while the first can contribute more to the mean hitting time than the one from (i), unless $\mu(m)/\text{cap}(m, J) \ll \mu(n)/\text{cap}(m, n)$.

Remark 8.14 The above arguments use that quantities like $\mu(m)/\mu(A(m))$ are not too small, i.e., the most massive points in a metastable set have a fairly large mass (compared to, say, ρ in Definition 8.2). The most restrictive contribution comes from case (iii), which is small only when $\rho|S|$ is small. Physically speaking, the latter avoids the situation where the time it takes for the dynamics to hit a target point in J after crossing the respective saddle is much longer than the time it takes to escape from the starting well.

Taking into account that $W_{n,J}(m)$ appears with the prefactor $\mu(m)/\mu(n)$ in (8.3.3), we see that contributions from case (ii) are always subdominant. In particular, when $J = \mathscr{M}\backslash n$, the term $m = n$ always gives the main contribution. The terms from case (iii) have a chance to contribute only when $\mu(m) \geq \mu(n)$. In subcase (iii1) they indeed contribute, and potentially dominate the sum, while in subcase (iii2) they may or may not contribute.

The estimates obtained in Lemma 8.13 can now be inserted into the sum in (8.3.3) and then into (8.1.6), to provide estimates on the mean hitting times $\mathbb{E}_n[\tau_J]$. We state the outcome in the special case when only the term involving the starting minimum contributes.

Theorem 8.15 (Mean metastable exit time) *Let $n \in \mathscr{M}$ and $J \subset \mathscr{M}\backslash n$ be such that for all $m \notin J \cup n$ $\mu(m) \ll \mu(n)$ or $\mathrm{cap}(m, J)/\mu(m) \gg \mathrm{cap}(m, n)/\mu(n)$. Then*

$$\mathbb{E}_n[\tau_J] = \frac{\mu(A(n))}{\mathrm{cap}(n, J)}[1 + \mathrm{error}], \quad 0 < \mathrm{error} \ll 1. \tag{8.3.26}$$

Proof The proof is straightforward from (8.3.3) and Lemmas 8.11 and 8.13. See the discussion before Remark 8.14. □

Note that the theorem covers in particular the case $J = \mathscr{M}_n$, where

$$\mathscr{M}_n = \left\{m \in \mathscr{M} : \mu(m) \geq \mu(n)\right\}. \tag{8.3.27}$$

We call $\mathbb{E}_n[\tau_{\mathscr{M}_n}]$ the *mean metastable exit time* from the metastable point n. This quantity plays an important rôle in Sect. 8.4 as well.

We will see in Parts IV–VIII that (8.3.26) is *the key formula for the computation of mean crossover times in metastable systems.*

8.4 Spectral characterisation of metastability

We now turn to the characterisation of metastability through spectral data. We will show that Definition 8.2 implies that the spectrum of the generator decomposes into a *cluster of $|\mathscr{M}|$ small real eigenvalues* that are *separated from the rest of the spectrum by a gap* (see Fig. 8.4). The associated eigenfunctions each live essentially in their own valley (see Fig. 8.5).

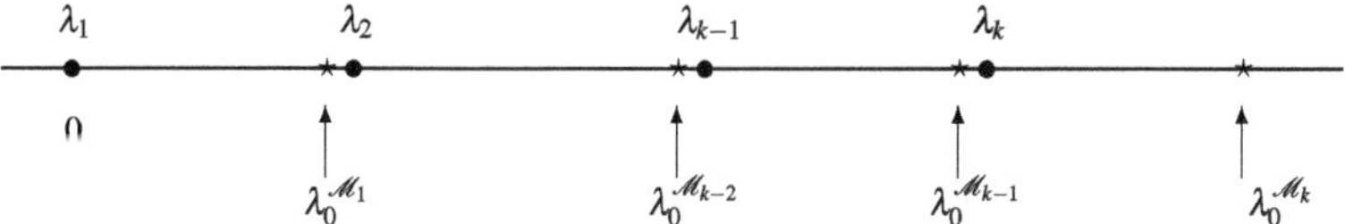

Fig. 8.4 Schematic picture of eigenvalues and Dirichlet eigenvalues when $|\mathscr{M}| = k$, $k \in \mathbb{N}\backslash\{1\}$. The eigenvalues $0 = \lambda_1 < \lambda_2 < \cdots < \lambda_{k-1} < \lambda_k$ are indicated by *dots*, the Dirichlet eigenvalues $\lambda_0^{\mathscr{M}_1} < \cdots < \lambda_0^{\mathscr{M}_{k-2}} < \lambda_0^{\mathscr{M}_{k-1}} < \lambda_0^{\mathscr{M}_k}$ are indicated by *stars*. The latter correspond to a nested sequence of subsets of $\mathscr{M}$, namely, $\mathscr{M}_\ell = \{x_1, \ldots, x_\ell\}$, $\ell = 1, \ldots, k$, with $\mathscr{M}_k = \mathscr{M}$, ordered according to the depths of the valleys (see (8.4.61) below). The distance between the two spectra is much smaller than the gaps within each of the two spectra

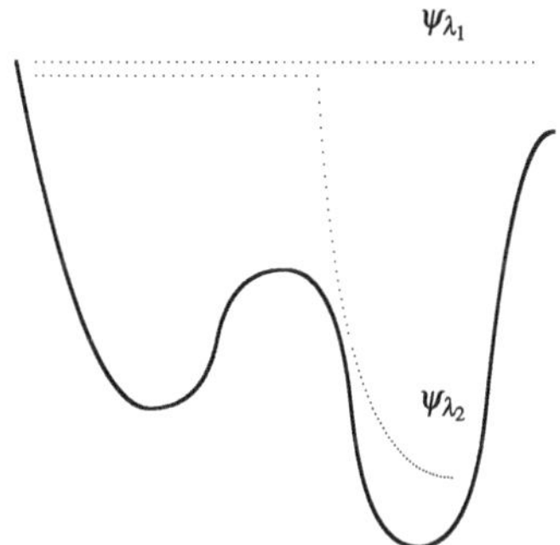

Fig. 8.5 Each eigenfunction essentially lives in its own valley

For the generator of a Markov process with countable state space, the eigenvalues of $-L$ are those values of λ for which

$$\big((-L-\lambda)\psi_\lambda\big)(x) = 0, \quad x \in S, \tag{8.4.1}$$

has a solution $\psi_\lambda \in \mathbb{R}^S$, called the eigenvector. For a subset $I \subset S$, we denote by L^I the *Dirichlet operator* with Dirichlet boundary conditions on I. The eigenvalues of $-L^I$ are denoted by λ^I and are those values of λ for which

$$\begin{aligned} \big((-L-\lambda)\psi_\lambda\big)(x) &= 0, \quad x \in S\backslash I, \\ \psi_\lambda(x) &= 0, \quad x \in I, \end{aligned} \tag{8.4.2}$$

has a solution. The smallest eigenvalue is called *principal eigenvalue*. The comparison of eigenvalues and Dirichlet eigenvalues will be an important tool in the analysis of the spectra of metastable Markov processes.

In Sect. 8.4.1 we derive rough bounds on the smallest eigenvalue $\lambda_0^{\mathscr{M}}$ of $-L^{\mathscr{M}}$. In Sect. 8.4.2 we characterise the eigenvalues smaller than $\lambda_0^{\mathscr{M}}$, in Sect. 8.4.3 we find the asymptotics of these eigenvalues, while in Sect. 8.4.4 we use this asymptotics to prove the exponential limit law of metastable exit times.

8.4.1 A priori bounds

The first step consists in getting a rough bound on the principal eigenvalue of $-L^{\mathscr{M}}$, with $\mathscr{M}$ the set of metastable points. This uses an important tool that is due to Donsker and Varadhan [94].

Lemma 8.16 (Lower bound on principal Dirichlet eigenvalues) *Let $I \subset S$, and let λ_0^I be the smallest eigenvalue of $-L^I$. Then*

$$\lambda_0^I \geq \frac{1}{\sup_{z\in S}\mathbb{E}_z[\tau_I]}. \tag{8.4.3}$$

Proof For any $\phi \in \mathbb{R}^S$, any $C > 0$ and any $x, y \in S$, we have

$$\phi(x)\phi(y) \leq \frac{1}{2}\left(C\phi(x)^2 + \frac{1}{C}\phi(y)^2\right). \tag{8.4.4}$$

Let $w \in \mathbb{R}^S$ be such that $w(x) > 0$ whenever $\phi(x) \neq 0$. Using (8.4.4) with $C = w(y)/w(x)$ within the Dirichlet form, we get

$$\mathscr{E}(\phi,\phi) \geq \sum_{x\in S}\mu(x)\phi(x)^2\left(\frac{(-Lw)(x)}{w(x)}\right). \tag{8.4.5}$$

Let $w(x) = \mathbb{E}_x[\tau_I]$, $x \in S\backslash I$, and let ϕ be an eigenvector of $-L^I$ with eigenvalue λ. Recalling that w solves the Dirichlet problem in (7.1.8), we get

$$\lambda\|\phi\|_{2,\mu}^2 \geq \sum_{x\in S\backslash I}\mu(x)\phi(x)^2\left(\frac{1}{w(x)}\right) \geq \frac{1}{\sup_{x\in S\backslash I} w(x)}\|\phi\|_{2,\mu}^2. \tag{8.4.6}$$

Since this holds for all eigenvalues of $-L^I$, it implies the assertion in (8.4.3). □

Lemma 8.16 links the time scale of the metastable dynamics to the smallest eigenvalue of the Dirichlet operator in a way that is intuitively plausible. The estimate sometimes needs improvement, but at least it shows the basic twist. Note that the bound is not very precise. We later derive a more precise relation for the cluster of $|\mathscr{M}|$ small real eigenvalues alluded to above.

Combining Lemma 8.16 with Lemma 8.10, we obtain the following.

Corollary 8.17 (Lower bound on principal Dirichlet eigenvalues) *For a reversible Markov process with a finite state space S and a set of metastable points $\mathscr{M}$,*

$$\lambda_0^{\mathscr{M}} \geq \inf_{y\notin\mathscr{M}} \frac{\mathrm{cap}(y,\mathscr{M})}{3\,|\mathscr{M}|\,|S|\,\mu(y)}. \tag{8.4.7}$$

8.4.2 Characterisation of small eigenvalues

We next obtain a representation formula for the eigenvalues that are smaller than $\lambda_0^{\mathscr{M}}$. We show that there are precisely $|\mathscr{M}|$ such eigenvalues. The idea is to use the fact that the solution of the Dirichlet problem

$$\begin{aligned} (-Lf)(x) - \lambda f(x) &= 0, \quad x \in \mathscr{M}^c, \\ f(x) &= \phi_x, \quad x \in \mathscr{M}, \end{aligned} \tag{8.4.8}$$

already solves the eigenvalue equation $-Lf = \lambda f$ on $\mathscr{M}^c$. The question is whether an appropriate choice of boundary conditions and of λ leads to a solution. The following observation is elementary.

Lemma 8.18 *Assume that $\lambda < \lambda_0^{\mathscr{M}}$ is an eigenvalue of $-L$ and that ϕ is the corresponding eigenfunction. Then the unique solution of (8.4.8) with $\phi_x = \phi(x)$, $x \in \mathscr{M}$, satisfies $f(y) = \phi(y)$ for all $y \in S$.*

Proof Inserting $f = \phi$ into (8.4.8), we see that the first line in (8.4.8) is satisfied because ϕ is an eigenfunction with eigenvalues λ. The second line holds by assumption. □

For any $\lambda < \lambda_0^{\mathscr{M}}$, the boundary value problem in (8.4.8) has a unique solution for any choice of boundary condition ϕ. Denote by $h^\lambda_{x,\mathscr{M}\backslash x} = h^\lambda_x$, $x \in \mathscr{M}$, the solutions for the special case

$$\begin{aligned} \big((-L-\lambda)h^\lambda_x\big)(y) &= 0, \quad y \in \mathscr{M}^c, \\ h^\lambda_x(x) &= 1, \\ h^\lambda_x(y) &= 0, \quad y \in \mathscr{M}\backslash x. \end{aligned} \tag{8.4.9}$$

Then the unique solution of (8.4.8) can be represented as

$$f(y) = \sum_{x\in\mathscr{M}} \phi_x h^\lambda_x(y), \quad y \in S. \tag{8.4.10}$$

Asking for f to be an eigenfunction therefore amounts to imposing the following $|\mathscr{M}|$ additional equations:

$$\sum_{x\in\mathscr{M}} \big((-L-\lambda)h^\lambda_x\big)(y)\phi_x = \sum_{x\in\mathscr{M}} \mathrm{e}^\lambda_{x,\mathscr{M}\backslash x}(y)\phi_x = 0, \quad y \in \mathscr{M}. \tag{8.4.11}$$

Here, $\mathrm{e}^\lambda_{x,\mathscr{M}\backslash x}(y) = ((-L-\lambda)h^\lambda_{x,\mathscr{M}\backslash x})(y)$ is the λ-analogue of the equilibrium measure. Thus, denoting by $\mathscr{E}_{\mathscr{M}}(\lambda)$ the $(|\mathscr{M}| \times |\mathscr{M}|)$-matrix with elements

$$\big(\mathscr{E}_{\mathscr{M}}(\lambda)\big)_{xy} = \mathrm{e}^\lambda_{x,\mathscr{M}\backslash x}(y), \quad x, y \in \mathscr{M}, \tag{8.4.12}$$

we get the following lemma.

Lemma 8.19 *A number $\lambda < \lambda_0^{\mathscr{M}}$ is an eigenvalue of the matrix $-L$ if and only if*

$$\det\big(\mathscr{E}_{\mathscr{M}}(\lambda)\big) = 0. \tag{8.4.13}$$

Remark 8.20 Equation (8.4.12) can be seen as a non-linear generalisation of the characteristic equation for eigenvalues. It in fact coincides with that equation if we replace $\mathscr{M}$ by S.

Anticipating that we are interested in small λ, we want to rewrite the matrix $\mathscr{E}_{\mathscr{M}}(\lambda)$ in a more convenient form. In order to do so, let us set

$$h_x^\lambda(y) = h_x(y) + \psi_x^\lambda(y), \tag{8.4.14}$$

where $h_x = h_x^0$ is the equilibrium potential $h_x = h_{x,\mathscr{M}\backslash x}$. Since $h_y(y) = 1$ for $y \in \mathscr{M}$, we have

$$\big((-L-\lambda)h_x^\lambda\big)(y) = h_y(y)\big((-L-\lambda)h_x^\lambda\big)(y), \quad x, y \in \mathscr{M}, \tag{8.4.15}$$

and since $h_y^\lambda(z) = 0$ for $z \neq y \in \mathscr{M}$, we have

$$h_y(z)\big((-L-\lambda)h_x^\lambda\big)(z) = 0, \quad z \neq y. \tag{8.4.16}$$

Hence, we get that

$$\begin{aligned} e_{x,\mathscr{M}\backslash x}^\lambda(y) &= \big((-L-\lambda)h_x^\lambda\big)(y) \\ &= \frac{1}{\mu(y)}\sum_{z\in S}\mu(z)h_y(z)\big((-L-\lambda)h_x^\lambda\big)(z) \\ &= \frac{1}{\mu(y)}\sum_{z\in S}\mu(z)h_y(z)\big((-L-\lambda)h_x\big)(z) \\ &\quad + \frac{1}{\mu(y)}\sum_{z\in S}\mu(z)h_y(z)\big((-L-\lambda)\psi_x^\lambda\big)(z). \end{aligned} \tag{8.4.17}$$

The first sum equals

$$\sum_{z\in S}\mu(z)h_y(z)\big((-L-\lambda)h_x\big)(z) = \mathscr{E}(h_y, h_x) - \lambda(h_y, h_x)_\mu. \tag{8.4.18}$$

The matrix with elements $\mathscr{E}(h_x, h_y)$ is referred to as the capacity matrix. It will turn out that the small eigenvalues of the generator are very close to the eigenvalues of the capacity matrix.

For the second sum, we use the symmetry of L, plus the fact that ψ_x^λ vanishes on $\mathscr{M}$ and $(-Lh)_y$ vanishes on $\mathscr{M}^c$, to write it as

$$\sum_{z\in S}\mu(z)h_y(z)\big((-L-\lambda)\psi_x^\lambda\big)(z)$$
$$= -\lambda\big(h_y,\psi_x^\lambda\big)_\mu + \sum_{z\in S}\mu(z)(-Lh_y)(z)\psi_x^\lambda(z) = -\lambda\big(h_y,\psi_x^\lambda\big)_\mu. \tag{8.4.19}$$

Hence $\mathrm{e}^\lambda_{x,\mathscr{M}\backslash x}(y)$ can be written in the form

$$\mathrm{e}^\lambda_{x,\mathscr{M}\backslash x}(y) = \frac{1}{\mu(y)}\big(\mathscr{E}(h_y,h_x) - \lambda(h_y,h_x)_\mu - \lambda\big(h_y,\psi_x^\lambda\big)_\mu\big). \tag{8.4.20}$$

The term involving ψ^λ is of order λ^2, and hence is a small perturbation. This is implied by the following lemma.

Lemma 8.21 (ℓ^2-bounds) *Let $\lambda_0^{\mathscr{M}}$ denote the principal eigenvalue of the operator $-L$ with Dirichlet boundary conditions in $\mathscr{M}$.*

(i) *If $\lambda < \lambda_0^{\mathscr{M}}$, then for all $x \in \mathscr{M}$,*

$$\|\psi_x^\lambda\|_{2,\mu} \leq \frac{\lambda}{\lambda_0^{\mathscr{M}} - \lambda}\|h_x\|_{2,\mu}. \tag{8.4.21}$$

(ii) *For all $x, y \in \mathscr{M}$,*

$$\big|\big(h_y,\psi_x^\lambda\big)_\mu\big| \leq \frac{\lambda}{(\lambda_0^{\mathscr{M}} - \lambda)}\|h_x\|_{2,\mu}\|h_y\|_{2,\mu}. \tag{8.4.22}$$

(iii) *For all $\lambda, \lambda' < \lambda_0^{\mathscr{M}}$ and $x \in \mathscr{M}$,*

$$\|\psi_x^\lambda - \psi_x^{\lambda'}\|_{2,\mu} \leq \frac{|\lambda-\lambda'|}{\lambda_0^{\mathscr{M}} - \lambda'}\|h_x^\lambda\|_{2,\mu}. \tag{8.4.23}$$

(iv) *For all $x, y \in \mathscr{M}$,*

$$\big|\big(\psi_x^\lambda - \psi_x^{\lambda'}, h_y\big)_\mu\big| \leq \frac{|\lambda-\lambda'|}{\lambda_0^{\mathscr{M}} - \lambda'}\|h_x^\lambda\|_{2,\mu}\|h_y\|_{2,\mu}. \tag{8.4.24}$$

Proof It is readily verified that ψ_x^λ solves the Dirichlet problem

$$\begin{aligned} \big((-L-\lambda)\psi_x^\lambda\big)(y) &= \lambda h_x(y), \quad y \in \mathscr{M}^c, \\ \psi_x^\lambda(y) &= 0, \quad y \in \mathscr{M}. \end{aligned} \tag{8.4.25}$$

But the Dirichlet operator $-L^{\mathcal{M}} - \lambda$ is invertible for $\lambda < \lambda_0^{\mathcal{M}}$, and its norm as an operator on $\ell^2(S, \mu)$ is bounded by $1/(\lambda_0^{\mathcal{M}} - \lambda)$. Hence

$$\|\psi_x^\lambda\|_{2,\mu} \le \frac{\lambda}{\lambda_0^{\mathcal{M}} - \lambda} \|h_x\|_{2,\mu}, \tag{8.4.26}$$

which proves (i). The assertion in (ii) follows from (i) together with the Cauchy-Schwarz inequality. Finally,

$$\left(-L\left(\psi_x^\lambda - \psi_x^{\lambda'}\right)\right)(z) = \left(\lambda - \lambda'\right) h_x^\lambda(z) + \lambda'\left(\psi_x^\lambda - \psi_x^{\lambda'}\right). \tag{8.4.27}$$

Hence

$$\left(\left(-L - \lambda'\right)\left(\psi_x^\lambda - \psi_x^{\lambda'}\right)\right)(z) = \left(\lambda - \lambda'\right) h_x^\lambda(z), \tag{8.4.28}$$

and so (8.4.23) follows in the same way as (8.4.21). Assertion (iv) follows again via the Cauchy-Schwarz inequality. □

The ℓ^2-bounds in Lemma 8.21 can be improved to ℓ^∞-bounds.

Lemma 8.22 (ℓ^∞-bounds) *With the notation of Lemma* 8.21, *the following bounds hold.*

(i) *For all* $\lambda < \lambda_0^{\mathcal{M}}$ *and* $x \in \mathcal{M}$,

$$\left\| \frac{\psi_x^\lambda}{h_{x,\mathcal{M}}} \right\|_\infty \le \frac{\lambda |S| \sup_{y \notin \mathcal{M}} \frac{\mu(y)}{\operatorname{cap}(y,\mathcal{M})}}{1 - \lambda |S| \sup_{y \notin \mathcal{M}} \frac{\mu(y)}{\operatorname{cap}(y,\mathcal{M})}}. \tag{8.4.29}$$

(ii) *For all* $\lambda, \lambda' < \lambda_0^{\mathcal{M}}$ *and* $x \in \mathcal{M}$,

$$\left\| \frac{\psi_x^\lambda - \psi_x^{\lambda'}}{h_{x,\mathcal{M}}} \right\|_\infty \le \frac{|\lambda - \lambda'| |S| \sup_{y \notin \mathcal{M}} \frac{\mu(y)}{\operatorname{cap}(y,\mathcal{M})}}{1 - |\lambda - \lambda'| |S| \sup_{y \notin \mathcal{M}} \frac{\mu(y)}{\operatorname{cap}(y,\mathcal{M})}}. \tag{8.4.30}$$

Proof Note that ψ_x^λ satisfies

$$\begin{aligned} \left(-L\psi_x^\lambda\right)(y) &= \lambda h_x^\lambda(y), \quad y \notin \mathcal{M}, \\ \psi_x^\lambda(z) &= 0, \quad z \in \mathcal{M}. \end{aligned} \tag{8.4.31}$$

Thus, with $G_{\mathcal{M}}$ the Green function with Dirichlet boundary conditions in $\mathcal{M}$, ψ_x^λ satisfies

$$\psi_x^\lambda(y) = \lambda \sum_{a \notin \mathcal{M}} G_{\mathcal{M}}(y, a) h_x^\lambda(a). \tag{8.4.32}$$

Dividing by $h_{x,\mathscr{M}\backslash x}(y)$, we get

$$\frac{\psi_x^\lambda(y)}{h_{x,\mathscr{M}\backslash x}(y)} - \lambda \sum_{a\notin\mathscr{M}} \frac{1}{h_{x,\mathscr{M}\backslash x}(y)} G_{\mathscr{M}}(y,a) h_{x,\mathscr{M}\backslash x}(a) \frac{h_x^\lambda(a)}{h_{x,\mathscr{M}\backslash x}(a)}$$

$$= \lambda \sum_{a\notin\mathscr{M}} G_{\mathscr{M}}^{h_x}(y,a) + \lambda \sum_{a\notin\mathscr{M}} G_{\mathscr{M}}^{h_x}(y,a) \frac{\psi_x^\lambda(a)}{h_{x,\mathscr{M}\backslash x}(a)}. \tag{8.4.33}$$

Here, $G_{\mathscr{M}}^{h_x}(y,a) = \frac{1}{h_{x,\mathscr{M}\backslash x}(y)} G_{\mathscr{M}}(y,a) h_{x,\mathscr{M}\backslash x}(a)$ is the Green function of the Doob-transformed generator L^{h_x}, which is the generator of the process *conditioned* to hit $\mathscr{M}$ in x (see Sect. 4.2.4). In particular,

$$\sum_{a\notin\mathscr{M}} G_{\mathscr{M}}^{h_x}(y,a) = \mathbb{E}_y[\tau_{\mathscr{M}} | \tau_x = \tau_{\mathscr{M}}]. \tag{8.4.34}$$

Using the representation of the Green function given in (7.1.23), we see that

$$G_{\mathscr{M}}^{h_x}(y,a) = \frac{1}{h_{x,\mathscr{M}\backslash x}(y)} \frac{h_{a,\mathscr{M}}(y)}{e_{a,\mathscr{M}}(a)} h_{x,\mathscr{M}\backslash x}(a). \tag{8.4.35}$$

But

$$\frac{h_{a,\mathscr{M}}(y) h_{x,\mathscr{M}\backslash x}(a)}{h_{x,\mathscr{M}\backslash x}(y)} = \frac{\mathbb{P}_y(\tau_a < \tau_{\mathscr{M}} \wedge \tau_x = \tau_{\mathscr{M}})}{\mathbb{P}_y(\tau_x = \tau_{\mathscr{M}})} = \mathbb{P}_y(\tau_a < \tau_{\mathscr{M}} | \tau_x = \tau_{\mathscr{M}}). \tag{8.4.36}$$

Hence

$$\mathbb{E}_y[\tau_{\mathscr{M}} | \tau_x = \tau_{\mathscr{M}}] \leq \sup_{a\in S\backslash\mathscr{M}} \frac{\mu(a)}{\mathrm{cap}(a,\mathscr{M})} |S|. \tag{8.4.37}$$

From (8.4.33–8.4.36) it follows that, for all $y \in S\backslash\mathscr{M}$,

$$\frac{\psi_x^\lambda(y)}{h_{x,\mathscr{M}\backslash x}(y)} \leq \lambda \mathbb{E}_y[\tau_{\mathscr{M}} | \tau_x = \tau_{\mathscr{M}}] \left(1 + \sup_{a\notin\mathscr{M}} \frac{\psi_x^\lambda(a)}{h_{x,\mathscr{M}\backslash x}(a)}\right). \tag{8.4.38}$$

Via the bound in (8.4.37) this implies (8.4.29). The bound in (8.4.30) is proven analogously. □

The main application of the bounds in Lemma 8.22 is the following improvement of (8.4.22) and (8.4.24).

Corollary 8.23 *For all* $x, y \in \mathscr{M}$,

$$\left|(\psi_x^\lambda, h_y)_\mu\right| \leq C \frac{\lambda |S| \sup_{y\notin\mathscr{M}} \frac{\mu(y)}{\mathrm{cap}(y,\mathscr{M})}}{1 - \lambda |S| \sup_{y\notin\mathscr{M}} \frac{\mu(y)}{\mathrm{cap}(y,\mathscr{M})}} (h_x, h_y)_\mu, \tag{8.4.39}$$

and

$$\left|\left(\psi_x^\lambda - \psi_x^{\lambda'}, h_y\right)_\mu\right| \le C \frac{|\lambda-\lambda'||S| \sup_{y\notin \mathcal{M}} \frac{\mu(y)}{\mathrm{cap}(y,\mathcal{M})}}{1-|\lambda-\lambda'||S| \sup_{y\notin \mathcal{M}} \frac{\mu(y)}{\mathrm{cap}(y,\mathcal{M})}} (h_x, h_y)_\mu. \tag{8.4.40}$$

Proof Follows from the estimates in Lemma 8.22. □

The next lemma controls the off-diagonal elements $(h_x, h_y)_\mu$.

Lemma 8.24 *There is a* $C < \infty$ *such that*

$$\sup_{\substack{x,y\in\mathcal{M}\\ x\neq y}} \frac{(h_x, h_y)_\mu}{\|h_x\|_{2,\mu}\|h_y\|_{2,\mu}} \le C\rho. \tag{8.4.41}$$

Proof To bound the numerator we can use the computations from the proof of Theorem 8.15. Analogously to (8.3.3), we write

$$\sum_{z\in S}\mu(z)h_x(z)h_y(z) = \sum_{m\in\mathcal{M}}\mu(m)\widetilde{W}_{x,y}(m), \quad \widetilde{W}_{x,y}(m) = \sum_{z\in A(m)} \frac{\mu(z)}{\mu(m)} h_x(z)h_y(z). \tag{8.4.42}$$

We need to distinguish the terms $m \in \{x, y\}$ from the rest. First,

$$\widetilde{W}_{x,y}(x) \le \sum_{z\in A(x)} \frac{\mu(z)}{\mu(x)} h_y(z) = W_{y,\mathcal{M}\setminus y}(x). \tag{8.4.43}$$

Hence, by Lemma 8.13(ii),

$$\widetilde{W}_{x,y}(x) \le C|\mathcal{M}|a^{-1}\frac{\mathrm{cap}(x,y)}{\mu(x)}\left|A(x)\right|, \tag{8.4.44}$$

and analogously

$$\widetilde{W}_{x,y}(y) \le C|\mathcal{M}|a^{-1}\frac{\mathrm{cap}(x,y)}{\mu(y)}\left|A(y)\right|. \tag{8.4.45}$$

For $m \notin \{x, y\}$, we use the bounds from Lemma 8.11(ii). This yields

$$\begin{aligned} \widetilde{W}_{x,y}(m) &= \sum_{z\in A(m)} \frac{\mu(z)}{\mu(m)} h_x(z)h_y(z) \\ &\le \sum_{z\in A(m)} 2|\mathcal{M}|a^{-1}\frac{\min(\mathrm{cap}(x,m), \mathrm{cap}(y,m))}{\mu(m)} \\ &\le |A(m)| 4|\mathcal{M}|a^{-1}\frac{\mathrm{cap}(x,y)}{\mu(m)}. \end{aligned} \tag{8.4.46}$$

This yields

$$\sum_{z\in S} \mu(z)h_x(z)h_y(z) \leq C|\mathscr{M}|a^{-1}|S|\,\mathrm{cap}(x,y). \tag{8.4.47}$$

The denominator in (8.4.41) is trivially bounded from below by $\sqrt{\mu(x)\mu(y)}$. Thus, the left-hand side of (8.4.41) is bounded from above by

$$\frac{\mathrm{cap}(x,y)}{\sqrt{\mu(x)\mu(y)}} \leq \max\left(\frac{\mathrm{cap}(x,y)}{\mu(x)}, \frac{\mathrm{cap}(x,y)}{\mu(y)}\right) \leq a\rho/|S|, \tag{8.4.48}$$

where the last inequality uses Definition 8.2. The assertion of the lemma now follows with $C = a/|S|$. □

Remark 8.25 The bounds can be improved. For instance, with a little more care we get

$$\|h_x\|_{2,\mu}\|h_y\|_{2,\mu} \geq \sqrt{\mu\big(A(x)\big)\mu\big(A(y)\big)}\,\big[1-O(\rho)\big]. \tag{8.4.49}$$

We are now in a position to relate the small eigenvalues of $-L$ to capacities.

8.4.3 Computation of small eigenvalues

We have a bound on $\lambda_0^{\mathscr{M}}$ and a characterisation of the eigenvalues of $-L$ that are smaller than $\lambda_0^{\mathscr{M}}$. We will show that there are $|\mathscr{M}|$ such eigenvalues. The strategy to compute them is as follows. First, compute the *largest* eigenvalue that is smaller than $\lambda_0^{\mathscr{M}}$. It will turn out that this eigenvalue is slightly larger than, but very close to, $\lambda_0^{\mathscr{M}\backslash x}$ for some $x \in \mathscr{M}$. Next, start all over with $\mathscr{M}$ replaced by $\mathscr{M}\backslash x$, i.e., compute the largest eigenvalue of $-L$ that is smaller than $\lambda_0^{\mathscr{M}\backslash x}$. Finally, repeat this procedure until the set $\mathscr{M}$ is exhausted and all $|\mathscr{M}|$ eigenvalues are computed. For this strategy to work, we need some non-degeneracy assumptions that will be stated below.

Let us note that principal eigenvalues are also characterised through the *Rayleigh-Ritz variational principle*. Namely, for any $I \subset S$,

$$\lambda_0^I = \inf_{\substack{f:f(x)=0,x\in I\\ \|f\|_{2,\mu}=1}} \mathscr{E}(f,f). \tag{8.4.50}$$

This immediately implies that, for $I \subset J \subset S$,

$$\lambda_0^I \leq \lambda_0^J. \tag{8.4.51}$$

We start by deriving a precise estimate on the principal eigenvalue $\lambda_0^{\mathscr{M}\backslash x}$ for $x \in \mathscr{M}$. The next theorem gives two characterisations of principal eigenvalues.

Theorem 8.26 (Capacity bounds for principal eigenvalues) *Let $\mathscr{N} \subset \mathscr{M}$ be non-empty, and let $x \in \mathscr{N}$. Then $-L^{\mathscr{N}\backslash x}$ has a unique eigenvalue $\lambda = \lambda_0^{\mathscr{N}\backslash x}$ smaller than $\lambda_0^{\mathscr{N}}$, given by the solution of the equation*

$$e^{\lambda}_{x,\mathscr{N}\backslash x}(x) = 0. \tag{8.4.52}$$

Moreover,

$$\frac{\mathrm{cap}(x,\mathscr{N}\backslash x)}{\|h_{x,\mathscr{N}\backslash x}\|^2_{2,\mu}}\big[1 - O(\delta)\big] \leq \lambda = \frac{\mathrm{cap}(x,\mathscr{N}\backslash x)}{\|h_{x,\mathscr{N}\backslash x}\|^2_{2,\mu}}, \tag{8.4.53}$$

where $\delta = \frac{\mathrm{cap}(x,\mathscr{N}\backslash x)}{\|h_{x,\mathscr{N}\backslash x}\|^2_{2,\mu}} / \lambda_0^{\mathscr{N}}$.

Proof The same argument leading to Lemma 8.19 shows that any eigenvalue of $-L^{\mathscr{N}\backslash x}$ smaller than $\lambda_0^{\mathscr{N}}$ must satisfy (8.4.52). To show (8.4.53), note that the principal eigenvector of $-L^{\mathscr{N}\backslash x}$ must be strictly positive on $(\mathscr{N}\backslash x)^c$. In particular, it must be positive at x. Hence we can reformulate the Rayleigh-Ritz variational principle in (8.4.50) as

$$\lambda_0^{\mathscr{N}\backslash x} = \inf_{f:\ f(z)=0, z\in\mathscr{N}\backslash x, f(x)=1} \frac{\mathscr{E}(f,f)}{\|f\|^2_{2,\mu}} \leq \frac{\mathscr{E}(h_{x,\mathscr{N}\backslash x}, h_{x,\mathscr{N}\backslash x})}{\|h_{x,\mathscr{N}\backslash x}\|^2_{2,\mu}}. \tag{8.4.54}$$

On the other hand, using (8.4.20) to write out (8.4.52), we see that this equation implies

$$\lambda_0^{\mathscr{N}\backslash x} = \frac{\mathscr{E}(h_{x,\mathscr{N}\backslash x}, h_{x,\mathscr{N}\backslash x})}{\|h_{x,\mathscr{N}\backslash x}\|^2_{2,\mu} + (h_{x,\mathscr{N}\backslash x}, \psi^{\lambda}_x)_\mu}. \tag{8.4.55}$$

Via the bound in (8.4.22) from Lemma 8.21, this implies that

$$\lambda_0^{\mathscr{N}\backslash x} \geq \frac{\mathscr{E}(h_{x,\mathscr{N}\backslash x}, h_{x,\mathscr{N}\backslash x})}{\|h_{x,\mathscr{N}\backslash x}\|^2_{2,\mu}} \frac{1}{1 + \lambda_0^{\mathscr{N}\backslash x}/(\lambda_0^{\mathscr{N}} - \lambda_0^{\mathscr{N}\backslash x})}. \tag{8.4.56}$$

Finally, using the upper bound on $\lambda_0^{\mathscr{N}\backslash x}$ from (8.4.54) and the definition of δ, we get

$$\frac{1}{1 + \lambda_0^{\mathscr{N}\backslash x}/(\lambda_0^{\mathscr{N}} - \lambda_0^{\mathscr{N}\backslash x})} \geq \frac{1}{1 + \frac{\delta}{1-\delta}} \geq 1 - O(\delta). \tag{8.4.57}$$

This concludes the proof of Theorem 8.26. □

We define a sequence of nested subsets of $\mathscr{M}$ as follows.

Definition 8.27 (Nested subsets of metastable sets) Let $\mathscr{M}$ be a set of metastable points, and let $|\mathscr{M}| = k$. Set

$$\mathscr{M}_k = \mathscr{M}. \tag{8.4.58}$$

For $\ell = k, \dots, 1$, set

$$x_\ell = \operatorname{argmax}\left(x \in \mathcal{M}_\ell : \frac{\operatorname{cap}(x, \mathcal{M}_\ell \backslash x)}{\mu(x)} \right) \tag{8.4.59}$$

and

$$\mathcal{M}_{\ell-1} = \mathcal{M}_\ell \backslash x_\ell. \tag{8.4.60}$$

We call the set $\mathcal{M}$ non-degenerate if for any $\ell = k, \dots, 2$ the set $\mathcal{M}_\ell$ is itself a set of metastable points in the sense of Definition 8.2.

What the recursive construction in Definition 8.27 does is to look for the minimum with the smallest stability level and remove it from the set of minima that are left over. The sequence of sets thus obtained has the form

$$\begin{aligned} \mathcal{M} &= \mathcal{M}_k \supset \mathcal{M}_{k-1} = \mathcal{M}_k \backslash x_k \supset \mathcal{M}_{k-2} = \mathcal{M}_{k-1} \backslash x_{k-1} \supset \cdots \supset \mathcal{M}_1 \\ &= \mathcal{M}_2 \backslash x_2 = x_1 \supset \emptyset. \end{aligned} \tag{8.4.61}$$

Note that the non-degeneracy condition implies that

$$\frac{\operatorname{cap}(x_\ell, \mathcal{M}_{\ell-1})}{\mu(x_\ell)} \leq \delta \frac{\operatorname{cap}(x_{\ell+1}, \mathcal{M}_\ell)}{\mu(x_{\ell+1})}, \tag{8.4.62}$$

with

$$|S|\delta \leq \rho. \tag{8.4.63}$$

We next establish some properties of the sets $\mathcal{M}_\ell$. We first obtain some information on the relation between the sets $\mathcal{M}_\ell$ and $\mathcal{M}_{x_\ell}$.

Lemma 8.28 *Decompose disjointly* $\mathcal{M}_{x_\ell} = \bigcup_j \mathcal{M}_{x_\ell}^{(j)}$ *such that* $\operatorname{cap}(x_\ell, z) \sim \operatorname{cap}(x_\ell, \mathcal{M}_{x_\ell}^{(j)})$ *for all* $z \in \mathcal{M}_{x_\ell}^{(j)}$. *Let* $x_*^{(j)}$ *be such that* $\mu(x_*^{(j)}) = \max(\mu(z) \colon z \in \mathcal{M}_{x_\ell}^{(j)})$. *Then*

(i) *For all* j, $x_*^{(j)} \in \mathcal{M}_{\ell-1}$.
(ii) *For all* $z \in \mathcal{M}_{x_\ell}^{(j)} \backslash \mathcal{M}_{\ell-1}$,

$$\frac{\operatorname{cap}(z, \mathcal{M}_{x_\ell}^{(j)})}{\mu(z)} > \frac{\operatorname{cap}(x_\ell, \mathcal{M}_{\ell-1})}{\mu(x_\ell)}. \tag{8.4.64}$$

Proof Throughout we use the approximate ultrametricity of capacity and the non-degeneracy assumptions. Consider a single set $\mathcal{M}_{x_\ell}^{(j)}$. If all points in this set are in $\mathcal{M}_{\ell-1}$, then there is nothing to prove. Otherwise, there is some ℓ' with $k \geq \ell' > \ell$ for which a first point $z \in \mathcal{M}_{x_\ell}^{(j)}$ is selected to be removed as $x_{\ell'}$. But then it must be that

$$\frac{\operatorname{cap}(z, \mathcal{M}_{\ell'} \backslash z)}{\mu(z)} \geq \frac{\operatorname{cap}(x_\ell, \mathcal{M}_{\ell'} \backslash x_\ell)}{\mu(x_\ell)} \geq \frac{\operatorname{cap}(x_\ell, \mathcal{M}_{\ell-1})}{\mu(x_\ell)}. \tag{8.4.65}$$

Assume first that $\{z\} = \mathscr{M}_{x_\ell}^{(j)}$. Then

$$\frac{\mathrm{cap}(z, \mathscr{M}_{\ell'} \backslash z)}{\mu(z)} < \frac{\mathrm{cap}(z, x_\ell)}{\mu(z)} < \frac{\mathrm{cap}(z, x_\ell)}{\mu(x_\ell)} \leq \frac{\mathrm{cap}(x_\ell, \mathscr{M}_{\ell'} \backslash x_\ell)}{\mu(x_\ell)}, \tag{8.4.66}$$

which contradicts (8.4.65). Hence, z is not selected before the ℓ-th step, and so $z \in \mathscr{M}_{\ell-1}$. Now let $\mathscr{M}_{x_\ell}^{(j)}$ contain several points. Then, for (8.4.65) to be satisfied, it must be a point $y \in \mathscr{M}_{x_\ell}^{(j)}$ such that $\mathrm{cap}(z, y) \sim \mathrm{cap}(z, \mathscr{M}_{\ell'} \backslash z)$, and then z is selected only if $\mu(y) > \mu(z)$. Otherwise, z cannot be selected and must be in $\mathscr{M}_{\ell-1}$. Continuing in this way, we must arrive at a point $x_*^{(j)} \in \mathscr{M}_{x_\ell}^{(j)}$ with maximal invariant mass that cannot be removed at any step $\ell' > \ell$, and thus must be in $\mathscr{M}_{\ell-1}$. This proves (i). Since now for any point in $\mathscr{M}_{x_\ell}^{(j)}$ there is a point y with $\mu(y) > \mu(z)$, and $\mathrm{cap}(z, y)/\mu(z) > \mathrm{cap}(x_\ell, \mathscr{M}_{\ell-1})/\mu(x_\ell)$, it follows that $\mathrm{cap}(z, y)/\mu(z) > \mathrm{cap}(z, x_*^{(j)})/\mu(z)$, which implies (ii). □

Corollary 8.29 *Under the assumptions of Lemma* 8.28,

$$\mathrm{cap}(x_\ell, \mathscr{M}_{x_\ell}) = \mathrm{cap}(x_\ell, \mathscr{M}_{\ell-1}) \left[1 + O(\delta)\right]. \tag{8.4.67}$$

Proof We have shown that each component $\mathscr{M}_{x_\ell}^{(j)}$ contains one point from $\mathscr{M}_{\ell-1}$. We need the following lemma.

Lemma 8.30 *Let $Y \subset X \subset S$, and let $x \in S \backslash X$. Then*

$$\mathrm{cap}(x, Y) \leq \mathrm{cap}(x, X) \leq \frac{\mathrm{cap}(x, Y)}{1 - \sup_{z \in X \backslash Y} \frac{\mathrm{cap}(z,x)}{\mathrm{cap}(z,Y)}}. \tag{8.4.68}$$

Proof The first inequality in (8.4.68) is trivial. For the second, note that

$$\begin{aligned} \mathbb{P}_x(\tau_X < \tau_x) &= \mathbb{P}_x(\tau_X < \tau_x \wedge \tau_Y < \tau_x) + \mathbb{P}_x(\tau_X < \tau_x \wedge \tau_Y > \tau_x) \\ &\leq \mathbb{P}_x(\tau_Y < \tau_x) + \sum_{z \in X \backslash Y} \mathbb{P}_x(\tau_z < \tau_{x \cup X \backslash z}) \mathbb{P}_z(\tau_x < \tau_Y) \\ &\leq \mathbb{P}_x(\tau_Y < \tau_x) + \sup_{z \in X \backslash Y} \mathbb{P}_z(\tau_x < \tau_Y) \mathbb{P}_x(\tau_{X \backslash Y} < \tau_{x \cup Y}) \\ &\leq \mathbb{P}_x(\tau_Y < \tau_x) + \sup_{z \in X \backslash Y} \frac{\mathrm{cap}(z, x)}{\mathrm{cap}(z, Y)} \mathbb{P}_x(\tau_X < \tau_x). \end{aligned} \tag{8.4.69}$$

Using that $\mathrm{cap}(x, Y) = \mu(x) \mathbb{P}_x(\tau_X < \tau_x)$ (see (7.1.19)), we get the upper bound in (8.4.68). □

At this point we have

$$\mathrm{cap}(x_\ell, \mathscr{M}_{x_\ell} \cap \mathscr{M}_{\ell-1}) \leq \mathrm{cap}(x_\ell, \mathscr{M}_{x_\ell}) \leq \frac{\mathrm{cap}(x_\ell, \mathscr{M}_{x_\ell} \cap \mathscr{M}_{\ell-1})}{1 - \sup_{z \in \mathscr{M}_{x_\ell} \backslash \mathscr{M}_{\ell-1}} \frac{\mathrm{cap}(z,x_\ell)}{\mathrm{cap}(z,\mathscr{M}_{x_\ell} \cap \mathscr{M}_{\ell-1})}}. \tag{8.4.70}$$

It follows from Lemma 8.28 and the non-degeneracy conditions that the ratios of the capacities in the denominators are at most δ. Next, we use the same reasoning to show that

$$\operatorname{cap}(x_\ell, \mathcal{M}_{x_\ell} \cap \mathcal{M}_{\ell-1}) \leq \operatorname{cap}(x_\ell, \mathcal{M}_{\ell-1}) \leq \frac{\operatorname{cap}(x_\ell, \mathcal{M}_{x_\ell} \cap \mathcal{M}_{\ell-1})}{1 - \sup_{z \in \mathcal{M}_{\ell-1} \setminus \mathcal{M}_{x_\ell}} \frac{\operatorname{cap}(z, x_\ell)}{\operatorname{cap}(z, \mathcal{M}_{x_\ell} \cap \mathcal{M}_{\ell-1})}}. \tag{8.4.71}$$

Here, again the ratios of the capacities in the denominator must all be smaller than δ, since if for some z in the supremum this is not true, then it leads to a contradiction with the definition of x_ℓ. □

The first important consequence is the following estimate on the ℓ^2-norms of the corresponding equilibrium potentials.

Lemma 8.31 *Let $\mathcal{M}$ be a non-degenerate set of metastable points and let $\mathcal{M}_\ell$, $\ell = k, \ldots, 1$ be defined in Definition 8.27. Then, for $\ell = k, \ldots, 2$,*

$$\|h_{x_\ell, \mathcal{M}_{\ell-1}}\|_{2,\mu}^2 = \mu\big(A(x_\ell)\big), \tag{8.4.72}$$

where the valley $A(x_\ell)$ is defined with respect to the original set $\mathcal{M}$.

Proof Use the estimates on equilibrium potentials in Lemma 8.11 and Lemma 8.28. □

Corollary 8.32 *Let $\mathcal{M}$ be a non-degenerate set of metastable points. Then*

$$\frac{\operatorname{cap}(x_\ell, \mathcal{M}_{\ell-1})}{\|h_{x_\ell, \mathcal{M}_{\ell-1}}\|_{2,\mu}^2} \leq \delta \frac{\operatorname{cap}(x_{\ell+1}, \mathcal{M}_\ell)}{\|h_{x_{\ell+1}, \mathcal{M}_\ell}\|_{2,\mu}^2}, \tag{8.4.73}$$

with $0 < \delta \ll 1$ as in (8.4.63).

Theorem 8.33 (Sharp bounds on principal eigenvalues) *Assume that $\mathcal{M}$ is a non-degenerate set of metastable points and let $|\mathcal{M}| = k$. Define the sequence of points $x_k, \ldots, x_1$ and the sequence of sets $\mathcal{M}_\ell$, $\ell = k, \ldots, 1$ as in Definition* 8.27. *Then, for all $\ell = 1, \ldots, k-1$,*

$$\frac{\operatorname{cap}(x_\ell, \mathcal{M}_{\ell-1})}{\|h_{x_\ell, \mathcal{M}_{\ell-1}}\|_{2,\mu}^2}\big[1 - O(\delta)\big] \leq \lambda_0^{\mathcal{M}_\ell} \leq \frac{\operatorname{cap}(x_\ell, \mathcal{M}_{\ell-1})}{\|h_{x_\ell, \mathcal{M}_{\ell-1}}\|_{2,\mu}^2}\big[1 + O(\delta)\big] \tag{8.4.74}$$

and $\lambda_0^{\mathcal{M}_{\ell-1}} \leq O(\delta)\lambda_0^{\mathcal{M}_\ell}$, and the sequence $\mathcal{M}_\ell$, $\ell = k, \ldots, 1$, realises the sequence defined in (8.4.61).

We will show that each of these principal Dirichlet eigenvalues is very close to one of the small eigenvalues of $-L$.

Theorem 8.34 (Sharp asymptotics of principal eigenvalues) *Assume that there exists an $x \in \mathscr{M}$ such that, for some $0 < \delta \ll 1$,*

$$\delta \frac{\mathrm{cap}(x, \mathscr{M}\backslash x)}{\|h_x\|_{2,\mu}^2} \geq \max_{z \in \mathscr{M}\backslash x} \frac{\mathrm{cap}(z, \mathscr{M}\backslash z)}{\|h_z\|_{2,\mu}^2}. \tag{8.4.75}$$

Then the largest eigenvalue of $-L$ smaller than $\lambda_0^{\mathscr{M}}$ is given by

$$\lambda_x = \frac{\mathrm{cap}(x, \mathscr{M}\backslash x)}{\|h_x\|_{2,\mu}^2} \big[1 + O(\delta)\big], \tag{8.4.76}$$

and all other eigenvalues λ of $-L$ satisfy

$$\lambda \leq C\delta\lambda_x. \tag{8.4.77}$$

Moreover, the eigenvector $\phi^{(x)}$ corresponding to λ_x, normalised such that $\phi^{(x)}(x) = 1$, satisfies $\phi^{(x)}(z) \leq C\delta$, $z \neq x$, for some constant $C < \infty$.

Proof Let $x = x_k \in \mathscr{M}$ be defined in Definition 8.27. We know from Theorem 8.26 that $\lambda_0^{\mathscr{M}\backslash x} \sim \mathrm{cap}(x, \mathscr{M}\backslash x)/\|h_x\|_{2,\mu}^2$.

Assume that there is an eigenvalue λ_x smaller than $\lambda_0^{\mathscr{M}}$. We try to compute the precise value of this eigenvalue, i.e., we look for a root of the determinant of $\mathscr{E}_{\mathscr{M}}(\lambda)$ that is of order $\mathrm{cap}(x, \mathscr{M}\backslash x)/\|h_x\|_{2,\mu}^2$.

The determinant of $\mathscr{E}_{\mathscr{M}}(\lambda)$ vanishes together with that of the matrix $\widehat{\mathscr{K}}$ with elements

$$\begin{aligned} \widehat{\mathscr{K}}_{xy}^{\lambda} &= \frac{\mu(x)}{\|h_x\|_{2,\mu}\|h_y\|_{2,\mu}} \big(\mathscr{E}_{\mathscr{M}}(\lambda)\big)_{xy} \\ &= \frac{\mathscr{E}(h_x, h_y)}{\|h_x\|_{2,\mu}\|h_y\|_{2,\mu}} - \lambda\left(\frac{(h_x, h_y)_\mu + (\psi_x^\lambda, h_y)_\mu}{\|h_x\|_{2,\mu}\|h_y\|_{2,\mu}}\right). \end{aligned} \tag{8.4.78}$$

Lemma 8.21, Corollary 8.23 and Lemma 8.24 already control the term involving ψ_x^λ and the scalar products $(h_x, h_y)_\mu$. The terms involving $\mathscr{E}(h_x, h_y)$, $x \neq y$, can be bounded using the Cauchy-Schwarz inequality,

$$\begin{aligned} \mathscr{E}(h_x, h_y) &= \left|\frac{1}{2} \sum_{z,z' \in S} \mu(z')p(z', z)\big[h_x(z') - h_x(z)\big]\big[h_y(z') - h_y(z)\big]\right| \\ &\leq \sqrt{\mathscr{E}(h_x, h_x)\mathscr{E}(h_y, h_y)}, \end{aligned} \tag{8.4.79}$$

and hence

$$\left|\frac{\mathscr{E}(h_x, h_y)}{\|h_x\|_{2,\mu}\|h_y\|_{2,\mu}}\right| \leq \sqrt{\frac{\mathscr{E}(h_x, h_x)}{\|h_x\|_{2,\mu}^2}} \sqrt{\frac{\mathscr{E}(h_y, h_y)}{\|h_y\|_{2,\mu}^2}}. \tag{8.4.80}$$

Therefore, by the assumption in Theorem 8.34, there exists an $x \in \mathcal{M}$ such that

$$\left| \frac{\mathcal{E}(h_x, h_y)}{\|h_x\|_{2,\mu} \|h_y\|_{2,\mu}} \right| \leq \sqrt{\delta} \frac{\mathcal{E}(h_x, h_x)}{\|h_x\|_{2,\mu}^2}. \tag{8.4.81}$$

Collecting estimates, we have the following, where we abbreviate

$$A_x = \frac{\mathcal{E}(h_x, h_x)}{\|h_x\|_{2,\mu}^2}. \tag{8.4.82}$$

Lemma 8.35

(i) *Let x be the point specified in the assumptions in Theorem* 8.34. *Then*

$$\widehat{\mathcal{K}_{xx}^{\lambda}} = A_x - \lambda\big[1 + O(\lambda)\big]. \tag{8.4.83}$$

(ii) *For $y \neq x$, the diagonal elements satisfy*

$$\widehat{\mathcal{K}_{yy}^{\lambda}} = A_x\, O(\delta) - \lambda\big[1 + O(\lambda)\big], \quad y \neq x. \tag{8.4.84}$$

(iii) *All off-diagonal elements satisfy*

$$|\widehat{\mathcal{K}_{uv}^{\lambda}}| \leq C(\sqrt{\delta} A_x + \lambda\rho), \quad u \neq v. \tag{8.4.85}$$

Recall that $\lambda < \lambda_0^{\mathcal{M}}$ is an eigenvalue of $-L$ if there is a non-zero solution to the equations

$$\sum_{y \in \mathcal{M}} \widetilde{\mathcal{K}}_{zy} c_y = 0, \quad z \in \mathcal{M}. \tag{8.4.86}$$

Trivially, we may choose the vector c in such a way that $\max_{z \in \mathcal{M}} |c_z| = 1$, and the component realising the maximum is equal to 1. Assume that, with this normalisation, $c_z = 1$ for $z \neq x$. Then the z-line of (8.4.86) reads

$$-\widetilde{\mathcal{K}}_{zz} = \sum_{y \neq z} \widetilde{\mathcal{K}}_{zy} c_y, \tag{8.4.87}$$

and inserting the estimates on the matrix elements, we find

$$\lambda \leq A_x C |\mathcal{M}| \sqrt{\delta} + \lambda C \rho \| \mathcal{M}|, \tag{8.4.88}$$

which implies that λ must be much smaller than A_x. Thus, such a c would not correspond to an eigenvalues that is larger than $\lambda_0^{\mathcal{M} \setminus x}$. Hence we may assume that $c_x = 1 \geq |c_y|$ for all $y \neq x$. Now, (8.4.86) with $z = x$,

$$\widetilde{\mathcal{K}}_{xx} = \sum_{y \neq x} \widetilde{\mathcal{K}}_{xy} c_y, \tag{8.4.89}$$

implies, in view of the bounds on $\widetilde{\mathscr{K}}_{xy}$ and the fact that $|c_y| \leq 1$,

$$\left|\widetilde{\mathscr{K}}_{xx}^{\lambda}\right| \leq C|\mathscr{M}|\left(\sqrt{\delta}A_x + \lambda^2/\lambda_0^{\mathscr{M}}\right), \tag{8.4.90}$$

i.e.,

$$|A_x - \lambda| \leq C|\mathscr{M}|\left(\sqrt{\delta}A_x + \lambda^2/\lambda_0^{\mathscr{M}}\right), \tag{8.4.91}$$

which in turn implies that

$$\lambda = A\left[1 + O(\sqrt{\delta} + \rho)\right]. \tag{8.4.92}$$

This bound can be improved if we consider the remaining equations in (8.4.86). Namely, for $z \neq x$,

$$-\widetilde{\mathscr{K}}_{zz}c_z = \sum_{y \neq z} \widetilde{\mathscr{K}}_{zy}c_y. \tag{8.4.93}$$

Solving for c_z, using (ii) and (iii), and employing $\lambda \sim A_x$, we see that

$$|c_z| \leq C|\mathscr{M}|(\sqrt{\delta} + \rho). \tag{8.4.94}$$

This allows us to improve (8.4.91) to

$$|A_x - \lambda| \leq C^2|\mathscr{M}|^2 A_x, \tag{8.4.95}$$

which is the first claim in Theorem 8.34. The assertion on the eigenvector follows from our estimates on the vector c.

It remains to show that a solution of (8.4.86) as specified above *exists*. This can be shown with the help of a fixed-point argument. Rearranging terms, we can cast (8.4.86) into the form

$$\begin{aligned} \lambda &= \Lambda(\lambda, c_1, \ldots, c_{k-1}), \\ c_\ell &= C_\ell(\lambda, c_1, \ldots, c_{k-1}), \quad \ell = 1, \ldots k-1. \end{aligned} \tag{8.4.96}$$

Explicitly, the maps Λ and C read (we abbreviate $\underline{c} = (c_1, \ldots, c_{k-1})$)

$$\begin{aligned} \Lambda(\lambda, \underline{c}) &= \frac{\mathscr{E}(h_x, h_x)}{\|h_x\|_{2,\mu}^2} - \lambda \frac{(\psi_x^\lambda, h_x)}{\|h_x\|_{2,\mu}^2} + \sum_{j=1}^{k-1} \mathscr{H}_{xx_j}^\lambda c_j, \\ C_\ell(\lambda, \underline{c}) &= \lambda^{-1}\left(\frac{\mathscr{E}(h_{x_\ell}, h_{x_\ell})}{\|h_{x_\ell}\|_{2,\mu}^2} c_\ell - \lambda \frac{(\psi_{x_\ell}^\lambda, h_{x_\ell})}{\|h_{x_\ell}\|_{2,\mu}^2} c_\ell + \sum_{j \neq \ell} \mathscr{H}_{x_\ell z_j}^\lambda c_j\right) \end{aligned} \tag{8.4.97}$$

for $\ell = 2, \ldots, k$. We want to construct a solution by the following iteration scheme.

Let $\lambda^{(0)} = A_x$ and $c_\ell^{(0)} = 0$, $\ell = 1, \ldots k-1$. For $n \in \mathbb{N}$, let $\lambda^{(n)}$ be the solution of $\lambda^{(n)} = \Lambda(\lambda^{(n)}, \underline{c}^{(n-1)})$, and let $\underline{c}^{(n)}$ be the solution of $\underline{c}^{(n)} = C(\lambda^{(n)}, \underline{c}^{(n)})$. We want to prove that the sequence $(\lambda^{(n)}, \underline{c}^{(n)})_{n \in \mathbb{N}}$ converges. To do this, we need the following facts.

(i) For $\underline{c}$ in a small neighbourhood of 0, the map $\Lambda(\cot, \underline{c})\colon \mathbb{R} \to \mathbb{R}$ is a contraction on a neighbourhood of A_x, and hence the steps $\lambda^{(n-1)} \to \lambda^{(n)}$ are well defined.
(ii) For λ in a small neighbourhood of A_x, the map $C(\lambda, \cdot)\colon \mathbb{R}^{k-1} \to \mathbb{R}^{k-1}$ is a contraction on a neighbourhood of 0, and hence the steps $\underline{c}^{(n-1)} \to \underline{c}^{(n)}$ are well defined.
(iii) On the respective sets, the solutions of $\lambda = \Lambda(\lambda, \underline{c})$ are Lipschitz in $\underline{c}$, and the solutions of $\underline{c} = C(\lambda, \underline{c})$ are Lipschitz in λ, with Lipschitz constants such that the composition of these maps yields a contraction.

In the following statements the assumptions of Theorem 8.34 are in place.

Lemma 8.36 *For any $\underline{c}$ with $\|\underline{c}\|_\infty \leq 1$, the map $\Lambda(\cdot, \underline{c})\colon (A_x/2, 3A_x/2) \to \mathbb{R}$ is a contraction. More precisely,*

$$\left|\Lambda(\lambda, \underline{c}) - \Lambda\big(\lambda', \underline{c}\big)\right| \leq \left|\lambda - \lambda'\right| C\rho, \tag{8.4.98}$$

where $C < \infty$ is independent of ρ.

Proof The estimate in (8.4.98) is straightforward from Lemma 8.21, Corollary 8.23 and Lemma 8.24, together with the assumption that all $\mathscr{M}_\ell$ are ρ-metastable sets. □

Corollary 8.37 *For any $\underline{c}$ with $\|\underline{c}\|_\infty \leq 1$,*

$$\lambda = \Lambda(\lambda, \underline{c}) \tag{8.4.99}$$

has a unique solution $\lambda(\underline{c}) \in (A_x/2, 3A_x/2)$.

Proof Set $\lambda^{(0)}(\underline{c}) = A_x$, and $\lambda^{(n)}(\underline{c}) = \Lambda(\lambda^{(n-1)}(\underline{c}), \underline{c})$. Then, as $n \to \infty$, $\lambda^{(n)}(\underline{c})$ converges to the unique fixed point of the map $\Lambda(\cdot, \underline{c})$ on $(A_x/2, 3A_x/2)$, which is the solution of (8.4.99). □

Lemma 8.38 *The solution of* (8.4.99) *from Corollary* 8.37 *is Lipschitz continuous with respect to the ℓ^1-norm in $\underline{c}$ with Lipschitz constant $C\sqrt{\delta}A_x$.*

Proof We show that, for fixed $\lambda \in (A_x/2, 3A_x/2)$, $\Lambda(\lambda, \underline{c})$ is Lipschitz in $\underline{c}$. Namely,

$$\begin{aligned} \left|\Lambda(\lambda, \underline{c}) - \Lambda\big(\lambda, \underline{c}'\big)\right| &\leq \sum_{\ell=1}^{k-1} \left|c_\ell - c'_\ell\right| \mathscr{H}^\lambda_{xx_\ell} \\ &= \sum_{\ell=1}^{k-1} \left|c_\ell - c'_\ell\right| (\sqrt{\delta}A_x + 3A_x C\rho), \end{aligned} \tag{8.4.100}$$

which is dominated by the $\sqrt{\delta}A_x$-term. This gives the Lipschitz bound in ℓ^1 and in ℓ^∞. Combining this with the bound (8.4.98), we get

$$\left|\lambda(\underline{c}) - \lambda(\underline{c}')\right| \leq \frac{C\sqrt{\delta}A_x}{1 - C\rho}, \tag{8.4.101}$$

which proves the lemma. □

Lemma 8.39 *For $\lambda \in (A_x/2, 3A_x/2)$, the map $C(\lambda, \cdot)\colon [-1,1]^{k-1} \to [-1,1]^{k-1}$ is a contraction. More precisely,*

$$\left|C(\lambda, \underline{c}) - C(\lambda, \underline{c}')\right| \leq C\delta\|\underline{c} - \underline{c}'\|_1. \tag{8.4.102}$$

Proof This is again elementary from what is already proven. □

Corollary 8.40 *For any $\lambda \in (A_x/2, 3A_x/2)$, the equation*

$$\underline{c} = C(\lambda, \underline{c}) \tag{8.4.103}$$

has a unique solution in $[-1,1]^{k-1}$.

Proof Same as the proof of Corollary 8.37. □

Lemma 8.41 *Let $\underline{c}(\lambda)$ denote the solution of* (8.4.103) *from Corollary* 8.40. *Then $\underline{c}(\lambda)$ is Lipschitz in λ. More precisely,*

$$\left|\underline{c}(\lambda) - \underline{c}(\lambda')\right| \leq C\frac{\sqrt{\delta}}{A_x}\left|\lambda - \lambda'\right|. \tag{8.4.104}$$

Proof The proof goes like the proof of Lemma 8.38. The fairly large bound on the Lipschitz constant comes from the term involving $\mathscr{H}^{\lambda}_{x_\ell x}$ that gives rise to a term

$$\left|\lambda^{-1} - \lambda'^{-1}\right| \frac{\mathscr{E}(h_{x_{k-1}}, h_x)}{\|h_{x_{k-1}}\|_{2,\mu}\|h_x\|_{2,\mu}}, \tag{8.4.105}$$

which cannot be bounded by less than what is claimed. □

Corollary 8.42 *The map $T\colon (A_x/2, 3A_x/2) \to (A_x/2, 3A_x/2)$ defined by $T(\lambda) = \lambda(\underline{c}(\lambda))$, where $\lambda(\underline{c})$ is the unique solution of $\lambda = \Lambda(\lambda, \underline{c})$ and $\underline{c}(\lambda)$ is the unique solution of $\underline{c} = C(\lambda, \underline{c})$, is a contraction. More precisely,*

$$\left|T(\lambda) - T(\lambda')\right| \leq C\sqrt{\delta}\left|\lambda - \lambda'\right|, \tag{8.4.106}$$

where $C < \infty$ is independent of δ.

Proof Using Lemmas 8.38 and 8.41, we have that

$$\begin{aligned} \left|\lambda(\underline{c}(\lambda)) - \lambda(\underline{c}(\lambda'))\right| &\leq C\sqrt{\delta}A_x \left\|\underline{c}(\lambda) - c(\lambda')\right\|_1 \\ &\leq C\sqrt{\delta}C\sqrt{\delta}A_x A_x^{-1}\left|\lambda - \lambda'\right| = C^2\delta\left|\lambda - \lambda'\right|, \end{aligned} \tag{8.4.107}$$

which proves the claim. □

Corollary 8.42 implies that there exists a unique $\lambda \in (A_x/2, 3A_x/2)$ such that $\lambda = \lambda(\underline{c}(\lambda))$. Hence $(\lambda, \underline{c}(\lambda))$ is the unique solution of $\lambda = \Lambda(\lambda, \underline{c})$ and $\underline{c} = C(\lambda, \underline{c})$ with λ near A_x. This proves the existence of the solution to (8.4.86) and concludes the proof of Theorem 8.34. □

At this point we have shown that $\lambda^{\mathcal{M}_0} > \lambda_x > \lambda_0^{\mathcal{M}\backslash x}$, where the last two eigenvalues are almost the same and are smaller by a factor at least δ than the first. This procedure can now be repeated with $\mathcal{M}$ replaced by $\mathcal{M}\backslash x$, provided $\mathcal{M}\backslash x$ satisfies the hypothesis of a set of metastable points.

Theorem 8.43 (Asymptotics of the spectrum and mean metastable exit times) *Let $|\mathcal{M}| = k \geq 2$, and let $\mathcal{M}_\ell$, $\ell = k, \ldots, 1$ be the sequence of sets defined in Theorem 8.33. Assume further that, for each $\ell = 1, \ldots, k$, $\mathcal{M}_\ell$ is a set of metastable points in the sense of Definition 8.2 (with the same parameter ρ). Then $-L$ has k eigenvalues $\lambda_1 < \lambda_2 < \cdots < \lambda_k < \lambda_0^{\mathcal{M}}$, where*

$$\lambda_1 = 0, \tag{8.4.108}$$

and

$$\lambda_\ell = \frac{\operatorname{cap}(x_\ell, \mathcal{M}_{\ell-1})}{\mu(A(x_\ell))}\left[1 + O(\delta)\right], \quad \ell = 2, \ldots, k. \tag{8.4.109}$$

Consequently,

$$\lambda_\ell = \frac{1}{\mathbb{E}_{x_\ell}[\tau_{\mathcal{M}_{\ell-1}}]}\left[1 + O(\delta)\right] = \frac{1}{\mathbb{E}_{x_\ell}[\tau_{\mathcal{M}_{x_\ell}}]}\left[1 + O(\delta)\right]. \tag{8.4.110}$$

The corresponding normalised eigenfunction has the form

$$\psi_\ell(y) = \frac{h_{x_\ell, \mathcal{M}_{\ell-1}}(y)}{\|h_{x_\ell, \mathcal{M}_{\ell-1}}\|_{2,\mu}} + \sum_{j=1}^{\ell-1} O(\delta) \frac{h_{x_\ell, \mathcal{M}_{j-1}}(y)}{\|h_{x_\ell, \mathcal{M}_{j-1}}\|_{2,\mu}}. \tag{8.4.111}$$

Proof Applying Theorem 8.34 to the sets $\mathcal{M}_\ell$, we successively show that $-L$ has $k-1$ eigenvalues below $\lambda_0^{\mathcal{M}}$ that satisfy

$$\lambda_{x_\ell} = \frac{\operatorname{cap}(x_\ell, \mathcal{M}_{\ell-1})}{\|h_{x_\ell, \mathcal{M}_{\ell-1}}\|_{2,\mu}^2}\left[1 + O(\delta)\right], \quad \ell = 2, \ldots, k. \tag{8.4.112}$$

Using the same arguments as in the proof of Lemma 8.24, we show that

$$\|h_{x_\ell,\mathcal{M}_{\ell-1}}\|_{2,\mu}^2 = \|h_{x_\ell,\mathcal{M}_{\ell-1}}\|_{1,\mu}\big[1+O(\rho)\big] = \|1_{A(x_\ell)}\|_{1,\mu}\big[1+O(\rho)\big]. \quad (8.4.113)$$

It remains to identify the right-hand side with the inverse mean hitting time of the set $\mathcal{M}_{x_\ell} = \{z \in \mathcal{M} : \mu(z) > \mu(x_\ell)\}$.

Lemma 8.44 *Under the assumptions of Lemma* 8.28,

$$\mathbb{E}_{x_\ell}[\tau_{\mathcal{M}_{x_\ell}}] = \mathbb{E}_{x_\ell}[\tau_{\mathcal{M}_{\ell-1}}]\big[1+O(\delta)\big]. \quad (8.4.114)$$

Proof First, from Theorem 8.15 we have that

$$\mathbb{E}_{x_\ell}[\tau_{\mathcal{M}_{x_\ell}}] = \frac{\mu(A(x_\ell))}{\mathrm{cap}(x_\ell,\mathcal{M}_{x_\ell})}\big[1+o(1)\big]. \quad (8.4.115)$$

To estimate $\mathbb{E}_{x_\ell}[\tau_{\mathcal{M}_{\ell-1}}]$, we use that, by assumption, $\mathcal{M}_\ell$ is a set of metastable points and so, by Theorem 8.15,

$$\mathbb{E}_{x_\ell}[\tau_{\mathcal{M}_{\ell-1}}] = \frac{1}{\mathrm{cap}(x_\ell,\mathcal{M}_{\ell-1})}\sum_{z\in\widehat{A}(x_\ell)}\mu(z)h_{x_\ell,\mathcal{M}_{\ell-1}}(z)(1+o(1)), \quad (8.4.116)$$

where $\widehat{A}(x_\ell)$ now refers to the set $\mathcal{M}_\ell$. However, by the construction of the sequence of sets $\mathcal{M}_\ell$, $\ell = k,\dots,1$, we have $\mu(\widehat{A}(x_\ell)) = \mu(A(x_\ell))$, $\ell = k,\dots,1$. By Corollary 8.29, the capacities in the two formulas are also equal up to a factor $1+O(\delta)$, which proves the lemma. □

This observation allows us to conclude that the k smallest eigenvalues of L are precisely the inverses of the mean exit times from the metastable points $\mathcal{M}$. The estimate on the eigenvectors is inherited from (8.4.94). □

8.4.4 Exponential law of the metastable exit times

There are different ways to prove that the distribution of metastable exit times is asymptotically exponential. The most robust argument is based on a renewal argument: Since the probability to reach the set $\mathcal{M}_x$ starting from $x \in \mathcal{M}$ *without* returning to x is very small, the process returns many times to x before a successful excursion happens. The number of such excursions is geometrically distributed, and the time of an unsuccessful recursion is $\mu(A(x))/\mu(x)$, by the ergodic theorem. Since this time is small compared to the number of excursions, the rescaled time until a successful excursion converges to an exponential distribution. Finally, the time of the last excursion is negligible compared to this time.

Theorem 8.45 (Exponential law of metastable exit times) *Under the non-degeneracy hypothesis of Theorem* 8.34 *with* δ *satisfying* (8.4.63), *for all* $t > 0$,

$$\lim_{\rho\downarrow 0}\mathbb{P}_x\big(\tau_{\mathscr{M}_x} > t\,\mathbb{E}_x[\tau_{\mathscr{M}_x}]\big) = \mathrm{e}^{-t}. \tag{8.4.117}$$

Proof To exploit the renewal structure it is convenient to use Laplace transforms. We set $\hat\tau_{\mathscr{M}_x} = \tau_{\mathscr{M}_x}/\mathbb{E}_x[\tau_{\mathscr{M}_x}]$ and $\hat\tau_x = \tau_x/\mathbb{E}_x[\tau_{\mathscr{M}_x}]$ and

$$R_x(\lambda) = \mathbb{E}_x\big[\exp(\lambda\hat\tau_{\mathscr{M}_x})\big]. \tag{8.4.118}$$

Note that $R_x(\lambda) < \infty$ for all $\lambda < 1$, due to the fact that the principle eigenvalue of the Dirichlet generator with Dirichlet conditions in $\mathscr{M}_x$ is essentially $1/\mathbb{E}_x[\tau_{\mathscr{M}_x}]$. Moreover, $R_x(\lambda)$ satisfies the following *renewal equation*.

Lemma 8.46 (Renewal equation for Laplace transforms) *For all* $\lambda < 1$,

$$R_x(\lambda) = \frac{\mathbb{E}_x[\mathrm{e}^{\lambda\hat\tau_{\mathscr{M}_x}}\mathbb{1}_{\tau_{\mathscr{M}_x}<\tau_x}]}{1-\mathbb{E}_x[\mathrm{e}^{\lambda\hat\tau_x}\mathbb{1}_{\tau_x<\tau_{\mathscr{M}_x}}]}. \tag{8.4.119}$$

Proof Noting that $1 = \mathbb{1}_{\tau_{\mathscr{M}_x}<\tau_x} + \mathbb{1}_{\tau_x<\tau_{\mathscr{M}_x}}$ and using the strong Markov property, we see that

$$R_x(\lambda) = \mathbb{E}_x\big[\mathrm{e}^{\lambda\hat\tau_{\mathscr{M}_x}}\mathbb{1}_{\tau_{\mathscr{M}_x}<\tau_x}\big] + \mathbb{E}_x\big[\mathrm{e}^{\lambda\hat\tau_x}\mathbb{1}_{\tau_x<\tau_{\mathscr{M}_x}}\big]\mathbb{E}_x\big[\mathrm{e}^{\lambda\hat\tau_{\mathscr{M}_x}}\big]. \tag{8.4.120}$$

Equation (8.4.119) is now immediate. □

As a result of the representation in (8.4.119), Theorem 8.45 is a consequence of the following lemma.

Lemma 8.47 *With the notation in Lemma* (8.46), *for all* $\lambda < 1$,

$$\lim_{\rho\downarrow 0}\frac{\mathbb{E}_x[\mathrm{e}^{\lambda\hat\tau_{\mathscr{M}_x}}\mathbb{1}_{\hat\tau_{\mathscr{M}_x}<\tau_x}]}{1-\mathbb{E}_x[\mathrm{e}^{\lambda\tau_x}\mathbb{1}_{\tau_x<\hat\tau_{\mathscr{M}_x}}]} = \frac{1}{1+\lambda}. \tag{8.4.121}$$

Proof The first step in the proof is the following crucial pointwise bound.

Lemma 8.48 *There exists a* $1 < C < \infty$ *such that*

$$\mathbb{E}_x[\tau_{\mathscr{M}_x}\mathbb{1}_{\tau_{\mathscr{M}_x}<\tau_x}] \leq \frac{C\rho}{1-C\rho}\mathbb{E}_x[\tau_x\mathbb{1}_{\tau_x<\tau_{\mathscr{M}_x}}]. \tag{8.4.122}$$

Proof Instead of proving (8.4.122) directly, we first show the (more natural) estimate

$$\mathbb{E}_x[\tau_{\mathscr{M}_x}\mathbb{1}_{\tau_{\mathscr{M}_x}<\tau_x}] \leq C\rho. \tag{8.4.123}$$

To do so, we use the fact that

$$\mathbb{E}_x[\tau_{x\cup\mathcal{M}_x}] = \mathbb{E}_x[\tau_x \mathbb{1}_{\tau_x<\tau_{\mathcal{M}_x}}] + \mathbb{E}_x[\tau_{\mathcal{M}_x} \mathbb{1}_{\tau_{\mathcal{M}_x}<\tau_x}]. \tag{8.4.124}$$

Define the function

$$w_{x,\mathcal{M}_x}(y) = \begin{cases} \mathbb{E}_y[\tau_x \mathbb{1}_{\tau_x<\tau_{\mathcal{M}_x}}], & \text{if } y \notin x \cup \mathcal{M}_x, \\ 0, & \text{else.} \end{cases} \tag{8.4.125}$$

Then $w_{x,\mathcal{M}_x}$ solves the Dirichlet problem

$$\begin{aligned} (-Lw_{x,\mathcal{M}_x})(y) = h_{x,\mathcal{M}_x}(y), \quad & y \notin x \cup \mathcal{M}_x, \\ w_{x,\mathcal{M}_x}(y) = 0, \qquad & y \in x \cup \mathcal{M}_x. \end{aligned} \tag{8.4.126}$$

Note that

$$\mathbb{E}_x[\tau_x \mathbb{1}_{\tau_x<\tau_{\mathcal{M}_x}}] = \mathbb{P}_x(\tau_x < \tau_{\mathcal{M}_x}) - (-Lw_{x,\mathcal{M}_x})(x). \tag{8.4.127}$$

Next, using reversibility, we have

$$\sum_y \mu(y) h_{x,\mathcal{M}_x}(y)(-Lw_{x,\mathcal{M}_x})(y) = \sum_y \mu(y)(-Lh_{x,\mathcal{M}_x})(y) w_{x,\mathcal{M}_x}(y). \tag{8.4.128}$$

By the properties of the functions $h_{x,\mathcal{M}_x}$ and $w_{x,\mathcal{M}_x}$, the right-hand side of (8.4.128) vanishes identically, while the left-hand side is equal to

$$\mu(x)(-Lw_{x,\mathcal{M}_x})(x) + \sum_{y\notin x\cup\mathcal{M}_x} \mu(y) h_{x,\mathcal{M}_x}(y)^2 = 0. \tag{8.4.129}$$

Hence, inserting (8.4.127), we get

$$\mu(x)\mathbb{E}_x[\tau_x \mathbb{1}_{\tau_x<\tau_{\mathcal{M}_x}}] = \mu(x)\mathbb{P}_x(\tau_x < \tau_{\mathcal{M}_x}) + \sum_{y\notin x\cup\mathcal{M}_x} \mu(y) h_{x,\mathcal{M}}(y)^2. \tag{8.4.130}$$

With a similar procedure we show that

$$\mu(x)\mathbb{E}_x[\tau_{x\cup\mathcal{M}_x}] = \mu(x) + \sum_{y\notin x\cup\mathcal{M}_x} \mu(y) h_{x,\mathcal{M}_x}(y). \tag{8.4.131}$$

Therefore, combining (8.4.124), (8.4.30) and (8.4.131), we get

$$\mathbb{E}_x[\tau_{\mathcal{M}_x} \mathbb{1}_{\tau_{\mathcal{M}_x}<\tau_x}] = \mathbb{P}_x(\tau_{\mathcal{M}_x} < \tau_x) + \frac{1}{\mu(x)} \sum_{y\notin x\cup\mathcal{M}_x} \mu(y) h_{x,\mathcal{M}_x}(y) h_{\mathcal{M}_x,x}(y). \tag{8.4.132}$$

The first term in the right-hand side is exponentially small. The second term is controlled in the same way as in Lemma 8.24 and is bounded by $C\rho$. Thus,

(8.4.123) holds. Finally, $\mathbb{E}_x[\tau_{x\cup\mathcal{M}_x}] \geq 1$ and so, via (8.4.124), it follows that $\mathbb{E}_x[\tau_x \mathbb{1}_{\tau_x<\tau_{\mathcal{M}_x}}] \geq 1 - C\rho$, and we deduce (8.4.122). □

Next, define $T = \mathbb{E}_x[\tau_{\mathcal{M}_x}]$. Then (8.4.122) implies that

$$\begin{aligned} T &= \mathbb{E}_x[\tau_{\mathcal{M}_x} \mathbb{1}_{\tau_{\mathcal{M}_x}<\tau_x}] + \mathbb{E}_x[\tau_{\mathcal{M}_x} \mathbb{1}_{\tau_x<\tau_{\mathcal{M}_x}}] \\ &= \mathbb{E}_x[\tau_{\mathcal{M}_x} \mathbb{1}_{\tau_{\mathcal{M}_x}<\tau_x}] + \mathbb{E}_x[\tau_x \mathbb{1}_{\tau_x<\hat{\tau}_{\mathcal{M}_x}}] + \mathbb{P}_x(\tau_x < \tau_{\mathcal{M}_x})\mathbb{E}_x[\tau_{\mathcal{M}_x}] \\ &= \mathbb{E}_x[\tau_{x\cup\mathcal{M}_x}] + \mathbb{P}_x(\tau_x < \tau_{\mathcal{M}_x})T. \end{aligned} \tag{8.4.133}$$

Hence we have

$$T = \frac{\mathbb{E}_x[\tau_{x\cup\mathcal{M}_x}]}{\mathbb{P}_x(\tau_{\mathcal{M}_x} < \tau_x)}. \tag{8.4.134}$$

Because $1 \geq \mathrm{e}^{-x} \geq 1 - x$, $x \geq 0$, it follows that the numerator in (8.4.119) satisfies, for $\lambda \leq 0$,

$$\mathbb{P}_x(\tau_{\mathcal{M}_x} < \tau_x) \geq \mathbb{E}_x\left[\mathrm{e}^{\lambda\hat{\tau}_{\mathcal{M}_x}} \mathbb{1}_{\hat{\tau}_{\mathcal{M}_x}<\tau_x}\right] \geq \mathbb{P}_x(\tau_{\mathcal{M}_x} < \tau_x)\left(1 + \lambda \frac{\mathbb{E}_x[\tau_{\mathcal{M}_x} \mathbb{1}_{\tau_{\mathcal{M}_x}<\tau_x}]}{\mathbb{E}_x[\tau_{x\cup\mathcal{M}_x}]}\right). \tag{8.4.135}$$

Let us now turn to the denominator in (8.4.119), which we rewrite as

$$\mathbb{P}_x(\tau_{\mathcal{M}_x} < \tau_x)\left(1 + \lambda \frac{\mathbb{E}_x[(1 - \mathrm{e}^{\lambda\hat{\tau}_x})\mathbb{1}_{\tau_x<\hat{\tau}_{\mathcal{M}_x}}]}{\lambda\mathbb{P}_x(\tau_{\mathcal{M}_x} < \tau_x)}\right). \tag{8.4.136}$$

Combining this with (8.4.135) we get that, for $\lambda \leq 0$,

$$\frac{1}{1 + \lambda \frac{\mathbb{E}_x[(1-\mathrm{e}^{\lambda\hat{\tau}_x})\mathbb{1}_{\tau_x<\hat{\tau}_{\mathcal{M}_x}}]}{\lambda\mathbb{P}_x(\tau_{\mathcal{M}_x}<\tau_x)}} \geq R_x(\lambda) \geq \frac{1 + \lambda \frac{\mathbb{E}_x[\tau_{\mathcal{M}_x} \mathbb{1}_{\tau_{\mathcal{M}_x}<\tau_x}]}{\mathbb{E}_x[\tau_{x\cup\mathcal{M}_x}]}}{1 + \lambda \frac{\mathbb{E}_x[(1-\mathrm{e}^{\lambda\hat{\tau}_x})\mathbb{1}_{\tau_x<\hat{\tau}_{\mathcal{M}_x}}]}{\lambda\mathbb{P}_x(\tau_{\mathcal{M}_x}<\tau_x)}}. \tag{8.4.137}$$

What is left to control is

$$\frac{\mathbb{E}_x[(1 - \mathrm{e}^{\lambda\hat{\tau}_x})\mathbb{1}_{\tau_x<\hat{\tau}_{\mathcal{M}_x}}]}{\lambda\mathbb{P}_x(\tau_{\mathcal{M}_x} < \tau_x)} = -\frac{\mathbb{E}_x[\hat{\tau}_x \mathbb{1}_{\tau_x<\hat{\tau}_{\mathcal{M}_x}}]}{\mathbb{P}_x(\tau_{\mathcal{M}_x} < \tau_x)} + \frac{\mathbb{E}_x[(1 + \lambda\hat{\tau}_x - \mathrm{e}^{\lambda\hat{\tau}_x})\mathbb{1}_{\tau_x<\hat{\tau}_{\mathcal{M}_x}}]}{\lambda\mathbb{P}_x(\tau_{\mathcal{M}_x} < \tau_x)}. \tag{8.4.138}$$

The first term is fine because

$$\frac{\mathbb{E}_x[\hat{\tau}_x \mathbb{1}_{\tau_x<\hat{\tau}_{\mathcal{M}_x}}]}{\mathbb{P}_x(\tau_{\mathcal{M}_x} < \tau_x)} = \frac{\mathbb{E}_x[\tau_x \mathbb{1}_{\tau_x<\hat{\tau}_{\mathcal{M}_x}}]}{\mathbb{E}_x[\tau_{x\cup\mathcal{M}_x})}, \tag{8.4.139}$$

which tends to one rapidly. To deal with the second term, we use that, for $u \leq 0$,

$$0 \geq 1 + u\tau_x - \mathrm{e}^{u\tau_x} \geq -\tfrac{1}{2}u^2\tau_x^2. \tag{8.4.140}$$

Next, we note that, for $u \le 0$,

$$0 \ge \mathbb{E}_x\big[\big(1 + u\tau_x - e^{u\tau_x}\big)\mathbb{1}_{\tau_x < \tau_{\mathcal{M}_x}}\big] \tag{8.4.141}$$
$$= \mathbb{E}_x\big[\big(1 + u\tau_{x\cup\mathcal{M}_x} - e^{u\tau_{x\cup\mathcal{M}_x}}\big)\mathbb{1}_{\tau_x < \tau_{\mathcal{M}_x}}\big] \ge \mathbb{E}_x\big[1 + u\tau_{x\cup\mathcal{M}_x} - e^{u\tau_{x\cup\mathcal{M}_x}}\big].$$

Thus we need to control the non-negative function

$$v_u(y) = \begin{cases} \mathbb{E}_y[e^{u\tau_x} - 1 - u\tau_x], & y \notin x \cup \mathcal{M}_x, \\ 0, & y \in x \cup \mathcal{M}_x, \end{cases} \tag{8.4.142}$$

for $u \sim 1/\mathbb{E}_x[\tau_{\mathcal{M}_x}]$. This goes in a similar fashion as the derivation of the ℓ^∞-bounds in Lemma 8.22.

Set $w(y) = \mathbb{E}_y[\tau_{x\cup\mathcal{M}_x}]$. We easily verify that v_u solves the Dirichlet problem

$$\begin{aligned} (-L - \bar{u})v_u(y) &= u\bar{u}w(y) + (\bar{u} - u), \quad y \notin x \cup \mathcal{M}_x, \\ v_u(y) &= 0, \quad y \in x \cup \mathcal{M}_x, \end{aligned} \tag{8.4.143}$$

where we set $\bar{u} = 1 - e^{-u}$. Note that $\bar{u} - u \le 0$ for $u \le 0$, and is $O(u^2)$ for small u. The Dirichlet problem in (8.4.143) has a unique solution for $\bar{u} < \lambda_0^{x\cup\mathcal{M}_x}$. Note also that

$$\mathbb{E}_x\big[e^{u\tau_{x\cup\mathcal{M}_x}}\big] = e^u\big((Lw_u)(x) + 1\big), \tag{8.4.144}$$

and

$$\mathbb{E}_x[\tau_{x\cup\mathcal{M}_x}] = (Lw)(x) + 1. \tag{8.4.145}$$

Proceeding as in the proof of Lemma 8.22, we obtain the following estimates.

Lemma 8.49 *Assume that $\bar{u} < \sup_{z\notin x\cup\mathcal{M}_x} \mathbb{E}_z[\tau_{x\cup\mathcal{M}_x}]$. Then, for any y and $u \le 0$,*

$$0 \le v_u(y) \le u\bar{u}\; \mathbb{E}_y[\tau_{x\cup\mathcal{M}_x}] \sup_{z\notin x\cup\mathcal{M}_x} \mathbb{E}_z[\tau_{x\cup\mathcal{M}_x}] + (\bar{u} - u)\,\mathbb{E}_y[\tau_{x\cup\mathcal{M}_x}]. \tag{8.4.146}$$

Proof Rewrite the first line of (8.4.143) as

$$(-Lv_u)(y) = \bar{u}\, v_u(y) + u\bar{u}\, w(y) + (\bar{u} - u), \tag{8.4.147}$$

and solve for v_u. This gives, for $u \le 0$,

$$\begin{aligned} v_u(y) &= \bar{u} \sum_{z\notin x\cup\mathcal{M}_x} G_{x\cup\mathcal{M}_x}(y,z)\big(v_u(z) + u\, w(z) + (1 - u/\bar{u}u)\big) \\ &\le u\bar{u} \sup_{z\notin x\cup\mathcal{M}_x} w(z)\mathbb{E}_y[\tau_{x\cup\mathcal{M}_x}] + (\bar{u} - u)\mathbb{E}_y[\tau_{x\cup\mathcal{M}_x}], \end{aligned} \tag{8.4.148}$$

which yields the claim. □

A rough estimate gives

$$\sup_{y\notin x\cup\mathcal{M}_x} \mathbb{E}_y[\tau_{x\cup\mathcal{M}_x}] \leq |S| \sup_{y\notin x\cup\mathcal{M}_x} \frac{\mu(y)}{\mathrm{cap}(y,\mathcal{M}_x\cup x)} \leq |S|/\lambda_0^{x\cup\mathcal{M}_x} \leq \rho/\lambda_0^{\cup\mathcal{M}_x}, \tag{8.4.149}$$

where the last inequality uses the assumption that $\lambda_0^{x\cup\mathcal{M}_x} \geq \delta\lambda_0^{\mathcal{M}_x}$ with $\delta|S| \leq \rho$. Finally, we recall that $\lambda_0^{\mathcal{M}_x} \sim 1/\mathbb{E}_x[\tau_{\mathcal{M}_x}]$. Hence, for $u = \lambda/\mathbb{E}_x[\tau_{\mathcal{M}_x}]$, $\lambda \leq 0$,

$$\left|\mathbb{E}_x\left[\left(1+\lambda\hat{\tau}_x - \mathrm{e}^{\lambda\hat{\tau}_x}\right)\mathbb{1}_{\tau_x<\hat{\tau}_{\mathcal{M}_x}}\right]\right| \leq C\lambda^2\rho\frac{\mathbb{E}_x[\tau_{x\cup\mathcal{M}_x}]}{\mathbb{E}_x[\tau_{\mathcal{M}_x}]}. \tag{8.4.150}$$

Inserting this bound into (8.4.138), we see that the second term in the right-hand side is bounded in absolute value by $\lambda C\rho$, which tends to zero as desired. But this implies the assertion of Lemma 8.47. □

This concludes the proof of Theorem 8.45. □

8.5 Metastability in uncountable state spaces

In Sects. 8.2–8.4 we have seen that the definition of metastability in Definition 8.2 through capacities leads to precise predictions on the connection between capacities, metastable exit times and small eigenvalues of the generator. These results relied on renewal arguments that are only available in the context of countable state spaces. For uncountable state spaces, however, renewal arguments are not possible because capacities of single points are either zero or are much smaller than capacities of small neighbourhoods around them. This means that the process may take a very long time to hit a point, even when it has already reached its neighbourhood. A proper definition of metastable sets must look like (8.1.3), but with a suitable and model-dependent choice of the sets B_i, $i \in I$.

In Chap. 11 we explain in detail, within the context of diffusion processes, how the theory developed above carries over, provided the sets B_i, $i \in I$, can be chosen in such a way that solutions of the Dirichlet problems we have encountered are almost constant on these sets. At present this is the only example where the full picture established in the present chapter has been carried over. For this reason, we do not formulate an abstract model-independent result, but rather provide the details in the specific context of Chap. 11.

An interesting setting where we would like to establish similar results are interacting particle systems with state spaces like $S = \{-1,+1\}^\Lambda$ with $\Lambda \subset \mathbb{Z}^d$. In this situation, only partial results in specific models are presently available. These are the random-field Curie Weiss model (Chap. 15), when $\Lambda = \{1,\ldots,N\}$ with $N \to \infty$, as well as the Ising model with Glauber dynamics (Chap. 19) and the lattice-gas model with Kawasaki dynamics (Chap. 20), both at low temperature, when the size of Λ diverges as the temperature tends to zero.

8.6 Bibliographical notes

1. The material in this chapter was developed by Bovier, Eckhoff, Gayrard and Klein [34]. The exposition here is streamlined.

2. A somewhat opposite approach, connecting metastability to spectral data, was initiated in a series of papers by Davies [70–74] and developed further by Gaveau and Schulman [124] and Gaveau and Moreau [123]. Here, the starting point is a set of assumptions on the properties of eigenvalues and eigenfunctions of the generator of the Markov process that *imply* certain metastable behaviour. Low-lying eigenvalues are related to metastable time scales and the corresponding eigenfunctions are related to metastable states. The problem is, however, that the computation of the spectrum of a generator and of its eigenfunctions is difficult, and even numerically prohibitive for large systems.

3. In much the same way as in Theorem 8.15, conditional mean hitting times such as $\mathbb{E}_x[\tau_J | \tau_J \leq \tau_I]$ can be computed (for an example see (8.4.37)). Formulas are given in Bovier, Eckhoff, Gayrard and Klein [33, 34].

4. The characterisation of eigenvalues given in Lemma 8.19 was first exploited in Bovier, Eckhoff, Gayrard and Klein [34], but similar ideas were put forward earlier in Wentzel [233, 234].

5. The non-degeneracy conditions of Theorem 8.34, ensuring the simplicity of eigenvalues, can be lifted at the expense of more work. Some examples have been worked out in the context of finite-state Markov processes with exponentially small transition probabilities by Berglund and Dutercq [20] and by Cirillo and Nardi [65].

6. A similar approach to metastability for continuous-time Markov processes, based on potential theory combined with martingale convergence theory, has been taken by Beltrán and Landim [16, 17].

7. A probabilistic proof of Lemma 8.5 (with slightly different bounds) was given in Bovier, Eckhoff, Gayrard and Klein [34]. An analytic proof can be found in Slowik [219].

8. The *transition path theory* championed by E and Vanden-Eijnden [100] analyses path properties of metastable systems via Doob transforms based on the assumption that the equilibrium potentials ("committors" in their terminology) are known. They use numerical methods to compute equilibrium potentials in concrete models.

9. The proof of the asymptotic exponential law for metastable exit times given in Sect. 8.4.4 can be extended to settings where only approximate renewal properties hold. See, for instance, Bianchi, Bovier and Ioffe [25]. An alternative proof in the discrete setting can be found in Bovier, Eckhoff, Gayrard and Klein [34]. An elegant earlier proof was given by Martinelli and Scoppola [177] and Martinelli, Olivieri and Scoppola [174, 175].

Chapter 9
Basic Techniques

Il y a peu de choses impossibles d'elles-mêmes; et l'application pour les faire réussir nous manque plus que les moyens.
(François de La Rochefoucauld, Réflexions)

This chapter collects techniques that are basic for the study of metastability and that will be used throughout Parts IV–VIII. Section 9.1 focusses on capacity estimates, and derives upper and lower bounds on capacities with the help of variational principles, namely, the Dirichlet principle and the Berman-Konsowa principle introduced in Sect. 7.3. We outline the strategies that are used to exploit these variational principles in an efficient way to derive matching upper and lower bounds. Section 9.2 introduces the notion of coarse-graining, which is particularly useful for mean-field systems. Section 9.3 states conditions under which the Markov property is preserved when states are lumped. Section 9.4 deals with regularity estimates for harmonic functions with the help of elliptic regularity theory and coupling.

9.1 Capacity estimates

We have seen in Chap. 8 that in metastable dynamics the computation of key quantities such as hitting probabilities and mean hitting times can be largely reduced to the computation of *capacities*. The usefulness of this reduction depends on how well we can compute capacities. While clearly the universality of our approach ends here, and model-specific properties have to come into the game, it is both surprising and rewarding that precise computations of capacities are possible in a multitude of specific systems, as we will come to appreciate as we go along.

In Sect. 9.1.1 we outline general strategies to obtain upper and lower bounds on capacities via the Dirichlet principle. In Sect. 9.1.2 we concentrate on lower bounds on capacities via the Berman-Konsowa principle.

A. Bovier, F. den Hollander, *Metastability*,
Grundlehren der mathematischen Wissenschaften 351,
DOI 10.1007/978-3-319-24777-9_9

9.1.1 General strategies

The key to success is the variational representation of capacities through the Dirichlet principle stated in Theorem 7.33. The Dirichlet principle immediately yields two avenues towards bounds:

(i) Upper bounds via judiciously chosen test potentials.
(ii) Lower bounds via monotonicity of the Dirichlet form in the transition probabilities.

These two avenues are well-known and give rise to what are called "Rayleigh's short-cut rules" in the language of electric networks (see e.g. Doyle and Snell [92] and references therein). In the context of metastable systems, their usefulness can be enhanced by an iterative method. Moreover, the lower bound can be tackled with the help of the Thomson principle or the Berman-Konsowa principle, of which the latter will turn out to be especially powerful.

The key idea of iteration is to get control of the minimiser in the Dirichlet principle, i.e., the equilibrium potential. In metastable systems, when we are interested in computing for instance the capacity $\mathrm{cap}(B_x, B_y)$ for a pair of metastable sets $B_x, B_y \subset S$, our initial goal will always be to identify domains where h_{B_x,B_y} is close to 0 or close to 1. This can be done with the help of the renewal estimate in Lemma 8.4. While this approach looks circular at first glance (namely, we need to know capacities to estimate equilibrium potentials, which we need to estimate capacities), it turns out to yields a tool to turn "poor" bounds into "less poor" bounds. Thus, the first step in the program is:

(i) Guess a test potential that produces a good upper bound based on proper intuition.
(ii) Simplify the state space to obtain a system that can be solved exactly for the lower bound. In many examples, this leads to a one-dimensional or quasi-one-dimensional system.
(iii) Insert the resulting bounds into (8.2.1) to obtain bounds on the harmonic function.

Using the bounds on the harmonic function, we can identify the sets

$$D_x = \left\{z \in S\colon h_{B_x,B_y}(z) < \delta\right\}, \quad \text{and} \quad D_y = \left\{z \in S\colon h_{B_x,B_y}(z) > 1-\delta\right\}, \tag{9.1.1}$$

for $0 < \delta \ll 1$ suitably chosen. If $S\backslash(D_x \cup D_y)$ contains no further metastable set, then we define

$$I = \left\{z \in S\backslash(D_x \cup D_y)\colon \mu(z) < \rho \sup_{w \in S\backslash(D_x \cup D_y)} \mu(w)\right\} \tag{9.1.2}$$

for $0 < \rho \ll 1$ suitably chosen, with μ the invariant measure on S. The idea is that the set I will be irrelevant for the value of the capacity, no matter what values h_{B_x,B_y} takes on I, and that the sets D_x and D_y give no contribution to the capacity to leading order. The only problem therefore is to find the equilibrium potential, or a reasonably good approximation of it, on the set $\mathscr{S} = S\backslash(D_x \cup D_y \cup I)$ (see Fig. 9.1). We return to this problem shortly.

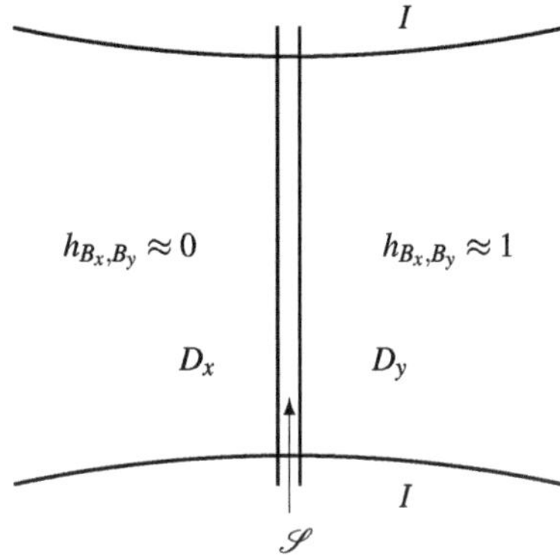

Fig. 9.1 Schematic picture of the harmonic function h_{B_x,B_y}: trivial on D_x and D_y, irrelevant on I, and nontrivial on $\mathscr{S}$. In many applications, $\mathscr{S}$ is small and well structured

Remark 9.1 Of course, the above approach can only make sense when the sets D_x and D_y are connected through $\mathscr{S}$. If that is not the case, then we will have to analyse the set $S\backslash(D_x \cup D_y)$ more carefully. If $D_x \cup D_y$ contains further metastable sets B_w, then it will be possible to identify domains $D_w \supset B_w$ on which h_{B_x,B_y} takes a constant value c_w (to be determined later). Note that this can be done again with the help of the renewal bounds we encountered in Chap. 8. The starting point is the observation that

$$\begin{aligned} h_{B_x,B_y}(z) &= \mathbb{P}_z(\tau_{B_x} < \tau_{B_y}) \\ &= \mathbb{P}_z(\tau_{B_x} < \tau_{B_y}, \tau_{B_w} < \tau_{B_x}) + \mathbb{P}_z(\tau_{B_x} < \tau_{B_y}, \tau_{B_w} \geq \tau_{B_x}) \\ &\sim \mathbb{P}_z(\tau_{B_w} < \tau_{B_x \cup B_y}) \mathbb{P}_w(\tau_{B_x} < \tau_{B_y}) \\ &= \mathbb{P}_z(\tau_{B_w} < \tau_{B_x \cup B_y}) c_w, \end{aligned} \tag{9.1.3}$$

where we anticipate that $\mathbb{P}_w(\tau_{B_x} < \tau_{B_y}) \sim \mathbb{P}_z(\tau_{B_x} < \tau_{B_y}) = c_w$ for all $z \in B_w$ (and later for all $z \in D_w$). Thus, the problem (to be solved with the help of the a priori bounds on the equilibrium potential in Lemma 8.4 and the a priori bounds on capacities in Lemma 8.5) is to determine the set of points z for which $\mathbb{P}_z(\tau_w < \tau_{B_x \cup B_y}) > 1 - \delta$. Once this is done, we proceed as before, after increasing the set $D_x \cup D_y$ in the definition of I to $D = D_x \cup D_y \cup D_{w_1} \cup \cdots \cup D_{w_k}$ when k such sets can be identified. It should then be the case that the set $D \cup \mathscr{S}$ is connected. The remaining problem consists in determining the equilibrium potential on the set $\mathscr{S}$ and the values $c_{w_1}, \ldots, c_{w_k}$. At this stage we can obtain upper and lower bounds in terms of variational formulas that involve only the set $\mathscr{S}$. To what extent these variational formulas can be solved depends on the situation at hand.

To obtain an upper bound, we choose a test function h^+ with the properties

$$\begin{aligned} h^+(z) &= 1, & z &\in D_x, \\ h^+(z) &= 0, & z &\in D_y, \\ h^+(z) &= c_{w_i}, & z &\in D_{w_i},\ i = 1, \ldots, k, \end{aligned} \tag{9.1.4}$$

where the constants c_{w_i} are determined later. On I the function h^+ can be chosen essentially arbitrarily, while on $\mathscr{S}$ it must be chosen such that it optimises the restriction of the Dirichlet form to $\mathscr{S}$ with boundary conditions implied by (9.1.4). Finally, the constants c_{w_i}, $i = 1, \dots, k$, are determined by minimising the outcome as a function of these constants.

A first strategy to obtain matching lower bounds that works well in many situations goes as follows. If h^* denotes the true minimiser in the Dirichlet form on S, then

$$\mathscr{E}(h^*, h^*) \geq \mathscr{E}_{\mathscr{S}}(h^*, h^*), \tag{9.1.5}$$

where $\mathscr{E}_{\mathscr{S}}$ is the restriction of the Dirichlet form to $\mathscr{S}$, i.e.,

$$\mathscr{E}_{\mathscr{S}}(h, h) = \tfrac{1}{2} \sum_{x \in \mathscr{S} \text{ and/or } y \in \mathscr{S}} \mu(x) p(x, y) \big[h(x) - h(y)\big]^2, \tag{9.1.6}$$

where for the sake of exposition we focus on the case of countable S. We minorise $\mathscr{E}_{\mathscr{S}}(h^*, h^*)$ by taking the infimum over all h on $\mathscr{S}$, with boundary conditions imposed by what we know a priori about the equilibrium potential. In particular, we know that these boundary conditions are close to constants on the different components of D. Of course, we do not know the constants c_{w_i}, $i = 1, \dots, k$, but by minimising the result over their possible values at the end we get a lower bound.

Thus, if we can show that the minimisers in the lower bound differ little from the minimisers with constant boundary conditions, then we get upper and lower bounds that coincide up to small error terms. In general it may be difficult to compute these minimisers. However, *in metastable systems typically the problem reduces in complexity* compared to the original problem, and in many instances can be solved explicitly.

9.1.2 Lower bounds via flows

The method to get lower bounds described above leaves open the problem of how to obtain a lower bound for the reduced Dirichlet form $\mathscr{E}_{\mathscr{S}}$. Sometimes this can be done by setting sufficiently many transition probabilities $p(x, y)$ equal to 0 until the remaining Dirichlet problem can be solved exactly by hand (for instance, because it corresponds to the one-dimensional chains discussed in Sects. 7.1 and 7.2).

A more versatile and systematic tool is provided by the variational principles of Thomson and Berman-Konsowa presented in Sect. 7.3. The latter appears to be particularly suitable, and we describe now the strategies used in exploiting it.

The Berman-Konsowa principle yields a lower bound whenever we insert some flow. However, guessing a flow is much harder than guessing a potential, due to the local constraints imposed by Kirchhoff's law. In principle we would like to guess a good approximation of the harmonic flow, but this is not easy in practice. A natural idea is to use the approximate harmonic function that was guessed in the derivation of the upper bound to produce an approximation of the harmonic flow. But since an approximate harmonic function is not harmonic, it is not straightforward how to get a flow out of it. It is useful to inspect the proof of the Berman-Konsowa principle in Theorem 7.43 to see what flexibility we have in playing with the test flow. The flow property is used only in the proof of Lemma 7.41, more precisely, in the derivation of (7.3.28). We therefore define the notion of a *defective flow*.

Definition 9.2 (Defective loop-free unit flow) Let $\Gamma = (S, E)$ be a graph with vertex set S and edge set E. Let $A, B \subset S$ be non-empty and disjoint. A map $f\colon E \to \mathbb{R}$ is called a *defective loop-free unit flow* from A to B if:

(i) There exists a defect function $\delta\colon S \to \mathbb{R}$ such that

$$\sum_{\substack{y\in S:\\(y,x)\in E}} f\big((y,x)\big) = \sum_{\substack{z\in S:\\(x,z)\in E}} f\big((x,z)\big) + \delta(x), \quad x \in S\backslash(A\cup B). \tag{9.1.7}$$

(ii) The total flow out of A is equal to 1, i.e.,

$$\sum_{x\in A}\sum_{\substack{z\in S:\\(x,z)\in E}} f\big((x,z)\big) = 1. \tag{9.1.8}$$

(iii) Any path γ from A to B such that $f(e) > 0$ for all $e \in \gamma$ is self-avoiding. In particular, if $f((x, y)) > 0$, then $f((y, x)) = 0$.

Given a defective loop-free unit flow f, we can construct a Markov chain with transition rates

$$q^f(x,y) = \frac{f((x,y))}{\mathscr{F}(x)}, \tag{9.1.9}$$

where $\mathscr{F}(x) = \sum_{y\in S} f((x, y))$, and with initial distribution

$$\mathbb{P}^f(X_0 = x) = \mathscr{F}(x)\mathbb{1}_{x\in A}. \tag{9.1.10}$$

Here, without loss of generality, we assume that $\mathscr{F}(x) > 0$ for all $x \in S$. We denote the law of this Markov chain by $\mathbb{P}^f$, and define the sets A_n, $n \in \mathbb{N}_0$, with $A_0 = A$, as in (7.3.31) in Sect. 7.3. Recall that $n^*(z)$ is the unique value of n such that $z \in A_n$. The elementary estimate that results is the following.

Lemma 9.3 *Let f be a defective loop-free unit flow from A to B. Then*

$$\mathbb{P}^f(\tau_z < \tau_B) \le \prod_{k=1}^{n^*(z)} \left(1 + \max_{y\in A_k} \frac{\delta(y)}{\mathscr{F}(y)}\right)_+ \mathscr{F}(z), \quad z \in S, \tag{9.1.11}$$

where $a_+ = \max(a, 1)$.

Proof The proof of this estimate proceeds by induction on n in exactly the same way as the proof of Lemma 7.41. We know that (9.1.11) holds for all $z \in A_0$ because $\mathscr{F} = 1$ on A_0. We assume that (9.1.11) holds for all $z \in A_k$ with $0 \le k \le n$. Then the recursion in (7.3.33) yields, for $z \in A_{n+1}$,

$$\begin{aligned} \mathbb{P}^f(\tau_z < \tau_B) &= \sum_{y \in A_0 \cup \cdots \cup A_n} \mathbb{P}^f(\tau_y < \tau_B) q^f\big((y,z)\big) \\ &\le \sum_{k=0}^{n} \prod_{\ell=1}^{k-1} \left(1 + \max_{y \in A_\ell} \frac{\delta(y)}{\mathscr{F}(y)}\right)_+ \sum_{y \in A_k} f\big((y,z)\big) \\ &\le \prod_{k=1}^{n-1} \left(1 + \max_{y \in A_k} \frac{\delta(y)}{\mathscr{F}(y)}\right)_+ \sum_{y \in S} f\big((y,z)\big) \\ &\le \prod_{k=1}^{n} \left(1 + \max_{y \in A_k} \frac{\delta(y)}{\mathscr{F}(y)}\right)_+ \mathscr{F}(z), \end{aligned} \tag{9.1.12}$$

where the first inequality uses (9.1.9) and the induction hypothesis, and the last inequality uses (9.1.7). □

Recalling (7.3.29) and using (9.1.9) and (9.1.11), we have

$$\mathbb{P}^f\big((x,y) \in \gamma\big) \le \prod_{k=1}^{n^*(x)-1} \left(1 + \max_{y \in A_k} \frac{\delta(y)}{\mathscr{F}(y)}\right)_+ f\big((x,y)\big), \tag{9.1.13}$$

and hence

$$\mathbb{1}_{f((x,y))>0} \ge \frac{\mathbb{P}^f((x,y) \in \gamma)}{f((x,y))} \prod_{k=1}^{n^*(x)-1} \left(1 + \max_{y \in A_k} \frac{\delta(y)}{\mathscr{F}(y)}\right)_+^{-1}. \tag{9.1.14}$$

Inserting this lower bound into the Dirichlet form in (7.3.38), we get

$$\mathscr{E}(h,h) \ge \prod_{k=1}^{M} \left(1 + \max_{y \in A_k} \frac{\delta(y)}{\mathscr{F}(y)}\right)_+^{-1} \mathbb{E}^f\left[\sum_{(x,y) \in \gamma} \frac{\mu(x)p(x,y)}{f((x,y))} \big[h(x) - h(y)\big]^2\right] \tag{9.1.15}$$

with $M = \max_{x \in S} n^*(x)$. Taking the infimum over h, we get the following lower bound on the capacity (recall (7.3.39)).

Lemma 9.4 (Berman-Konsowa defective flow bound) *Let f be a defective loop-free unit flow from A to B with defect function δ. Then*

$$\operatorname{cap}(A,B) \ge \prod_{k=1}^{M} \left(1 + \max_{y \in A_k} \frac{\delta(y)}{\mathscr{F}(y)}\right)_+^{-1} \mathbb{E}^f\left[\left(\sum_{(x,y) \in \gamma} \frac{f((x,y))}{\mu(x)p(x,y)}\right)^{-1}\right]. \tag{9.1.16}$$

Situations where we may want to apply Lemma 9.4 arise when an approximate harmonic function has been guessed in the upper bound. Given such a function, say g, we may define

$$f\big((x,y)\big) = N(g)^{-1}\mu(x)p(x,y)\big[g(y)-g(x)\big]_+, \tag{9.1.17}$$

where $N(g)$ is a normalising constant that fixes the total outgoing flow from A to be 1, and we may write, using reversibility,

$$\begin{aligned}
\sum_{z\in S} f\big((z,w)\big) &= N(g)^{-1} \sum_{z\in S,(z,w)\in E, g(w)\geq g(z)} \mu(z)p(z,w)\big(g(w)-g(z)\big) \qquad (9.1.18)\\
&= N(g)^{-1} \sum_{z\in S,(z,w)\in E} \mu(w)p(w,z)\big(g(w)-g(z)\big)\\
&\quad + N(g)^{-1} \sum_{z\in S,(z,w)\in E, g(w)<g(z)} \mu(w)p(w,z)\big(g(z)-g(w)\big)\\
&= N(g)^{-1}\mu(w)(\mathscr{L}g)(w) + \mathscr{F}(w),
\end{aligned}$$

i.e., the defect function is given by $\delta(w) = N(g)^{-1}\mu(w)(\mathscr{L}g)(w)$. In Chap. 10 we will encounter a set-up where this strategy works nicely. In general, however, there is quite a bit of artistry involved in working out good test flows. We will see examples in Chaps. 15, 19 and 20.

9.2 Coarse-graining

One of the heuristic ideas in physics is that of *renormalisation* or *coarse-graining*. Here, the hope is that a system defined on the level of *microscopic* variables (such as particles or spins), can be effectively described by a lower-dimensional system that captures only the *mesoscopic* variables (such as block-densities or block-magnetisations). If we want to understand the behaviour of such systems at moderate temperatures, then a coarse-grained description becomes imperative, since on the microscopic level the competition between energy and entropy does not allow for a proper understanding of metastable states and their transition paths.

The paradigmatic example is the Curie-Weiss model (to be discussed in Chap. 13). This offers the first link between the dynamics of a spin system and the Kramers diffusion mentioned in Sect. 2.1.1. The microscopic model is a spin system with state space $\{-1,+1\}^N$ whose state $\sigma(t) = \{\sigma_i(t)\}_{i=1}^N$ at time t evolves as a Markov process with transition rates that depend only on the total magnetisation $m_N(\sigma(t)) = \frac{1}{N}\sum_{i=1}^N \sigma_i(t)$. The dynamics is chosen to be reversible with respect to a Gibbs measure $\mu_{N,\beta}(\sigma) = Z_N^{-1}\exp[-\beta H_N(\sigma)]$, where the Hamiltonian

$$H_N(\sigma) = -\tfrac{1}{2}Nm_N(\sigma)^2 - hNm_N(\sigma) \tag{9.2.1}$$

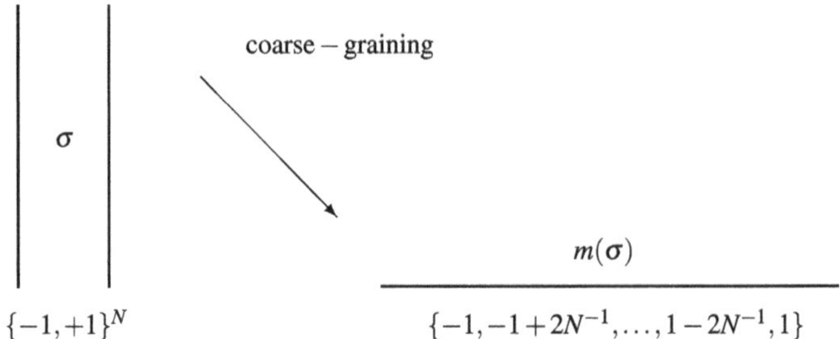

Fig. 9.2 Coarse-graining from spin configuration to magnetisation

is a function of the total magnetisation. In this case it is easy to verify that

$$m_N(t) = m_N\big(\sigma(t)\big) \tag{9.2.2}$$

is again a Markov process, this time on the much smaller state space $\{-1, -1 + 2N^{-1}, \dots, 1 - 2N^{-1}, 1\}$ (see Fig. 9.2), and is reversible with respect to the measure $\exp[-Nf_{\beta,N}(m)]$, where $f_{\beta,N}$ is a *double-well potential* for $\beta > 1$ and h small enough (for more details, see Fig. 13.1). This Markov process is a nearest-neighbour random walk that is attracted to the local minima of the function $f_{\beta,N}$ and behaves similarly to the Kramers diffusion. The key point here is that the effective inverse temperature in the *coarse-grained* model is of order N, i.e., entropic effects on the mesoscopic scale have become marginal compared to the original model, and can be ignored in the limit as $N \to \infty$.

We would like to consider a similar mapping down to mesoscopic variables in similar situations. This is called *lumping* in the theory of Markov chains, and is extensively discussed in the monograph by Kemeny and Snell [151]. Let us briefly state the main results. The technique is still on the level of an art, and will be illustrated in the example of the random-field Curie-Weiss model treated in Chaps. 14 and 15.

9.3 Lumping

Consider a Markov process $X = (X_t)_{t\in\mathbb{R}_+}$ on some state space S. Let T be some other state space, and let f be a map from S to T. Then $Y = (Y_t)_{t\in\mathbb{R}_+}$ with $Y_t = f(X_t)$ is again a stochastic process. Typically, Y is not Markov, but there are easy conditions under which it is (see Burke and Rosenblatt [43]).

Theorem 9.5 (Preservation of Markov property under lumping) *Let $\mathbb{P}$ be the law of a Markov process $X = (X_t)_{t\in\mathbb{R}_+}$ with state space $(S, \mathscr{B}(S))$ and stationary transition kernels P_t, $t \geq 0$. Let $\mathscr{F}_t = \sigma(X_s, 0 \leq s \leq t)$, $t \in \mathbb{R}_+$, be the σ-algebra generated by X up to time t. Let $(T, \mathscr{B}(T))$ be a measurable space and f a measurable*

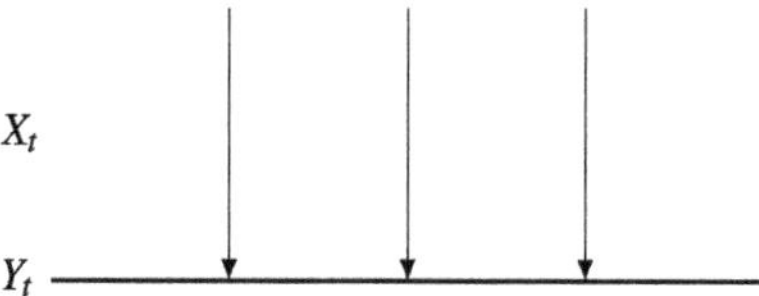

Fig. 9.3 Lumping: map with symmetry

map from S to T. Then $Y = (Y_t)_{t\in\mathbb{R}_+}$ with $Y_t = f(X_t)$ is a Markov process when for every $t \geq 0$ and $B \in \mathscr{B}(T)$ the maps

$$\mathbb{P}\big(f(X_t) \in B \mid \mathscr{F}_s\big), \quad 0 \leq s < t, \tag{9.3.1}$$

are measurable with respect to the σ-algebra generated by Y up to time s. In that case, the transition kernels R_t of the Markov process Y are given by

$$R_t(B, y) = P_t\big(f^{-1}(B), x\big), \quad B \in \mathbb{B}(T),\ y \in T, \quad \forall x \in S\colon\ f(x) = y. \tag{9.3.2}$$

In words, the image process Y is Markov when the original process X is Markov and has a high degree of symmetry (see Fig. 9.3).

In the case of a countable state space, the conditions of Theorem 9.5 can be restated as saying that, if $p(x, x')$, $x, x' \in S$, are the transition probabilities (or transition rates) of X, then for any $y, y' \in T$ and $x \in S$ such that $f(x) = y$ the formula

$$r\big(y, y'\big) = \sum_{\substack{x'\in S:\\ f(x')=y'}} p\big(x, x'\big), \tag{9.3.3}$$

is independent of the specific choice of x. Clearly, if X has invariant measure μ, then Y has invariant measure $\nu = \mu \circ f^{-1}$. Also, if X is reversible with respect to μ, then Y is reversible with respect to ν.

Remark 9.6 Note that we can alternatively define transition probabilities

$$r\big(y, y'\big) = \sum_{\substack{x\in S:\\ f(x)=y}} \frac{\mu(x)}{\nu(y)} \sum_{\substack{x'\in S:\\ f(x')=y'}} p\big(x, x'\big). \tag{9.3.4}$$

If p is reversible with respect to μ, then r is reversible with respect to ν, and we may hope that r generates a Markov chain that is a good approximation of Y.

The following theorem states some consequences of lumpability.

Theorem 9.7 (Consequences of lumpability) *Let X, Y and f be as in Theorem 9.5. Let $A, B \subset S$ be such that there exist $a, b \subset T$ such that $A = f^{-1}(a)$, $B = f^{-1}(b)$. Then*

(i)

$$h^X_{A,B}(x) = h^Y_{a,b}\big(f(x)\big), \quad x \in S, \tag{9.3.5}$$

where h^X, h^Y denote the equilibrium potentials with respect to X and Y.

(ii)

$$\mathrm{cap}_X(A,B) = \mathrm{cap}_Y(a,b), \tag{9.3.6}$$

where $\mathrm{cap}_X, \mathrm{cap}_Y$ *denote the capacities with respect to X and Y.*

Proof Property (i) is immediate from (9.3.1) in combination with the representations $h^X_{A,B}(x) = \mathbb{P}(\tau^X_A < \tau^X_B \mid X_0 = x)$, $x \in S$, and $h^Y_{a,b}(y) = \mathbb{P}(\tau^Y_a < \tau^Y_b \mid Y_0 = y)$, $y \in T$, where τ^X_A is the first hitting time of A for X and τ^Y_a is the first hitting time of a for Y. Property (ii) is immediate from (9.3.1) and the representations $\mathrm{cap}_X(A,B) = \mathscr{E}^X(h^X_{A,B}, h^X_{A,B})$ and $\mathrm{cap}_Y(a,b) = \mathscr{E}^Y(h^Y_{a,b}, h^Y_{a,b})$, where $\mathscr{E}^X, \mathscr{E}^Y$ are the Dirichlet forms associated with X, Y (recall Lemmas 7.12 and 7.26). □

9.4 Regularity estimates

For models with an uncountable state space, in order to carry over the general formalism discussed in Chap. 8 we need some a priori control on the behaviour of harmonic functions and other solutions of relevant Dirichlet problems. There are two methods that can work in different cases: elliptic regularity theory (Sect. 9.4.1) and coupling methods (Sect. 9.4.2).

9.4.1 Elliptic regularity theory

In the case of Markov processes with a state space that is a subset of $\mathbb{R}^d$ and with a generator given by (the closure of) an elliptic operator of the form (7.2.15), there is a well developed analytic theory that provides quantitative control on the regularity of solutions of homogeneous and inhomogeneous Dirichlet problems associated with these operators. The following two key lemmas are taken from Gilbarg and Trudinger [126, Corollaries 9.24–9.25], and concern second-order elliptic operators

$$\mathscr{L} = \sum_{i,j} a_{ij}(x)\frac{\partial^2}{\partial x_i \partial x_j} + \sum_i b_i(x)\frac{\partial}{\partial x_j} + d(x) \tag{9.4.1}$$

defined on some domain $\Omega \subset \mathbb{R}^d$, where $a_{ij} \in C^0(\Omega)$, $b_i, d \in L^\infty(\Omega)$. Assume that, for two numbers $0 < c \le C < \infty$,

$$C(\xi,\xi) \ge \big(\xi, a(x)\xi\big) \ge c(\xi,\xi) > 0 \quad \forall \xi \in \mathbb{R}^d. \tag{9.4.2}$$

Let $\gamma = C/c$, and choose ν such that $(\|b\|_\infty/c)^2 \leq \nu$ and $\|b\|_\infty \leq \nu$. For $n \in \mathbb{N}$, let $W^{2,n}(\Omega)$ denote the Sobolev spaces of twice (weakly) differentiable functions on Ω whose derivatives of order ≤ 2 are in $L^n(\Omega)$. Let $B_R(x)$ denote the ball of radius R centred at x.

Lemma 9.8 *If $u \in W^{2,n}(\Omega)$ is positive and satisfies $\mathscr{L}u = 0$ in Ω, then for any $x \in \Omega$ and $R > 0$ such that $B_{2R}(x) \subset \Omega$,*

$$\sup_{z \in B_R(x)} u(z) \leq C \inf_{z \in B_R(x)} u(z), \tag{9.4.3}$$

where $C = C(n, \gamma, \nu R^2) < \infty$.

Lemma 9.9 *If $u \in W^{2,n}(\Omega)$ is positive and satisfies $\mathscr{L}u = f$ in $B_{R_0}(x)$, then for any $0 < R \leq R_0$,*

$$\operatorname{osc}_{B_R(x)} u \leq C \left(\frac{R}{R_0}\right)^\alpha \bigl(\operatorname{osc}_{B_{R_0}(x)} u + R_0 \|f - cu\|_{n, B_{R_0}(x)}\bigr), \tag{9.4.4}$$

where $\operatorname{osc}_A u = \sup_A u - \inf_A u$, $\alpha = \alpha(n, \gamma, \nu R_0^2) > 0$ and $C = C(n, \gamma, \nu R_0^2) < \infty$.

In the context of reversible diffusions with small noise, the term involving second derivatives is scaled by the small parameter ε, i.e., we deal with operators of the form

$$\begin{aligned} \mathscr{L}_\varepsilon &= \varepsilon \sum_{i,j} e^{F(x)/\varepsilon} \frac{\partial}{\partial x_i} a_{ij}(x)\, e^{-F(x)/\varepsilon} \frac{\partial}{\partial x_j} \\ &= \varepsilon \sum_{i,j} a_{ij}(x) \frac{\partial^2}{\partial x_i \partial x_j} + \sum_{i,j} \left(\varepsilon \frac{\partial a_{ij}(x)}{\partial x_i} - a_{ij}(x) \frac{\partial F(x)}{\partial x_i}\right) \frac{\partial}{\partial x_j}. \end{aligned} \tag{9.4.5}$$

This means, in particular, that the ellipticity constant scales with ε. The way we will use Lemmas 9.8–9.9 is to consider a family of domains depending on ε, chosen in such a way that the numerical constants C and α are independent of ε. For the operator $\mathscr{L}_\varepsilon$, both c and C are proportional to ε, $\gamma = O(1)$, and we can choose $\nu \sim \varepsilon^{-2} \sup_{y \in \Omega} \|\nabla F(y)\|_\infty^2$.

An important application of the regularity estimates in Lemmas 9.8–9.9 is to obtain bounds on harmonic functions. The basic tool used throughout Chap. 8 for countable state spaces was Lemma 8.4, which was based on a simple renewal argument contained in the *renewal equation* (recall (8.2.2))

$$\mathbb{P}_x(\tau_A < \tau_B) = \frac{\mathbb{P}_x(\tau_A < \tau_{B \cup x})}{\mathbb{P}_x(\tau_{A \cup B} < \tau_x)}. \tag{9.4.6}$$

While this formula remains true in the diffusion setting, it is useless for $d > 1$ because the denominator equals 1 and so the numerator equals the left-hand side. Fortunately, it is easy to obtain a useful analogue of (9.4.6) by purely analytic considerations, contained in the following theorem for operators of the form (9.4.5).

Theorem 9.10 (Upper bound on harmonic function) *Let $A, B \subset \mathbb{R}^d$ be disjoint closed sets whose complement is regular, and let $x \in (A \cup B)^c$ be such that* $\mathrm{dist}(x, A \cup B) > c\varepsilon$. *Then, for any $\rho \leq c\varepsilon$, with $c \in \mathbb{R}_+$, there exists a $C \in (0, \infty)$ (depending only on c and on the value of $\|\nabla F(x)\|_\infty$) such that*

$$h_{A,B}(x) \leq C \frac{\mathrm{cap}(B_\rho(x), A)}{\mathrm{cap}(B_\rho(x), B)}. \tag{9.4.7}$$

Proof We begin by proving the following lemma.

Lemma 9.11 *With the notation of Theorem* 9.10,

$$\begin{aligned} &\sup_{z \in \partial B_\rho(x)} G_{(A\cup B)^c}(z, x)\, \mathrm{e}^{F(x)/\varepsilon} \int_{\partial B_\rho(x)} \mathrm{e}^{-F(y)/\varepsilon} e_{B \cup B_\rho(x), A}(dy), \\ &\geq h_{A,B}(x) \\ &\geq \inf_{z \in \partial B_\rho(x)} G_{(A\cup B)^c}(z, x)\, \mathrm{e}^{F(x)/\varepsilon} \int_{\partial B_\rho(x)} \mathrm{e}^{-F(y)/\varepsilon} e_{B \cup B_\rho(x), A}(dy), \end{aligned} \tag{9.4.8}$$

where $e_{B \cup B_\rho(x), A}$ is the equilibrium measure defined in (7.2.56).

Proof Let $\Omega \subset \mathbb{R}^d$ be a regular domain, and let f be a function defined on $\partial\Omega$. Recall that the Poisson kernel $H_\Omega = H_\Omega^{\lambda=0}$ defined in (7.2.6) maps a function f defined on $\partial\Omega$ to a harmonic function on Ω, called its *harmonic extension*. Choosing $\Omega = (A \cup B)^c$, we see that the equilibrium potential $h_{A,B}$ satisfies the mean-value property

$$h_{A,B}(x) = (H_{(A\cup B)^c} h_{A,B})(x). \tag{9.4.9}$$

Let $C \subset (A \cup B)^c$ be a regular neighbourhood of x. Since $h_{A,B\cup C}$ and $h_{A,B}$ coincide on $\partial(A \cup B)$, it is obvious that

$$h_{A,B}(z) = (H_{(A\cup B)^c} h_{A,B\cup C})(z) \quad \forall z \in (A \cup B \cup C)^c. \tag{9.4.10}$$

Using the first Green identity in (7.2.49) for $\Omega = \Gamma = (A \cup B \cup C)^c$, $f = G_{(A\cup B)^c}(x, \cdot)$ and $g = h_{A,B\cup C}$, we get

$$\begin{aligned} &(H_{(A\cup B)^c} h_{A,B\cup C})(x) \\ &= -\varepsilon \int_{\partial(A\cup B)} \mathrm{e}^{[F(x)-F(y)]/\varepsilon} h_{A,B\cup C}(y)\, \partial_{n(y)}\, G_{(A\cup B)^c}(y, x)\, d\sigma_{A\cup B}(y) \\ &= -\varepsilon \int_{\partial C} \mathrm{e}^{[F(x)-F(y)]/\varepsilon} G_{(A\cup B)^c}(y, x)\, \partial_{n(y)}\, h_{A,B\cup C}(y)\, d\sigma_C(y) \\ &= -\int_{\partial C} \mathrm{e}^{[F(x)-F(y)]/\varepsilon} G_{(A\cup B)^c}(y, x)\, e_{A,B\cup C}(dy), \end{aligned} \tag{9.4.11}$$

where $\partial_{n(y)}$ is the normal derivative defined in (7.2.51). We use that $h_{A,B\cup C}(y)=0$ when $y\in\partial C$ and $G_{A\cup B}(y,x)=0$ when $x\in\partial(A\cup B)$. The last equality follows from (7.2.56). (Note that the factor ε appears because the definition of the normal derivative does not include the factor ε.)

Now choose $C=B_\rho(x)$. If we could replace $G_{(A\cup B)^c}(y,x)$ by a constant for $y\in\partial B_\rho(x)$, then we could extract this constant from the integral, and the remaining integral would be some partial capacity. In fact, on a countable state space instead of the ball $B_\rho(x)$ we could choose the point x, in which case the problem would be absent and we would readily get (9.4.6). In the present setting, by combining (9.4.10)–(9.4.11), we still get two bounds, namely,

$$\begin{aligned} h_{A,B}(x) &\geq -\sup_{z\in\partial B_\rho(x)} G_{(A\cup B)^c}(z,x)\,\mathrm{e}^{F(x)/\varepsilon}\int_{\partial B_\rho(x)} \mathrm{e}^{-F(y)/\varepsilon} e_{A,B\cup B_\rho(x)}(dy), \\ h_{A,B}(x) &\leq -\inf_{z\in\partial B_\rho(x)} G_{(A\cup B)^c}(z,x)\,\mathrm{e}^{F(x)/\varepsilon}\int_{\partial B_\rho(x)} \mathrm{e}^{-F(y)/\varepsilon} e_{A,B\cup B_\rho(x)}(dy). \end{aligned} \tag{9.4.12}$$

But, trivially, $-e_{A\cup B,B_\rho(x)}=e_{B_\rho(x),A\cup B}$, which implies (9.4.8). □

At this point it is clear that we need to be able to control the Green function near the diagonal. Before turning to estimates, we bring (9.4.12) into a more suitable form.

Lemma 9.12 *Within the setting of Lemma* 9.11,

$$h_{A,B}(x)\leq \sup_{z\in\partial B_\rho(x)} G_{(A\cup B)^c}(z,x)\,\mathrm{e}^{F(x)/\varepsilon}\operatorname{cap}\big(B_\rho(x),A\big). \tag{9.4.13}$$

Proof By the representation in (7.2.56), we have $e_{B\cup B_\rho(x),A}(dy)\leq e_{B_\rho(x),A}(dy)$. Hence, by (7.2.57),

$$\int_{\partial B_\rho(x)} \mathrm{e}^{-F(y)/\varepsilon} e_{B\cup B_\rho(x),A}(dy)\leq \int_{\partial B_\rho(x)} \mathrm{e}^{-F(x)/\varepsilon} e_{B_\rho(x),A}(dy)=\operatorname{cap}\big(B_\rho(x),A\big). \tag{9.4.14}$$

Thus, the upper bound in (9.4.8) implies the upper bound in (9.4.13). □

We want to express the Green function in Lemma 9.11 in terms of capacity. Using the symmetry of the Green function and the fundamental relation between the Green function, the equilibrium measure and the equilibrium potential in Theorem 7.28, we get

$$\begin{aligned} &\mathrm{e}^{F(x)/\varepsilon}\int_{\partial B_\rho(x)} \mathrm{e}^{-F(z)/\varepsilon} G_{(A\cup B)^c}(x,z) e_{B_\rho(x),A\cup B}(dz) \\ &\quad=\int_{\partial B_\rho(x)} G_{(A\cup B)^c}(z,x) e_{B_\rho(x),A\cup B}(dz) \\ &\quad=h_{B_\rho(x),A\cup B}(x)=1. \end{aligned} \tag{9.4.15}$$

This implies that

$$\begin{aligned}1 &\geq e^{F(x)/\varepsilon}\Big(\inf_{z\in B_\rho(x)} G_{(A\cup B)^c}(x,z)\Big)\int_{\partial B_\rho(x)} e^{-F(z)/\varepsilon}\, e_{B_\rho(x),A\cup B}(dz)\\ &= e^{F(x)/\varepsilon}\Big(\inf_{z\in B_\rho(x)} G_{(A\cup B)^c}(x,z)\Big)\mathrm{cap}\big(B_\rho(x),A\cup B\big),\end{aligned} \tag{9.4.16}$$

i.e.,

$$e^{F(x)/\varepsilon}\inf_{z\in B_\rho(x)} G_{(A\cup B)^c}(x,z) \leq \frac{1}{\mathrm{cap}(B_\rho(x),A\cup B)}. \tag{9.4.17}$$

It is clear at this point that we cannot continue unless we can compare the infimum and the supremum of $G_{(A\cup B)^c}(z,x)$ with $z\in B_\rho(x)$. Such a comparison is provided by the *Harnack inequalities*.

Lemma 9.13 (Harnack inequality for the Green function) *If $\rho = c\varepsilon$ for some $c<\infty$, then there exists a constant C, depending on c only, such that*

$$\sup_{z\in B_\rho(x)} G_{(A\cup B)^c}(z,x) \leq C \inf_{z\in B_\rho(x)} G_{(A\cup B)^c}(z,x), \quad x\in(A\cup B)^c. \tag{9.4.18}$$

Proof We will apply Lemma 9.8. If we choose $R\leq\varepsilon$, then we can use (9.4.3) with a constant that does not depend on ε. (If x is a quadratic critical point of F, then we can even choose $R=\varepsilon^{1/2}$.) Let $u(z)=G_{(A\cup B)^c}(z,x)$, $z\in B_\rho(x)$. Then u is harmonic in $(A\cup B)^c\backslash x$. Therefore, if $\rho>2R$, then u is harmonic in $B_{2R}(y)$ for every $y\in\partial B_\rho(x)$. Let $a,b\in\partial B_\rho(x)$ be such that $\sup_{z\in\partial B_\rho(x)} u(z)=u(a)$ and $\inf_{z\in\partial B_\rho(x)} u(z)=u(b)$. Then we can find k points $x_1,\dots,x_k\in\partial B_\rho(x)$, with $k\leq \pi\rho/R$, such that $x_1=a$, $b\in B_R(x_k)$ and $B_R(x_i)\cap B_R(x_{i+1})\neq\emptyset$. Clearly,

$$\begin{aligned}u(a) &\leq C\inf_{z\in B_R(a)} u(z) \leq C \inf_{z\in B_R(a)\cap B_R(x_2)} u(z) \leq C\sup_{z\in B_R(x_2)} u(z)\\ &\leq C^2 \inf_{z\in B_R(x_2)} u(z) \leq \cdots \leq C^{k-1}\sup_{z\in B_R(x_k)} u(z) \leq C^k \inf_{z\in B_R(x_k)} u(z) = u(b).\end{aligned} \tag{9.4.19}$$

Thus, $u(a)\leq C^{\rho/R}u(b)$, and so if $\rho=c\varepsilon$ and $R=\varepsilon$, then the supremum and the infimum are related by at most a finite ε-independent constant. □

Combining (9.4.16) with Lemmas 9.12–9.13, we arrive at the assertion in (9.4.7). □

9.4.2 Coupling methods

An alternative way to obtain regularity estimates that are suitable for metastable systems is via *coupling*. The basic idea is as follows. Take some function depending

Fig. 9.4 Coupling of two trajectories starting from two different initial values in a certain neighbourhood

on the initial value x of the process, for instance, the expected hitting time $x \mapsto \mathbb{E}_x[\tau_D]$ of a set D. We want to show that in a certain uniform sense this function is continuous in x on a certain neighbourhood. To do so, we start two copies of the process in two different initial values, say x and y, *coupled* in such a way as to favour convergence of the trajectories over time. If the trajectories meet before D is hit, then the processes realise the hit together. If this happens with large probability after a small time, then the difference between $\mathbb{E}_x[\tau_D]$ and $\mathbb{E}_y[\tau_D]$ must be small. See Fig. 9.4 for an illustration.

To exemplify this technique, we present its application in finite- and infinite-dimensional diffusion processes first used by Martinelli et al. [174–177]. To that end we place ourselves in the setting of the SDEs and SPDEs discussed in Sects. 5.6 and 5.7, respectively, 6.2 and 6.3, where the noise is scaled by a small parameter ε. As pointed out in Sect. 6.5, there may be several minima of the action functional that are *metastable states* in the sense of Freidlin-Wentzell theory. The theory therefore yields the exponential asymptotics of transition times between such states.

In the potential-theoretic approach, Corollary 7.30 gives us a formula for mean transition times when the system is started in a specific initial distribution on a ball around a metastable state, namely, the last-exit biased distribution. The desired result, however, is that the mean transition times do not really depend on the initial distribution, i.e., are essentially the same no matter where in the ball the system starts.

We limit our discussion to the reversible setting, although the results of Martinelli et al. apply also to non-reversible processes. The main conditions formulated below guarantee that the deterministic dynamics is attractive in the proper sense. If we are looking at an S(P)DE of the form

$$dX_t = -\nabla F(X_t)\,dt + \sqrt{2\varepsilon}\,dB_t, \tag{9.4.20}$$

(where X is a finite-dimensional vector or a function, and B is d-dimensional Brownian motion or the derivative of a Brownian sheet), then we assume that

(i) F is a Morse function with finitely many local minima, and the Hessian matrix of F is non-degenerate at all critical points.
(ii) For some $R < \infty$ and all $x \notin B_R(0)$, the gradient $\nabla F(x)$ is *pointing inward*, i.e., $(\nabla F(x), n(x)) \leq b < 0$, where $n(x)$ is the normal vector at x.
(iii) The second derivatives of F are locally bounded.

(See Chap. 12 for more precise statements for SPDEs.)

The finitely many local minima of F correspond to metastable states in the Freidlin-Wentzell theory. We denote this set by $\mathcal{M}$. For $m_\ell \in \mathcal{M}$ we denote by $\mathcal{M}_\ell \subset \mathcal{M}$ the set of local minima of F below $F(m_\ell)$.

Theorem 9.14 *For any $m_\ell \in \mathcal{M}$ there exist $\rho_0 > 0$, $\eta > 0$ and $\rho_0 > \rho > 0$ such that, for any $\delta > 0$ and for ε small enough,*

$$\sup_{\|z-m_\ell\|_\infty<\rho_0} \frac{|\mathbb{E}_{m_\ell}[\tau(\mathcal{B}_\delta(\mathcal{M}_\ell))] - \mathbb{E}_z[\tau(\mathcal{B}_\delta(\mathcal{M}_\ell))]|}{\mathbb{E}_{m_\ell}[\tau(\mathcal{B}_\delta(\mathcal{M}_\ell))]} \leq \mathrm{e}^{-\eta/\varepsilon}. \tag{9.4.21}$$

Moreover, for any $m_\ell \in \mathcal{M}$,

$$\begin{aligned} \sup_{\|z-m_\ell\|_\infty<\rho_0} \Big| & \mathbb{P}_{m_\ell}\big[\tau\big(\mathcal{B}_\delta(m_\ell)\big) < \tau\big(\mathcal{B}_\delta(\mathcal{M}_\ell)\big)\big] \\ & - \mathbb{P}_z\big[\tau\big(\mathcal{B}_\delta(m_\ell)\big) < \tau\big(\mathcal{B}_\delta(\mathcal{M}_\ell)\big)\big]\Big| \leq \mathrm{e}^{-\eta/\varepsilon}. \end{aligned} \tag{9.4.22}$$

Remark 9.15 This result applies both for finite-dimensional diffusions and for SPDEs under the conditions stated in Sect. 5.7. For sequences of N-dimensional discretisations of SPDEs as described in Sect. 5.7, the corresponding estimates hold uniformly in N, as was shown in Barret [11].

The proof makes essential use of the attractive nature of the deterministic ($\varepsilon = 0$) equation. The key estimate is the following bound on solutions of (9.4.20). This holds both for SDEs and SPDEs.

Lemma 9.16 *Denote by X_z^ε the solution of* (9.4.20) *starting in z. Let m_ℓ be a minimum of F. Then there exist $k, C > 0$ and $\varepsilon_0, \rho_0 > 0$ such that, for $\varepsilon_0 > \varepsilon > 0$,*

$$\begin{aligned} \mathbb{P}\Big[\sup_{\|z-m_\ell\|_\infty<\rho_0} \big\|X_z^\varepsilon(t) - X_{m_\ell}(t)\big\|_\infty \leq \mathrm{e}^{-kt}\big\|X_z^\varepsilon(0) - X_{m_\ell}^\varepsilon(0)\big\|_\infty, \forall t > 0\Big] \\ \geq 1 - \mathrm{e}^{-C/\varepsilon}. \end{aligned} \tag{9.4.23}$$

The proof of this contraction result is tedious but relies on large deviation estimates only. These are used to show that solutions cannot spend a substantial fraction of time away from local minima. Two solutions driven by the same Brownian motion approach each other when they are in the neighbourhood of a minimum. Careful book-keeping yields the result. For details see, in particular, the elegant proof given in Martinelli, Sbano and Scoppola [176].

9.5 Bibliographical notes

1. The first reference to lumping appears to be Burke and Rosenblatt [43]. In this paper, a necessary and sufficient criterion is given for a function of a Markov process to be Markovian. Kemeny, Snell and Laurie [151] introduced the notion of lumping and of a lumpable Markov process. A systematic presentation of conditions for lumpability on terms of symmetries of the transition rates is given in Baake, Baake, Bovier and Klein [9]. Liggett [166] discusses lumping in connection with capacities. In the context of metastability, lumping was used heavily in Bovier, Eckhoff, Gayrard and Klein [33]. Sharp bounds through coarse-graining in non-lumpable models have been obtained in Bianchi, Bovier and Ioffe [24] for the random-field Curie-Weiss model, and by Slowik [219] for the Potts version of this model.

2. Coupling methods to prove regularity were used by Martinelli, Olivieri and Scoppola [174, 175], and in an improved form by Martinelli, Sbano and Scoppola [176], to prove exponential convergence of exit times for finite- and infinite-dimensional diffusions. They were applied to discretisations of SPDEs by Barret [11].

3. Coupling techniques were used to obtain similar bounds as in Theorem 9.14 for Glauber dynamics of the random-field Curie-Weiss model by Bianchi, Bovier and Ioffe in [25]. The coupling used was an extension of a coupling constructed for the Glauber dynamics of the Curie-Weiss model by Levin, Luczak and Peres in [164].

4. Bianchi and Gaudillière [26] consider families of finite-state Markov processes in a setting where the size of the state space tends to infinity. With the help of potential theory they are able to prove that metastable transition times depend only weakly on the starting distribution, provided this has a support that lies in a small neighbourhood of a metastable state. An immediate consequence of their result is that metastable transition times are asymptotically exponential. The key idea is to use Corollary 7.11, not for S, but for $S \cup (A' \cup B')$ with A', B' copies of A, B, and to allow transitions $A \leftrightarrow A'$ and $B \leftrightarrow B'$ between mirror sites at non-zero rates. The effect of this extension is that the Markov process can start in A', move to A, run around A for awhile so as to approach a quasi-stationary distribution on A before exiting A, then move to B, run around B for awhile so as to approach a quasi-stationary distribution on B before exiting B, and finally enter B'. It is shown that, under certain conditions, the mean metastable transition time from A to B starting from the quasi-stationary distribution on A is close to the mean metastable transition time from A' to B' starting from the last-exit based distribution on A'. For the latter the formula in Corollary 7.11 is available and the techniques outlined in Chaps. 8–9 can be used.

Part IV
Applications: Diffusions with Small Noise

Parts IV–VIII bring the general theory outlined in Part III to bear on a number of selected examples.

In Part IV we study diffusions with small noise. Chapter 10 deals with diffusions on a lattice with a vanishing spacing. Chapter 11 looks at finite-dimensional diffusions on subsets of $\mathbb{R}^d$ and sharpens the results of Freidlin and Wentzell by using the potential-theoretic tools introduced in Part III. Chapter 12 looks at stochastic partial differential equations, which are the infinite-dimensional analogues of the diffusions dealt with in Chap. 11.

Chapter 10
Discrete Reversible Diffusions

I noticed an unlighted cigar and an open box of cigar-lights: all things betokened that the Doctor, usually so methodical and so self-contained, had been trying every form of occupation and could settle to none.

(Lewis Carroll, Sylvie and Bruno Concluded)

One of the simplest settings in which the general theory of metastability outlined in Part III can be applied is that of *discrete diffusions*. By this we understand discrete-time or continuous-time (nearest-neighbour) random walks on d-dimensional lattices of spacing $\varepsilon > 0$ subject to a drift field derived from a potential F that may have several local minima. One of the motivations for studying discrete reversible diffusions is that they appear as coarse-grained versions of mean-field spin systems. The results of this chapter will be used in Part V.

In Sect. 10.1 we define the setting and state the necessary assumptions on the potential. In Sects. 10.2 and 10.3 we derive upper and lower bounds on the relevant capacities.

10.1 Definitions

For simplicity, we focus on the discrete-time setting (the extension to continuous time is trivial). The transitions take place in subsets S_ε of the lattice $(\varepsilon\mathbb{Z})^d$, $\varepsilon > 0$. In particular, we assume that $S_\varepsilon = \Omega \cap (\varepsilon\mathbb{Z})^d$ for some fixed connected open set $\Omega \subset \mathbb{R}^d$. We write $x \sim y$ when x and y are nearest-neighbour sites in $(\varepsilon\mathbb{Z})^d$.

We denote by X^ε the time-homogeneous nearest-neighbour random walk on S_ε with transition matrix $p_\varepsilon(x, y)$, $x, y \in S_\varepsilon$, and assume reversibility with respect to a probability distribution μ_ε, i.e., $\mu_\varepsilon(x)p_\varepsilon(x, y) = \mu_\varepsilon(y)p_\varepsilon(y, x)$ for all $x, y \in S_\varepsilon$. We write L_ε for the generator of X^ε. To avoid notational complications, we assume that

$$\mu_\varepsilon(x) = \exp\left[-F(x)/\varepsilon\right] \tag{10.1.1}$$

for some $F : \mathbb{R}^d \to \mathbb{R}$. In many applications it is necessary to allow F to depend on ε as well, but this poses no additional difficulties. We ignore the issue of normalisation

A. Bovier, F. den Hollander, *Metastability*,
Grundlehren der mathematischen Wissenschaften 351,
DOI 10.1007/978-3-319-24777-9_10

of μ_ε, which is of no consequence for what follows. Metastability occurs when F has at least two local minima.

We also assume that the transition probabilities $p_\varepsilon(x,y)$ depend smoothly on x. A possible choice is

$$p_\varepsilon(x,y)=\begin{cases} r_\ell \mathrm{e}^{-[F(y)-F(x)]_+/\varepsilon}, & \text{if } y=x\pm\varepsilon \mathrm{e}_\ell,\ \ell=1,\dots,d,\\ 1-\Sigma(x), & \text{if } y=x,\\ 0, & \text{otherwise,}\end{cases} \tag{10.1.2}$$

where e_ℓ denotes the ℓ-th basis vector of the lattice $\mathbb{Z}^d$, $r_\ell\in(0,1)$ is an isotropy in the direction of e_ℓ such that $\sum_{l=1}^d r_\ell\le 1$, and $\Sigma(x)$ is chosen such that $\sum_{y\in S_\varepsilon} p_\varepsilon(x,y)=1$. We assume that:

Assumption 10.1 $F\in C^3(\Omega,\mathbb{R})$.

Definition 10.2 (Communication heights, communication level sets, optimal paths, gates) Given two non-empty subsets $A,B\subset\Omega$:

(a) The *communication height* between A and B is

$$\Phi(A,B)=\inf_{\substack{\gamma\in C([0,1],\Omega)\\ \gamma(0)\in A,\gamma(1)\in B}}\ \sup_{t\in[0,1]} F\big(\gamma(t)\big), \tag{10.1.3}$$

where the infimum runs over all continuous paths γ in Ω. The *communication level set* between A and B is

$$\mathscr{S}(A,B)=\big\{z\in\mathbb{R}^d:\ F(z)=\Phi(A,B)\big\}. \tag{10.1.4}$$

(b) The set of *optimal paths* from A to B is

$$(A\to B)_{\mathrm{opt}} =\Big\{\gamma\in C\big([0,1],\Omega\big):\ \gamma(0)\in A,\gamma(1)\in B,\ \sup_{t\in[0,1]} F\big(\gamma(t)\big)=\Phi(A,B)\Big\}. \tag{10.1.5}$$

(c) A subset $\mathscr{W}\subseteq\mathscr{S}(A,B)$ is a *gate* if it is a minimal subset with the property that all optimal paths intersect $\mathscr{W}$. A priori there may be several (not necessarily disjoint) gates. Their union is denoted by $\mathscr{G}(A,B)$ and is called the *essential gate*.

We make the following simplifying assumptions on F:

Assumption 10.3

(i) The set $\mathscr{M}$ of local minima of F is finite, and for all pairs $x,y\in\mathscr{M}$ there is a unique essential gate $\mathscr{G}(x,y)$ consisting of a finite collection of isolated *saddle points* $z_k^*(x,y)$, $k\in I(x,y)$, with $I(x,y)$ an index set.

(ii) At all local minima $x \in \mathcal{M}$ and all saddle points $z_k^*(x, y)$, $x, y \in \mathcal{M}$, $k \in I(x, y)$, the Hessian matrix of F, denoted by $\mathbb{A}(x)$ and $\mathbb{A}(z_k^*(x, y))$, is non-degenerate (i.e., has only non-zero eigenvalues).

Remark 10.4 Assumption 10.3 amounts to saying that F is a *Morse function*. Under this assumption, the saddle points $z_k^*(x, y)$ are the critical points where $\mathbb{A}(z_k^*(x, y))$ has exactly one negative eigenvalue.

We may encounter situations where saddle points in $\partial\Omega$ are relevant. While this does not necessarily lead to problems, there are many instances where the formulation of general results becomes somewhat cumbersome. In order to avoid these complications we exclusively deal with situations where $\partial\Omega$ is never reached by X^ε:

Assumption 10.5 $\lim_{i\to\infty} F(x_i) = \infty$ for any sequence of points $(x_i)_{i\in\mathbb{N}}$ in Ω such that $\lim_{i\to\infty} x_i = x \in \partial\Omega$.

In the setting described above, the general theory of metastability for Markov chains on countable state spaces described in Chap. 8 applies.

Theorem 10.6 (Metastable set) *Let $\mathcal{M}_\varepsilon \subset S_\varepsilon$ be a set of best lattice approximations of the points in $\mathcal{M}$. Then $\mathcal{M}_\varepsilon$ is a set of metastable points in the sense of Definition* 8.2, *with* $\rho = \exp(-c/\varepsilon)$ *for some* $c > 0$ *depending on* F.

Proof The proof of this fact is easy. As we will see later in full detail, if $x, y \in \mathcal{M}$, then $\mathrm{cap}(x, \mathcal{M}\backslash x)/\mu_\varepsilon(x) \sim \exp(\varepsilon^{-1}[F(x) - F(z^*(x, y))])$, which is of order $\exp(-C/\varepsilon)$ with $C > 0$. If $z \notin \mathcal{M}$, then there is a path $\gamma = (\gamma_0, \dots, \gamma_n)$ from z to $\mathcal{M}$ along which F is decreasing. As pointed out in Sect. 9.1.2, it follows that $\mathrm{cap}(z, \mathcal{M}) \geq \mathrm{cap}_\gamma(\gamma(0), \gamma(n))$, where cap_γ is the capacity of the Markov process in which all connections except those on γ are removed. The lower bound can be computed explicitly (see Sect. 7.1.4). This leads to the estimate $\mathrm{cap}(z, \mathcal{M})/\mu_\varepsilon(z) \geq O(\varepsilon^p)$ for some dimension-dependent $p < \infty$, which in turn implies that the conditions of Definition 8.2 are satisfied. □

Remark 10.7 Note that when S_ε is infinite, our assumptions on F imply that there exists a subset $S_{\varepsilon,0} = \Lambda \cap (\varepsilon\mathbb{Z})^d$, with Λ some finite box, satisfying the hypothesis on the subset S_0 mentioned in the remark following Definition 8.2. We just need to take Λ such that it contains all local minima of F and such that, outside Λ, F is large enough.

In view of Theorem 10.6, all that is required is to compute capacities between the single points in $\mathcal{M}_\varepsilon$. To further simplify the presentation, we assume that *all gates consist of a single saddle point*. The general case is obtained simply by adding up the contributions to the capacity coming from the different saddle points. Our goal is to apply Theorem 8.15, which expresses the metastable crossover times between the local minima in $\mathscr{M}_\varepsilon$ in terms of capacities and the invariant measure. Theorem 8.43 automatically yields the associated spectral estimates.

Let A, B be disjoint non-empty subsets of $\mathscr{M}_\varepsilon$ connected through a unique saddle point $z^*(A, B)$, i.e., A and B are contained in two different connected components of the level set $\{y \in S_\varepsilon\colon\ F(y) < \Phi(A, B)\}$. For $z^* \in \mathscr{M}_\varepsilon$, let $\mathbb{B}(z^*)$ be the matrix with elements

$$\mathbb{B}(z^*)_{\ell,k} = \sqrt{r_\ell}\,\mathbb{A}(z^*)_{\ell,k}\sqrt{r_k}, \tag{10.1.6}$$

and let $\hat{\gamma}_1(z^*)$ be the unique negative eigenvalue of $\mathbb{B}(z^*)$. For $x \in \mathscr{M}_\varepsilon$, write $\mathscr{M}_{\varepsilon,x} = \{y \in \mathscr{M}_\varepsilon\colon\ F(y) < F(x)\}$ and let $z^*(x, \mathscr{M}_{\varepsilon,x})$ be the unique saddle point connecting x and $\mathscr{M}_{\varepsilon,x}$.

Our main results in this chapter are the following.

Theorem 10.8 (Sharp asymptotics for capacities) *For A, B as above, as $\varepsilon \downarrow 0$,*

$$\operatorname{cap}(A, B) = \mathrm{e}^{-\Phi(A,B)/\varepsilon}\frac{\varepsilon[-\hat{\gamma}_1(z^*)]}{2\pi\sqrt{-\det\mathbb{A}(z^*)}}\left(\frac{2\pi}{\varepsilon}\right)^{d/2}\left(1 + O\left(\sqrt{\varepsilon\left[\ln(1/\varepsilon)\right]^3}\right)\right), \tag{10.1.7}$$

where $z^* = z^*(A, B)$.

Theorem 10.9 (Mean metastable exit time) *For every $x \in \mathscr{M}_\varepsilon$, as $\varepsilon \downarrow 0$,*

$$\mathbb{E}_x[\tau_{\mathscr{M}_{\varepsilon,x}}] = \mathrm{e}^{[\Phi(x,\mathscr{M}_{\varepsilon,x})-F(x)]/\varepsilon}\frac{2\pi}{\varepsilon[-\hat{\gamma}_1(z^*)]}\sqrt{\frac{-\det\mathbb{A}(z^*)}{\det\mathbb{A}(x)}}\left(1+O\left(\sqrt{\varepsilon\left[\ln(1/\varepsilon)\right]^3}\right)\right), \tag{10.1.8}$$

where $z^* = z^*(x, \mathscr{M}_{\varepsilon,x})$.

Theorem 10.10 (Link between spectrum and metastable exit times) *Suppose that there exists a $\theta > 0$ such that the elements of $\mathscr{M}_\varepsilon$ can be labeled in such a way that*

$$\Phi(x_k, \mathscr{M}_{k-1}) - F(x_k) \le \min_{1\le l<k}\left[\Phi(x_l, \mathscr{M}_k\backslash x_l) - F(x_l)\right] - \theta, \quad k = 2, \dots, |\mathscr{M}_\varepsilon|, \tag{10.1.9}$$

where $\mathscr{M}_k = \{x_1, \dots, x_k\}$, $k = 1, \dots, |\mathscr{M}_\varepsilon|$. As $\varepsilon \downarrow 0$,

$$\lambda_k = \frac{1}{\mathbb{E}_{x_k}[\tau_{\mathscr{M}_k}]}\left(1 + O\left(\mathrm{e}^{-\delta/\varepsilon}\right)\right), \quad k = 2, \dots, |\mathscr{M}_\varepsilon|, \tag{10.1.10}$$

for some $\delta = \delta(\theta) > 0$, where λ_k is the k-th eigenvalue of $-L_\varepsilon$ (in increasing order), and $\lambda_1 = 0$.

Theorem 10.11 (Exponential law of the metastable exit time) *Under the assumptions of Theorem 10.10, for $k = 1, \dots, |\mathscr{M}_\varepsilon|$,*

$$\lim_{\varepsilon\downarrow 0}\mathbb{P}_{x_k}\left(\tau_{\mathscr{M}_k}/\mathbb{E}_{x_k}[\tau_{\mathscr{M}_k}] > t\right) = \mathrm{e}^{-t}, \quad t \ge 0. \tag{10.1.11}$$

The proof of Theorem 10.8 is given in Sects. 10.2–10.3. Theorems 10.9–10.11 follow from Theorems 8.15, 8.43 and 8.45, together with Theorem 10.8 and the following lemma.

Lemma 10.12 *For every $x \in \mathscr{M}_\varepsilon$, as $\varepsilon \downarrow 0$,*

$$\mu_\varepsilon\big(A(x)\big) = e^{-F(x)/\varepsilon}\left(\frac{2\pi}{\varepsilon}\right)^{d/2}\frac{1}{\sqrt{\det \mathbb{A}(x)}}\Big(1+O\Big(\sqrt{\varepsilon\big[\ln(1/\varepsilon)\big]^3}\Big)\Big), \quad (10.1.12)$$

where $A(x)$ is the valley around x defined in (8.2.10).

Proof The lemma states that, up to the error given, the mass of $A(x)$ is the same as if the potential were of the form $F(y) = F(x) + \frac{1}{2}((y-x), \mathbb{A}(x)(y-x))$. In that case the discrete sum over lattice points with spacing ε is well approximated by the corresponding Gaussian integral. Furthermore, the contributions to both the Gaussian integral and the original sum coming from the region where $\|y-x\|_\infty \geq C\sqrt{\varepsilon\ln(1/\varepsilon)}$ are by a factor of order ε^C smaller than the main contribution and can be neglected. On the remaining set, by Taylor expansion,

$$\varepsilon^{-1}\big|F(y) - F(x) - \tfrac{1}{2}((y-x), \mathbb{A}(x)(y-x))\big| \leq C'\sqrt{\varepsilon\big[\ln(1/\varepsilon)\big]^3}. \quad (10.1.13)$$

This results in the error term in (10.1.12). More details can be found in the computation of capacities carried out in Sects. 10.2–10.3, which uses very similar approximations. □

10.2 Upper bounds on capacities

In this section we derive upper bounds on capacities between two local minima. For this we use the Dirichlet principle. We only need to produce a good test function.

We want to estimate

$$\mathrm{cap}(A,B) = \inf_{h\in\mathscr{H}_{A,B}} \mathscr{E}(h,h), \quad (10.2.1)$$

where the Dirichlet form is given by

$$\mathscr{E}(h,h) = \tfrac{1}{2}\sum_{x,y\in S_\varepsilon} e^{-F(x)/\varepsilon} p_\varepsilon(x,y)\big[h(x)-h(y)\big]^2, \quad (10.2.2)$$

and $\mathscr{H}_{A,B} = \{h\colon S_\varepsilon \to \mathbb{R}\colon \mathscr{E}(h,h) < \infty, h|_A \geq 1, h|_B \leq 0\}$. The general strategy to construct a test function is the following. We choose a strip W_0 of width $C\sqrt{\varepsilon\ln(1/\varepsilon)}$ with the following properties (see Fig. 10.1):

(i) The complement of W_0 in S_ε consists of two parts: W_1 containing A and W_2 containing B.

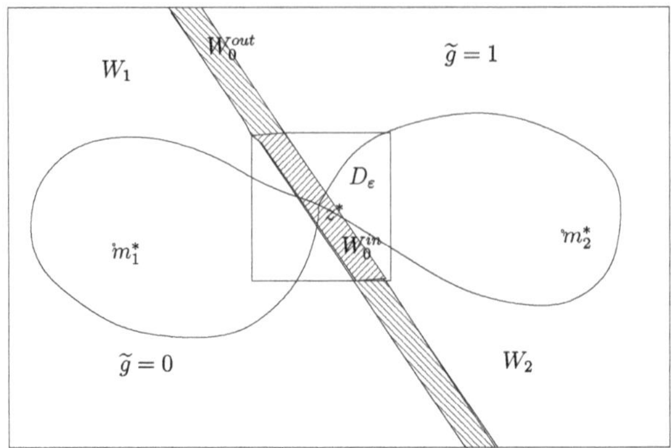

Fig. 10.1 Domains for the construction of the test function in (10.2.3), with $m_1^* \in A$ and $m_2^* \in B$

(ii) W_0 contains z^*, and for a cube D_ε of linear size $C'\sqrt{\varepsilon \ln(1/\varepsilon)}$ centered at z^*, with C' large enough, $W_0 \cap D_\varepsilon$ is contained in the set $\{x \in S_\varepsilon : F(x) > F(z^*) + c\varepsilon \ln(1/\varepsilon)\}$ for a suitably chosen $c > 1$.

A test function $\widetilde{g}$ is taken of the form

$$\widetilde{g}(x) = \begin{cases} 0, & \text{if } x \in W_1, \\ 1, & \text{if } x \in W_2, \\ g(x), & \text{if } x \in W_0 \cap D_\varepsilon = W_0^{in}, \\ 0, & \text{if } x \in W_0 \cap D_\varepsilon^c = W_0^{out}, \end{cases} \tag{10.2.3}$$

where $g(x)$ has to be chosen carefully as an approximately harmonic function. It should be clear that the non-negligible contributions to the Dirichlet form come from the region $W_0 \cap D_\varepsilon$. The advantage is that, within the small region D_ε, the Dirichlet form can be approximated by a simplified form for which a harmonic function is readily found.

The proof proceeds in three steps: cleaning of the Dirichlet form (Sect. 10.2.1); construction of approximate harmonic functions (Sect. 10.2.2); final estimate (Sect. 10.2.3).

10.2.1 Cleaning of the Dirichlet form

The following lemma provides a continuity property for capacities.

Lemma 10.13 *Let $\mathscr{E}, \widetilde{\mathscr{E}}$ be two Dirichlet forms defined for the same state space S_ε, corresponding to reversible measures $\mu_\varepsilon, \widetilde{\mu}_\varepsilon$ and transition matrices $p_\varepsilon, \tilde{p}_\varepsilon$,* *respectively. Assume that, for all* $x, x' \in S_\varepsilon$,

$$\left|\frac{\mu_\varepsilon(x)}{\widetilde{\mu}_\varepsilon(x)} - 1\right| \leq \delta, \qquad \left|\frac{p_\varepsilon(x,x')}{\widetilde{p}_\varepsilon(x,x')} - 1\right| \leq \delta. \tag{10.2.4}$$

Then for any disjoint and non-empty $A, B \subset S_\varepsilon$,

$$(1-\delta)^2 \leq \frac{\mathrm{cap}(A,B)}{\widetilde{\mathrm{cap}}(A,B)} \leq (1-\delta)^{-2}, \tag{10.2.5}$$

where $\mathrm{cap}(A,B) = \inf_{u\in\mathscr{H}_{A,B}} \mathscr{E}(u,u)$ *and* $\widetilde{\mathrm{cap}}(A,B) = \inf_{u\in\mathscr{H}_{A,B}} \widetilde{\mathscr{E}}(u,u)$.

Proof Note that there exists an h^* such that

$$\begin{aligned}
\mathrm{cap}(A,B) &= \mathscr{E}\big(h^*,h^*\big) \\
&= \tfrac{1}{2}\sum_{x,x'\in S_\varepsilon} \widetilde{\mu}_\varepsilon(x)\frac{\mu_\varepsilon(x)}{\widetilde{\mu}_\varepsilon(x)}\widetilde{p}_\varepsilon\big(x,x'\big)\frac{p_\varepsilon(x,x')}{\widetilde{p}_\varepsilon(x,x')}\big[h^*(x)-h^*\big(x'\big)\big]^2 \\
&\geq \tfrac{1}{2}\sum_{x,x'\in S_\varepsilon} \widetilde{\mu}_\varepsilon(x)(1-\delta)\widetilde{p}_\varepsilon\big(x,x'\big)(1-\delta)\big[h^*(x)-h^*\big(x'\big)\big]^2 \\
&\geq (1-\delta)^2 \inf_{u\in\mathscr{H}_{A,B}} \tfrac{1}{2}\sum_{x,x'\in S_\varepsilon} \widetilde{\mu}_\varepsilon(x)\widetilde{p}_\varepsilon\big(x,x'\big)\big[h(x)-h\big(x'\big)\big]^2 \\
&= (1-\delta)^2\,\widetilde{\mathrm{cap}}(A,B).
\end{aligned} \tag{10.2.6}$$

Reversing the rôles of $\mathscr{E}$ and $\widetilde{\mathscr{E}}$, we also get

$$\widetilde{\mathrm{cap}}(A,B) \geq (1-\delta)^2\,\mathrm{cap}(A,B), \tag{10.2.7}$$

and the claim follows. □

We use Lemma 10.13 for our Dirichlet forms restricted to neighbourhoods of the saddle point z^*. For $\rho = \rho(\varepsilon) = C\sqrt{\varepsilon \ln(1/\varepsilon)}$ with $C < \infty$, define

$$D_\varepsilon(\rho) = \big\{x \in S_\varepsilon\colon \big|z^*_\ell - x_\ell\big| \leq \rho \ \forall \ell = 1,\dots,d\big\}. \tag{10.2.8}$$

We need to control the transition probabilities $p_\varepsilon(x,x')$ and the reversible measure $\mu_\varepsilon(x) = \exp(-F(x)/\varepsilon)$ in terms of suitable modifications. Let $\mathbb{A}(z^*) = \mathbb{A}$ be the Hessian matrix of F at the saddle point z^*, and set

$$\tilde{\mu}_\varepsilon(x) = \exp\big[-\tfrac{1}{2}\big(x-z^*\big)\mathbb{A}\big(x-z^*\big)/\varepsilon\big]. \tag{10.2.9}$$

Then, by Taylor expansion, for some $K < \infty$,

$$\big|\mu_\varepsilon(x)/\tilde{\mu}_\varepsilon(x) - 1\big| \leq K\rho^3/\varepsilon, \quad x \in D_\varepsilon(\rho). \tag{10.2.10}$$

By (10.1.2) and Assumption 10.1, the transition probabilities are C^1-functions so that, for some $C'' < \infty$,

$$\left| \frac{p_\varepsilon(x, x + \varepsilon e_\ell)}{r_\ell} - 1 \right| \le C'' \sqrt{\varepsilon \ln(1/\varepsilon)}, \quad x \in D_\varepsilon(\rho). \tag{10.2.11}$$

With this in mind, we let $\widetilde{L}_\varepsilon$ be the generator of the dynamics on $D_\varepsilon(\rho)$ with transition probabilities $\tilde{r}(x, y)$ given by

$$\begin{aligned} \widetilde{r}(x, x + \varepsilon e_\ell) &= r_\ell, \\ \widetilde{r}(x + \varepsilon e_\ell, x) &= r_\ell \widetilde{\mu}_\varepsilon(x) / \widetilde{\mu}_\varepsilon(x + \varepsilon e_\ell). \end{aligned} \tag{10.2.12}$$

The fact that we choose the transition probabilities in the directions $+e_\ell$ to be constant is arbitrary. In the directions $-e_\ell$ we must choose them such that reversibility with respect to the modified reversible measure $\tilde{\mu}_\varepsilon$ is satisfied.

For $u \in \mathscr{H}_{A,B}$, we write the corresponding Dirichlet form as

$$\widetilde{\mathscr{E}}_{D_\varepsilon}(u, u) = \sum_{x \in D_\varepsilon(\rho)} \sum_{\ell=1}^{d} r_\ell \tilde{\mu}_\varepsilon(x) \big[u(x) - u(x + \varepsilon e_\ell) \big]^2, \tag{10.2.13}$$

where we note that $\tilde{\mu}_\varepsilon(z^*) = 1$.

10.2.2 Construction of an approximate harmonic function

In this section we construct a function that is almost harmonic with respect to the Dirichlet form $\widetilde{\mathscr{E}}_{D_\varepsilon}$.

Recall the matrix $\mathbb{B}(z^*) = \mathbb{B}$ defined in (10.1.6). Let $\hat{v}^{(i)}$, $i = 1, \dots, d$, be the normalized eigenvectors of $\mathbb{B}$, and $\hat{\gamma}_i$ the corresponding eigenvalues. Denote by $\hat{\gamma}_1$ the unique negative eigenvalue of $\mathbb{B}$. Define vectors $v^{(i)}$ by

$$v_\ell^{(i)} = \hat{v}_\ell^{(i)} / \sqrt{r_\ell}, \quad \ell = 1, \dots, d, \tag{10.2.14}$$

and vectors $\check{v}^{(i)}$ by

$$\check{v}_\ell^{(i)} = \hat{v}_\ell^{(i)} \sqrt{r_\ell} = r_\ell v_\ell^{(i)}, \quad \ell = 1, \dots, d. \tag{10.2.15}$$

The important fact about these vectors is that

$$\mathbb{A} \check{v}^{(i)} = \hat{\gamma}_i v^{(i)} \tag{10.2.16}$$

and

$$\big(\check{v}^{(i)}, v^{(j)} \big) = \delta_{ij}. \tag{10.2.17}$$

This implies the following non-orthogonal decomposition of the quadratic form $\mathbb{A}$:

$$(y, \mathbb{A}x) = \sum_{i=1}^{d} \hat{\gamma}_i \big(y, v^{(i)}\big)\big(x, v^{(i)}\big). \tag{10.2.18}$$

Define the function $f : \mathbb{R} \to [0, 1]$ by

$$f(a) = \frac{\int_{-\infty}^{a} e^{-|\hat{\gamma}_1|u^2/2\varepsilon} du}{\int_{-\infty}^{\infty} e^{-|\hat{\gamma}_1|u^2/2\varepsilon} du} = \sqrt{\frac{|\hat{\gamma}_1|}{2\pi\varepsilon}} \int_{-\infty}^{a} e^{-|\hat{\gamma}_1|u^2/2\varepsilon} du. \tag{10.2.19}$$

Finally, we single out the vectors $v = v^{(1)}$, $\check{v} = \check{v}^{(1)}$, $\hat{v} = \hat{v}^{(1)}$ and set

$$g(x) = f((v, x)), \tag{10.2.20}$$

which is our choice for the approximately harmonic function in the definition of $\widetilde{g}$ in (10.2.3). Note that $g(x)$ only varies in the direction of the vector v, and that it is close to 0 when $(v, x) \leq -\rho$, and close to 1 when $(v, x) \geq \rho$. Moreover, the following estimate holds.

Lemma 10.14 *Let g be as in* (10.2.20), *and let $\widetilde{L}_\varepsilon$ be the generator defined after* (10.2.11). *Then, for all $x \in D_\varepsilon(\rho)$, there exists a constant $c < \infty$ such that*

$$\big|(\widetilde{L}_\varepsilon g)(x)\big| \leq \left(\sqrt{\frac{\varepsilon|\hat{\gamma}_1|}{2\pi}} e^{-|\hat{\gamma}_1|(v,x)^2/2\varepsilon} \sum_{\ell=1}^{d} r_\ell v_\ell\right) O(\rho^2). \tag{10.2.21}$$

Proof We choose coordinates such that $z^* = 0$, and set $\mathbb{A} = \mathbb{A}(z^*)$. Using reversibility, we get

$$\begin{aligned} \widetilde{r}(x, x - \varepsilon e_\ell) &= \exp\big[-\tfrac{1}{2}\varepsilon^{-1}\big[(x, \mathbb{A}x) - \big((x - \varepsilon e_\ell), \mathbb{A}(x - \varepsilon e_\ell)\big)\big]\big] r_\ell \\ &= \exp\big[-(e_\ell, \mathbb{A}x)\big]\big[1 + O(\varepsilon)\big] r_\ell. \end{aligned} \tag{10.2.22}$$

Therefore

$$\begin{aligned} (\widetilde{L}_\varepsilon g)(x) &= \sum_{\ell=1}^{d} r_\ell \big[g(x + \varepsilon e_\ell) - g(x)\big] \\ &\quad \times \left(1 - \exp\big[-(e_\ell, \mathbb{A}x)\big] \frac{g(x) - g(x - \varepsilon e_\ell)}{g(x + \varepsilon e_\ell) - g(x)} \big[1 + O(\varepsilon)\big]\right). \end{aligned} \tag{10.2.23}$$

Next, we use the explicit form of g given in (10.2.20) to obtain, by Taylor expansion, that for some $\tilde{x} \in [x, x + \varepsilon e_\ell]$,

$$\begin{aligned} g(x+\varepsilon e_\ell)-g(x) &= f\big((v,x)+\varepsilon v_\ell\big)-f\big((v,x)\big) \\ &= v_\ell \varepsilon f'\big((v,x)\big)+\tfrac{1}{2}v_\ell^2\varepsilon^2 f''(v,x)+\tfrac{1}{6}v_\ell^3\varepsilon^3 f'''\big((v,\tilde{x})\big) \\ &= v_\ell\sqrt{\frac{\varepsilon|\hat{\gamma}_1|}{2\pi}}\mathrm{e}^{-|\hat{\gamma}_1|(v,x)^2/2\varepsilon}\big[1-v_\ell|\hat{\gamma}_1|(v,x)+O\big(\rho^2\big)\big], \end{aligned} \tag{10.2.24}$$

where we use (10.2.19) and $\rho = C\sqrt{\varepsilon \ln(1/\varepsilon)}$. In particular, we get that

$$\begin{aligned} \frac{g(x)-g(x-\varepsilon e_\ell)}{g(x+\varepsilon e_\ell)-g(x)} &= \exp\big[-|\hat{\gamma}_1|\big[(v,x-\varepsilon e_\ell)^2-(v,x)^2\big]/2\varepsilon\big] \\ &\quad\times\frac{1-v_\ell|\hat{\gamma}_1|[(v,x)-v_\ell\varepsilon]+O(\rho^2)}{1-v_\ell|\hat{\gamma}_1|(v,x)+O(\rho^2)} \\ &= \exp\big[-|\hat{\gamma}_1|v_\ell(v,x)\big]\left(1+\frac{v_\ell^2\varepsilon|\hat{\gamma}_1|+O(\rho^2)}{1-v_\ell|\hat{\gamma}_1|(v,x)+O(\rho^2)}\right) \\ &= \exp\big[-|\hat{\gamma}_1|v_\ell(v,x)\big]\big[1+O\big(\rho^2\big)\big]. \end{aligned} \tag{10.2.25}$$

Inserting these equations into (10.2.23), we get

$$\begin{aligned} (\widetilde{L}_\varepsilon g)(x) &= \sqrt{\frac{\varepsilon|\hat{\gamma}_1|}{2\pi}}\mathrm{e}^{-|\hat{\gamma}_1|(v,x)^2/2\varepsilon}\sum_{\ell=1}^{d} r_\ell v_\ell\big[1-v_\ell|\hat{\gamma}_1|(v,x)+O\big(\rho^2\big)\big] \\ &\quad\times\big(1-\exp\big[-(e_\ell,\mathbb{A}x)-|\hat{\gamma}_1|v_\ell(v,x)\big]\big[1+O\big(\rho^2\big)\big]\big). \end{aligned} \tag{10.2.26}$$

Now

$$\begin{aligned} &1-\exp\big[-(e_\ell,\mathbb{A}x)-|\hat{\gamma}_1|v_\ell(v,x)\big]\big[1+O\big(\rho^2\big)\big] \\ &\quad=(e_\ell,\mathbb{A}x)+|\hat{\gamma}_1|v_\ell(v,x)+O\big(\rho^2\big). \end{aligned} \tag{10.2.27}$$

Using this fact, and collecting the leading order terms, we get

$$\begin{aligned} (\widetilde{L}_\varepsilon g)(x) &= \sqrt{\frac{\varepsilon|\hat{\gamma}_1|}{2\pi}}\mathrm{e}^{-|\hat{\gamma}_1|(v,x)^2/2\varepsilon} \\ &\quad\times\sum_{\ell=1}^{d} r_\ell v_\ell\big\{\big[(e_\ell,\mathbb{A}x)+|\hat{\gamma}_1|v_\ell(v,x)\big]+O\big(\rho^2\big)\big\}. \end{aligned} \tag{10.2.28}$$

Thus, since $\hat{\gamma}_1 < 0$, we have proved the claim, provided

$$\sum_{\ell=1}^{d} r_\ell v_\ell\big[(e_\ell,\mathbb{A}x)-\hat{\gamma}_1 v_\ell(v,x)\big]=0. \tag{10.2.29}$$

But from (10.2.18) we obtain

$$(\mathrm{e}_\ell, \mathbb{A}x) - \hat{\gamma}_1 v_\ell(v, x) = \sum_{j=2}^{d} \hat{\gamma}_j v_\ell^{(j)}\big(v^{(j)}, x\big). \tag{10.2.30}$$

Hence, recalling that $r_\ell v_\ell = \breve{v}_\ell^{(1)}$ by (10.2.15) and that $\breve{v}^{(1)}$ is orthogonal to $v^{(j)}$ for $j \geq 2$ by (10.2.17), we see that (10.2.29) holds. □

10.2.3 Final estimate

Lemma 10.14 justifies the choice of g and will play an important rôle also in the derivation of the *lower bound* in Sect. 10.3. But first we state the *upper bound* that follows from it.

Proposition 10.15 *With the notation introduced above,*

$$\operatorname{cap}(A, B) \leq \mu_\varepsilon\big(z^*\big)\frac{|\hat{\gamma}_1 \varepsilon|}{2\pi}\left(\frac{2\pi}{\varepsilon}\right)^{d/2}\frac{1}{\sqrt{-\det \mathbb{A}(z^*)}}\Big(1 + O\Big(\sqrt{\varepsilon\big[\ln(1/\varepsilon)\big]^3}\Big)\Big). \tag{10.2.31}$$

Proof Return to Fig. 10.1. We first estimate the contribution of the set $D_\varepsilon \cap W_0$. By Lemma 10.13, this can be controlled in terms of the modified Dirichlet form $\widetilde{\mathscr{E}}_{D_\varepsilon}$ in (10.2.13). Thus, let g be the function defined in (10.2.20), and choose coordinates such that $z^* = 0$ and $F(z^*) = 0$. Then, by (10.2.9) and (10.2.24),

$$\begin{aligned}
\widetilde{\mathscr{E}}_{D_\varepsilon}(g, g) &= \sum_{x \in D_\varepsilon}\sum_{\ell=1}^{d} \mathrm{e}^{-(x,\mathbb{A}x)/2\varepsilon} r_\ell\big[g(x + \mathrm{e}_\ell) - g(x)\big]^2 \qquad (10.2.32)\\
&= \frac{|\hat{\gamma}_1|\varepsilon}{2\pi} \sum_{x \in D_\varepsilon} \mathrm{e}^{-|\hat{\gamma}_1|(v,x)^2/\varepsilon}\mathrm{e}^{-(x,\mathbb{A}x)/2\varepsilon}\sum_{\ell=1}^{d} r_\ell v_\ell^2\\
&\quad \times \big[1 - v_\ell|\hat{\gamma}_1|(v, x) + O\big(\rho^2\big)\big]^2\\
&= \frac{|\hat{\gamma}_1|\varepsilon}{2\pi}\big[1 + O\big(\sqrt{\varepsilon}\ln(1/\varepsilon)\big)\big] \sum_{x \in D_\varepsilon} \mathrm{e}^{-|\hat{\gamma}_1|(v,x)^2/\varepsilon}\mathrm{e}^{-(x,\mathbb{A}x)/2\varepsilon},
\end{aligned}$$

where we use that $\sum_{\ell=1}^{d} r_\ell v_\ell^2 = \sum_{\ell=1}^{d} \hat{v}_\ell^2 = 1$. It remains to compute the sum over x. Via a standard approximation of the sum by an integral we get, by (10.1.6), (10.2.14) and (10.2.19),

$$\begin{aligned}
&\sum_{x\in D_\varepsilon} \exp\big[-|\hat\gamma_1|(v,x)^2/\varepsilon-(x,\mathbb{A}x)/2\varepsilon\big]\\
&\quad=\varepsilon^{-d}\big[1+O(\rho)\big]\int_{D_\varepsilon} dx\,\exp\big[-|\hat\gamma_1|(v,x)^2/\varepsilon-(x,\mathbb{A}x)/2\varepsilon\big]\\
&\quad=\big[1+O(\rho)\big]\prod_{\ell=1}^{d}\frac{\sqrt{r_\ell}}{\varepsilon}\int_{\bar D_\varepsilon} dy\,\exp\big[-|\hat\gamma_1|(y,\hat v)^2/\varepsilon-(y,\mathbb{B}y)/2\varepsilon\big]\\
&\quad=\big[1+O(\rho)\big]\prod_{\ell=1}^{d}\frac{\sqrt{r_\ell}}{\varepsilon}\int_{\bar D_\varepsilon} dy\,\exp\left[-|\hat\gamma_1|(y,\hat v)^2/\varepsilon-\sum_{j=1}^{d}\hat\gamma_j\big(\hat v^{(j)},y\big)^2/2\varepsilon\right]\\
&\quad=\big[1+O(\rho)\big]\prod_{\ell=1}^{d}\frac{\sqrt{r_\ell}}{\varepsilon}\int_{\bar D_\varepsilon} dy\,\exp\left[-\sum_{j=1}^{d}|\hat\gamma_j|\big(\hat v^{(j)},y\big)^2/2\varepsilon\right]\\
&\quad=\big[1+O(\rho)\big]\left(\frac{2\pi}{\varepsilon}\right)^{d/2}\prod_{\ell=1}^{d}\sqrt{\frac{r_\ell}{|\hat\gamma_\ell|}}\\
&\quad=\big[1+O(\rho)\big]\left(\frac{2\pi}{\varepsilon}\right)^{d/2}\big[-\det\mathbb{A}\big]^{-1/2}.
\end{aligned}\tag{10.2.33}$$

Here, $\bar D_\varepsilon$ is the image of D_ε under the change of variables $y_\ell = x_\ell/\sqrt{r_\ell}$, and in the last equality we use the fact that Gaussian integrals over intervals of length $\rho = C\sqrt{\varepsilon\ln(1/\varepsilon)}$ are equal to integrals over $\mathbb{R}$ up to errors of order ε^C.

Inserting (10.2.33) into (10.2.32), we see that the left-hand side of (10.2.32) is equal to the right-hand side of (10.2.31) up to error terms. It therefore remains to show that the sum outside D_ε in the Dirichlet form does not contribute significantly to the capacity. But we can always choose D_ε and W_0 in such a way that the following hold:

(i) For $x \in W_0 \cap D_\varepsilon^c$, $\mu_\varepsilon(x) \le \mu_\varepsilon(z^*)\varepsilon^K$ with K as large as desired.
(ii) If $x \in W_0 \cap D_\varepsilon$ and $y \in W_1$, with $p_\varepsilon(x,y) > 0$, then $g(x)^2\mu_\varepsilon(x)/\mu_\varepsilon(z^*) \le \varepsilon^K$. Similarly, if $x \in W_0 \cap D_\varepsilon$ and $y \in W_2$, with $p_\varepsilon(x,y) > 0$, then $(g(x)-1)^2\mu_\varepsilon(x)/\mu_\varepsilon(z^*) \le \varepsilon^K$. Both facts follow from the explicit form of g and the fact that F is close to its quadratic approximation on D_ε.

From these observations we easily derive that the contribution to the Dirichlet form from $(W_0 \cap D_\varepsilon)^c$ is negligible compared to the contribution from $W_0 \cap D_\varepsilon$. This yields Proposition 10.15. □

10.3 Lower bounds on capacities

To obtain sharp lower bounds on capacities we use the *Berman-Konsowa* principle from Sect. 7.3 and the ideas presented in Sect. 9.1, in particular, Lemma 9.4. We prove the following counterpart of Proposition 10.15.

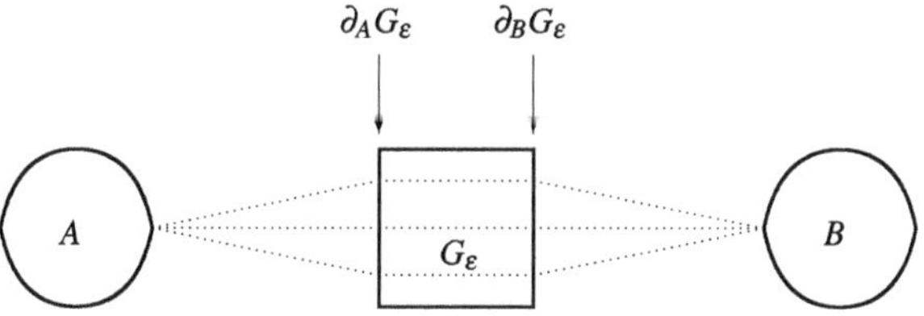

Fig. 10.2 The defective flow from A to B

Proposition 10.16 *With the notation from Proposition* 10.15,

$$\mathrm{cap}(A,B) \geq \mu_\varepsilon(z^*)\frac{|\hat{\gamma}_1|\varepsilon}{2\pi}\left(\frac{2\pi}{\varepsilon}\right)^{d/2}\frac{1}{\sqrt{-\det \mathbb{A}(z^*)}}\left(1+O\left(\sqrt{\varepsilon[\ln(1/\varepsilon)]^3}\right)\right). \tag{10.3.1}$$

Proof We have to construct a defective unit flow $f_{A,B}$ from A to B that reproduces the upper bound from Proposition 10.15. The construction of the flow is a bit more artistic than the construction of the approximate harmonic function. The idea is to channel the flow through a certain neighbourhood G_ε of z^* that plays a rôle similar to that of the cube D_ε in Sect. 10.2 (recall Fig. 10.2). We will construct the flow from three pieces, f_A, f, f_B. Here, f_A is a unit flow from A to $\partial_A G_\varepsilon$, f_B is a unit flow from $\partial_B G_\varepsilon$ to B, and f is a *defective unit flow* from $\partial_A G_\varepsilon$ to $\partial_B G_\varepsilon$ associated with the approximate harmonic function g that we used in the upper bound (recall Fig. 10.1 and see Fig. 10.2). In fact, for $x \in G_\varepsilon$ we set

$$f\big((x,x+\varepsilon e_\ell)\big) = \frac{\tilde{\mu}_\varepsilon(x) r_\ell[g(x+\varepsilon e_\ell)-g(x)]}{N(g)}, \tag{10.3.2}$$

where $N(g)$ is the normalising constant so that the total flow out of $\partial_A G_\varepsilon$ equals 1, which is given by

$$N(g) = \sum_{x\in\partial_A G_\varepsilon}\sum_{\substack{1\leq \ell\leq d:\\ x+\varepsilon e_\ell \in G_\varepsilon}} \tilde{\mu}_\varepsilon(x) r_\ell\big[g(x+\varepsilon e_\ell)-g(x)\big]. \tag{10.3.3}$$

Recall from (10.2.24) that

$$g(x+\varepsilon e_\ell)-g(x) = v_\ell\sqrt{\frac{\varepsilon|\hat{\gamma}_1|}{2\pi}}\exp\big[-|\hat{\gamma}_1|(v,x)^2/2\varepsilon\big]\big[1+O(\rho)\big] \tag{10.3.4}$$

uniformly in G_ε. Recall the definition of the transition probabilities q^f in (7.3.26). Substitution of (10.3.4) into (10.3.2) yields, because $r_\ell v_\ell = \check{v}_\ell$,

$$q^f\big((x,x+\varepsilon e_\ell)\big) = \frac{f((x,x+\varepsilon e_\ell))}{\sum_{k=1}^d f((x,x+\varepsilon e_k))} = \frac{\check{v}_\ell}{\sum_{k=1}^d \check{v}_k}\big[1+O(\rho)\big]. \tag{10.3.5}$$

This is essentially a directed nearest-neighbour random walk with drift in the direction of $\check{v}$. Recall that $\check{v}$ is the direction of steepest descent of F at the saddle point z^*. This implies, in particular, that with probability tending to one paths starting in $\partial_A G_\varepsilon$ stay within a larger cylinder in the direction of $\check{v}$ with base containing $\partial_A G_\varepsilon$ before leaving D_ε.

The choice of the flow into $\partial_A G_\varepsilon$ is rather arbitrary. Ideally, we would like to take disjoint paths from A to each point in $\partial_A G_\varepsilon$ and send the flow through this path. Of course, this is not possible when e.g. A is a single point, since in that case paths need to merge. However, this is not really a problem because these parts of the paths will not give any relevant contributions to the capacity anyway. Likewise, the flow arriving in $\partial_B G_\varepsilon$ will be channeled into B along coalescing paths. Figure 10.2 depicts this choice.

We will only consider paths from A to B that enter G_ε through $\partial_A G_\varepsilon$ and exit G_ε through $\partial_B G_\varepsilon$. By construction, this set of paths has $\mathbb{P}^f$-probability at least $1-o(1)$. Any such path consists of three pieces: $\gamma_1\colon A \to \partial_A G_\varepsilon$, $\gamma_2\colon \partial_A G_\varepsilon \to \partial_B G_\varepsilon$ and $\gamma_3\colon \partial_B G_\varepsilon \to B$. Consequently,

$$\sum_{(x,y)\in\gamma} \frac{f_{A,B}((x,y))}{\tilde{\mu}_\varepsilon(x)p_\varepsilon(x,y)} = \sum_{i=1}^{3} \sum_{(x,y)\in\gamma_i} \frac{f_{A,B}((x,y))}{\tilde{\mu}_\varepsilon(x)p_\varepsilon(x,y)} = K_1+K_2+K_3. \tag{10.3.6}$$

The term K_2 gives the desired contribution (recall (10.3.2)). Using (10.2.10–10.2.11) and the explicit form of g in (10.2.19–10.2.20), we get (recall (10.1.2))

$$\begin{aligned} K_2 = \sum_{(x,y)\in\gamma_2} \frac{f((x,y))}{\tilde{\mu}_\varepsilon(x)p_\varepsilon(x,y)} &= \frac{1}{N(g)} \sum_{(x,y)\in\gamma_2} \big[g(x)-g(y)\big] \\ &= \frac{1}{N(g)}\big[g\big(\gamma_2(|\gamma_2|)\big) - g\big(\gamma_2(0)\big)\big] \\ &= \frac{1}{N(g)}\sqrt{\frac{|\hat{\gamma}_1|}{2\pi\varepsilon}} \int_{-C\sqrt{\varepsilon\ln(1/\varepsilon)}}^{C\sqrt{\varepsilon\ln(1/\varepsilon)}} \mathrm{e}^{-|\hat{\gamma}_1|u^2/2\varepsilon}du \\ &= \frac{1}{N(g)}\big[1-O\big(\varepsilon^C\big)\big]. \end{aligned} \tag{10.3.7}$$

The terms K_1 and K_3 are negligible, provided the paths γ_1, γ_3 stay within the level set $\{x \in S_\varepsilon\colon F(x) \le F(z^*) - C'\varepsilon\ln(1/\varepsilon)\}$, which can always be achieved due to our assumptions on the function F. Namely, even the crudest possible bound $f((x,y)) \le 1$ implies that (recall (10.1.1)–(10.1.2))

$$\begin{aligned} K_1 = \sum_{(x,y)\in\gamma_1} \frac{f_A((x,y))}{\mu_\varepsilon(x)p_\varepsilon(x,y)} &\le \sum_{(x,y)\in\gamma_1} \frac{1}{\mu_\varepsilon(x)p_\varepsilon(x,y)} \\ &\le C|\gamma_1|\mathrm{e}^{F(z^*)/2\varepsilon - C'\ln(1/\varepsilon)} \ll \frac{1}{N(g)}, \end{aligned} \tag{10.3.8}$$

provided C' is large enough. The last estimate in (10.3.8) will be shown in (10.3.18–10.3.19) below. The same argument applies to K_3 and f_B. Hence, we obtain with the help of Lemma 10.13 that

$$\mathbb{E}^f\biggl(\biggl[\sum_{(x,y)\in\gamma}\frac{f_{A,B}((x,y))}{\mu_\varepsilon(x)p_\varepsilon(x,y)}\biggr]^{-1}\biggr)$$
$$\geq N(g)\bigl[1-o(1)\bigr]\mathbb{P}^f(\tau_{\partial_B G_\varepsilon}<\tau_{\partial G_\varepsilon\backslash\partial_B G_\varepsilon})=N(g)\bigl[1-o(1)\bigr]. \tag{10.3.9}$$

In order to apply Lemma 9.4, we must ensure that the accumulated defect is negligible. Recall from Lemma 9.4 that the error factor is bounded by

$$\prod_{k=1}^{M}\biggl(1+\max_{y\in A_k}\frac{\delta(y)}{\mathscr{F}(y)}\biggr)^{-1}, \tag{10.3.10}$$

where M is the length of the path and A_k, $k=1,\dots,M$ are the sets defined in (7.3.31). For our choice of the flow,

$$\frac{\delta(y)}{\mathscr{F}(y)}=\frac{(\widetilde{L}_\varepsilon g)(y)}{\sum_{z\in G_\varepsilon:\, g(y)<g(z)}\mu_\varepsilon(y)p_\varepsilon(y,z)[g(z)-g(y)]}. \tag{10.3.11}$$

By Lemma 10.14 and (10.2.24), for all $y\in G_\varepsilon$,

$$\frac{(\widetilde{L}_\varepsilon g)(y)}{\sum_{z\in G_\varepsilon:\, g(y)<g(z)}\mu_\varepsilon(y)p_\varepsilon(y,z)[g(z)-g(y)]}\leq O(\rho^2). \tag{10.3.12}$$

On the other hand, the paths in G_ε have length at most ρ/ε, so that

$$\prod_{k=1}^{M}\biggl(1+\max_{y\in A_k}\frac{\delta(y)}{\mathscr{F}(y)}\biggr)^{-1}\geq\bigl(1+O(\rho^2)\bigr)^{-\rho/\varepsilon}$$
$$\geq 1-O\Bigl(\sqrt{\varepsilon\bigl[\ln(1/\varepsilon)\bigr]^3}\Bigr), \tag{10.3.13}$$

which controls the error factor.

Finally, we must show that, with the right choice of G_ε, the normalisation $N(g)$ is essentially equal to $\widetilde{\mathscr{E}}_{D_\varepsilon}(g,g)$. Let G_ε be the cylinder with axis $\hat{v}$, radius ρ and length $\rho'=C'\sqrt{\varepsilon\ln(1/\varepsilon)}$, centred at z^*. The constants C,C' will be chosen such that, for all x in the front and end bases of the cylinder, $F(x)\leq F(z^*)-c\varepsilon\ln(1/\varepsilon)$ for some $c>0$. This is possible because, to leading order for $x=\sum_{i=1}^d a_i v^{(i)}$,

$$F(x)=-\tfrac{1}{2}|\check{\gamma}_1|a_1^2+\tfrac{1}{2}\sum_{i=2}^{d}\check{\gamma}_i a_i^2+O(\varepsilon^{3/2}), \tag{10.3.14}$$

where we recall (10.1.13) and (10.2.16). For points in the front base, $a_1=C'\sqrt{\varepsilon\ln(1/\varepsilon)}$ and $a_i\leq C\sqrt{\varepsilon\ln(1/\varepsilon)}$, $i=2,\dots,d$, so by making C' large we can achieve our objective.

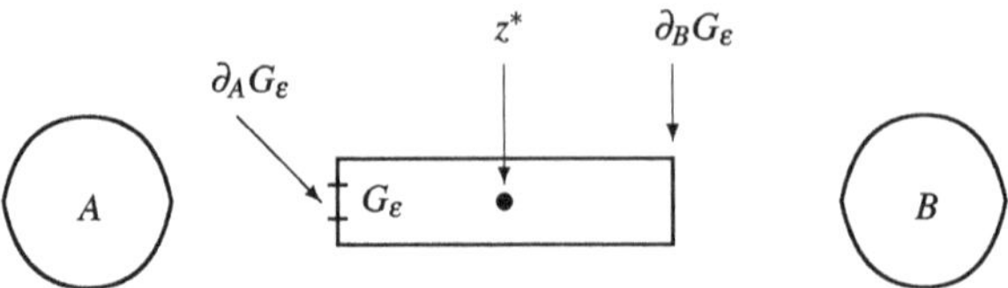

Fig. 10.3 The cylinder G_ε, and the pieces $\partial_A G_\varepsilon$ and $\partial_B G_\varepsilon$ of ∂G_ε at the front and end base of the cylinder

Let $\partial_B G_\varepsilon$ be the end base of the cylinder, in the direction of B. Let $\partial_A G_\varepsilon$ be the central part of radius $C''\sqrt{\varepsilon \ln(1/\varepsilon)}$ with $C'' < C$ of the front base of the cylinder, in the direction of A. Any choice of $C'' < C$ will actually be fine. The cylinder G_ε is depicted in Fig. 10.3.

By Lemma 10.13, inside G_ε we may work with the modified Dirichlet form given in (10.2.13). The boundary ∂G_ε of G_ε consists of three disjoint pieces, $\partial G_\varepsilon = \partial_A G_\varepsilon \cup \partial_B G_\varepsilon \cup \partial_r G_\varepsilon$, where $\partial_r G_\varepsilon$ is simply what is left over after the other two pieces are removed. Let g be the approximate harmonic function defined in (10.2.3) and (10.2.20). Proceeding along the lines of (10.2.32)–(10.2.33), we see that

$$\mathscr{E}_{G_\varepsilon}(g,g) = \big[1+o(1)\big] \sum_{x\in G_\varepsilon} \tilde{\mu}_\varepsilon(x) \sum_{\ell=1}^{d} r_\ell \big[g(x+\varepsilon e_\ell) - g(x)\big]^2. \tag{10.3.15}$$

Now use the first Green identity in (10.3.15), to get

$$\begin{aligned} &\sum_{x\in G_\varepsilon} \tilde{\mu}_\varepsilon(x) \sum_{\ell=1}^{d} r_\ell \big[g(x+\varepsilon e_\ell) - g(x)\big]^2 \\ &\quad = -\sum_{x\in G_\varepsilon} \tilde{\mu}_\varepsilon(x) g(x) (\widetilde{L}_\varepsilon g)(x) \\ &\qquad + \sum_{x\in\partial G_\varepsilon} \sum_{\substack{1\le\ell\le d:\\ x+\varepsilon e_\ell\in G_\varepsilon}} r_\ell \tilde{\mu}_\varepsilon(x) g(x) \big[g(x) - g(x+\varepsilon e_\ell)\big]. \end{aligned} \tag{10.3.16}$$

Using the bound in (10.2.21) from Lemma 10.14, we get

$$\Big|\sum_{x\in G_\varepsilon} \tilde{\mu}_\varepsilon(x) g(x) (\widetilde{L}_\varepsilon g)(x)\Big| \le O\big((\rho/\varepsilon)^d \sqrt{\varepsilon}\rho^2\big) = O\big(\varepsilon^{3/2-d/2}\big[\ln(1/\varepsilon)\big]^{-1+d/2}\big). \tag{10.3.17}$$

But this is much smaller than $\tilde{\mathscr{E}}_{G_\varepsilon}(g,g)$, which we know to be of order $\varepsilon^{1-d/2}$ by (10.2.32–10.2.33). Finally, in view of the fact that $g(x) \le \varepsilon^{C_*}$, $x \in \partial_B G_\varepsilon$ (where C_* can be taken as large as desired by taking C large enough) and that $\tilde{\mu}_\varepsilon(x)$ decays as

$(x, \hat{v})^2$ increases (see (10.3.14), we get that

$$\sum_{x\in\partial G_\varepsilon}\sum_{\substack{1\le\ell\le d:\\ x+\varepsilon e_\ell\in G_\varepsilon}} r_\ell\tilde{\mu}_\varepsilon(x)g(x)\big[g(x)-g(x+\varepsilon e_\ell)\big]$$
$$=\big[1+o(1)\big]\sum_{x\in\partial_A G_\varepsilon}\sum_{\substack{1\le\ell\le d:\\ x+\varepsilon e_\ell\in G_\varepsilon}} r_\ell\tilde{\mu}_\varepsilon(x)\big[g(x)-g(x+\varepsilon e_\ell)\big]. \tag{10.3.18}$$

Hence

$$\tilde{\mathscr{E}}_{G_\varepsilon}(g,g)=\big[1+o(1)\big]\sum_{x\in\partial_A G_\varepsilon}\sum_{\substack{1\le\ell\le d\\ x+\varepsilon e_\ell\in G_\varepsilon}} r_\ell\tilde{\mu}_\varepsilon(x)\big[g(x)-g(x+\varepsilon e_\ell)\big]$$
$$=\big[1+o(1)\big]N(g), \tag{10.3.19}$$

where we recall (10.3.3). This completes the proof of Proposition 10.16. □

In conclusion, we have achieved our goal to derive asymptotically coinciding upper and lower bounds for capacities of disjoint sets that are neighbourhoods of two local minima of F. In the current formulation, we assumed that there is only one relevant saddle point connecting these minima, but the extension to more complicated situations is straightforward (see the general discussion in Sect. 9.1).

10.4 Bibliographical notes

1. The class of models treated in this chapter served as the starting point of the potential-theoretic approach to metastability initiated in Bovier, Eckhoff, Gayrard and Klein [33]. This paper was leaning heavily on renewal ideas, and identified prefactors of capacities and mean hitting times only up to constants.

2. The method that allows us to identify the prefactors up to a multiplicative error $1+o(1)$ is the Berman-Konsowa principle for lower bounds on capacities, introduced in Bianchi, Bovier and Ioffe [24] for the analysis of the random-field Curie-Weiss model with a continuous distribution of the random magnetic field, which is treated in Chap. 15. The proof using defective flows given here is new.

Chapter 11
Diffusion Processes with Gradient Drift

> *"Ignorance of Axioms", the Lecturer continued, "is a great drawback in life. It wastes so much time to have to say them over and over again. For instance, take the Axiom* "Nothing is greater than itself", *that is,* "Nothing can contain itself". *How often do you hear people say "He was so excited, he was quite unable to contain himself". Why,* of course *he was unable! The excitement had nothing to do with it".*
>
> (Lewis Carroll, Sylvie and Bruno Concluded)

The first steps towards describing metastability for models with a non-discrete state space lead to *finite-dimensional diffusions*. These are the processes originally studied by Freidlin and Wentzell [115]. In the case of *gradient drifts* we are able to recover the heuristic predictions by Eyring and Kramers explained in Sect. 2.1.1. The presentation below contains two main parts. After describing the setting in Sect. 11.1, we derive sharp estimates on average hitting times in Sect. 11.2. This requires the use of sharp estimates on capacities, together with the regularity estimates that were presented in Sect. 9.4. In Sect. 11.3 we compute the low-lying spectrum of the generator of the diffusion. The result is completely analogous to the spectral result described in Chap. 8 for Markov processes with a discrete state space. The main message of this chapter is that all the results about metastable systems that were obtained in Chap. 8 in the setting of a discrete state space carry over to diffusions with the help of some regularity theory.

11.1 The setting

In this chapter we investigate metastability in the context of reversible diffusion processes $X_\varepsilon = (X_\varepsilon(t))_{t\geq 0}$ that were discussed in Sect. 5.6. We limit ourselves to the simplest case of SDEs of the form

$$dX_\varepsilon(t) = -\nabla F\big(X_\varepsilon(t)\big)\,dt + \sqrt{2\varepsilon}\,dB(t) \tag{11.1.1}$$

on a regular domain $\Omega \subseteq \mathbb{R}^d$, where the drift ∇F is generated by a potential function F that is sufficiently regular. The parameter ε scales the strength of the noise. Metastability occurs when F has one or more local minima that are not global min-

A. Bovier, F. den Hollander, *Metastability*,
Grundlehren der mathematischen Wissenschaften 351,
DOI 10.1007/978-3-319-24777-9_11

ima, and when ε is *small*. We assume that X_ε is killed as soon as it exits Ω. This is the natural extension of the paradigmatic double-well model of Kramers to the multi-dimensional setting. We will show that the formalism outlined in Chap. 8 is well suited to study the metastable behaviour of X_ε in the limit as $\varepsilon \downarrow 0$, and yields sharp results in a relatively simple manner. Recall that the process X_ε is reversible with respect to the measure

$$\mu_\varepsilon(dx) = \exp\bigl(-F(x)/\varepsilon\bigr)dx. \tag{11.1.2}$$

Recall that the generator $\mathscr{L}_\varepsilon$ of this process acts on smooth functions f as

$$(-\mathscr{L}_\varepsilon f)(x) = \varepsilon \Delta f(x) - \nabla F(x) \cdot \nabla f(x). \tag{11.1.3}$$

The Dirichlet form is given by

$$\mathscr{E}(f,g) = \varepsilon \int_\Omega \mathrm{e}^{-F(x)/\varepsilon}\bigl(\nabla f(x), \nabla g(x)\bigr)dx. \tag{11.1.4}$$

We begin by formulating assumptions on Ω and F.

Assumption 11.1

(i) $\Omega \subseteq \mathbb{R}^d$ is open and connected, and $F \in C^3(\Omega)$.
(ii) If Ω is unbounded, then

  (ii.1) $\lim_{\|x\|\to\infty} |(\frac{x}{\|x\|}, \nabla F(x))| = \infty$,
  (ii.2) $\lim_{\|x\|\to\infty} [\|\nabla F(x)\| - 2\Delta F(x)] = \infty$.

Assumption 11.1 ensures that the resolvent of the generator $-\mathscr{L}_\varepsilon$ of X_ε is compact for ε sufficiently small. Moreover, it implies that F has exponentially tight level sets, i.e.,

$$\int_{\{y\in\mathbb{R}^d:\ F(y)\geq a\}} \mathrm{e}^{-F(y)/\varepsilon} dy \leq C\mathrm{e}^{-a/\varepsilon} \quad \forall a > 0, \tag{11.1.5}$$

where $C = C(a) < \infty$ is uniform in $0 < \varepsilon \leq 1$.

Recall Definition 10.2. Throughout the sequel, Assumption 11.1 is in force, as well as Assumptions 10.3 and 10.5 in Sect. 10. In this situation, it can be shown easily that (11.1.1) has a global strong solution (Bauer [14]).

11.2 Capacity estimates and mean hitting times

Our main interest in this section is the mean of the hitting time

$$\tau_A = \inf\bigl\{t > 0:\ X(t) \in A\bigr\}, \quad A \subset \mathbb{R}^d, \tag{11.2.1}$$

for X starting in a minimum of F, say $x \in \mathscr{M}$, when $A = B_\rho(y)$ is a small ball of radius ρ around another minimum of F, say $y \in \mathscr{M}$. It will become apparent

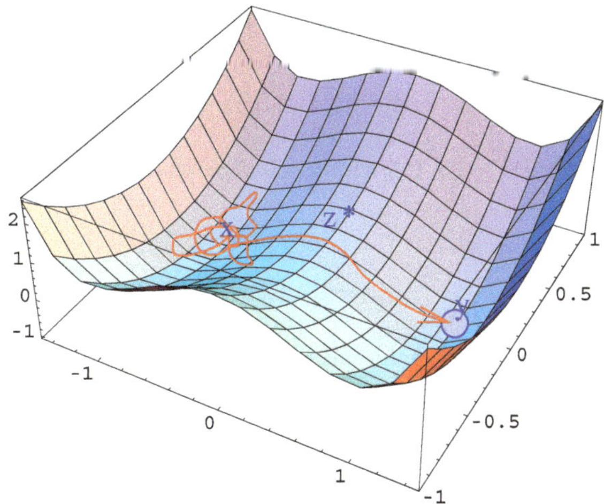

Fig. 11.1 Motion in a potential with two wells

that the precise choice of the hitting set is not important, and that the problem of computing τ_A is virtually equivalent to computing the escape time from a suitably chosen neighbourhood of x, provided this neighbourhood contains the relevant *saddle points* connecting x and y. Figure 11.1 schematically depicts the motion of a particle in a two-well potential.

11.2.1 Main results

The basis for the success of the potential-theoretic approach to metastability is the fact that capacities can be estimated sharply. Recall the definition of the communication height $\Phi(A,B)$ and the essential gate $\mathscr{G}(A,B)$ and between two disjoint sets A, B, as introduced in Definition 10.2.

Theorem 11.2 (Capacity asymptotics) *Assume that $A, B \subset \mathbb{R}^d$ are closed and disjoint such that*

(i) $\operatorname{dist}(\mathscr{G}(A,B), A \cup B) \geq \delta > 0$ *for some δ independent of ε.*
(ii) *Both A and B contain a closed ball of radius at least ε.*

If $\mathscr{G}(A,B) = \{z_1^*, \dots, z_n^*\}$, then

$$\operatorname{cap}(A,B) = \mathrm{e}^{-\Phi(A,B)/\varepsilon} \frac{(2\pi\varepsilon)^{d/2}}{2\pi} \sum_{i=1}^{n} \frac{[-\lambda_1^*(z_i^*)]}{\sqrt{-\det(\nabla^2 F(z_i^*))}} \left[1 + O\left(\sqrt{\varepsilon[\ln(1/\varepsilon)]^3}\right)\right], \tag{11.2.2}$$

where $\lambda_1^*(z_i^*)$ denotes the negative eigenvalue of the Hessian of F at z_i^*.

Theorem 11.3 (Mean metastable exit times) *Let $x_i \in \mathscr{M}$ be a minimum of F and let $D \subset \mathbb{R}^d$ be a closed set such that*:

(i) $\bigcup_{j=1}^{k(i)} B_\varepsilon(y_j) \subset D$, *where* $\mathscr{M}_i = \{y_1, \dots, y_{k(i)}\} \subset \mathscr{M}$ *enumerates all the minima of F with* $F(y_j) \leq F(x_i)$, $j = 1, \dots, k(i)$.
(ii) $\operatorname{dist}(\mathscr{G}(x_i, \mathscr{M}_i), D) \geq \delta > 0$ *for some δ independent of ε.*

Then

$$\mathbb{E}_{x_i}[\tau_D]$$
$$= \frac{2\pi e^{[\Phi(x_i, \mathscr{M}_i) - F(x_i)]/\varepsilon}}{\sqrt{\det(\nabla^2 F(x_i))}} \left(\sum_{j=1}^{k(i)} \frac{[-\lambda_1^*(z_j^*)]}{\sqrt{-\det(\nabla^2 F(z_j^*))}} \right)^{-1} \left[1 + O\left(\sqrt{\varepsilon[\ln(1/\varepsilon)]^3}\right)\right]. \tag{11.2.3}$$

In the special case where there is only one saddle point z^*, (11.2.3) reduces to the classical Eyring-Kramers formula in (2.1.2):

$$\mathbb{E}_{x_i}[\tau_D] = \frac{2\pi e^{[F(z^*) - F(x_i)]/\varepsilon}}{[-\lambda_1^*(z^*)]} \frac{\sqrt{-\det(\nabla^2 F(z^*))}}{\sqrt{\det(\nabla^2 F(x_i))}} \left[1 + O\left(\sqrt{\varepsilon[\ln(1/\varepsilon)]^3}\right)\right]. \tag{11.2.4}$$

11.2.2 Rough estimates on capacities and harmonic functions

To prove Theorems 11.2–11.3, we follow the general strategy outlined in Sect. 9.1. In the present section we derive rough estimates on capacities. Via the renewal estimate in (9.4.6) these lead to rough estimates on harmonic functions. These will in turn lead to sharp estimates on capacities and the equilibrium potential.

Lemma 11.4 *Let $D \subset \mathbb{R}^d$ be closed, and let $x \in D^c$ be such that $d(x, D) \geq \rho$ for some $0 < \rho \leq \varepsilon$. Let $z^* = z^*(x, D)$ be any point in $\mathscr{G}(x, D)$. Then there are constants $C_\ell, C_u > 0$ independent of ε such that, for ε small enough,*

$$C_\ell \rho^{d-1} e^{-F(z^*)/\varepsilon} \leq \operatorname{cap}\big(B_\rho(x), D\big) \leq C_u \varepsilon \rho^{-1} e^{-F(z^*)/\varepsilon}. \tag{11.2.5}$$

Proof To prove the lower bound, we use the Dirichlet principle (recall Theorem 7.33) and monotonicity. We begin by choosing a smooth path ω from x to D remaining in the level set $\{z \in \mathbb{R}^d : F(z) \leq F(z^*)\}$ and reaching the value $F(z^*)$ only when passing through z^*. (The canonical path can be constructed by using pieces of the deterministic trajectory of the unperturbed equation $dX_\varepsilon(t) = -\nabla F(X_\varepsilon(t))dt$ in a rather obvious manner, but this is not important.) Given this path, we parametrise it by arc-length, so that $\|\dot\omega(t)\|_2 = 1$ for all t.

Given $\omega(t)$, we consider the tube of width ρ around $\omega(t)$:

$$\omega^\rho = \big\{z \in \mathbb{R}^d : \exists t \in \big[0, |\omega|\big] \text{ such that } \big\|\omega(t) - z\big\|_2 \leq \rho\big\}. \tag{11.2.6}$$

Let D_ρ denote the $(d-1)$-dimensional disk of radius ρ centred at the origin. The important fact to note is that, for any h,

$$\left\|\nabla h\big(\omega(t)+z_\perp\big)\right\|_2^2 \geq \left[\frac{d}{dt}h\big(\omega(t)+z_\perp\big)\right]^2. \tag{11.2.7}$$

Here $z_\perp$ denotes a vector in the subspace orthogonal to ω at $\omega(t)$.

We may therefore bound the Dirichlet form in (11.1.4) as

$$\mathscr{E}(h,h) \geq \varepsilon \int_{D_\rho} dz_\perp \int_0^{|\omega|} dt\, \mathrm{e}^{-F(\omega(t)+z_\perp)/\varepsilon}\left[\frac{d}{dt}h\big(\omega(t)+z_\perp\big)\right]^2. \tag{11.2.8}$$

The minimisation problem is now trivial, i.e., it decomposes for each fixed $z_\perp$ into a one-dimensional problem whose solution is well known. In fact, the minimiser $h_{z_\perp}(t)$ is the solution of the 1-dimensional Dirichlet problem

$$\begin{aligned} \left[-\varepsilon\frac{d}{dt}+\frac{d}{dt}F\big(\omega(t)+z_\perp\big)\right]\frac{d}{dt}h_{z_\perp}(t) &= 0,\\ h_{z_\perp}(0) &= 1,\\ h_{z_\perp}(|\omega|) &= 0, \end{aligned} \tag{11.2.9}$$

whose solution is readily found to be

$$h_{z_\perp}(t) = \frac{\int_t^{|\omega|} ds\, \mathrm{e}^{F(\omega(s)+z_\perp)/\varepsilon}}{\int_0^{|\omega|} ds\, \mathrm{e}^{F(\omega(s)+z_\perp)/\varepsilon}}. \tag{11.2.10}$$

Inserting this solution into the lower bound (11.2.8), we get

$$\mathrm{cap}\big(B_\rho(x), D\big) \geq \varepsilon \int_{D_\rho} dz_\perp \left[\int_0^{|\omega|} dt\, \mathrm{e}^{F(\omega(t)+z_\perp)/\varepsilon}\right]^{-1}. \tag{11.2.11}$$

From here the lower bound in (11.2.5) follows from simple saddle point evaluations of the integral in the denominator.

The upper bound is obtained by choosing a suitable test function that changes from zero to one over a distance ρ along a plane separating $B_\rho(x)$ and D passing at distance ρ from x. □

Corollary 11.5 *Let $A, D \subset \mathbb{R}^d$ be disjoint sets. Let $x \in (A \cup D)^c$. Then, for $0 < \rho \leq \varepsilon$, there exists a constant C such that, for ε small enough,*

$$h_{A,D}(x) \leq C\rho^{-d}\mathrm{e}^{-F(z^*)/\varepsilon}\,\mathrm{cap}\big(B_\rho(x), A\big) \tag{11.2.12}$$

with z^* the same as in Lemma 11.4.

Proof Combine Lemma 11.4 with Lemma 9.4.7. □

11.2.3 Sharp estimates on capacities

In this section we prove Theorem 11.2. The proof is similar to the one for discrete diffusions in Chap. 10.

Proof The capacity $\mathrm{cap}(A, B)$ satisfies the Dirichlet principle in Theorem 7.33,

$$\mathrm{cap}(A,B) = \inf_{h \in \mathscr{H}_B^A} \mathscr{E}(h,h), \tag{11.2.13}$$

where

$$\mathscr{H}_B^A = \{h \in W^{1,2}(\mathbb{R}^d, \mathbb{Q}(dx)) \colon h(z) \in [0,1], \forall z \in \mathbb{R}^d, h_{|A} = 1, h_{|B} = 0\} \tag{11.2.14}$$

with $\mathbb{Q}(dx) = \mathrm{e}^{-F(x)/\varepsilon} dx$. For simplicity we only consider the case of a single saddle point z^* (i.e., $n = 1$ in (11.2.2)).We may assume without loss of generality that $F(z^*) = 0$ and $z^* = 0$. We choose coordinates that diagonalise the Hessian of F at z^*, so that (with $\lambda_i^* = \lambda_i^*(z^*)$)

$$F(z) = \tfrac{1}{2}\lambda_1^* z_1^2 + \sum_{i=2}^{d} \tfrac{1}{2}\lambda_i^* z_i^2 + O(\|z\|_2^3), \quad \|z\| \downarrow 0. \tag{11.2.15}$$

Define a neighbourhood of zero

$$C_\delta = \Big[-\delta/\sqrt{[-\lambda_1^*]}, \delta/\sqrt{[-\lambda_1^*]}\Big] \bigotimes_{i=2}^{d} \Big[-2\delta/\sqrt{\lambda_i^*}, 2\delta/\sqrt{\lambda_i^*}\Big]. \tag{11.2.16}$$

Since we have assumed that there is a single saddle point at the communication height between A and B, it is possible to choose $\delta > 0$ so small that there exists a strip $\mathscr{S}_\delta$ of width $2\delta/\sqrt{[-\lambda_1^*]}$, containing 0, separating A and B in the sense that any path connecting them must cross $\mathscr{S}_\delta$, and such that $F(z) \geq \delta^2$ for all $z \in \mathscr{S}_\delta \backslash C_\delta$. Let D_A and D_B be the connected components of $\mathbb{R}^d \backslash \mathscr{S}_\delta$ containing A and B, respectively.

Upper bound

To prove an upper bound on the capacity we simply choose a convenient function h^+ (see Fig. 11.2):

$$h^+(z) = \begin{cases} 1, & \text{if } z \in D_A, \\ 0, & \text{if } z \in D_B, \\ \text{arbitrary in } [0,1], & \text{if } z \in \mathscr{S}_\delta \backslash C_\delta \text{ with } \|\nabla h^+\|_{2,\mu_\varepsilon} \leq c\sqrt{[-\lambda_1^*]}/\delta, \\ f(z_1), & \text{if } z \in C_\delta, \end{cases} \tag{11.2.17}$$

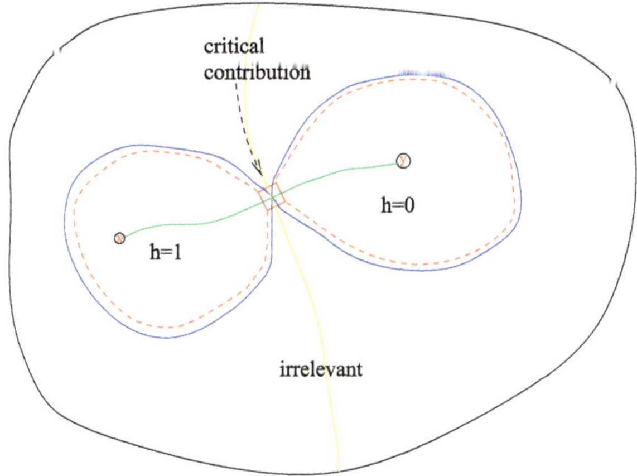

Fig. 11.2 Schematic construction of the test function

where f is the solution of the one-dimensional Dirichlet problem

$$\begin{aligned}\left(-\varepsilon\frac{d}{dz_1}+\frac{d}{dz_1}F(z_1,0)\right)\frac{d}{dz_1}f(z_1)&=0,\\ f\left(-\delta/\sqrt{[-\lambda_1^*]}\right)&=1,\\ f\left(+\delta/\sqrt{[-\lambda_1^*]}\right)&=0.\end{aligned} \tag{11.2.18}$$

The solution of this problem is obviously

$$f(z_1)=\frac{\int_{z_1}^{\delta/\sqrt{[-\lambda_1^*]}}\mathrm{e}^{F(t,0)/\varepsilon}dt}{\int_{-\delta/\sqrt{[-\lambda_1^*]}}^{\delta/\sqrt{[-\lambda_1^*]}}\mathrm{e}^{F(t,0)/\varepsilon}dt}. \tag{11.2.19}$$

Inserting (11.2.17) into (11.2.13), we see that

$$\begin{aligned}\operatorname{cap}(A,B)\le\varepsilon&\int_{-2\delta/\sqrt{\lambda_2^*}}^{2\delta/\sqrt{\lambda_2^*}}dz_2\ldots\int_{-2\delta/\sqrt{\lambda_d^*}}^{2\delta/\sqrt{\lambda_d^*}}dz_d\\ &\times\left(\int_{-\delta/\sqrt{[-\lambda_1^*]}}^{\delta/\sqrt{[-\lambda_1^*]}}dz_1\,\mathrm{e}^{-F(z)/\varepsilon}\left|f'(z_1)\right|^2\right)\\ &+\varepsilon c^2[-\lambda_1^*]\delta^{-2}\int_{\mathscr{S}_\delta\setminus C_\delta}dz\,\mathrm{e}^{-F(z)/\varepsilon}.\end{aligned} \tag{11.2.20}$$

The second term in (11.2.20) is bounded by a constant times $\varepsilon\delta^{-2}\mathrm{e}^{-\delta^2/\varepsilon}$, because of our assumption on F. The first term equals

$$\mathscr{E}_{C_\delta}\big(h^+, h^+\big) = \varepsilon \frac{\int_{C_\delta} dz\, \mathrm{e}^{-F(z)/\varepsilon}\mathrm{e}^{2F(z_1,0)/\varepsilon}}{\Big(\int_{-\delta/\sqrt{[-\lambda_1^*]}}^{\delta/\sqrt{[-\lambda_1^*]}} dt\, \mathrm{e}^{F(t,0)/\varepsilon}\Big)^2}. \tag{11.2.21}$$

Now, on the set C_δ we have

$$F(z) = F(0) + \frac{\lambda_1^* z_1^2 + \lambda_2^* z_2^2 + \cdots + \lambda_d^* z_d^2}{2} + O\big(\|z\|_2^3\big) \tag{11.2.22}$$

and hence

$$F(z) - 2F(z_1, 0) = -F(0) + \frac{-\lambda_1^* z_1^2 + \lambda_2^* z_2^2 \cdots + \lambda_d^* z_d^2}{2} + O\big(\|z\|_2^3\big). \tag{11.2.23}$$

But $\|z\|_2 \leq C'\delta$ on C_δ for some so if we choose $\delta = C''\sqrt{\varepsilon \ln(1/\varepsilon)}$ for some C'', for some $C'' < \infty$, then the numerator in (11.2.21) satisfies the bound

$$\begin{aligned}
&\int_C dz\, \mathrm{e}^{-F(z)/\varepsilon}\mathrm{e}^{2F(z_1,0)/\varepsilon} \\
&\quad \leq \mathrm{e}^{-F(0)/\varepsilon}\mathrm{e}^{O(\varepsilon^{1/2}[\ln(1/\varepsilon)]^{3/2})} \int_{\mathbb{R}^d} dz\, \exp\Big(-\frac{-\lambda_1^* z_1^2 + \cdots + \lambda_d^* z_d^2}{2\varepsilon}\Big) \\
&\quad = \mathrm{e}^{-F(0)/\varepsilon} \frac{(2\pi\varepsilon)^{d/2}}{\sqrt{[-\lambda_1^*]}\prod_{i=2}^d \sqrt{\lambda_i^*}} \big[1 + O\big(\varepsilon^{1/2}\big[\ln(1/\varepsilon)\big]^{3/2}\big)\big].
\end{aligned} \tag{11.2.24}$$

Similarly, the integral in the denominator in (11.2.21) is bounded from below by

$$\begin{aligned}
&\int_{-\delta/\sqrt{[-\lambda_1^*]}}^{\delta/\sqrt{[-\lambda_1^*]}} dt\, \mathrm{e}^{F(t,0)/\varepsilon} \\
&\quad \geq \mathrm{e}^{O(\varepsilon^{1/2}[\ln(1/\varepsilon)]^{3/2})}\mathrm{e}^{F(0)/\varepsilon}\left(\frac{(2\pi\varepsilon)^{1/2}}{\sqrt{[-\lambda_1^*]}} - 2\int_{\delta/\sqrt{[-\lambda_1^*]}}^{\infty} dt\, \mathrm{e}^{\lambda_1^* t^2/\varepsilon}\right) \\
&\quad \geq \mathrm{e}^{O(\varepsilon^{1/2}[\ln(1/\varepsilon)]^{3/2})}\mathrm{e}^{F(0)/\varepsilon}\left(\frac{(2\pi\varepsilon)^{1/2}}{\sqrt{[-\lambda_1^*]}} - \frac{\mathrm{e}^{-\delta^2/\varepsilon}}{\delta\varepsilon^{-1/2}}\right) \\
&\quad = \mathrm{e}^{F(0)/\varepsilon}\frac{(2\pi\varepsilon)^{1/2}}{\sqrt{[-\lambda_1^*]}}\Big[1 + O\big(\varepsilon^{1/2}\big[\ln(1/\varepsilon)\big]^{3/2}\big)\Big].
\end{aligned} \tag{11.2.25}$$

Combining the estimates in (11.2.20) and (11.2.24)–(11.2.25), we arrive at the upper bound

$$\mathscr{E}_{C_\delta}\big(h^+, h^+\big) \leq \mathrm{e}^{-F(0)/\varepsilon}(2\pi\varepsilon)^{d/2} \frac{[-\lambda_1^*]}{2\pi\sqrt{-\det(\nabla^2 F(0))}}\Big[1 + O\big(\varepsilon^{1/2}\big[\ln(1/\varepsilon)\big]^{3/2}\big)\Big]. \tag{11.2.26}$$

Lower bound

For the lower bound we consider a different domain, namely,

$$\begin{aligned}\widehat{C}_\delta &= \Big[-2\delta/\sqrt{[-\lambda_1^*]}, 2\delta/\sqrt{[-\lambda_1^*]}\Big] \bigotimes_{i=2}^{d} \Big[-\delta/\sqrt{(d-1)\lambda_i^*}, \delta/\sqrt{(d-1)\lambda_i^*}\Big] \\ &= \Big[-2\delta/\sqrt{[-\lambda_1^*]}, 2\delta/\sqrt{[-\lambda_1^*]}\Big] \otimes \widehat{C}_\delta^\perp. \end{aligned} \tag{11.2.27}$$

Let h^* denote the minimiser of the variational problem in (11.2.13), i.e., the equilibrium potential of the capacitor (A, B). Then

$$\inf_{h \in \mathscr{H}_B^A} \mathscr{E}(h,h) = \mathscr{E}\big(h^*, h^*\big) \geq \mathscr{E}_{\widehat{C}_\delta}\big(h^*, h^*\big), \tag{11.2.28}$$

where $\mathscr{E}_{\widehat{C}_\delta}$ is the restriction of the Dirichlet form to the domain $\widehat{C}_\delta$. Obviously,

$$\begin{aligned}\mathscr{E}_{\widehat{C}_\delta}(h,h) \geq \bar{\mathscr{E}}_{\widehat{C}_\delta}(h,h) &= \varepsilon \int_{\widehat{C}_\delta} dz\, \mathrm{e}^{-F(z)/\varepsilon} \left(\frac{\partial h(z)}{\partial z_1}\right)^2 \\ &= \varepsilon \int_{\widehat{C}_\delta^\perp} dz_\perp \left(\int_{-2\delta/\sqrt{[-\lambda_1^*]}}^{2\delta/\sqrt{[-\lambda_1^*]}} dz_1\, \mathrm{e}^{-F(z)/\varepsilon} \left| \frac{\partial h(z_1, z_\perp)}{\partial z_1} \right|^2 \right) \\ &\geq \varepsilon \int_{\widehat{C}_\delta^\perp} dz_\perp \Bigg(\inf_{f\colon f(\pm\delta/\sqrt{[-\lambda_1^*]}) = h^*(\pm\delta/\sqrt{[-\lambda_1^*]})} \\ &\qquad \int_{-2\delta/\sqrt{[-\lambda_1^*]}}^{2\delta/\sqrt{[-\lambda_1^*]}} dz_1\, \mathrm{e}^{-F(z)/\varepsilon} \left| f'(z_1) \right|^2 \Bigg). \end{aligned} \tag{11.2.29}$$

The minimisation problem for fixed values of $z_\perp$ is the solution of the Dirichlet problem

$$\begin{aligned} &\left(-\varepsilon \frac{d}{dz_1} + \frac{d}{dz_1} F(z_1, z_\perp)\right) \frac{d}{dz_1} f(z_1) = 0, \\ &\qquad f\big(-2\delta/\sqrt{[-\lambda_1^*]}\big) = h^*\big(-2\delta/\sqrt{[-\lambda_1^*]}, z_\perp\big), \\ &\qquad f\big(+2\delta/\sqrt{[-\lambda_1^*]}\big) = h^*\big(2\delta/\sqrt{[-\lambda_1^*]}, z_\perp\big). \end{aligned} \tag{11.2.30}$$

Set $a = h^*(-2\delta/\sqrt{[-\lambda_1^*]}, z_\perp)$, $b = h^*(2\delta/\sqrt{[-\lambda_1^*]}, z_\perp)$ and $g(z_1) = F(z_1, z_\perp)$. Then the general solution of the differential equation in (11.2.30) is

$$f(z_1) = c \int_{z_1}^{s} \mathrm{e}^{g(t)/\varepsilon} dt, \tag{11.2.31}$$

where the constants c and s are determined by the boundary conditions, i.e.,

$$
\begin{aligned}
c\int_{-2\delta/\sqrt{[-\lambda_1^*]}}^{s} e^{g(t)/\varepsilon}dt &= a,\\
c\int_{2\delta/\sqrt{[-\lambda_1^*]}}^{s} e^{g(t)/\varepsilon}dt &= b,
\end{aligned}
\tag{11.2.32}
$$

from which we get

$$
c = \frac{a}{\int_{-2\delta/\sqrt{[-\lambda_1^*]}}^{s} e^{g(t)/\varepsilon}dt}. \tag{11.2.33}
$$

The value of s is determined through the equation

$$
\int_{-2\delta/\sqrt{[-\lambda_1^*]}}^{s} e^{g(t)/\varepsilon}dt = \frac{a}{a-b}\int_{-2\delta/\sqrt{[-\lambda_1^*]}}^{2\delta/\sqrt{[-\lambda_1^*]}} e^{g(t)/\varepsilon}dt. \tag{11.2.34}
$$

Inserting this solution into (11.2.29) and recalling the definition of g, we obtain

$$
\begin{aligned}
&\mathscr{E}_{\widehat{C}_\delta}(h^*,h^*)\\
&\geq \varepsilon\int_{\widehat{C}_\delta^\perp} dz_\perp \int_{-2\delta/\sqrt{[-\lambda_1^*]}}^{2\delta/\sqrt{[-\lambda_1^*]}} dz_1\, \frac{e^{-F(z_1,z_\perp)/\varepsilon}[h^*(-2\delta/\sqrt{[-\lambda_1^*]},z_\perp)]^2 e^{2F(z_1,z_\perp)/\varepsilon}}{\left(\int_{-2\delta/\sqrt{[-\lambda_1^*]}}^{s(z_\perp)} e^{F(t,z_\perp)/\varepsilon}dt\right)^2}\\
&= \varepsilon\int_{\widehat{C}_\delta^\perp} dz_\perp \frac{[h^*(-2\delta/\sqrt{[-\lambda_1^*]},z_\perp) - h^*(2\delta/\sqrt{[-\lambda_1^*]},z_\perp)]^2}{\int_{-2\delta/\sqrt{[-\lambda_1^*]}}^{2\delta/\sqrt{[-\lambda_1^*]}} e^{F(t,z_\perp)/\varepsilon}dt}.
\end{aligned}
\tag{11.2.35}
$$

Using (11.2.15) again, we see that

$$
\begin{aligned}
\int_{-2\delta/\sqrt{[-\lambda_1^*]}}^{2\delta/\sqrt{[-\lambda_1^*]}} e^{F(t,z_\perp)/\varepsilon}dt &= \left(e^{\sum_{i=2}^d \frac{\lambda_i^* z_i^2}{2\varepsilon} + O(\delta^3/\varepsilon)}\right)\int_{-2\delta/\sqrt{[-\lambda_1^*]}}^{2\delta/\sqrt{[-\lambda_1^*]}} dt\, e^{-\frac{[-\lambda_1^*]t^2}{2\varepsilon}}\\
&\leq \frac{\sqrt{2\pi\varepsilon}}{\sqrt{[-\lambda_1^*]}}\left(e^{\sum_{i=2}^d \frac{\lambda_i^* z_i^2}{2\varepsilon} + O(\delta^3/\varepsilon)}\right),
\end{aligned}
\tag{11.2.36}
$$

and so

$$
\begin{aligned}
\mathscr{E}_{\widehat{C}_\delta}(h^*,h^*) \geq{}& \frac{\sqrt{\varepsilon[-\lambda_1^*]}}{\sqrt{2\pi}}\int_{\widehat{C}_\delta^\perp} dz_\perp \exp\left[-\sum_{i=2}^d \frac{\lambda_i^* z_i^2}{2\varepsilon} + O(\delta^3/\varepsilon)\right]\\
&\times\left(h^*\left(-2\delta/\sqrt{[-\lambda_1^*]},z_\perp\right) - h^*\left(2\delta/\sqrt{[-\lambda_1^*]},z_\perp\right)\right)^2.
\end{aligned}
\tag{11.2.37}
$$

The following lemma shows how close the values of h^* appearing in (11.2.37) are to 0 and 1, respectively.

Lemma 11.6 *Uniformly in* $z_\perp \in \widehat{C}_\delta^\perp$,

$$1 - h^*\big(-2\delta/\sqrt{[-\lambda_1^*]}, z_\perp\big) \le C\varepsilon^{-d/2} e^{-\delta^2/4\varepsilon},$$
$$h^*\big(2\delta/\sqrt{[-\lambda_1^*]}, z_\perp\big) \le C\varepsilon^{-d/2} e^{-\delta^2/4\varepsilon}. \tag{11.2.38}$$

Proof Use that $h^*(z) = \mathbb{P}_z[\tau_A < \tau_B] = h_{A,B}(z)$ in combination with Corollary 11.5. □

Inserting these bounds into (11.2.37), we arrive at

$$\mathscr{E}_{\widehat{C}_\delta}(h^*, h^*) \ge e^{-F(0)/\varepsilon} \frac{(2\pi\varepsilon)^{d/2}}{2\pi} \frac{\sqrt{[-\lambda_1^*]}}{\prod_{i=2}^d \sqrt{\lambda_i^*}} \times \big(1 - C\varepsilon^{-d/2} e^{-\delta^2/4\varepsilon}\big)^2 e^{-O(\delta^3/\varepsilon)} \left(1 - \frac{\sqrt{\varepsilon(d-1)}}{\delta} e^{-\frac{\delta^2}{2(d-1)\varepsilon}}\right)^{d-1}. \tag{11.2.39}$$

Choosing, as before, $\delta^2 = C\varepsilon \ln(1/\varepsilon)$, we see that to leading order (11.2.39) coincides with the upper bound (11.2.26), which proves Theorem 11.2 for the case $n = 1$ saddle points in the essential gate.

The generalisation to the case of several saddle points in the essential gate is straightforward and is left to the reader. □

11.2.4 Metastable exit times and capacities

In Propositions 11.7–11.8 below we compute the mean value of certain *metastable exit times* in terms of capacities. At the end we use these propositions to prove Theorem 11.2.

Proposition 11.7 *Let x be a (non-degenerate quadratic) critical point of F, and let $A, D \subset \mathbb{R}^d$ be disjoint closed sets. Then there exists an $\alpha > 0$ such that*

$$\mathbb{E}_x[\tau_D] = \frac{\int_{D^c} dy\, e^{-F(y)/\varepsilon} h_{B_\varepsilon(x),D}(y)}{\mathrm{cap}(B_\varepsilon(x), D)} \big[1 + O(\varepsilon^{\alpha/2})\big] \tag{11.2.40}$$

and

$$\mathbb{E}_x[\tau_D \mathbb{1}_{\tau_D < \tau_A}] = \frac{\int_{(A\cup D)^c} dy\, e^{-F(y)/\varepsilon} h_{B_\varepsilon(x),A\cup D}(y) h_{D,A}(y)}{\mathrm{cap}(B_\varepsilon(x), A \cup D)} \big[1 + O(\varepsilon^{\alpha/2})\big]. \tag{11.2.41}$$

Proof The proofs of (11.2.40) and (11.2.41) are analogous, and therefore we will only give the former. The strategy is to first use Corollary 7.30 with A a small ball around x, and then to use that the average exit time does not vary much as a function of the starting point on this ball. In fact, by Corollary 7.30,

$$\int \nu_{B_\varepsilon(x),D}(dy)\mathbb{E}_y[\tau_D \mathbb{1}_{\tau_D<\tau_A}] = \frac{\int_{(A\cup D)^c} dy\, \mathrm{e}^{-F(y)/\varepsilon} h_{B_\varepsilon(x),A\cup D}(y) h_{D,A}(y)}{\mathrm{cap}(B_\varepsilon(x), A\cup D)}. \tag{11.2.42}$$

Recall that $w_D(y) = \mathbb{E}_y[\tau_D]$, $y \notin D$, solves the Dirichlet problem in (7.2.74) with $g = 1$. Let $B_{R_0}(x)$ be the ball of radius R_0 centred at x, where x is a critical point of F. Then there is a $K < \infty$ such that $\sup_{y\in B_{R_0}(x)} \|\nabla F(y)\|_\infty \leq K R_0$ for R_0 small enough. Hence Lemmas 9.8–9.9 applied to this function have uniform constants when $R_0 \leq \sqrt{\varepsilon}$. Thus, by Lemma 9.13, w_D inherits from Lemma 9.8 the uniform Harnack bound

$$\sup_{y\in B_{\sqrt{\varepsilon}}(x)} w_D(y) \leq C \inf_{y\in B_{\sqrt{\varepsilon}}(x)} w_D(y). \tag{11.2.43}$$

Now use Lemma 9.9 with $R = \varepsilon$. This yields

$$\mathrm{osc}_{B_\varepsilon(x)} w_D \leq C\varepsilon^{\alpha/2}\Big(\sup_{y\in B_{R_0}(x)} w_D(y) + R_0\Big) \tag{11.2.44}$$

and implies

$$\begin{aligned} \sup_{y\in B_\varepsilon(x)} w_D(y) &\leq w_D(x) + C^2\mathrm{e}^{\alpha/2} w_D(x) + C\varepsilon^{1/2+\alpha/2}, \\ \inf_{y\in B_\varepsilon(x)} w_D(y) &\geq w_D(x) - C^2\mathrm{e}^{\alpha/2} w_D(x) - C\varepsilon^{1/2+\alpha/2}. \end{aligned} \tag{11.2.45}$$

Using these estimates in (7.2.78) of Corollary 7.30 with $\rho = \varepsilon$ and $A = B_\varepsilon(x)$, we get the claim in (11.2.40). □

Proposition 11.8 *Let x_j, $j = 1,\dots,n$, be the local minima of F. Let $\mathscr{S}_k = \cup_{i=1}^k B_\varepsilon(x_i)$, $k = 1,\dots,n$, be the collection of disjoint balls around the first k minima such that no ball contains a saddle point of F. Assume that, for all $j > k$, $F(x_i) > F(x_j)$ for all $i > k$ with $i \neq j$. Then, for all $j > k$*

$$\mathbb{E}_{x_j}[\tau_{\mathscr{S}_k}] = \frac{1}{\mathrm{cap}(B_\varepsilon(x_j), \mathscr{S}_k)} \frac{(2\pi\varepsilon)^{d/2}}{\sqrt{\det(\nabla^2 F(x_j))}} \mathrm{e}^{-F(x_j)/\varepsilon} \times \Big[1 + O\Big(\sqrt{\varepsilon[\ln(1/\varepsilon)]^3}, \varepsilon^{\alpha/2}\Big)\Big]. \tag{11.2.46}$$

Proof Fix $j > k$. Consider the set $\Gamma_j = \{y \in \Omega\colon\ F(y) \leq \Phi(x_j, \mathscr{S}_k) + \delta\}$ for $\delta > 0$ sufficiently small. Decompose Γ_j into its connected components: $\Gamma_j = \cup_{\tilde{\imath}} \Gamma_j(\tilde{\imath})$.

Write

$$\begin{aligned}&\int_{\Omega} dy\, \mathrm{e}^{-F(y)/\varepsilon} h_{B_\varepsilon(x_j),\mathscr{S}_k}(y)\\ &\quad = \int_{\Gamma_j^c} dy\, \mathrm{e}^{-F(y)/\varepsilon} h_{B_\varepsilon(x_j),\mathscr{S}_k}(y) + \sum_{\tilde{\imath}} \int_{\Gamma_j(\tilde{\imath})} dy\, \mathrm{e}^{-F(y)/\varepsilon} h_{B_\varepsilon(x_j),\mathscr{S}_k}(y).\end{aligned} \tag{11.2.47}$$

The first integral is bounded from above by $C\exp(-[\Phi(x_j,\mathscr{S}_k)+\delta]/\varepsilon)$ and is therefore negligible. The sum over $\tilde{\imath}$ can be split into $\tilde{\imath}\in L$ and $\tilde{\imath}\in R$ with $L=\{\tilde{\imath}\colon \Phi(x_i,\mathscr{S}_k)>\Phi(x_i,x_j)\ \forall x_i\in\Gamma_j(\tilde{\imath})\}$ and R the remaining $\tilde{\imath}$'s. The point of this decomposition is that $h_{B_\varepsilon(x_j),\mathscr{S}_k}(y)$ is close to 1 for $\tilde{\imath}\in L$ and $y\in\Gamma_j(\tilde{\imath})$, while $h_{B_\varepsilon(x_j),\mathscr{S}_k}(y)$ is close to 0 otherwise. (Here we make use of the fact that if $y,x_i\in\Gamma_j(\tilde{\imath})$ and $\Phi(x_i,\mathscr{S}_k)>\Phi(x_i,x_j)$, then $\Phi(y,\mathscr{S}_k)=\Phi(\tilde{\imath},\mathscr{S}_k)$.) We have

$$\sum_{\tilde{\imath}\in L}\int_{\Gamma_j(\tilde{\imath})} dy\, \mathrm{e}^{-F(y)/\varepsilon} h_{B_\varepsilon(x_j),\mathscr{S}_k}(y) = \sum_{\tilde{\imath}\in L}\int_{\Gamma_j(\tilde{\imath})} dy\, \mathrm{e}^{-F(y)/\varepsilon}\big(1-h_{\mathscr{S}_k,B_\varepsilon(x_j)}(y)\big). \tag{11.2.48}$$

By Corollary 11.5 and the upper bound on $\mathrm{cap}(B_\varepsilon(x_i),\mathscr{S}_k)$ provided by Theorem 11.2, we get, for $\tilde{\imath}\in L$ and $y\in\Gamma_j(\tilde{\imath})$,

$$0\le h_{\mathscr{S}_k,B_\varepsilon(x_j)}(y)\le C\varepsilon^{-d/2}\mathrm{e}^{-[\Phi(x_{\tilde{\imath}},\mathscr{S}_k)-\Phi(x_{\tilde{\imath}},x_j)]/\varepsilon}, \tag{11.2.49}$$

which is exponentially small. On the other hand, if $x_{\tilde{\imath}}$ denotes the absolute minimum of F within $\Gamma_j(\tilde{\imath})$ and the Hessian $\nabla^2 F(x_{\tilde{\imath}})$ at this minimum is non-degenerate, then

$$\int_{\Gamma_j(\tilde{\imath})\backslash\mathscr{S}_k} dy\, \mathrm{e}^{-F(y)/\varepsilon} = \frac{(2\pi\varepsilon)^{d/2}}{\sqrt{\det(\nabla^2 F(x_{\tilde{\imath}}))}}\,\mathrm{e}^{-F(x_{\tilde{\imath}})/\varepsilon}\Big[1+O\Big(\sqrt{\varepsilon[\ln(1/\varepsilon)]^3}\Big)\Big] \tag{11.2.50}$$

by standard Laplace asymptotics. Combining (11.2.48)–(11.2.50), we get

$$\begin{aligned}&\sum_{\tilde{\imath}\in L}\int_{\Gamma_j(\tilde{\imath})} dy\, \mathrm{e}^{-F(y)/\varepsilon} h_{B_\varepsilon(x_j),\mathscr{S}_k}(y)\\ &\quad = \sum_{\tilde{\imath}\in L}\frac{(2\pi\varepsilon)^{d/2}}{\sqrt{\det(\nabla^2 F(x_{\tilde{\imath}}))}}\,\mathrm{e}^{-F(x_{\tilde{\imath}})/\varepsilon}\Big[1+O\Big(\sqrt{\varepsilon[\ln(1/\varepsilon)]^3}\Big)\Big].\end{aligned} \tag{11.2.51}$$

Note that, under our assumptions on F, x_j is the unique value in the sum over $\tilde{\imath}$ for which $F(x_{\tilde{\imath}})$ takes its minimal value, and hence the sum is dominated by this single term.

The terms with $\tilde{\imath}\in R$ cannot be computed as precisely, but they are negligible. Indeed, note that, under our assumptions, all components $\Gamma_j(\tilde{\imath})$ that do not intersect $\mathscr{S}_k$ give a contribution that is smaller than $\exp(-F(x_j)/\varepsilon)$ times an exponentially small factor, and hence are negligible compared to what we get from (11.2.51).

Again using Corollary 11.5 when $\tilde{\iota} \in R$, we get

$$\int_{\Gamma_j(\tilde{\iota})} dy\, e^{-F(y)/\varepsilon} h_{B_\varepsilon(x_j),\mathscr{S}_k}(y) \le C\varepsilon^{-d} \int_{\Gamma_j(\tilde{\iota})\setminus\mathscr{S}_k} dy\, e^{-F(y)/\varepsilon} e^{-[\Phi(y,x_j)-\Phi(y,\mathscr{S}_k)]/\varepsilon}. \tag{11.2.52}$$

There are two possibilities. Either y is such that $\Phi(y,\mathscr{S}_k) = F(y)$. For those terms the integrand is bounded from above by $\exp(-\Phi(y,x_j)/\varepsilon) = \exp(-F(\mathscr{S}_k,x_j)/\varepsilon)$, which is exponentially small. The integral over all such y is therefore bounded by this small factor times the cardinality of the set $\Gamma_j(\tilde{\iota})$. All other y must lie in the valley A_i of a minimum x_i with $i > k$, and the contribution of such a valley can be at most of order $\exp(-F(x_j)/\varepsilon)$, which again under our assumptions is smaller than the main contribution from (11.2.51) times an exponentially small factor. Thus, in fact, all contributions coming from the terms with $\tilde{\iota} \in R$ are smaller than the main term in (11.2.51) by an exponentially small factor. Hence (11.2.46) holds. □

We are finally in a position to prove Theorem 11.3.

Proof The proof is immediate by inserting the formula for the capacity in Theorem 11.2 into (11.2.46), except for the error terms of order $\varepsilon^{\alpha/2}$, which we will show can be removed. Namely, note that nothing changes in the proof of Proposition 11.8 when we replace the starting point x_j by some point $x \in B_{\sqrt{\varepsilon}}(y)$. Also, inspection of the proof of Theorem 11.2 shows that the difference between $\text{cap}(B_\varepsilon(x_j),\mathscr{S}_k)$ and $\text{cap}(B_\varepsilon(x),\mathscr{S}_k)$ for $x \in B_{\sqrt{\varepsilon}}(y)$ is in fact much smaller than the error terms. Thus, we get

$$\text{osc}_{x\in B_{\sqrt{\varepsilon}}(x_j)} \mathbb{E}_x[\tau_{\mathscr{S}_k}] \le C\Big[\varepsilon^{\alpha/2} + \sqrt{\varepsilon\big[\ln(1/\varepsilon)\big]^3}\Big]\mathbb{E}_{x_j}[\tau_{\mathscr{S}_k}], \tag{11.2.53}$$

which improves the input in the Hölder estimate by a factor $\varepsilon^{\alpha/2}$, which in turn allows us to improve the error estimates in Proposition 11.8 from $\varepsilon^{\alpha/2}$ to ε^α. Iterating this procedure, we can reduce these errors until they are of the same order as $\sqrt{\varepsilon[\ln(1/\varepsilon)]^3}$. □

11.3 Spectral theory

In this section we turn to the analysis of the low-lying spectrum of the generator (11.1.3) with Dirichlet boundary conditions on Ω^c (when $\Omega \neq \mathbb{R}^d$). The strategy we follow is similar to that outlined in Sect. 8.4 in the context of discrete state spaces. The additional input that is needed is again the regularity estimates.

Assumption 11.1 on F ensures that the spectrum of $\mathscr{L}_\varepsilon$ is discrete. Moreover, it is well known from Wentzell-Freidlin theory [115] that the spectrum has precisely one exponentially small eigenvalue for each local minimum of the function F. We show how to get *sharp* estimates as $\varepsilon \downarrow 0$.

In Sect. 11.3.1 we state our main results. In Sect. 11.3.2 we derive a priori lower bounds on the spectrum. In Sect. 11.3.3 we look at the principal Dirichlet eigenvalue, in Sect. 11.3.4 at the small eigenvalues. In Sect. 11.3.5 we derive improved error estimates. In Sect. 11.3.6 we show that the exit times are asymptotically exponentially distributed.

11.3.1 Main results

Theorem 11.9 (Small eigenvalues) *Suppose that F has n local minima $x_1, \dots, x_n$, and that for some $\theta > 0$ these minima can be labeled in such a way that*

$$\Phi(x_k, \mathscr{M}_{k-1}) - F(x_k) \leq \min_{1 \leq l < k} \left[\Phi(x_l, \mathscr{M}_k \backslash x_l) - F(x_l)\right] - \theta, \quad k = 1, \dots, n, \tag{11.3.1}$$

where $\mathscr{M}_0 = \Omega^c$ and $\mathscr{M}_k = \{x_1, \dots, x_k\}$, $k = 1, \dots, n$. Set $B_k = B_\varepsilon(x_k)$ and $\mathscr{S}_k = \bigcup_{l=1}^k B_l$ and $h_k(y) = h_{B_k, \mathscr{S}_{k-1}}(y)$. Suppose that all essential gates $\mathscr{G}(x_k, \mathscr{M}_{k-1})$ consist of a single saddle point $z^*(x_k, \mathscr{M}_{k-1})$, and that the Hessian of F is non-degenerate at all the saddle points and all the local minima. Then there exists a $\delta > 0$ such that the n exponentially small eigenvalues $0 \leq \lambda_1 < \lambda_2 < \cdots < \lambda_n$ of $-\mathscr{L}_\varepsilon$ satisfy

$$\lambda_1 = 0 \tag{11.3.2}$$

and

$$\begin{aligned}
\lambda_k &= \frac{\operatorname{cap}(B_k, \mathscr{S}_{k-1})}{\|h_k\|_{2,\mu_\varepsilon}^2} \left[1 + O\left(\mathrm{e}^{-\delta/\varepsilon}\right)\right] \\
&= \frac{1}{\mathbb{E}_{x_k}[\tau_{\mathscr{S}_{k-1}}]} \left[1 + O\left(\mathrm{e}^{-\delta/\varepsilon}\right)\right] \\
&= \frac{[-\lambda_1^*(z^*(x_k, \mathscr{M}_{k-1}))]}{2\pi} \sqrt{\frac{\det(\nabla^2 F(x_k))}{-\det(\nabla^2 F(z^*(x_k, \mathscr{M}_{k-1})))}} \\
&\quad \times \mathrm{e}^{-[\Phi(x_k, \mathscr{M}_{k-1}) - F(x_k)]/\varepsilon} \\
&\quad \times \left[1 + O\left(\varepsilon^{1/2} \left[\ln(1/\varepsilon)\right]^{3/2}\right)\right], \quad k = 2, \dots, n,
\end{aligned} \tag{11.3.3}$$

where $\lambda_1^*(z^*)$ denotes the unique negative eigenvalue of the Hessian of F at the saddle point z^*.

The conditions in (11.3.1) state that *"all valleys of F have different depth"*, which is the generic situation. This is analogous to the condition that $\mathscr{M}$ be a regular set of metastable points made in Chap. 8.

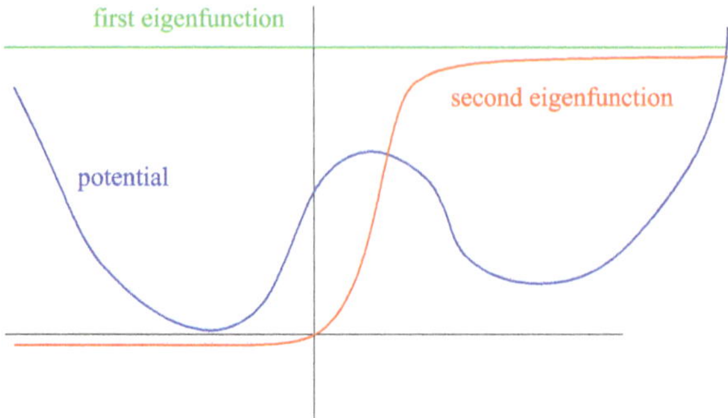

Fig. 11.3 First and second eigenfunction in a two-well potential

Remark 11.10 The Wentzell-Freidlin theory of metastability also provides estimates on the small eigenvalues, however, with less precise error estimates, namely,

$$\lim_{\varepsilon \downarrow 0} \ln \varepsilon \lambda_k = F(x_k) - \Phi(x_k, \mathscr{M}_{k-1}). \tag{11.3.4}$$

In the course of the proof of Theorem 11.9 we also obtain detailed control on the eigenfunctions of $-\mathscr{L}_\varepsilon$ corresponding to the small eigenvalues (see Fig. 11.3 for a schematic representation of the first two eigenfunctions in a double-well potential in one dimension).

Theorem 11.11 (Properties of eigenfunctions) *Under the assumptions of Theorem* 11.9, *if ϕ_k denotes the normalised eigenfunction corresponding to the eigenvalue λ_k, then there exists a $\delta > 0$ such that*

$$\phi_k(y) = \frac{h_k(y)}{\|h_k\|_{2,\mu_\varepsilon}} + O\big(\mathrm{e}^{-\delta/\varepsilon}\big), \quad k = 1, \dots, n. \tag{11.3.5}$$

Finally, metastable exit times are asymptotically exponentially distributed when appropriate non-degeneracy conditions are met.

Theorem 11.12 (Exponential law of metastable exit times) *Suppose that the assumptions of Theorem* 11.9 *are satisfied. Let $D \subset \mathbb{R}^d$ be a closed subset such that:*

(i) *If $\mathscr{M}_k = \{y_1, \dots, y_l\} \subset \mathscr{M}$ enumerates all the minima of F such that $F(y_l) \leq F(x_k)$, then $\bigcup_{l=1}^k B_\varepsilon(y_l) \subset D$.*
(ii) $\operatorname{dist}(z^*(x_i, \mathscr{M}_i), D) \geq \delta > 0$ *for some $\delta > 0$ independent of ε.*

Then there exists a $\delta > 0$, independent of ε and t, such that, for all $t > 0$,

$$\begin{aligned} \mathbb{P}_{x_k}\big(\tau_D > t\,\mathbb{E}_{x_k}[\tau_D]\big) &= \big[1 + O\big(\mathrm{e}^{-\delta/\varepsilon}\big)\big]\mathrm{e}^{-t[1+O(\mathrm{e}^{-\delta/\varepsilon})]} \\ &\quad \times O\big(\mathrm{e}^{-\delta/\varepsilon}\big) \sum_{l>k} \mathrm{e}^{-t\lambda_l \mathbb{E}_{x_k}[\tau_D]} + O(1)\,\mathrm{e}^{-t\,O(\varepsilon^{d-1})\,\mathbb{E}_{x_k}[\tau_D]}. \end{aligned} \tag{11.3.6}$$

11.3.2 A priori spectral estimates

In this section we derive a priori lower bounds on principal eigenvalues for the Dirichlet problem in regular open sets $D \subset \Omega \subseteq \mathbb{R}^d$. The closure of D is denoted by $\bar{D}$, the complement by D^c, and the boundary by ∂D. We denote by $\lambda_0^{D^c}$ the principal (= smallest) eigenvalue of the Dirichlet problem

$$\begin{aligned}(-\mathscr{L}_\varepsilon - \lambda) f(x) &= 0, \quad x \in D,\\ f(x) &= 0, \quad x \in D^c.\end{aligned} \tag{11.3.7}$$

We sometimes use the notation $\mathscr{L}_\varepsilon^D$ to indicate the Dirichlet operator corresponding to (11.3.7).

The following lemma improves the Donsker-Varadhan estimate in Lemma 8.16 when D is unbounded.

Lemma 11.13 *Let ϕ_D denote the normalised eigenfunction corresponding to the principal eigenvalue of $-\mathscr{L}_\varepsilon^D$, and let $A \subset D$ be a compact set. Then*

$$\lambda_0^{D^c} \geq \frac{1}{\sup_{x\in A} \mathbb{E}_x[\tau_{D^c}]} \left(1 - \int_{D\setminus A} dy\, \mathrm{e}^{-F(y)/\varepsilon} \phi_D(y)^2\right). \tag{11.3.8}$$

Moreover, for any $\delta > 0$, there exists a bounded set $A \subset D$, independent of ε, such that

$$\lambda_0^{D^c} \geq \frac{1-\delta}{\sup_{x\in A} \mathbb{E}_x[\tau_{D^c}]}. \tag{11.3.9}$$

For $B \subset D$,

$$\lambda_0^{D^c\cup B} \geq \frac{1}{\sup_{x\in A} \mathbb{E}_x[\tau_B | \tau_B \leq \tau_{D^c}]} \left(1 - \int_{D\setminus A} dy\, \mathrm{e}^{-F(y)/\varepsilon} \phi_{D\setminus B}(y)^2\right). \tag{11.3.10}$$

Proof Let $w(x)$ denote the solution of the Dirichlet problem

$$\begin{aligned}(-\mathscr{L}_\varepsilon w)(x) &= 1, \quad x \in D,\\ w(x) &= 0, \quad x \in D^c.\end{aligned} \tag{11.3.11}$$

Note that $w(x) = \mathbb{E}_x[\tau_{D^c}]$. Moreover,

$$\begin{aligned}&\int_D dx\, \mathrm{e}^{-F(x)/\varepsilon} \phi(x)(-\mathscr{L}_\varepsilon \phi)(x)\\ &= \int_D dx\, \mathrm{e}^{-F(x)/\varepsilon}\, \nabla\phi(x)\cdot\nabla\phi(x)\\ &= \int_D dx\, \mathrm{e}^{-F(x)/\varepsilon} \lim_{h\downarrow 0} h^{-2} \sum_{i=1}^d [\phi(x+he_i) - \phi(x)]^2.\end{aligned} \tag{11.3.12}$$

Using that, for any $a, b \in \mathbb{R}$ and $C > 0$, $ab \leq \frac{1}{2}(Ca^2 + b^2/C)$, and picking $a = \phi(x + he_i)$, $b = \phi(x)$ and $C = w(x)/w(x + he_i)$, we have

$$\big[\phi(x+he_i) - \phi(x)\big]^2 \geq \left(\frac{\phi(x+he_i)^2}{w(x+he_i)} - \frac{\phi(x)^2}{w(x)}\right)\big[w(x+he_i) - w(x)\big]. \tag{11.3.13}$$

Inserting this inequality into (11.3.12), we obtain

$$\begin{aligned}\int_D dx\, \mathrm{e}^{-F(x)/\varepsilon}\phi(x)(-\mathscr{L}_\varepsilon\phi)(x) &\geq \int_D dx\, \mathrm{e}^{-F(x)/\varepsilon}\frac{\phi(x)^2}{w(x)}(-\mathscr{L}_\varepsilon w)(x)\\ &= \int_D dx\, \mathrm{e}^{-F(x)/\varepsilon}\frac{\phi(x)}{w(x)}\phi(x)\\ &\geq \frac{1}{\sup_{x\in A} w(x)}\int_A dx\, \mathrm{e}^{-F(x)/\varepsilon}\phi(x)^2.\end{aligned} \tag{11.3.14}$$

Choosing ϕ as the normalised eigenfunction of $-\mathscr{L}_\varepsilon^D$ with maximal eigenvalue, we arrive at (11.3.8).

Next we claim that, for any $\gamma > 0$,

$$\int_D dy\, \mathrm{e}^{-\gamma\widetilde{F}(y)/\varepsilon}\phi_D(y)^2 < C_\gamma < \infty, \tag{11.3.15}$$

where $\widetilde{F}(y) = \min_{x\in\mathscr{M}}[F(y) - F(x)]$. Clearly this implies (11.3.9). To see why (11.3.15) is true, set $v(y) = \mathrm{e}^{-F(y)/2\varepsilon}\phi_D(y)$, which is the corresponding ground-state eigenfunction of the operator

$$-\big(\mathrm{e}^{-F/2\varepsilon}\mathscr{L}_\varepsilon \mathrm{e}^{F/2\varepsilon}\big)(x) = -\varepsilon\Delta + \frac{1}{4\varepsilon}\big|\nabla F(x)\big|^2 - \frac{1}{2}\Delta F(x), \tag{11.3.16}$$

which is a symmetric operator on $L^2(D, dy)$. A semi-classical Agmon estimate for the ground-state eigenfunction v that can be found in Helffer and Sjöstrand [138] yields

$$\int_D dy\, \mathrm{e}^{(1-\gamma)\widetilde{F}(y)/\varepsilon}v(y)^2 < C_\gamma < \infty, \tag{11.3.17}$$

which in turn implies (11.3.15). To obtain (11.3.10), note that $w_{B,D}(x) = \mathbb{E}_x[\tau_B|\tau_B \leq \tau_{D^c}]$, $x \in D\backslash B$, solves the Dirichlet problem

$$\begin{aligned}(-\mathscr{L}_\varepsilon w_{B,D})(x) &= h_{B,D^c}(x), \quad x \in D\backslash B,\\ w_{B,D}(x) &= 0, \quad x \in B \cup D^c.\end{aligned} \tag{11.3.18}$$

Rerunning the proof of (11.3.8) with w replaced by $w_{B,D}$, we obtain (11.3.10). □

We will next establish that $\lambda_0^{D^c}$ is at most polynomially small in ε when D does not contain local minima. Define

$$\mathscr{M}_\varepsilon = \big\{z \in \Omega:\ \mathrm{dist}(z, \mathscr{M}) \leq \varepsilon\big\}. \tag{11.3.19}$$

Lemma 11.14 *Assume that $D\cap\mathcal{M}_{2\varepsilon}=\emptyset$. Then there is a finite positive constant C, independent of ε, such that*

$$\sup_{x\in D}\mathbb{E}_x[\tau_{D^c}]\leq C\varepsilon^{-2d+2}\sup_{x\in D}\int_\Omega \mathbb{1}_{F(y)\leq F(x)}dy. \tag{11.3.20}$$

Proof We start from the relation

$$\int_D dy\,\mathrm{e}^{-F(y)/\varepsilon}h_{B_\varepsilon(x),D^c}(y)\geq\Big(\inf_{z\in\partial B_\varepsilon(x)}\mathbb{E}_z[\tau_{D^c}]\Big)\,\mathrm{cap}\big(B_\varepsilon(x),D^c\big), \tag{11.3.21}$$

which is an immediate consequence of Corollary 7.30. The Harnack inequality in Lemma 9.13 and the representation formula (7.2.47) give

$$\sup_{z\in\partial B_\varepsilon(x)}\mathbb{E}_z[\tau_{D^c}]\leq C\inf_{z\in\partial B_\varepsilon(x)}\mathbb{E}_z[\tau_{D^c}]. \tag{11.3.22}$$

Combining this with (11.3.21), we get

$$\sup_{z\in\partial B_\varepsilon(x)}\mathbb{E}_z[\tau_{D^c}]\leq C\frac{\int_D dy\,\mathrm{e}^{-F(y)/\varepsilon}h_{B_\varepsilon(x),D^c}(y)}{\mathrm{cap}(B_\varepsilon(x),D^c)}. \tag{11.3.23}$$

We distinguish between the regions $\{y\in D\colon\ F(y)>F(x)\}$ and $\{y\in D\colon\ F(y)\leq F(x)\}$ in the integral. In the former, we use that $h_{B_\varepsilon(x),D^c}(y)\leq 1$, while in the latter we use the upper bound in Theorem 9.10. This gives

$$\begin{aligned}\sup_{z\in\partial B_\varepsilon(x)}&\mathbb{E}_z[\tau_{D^c}]\\ &\leq C\frac{\int_{\{y\in D\colon\ F(y)>F(x)\}}dy\,\mathrm{e}^{-F(y)/\varepsilon}}{\mathrm{cap}(B_\varepsilon(x),D^c)}\\ &\quad+C\frac{1}{\mathrm{cap}(B_\varepsilon(x),D^c)}\int_{\{y\in D\colon\ F(y)\leq F(x)\}}dy\,\mathrm{e}^{-F(y)/\varepsilon}\frac{\mathrm{cap}(B_\varepsilon(y),B_\varepsilon(x))}{\mathrm{cap}(B_\varepsilon(y),D^c)}.\end{aligned} \tag{11.3.24}$$

Using the bounds on capacities given in Proposition 11.4, with $\rho=\varepsilon$, we get

$$\begin{aligned}\sup_{z\in\partial B_\varepsilon(x)}\mathbb{E}_z[\tau_{D^c}]\leq C'\varepsilon^{-d+1}\mathrm{e}^{F(x)/\varepsilon}&\int_{\{y\in D\colon\ F(y)>F(x)\}}dy\,\mathrm{e}^{-F(y)/\varepsilon}\\ +C'\varepsilon^{-2d+2}&\int_{\{y\in D\colon\ F(y)\leq F(x)\}}dy.\end{aligned} \tag{11.3.25}$$

By our assumption on F, the first integral is bounded by a constant times $\mathrm{e}^{-F(x)/\varepsilon}$ and the second integral is equal to the volume of the level set $\{y\in D\colon\ F(y)\leq F(x)\}$. The second term in (11.3.25) is dominant. □

Corollary 11.15 *If $D \cap \mathscr{M}_{2\varepsilon} = \emptyset$, then there exists a finite positive constant C, independent of ε, such that*

$$\lambda_0^{D^c} \geq C\varepsilon^{2d-2}. \tag{11.3.26}$$

We can generalise the bounds obtained so far to sets D containing some of the local minima of F. Let $\mathscr{N} \subset \mathscr{M}$ be non-empty, and let

$$\mathscr{N}_\varepsilon = \left\{y \in \mathbb{R}^d \colon \operatorname{dist}(y, \mathscr{N}) \leq \varepsilon\right\}. \tag{11.3.27}$$

Assume that $D \supset \mathscr{N}_\varepsilon$, and set

$$A(x) = \left\{y \in D \colon h_{B_\varepsilon(x), D^c \backslash B_\varepsilon(x)}(y) = \max_{z \in \mathscr{M}} h_{B_\varepsilon(x), D^c \backslash B_\varepsilon(x)}(z)\right\}. \tag{11.3.28}$$

Lemma 11.16 *Under the assumptions of Lemma* 11.13,

$$\frac{1}{\lambda_0^{D^c}} \leq \sum_{k \colon x_k \in \mathscr{N}_\varepsilon} \frac{\int_{A(x_k)} \mathrm{e}^{-F(y)/\varepsilon} dy}{\operatorname{cap}(B_\varepsilon(x_k), D \backslash B_\varepsilon(x_k))}. \tag{11.3.29}$$

Proof The proof is similar to that of Lemma 11.14 when combined with the estimate on mean exit times given in Proposition 11.8. We leave the details to the reader. □

Lemma 11.16 and Theorem 11.2 imply, under the assumptions of Theorem 11.9, that

$$\lambda_0^{D^c} \geq C_\varepsilon \min_{k \colon x_k \in \mathscr{N}_\varepsilon} \mathrm{e}^{-[\Phi(x_k, \mathscr{M}_k) - F(x_k)]/\varepsilon}, \tag{11.3.30}$$

where C_ε is polynomially bounded in ε. This rough bound will be made more precise in the next section.

11.3.3 Principal Dirichlet eigenvalues

We now give a precise characterisation of Dirichlet eigenvalues.

General strategy

Let $x_i, i = 1, \dots, n$, be the local minima of F labelled as in Theorem 11.9. Let $B_i = B_\varepsilon(x_i) \subset \Omega, i = 1, \dots, n$, be ε-balls around them. For $n \geq k \geq 1$, set $\mathscr{S}_k = \bigcup_{i=1}^k B_i$ and let $\bar{\lambda}_k = \lambda_0^{\mathscr{S}_k}$ denote the principal eigenvalue of the Dirichlet operator $\mathscr{L}_\varepsilon$ with Dirichlet boundary conditions on $\partial \mathscr{S}_k = \bigcup_{i=1}^k \partial B_i$ (and on $\partial \Omega$ when this is not empty). For $\lambda < \bar{\lambda}_k$, consider the Dirichlet problem

$$
\begin{aligned}
(-\mathscr{L}_\varepsilon - \lambda) f^\lambda(x) &= 0, & x &\in \Omega \backslash \partial \mathscr{S}_k, \\
f^\lambda(x) &= \phi(x), & x &\in \partial \mathscr{S}_k, \\
f^\lambda(x) &= 0, & x &\in \partial \Omega.
\end{aligned}
\tag{11.3.31}
$$

This is the Dirichlet problem in the exterior and the interior of the balls simultaneously (note that the principal eigenvalue of $-\mathscr{L}_\varepsilon$ within a ball is larger than $\bar{\lambda}_k$, and therefore plays no rôle). In the sequel, when we specify Dirichlet problems, the vanishing of the solution on the boundary of Ω will always be understood and will not be mentioned anymore.

The basic idea is to construct an eigenfunction of the full operator $-\mathscr{L}_\varepsilon$ as a solution of the problem in (11.3.31) with a suitably chosen ϕ. Namely, if $\lambda \leq \bar{l}$, then (11.3.31) has a unique solution. Suppose that there is an eigenfunction ϕ^λ with eigenvalue λ. If we choose $\phi(x) = \phi^\lambda(x)$ as boundary condition, then ϕ^λ solves (11.3.31), which shows that eigenfunctions can be obtained as solutions of this Dirichlet problem for suitable boundary conditions. On the other hand, when we want to check whether λ is an eigenvalue, we just need to verify whether there is a function ϕ on $\partial \mathscr{S}_k$ such that the corresponding solution f^λ of (11.3.31) also verifies

$$
(-\mathscr{L}_\varepsilon - \lambda) f^\lambda(x) = 0, \quad \forall x \in \partial \mathscr{S}_k. \tag{11.3.32}
$$

In fact, $(-\mathscr{L}_\varepsilon - \lambda) f^\lambda$ in general is a measure concentrated on the surface $\partial \mathscr{S}_k$. Demanding that this surface measure be zero, we are led to an integral equation for ϕ on $\partial \mathscr{S}_k$ that is not particularly easy to handle.

This procedure is completely analogous to that in Chap. 8 for Markov processes with countable state space. There, instead of balls B_i we just had points x_i. The equations in (11.3.32) were just k equations, and the boundary condition reduced to the k numbers $\phi(x_i)$. This led to a set of *linear equations* for the unknown vector $\phi(x_i)$, $i = 1, \dots, k$. The condition for λ to be an eigenvalue reduced to the vanishing of a certain determinant. It would be nice if in the present setting we could reduce the computation to a similarly simple condition. Indeed, this will be almost the case, due to the fact that the eigenfunctions are very close to constants on the balls B_i.

We begin the program in this section with the somewhat simpler problem of the computation of the principal eigenvalues in domains $D \subset \Omega$. This problem is considerably simpler because principal eigenfunctions are positive. Later the main application will be to the case where D equals Ω with some small balls around local minima of F removed.

Regularity properties of eigenfunctions

We first state a simple application of the Harnack and Hölder inequalities in Lemmas 9.8–9.9.

Lemma 11.17 *Assume that x is a local minimum of F. Let ϕ be a positive strong solution of $(-\mathscr{L}_\varepsilon - \lambda)\phi = 0$, $|\lambda| \leq 1$, on the ball $B_{4\sqrt{\varepsilon}}(x)$. Then there exist* $0 < C <$

∞ *and* $\alpha > 0$, *both independent of* ε, *such that*

$$\operatorname{osc}_{y\in B_\varepsilon(x)}\phi(y) \leq C\varepsilon^{\alpha/2} \min_{y\in B_\varepsilon(x)} \phi(x). \tag{11.3.33}$$

Sharp estimates on principal eigenvalues

We want to improve the estimates on principal eigenvalues $\lambda_0^{D^c}$ obtained in Sect. 11.3.2 when D contains a local minimum of F.

Proposition 11.18 *Assume that D contains l local minima of the function F and that there is a single minimum* $x \in D$ *that realises*

$$\Phi\big(x, D^c\big) - F(x) = \max_{1\leq i\leq l}\big[\Phi\big(x_i, D^c\big) - F(x_i)\big]. \tag{11.3.34}$$

Write $B = B_\varepsilon(x)$. *Then there exist* $\alpha > 0$, $C < \infty$ *and* $\delta > 0$, *independent of* ε, *such that the principal eigenvalue* $\lambda_0^{D^c}$ *of the Dirichlet problem on D satisfies*

$$\frac{\operatorname{cap}(B, D^c)}{\|h_{B,D^c}\|_{2,\mu_\varepsilon}^2}\big(1 - C\varepsilon^{\alpha/2}\big)\big(1 - \mathrm{e}^{-\delta/\varepsilon}\big) \leq \lambda_0^{D^c} \leq \frac{\operatorname{cap}(B, D^c)}{\|h_{B,D^c}\|_{2,\mu_\varepsilon}^2}\big(1 + C\varepsilon^{\alpha/2}\big)\big(1 + \mathrm{e}^{-\delta/\varepsilon}\big), \tag{11.3.35}$$

where $\|\cdot\|_{2,\mu_\varepsilon}$ *denotes the* L^2*-norm with respect to the measure* $\mu_\varepsilon(dy) = \mathrm{e}^{-F(y)/\varepsilon}dy$. *In particular,*

$$\begin{aligned}&\frac{\operatorname{cap}(B_k, \mathscr{S}_{k-1})}{\|h_{B_k,\mathscr{S}_{k-1}}\|_{2,\mu_\varepsilon}^2}\big(1 - C\varepsilon^{\alpha/2}\big)\big(1 - \mathrm{e}^{-\delta/\varepsilon}\big)\\ &\quad\leq \bar{\lambda}_k \leq \frac{\operatorname{cap}(B_k, \mathscr{S}_{k-1})}{\|h_{B_k,\mathscr{S}_{k-1}}\|_{2,\mu_\varepsilon}^2}\big(1 + C\varepsilon^{\alpha/2}\big)\big(1 + \mathrm{e}^{-\delta/\varepsilon}\big).\end{aligned} \tag{11.3.36}$$

Proof We know by Lemma 11.16 that

$$\lambda_0^{D^c\cup B} \geq \mathrm{e}^{-[F(z^*(x,D^c))-F(x)]/\varepsilon}\mathrm{e}^{\delta/\varepsilon}. \tag{11.3.37}$$

We also know that $\lambda_0^{D^c} < \lambda_0^{D^c\cup B}$ (and expect $\lambda_0^{D^c} \approx \mathrm{e}^{-[F(z^*(x,D^c))-F(x)]/\varepsilon}$, which is much smaller than the lower bound in (11.3.37)). By the philosophy outlined above, we know that the principal eigenfunction can be represented as the solution of the Dirichlet problem (both inside B and outside B)

$$\begin{aligned}(-\mathscr{L}_\varepsilon - \lambda)f^\lambda(y) &= 0, &\quad y &\in D\backslash\partial B,\\ f^\lambda(y) &= \phi_D(y), & y &\in \partial B,\\ f^\lambda(y) &= 0, & y &\in D^c,\end{aligned} \tag{11.3.38}$$

where the boundary conditions ϕ_D are given by the actual principal eigenfunction. We assume that $\operatorname{dist}(x, D^c) \geq \delta > 0$, with δ independent of ε. Then $B_{4\sqrt{\varepsilon}}(x) \subset$

D, and since ϕ_D is the principal eigenfunction, it may be chosen positive on D. Therefore Lemma 11.17 applies and shows that

$$\inf_{y\in\partial B}\phi_D(y)=c\le\sup_{y\in\partial B}\phi_D(y)\le\big(1+C\varepsilon^{\alpha/2}\big)c. \tag{11.3.39}$$

We normalise the eigenfunction such that $c=1$. Then $f^\lambda(x)=h^\lambda_{B,D^c}(x)+\chi^\lambda_{B,D^c}(x)$, where h^λ_{B,D^c} is the λ-equilibrium potential that solves

$$\begin{aligned}(-\mathscr{L}_\varepsilon-\lambda)h^\lambda_{B,D^c}(y)&=0, & y&\in D\backslash\partial B,\\ h^\lambda_{B,D^c}(y)&=1, & y&\in\partial B,\\ h^\lambda_{B,D^c}(y)&=0, & y&\in D^c,\end{aligned} \tag{11.3.40}$$

while χ^λ_{B,D^c} solves

$$\begin{aligned}(-\mathscr{L}_\varepsilon-\lambda)\chi^\lambda_{B,D^c}(y)&=0, & y&\in D\backslash\partial B,\\ \chi^\lambda_{B,D^c}(y)&=\phi_D(y)-1, & y&\in\partial B,\\ \chi^\lambda_{B,D^c}(y)&=0, & y&\in D^c.\end{aligned} \tag{11.3.41}$$

We want that $(-\mathscr{L}_\varepsilon-\lambda)f^\lambda$ vanishes also as a surface measure on ∂B. This requires that there is no discontinuity in the derivative of f^λ normal to ∂B, which we can express as saying that, for g a smooth test function that vanishes on D^c,

$$\int_{\partial B}\mathrm{e}^{-F(y)/\varepsilon}\big[g(y)\partial_{n(y)}f^\lambda(y)+g(y)\partial_{-n(y)}f^\lambda(y)\big]d\sigma_B(y)=0, \tag{11.3.42}$$

where $d\sigma_B(y)$ denotes the Euclidean surface measure on ∂B, and $\partial_{\pm n(y)}$ denotes the normal derivative at $y\in\partial B$ from the exterior and interior of B, respectively. As we will see, it already suffices to require that this equation hold for functions g that are equal to 1 on ∂B. In fact, we will choose $g=h_{B,D^c}$. To evaluate this expression, it will be convenient to observe that $h_{B,D^c}(y)=1$ for $y\in\partial B$. Moreover, $h_{B,D^c}(y)=1$ on B, so that $\partial_{-n(y)}h_{B,D^c}(y)$ vanishes on ∂B. Using these facts, together with the second Green identity, we get from (11.3.42) the condition

$$\begin{aligned}0&=\int_{\partial B}\mathrm{e}^{-F(y)/\varepsilon}\partial_{n(y)}h_{B,D^c}(y)f^\lambda(y)d\sigma_B(y)-\frac{\lambda}{\varepsilon}\int_D dy\,\mathrm{e}^{-F(y)/\varepsilon}h_{B,D^c}(y)f^\lambda(y)\\ &=\int_{\partial B}\mathrm{e}^{-F(y)/\varepsilon}\partial_{n(y)}h_{B,D^c}(y)d\sigma_B(y)-\frac{\lambda}{\varepsilon}\int_D dy\,\mathrm{e}^{-F(y)/\varepsilon}h_{B,D^c}(y)h^\lambda_{B,D^c}(y)\\ &\quad+\int_{\partial B}\mathrm{e}^{-F(y)/\varepsilon}\partial_{n(y)}h_{B,D^c}(y)\chi^\lambda_{B,D^c}(y)d\sigma_B(y)\\ &\quad-\frac{\lambda}{\varepsilon}\int_D dy\,\mathrm{e}^{-F(y)/\varepsilon}h_{B,D^c}(y)\chi^\lambda_{B,D^c}(y).\end{aligned} \tag{11.3.43}$$

(Note that the derivative $\partial_{n(y)}$ is in the direction of the interior of B.) The two terms involving χ^λ_{B,D^c} will be naturally treated as error terms. Since $\partial_{n(y)} h_{B,D^c}(y) > 0$, we get via Lemma 11.17 that

$$0 \leq \int_{\partial B} \mathrm{e}^{-F(y)/\varepsilon} \partial_{n(y)} h_{B,D^c}(y) \chi^\lambda_{B,D^c}(y) \leq C\varepsilon^{\alpha/2} \int_{\partial B} \mathrm{e}^{-F(y)/\varepsilon} \partial_{n(y)} h_{B,D^c}(y). \tag{11.3.44}$$

Defining $\delta\chi^\lambda_{B,D^c} = \chi^\lambda_{B,D^c} - \chi^0_{B,D^c}$, we see that $\delta\chi^\lambda_{B,D^c}$ solves the Dirichlet problem

$$\begin{aligned}(-\mathscr{L}_\varepsilon - \lambda)\delta\chi^\lambda_{B,D^c}(y) = \lambda\chi^0_{B,D^c}(y), &\quad y \in D\backslash\partial B,\\ \delta\chi^\lambda_{B,D^c}(y) = 0, &\quad y \in \partial B \cup D^c.\\ \delta\chi^\lambda_{B,D^c}(y) = 0, &\quad y \in D^c.\end{aligned} \tag{11.3.45}$$

In complete analogy with Lemma 8.21, we get the following $L^2(\mu_\varepsilon)$-estimates.

Lemma 11.19

(i)

$$\|\delta\chi^\lambda_{B,D^c}\|_{2,\mu_\varepsilon} \leq \frac{\lambda}{\lambda_0^{D^c\cup B} - \lambda} \|\chi^0_{B,D^c}\|_{2,\mu_\varepsilon}. \tag{11.3.46}$$

(ii)

$$\|h^\lambda_{B,D^c} - h_{B,D^c}\|_{2,\mu_\varepsilon} \leq \frac{\lambda}{\lambda_0^{D^c\cup B} - \lambda} \|h_{B,D^c}\|_{2,\mu_\varepsilon}. \tag{11.3.47}$$

(iii) *For all* $z \in D\backslash B$,

$$0 \leq \chi^0_{B,D^c}(z) \leq C\varepsilon^{\alpha/2} h_{B,D^c}(z). \tag{11.3.48}$$

Proof Items (i) and (ii) are the standard L^2-bounds as used in Lemma 8.21. Item (iii) follows from the Poisson kernel representation of χ^0_{B,D^c},

$$\chi^0_{B,D^c}(x) = -\varepsilon \int_{\partial B} (\phi_D(y) - 1)\partial_{n(y)} G_{D\backslash B}(x,y) d\sigma_B(y). \tag{11.3.49}$$

Since the normal derivative of the Green function $G_{D\backslash B}(x,y)$ is negative on ∂B and $\phi_D(y) \geq 1$ on ∂B, we get (11.3.48). □

Using the estimates above, together with

$$\varepsilon \int_{\partial B} \mathrm{e}^{-F(y)/\varepsilon} \partial_{n(y)} h_{B,D^c}(y) = \mathrm{cap}\big(B, D^c\big), \tag{11.3.50}$$

we see that (11.3.43) implies that

$$\left|\lambda - \frac{\mathrm{cap}(B, D^c)}{\|h_{B,D^c}\|^2_{2,\mu_\varepsilon}}\right| \leq C\varepsilon^{\alpha/2} \frac{\mathrm{cap}(B, D^c)}{\|h_{B,D^c}\|^2_{2,\mu_\varepsilon}} + \frac{\lambda}{\lambda_0^{D^c\cup B} - \lambda}. \tag{11.3.51}$$

This yields the bounds on $\lambda_0^{D^c}$ in (11.3.35). Note that, while we have only used a necessary condition for $\lambda_0^{D^c}$, the fact that there must be such an eigenvalue implies that it actually lies between the bounds given by (11.3.51). □

Uniform estimates on principal eigenfunctions

In complete analogy with Lemma 8.22 we can improve the L^2-estimates to uniform estimates.

Lemma 11.20 *With the notation above, the following estimates hold for all* ε *small enough*:

(i) *For all* $z \in D$,

$$\left|\chi^{\lambda}_{B,D^c}(z)\right| \le 2\frac{\lambda}{\lambda_0^{D^c\cup B}-\lambda}\left|\chi^0_{B,D^c}(z)\right|. \tag{11.3.52}$$

(ii) *For all* $z \in D\backslash B$,

$$\left|h^{\lambda}_{B,D^c}(z) - h_{B,D^c}(z)\right| \le \frac{\lambda}{a(D,B)-\lambda}\left|h_{B,D^c}(z)\right|, \tag{11.3.53}$$

where $a(D,B) = \inf_{y\in D\backslash B} \frac{1}{\mathbb{E}[\tau_B|\tau_B\le\tau_{D^c}]}$.

(iii) *For all* $z \in B$, $\chi^0_{B,D^c}(z) \le C\varepsilon^{\alpha/2}$.

(iv) *Consequently, the eigenfunction* ϕ_D, *normalised such that* $\inf_{y\in\partial B}\phi_D(y) = 1$, *satisfies, for all* $z \in D$,

$$h_{B,D^c}(y) \le \phi_D(y) \le h_{B,D^c}(y)\left(1 + C\varepsilon^{\alpha/2}\right)\left(1 + \mathrm{e}^{-\delta/\varepsilon}\right). \tag{11.3.54}$$

Proof Items (i) and (ii) follow from the same arguments that were used in the proof of Lemma 8.4.21. Item (iii) follows from the maximum principle. Combine these estimates to get (iv). □

Remark 11.21 Note that $a(D,B) = 1/\lambda_0^{D^c\cup B}[1 + o(1)]$ for sets $D\backslash B$ that do contain a local minimum of F.

11.3.4 Exponentially small eigenvalues and their eigenfunctions

The goal of this section is to generalise the analysis in Sect. 11.3.3 to all small eigenvalues of $-\mathscr{L}_\varepsilon$. To do this, we need to first establish some a priori estimates on the behaviour of eigenfunctions near the local minima of F.

A priori estimates on eigenfunctions near local minima

For the analysis of harmonic functions that are not necessarily positive, we need an estimate for sub-harmonic functions that allows us to relate the oscillation to the L^2-norm.

Lemma 11.22 *Let ϕ be a strong solution of $(-\mathscr{L}_\varepsilon - \lambda)\phi = 0$ on the ball $B_{c\sqrt{\varepsilon}}(x)$. Then there exist a $C < \infty$ independent of ε such that*

$$\operatorname{osc}_{B_{c\sqrt{\varepsilon}}} \phi \leq C\varepsilon^{-d/4} \left(\int_{B_{2c\sqrt{\varepsilon}}} \phi(x)^2 dx \right)^{1/2}. \tag{11.3.55}$$

Proof This is just a specialisation of Gilbar and Trudinger [126, Theorem 9.20] (which gives upper bounds on suprema of sub-harmonic functions in terms of L^p-norms), and is obtained after choosing the balls in such a way that the constants are uniform in ε. □

We want to show that in $\sqrt{\varepsilon}$-neighbourhoods the eigenfunctions corresponding to the exponentially small eigenvalues of $-\mathscr{L}_\varepsilon$ either have a constant sign or are irrelevantly small. This property is suggested by the following result.

Lemma 11.23 *Let ϕ be a normalised eigenfunction of $-\mathscr{L}_\varepsilon$ corresponding to one of the $|\mathscr{M}|$ smallest eigenvalues. Let $\gamma < \hat{\gamma} = \min_{x,y \in \mathscr{M}} [\Phi(x,y) - F(y)]$. For $i = 1, \dots, n$, let D_i be the set of points in $y \in \Omega$ such that the solution of the differential equation $\frac{d}{dt} y(t) = -\nabla F(y(t))$ with initial condition $y(0) = y$ converges to $x_i \in \mathscr{M}$. Then there exist constants c_i, $i = 1, \dots, n$, such that*

$$\left\| \phi - \sum_{i=1}^n c_i \mathbb{1}_{D_i} \right\|_{2,\mu_\varepsilon} \leq C e^{-\gamma/\varepsilon}, \tag{11.3.56}$$

for some $C = C_\gamma < \infty$.

Proof This proposition is stated and proved in Kolokoltsov [154] for smooth F, but it is easy to check that the proof carries through for $F \in C^3(\Omega)$. □

Unfortunately Lemma 11.23 is not quite enough to conclude that Φ is not changing sign near any minimum. We will, however, show that this is the case when the contribution of ϕ coming from a neighbourhood of a given minimum is significant. To that end, for $D \subset \Omega$ set

$$\|f\|_{2,\mu_\varepsilon,D} = \left(\int_D f(x)^2 \mu_\varepsilon(dx) \right)^{1/2}. \tag{11.3.57}$$

For a given eigenfunction ϕ, define the set

$$J = \left\{ 1 \leq j \leq n \colon \|\phi\|_{2,\mu_\varepsilon,D_j} \geq e^{-\gamma/2\varepsilon} \right\}, \tag{11.3.58}$$

where D_j are the sets defined in Lemma 11.23.

Lemma 11.24 *If ϕ is one of the eigenfunctions of Lemma 11.23 and $j \in J$, then there exist positive and finite constants c_j, C, a, independent of ε, such that $|\phi(x) - c_j| \leq C\varepsilon^{\alpha/2} c_j$, for all $x \in B_{\sqrt{\varepsilon}}(x_j)$.*

Proof We will first show that the weighted L^2-estimate on the deviation of ϕ from a constant implies a local unweighted L^2-estimate on balls of radius $r = \sqrt{\varepsilon}$ near the minima x_j, $j \in J$. To that end, note that (11.3.56) implies that

$$\|\phi - c_j\|_{2,\mu_\varepsilon,D_j} \leq C\mathrm{e}^{-\gamma/\varepsilon}. \tag{11.3.59}$$

Set $\widehat{\phi}(x) = \phi(x)/\|\phi\|_{2,\mu_\varepsilon,D_j}$ and $\hat{c}_j = c_j/\|\phi\|_{2,\mu_\varepsilon,D_j}$. Then, by the definition of J, this locally normalised function satisfies the estimate

$$\|\widehat{\phi} - \hat{c}_j\|_{2,\mu_\varepsilon,D_j} \leq C\mathrm{e}^{-\gamma/2\varepsilon}. \tag{11.3.60}$$

This estimate does not change if we add a constant to $F(x)$. Thus, we can pretend that $F(x_i) = 0$. Let $R > 0$ be such that $B_R(x_j) \in D_j$. Since x_j is a quadratic minimum, there exists a positive and finite constant b such that $F(x) \leq b(x - x_j)^2$ for $x \in B_R(x_j)$. Hence (11.3.60) implies, in particular, that

$$\int_{B_R(x_j)} \left(\widehat{\phi}(x) - \hat{c}_j\right)^2 dx \leq C\mathrm{e}^{bR^2/\varepsilon}\mathrm{e}^{-\gamma/2\varepsilon}. \tag{11.3.61}$$

Note that also

$$\int_{B_R(x_j)} \widehat{\phi}(x)^2\, dx \leq \mathrm{e}^{bR^2/\varepsilon}\|\widehat{\phi}\|_{2,\mu_\varepsilon,D_j} = \mathrm{e}^{bR^2/\varepsilon}. \tag{11.3.62}$$

Let $x \in B_{\sqrt{\varepsilon}}(x_j)$. Then Lemma 11.22 implies

$$\operatorname{osc}_{D_{2\sqrt{\varepsilon}}} \widehat{\phi} \leq C\varepsilon^{-d/4} \tag{11.3.63}$$

for a new positive, finite and ε-independent constant C. Now we can use the Hölder estimate in Lemma 9.9 to obtain that, for $r < \sqrt{\varepsilon}$,

$$\operatorname{osc}_{B_r} \widehat{\phi} \leq C\varepsilon^{-d/4}\left(\frac{r}{\varepsilon^{1/2}}\right)^{\alpha}, \tag{11.3.64}$$

for a new constant C and $\alpha > 0$ independent of ε. If we choose $r = \varepsilon^{\frac{d}{4\alpha}+1}$, then we can achieve that $\operatorname{osc}_{B_r(x)} \leq C\varepsilon^{\alpha/2} < \hat{c}_i/2$ for ε small enough by the estimate (11.3.61), it then follows that $\widehat{\phi}$ must be close to c_j, uniformly on $B_r(x)$. Since this argument holds for all $x \in B_{\sqrt{\varepsilon}}(x_j)$, we have $|\widehat{\phi} - \hat{c}_j| \leq C\varepsilon^{\alpha/2}$ on this ball. □

We will later see that Lemma 11.24 overestimates the fluctuations of ϕ.

Lemma 11.22 is also the appropriate tool to show that near the minima where the L^2-norm is small a similar estimate holds uniformly.

Lemma 11.25 *Let $x_i \in \mathcal{M}$, $i \notin J$. Then any eigenfunction ϕ of $-\mathcal{L}_\varepsilon$ corresponding to one of the $|\mathcal{M}|$ smallest eigenvalues satisfies*

$$\sup_{x\in B_{\sqrt{\varepsilon}}(x_i)} |\phi(x)| \leq C\varepsilon^{-d/4} \mathrm{e}^{-\gamma/2\varepsilon} \mathrm{e}^{F(x_i)/2\varepsilon}. \tag{11.3.65}$$

Proof Since $i \notin J$ we may assume that ϕ changes sign on $B_{\sqrt{e}}(x_i)$. Hence its absolute value is bounded by its oscillation, and so, by Lemma 11.22,

$$\begin{aligned} \sup_{x\in B_{\sqrt{\varepsilon}}(x_i)} |\phi(x)| &\leq C\varepsilon^{-d/4} \|\phi\|_{2,dx,B_{2\sqrt{\varepsilon}}(x_j)} \\ &\leq C'\varepsilon^{-d/4} \mathrm{e}^{F(x_i)/2\varepsilon} \|\phi\|_{2,\mu_\varepsilon,B_{2\sqrt{\varepsilon}}(x_j)} \\ &\leq C'\varepsilon^{-d/4} \mathrm{e}^{F(x_i)/2\varepsilon} \|\phi\|_{2,\mu_\varepsilon,D_j} \\ &\leq C'\varepsilon^{-d/4} \mathrm{e}^{-\gamma/2\varepsilon} \mathrm{e}^{F(x_i)/2\varepsilon}. \end{aligned} \tag{11.3.66}$$

This is the claimed bound. □

Characterisation of the eigenvalues

Recall that we are working under the assumption stated in Theorem 11.9. Suppose that we want to compute eigenvalues below $\lambda_0^{\mathcal{S}_k} = \bar{\lambda}_k$. We know that if ϕ^λ is an eigenfunction with $\lambda < \bar{\lambda}_k$, then it can be represented as the solution of the Dirichlet problem

$$\begin{aligned} (-\mathcal{L}_\varepsilon - \lambda) f^\lambda(y) &= 0, & y &\in \Omega \backslash \partial\mathcal{S}_k, \\ f^\lambda(y) &= \phi^\lambda(y), & y &\in \partial\mathcal{S}_k. \end{aligned} \tag{11.3.67}$$

As in the analysis of principal eigenvalues, the condition on λ will be the existence of a non-trivial ϕ^λ on $\partial\mathcal{S}_k$ such that the surface measure

$$dy\,\mathrm{e}^{-F(y)/\varepsilon}(-\mathcal{L}_\varepsilon - \lambda) f^\lambda(y) = \mathrm{e}^{-F(y)/\varepsilon}\big[\partial_{n(y)} f^\lambda(y) + \partial_{-n(y)} f^\lambda(y)\big] d\sigma_{\mathcal{S}_k}(y) \tag{11.3.68}$$

vanishes. A necessary condition for this to happen is the vanishing of the total mass on each of the surfaces ∂B_i, $1 \leq i \leq k$, i.e.,

$$\int_{\partial B_i} \mathrm{e}^{-F(y)/\varepsilon}\big[\partial_{n(y)} f^\lambda(y) + \partial_{-n(y)} f^\lambda(y)\big] d\sigma_{\mathcal{S}_k}(y) = 0. \tag{11.3.69}$$

Let $c_i = \inf_{y\in B_i} \phi^\lambda(y)$. In view of Lemmas 11.24 and 11.25, either of the following two properties holds:

(i) $\sup_{y\in B_i} |\phi^\lambda(y)/c_i - 1| \leq C\varepsilon^{\alpha/2}$.
(ii) $\sup_{y\in B_i} |\phi^\lambda(y)| \leq C\varepsilon^{-d/4}\mathrm{e}^{-\gamma/2\varepsilon}\mathrm{e}^{F(x_i)/2\varepsilon}$.

In what follows we analyse all possible cases. Let $J \subset \{1,\dots,k\}$ be the set of indices where (i) holds and $J^c = \{1,\dots,k\}\backslash J$ the set of indices where (ii) holds. Given this partition, set

$$f^\lambda = \sum_{j\in J} c_j\big(h^\lambda_{B_j,\mathscr{S}_k\backslash B_j} + \chi^\lambda_{B_j,\mathscr{S}_k\backslash B_j}\big) + \sum_{j\in J^c} \chi^\lambda_{B_j,\mathscr{S}_k\backslash B_j}. \tag{11.3.70}$$

To lighten the notation we set $h^\lambda_j = h^\lambda_{B_j,\mathscr{S}_k\backslash B_j}$ and $\chi^\lambda_j = \chi^\lambda_{B_j,\mathscr{S}_k\backslash B_j}$, etc. in the sequel. For $j \in J$, χ^λ_j is the solution of

$$\begin{aligned} (-\mathscr{L}_\varepsilon - \lambda)\chi^\lambda_j(y) &= 0, && y \in \Omega\backslash\partial\mathscr{S}_k,\\ \chi^\lambda_j(y) &= \phi^\lambda(y)/c_j - 1, && y \in \partial B_j,\\ \chi^\lambda_j(y) &= 0, && y \in \partial B_i, i \neq j, \end{aligned} \tag{11.3.71}$$

whereas, for $j \in J^c$, χ^λ_j is the solution of

$$\begin{aligned} (-\mathscr{L}_\varepsilon - \lambda)\chi^\lambda_j(y) &= 0, && y \in \Omega\backslash\partial\mathscr{S}_k,\\ \chi^\lambda_j(y) &= \phi^\lambda(y), && y \in \partial B_j,\\ \chi^\lambda_j(y) &= 0, && y \in \partial B_i, i \neq j. \end{aligned} \tag{11.3.72}$$

We now proceed as in the analysis of principal eigenvalues, i.e., as necessary condition for λ to be an eigenvalue we require that, for all $i = 1,\dots,k$,

$$\begin{aligned} 0 &= \int_{\partial B_i} e^{-F(y)/\varepsilon} h_i(y)\big(\partial_{n(y)} f^\lambda(y) + \partial_{-n(y)} f^\lambda(y)\big) d\sigma_{\partial\mathscr{S}_k}(y)\\ &= \int_{\partial\mathscr{S}_k} e^{-F(y)/\varepsilon}\partial_{n(y)} h_i(y) f^\lambda(y)\, d\sigma_{\partial\mathscr{S}_k}(y) - \frac{\lambda}{\varepsilon}\int_\Omega dy\, e^{-F(y)/\varepsilon} h_i(y) f^\lambda(y)\\ &= \sum_{j\in J} c_j \Big[\int_{\partial B_j} e^{-F(y)/\varepsilon}\partial_{n(y)} h_i(y)\big[h^\lambda_j(y) + \chi^\lambda_j(y)\big] d\sigma_{\partial\mathscr{S}_k}(y)\\ &\quad - \frac{\lambda}{\varepsilon}\int_\Omega dy\, e^{-F(y)/\varepsilon} h_i(y)\big[h^\lambda_j(y) + \chi^\lambda_j(y)\big]\Big]\\ &\quad + \sum_{j\in J^c}\Big[\int_{\partial B_j} e^{-F(y)/\varepsilon}\partial_{n(y)} h_i(y)\chi^\lambda_j(y) d\sigma_{\partial\mathscr{S}_k}(y)\\ &\quad - \frac{\lambda}{\varepsilon}\int_\Omega dy\, e^{-F(y)/\varepsilon} h_i(y)\chi^\lambda_j(y)\Big]. \end{aligned} \tag{11.3.73}$$

By the bounds in (i) and (ii) we have, for $j \in J$,

$$\left|\int_{\partial B_j} e^{-F(y)/\varepsilon} \partial_{n(y)} h_i(y) \chi_j^\lambda(y)\, d\sigma_{\partial \mathscr{S}_k}(y)\right| \\ \leq C\varepsilon^{\alpha/2} \left|\int_{\partial B_j} e^{-F(y)/\varepsilon} \partial_{n(y)} h_i(y)\, d\sigma_{\partial \mathscr{S}_k}(y)\right|, \tag{11.3.74}$$

and, for $j \in J^c$,

$$\left|\int_{\partial B_j} e^{-F(y)/\varepsilon} \partial_{n(y)} h_i(y) \chi_j^\lambda(y)\, d\sigma_{\partial \mathscr{S}_k}(y)\right| \\ \leq C\varepsilon^{-d/4} e^{-\gamma/2\varepsilon} e^{F(x_j)/2\varepsilon} \left|\int_{\partial B_j} e^{-F(y)/\varepsilon} \partial_{n(y)} h_i(y)\, d\sigma_{\partial \mathscr{S}_k}(y)\right|. \tag{11.3.75}$$

Since the h_i are harmonic, the first Green formula (7.23) implies that, for $i \neq j$,

$$\left|\int_{\partial B_j} e^{-F(y)/\varepsilon} \partial_{n(y)} h_i(y)\, d\sigma_{B_j}(y)\right| = \left|\int_{\partial B_j} e^{-F(y)/\varepsilon} h_j(y) \partial_{n(y)} h_i(y)\, d\sigma_{B_j}(y)\right| \\ = \varepsilon^{-1} \left|\int_{\mathscr{S}_k^c} dy\, e^{-F(y)/\varepsilon} \big(\nabla h_j(y), \nabla h_i(y)\big)\right| \\ \leq \varepsilon^{-1} \sqrt{\operatorname{cap}(B_i, \mathscr{S}_k \backslash B_i) \operatorname{cap}(B_j, \mathscr{S}_k \backslash B_j)}, \tag{11.3.76}$$

where the last inequality uses the Cauchy-Schwarz inequality. Thus, for $j \in J\backslash i$,

$$\left|\int_{\partial B_j} e^{-F(y)/\varepsilon} \partial_{n(y)} h_i(y) \chi_j^\lambda\, d\sigma_{\partial \mathscr{S}_k}(y)\right| \\ \leq C\varepsilon^{\alpha/2} \varepsilon^{-1} \sqrt{\operatorname{cap}(B_i, \mathscr{S}_k \backslash B_i) \operatorname{cap}(B_j, \mathscr{S}_k \backslash B_j)}. \tag{11.3.77}$$

and, for $j \in J^c\backslash i$,

$$\left|\int_{\partial B_j} e^{-F(y)/\varepsilon} \partial_{n(y)} h_i(y) \chi_j^\lambda\, d\sigma_{\partial \mathscr{S}_k}(y)\right| \\ \leq C\varepsilon^{-d/4} e^{-\gamma/2\varepsilon} e^{F(x_j)/2\varepsilon} \varepsilon^{-1} \sqrt{\operatorname{cap}(B_i, \mathscr{S}_k \backslash B_i) \operatorname{cap}(B_j, \mathscr{S}_k \backslash B_j)}. \tag{11.3.78}$$

For the diagonal terms $i = j \in J$ this simplifies to

$$\left|\int_{\partial B_j} e^{-F(y)/\varepsilon} \partial_{n(y)} h_i(y) \chi_j^\lambda d\sigma_{\mathscr{S}_k}(y)\right| \leq C\varepsilon^{\alpha/2} \operatorname{cap}(B_j, \mathscr{S}_k \backslash B_j). \tag{11.3.79}$$

For the remaining terms involving χ^λ in (11.3.73) we obtain, in complete analogy with the derivation of the bounds in Lemma 11.19, that, for $j \in J$,

$$\int_\Omega dy\, \mathrm{e}^{-F(y)/\varepsilon} h_i(y)\big[h_j^\lambda(y) - h_j(y) + \chi_j^\lambda(y)\big]$$
$$= O\big(\varepsilon^{\alpha/2}\big)\big[1 + O\big(\mathrm{e}^{-\delta/\varepsilon}\big)\big] \int_\Omega dy\, \mathrm{e}^{-F(y)/\varepsilon} h_i(y) h_j(y) \qquad (11.3.80)$$

and, for $j \in J^c$,

$$\int_\Omega dy\, \mathrm{e}^{-F(y)/\varepsilon} h_i(y) \chi_j^\lambda(y)$$
$$= O\big(\varepsilon^{-d/4} \mathrm{e}^{-\gamma/2\varepsilon} \mathrm{e}^{F(x_j)/2\varepsilon}\big) \int_\Omega dy\, \mathrm{e}^{-F(y)/\varepsilon} h_i(y) h_j(y). \qquad (11.3.81)$$

To control the off-diagonal terms we need to show that the normalised functions h_i and h_j are almost orthogonal.

Lemma 11.26

(i) *There is a constant $C < \infty$ such that, for $i \neq j$,*

$$(h_i, h_j)_{\mu_\varepsilon} = \int_\Omega dy\, \mathrm{e}^{-F(y)/\varepsilon} h_j(y) h_i(y) \le C \varepsilon^{-2d} \mathrm{e}^{-\Phi(x_i,x_j)/\varepsilon}. \qquad (11.3.82)$$

(ii) *For all j,*

$$\|h_j\|_{2,\mu_\varepsilon}^2 \ge C \varepsilon^{d/2} \mathrm{e}^{-F(x_j)/\varepsilon}. \qquad (11.3.83)$$

(iii) *There is a constant $C < \infty$ such that, for $i \neq j$,*

$$\frac{(h_i, h_j)_{\mu_\varepsilon}}{\|h_i\|_{2,\mu_\varepsilon} \|h_j\|_{2,\mu_\varepsilon}} \le C \varepsilon^{-3d} \max\big(\mathrm{e}^{-(\Phi(x_i,x_j) - F(x_i))/\varepsilon}, \mathrm{e}^{-(\Phi(x_i,x_j) - F(x_j))/\varepsilon}\big). \qquad (11.3.84)$$

Proof The proof goes in the same way as that of Lemma 8.24 in Chap. 8, and uses the bounds in (11.2.12) on harmonic functions and the bounds in (11.2.5) on capacities. □

Finally, we note that by Lemma 11.25 and Lemma 11.20, for $j \notin J$,

$$\big|\chi_j^\lambda(z)\big| \le C \varepsilon^{-d/4} \mathrm{e}^{-\gamma/2\varepsilon} \frac{h_j(z)}{\|h_j\|_{2,\mu_\varepsilon}}. \qquad (11.3.85)$$

Computation of small eigenvalues

The matrix $\mathscr{C}$ with elements given by

$$\mathscr{C}_{ij} = \mathscr{C}_{ij}^{(k)} = \varepsilon \int_{\partial B_j} e^{-F(y)/\varepsilon} h_j(y) \partial_{n(y)} h_i(y)\, d\sigma_{B_j}(y), \quad i, j = 1, \dots, k, \tag{11.3.86}$$

is the analogue of the *capacity matrix* that we encountered in Chap. 8. We also use its normalised version

$$\mathscr{K}_{ij} = \mathscr{K}_{ij}^{(k)} = \frac{\mathscr{C}_{ij}^{(k)}}{\|h_i\|_{2,\mu_\varepsilon} \|h_j\|_{2,\mu_\varepsilon}}. \tag{11.3.87}$$

Note that this matrix is symmetric and, by (11.3.76), satisfies

$$\mathscr{K}_{ij} \le \sqrt{\mathscr{K}_{ii} \mathscr{K}_{jj}}. \tag{11.3.88}$$

Also introduce the matrices

$$A_{ij} = \frac{\varepsilon \int_{\partial B_i} e^{-F(y)/\varepsilon} \partial_{n(y)} h_i(y) \chi_j^\lambda(y)\, d\sigma_{\partial \mathscr{S}_k}(y)}{\|h_i\|_{2,\mu_\varepsilon} \|h_j\|_{2,\mu_\varepsilon}}, \tag{11.3.89}$$

$$B_{ij} = \begin{cases} (1-\delta_{ij}) \frac{\int_\Omega dy\, e^{-F(y)/\varepsilon} h_i(y)[h_j^\lambda(y) - h_j(y) + \chi_j^\lambda(y)]}{\|h_i\|_{2,\mu_\varepsilon} \|h_j\|_{2,\mu_\varepsilon}}, & j \in J, \\ (1-\delta_{ij}) \frac{\int_\Omega dy\, e^{-F(y)/\varepsilon} h_i(y) \chi_j^\lambda(y)}{\|h_i\|_{2,\mu_\varepsilon} \|h_j\|_{2,\mu_\varepsilon}}, & j \in J^c, \end{cases} \tag{11.3.90}$$

$$D_{ij} = \delta_{ij} \frac{\int_\Omega dy\, e^{-F(y)/\varepsilon} h_j(y)[h_j^\lambda(y) - h_j(y) + \chi_j^\lambda]}{\|h_j\|_{2,\mu_\varepsilon}^2}. \tag{11.3.91}$$

Then (11.3.73) can be rewritten, for $i \in J$,

$$\sum_{j\in J} \hat{c}_j \big(\mathscr{K}_{ij} - \lambda\delta_{ij} + A_{ij} - \lambda(D_{ij} + B_{ij})\big) + \sum_{j \in J^c} (A_{ij} + \lambda B_{ij}) \|h_j\|_{2,\mu_\varepsilon} = 0, \tag{11.3.92}$$

where $\hat{c}_j = \|h_j\|_{2,\mu_\varepsilon} c_j$, $j = 1, \dots, k$. These equations are the analogues of (8.4.86) for countable state space. The following lemma, which is the analogue of Lemma 8.35, collects the estimates needed to analyse the solution of these equations.

Lemma 11.27 *The following bounds hold*:

(i) *For all* $i, j \in 1, \dots, k$,

$$|A_{ij}| \le |\mathscr{K}_{ij}| C \varepsilon^{\alpha/2}. \tag{11.3.93}$$

(ii) *For* $i \ne j \in J$,

$$|B_{ij}| \le O\big(\varepsilon^\alpha + e^{-\delta/\varepsilon}\big) \sqrt{\mathscr{K}_{ii} \mathscr{K}_{jj}}. \tag{11.3.94}$$

(iii) *For* $j \in J$,

$$|D_{jj}| \leq C\varepsilon^{\alpha/2}. \tag{11.3.95}$$

(iv) *For* $i \neq j \in J^c$,

$$\|h_j\|_{2,\mu_\varepsilon} |A_{ij}| \leq C\varepsilon^{-3d/4} \mathrm{e}^{-\gamma/2\varepsilon} |\mathscr{K}_{ii}|, \tag{11.3.96}$$

and

$$\|h_j\|_{2,\mu_\varepsilon} |B_{ij}| \leq C\varepsilon^{-d} \mathrm{e}^{-\gamma/\varepsilon} \sqrt{\mathscr{K}_{ii}\mathscr{K}_{jj}}. \tag{11.3.97}$$

Proof The bound in (11.3.93) is (11.3.77). The bounds in (11.3.94) and (11.3.95) follow from (11.3.80) and (11.3.88). The bound in (11.3.96) is a consequence of (11.3.78), while (11.3.97) follows from (11.3.81). □

From here on the analysis of the solutions of (11.3.92) is very similar to that of (8.4.86). Let us summarise the situation so far.

Theorem 11.28 *Let* $\mathscr{S}_k = \bigcup_{i=1}^k B_\varepsilon(x_i)$, *and let* $\bar{\lambda}_k$ *denote the principal eigenvalue of the operator* $-\mathscr{L}_\varepsilon$ *with Dirichlet boundary conditions on* $\partial\mathscr{S}_k$ *(and* $\partial\Omega$*). Then a necessary condition for a number* $\lambda < \bar{\lambda}_k$ *to be an eigenvalue of the operator* $-\mathscr{L}_\varepsilon$ *is that there exist a non-empty set* $J \subset \{1,\dots,k\}$ *and constants* $\hat{c}_j$, $j \in J$, *with* $\sum_{j\in J} \hat{c}_J^2 = 1$, *such that* (11.3.92) *holds for all* $i \in J$.

We expect that all the solutions of (11.3.92) are close to an eigenvalue of $\mathscr{K}$.

Lemma 11.29 *Let* $(\mathscr{K}_{ij})_{1\leq i,j\leq n}$ *be the normalised capacity matrix and assume that*

$$\max_{1\leq i<k} \mathscr{K}_{ii} \leq \mathrm{e}^{-\delta/\varepsilon} \mathscr{K}_{kk}. \tag{11.3.98}$$

Then $J \ni k$, *and the largest eigenvalue* μ_k *of* $\mathscr{K}$ *satisfies*

$$\mu_k = \mathscr{K}_{kk}\big[1 + O\big(\mathrm{e}^{-\delta/2\varepsilon}\big)\big], \tag{11.3.99}$$

while all other eigenvalues are smaller than $C\mathrm{e}^{-\delta/\varepsilon}\lambda_k$. *Moreover, the eigenvector* $v = (v_1,\dots,v_k)$ *corresponding to the largest eigenvalue normalised such that* $v_k = 1$ *satisfies* $|v_i| \leq C\mathrm{e}^{-\delta/\varepsilon}$ *for* $1 \leq i < k$.

Proof The proof is a simple perturbation argument. Note that we can write

$$\mathscr{K} = \hat{\mathscr{K}} + \check{\mathscr{K}}, \tag{11.3.100}$$

where $\hat{\mathscr{K}}_{ij} = \mathscr{K}_{kk}\delta_{ik}\delta_{jk}$. Estimate the norm of $\check{\mathscr{K}}$ as in the proof of Lemma 11.26. Recall that

$$|\mathscr{K}_{ij}| \leq \mathscr{K}_{ii}\mathscr{K}_{jj}. \tag{11.3.101}$$

Hence, by assumption (11.3.98),

$$\|\check{\mathscr{K}}\| \leq \mathscr{K}_{kk}\sqrt{\mathrm{e}^{-\delta/\varepsilon}k + \mathrm{e}^{-\delta^2/\varepsilon^2}k^2}. \tag{11.3.102}$$

Since $\hat{\mathscr{K}}$ has one eigenvalue $\mathscr{K}_{kk}$ with the obvious eigenvector and all other eigenvalues are zero, the claim follows from standard perturbation theory. □

Since $\mathscr{K}_{kk} = \mathrm{cap}(B_k, \mathscr{S}_{k-1})/\|h_k\|^2_{2,\mu_\varepsilon} \approx \bar{\lambda}_{k-1}$ (i.e., is equal to $\bar{\lambda}_{k-1}$ up to polynomial terms in ε), Lemma 11.29 tells us that $\mu_k \approx \bar{\lambda}_{k-1}$, which is precisely the value we expect.

Corollary 11.30 *Under the hypothesis in* (11.3.98) *the following hold*:

(i) *If there exists an eigenvalue λ_k of $-\mathscr{L}_\varepsilon$ in the interval $(\bar{\lambda}_k, \bar{\lambda}_{k-1}]$, then*

$$\lambda_k = \frac{\mathrm{cap}(B_k, \mathscr{S}_{k-1})}{\|h_k\|^2_{2,\mu_\varepsilon}}\left[1 + O\left(\varepsilon^{\alpha/2}, \mathrm{e}^{-\delta/\varepsilon}\right)\right]. \tag{11.3.103}$$

(ii) *The eigenvalue λ_k is simple and the corresponding eigenfunction f_k^λ can be written as*

$$\phi_k^\lambda(y) = \frac{h_k(y)}{\|h_k\|_{2,\mu_\varepsilon}}\left[1 + O\left(\varepsilon^{\alpha/2}\right)\right] + \sum_{j=1}^{k-1} d_j(y)\frac{h_j(y)}{\|h_j\|_{2,\mu_\varepsilon}}, \tag{11.3.104}$$

where $|d_j(y)| \leq \mathrm{e}^{-\delta/\varepsilon}$ for some $\delta > 0$ (uniformly on compact subsets when Ω is unbounded).

Proof (i) We already know (see Remark 11.3.4) that $\lambda = \frac{\mathrm{cap}(B_k, \mathscr{S}_{k-1})}{\|h_k\|^2_{2,\mu_\varepsilon}}\mathrm{e}^{o(1)/\varepsilon}$. But the only coefficient in (11.3.92) that is of this order is $\mathscr{K}_{kk}$, *if* $J \ni k$. So if $k \notin J$, then all $\hat{c}_i, i \in J$, would need to be exponentially small, contradicting the normalisation condition. Hence we may assume $k \in J$. By considering all the equations with $i \neq k$, we see that the same argument as before shows that $|\hat{c}_i| \leq C\mathrm{e}^{-\delta/2\varepsilon}$, and hence $\hat{c}_k \approx 1$, so the equation labelled k implies that

$$\left|(\mathscr{K}_{kk} + A_{kk} - \lambda_k)\right| \leq C|\mathscr{K}_{kk}|\mathrm{e}^{-\delta/\varepsilon}\varepsilon^{\alpha/2}, \tag{11.3.105}$$

and since also $|A_{kk}| \leq C\varepsilon^{\alpha/2}\mathscr{K}_{kk}$, (11.3.103) follows.

(ii) We have just seen that a solution of (11.3.92) with $\hat{c}_k = 1$ must satisfy $|\hat{c}_j| \leq \mathrm{e}^{-\delta/\varepsilon}$ for all $j \neq k$. Hence, by (11.3.70),

$$\phi_k^\lambda(y) = \frac{h_k^\lambda(y) + \chi_k^\lambda(y)}{\|h_k\|_{2,\mu_\varepsilon}} + \sum_{j \in J\setminus k} \hat{c}_j \frac{h_j^\lambda(y) + \chi_j^\lambda(y)}{\|h_j\|_{2,\mu_\varepsilon}} + \sum_{j \in J^c} \chi_j^\lambda(y). \tag{11.3.106}$$

Using the same arguments as in the proof of Lemma 11.20, and the bounds on $\phi^\lambda - c_j$ on the boundaries ∂B_j, we get that, for $j \in J$,

$$\frac{|\chi_j^\lambda(y)|}{\|h_j\|_{2,\mu_\varepsilon}} \le C\varepsilon^{\alpha/2}\frac{h_j(y)}{\|h_j\|_{2,\mu_\varepsilon}}. \tag{11.3.107}$$

Combining these estimates we arrive at (11.3.104). Note that this final estimate does not depend on the choice of J. □

At this point we can further explore the eigenvalues below $\bar{\lambda}_{k-1}$, etc., with the same result. Thus, at the end of the procedure, we arrive at the conclusion that $-\mathscr{L}_\varepsilon$ can have at most the n simple eigenvalues given by the values in Corollary 11.30 below $C\varepsilon^{d-1}$. However, since we know that there must be n such eigenvalues, we conclude that all these candidate eigenvalues are in fact the true eigenvalues, which yields the following proposition.

Proposition 11.31 *Under the assumptions of Theorem* 11.9, *the spectrum of* $-\mathscr{L}_\varepsilon$ *below* $C\varepsilon^{d-1}$ *consists of* n *simple eigenvalues that satisfy*:

$$\begin{aligned}\lambda_k &= \frac{\operatorname{cap}(B_k, \mathscr{S}_{k-1})}{\|h_k\|_{2,\mu_\varepsilon}^2}\left[1 + O\left(\varepsilon^{\alpha/2}, \mathrm{e}^{-\delta/\varepsilon}\right)\right] \\ &= \operatorname{cap}(B_k, \mathscr{S}_{k-1})\frac{\sqrt{\det(\nabla^2 F(x_k))}}{(\sqrt{2\pi\varepsilon})^d}\mathrm{e}^{F(x_k)/\varepsilon}\left[1 + O\left(\varepsilon^{1/2}\ln(1/\varepsilon), \varepsilon^{\alpha/2}, \mathrm{e}^{-\delta/\varepsilon}\right)\right] \\ &= \frac{1}{\mathbb{E}_{x_k}[\tau_{\mathscr{S}_{k-1}}]}\left[1 + O\left(\varepsilon^{\alpha/2}, \mathrm{e}^{-\delta/\varepsilon}\right)\right], \quad k = 1, \ldots, n.\end{aligned} \tag{11.3.108}$$

The corresponding eigenfunctions satisfy (11.3.104).

Proof We have seen that $\lambda_k = \mathscr{K}_{kk}^{(k)}(1 + O(\mathrm{e}^{-\theta/\varepsilon}, \varepsilon^{\alpha/2}))$, which proves the first assertion. It remains to identify the eigenvalues with the inverse mean times. The argument is essentially the same as in the proof of Theorem 8.43.

By virtue of Theorem 11.2 we need to show that

$$\int_\Omega dy\, \mathrm{e}^{-F(y)/\varepsilon} h_k^2(y) \sim \int_\Omega dy\, \mathrm{e}^{-F(y)/\varepsilon} h_k(y). \tag{11.3.109}$$

In fact, we will show more, namely, that both sides of (11.3.109) are asymptotically equal to

$$\mathrm{e}^{-F(x_k)/\varepsilon}\frac{(\sqrt{2\pi\varepsilon})^d}{\sqrt{\det(\nabla^2 F(x_k))}}. \tag{11.3.110}$$

We must show that the main contribution of the integrals comes from a small neighbourhood of x_k, which yields the contribution in (11.3.110). It is clear that all contributions from the set $\{y \in \Omega\colon\ F(y) > F(x_k) + \varepsilon\ln(1/\varepsilon)\}$ give only sub-leading corrections. To treat the complement of this set, we use the bounds on the equilibrium potential in (11.2.12). Up to polynomial factors in ε, these imply that, on the connected component of the level set that does not contain x_k, the integrand in the right-hand side of (11.3.109) (and a fortiori in the left-hand side) is smaller than

$$e^{-[F(y)+\Phi(y,B_k)-\Phi(y,\mathscr{S}_{k-1})]/\varepsilon}. \tag{11.3.111}$$

If y is in the component of the level set that contains the minimum x_j, $1 \leq j \leq k$, then we see that the latter is equal to

$$e^{-\Phi(x_j,B_k)/\varepsilon}, \tag{11.3.112}$$

which is exponentially smaller than $\exp(-F(x_k)/\varepsilon)$, independently of y. If $j > k$, then we still get the same result when $F(y) \geq \Phi(x_j, \mathscr{S}_{k-1})$. Otherwise, we can write (11.3.111) as

$$e^{-[F(y)-F(x_j)]/\varepsilon} e^{-\{F(x_k)+[\Phi(x_j,B_k)-F(x_k)]-[\Phi(x_j,\mathscr{S}_{k-1})-F(x_j)]\}/\varepsilon}. \tag{11.3.113}$$

We will argue that

$$\Phi(x_j, B_k) - F(x_k) > \Phi(x_j, \mathscr{S}_{k-1}) - F(x_j). \tag{11.3.114}$$

Suppose that the contrary holds. Trivially,

$$\Phi(x_j, \mathscr{S}_{k-1}) \geq \Phi(x_j, \mathscr{S}_{j-1}), \tag{11.3.115}$$

while

$$\Phi(x_j, B_k) = \Phi(x_k, B_j) \leq \Phi(x_k, \mathscr{S}_j \backslash B_k). \tag{11.3.116}$$

Therefore, our supposition implies that

$$\Phi(x_j, \mathscr{S}_{j-1}) - F(x_j) \leq \Phi(x_k, \mathscr{S}_j \backslash B_k) - F(x_k), \tag{11.3.117}$$

which contradicts the conditions in (11.3.1) at stage j. In other words, if our supposition were true, then the set B_k would have to yield the largest eigenvalue at stage j, i.e., it would have to be labelled B_j. Hence (11.3.114) must hold.

Since, by assumption, the inequalities are strict (which is more than we need), it indeed follows that

$$\int_{\Omega} dy\, e^{-F(y)/\varepsilon} h_k(y) = e^{-F(x_k)/\varepsilon} \frac{(\sqrt{2\pi\varepsilon})^d}{\sqrt{\det(\nabla^2 F(x_k))}} \left[1 + O\left(\varepsilon^{1/2} \left[\ln(1/\varepsilon)\right]^{3/2}\right)\right], \tag{11.3.118}$$

and the same bound holds when h_k is replaced by h_k^2. □

11.3.5 Improved error estimates

To conclude the proofs of Theorems 11.9 and 11.11, we only need to improve the error estimates. So far the proofs have produced error terms from two sources: the exponentially small errors resulting from the perturbation around $\lambda = 0$ and from the imperfect orthogonality of the functions h_i, $i = 1, \dots, n$, and the much larger errors of order $\varepsilon^{\alpha/2}$ resulting from the a priori control on the regularity of the eigen functions obtained from the Hölder estimate of Lemma 11.24. In the light of the estimates obtained on the eigenfunctions, these can now be improved successively.

First, note that the eigenfunction corresponding to the minimum x_k is small enough at all the minima x_l, $1 \leq l < k$, so that we can actually take $J = \{k\}$ and $J_k = \{1, \dots, k-1\}$ in (11.3.71) and (11.3.73). Then we know from Corollary 11.30 that

$$\operatorname{osc}_{y \in B_{4\sqrt{\varepsilon}}(x_k)} \phi_k(y) \leq C\varepsilon^{\alpha/2} \sup_{y \in B_{4\sqrt{\varepsilon}}(x_k)} \phi_k(y), \tag{11.3.119}$$

which improves the a priori estimate in (11.3.33).

Next, the Hölder estimate in Lemma 9.9 gives the improvement

$$\begin{aligned} \operatorname{osc}_{y \in B_\varepsilon(x_k)} \phi_k(y) &\leq C\varepsilon^{\alpha/2}\big(C\varepsilon^{\alpha/2} + \lambda_k \varepsilon^{(d+1)/2}\big) \sup_{y \in B_{4\sqrt{\varepsilon}}(x_k)} \phi_k(y) \\ &\leq C\varepsilon^{\alpha} \sup_{y \in B_{4\sqrt{\varepsilon}}(x_k)} \phi_k(y) \end{aligned} \tag{11.3.120}$$

over the estimate in (11.3.33). This allows us to replace all errors of order $\varepsilon^{\alpha/2}$ by errors of order ε^{α}. This procedure can be iterated m times to get errors of order $\varepsilon^{m\alpha/2}$, which for m of order $\ln(1/\varepsilon)$ is as small as the exponentially small errors.

Finally, we want to improve the precision with which we relate the eigenvalues to the inverse of the mean exit times. This precision is so far limited by the precision with which

$$\mathbb{E}_{x_k}[\tau_{\mathscr{S}_{k-1}}] \approx \frac{\operatorname{cap}(B_\varepsilon(x_k), \mathscr{S}_{k-1})}{\|h_k\|_{2,\mu_\varepsilon}}. \tag{11.3.121}$$

From Proposition 11.7 we know that this precision is limited only by the variation of $\mathbb{E}_x[\tau_{\mathscr{S}_{k-1}}]$ on $B_\varepsilon(x_k)$. To improve (11.3.121), we need to control ($h_k = h_{B_\varepsilon(x_k), \mathscr{S}_{k-1}}$)

$$\frac{\operatorname{cap}(B_\varepsilon(x_k), \mathscr{S}_{k-1})}{\|h_k\|_{2,\mu_\varepsilon}} - \frac{\operatorname{cap}(B_\varepsilon(x), \mathscr{S}_{k-1})}{\|h_{B_\varepsilon(x), \mathscr{S}_{k-1}}\|_{2,\mu_\varepsilon}}, \quad x \in B_{\sqrt{\varepsilon}}(x_k). \tag{11.3.122}$$

Now, it is easy to see that if $x \in B_{\sqrt{\varepsilon}}(x_k)$, then

$$\big|h_{B_\varepsilon(x), \mathscr{S}_{k-1}}(y) - h_k(y)\big| \leq \mathrm{e}^{-\delta/\varepsilon} h_k(y). \tag{11.3.123}$$

Namely,

$$\begin{aligned} &\big|h_{B_\varepsilon(x), \mathscr{S}_{k-1}}(y) - h_k(y)\big| \\ &\quad \leq \mathbb{P}_y(\tau_{B_\varepsilon(x_k)} < \tau_{\mathscr{S}_{k-1}} < \tau_{B_\varepsilon(x)}) + \mathbb{P}_y(\tau_{B_\varepsilon(x)} < \tau_{\mathscr{S}_{k-1}} < \tau_{B_\varepsilon(x_k)}). \end{aligned} \tag{11.3.124}$$

By the Markov property, the first term in (11.3.124) is bounded as

$$\begin{aligned}&\mathbb{P}_y(\tau_{B_\varepsilon(x_k)} < \tau_{\mathscr{S}_{k-1}} < \tau_{B_\varepsilon(x)})\\&\leq \mathbb{P}_y(\tau_{B_\varepsilon(x_k)} < \tau_{\mathscr{S}_{k-1}}) \max_{z\in B_\varepsilon(x_k)} \mathbb{P}_z(\tau_{\mathscr{S}_{k-1}} < \tau_{B_\varepsilon(x)})\\&\leq \mathrm{e}^{-\delta/\varepsilon}\mathbb{P}_y(\tau_{B_\varepsilon(x_k)} < \tau_{\mathscr{S}_{k-1}}).\end{aligned} \tag{11.3.125}$$

The second term in (11.3.124) is bounded in the same way. This in turn implies that

$$\|h_{B_\varepsilon(x),\mathscr{S}_{k-1}}\|_{2,\mu_\varepsilon} - \|h_k\|_{2,\mu_\varepsilon} \leq \mathrm{e}^{-\delta/\varepsilon}\|h_k\|_{2,\mu_\varepsilon}. \tag{11.3.126}$$

To get an analogous estimate for capacities, we take advantage of the fact that, as long as $\lambda_0^{B_\varepsilon(x)\cup\mathscr{S}_{k-1}} \gg \lambda_k$, we can replace $B_\varepsilon(x_k)$ by $B_\varepsilon(x)$ in the proof of Proposition 11.31 without further changes. Thus

$$\lambda_k = \frac{\mathrm{cap}(B_\varepsilon(x),\mathscr{S}_{k-1})}{\|h_{B_\varepsilon(x),\mathscr{S}_{k-1}}\|_{2,\mu_\varepsilon}^2}\left[1+O\big(\mathrm{e}^{-\delta/\varepsilon}\big)\right] = \frac{\mathrm{cap}(B_\varepsilon(x_k),\mathscr{S}_{k-1})}{\|h_k\|_{2,\mu_\varepsilon}^2}\left[1+O\big(\mathrm{e}^{-\delta/\varepsilon}\big)\right], \tag{11.3.127}$$

which together with (11.3.126) implies that

$$\left|\mathrm{cap}\big(B_\varepsilon(x),\mathscr{S}_{k-1}\big) - \mathrm{cap}\big(B_\varepsilon(x_k),\mathscr{S}_{k-1}\big)\right| \leq \mathrm{e}^{-\delta/\varepsilon}\,\mathrm{cap}\big(B_\varepsilon(x_k),\mathscr{S}_{k-1}\big). \tag{11.3.128}$$

Based on (11.3.123) and (11.3.128), we can improve Proposition 11.7 iteratively as above to obtain

$$\mathbb{E}_{x_k}[\tau_{\mathscr{S}_k}] = \frac{\mathrm{cap}(B_\varepsilon(x_k),\mathscr{S}_{k-1})}{\|h_k\|_{2,\mu_\varepsilon}}\left[1+O\big(\mathrm{e}^{-\delta/\varepsilon}\big)\right], \tag{11.3.129}$$

which, together with the capacity estimate given in Theorem 11.2, implies the first equality in Theorem 11.9. Thus, all error terms of order $\varepsilon^{\alpha/2}$ can be removed from (11.3.104) and (11.3.108), which completes the proofs of Theorems 11.9 and Theorem 11.11. □

11.3.6 Exponential distribution of metastable exit times

The last assertion of Theorem 11.12, the asymptotic exponential distribution of the metastable exit time, follows from the spectral estimates above exactly as in the discrete case (see the proof of Theorem 8.45). This result can also be obtained via the coupling method of Martinelli et al. [174, 177].

11.4 Bibliographical notes

1. The material presented in this chapter is based on Bovier, Eckhoff, Gayrard and Klein [35] and Bovier, Gayrard and Klein [38], with some corrections taken from the Diploma Thesis of Erich Bauer [14]. Assumptions 10.3, 10.5 and 11.1 can be relaxed. In particular, we may take $F = F_\varepsilon$ depending on ε, or F with infinitely many local minima. See e.g. Berglund and Gentz [21].

2. A proof of the Eyring-Kramers formula for the special case when all minima of the potential are at the same level was given in two little-noticed papers by Sugiura [224, 225]. The approach used there runs via a direct variational control on principal eigenvalues.

3. If Assumption 10.3 fails, then the asymptotics in Theorem 11.2 becomes more complicated. Berglund and Gentz [21] classify various cases where the saddle point is not quadratic.

4. Rough estimates of the small eigenvalues λ_i associated with the local minima x_i of F were derived in Freidlin and Wentzell [115], Mathieu [179] and Miclo [185]. Wentzell [234] and Freidlin and Wentzell [115] obtained estimates for the exponential rate $\lim_{\varepsilon\downarrow 0} \varepsilon \ln \lambda_k(\varepsilon)$ with the help of large deviation methods. Sharper estimates, with multiplicative errors of order $\varepsilon^{\pm kd}$, were obtained for principal eigenvalues by Holley, Kusuoka and Strook [140] with the help of variational principles. These methods were extended to the full set of exponentially small eigenvalues in Miclo [185] and Mathieu [179].

5. For a long time sharp spectral estimates were known only in the one-dimensional case (see e.g. Buslov and Makarov [44, 45] and references therein), whereas in the multi-dimensional case only heuristic results based on formal power series expansions of the so-called WKB-type existed (see e.g. Kolokoltsov [154]). The proof in Sect. 11.3, which is based on potential theory, follows Bovier, Gayrard and Klein [38] and uses ideas that appeared already in Wentzell [233, 234]. More recently, a full analytic proof of the asymptotic expansion for these eigenvalues was given by Helffer, Klein and Nier [136], and Helffer and Nier [137], using a microlocal analysis of the so-called Witten complex. They show, in particular, that the error bounds in Theorem 11.9 can be improved to $O(\varepsilon)$. Moreover, they show that, under the assumption that F is C^∞, a full asymptotic expansion in ε for the eigenvalues can be computed.

6. There is considerable interest in the knowledge of eigenfunctions in the context of numerical schemes designed to recover metastable sets from the computation of eigenfunctions (see, in particular, Schütte, Huisinga and Meyn [216]). Using the bounds on equilibrium potentials obtained in Bovier, Eckhoff, Gayrard and Klein [35, Corollary 4.8], we can show that the result in Theorem 11.11 implies that the eigenfunction ϕ_k corresponding to the local minimum x_k of F is exponentially close to a constant, i.e., $\sim \mathrm{e}^{F(x_k)/2\varepsilon}$, in the connected component of the level set $\{y \in D\colon\ F(y) < F(z^*(x_k, \mathscr{M}_{k-1}))\}$ that contains x_k (i.e., in the valley below the saddle point that connects x_k to the set below the level of x_k), while it drops exponentially fast in the other connected components of the level set of this saddle, and below the level of x_k is exponentially small in absolute terms. Note that this implies that the zeros of ϕ_k are generally not in the neighbourhood of the saddle points, but close to the minima in $\mathcal{M}_{k-1}$. This fact was observed in Schütte, Huisinga and Meyn [216]. We would like to stress that the fact that the eigenfunctions drop sharply at the saddle points makes them very good indicators of the actual valley structure of F, i.e., they are excellent approximations of the indicator functions of the metastable sets corresponding to the metastable exit time $1/\lambda_k$.

7. An interesting approach to the characterisation of sharp Poincaré inequalities that allows for a derivation of the Kramers formula based on the theory of optimal transport was developed by Menz and Schlichtung [183, 210].

Chapter 12
Stochastic Partial Differential Equations

Il y a des faussetés déguisées qui représentes si bien la vérité que ce serait mal juger que de ne s'y pas laisser tromper.
(Francois de la Rochefoucauld, Réflexions)

A natural generalisation of the finite-dimensional diffusions considered Chap. 11 are *stochastic partial differential equations*. In this chapter we focus on the Allen-Cahn equation introduced in Sect. 5.7. Section 12.1 gives the main theorem and a rough outline of its proof. Section 12.2 lists some approximation properties for the potential that are crucial for the proof. Section 12.3 provides estimates on the relevant capacities, Sect. 12.4 on the equilibrium potential. The results are collected in Sect. 12.5 to complete the argument.

12.1 Definitions, main theorem and outline of proof

We return to the SPDE in (6.3.1). Let F be the functional defined in (5.7.5). The first and second *Frechét derivatives* D_ϕ and D^2_ϕ are defined through the requirement that F has a Taylor expansion up to second order in h,

$$F(\phi+h)=F(\phi)+(D_\phi F)(h)+\tfrac{1}{2}\big(D^2_\phi F\big)(h,h)+o\big(\|h\|^2_{C^2}\big), \qquad (12.1.1)$$

where $\|h\|_{C^2}=\|h\|_\infty+\|h'\|_\infty+\|h''\|_\infty$. The differentials $D_\phi F$ and $D^2_\phi F$ can be computed explicitly, namely,

$$(D_\phi F)(h)(x)=-Dh'(x)+V'\big(\phi(x)\big)h(x), \qquad (12.1.2)$$

while $(D^2_\phi F)(h,h)$ is the quadratic form associated with the Hessian operator $\mathscr{H}_\phi F$ given by

$$(\mathscr{H}_\phi F)(h)(x)=-Dh''(x)+V''\big(\phi(x)\big)h(x). \qquad (12.1.3)$$

Note that $\mathscr{H}_\phi F$ a Sturm-Liouville operator (see Coddington and Levinson [66]).

We say that ϕ is a *stationary point* of F when ϕ is a solution of the non-linear differential equation

$$-D\phi''+V'(\phi)=0. \qquad (12.1.4)$$

A. Bovier, F. den Hollander, *Metastability*,
Grundlehren der mathematischen Wissenschaften 351,
DOI 10.1007/978-3-319-24777-9_12

The notion of saddle points, communications heights, and gates are defined as in the finite-dimensional setting.

The theory can be developed under assumptions that are analogous to those used in the finite-dimensional setting:

Assumption 12.1

(i) F has finitely many local minima and saddle points.
(ii) All local minima and saddle points of F are non-degenerate: at each point the Hessian operator has only non-zero eigenvalues.

However, in this chapter we will only do computations for the simplest non-trivial case, namely,

$$V(s) = -\tfrac{1}{2}s^2 + \tfrac{1}{4}s^4 + bs, \quad s \in \mathbb{R}, \tag{12.1.5}$$

with $b \geq 0$ small enough so that the equation

$$s - s^3 - b = 0 \tag{12.1.6}$$

has three roots,

$$z_b^{-,*} < 0 < z_b^* < z_b^{+,*}, \tag{12.1.7}$$

corresponding to two minima and one saddle point of V. In this case, there are only two local minima of F, namely, the constant functions $I^{\pm}$ given by $I^{\pm}(x) = \pm z_b^{\pm,*}$, $x \in [0, 1]$.

Assumption 12.2 $\mathscr{S}(I^-, I^+)$, the communication level set of F between I^- and I^+ (recall Definition 10.2), consists of a single saddle point $O = O_b$.

This assumption holds when $D > \pi^{-2}$. The saddle point is the function $O_b(x) = z_b^*$, $x \in [0, 1]$. If $b = 0$, then $z_b^* = 0$.

For Sturm-Liouville operators, the notion of a determinant can be defined in the following way. For $\phi \in C([0, 1])$, let f be the solution of the initial value problem

$$(\mathscr{H}_\phi F)(f) = 0, \quad f(0) = 1, \quad f'(0) = 0. \tag{12.1.8}$$

Define $\mathrm{Det}(\mathscr{H}_\phi F) = f(1)$. Note that, as a regular Sturm-Liouville operator, $\mathscr{H}_\phi F$ has a countable number of real eigenvalues $(\lambda_k(\phi))_{k\in\mathbb{N}}$. The definition of $\mathrm{Det}(\mathscr{H}_\phi F)$ is justified by the following standard result from Sturm-Liouville theory (see Levit and Smilansky [165]).

Lemma 12.3 *For any ϕ and ψ with non-degenerate Hessian operator, the infinite product*

$$\prod_{k\in\mathbb{N}} \frac{\lambda_k(\phi)}{\lambda_k(\psi)} = \frac{\mathrm{Det}(\mathscr{H}_\phi F)}{\mathrm{Det}(\mathscr{H}_\psi F)} \tag{12.1.9}$$

is convergent.

For $\phi \in H^1_{bc}$, let

$$\mathscr{B}_\rho(\phi) = \{\sigma \in H^1_{bc} \colon \|\sigma - \psi\|_{L^2} \leq \rho\}. \tag{12.1.10}$$

Theorem 12.4 (Mean metastable exit time) *Suppose that Assumptions* 12.1 *and* 12.2 *are satisfied. Then there exists a* $\rho_0 \in (0, \infty)$ *such that, for any* $\rho \in (0, \rho_0)$,

$$\mathbb{E}_{I^+}\big[\tau\big(\mathscr{B}_\rho(I^-)\big)\big] = \frac{2\pi}{[-\lambda^-(O)]}\sqrt{\frac{[-\mathrm{Det}(\mathscr{H}_O F)]}{\mathrm{Det}(\mathscr{H}_{I^+} F)}}\, \mathrm{e}^{[F(O)-F(I^+)]/\varepsilon}\big[1 + \Psi(\varepsilon)\big], \tag{12.1.11}$$

where $\lambda^-(O)$ *is the unique negative eigenvalue of* $\mathscr{H}_O F$, *and the error term satisfies* $\Psi(\varepsilon) = O(\sqrt{\varepsilon}[\ln(1/\varepsilon)]^3)$.

The main idea behind the proof of Theorem 12.4 is the use of the space-discretisation introduced in Sect. 5.7.2. The proof comes in three steps:

(1) Let F_N be the space discretisation of F defined in (5.7.24). According to Theorem 5.70, given $\varepsilon > 0$ and sequences $I^{\pm,N}$, $N \in \mathbb{N}$, converging to $I^\pm$, we have

$$\lim_{N\to\infty} \mathbb{E}_{I^{+,N}}\big[\tau^N_\varepsilon\big(\mathscr{B}_\rho(I^{-,N})\big)\big] = \mathbb{E}_{I^+}\big[\tau\big(\mathscr{B}_\rho(I^-)\big)\big]. \tag{12.1.12}$$

(2) For fixed N, we compute the asymptotics of the transition time. This produces a prefactor $a_N(\varepsilon)$ such that

$$\left|\frac{1}{a_N(\varepsilon)}\mathbb{E}_{I^{+,N}}\big[\tau^N_\varepsilon\big(\mathscr{B}_\rho(I^{-,N})\big)\big] - 1\right| = \psi(\varepsilon, N). \tag{12.1.13}$$

We show that $\psi(\varepsilon, N) \leq \Psi(\varepsilon) = O(\sqrt{\varepsilon}[\ln(1/\varepsilon)]^3)$ for all N. This estimate is first shown for the process starting in the last-exit biased distribution and then transferred to a pointwise estimate with the help of a coupling argument as explained in Sect. 9.4.2.

(3) We show that $a_N(\varepsilon)$ converges to the explicit expression given in (12.1.11) as $N \to \infty$.

12.2 Approximation properties of the potential

In this section we collect some approximation properties of the potential and related quantities.

Recall Sect. 5.7.2. We identify $u^N = (u^N_1, \ldots, u^N_N) \in \mathbb{R}^N$ with the linear interpolation between the points $(i/N, u^N_i)$. We say that $u^N \in \mathbb{R}^N$ converges to $u \in H^1$ when the linear interpolation associated with u^N converges to u in the H^1-norm. The proof of the following lemma is elementary.

Lemma 12.5 *For any* $u^N \in \mathbb{R}^N$, $N \in \mathbb{N}$, *converging to* $u \in H^1$ *the following hold*:

(a) $\lim_{N\to\infty} F_N(u^N) = F(u) < \infty$.
(b) $\lim_{N\to\infty} \nabla F_N(u^N) \cdot h^N = (D_u F)(h)$ *for any* h^N, $N \in \mathbb{N}$, *converging to* h.
(c) $\lim_{N\to\infty} (\mathscr{H} F_N)(u^N)(h^N, k^N) = (D_u^2 F)(h, k)$ *for any* h^N, k^N, $N \in \mathbb{N}$, *converging to* h, k.

Let Δ^N denote the discrete Laplacian defined by

$$\big(\Delta^N u\big)(x) = N^2\big[u\big(x + N^{-1}\big) - 2u(x) + u\big(x - N^{-1}\big)\big]. \tag{12.2.1}$$

Let $(\lambda^0_{k,N})_{1\leq k\leq N}$ be the eigenvalues of $D\Delta^N$ and $(\lambda^0_k)_{k\in\mathbb{N}}$ the eigenvalues of $D\Delta$, in increasing order. In the case of periodic boundary conditions on $[0, 1]$, we have

$$\lambda^0_{k,N} = D\,(2N)^2 \sin^2\left(\frac{k\pi}{2N}\right), \qquad \lambda^0_k = D k^2\pi^2, \quad k \in \mathbb{N}. \tag{12.2.2}$$

Set

$$e_{k,N} = \lambda^0_{k,N} - \lambda^0_k. \tag{12.2.3}$$

Note that $\lim_{N\to\infty} e_{k,N} = 0$ for fixed k, but there is no convergence uniformly in k.

Fix $u^N \in \mathbb{R}^N$, $N \in \mathbb{N}$, converging to $u \in H^1$. Let $(\lambda_{k,N}(u^N))_{1\leq k\leq N}$ be the eigenvalues of $N(\mathscr{H} F_N)(u^N)$ and $(\lambda_k(u))_{k\in\mathbb{N}}$ the eigenvalues of $(\mathscr{H} F)(u)$. We would like to show that $(\lambda_{k,N}(u^N))_{1\leq k\leq N}$ converges to $(\lambda_k(u))_{k\in\mathbb{N}}$ in some appropriate sense. Since (recall (5.7.24))

$$N(\mathscr{H} F_N)\big(u^N\big) = -\tfrac{1}{2} D\Delta^N + V''\big(u^N\big) \tag{12.2.4}$$

and $V''(u)$ is bounded for any u fixed, we have the following estimates.

Lemma 12.6 *There is a constant C such that, for all $k, N, \phi^N, \psi^N, \phi, \psi$,*

$$\begin{aligned} \big|\lambda_{k,N}\big(\phi^N\big) - \lambda_{k,N}\big(\psi^N\big)\big| &\leq C, & \big|\lambda_k(\phi) - \lambda_k(\psi)\big| &\leq C, \\ \big|\lambda_{k,N}\big(\phi^N\big) - \lambda^0_{k,N}\big| &\leq C, & \big|\lambda_k(\phi) - \lambda^0_k\big| &\leq C. \end{aligned} \tag{12.2.5}$$

The following lemma, adapted from de Hoog and Anderssen [76], gives us tighter control under stronger assumptions.

Lemma 12.7 *Consider a sequence $u^N \in \mathbb{R}^N$, $N \in \mathbb{N}$, converging to $u \in C^2([0, 1])$ such that $\|u^N - u\|_\infty = O(N^{-2})$.*

(a) *For every $\alpha \in (0, 1)$ there is a constant C_1 such that, for all N and all $1 \leq k \leq \alpha N$,*

$$\big|\lambda_{k,N}\big(u^N\big) - \lambda_k(u) - e_{k,N}\big| \leq \frac{C_1}{N^2}. \tag{12.2.6}$$

(b) *There exists a constant C_2 such that*

$$|e_{k,N}| \leq \frac{C_2 k^4}{N^2}. \tag{12.2.7}$$

(c) *For fixed* $1 \le k \le N$, *the* H^1*-normalised eigenvector* $\phi_{k,N}$ *of* $(\mathscr{H}F_N)(u^N)$ *associated with* $\lambda_{k,N}(u^N)$ *converges in* H^1 *to the eigenvector* $\phi_k(u)$ *of* $(\mathscr{H}F)(u)$ *associated with* $\lambda_k(u)$, *and*

$$\frac{\|\phi_{k,N}\|_\infty}{\|\phi_{k,N}\|_2} \le \frac{C}{\sqrt{N}}. \tag{12.2.8}$$

Lemmas 12.6–12.7 imply that

$$\lim_{N\to\infty} \prod_{k=0}^{N-1} \frac{\lambda_k(\phi)}{\lambda_k(\psi)} = \prod_{k\in\mathbb{N}_0} \frac{\lambda_k(\phi)}{\lambda_k(\psi)}. \tag{12.2.9}$$

Indeed, this convergence holds because

$$\left|\frac{\lambda_k(\phi) - \lambda_k(\psi)}{\lambda_k(\psi)}\right| \le \frac{C}{k^2}. \tag{12.2.10}$$

Proposition 12.8 *For any* ϕ^N, ψ^N, $N \in \mathbb{N}$, *converging in* H^1 *to* ϕ, ψ *such that* $(\mathscr{H}F)(\psi)$ *and* $(\mathscr{H}F)(\phi)$ *do not have a zero eigenvalue and*

$$\left\|\phi^N - \phi\right\|_\infty \vee \left\|\psi^N - \psi\right\|_\infty \le \frac{C}{N^2}, \tag{12.2.11}$$

the following convergence holds:

$$\lim_{N\to\infty} \frac{\det[(\mathscr{H}F_N)(\phi^N)]}{\det[(\mathscr{H}F_N)(\psi^N)]} - \prod_{k\in\mathbb{N}} \frac{\lambda_k(\phi)}{\lambda_k(\psi)}. \tag{12.2.12}$$

Proof For $1 \le k \le \alpha N$ we proceed as follows. Put

$$\theta_{k,N}(\phi) = \lambda_k(\phi)^{-1}\left[\lambda_{k,N}\left(\phi^N\right) - \lambda_k(\phi)\right]. \tag{12.2.13}$$

Then

$$\frac{\lambda_{k,N}(\phi^N)}{\lambda_k(\phi)} \frac{\lambda_k(\psi)}{\lambda_{k,N}(\psi^N)} = \frac{1+\theta_{k,N}(\phi)}{1+\theta_{k,N}(\psi)} = 1 + \frac{\theta_{k,N}(\phi) - \theta_{k,N}(\psi)}{1+\theta_{k,N}(\psi)}. \tag{12.2.14}$$

By Lemmas 12.6–12.7, we have

$$\left|\theta_{k,N}(\psi)\right| \le \frac{C'}{k^2}\left(\frac{C_1}{N^2} + \frac{C_2 k^4}{N^2}\right) \le C''\left(\alpha^2 + \frac{1}{N^2}\right). \tag{12.2.15}$$

For α small enough and N large enough this gives $|\theta_{k,N}(\psi)| \le \frac{1}{2}$, and hence

$$\left|\ln \prod_{k=0}^{\alpha N} \frac{\lambda_{k,N}(\phi^N)}{\lambda_k(\phi)} \frac{\lambda_k(\psi)}{\lambda_{k,N}(\psi^N)}\right| \le 2\sum_{k=0}^{\alpha N}\left|\theta_{k,N}(\phi) - \theta_{k,N}(\psi)\right| \le \frac{2C\alpha}{N}, \tag{12.2.16}$$

where we use Lemma 12.6 to estimate $|\theta_{k,N}(\phi) - \theta_{k,N}(\psi)| \le C/N^2$.

For $k > \alpha N$ we proceed similarly. Put

$$\begin{aligned} \theta'_{k,N} &= \lambda_{k,N}(\psi)^{-1}\big[\lambda_{k,N}(\phi^N) - \lambda_{k,N}(\psi^N)\big], \\ \theta'_k &= \lambda_k(\psi)^{-1}\big[\lambda_k(\phi) - \lambda_k(\psi)\big]. \end{aligned} \tag{12.2.17}$$

Then

$$\frac{\lambda_{k,N}(\phi^N)}{\lambda_k(\phi)} \frac{\lambda_k(\psi)}{\lambda_{k,N}(\psi^N)} = \frac{1+\theta'_{k,N}}{1+\theta'_k} = 1 + \frac{\theta'_{k,N} - \theta'_k}{1+\theta'_k}, \tag{12.2.18}$$

and similarly for θ'_k. By Lemma 12.6, we have

$$|\theta'_k| \le \frac{C}{k^2}. \tag{12.2.19}$$

For α fixed and N large enough this gives $|\theta'_k| \le \frac{1}{2}$, and hence

$$\left| \ln \prod_{k=\alpha N}^{N-1} \frac{\lambda_{k,N}(\phi^N)}{\lambda_k(\phi)} \frac{\lambda_k(\psi)}{\lambda_{k,N}(\psi^N)} \right| \le 2 \sum_{k=\alpha N}^{N-1} |\theta'_{k,N} - \theta'_k| \le \sum_{k=\alpha N}^{N-1} \frac{2C}{k^2} \le \frac{2C}{\alpha N}, \tag{12.2.20}$$

where we use Lemma 12.6 to estimate $|\theta'_{k,N} - \theta'_k| \le C/k^2$. Combining (12.2.16) and (12.2.20), and recalling that

$$\frac{\det[(\mathscr{H} F_N)(\phi^N)]}{\det[(\mathscr{H} F_N)(\psi^N)]} = \prod_{k=0}^{N-1} \frac{\lambda_{k,N}(\phi^N)}{\lambda_{k,N}(\psi^N)}, \tag{12.2.21}$$

we get the claim. □

It can be shown that the conclusion of Proposition 12.8 holds when the condition in (12.2.11) is replaced by

$$\|\phi^N - \phi\|_{L^2} \vee \|\psi^N - \psi\|_{L^2} \le \frac{C}{N}. \tag{12.2.22}$$

The next lemma shows that every stationary point of F can be approximated by a sequence of stationary points of F_N, $N \in \mathbb{N}$, in the sense of (12.2.22). The proof is elementary.

Lemma 12.9 *There exist C, N_0 such that for all $N > N_0$ and all stationary points ϕ of F there is a stationary point ϕ^N of F_N such that*

$$\|\phi - \phi^N\|_{L^2} \le \frac{C}{N}. \tag{12.2.23}$$

12.3 Estimate of the capacity

In this section we compute the relevant capacities for the discretised process. This can be taken from Chap. 11, except that we have to take care of the N-dependence of the error terms.

Recall from (5.7.27) that, after proper rescaling, we are considering the N-dimensional diffusion

$$dX_t^N = -N\nabla F_N\big(X_t^N\big)\,dt + \sqrt{2\varepsilon N}\,dB_t. \tag{12.3.1}$$

Denote by μ_ε^N the invariant measure for the process $X^N = (X_t^N)_{t\in\mathbb{R}_+}$:

$$\mu_\varepsilon^N(dx) = \mathrm{e}^{-F_N(x)/\varepsilon}dx. \tag{12.3.2}$$

The Dirichlet form for this process is given by

$$\mathscr{E}^N(h,h) = \varepsilon N\int_{\mathbb{R}^N}\|\nabla h(x)\|_2^2\,\mu_\varepsilon^N(dx). \tag{12.3.3}$$

Let $\mathscr{B}_\rho^N(x)$ denote the Euclidean ball of radius ρ around $x\in\mathbb{R}^N$. Write $I^{+,N}$, $I^{-,N}$, O^N, $\lambda_N^-(O^N)$ to denote the analogues of I^+, I^-, O defined prior to Theorem 12.4. The following proposition is the desired estimate for the capacity with an error term that is uniform in N.

Proposition 12.10 *For all* $0<\varepsilon<\varepsilon_0$ *and* $\rho>0$,

$$\begin{aligned}&\mathrm{cap}\big(\mathscr{B}_\rho^N\big(I^{+,N}\big),\mathscr{B}_\rho^N\big(I^{-,N}\big)\big)\\&\quad=\varepsilon(2\pi\varepsilon)^{(N-2)/2}\frac{[-\lambda_N^-(O^N)]\,\mathrm{e}^{-F_N(O^N)/\varepsilon}}{\sqrt{-\det[(\mathscr{H}F_N)(O^N)]}}\big[1+\psi_1(\varepsilon,N)\big],\end{aligned} \tag{12.3.4}$$

where $\limsup_{N\to\infty}|\psi_1(\varepsilon,N)|\le C\sqrt{\varepsilon[\ln(1/\varepsilon)]^3}$ *for some constant* C.

The proof of this proposition is given in Sects. 12.3.1–12.3.3.

12.3.1 Properties of the potential

We need to control the potentials F_N globally and near their critical points. It is very convenient that in our setting the Hessians at all the three stationary points are diagonal in the same basis, namely,

$$v_l^k = \omega^{kl}/\sqrt{N},\quad k\in\{0,\dots,N-1\},\ l\in\{1,\dots,N\}, \tag{12.3.5}$$

with $\omega = e^{2\pi i/N}$. This allows us to choose global coordinates for which all relevant Hessians are diagonal. For $y \in \mathbb{R}^N$, define

$$\hat{y}_k = \hat{y}_k(y) = \frac{1}{\sqrt{N}} \sum_{l=1}^{N} v_l^k y_l, \tag{12.3.6}$$

and the inverse

$$y_l = y_l(\hat{y}) = \sqrt{N} \sum_{k=0}^{N-1} v_l^k \hat{y}_k. \tag{12.3.7}$$

Recall that the explicit form of F_N in the old coordinates is (recall (5.7.23))

$$F_N(y) = N^{-1} \sum_{l=1}^{N} V(y_l) + \tfrac{1}{4} N D \sum_{l=1}^{N} (y_l - y_{l+1})^2. \tag{12.3.8}$$

In the new coordinates this takes the form (recall (12.2.2))

$$\begin{aligned} F_N\big(y(\hat{y})\big) &= \tfrac{1}{2} \sum_{k=0}^{N-1} \lambda_{k,N}^0 \hat{y}_k^2 - \tfrac{1}{2} \hat{y}_0^2 + \frac{1}{4N} \big\| y(\hat{y}) \big\|_4^4 \\ &= \tfrac{1}{2} \sum_{k=0}^{N-1} \lambda_{k,N} \hat{y}_k^2 + \frac{1}{4N} \big\| y(\hat{y}) \big\|_4^4 \end{aligned} \tag{12.3.9}$$

where $\lambda_{k,N}^0$, $k = 0, \ldots, N-1$, are the eigenvalues of D times the discrete Laplacian ($\lambda_{0,N}^0 = 0$), and we put

$$\lambda_{0,N} = -1, \qquad \lambda_{k,N} = \lambda_{k,N}^0, \quad k = 1, \ldots, N-1. \tag{12.3.10}$$

The critical points in the new coordinates are

$$I^{\pm,N}(\hat{y}) = (\pm 1, 0, \ldots, 0), \qquad O^N(\hat{y}) = (0, 0, \ldots, 0). \tag{12.3.11}$$

Since these all lie on the line $\hat{y}_1 = \hat{y}_2 = \cdots = \hat{y}_{N-1} = 0$, it is useful to single out the 0-th coordinate. Note that

$$\frac{1}{N} \sum_{l=1}^{N} \big| y_l(\hat{y}) \big|^4 = \frac{1}{N} \sum_{l=1}^{N} \left| \hat{y}_0 + \sum_{k=1}^{N-1} \omega^{lk} \hat{y}_k \right|^4 = \frac{1}{N} \sum_{l=1}^{N} \big| \hat{y}_0 + w_l(\hat{y}) \big|^4, \tag{12.3.12}$$

where $w_l = w_l(\hat{y}) = \sum_{k=1}^{N-1} \omega^{kl} \hat{y}_k$. The important point is that $\sum_{l=1}^{N} \omega^{kl} = 0$ for all $k \neq 0$. Hence, expanding to fourth power, we get the following.

Lemma 12.11 *The potential F_N expressed in the new coordinates $\hat{y}$ satisfies*:

(i)

$$\left|F_N(y(\hat{y})) + \tfrac{1}{2}\hat{y}_0^2(1+3\|w\|_2^2) - \tfrac{1}{4}\hat{y}_0^4 - \tfrac{1}{2}\sum_{k=1}^{N-1}\lambda_{k,N}\hat{y}_k^2\right|$$
$$\leq \frac{1}{4N}\left(4|\hat{y}_0|\|w\|_3^3 + \|w\|_4^4\right). \tag{12.3.13}$$

(ii)

$$F_N(y(\hat{y})) \geq \tfrac{1}{2}\sum_{k=1}^{N-1}\lambda_{k,N}\hat{y}_k^2 - \tfrac{1}{2}\hat{y}_0^2. \tag{12.3.14}$$

Proof Item (i) follows from (12.3.9) and (12.3.13). Item (ii) follows trivially because the quartic term in (12.3.9) is non-negative. □

The following facts are crucial.

Lemma 12.12 *With the norms and maps defined above:*

(i) *The Parseval identity holds, i.e.,*

$$N^{-1/2}\|y(\hat{y})\|_2 = \|\hat{y}\|_2. \tag{12.3.15}$$

(ii) *The Hausdorff-Young inequality holds, i.e., for any $p \geq 2$ and for $q = p/(p-1)$ there exists a constant C_q such that*

$$N^{-1/p}\|y(\hat{y})\|_p \leq C_p\|\hat{y}\|_q. \tag{12.3.16}$$

Proof The Parseval identity is checked easily. The Hausdorff-Young inequality is proven as a consequence of the Riesz-Thorin interpolation theorem. Namely, since the components of the vectors v^k are bounded in absolute value by $C/\sqrt{N}$, $\hat{y} \mapsto y(\hat{y})$ is bounded as a map from L^1 to L^∞, i.e.,

$$\|y(\hat{y})\|_\infty \leq C\|\hat{y}\|_1. \tag{12.3.17}$$

Together with the Parseval identity, this provides the input to obtain (12.3.16) from the Riesz-Thorin interpolation theorem. See Reed and Simon [204, p. 328]. □

Define, for $\delta_0 > 0$ and for constants $r_{k,N}$, $k = 1, \ldots, N-1$, to be chosen later, the sets

$$C_\delta^{N,\perp} = \left\{\hat{y} \in \mathbb{R}^{N-1} : |\hat{y}_k| \leq \delta r_{k,N}/\sqrt{|\lambda_{k,N}|},\ 1 \leq k \leq N-1\right\}. \tag{12.3.18}$$

Then, for $\hat{y} \in C_\delta^{N,\perp}$,

$$\|w(\hat{y})\|_p \leq \delta^p \left(\sum_{k=1}^{N-1}\frac{r_{k,N}^q}{(\lambda_{k,N})^{q/2}}\right)^{p/q}. \tag{12.3.19}$$

From the explicit form of the eigenvalues of the discrete Laplacian in (12.2.2) and the relation in (12.3.10), we see that $\lambda_{k,N} = \lambda_{N-k,N}$, $k = 1, \dots, N-1$. Using that, for $0 \le t \le \frac{\pi}{2}$,

$$0 < t^2\left(1 - \tfrac{1}{3}t^2\right) \le \sin^2 t \le t^2, \tag{12.3.20}$$

we see that, for $1 \le k \le \lfloor \frac{N}{2} \rfloor$,

$$\lambda_{k,N} \ge k^2 \tfrac{1}{8} D\pi^2\left(1 - \tfrac{1}{12}\pi^2\right). \tag{12.3.21}$$

The constants $r_{k,N}$ are constructed as follows. For an increasing sequence $(\rho_k)_{k\in\mathbb{N}}$ set

$$r_{k,N} = r_{N-k,N} = \rho_k, \quad 1 \le k \le \left\lfloor \frac{N}{2} \right\rfloor. \tag{12.3.22}$$

Pick $\rho_k = k^\alpha$ with $\alpha > 0$ such that, for $q = \frac{3}{2}, \frac{4}{3}$,

$$\sum_{k\in\mathbb{N}} \frac{\rho_k^q}{k^q} = B_q < \infty. \tag{12.3.23}$$

This yields the following estimates.

Lemma 12.13 *There is a choice of $\alpha > 0$ such that for $\rho_k = k^\alpha$ and $p = 2, 3, 4$ there are constants $C_p < \infty$ (independent of N) such that*

(i) *For $\hat{y} \in C_\delta^{N,\perp}$,*

$$N^{-1}\left\|w(\hat{y})\right\|_p^p \le \delta^p C_p. \tag{12.3.24}$$

(ii) *For $\hat{y} \in C_\delta^{N,\perp}$ with $\hat{y}_0 \le \delta$,*

$$N^{-1}\left\|y(\hat{y})\right\|_4^4 \le \delta^4 C_4. \tag{12.3.25}$$

Proof Collect the estimates above. □

This uniform control on the quadratic approximation of F_N is the main ingredient needed to extend the analysis of capacities in Chap. 11 to the SPDE setting.

12.3.2 Upper bound

The strategy for the upper bound is the same as in the proof of Theorem 11.2.

Proof Define the following neighbourhood of the saddle point O^N:

$$C_\delta^N = C_\delta^N\left(O^N\right) = \left\{y(\hat{y}) \in \mathbb{R}^N : \ |\hat{y}_0| \le c_0\delta, \hat{y} \in C_\delta^{N,\perp}\right\}, \tag{12.3.26}$$

where $C_\delta^{N,\perp}$ is defined in (12.3.18) and $c_0 < \infty$ is a constant to be chosen. For the upper bound it is enough to replace F_N by its lower bound in (12.3.14). Define the set

$$\mathscr{U}_\delta^N = \{y(\hat{y}) \in \mathbb{R}^N : |\hat{y}_0| \le c_0\delta\}. \tag{12.3.27}$$

Choose a test function h^+ in the Dirichlet principle for the Dirichlet form in (12.3.3) to obtain an upper bound on the capacity of interest. The set $(\mathscr{U}_\delta^N)^c$ decomposes into two disjoint connected components, one of which contains $I^{+,N}$. We set $h^+(y) = 1$ on the latter component and $h^+(y) = 0$ on the other component. On $\mathscr{U}_\delta^N$ we choose h^+ as $h^+(y) = f(\hat{y}_0)$, where (recall that $\lambda_{0,N} = -1$)

$$f(s) = \frac{\int_s^{c_0\delta} e^{\lambda_{0,N}t^2/2\varepsilon}\,dt}{\int_{-c_0\delta}^{c_0\delta} e^{\lambda_{0,N}t^2/2\varepsilon}\,dt}. \tag{12.3.28}$$

The Dirichlet form evaluated on this test function then reduces to

$$\begin{aligned}
&\mathscr{E}^N(h^+,h^+)\\
&= \varepsilon N\sqrt{N}^N \int_{\mathscr{U}_\delta^N} d\hat{y}\, \|\nabla h^+(y(\hat{y}))\|_2^2\, e^{-F_N(y(\hat{y}))/\varepsilon}\\
&\le \varepsilon\sqrt{N}^N \int_{-c_0\delta}^{c_0\delta} d\hat{y}_0\, e^{-\lambda_{0,N}\hat{y}_0^2/2\varepsilon} |f'(\hat{y}_0)|^2 \int_{\mathbb{R}^{N-1}} d\hat{y}_1\dots d\hat{y}_{N-1}\, e^{-\sum_{k=1}^{N-1}\lambda_{k,N}\hat{y}_k^2/2\varepsilon}\\
&= \varepsilon\sqrt{N}^N \frac{1}{\int_{-c_0\delta}^{c_0\delta} e^{\lambda_{0,N}s^2/2\varepsilon}\,ds} \prod_{k=1}^{N-1}\sqrt{\frac{2\pi\varepsilon}{\lambda_{k,N}}}\\
&= \varepsilon\sqrt{N}^N \sqrt{\frac{-\lambda_{0,N}}{2\pi\varepsilon}} \prod_{k=1}^{N-1}\sqrt{\frac{2\pi\varepsilon}{\lambda_{k,N}}}\,[1 + O(e^{c_0^2\delta^2\lambda_{0,N}/2\varepsilon})].
\end{aligned} \tag{12.3.29}$$

In the first and second equality, the change of variable $y \to \hat{y}$ gives rise to the factor $\sqrt{N}^N$ and the relation $\|\nabla h^+(y(\hat{y}))\|_2^2 = N^{-1}|f'(\hat{y}_0)|^2$. Taking $\delta = K\sqrt{\varepsilon \ln(1/\varepsilon)}$, as in Chap. 11, we see that the right-hand side has the desired asymptotics. Thus we obtain that, for N large enough,

$$\begin{aligned}
&\mathrm{cap}(\mathscr{B}_\rho^N(I^{+,N}), \mathscr{B}_\rho^N(I^{-,N}))\\
&\le \varepsilon(2\pi\varepsilon)^{(N-2)/2} \frac{[-\lambda_{0,N}]\, e^{-F_N(O^N)/\varepsilon}}{\sqrt{-\det[(\mathscr{H}F_N)(O^N)]}} \left[1 + \varepsilon^{-c_0^2K^2/2}\right],
\end{aligned} \tag{12.3.30}$$

where we use that $\det[(\mathscr{H}F_N)(O^N)] = \prod_{k=0}^{N-1}\lambda_{k,N}$, and recall that $F_N(O^N) = 0$ (see (12.1.5), (12.3.8) and (12.3.11)). This is the upper bound with a better error estimate than in (12.3.4). □

Remark 12.14 Note that, due to the particularly simple form of the potential in (12.1.5), we did not need to use the fact that F_N is well approximated by a quadratic function in a suitable neighbourhood of the saddle point. In more general settings, however, this would be needed, together with an estimate showing that the contributions coming from outside this neighbourhood are negligible, as in Chap. 11.

12.3.3 Lower bound

We next turn to the proof of the complementing lower bound.

Proof Around the saddle point O^N we take a narrow corridor from one local minimum to the other, and minimise the Dirichlet form on this corridor. We use the same notation as in the proof of the upper bound.

We bound the capacity from below by

$$\begin{aligned} &\operatorname{cap}\big(\mathscr{B}_\rho^N(I^{+,N}),\mathscr{B}_\rho^N(I^{-,N})\big) \qquad (12.3.31)\\ &\geq \inf_{\substack{h:\,h(x)=1\,\forall x\in\mathscr{B}_\rho^N(I^{+,N})\\ h(x)=0\,\forall x\in\mathscr{B}_\rho^N(I^{-,N})}} \varepsilon N\sqrt{N}^N\int_{C_\delta^{N,\perp}}\big\|\nabla h\big(y(\hat y)\big)\big\|_2^2\,\mathrm{e}^{-F_N(y(\hat y))/\varepsilon}\,d\hat y\\ &\geq \inf_{\substack{h:\,h(x)=1\,\forall x\in\mathscr{B}_\rho^N(I^{+,N})\\ h(x)=0\,\forall x\in\mathscr{B}_\rho^N(I^{-,N})}} \varepsilon N\sqrt{N}^N\int_{C_\delta^{N,\perp}}\left|\frac{d}{d\hat y_0}h\big(y(\hat y)\big)\right|^2\mathrm{e}^{-F_N(y(\hat y))/\varepsilon}\,d\hat y. \end{aligned}$$

The infimum can now be performed for each value of the orthogonal coordinates $\hat y^\perp=(\hat y_1,\ldots,\hat y_{N-1})$ separately, i.e., the right-hand side of (12.3.31) is larger than or equal to

$$\begin{aligned} &\varepsilon\sqrt{N}^N\int_{C_\delta^{N,\perp}}d\hat y^\perp\sup_{f:\,f(1)=1,f(-1)=0}\int_{-1}^1 d\hat y_0\,\big|f'(\hat y_0)\big|^2\,\mathrm{e}^{-F_N(y(\hat y_0,\hat y^\perp))/\varepsilon}\\ &=\varepsilon\sqrt{N}^N\int_{C_\delta^{N,\perp}}d\hat y^\perp\left[\int_{-1}^1 d\hat y_0\,\mathrm{e}^{F_N(y(\hat y_0,\hat y^\perp))/\varepsilon}\right]^{-1}, \qquad (12.3.32) \end{aligned}$$

where we use that we already know how to solve the one-dimensional variational problem.

To conclude, we need to bound the second integral in (12.3.32) from above. Using the upper bound from Lemma 12.11 and bounding the norms of w appearing there with the help of Lemma 12.13, we obtain

$$\int_{-1}^1 d\hat y_0\,\mathrm{e}^{F_N(y(\hat y_0,\hat y^\perp))/\varepsilon}\leq \mathrm{e}^{\frac12\sum_{k=1}^{N-1}\lambda_{k,N}\hat y_k^2/\varepsilon+O(\delta^3)/\varepsilon}\int_{-1}^1 d\hat y_0\,\mathrm{e}^{(\frac12\lambda_{0,N}\hat y_0^2[1+O(\delta^2)]+\frac14\hat y_0^4)/\varepsilon} \qquad (12.3.33)$$

when $y_\perp\in C_\delta^{N,\perp}(O^N)$. We again choose $\delta=K\sqrt{\varepsilon\ln(1/\varepsilon)}$ for some sufficiently large K, and recall that $\lambda_{0,N}=-1$. Hence the exponent in the integrand in the right-hand side of (12.3.33) without the error term achieves its unique maximum at $-1/4\varepsilon$. It is therefore easy to see that

$$\int_{-1}^{1} d\hat{y}_0\, \mathrm{e}^{(\frac{1}{2}\lambda_{0,N}\hat{y}_0^2[1+O(\delta^2)]+\frac{1}{4}\hat{y}_0^4)/\varepsilon} = \sqrt{2\pi\varepsilon}\,(-\lambda_{0,N})^{-1/2}\big[1+O\big(\varepsilon\ln(1/\varepsilon)\big)\big]. \tag{12.3.34}$$

Inserting this bound into (12.3.32), we can now carry out all the integrals over the $\hat{y}_k$, $1\le k\le N-1$. It is again elementary to show that

$$\begin{aligned}
&\int_{C_\delta^{N,\perp}} d\hat{y}^\perp \mathrm{e}^{-\frac{1}{2}\sum_{k=1}^{N-1}\lambda_{k,N}\hat{y}_k^2/\varepsilon}\\
&\ge \int_{\mathbb{R}^{N-1}} d\hat{y}^\perp \mathrm{e}^{-\frac{1}{2}\sum_{k=1}^{N-1}\lambda_{k,N}\hat{y}_k^2/\varepsilon}\left(1-\sum_{k=1}^{N-1}\sqrt{\frac{\lambda_{k,N}}{2\pi\varepsilon}}\int_{|\hat{y}_k|\ge\delta r_{k,N}/\sqrt{\lambda_{k,N}}} d\hat{y}_k\,\mathrm{e}^{-\frac{1}{2}\lambda_{k,N}\hat{y}_k^2/\varepsilon}\right)\\
&\ge \sqrt{2\pi\varepsilon}^{\,N-1}\prod_{k=1}^{N-1}\frac{1}{\sqrt{\lambda_{k,N}}}\left(1-\sum_{k=1}^{N-1} r_{k,N}^{-1}\mathrm{e}^{-\frac{1}{2}K\ln(1/\varepsilon)r_{k,N}^2}\right).
\end{aligned} \tag{12.3.35}$$

If we choose $r_{k,N}$ as in (12.3.22), with $\rho_k=k^\alpha$ for some $\alpha>0$, and choose K large enough, then we can arrange that

$$1-\sum_{k=1}^{N-1} r_{k,N}^{-1}\mathrm{e}^{-\frac{1}{2}K\ln(1/\varepsilon)r_{k,N}^2}\ge 1-\widehat{K}\varepsilon^{\widetilde{K}} \tag{12.3.36}$$

for some $\widehat{K}<\infty$ for $\widetilde{K}>1$ as large as desired (uniformly in N). From here we get the desired lower bound

$$\begin{aligned}
&\operatorname{cap}\big(\mathscr{B}_\rho^N(I^{+,N}),\mathscr{B}_\rho^N(I^{-,N})\big)\\
&\quad\ge \varepsilon\sqrt{2\pi\varepsilon}^{\,N-2}\,\frac{[-\lambda_{0,N}]\mathrm{e}^{-F_N(O^N)/\varepsilon}}{\sqrt{-\det[(\mathscr{H}F_N)(O^N)]}}\Big[1-C\sqrt{\varepsilon\ln(1/\varepsilon)^3}\Big]
\end{aligned} \tag{12.3.37}$$

for some constant C that is independent of N. This is the claimed lower bound and concludes the proof of Proposition 12.10. □

12.4 Estimate of the equilibrium potential

Recall from Corollary 7.30 that we have the formula

$$\mathbb{E}_{\nu_{\mathscr{B}_\rho^N(I^{+,N}),\mathscr{B}_\rho^N(I^{-,N})}}[\tau_{\mathscr{B}_\rho^N(I^{-,N})}]=\frac{\int_{\mathbb{R}^N} h_{B_\rho^N(I^{+,N}),B_\rho^N(I^{-,N})}(x)\,d\mu_\varepsilon^N(x)}{\operatorname{cap}(\mathscr{B}_\rho^N(I^{+,N}),\mathscr{B}_\rho^N(I^{-,N}))}. \tag{12.4.1}$$

In Sect. 12.3 we derived upper and lower bounds on the denominator in (12.4.1) We next derive estimates on the numerator of (12.4.1). The point is to show that this is essentially the mass of a small neighbourhood of the starting minimum $I^{+,N}$.

Proposition 12.15 *For all* $0 < \varepsilon < \varepsilon_0$ *and* $\rho > 0$ *small enough,*

$$\int_{\mathbb{R}^N} h_{B_\rho^N(I^{+,N}),B_\rho^N(I^{-,N})}(x)\,d\mu_\varepsilon^N(x) = \frac{(2\pi\varepsilon)^N}{\sqrt{\det[(\mathscr{H}F_N)(I^{+,N})]}}\mathrm{e}^{-F_N(O^N)/\varepsilon}\big[1+\psi_2(\varepsilon,N)\big], \tag{12.4.2}$$

where $\limsup_{N\to\infty}|\psi_2(\varepsilon,N)| \le C\sqrt{\varepsilon[\ln(1/\varepsilon)]^3}$.

Proof We first consider the symmetric case $b = 0$ in (12.1.5). As in the previous section, we define around the minimum $I^{\pm,N} \in \mathbb{R}^N$ a neighbourhood $C_\delta^N(I^{\pm,N})$ by

$$C_\delta^N\big(I^{\pm,N}\big) = \big\{y(\hat{y}) \in \mathbb{R}^N:\ |\hat{y}_0 \mp 1| \le \delta,\ \hat{y} \in C_\delta^{N,\perp}\big\}. \tag{12.4.3}$$

To estimate the left-hand side of (12.4.2) we need yet another lower bound on the non-quadratic terms in F_N. This time we write

$$\|y(\hat{y})\|_{4,N}^4 = \hat{y}_0^4 + \|y(\hat{y})\|_{4,N}^4 - \hat{y}_0^4 \le \hat{y}_0^4, \tag{12.4.4}$$

where in the last inequality we use that, by the Cauchy-Schwarz inequality,

$$\hat{y}_0^4 = \left(N^{-1}\sum_{l=1}^N y_l\right)^4 \le N^{-2}\left(\sum_{l=1}^N y_l^2\right)^2 \le N^{-1}\sum_{l=1}^N y_l^4. \tag{12.4.5}$$

Inserting (12.4.4) into (12.3.9), we get

$$F_N\big(y(\hat{y})\big) \ge \sum_{k=0}^{N-1}\lambda_{k,N}\hat{y}_k^2 + \tfrac{1}{4}\hat{y}_0^4. \tag{12.4.6}$$

Note, moreover, that the coordinates of the two local minima are

$$\hat{y}\big(I^{\pm,N}\big)_k = \pm\delta_{k,0}, \tag{12.4.7}$$

and in the C_δ^N-neighbourhoods of these local minima the quadratic approximation is good. Finally, the sets $C_\delta^N(I^{\pm,N})$ are subsets of $B_\rho^N(I^{\pm,N})$, so that the integrand is equal to 1 on the set $C_\delta^N(I^{+,N})$ and equal to 0 on the set $C_\delta^N(I^{-,N})$. The claimed estimate on the integral is now straightforward.

Most of the analysis above carries over unchanged when $b > 0$. The saddle points remain the same, while the positions of the minima are shifted. More importantly, the value of F_N is now smaller by bs on the negative side. To show that nonetheless there is no contribution from the target valley, we need a bound on the equilibrium potential. Let

$$\mathscr{A} = \big\{x \in \mathbb{R}^N:\ F_N(x) \le F_N\big(I_+^N\big) + \delta\big\} \tag{12.4.8}$$

for some $\delta > 0$ small enough.

Lemma 12.16 *For all $\eta > 0$ there exist $\rho_0 > 0, \delta_0 > 0$ and $\varepsilon_0 > 0$ such that for all $0 < \rho < \rho_0, 0 < \delta < \delta_0, 0 < \varepsilon < \varepsilon_0$ and $x \in \mathscr{A}$,*

$$h_{B_\rho^N(I^{+,N}),B_\rho^N(I^{-,N}l)}(x) \leq \mathrm{e}^{-(F_N(O^N)-F_N(x)-c\delta^2\eta)/\varepsilon}. \tag{12.4.9}$$

Proof By the definition of the set $\mathscr{A}$, all paths from $x \in \mathscr{A}$ to $I^{+,N}$ must attain a height at least $F_N(O^N)$. Therefore it follows from the large deviation principle and the discussion on the exit problem (see Sect. 6.5.2) that for any $T < \infty$ fixed and all $x \in \mathscr{A}$,

$$\mathbb{P}_x(\tau_{B_\rho^N(I^{+,N})} < T) \leq \mathrm{e}^{-(F_N(O^N)-F_N(x)-\eta)/\varepsilon}. \tag{12.4.10}$$

On the other hand, for all $x \in \mathscr{A}$ there is a zero-action path from x to one of the minima in $B_\rho^N(I^{-,N})$ that takes only a finite time T_0. All zero-action paths must lead to $B_\rho^N(I^{-,N})$ in finite time. Therefore, to stay away from this set for a time T requires the path *not to follow* a minimiser of the action integral for time $T - T_0$. This costs a total action of order Ta for some $a > 0$, and thus the probability of this event is of order $\exp(-Ta/\varepsilon)$, which can be made as small as desired by choosing T large enough. In particular, it can be made much smaller than the probability in (12.4.10). Now the simple bound

$$\mathbb{P}_x(\tau_{B_\rho^N(I^{+,N})} < \tau_{B_\rho^N(l^{-,N})}) \leq \mathbb{P}_x(\tau_{B_\rho^N(I^{+,N})} < T) + \mathbb{P}_x(\tau_{B_\rho^N(I^{-,N})} > T) \tag{12.4.11}$$

yields the desired estimate. □

Using the bound in Lemma 12.16, we see that the results for the symmetric case $b = 0$ carry over to $b > 0$. This completes the proof of Proposition 12.15. □

Remark 12.17 In more complicated situations, i.e., in the presence of multiple stationary points, the argument gets a little more involved. In that case, the process may reach a small neighbourhood of some other stationary point before reaching its final destination, and in this neighbourhood it could spend a large amount of time without penalty. The probabilities to first reach the various stationary points are easily computed with the help of large deviations, and by continuing the analysis step for step from these new points as starting points we can show that this does not affect the ultimate estimate on the harmonic function. This type of analysis is the basis of the Freidlin-Wentzell theory [115]. All estimates involve only the potentials F_N, and since these converge to F as discussed earlier, the control that is obtained in this way is uniform in N.

12.5 Proof of the main theorem

Proof By putting all the estimates together, we obtain the following result on the mean metastable exit time.

Proposition 12.18 *Uniformly in N,*

$$\mathbb{E}_{I^{+,N}}\left[\tau^N\left(\mathscr{B}_\rho^N\left(I^{-,N}\right)\right)\right]$$
$$= \frac{2\pi\, \mathrm{e}^{(F_N(O^N)-F_N(I^{+,N}))/\varepsilon}}{[-\lambda_{0,N}]} \frac{\sqrt{-\det[(\mathscr{H}F_N)(O^N)]}}{\sqrt{\det(\mathscr{H}F_N)(I^{+,N})}}\left[1+\Psi(\varepsilon,N)\right], \quad (12.5.1)$$

where the error term satisfies

$$\limsup_{N\to\infty}\left|\Psi(\varepsilon,N)\right| \le C\sqrt{\varepsilon\left[\ln(1/\varepsilon)\right]^3}. \quad (12.5.2)$$

Proof Inserting the estimates for the denominator (Proposition 12.10) and the numerator (Proposition 12.15) into (12.4.1), we get that $\mathbb{E}_{\nu^N}[\tau_\varepsilon^N]$ is equal to the right-hand side of (12.5.1), where

$$\nu^N = \nu^N_{\mathscr{B}_\rho^N(I^{+,N}),\mathscr{B}_\rho^N(I^{-,N})} \quad (12.5.3)$$

is the last-exit biased distribution on $\mathscr{B}_\rho^N(I^{+,N})$. Then use Theorem 9.14 to replace ν^N by the point $\mathscr{B}_\rho^N(I^{+,N})$. □

The assertion of Theorem 12.4 follows from Proposition 12.18 and the convergence results established in Sect. 5.7, in particular, Theorem 5.70. □

12.6 Bibliographical notes

1. The system in (5.7.1) and its metastable behaviour have been studied for thirty years. The main techniques employed in the literature are based on large deviation principles and comparison estimates between the deterministic process ((5.7.1) with $\varepsilon = 0$) and the stochastic process ((5.7.1) with $\varepsilon > 0$). Faris and Jona-Lasinio [107] analysed (5.7.1) for the quartic double-well potential we considered here. Cassandro, Olivieri and Picco [52] obtained similar asymptotics as in [107] when the space interval [0, 1] is not fixed but tends to infinity as $\varepsilon \downarrow 0$ (sufficiently slowly). These results established the existence of a suitable exponential time scale on which the process undergoes a transition. For (6.3.1), Martinelli, Olivieri and Scoppola [175] obtained the asymptotic exponential law of the transition times. Brassesco [41] proved that the trajectories exhibit characteristics of metastable behaviour: the escape from the basin of attraction of the minimum occurs through the lowest saddle points and the process starting from this minimum spends most of its time before the transition near this minimum.

2. As in the finite-dimensional setting, local minima and saddle points play a key role in understanding metastability. In the infinite-dimensional setting, identifying the critical points is already a difficult task in itself. Fortunately, elegant methods are available to do so: see e.g. Fiedler and Rocha [113] and Wolfrum [237].

3. To analyse metastability for the infinite-dimensional diffusion, we used a spatial discretisation that brings us back to the case of finite-dimensional diffusions studied in Chap. 11. The use of spatial finite-difference approximation is natural. Berglund and Gentz [22] use a Galerkin approximation and obtain analogous results in a more general setting. Our main objective has been to derive the infinite-dimensional analogue of Kramer's formula for average metastable exit times. Such a formula was conjectured by Maier and Stein [170] (see also Vanden-Eijnden and Westdickenberg [230]) as a formal limit of the finite-dimensional systems. For the setting described in this chapter, this limit was justified rigorously in Barret, Bovier and Méléard [12]. Extensions to more general settings were studied by Barret [11], and Berglund and Gentz [22]. Berglund and Gentz also consider cases where the Hessian matrix is degenerate.

Part V
Applications: Coarse-Graining in Large Volumes at Positive Temperatures

Part V deals with Markov processes that allow for *coarse-graining*, i.e., a lumping of states that leads to a simpler Markov process on a reduced state space. For instance, the reduction of the state space of a high-dimensional spin system to that of a low-dimensional spin system, whenever possible, is a powerful tool for the analysis of its dynamics. Some mean-field models allow for such a reduction.

Chapter 13 looks at the Curie-Weiss model, Chaps. 14–15 at the random-field Curie-Weiss model.

Chapter 13
The Curie-Weiss Model

La simplicité affectée est une imposture délicate.
(François de La Rochefoucauld, Réflexions)

Most systems of interest in statistical physics are extremely high-dimensional, and become infinite-dimensional in the thermodynamic limit. Unlike in the diffusion-type models discussed in Part IV, their metastable behaviour cannot be read off from the energy of paths alone, because a true interplay between energy and entropy of paths takes place. This makes the analysis of such systems hard. A promising strategy is the reduction of this complexity via a mapping to a low-dimensional state space in the spirit of the coarse-graining and lumping explained in Sects. 9.2–9.3. In this chapter we deal with the Curie-Weiss model. Section 13.1 defines the model and introduces the coarse-graining. Section 13.2 solves the coarse-grained model and proves the theorems describing its metastable behaviour.

13.1 The Curie-Weiss model

The toy model where coarse-graining works perfectly well is the *Curie-Weiss model* of a ferromagnet. The state space is $S_\Lambda = \{-1,+1\}^\Lambda$ with $\Lambda = \{1,\dots,N\}$, $N \in \mathbb{N}$. The Hamiltonian is given by

$$H_N(\sigma) = -\frac{1}{2N}\sum_{i,j\in\Lambda}\sigma_i\sigma_j - h\sum_{i\in\Lambda}\sigma_i, \quad \sigma \in S_\Lambda, \tag{13.1.1}$$

with $h \in \mathbb{R}$ the magnetic field. The fact that this is a *mean-field model* is expressed by the fact that $H_N(\sigma)$ depends on σ only through the *empirical magnetisation*

$$m_N(\sigma) = \frac{1}{N}\sum_{i\in\Lambda}\sigma_i, \tag{13.1.2}$$

namely,

$$H_N(\sigma) = -N\big(\tfrac{1}{2}m_N^2(\sigma) + hm_N(\sigma)\big) = NE\big(m_N(\sigma)\big). \tag{13.1.3}$$

A. Bovier, F. den Hollander, *Metastability*,
Grundlehren der mathematischen Wissenschaften 351,
DOI 10.1007/978-3-319-24777-9_13

We choose a discrete-time dynamics $\sigma_N = (\sigma(n))_{n\in\mathbb{N}_0}$ on S_Λ with *Metropolis transition probabilities*

$$p(\sigma,\sigma') = \begin{cases} N^{-1}\exp[-\beta[H_N(\sigma') - H_N(\sigma)]_+], & \text{if } \|\sigma-\sigma'\|_1 = 2,\\ 0, & \text{if } \|\sigma-\sigma'\|_1 > 2,\\ 1-\sum_{\eta\neq\sigma} p(\sigma,\eta), & \text{if } \sigma=\sigma', \end{cases} \tag{13.1.4}$$

where $\|\cdot\|_1$ is the ℓ^1-norm on S_Λ, and the last line is put in to obtain a proper normalisation. This dynamics is reversible w.r.t. the Gibbs measure

$$\mu_{\beta,N}(\sigma) = \frac{1}{Z_{\beta,N}}\, e^{-\beta H_N(\sigma)}\, 2^{-N}, \quad \sigma\in S_\Lambda, \tag{13.1.5}$$

with $Z_{\beta,N}$ the normalising partition function and β the inverse temperature.

Let us look at the evolution of the magnetisation $m_N(n) = m_N(\sigma(n))$ at time $n\in\mathbb{N}_0$. Clearly, this quantity can only increase or decrease by $2N^{-1}$, and the probability of doing so only depends on the number of -1's and $+1$'s present in the configuration $\sigma(n)$, i.e., on $m_N(\sigma(n))$. In other words, with $\mathscr{F}_n$ denoting the σ-algebra up to time n,

$$\mathbb{P}\big(m_N(n+1) = m' \mid \mathscr{F}_n\big) = r_{\beta,N}\big(m_N(n), m'\big), \quad n\in\mathbb{N}_0, \tag{13.1.6}$$

is a function of $m_N(n)$ only, so that Theorem 9.5 applies and the image Markov process has transition probabilities (recall (9.3.3))

$$r_{\beta,N}\big(m,m'\big) = \begin{cases} \frac{1-m}{2}\exp[-\beta N[E(m') - E(m)]_+], & \text{if } m' = m+2N^{-1},\\ \frac{1+m}{2}\exp[-\beta N[E(m') - E(m)]_+], & \text{if } m' = m-2N^{-1}, \end{cases} \tag{13.1.7}$$

on the state space

$$\Gamma_N = \big\{-1, -1+2N^{-1}, \dots, 1-2N^{-1}, 1\big\}. \tag{13.1.8}$$

Moreover, this Markov process is reversible with respect to the image Gibbs measure

$$\nu_{\beta,N}(m) = \frac{1}{Z_{\beta,N}}\, e^{-\beta N E(m)} \binom{N}{\frac{1+m}{2}N} 2^{-N}, \quad m\in\Gamma_N. \tag{13.1.9}$$

In exponential form the latter can be written as

$$\nu_{\beta,N}(m) = \frac{1}{Z_{\beta,N}} e^{-\beta N f_{\beta,N}(m)}, \tag{13.1.10}$$

where

$$f_{\beta,N}(m) = -\tfrac{1}{2}m^2 - hm + \beta^{-1} I_N(m), \tag{13.1.11}$$

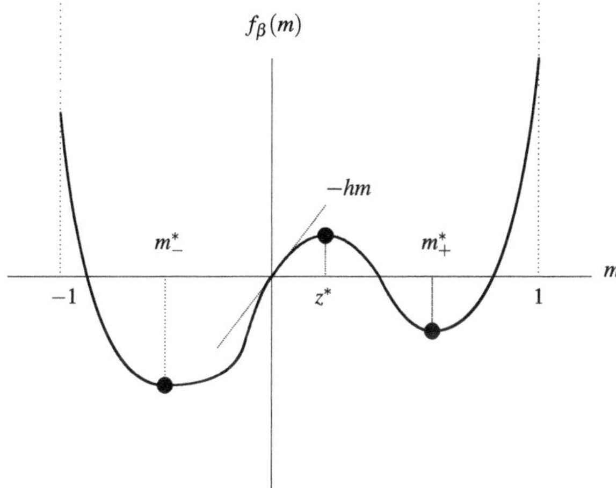

Fig. 13.1 Plot of $m \mapsto f_\beta(m)$ on $[-1, 1]$ when $\beta > 1$ and $h < 0$

with

$$-I_N(m) = \frac{1}{N} \ln\left[\binom{N}{\frac{1+m}{2}N} 2^{-N}\right]. \tag{13.1.12}$$

In the limit as $N \to \infty$, Γ_N lies dense in $[-1, 1]$ and

$$\lim_{N\to\infty} f_{\beta,N}(m) = f_\beta(m), \qquad \lim_{N\to\infty} I_N(m) = I(m), \tag{13.1.13}$$

with

$$\begin{aligned} f_\beta(m) &= -\tfrac{1}{2}m^2 - hm + \beta^{-1} I(m), \\ I(m) &= \tfrac{1}{2}(1+m)\ln(1+m) + \tfrac{1}{2}(1-m)\ln(1-m). \end{aligned} \tag{13.1.14}$$

The latter is the Cramér rate function for coin tossing (recall Sect. 6.1). Since $I(m) = I(-m)$ and $I(m) \sim \frac{1}{2}m^2$ as $m \to 0$, we see from (13.1.11) that $m \mapsto f_\beta(m)$ is a double well when $\beta > 1$ and $|h|$ is small enough (see Fig. 13.1). The stationary points of f_β are the solutions of the equation

$$m = \tanh\bigl[\beta(m+h)\bigr]. \tag{13.1.15}$$

The above observations show that $m_N = (m_N(n))_{n\in\mathbb{N}_0}$ is a random walk on $\Gamma_N \subset [-1, 1]$ with a reversible invariant measure that is close to $\exp[-\beta N f_\beta(m)]$ (modulo normalisation) for large N. Clearly, this bring us to a situation where we can obtain an exact solution, as was explained in Sect. 7.1.4. Moreover, since Γ_N is a lattice with spacing $2N^{-1}$, in the limit as $N \to \infty$ sums appearing in the exact solution can be approximated by integrals with the help of saddle-point techniques.

13.2 Metastable behaviour

The random walk m_N is close to a diffusion on $[-1, 1]$ given by the Kramers diffusion equation in (2.1.1) with $W(x) = \beta f_\beta(x)$ and $\varepsilon = N^{-1}$. In other words, for large N the dynamics of the magnetisation in the Curie-Weiss model can be approximated by a Brownian motion in a potential as encountered in Sect. 2.1. For $\beta > 1$ and $|h|$ small enough this potential is a double well and the diffusion exhibits metastable behaviour.

Let $m_-^* < m_+^*$ be the two local minima of $m \mapsto f_\beta(m)$, and z^* the saddle point in between. Let $m_-^*(N), m_+^*(N)$ denote the points in Γ_N that are closest to m_-^*, m_+^*. These points form a metastable set in the sense of Definition 8.2. In the setting of Fig. 13.1, we have $f_\beta(m_+) > f_\beta(m_-)$, so $m_+^*(N)$ is the metastable state and $m_-^*(N)$ is the stable state. Let $\mathbb{E}_{m_+^*(N)}$ denote expectation w.r.t. the Markov process starting in $m_+^*(N)$ and $\tau_{m_-^*(N)}$ the first hitting time of $m_-^*(N)$.

Theorem 13.1 (Mean metastable crossover time) *As $N \to \infty$,*

$$\begin{aligned}\mathbb{E}_{m_+^*(N)}[\tau_{m_-^*(N)}] &= \exp\big[\beta N\big[f_\beta(z^*) - f_\beta(m_+^*)\big]\big] \\ &\quad \times \big[1 + o(1)\big] \frac{2}{1 - z^*} \sqrt{\frac{1 - z^{*2}}{1 - m_+^{*2}}} \frac{2\pi N/4}{\beta\sqrt{[-f_\beta''(z^*)] f_\beta''(m_+^*)}}.\end{aligned} \tag{13.2.1}$$

Proof Since our Markov process m_N is a nearest-neighbour random walk on Γ_N, we can use the computations in Sect. 7.1.4. According to (7.1.61), we have

$$\mathbb{E}_{m_+^*(N)}[\tau_{m_-^*(N)}] = \sum_{\substack{m,m' \in \Gamma_N,\, m \le m' \\ m_-^*(N) < m \le m_+^*(N)}} \frac{\nu_{\beta,N}(m')}{\nu_{\beta,N}(m)} \frac{1}{r_{\beta,N}(m, m - 2N^{-1})}. \tag{13.2.2}$$

By (13.1.7) and (13.1.10), we have

$$\begin{aligned} r_{\beta,N}\big(m, m - 2N^{-1}\big) &= \frac{1+m}{2} \mathrm{e}^{-\beta N[E(m-2N^{-1}) - E(m)]_+}, \\ \frac{\nu_{\beta,N}(m')}{\nu_{\beta,N}(m)} &= \mathrm{e}^{\beta N[f_{\beta,N}(m) - f_{\beta,N}(m')]}. \end{aligned} \tag{13.2.3}$$

In the limit as $N \to \infty$, the sums in (13.2.2) are dominated by the terms with $m \to z^*$ and $m' \to m_+^*$, since for these terms $f_{\beta,N}(m) - f_{\beta,N}(m')$ is maximal. This explains the exponential factor in (13.2.1). To get the prefactor in (13.2.1), we need to look a bit more closely.

Note that $[E(m - 2N^{-1}) - E(m)]_+ = 2N^{-1}[(m + h) - N^{-1}]_+$. For $m \to z^*$, the first line of (13.2.3) converges to $\frac{1+z^*}{2} \exp(-2\beta[z^* + h]_+)$. In the situation depicted in Fig. 13.1, we have $z^* > 0$. But z^* is a solution of (13.1.15), and so we have $\exp(2\beta[z^* + h]) = (1 + z^*)/(1 - z^*) > 1$. Therefore (13.2.2)–(13.2.3) imply that, for any $\varepsilon > 0$,

$$\mathbb{E}_{m_+^*(N)}[\tau_{m_-^*(N)}] = \mathrm{e}^{\beta N[f_{\beta,N}(z^*)-f_{\beta,N}(m_+^*)]} \frac{2}{1-z^*}\big[1+o(1)\big] \times \sum_{\substack{m,m'\in\Gamma_N \\ |m-z^*|<\varepsilon,\, |m'-m_+^*|<\varepsilon}} \mathrm{e}^{\beta N[f_{\beta,N}(m)-f_{\beta,N}(z^*)]-\beta N[f_{\beta,N}(m')-f_{\beta,N}(m_+^*)]}. \tag{13.2.4}$$

It follows from (13.1.11)–(13.1.14) and Stirling's formula that

$$I_N(m) - I(m) = \big[1+o(1)\big]\frac{1}{2N}\ln\left(\frac{\pi N(1-m^2)}{2}\right) \tag{13.2.5}$$

and hence

$$\mathrm{e}^{\beta N[f_{\beta,N}(m)-f_\beta(m)]} = \big[1+o(1)\big]\sqrt{\frac{\pi N(1-m^2)}{2}}. \tag{13.2.6}$$

Consequently,

$$\mathrm{e}^{\beta N[f_{\beta,N}(z^*)-f_{\beta,N}(m_+^*)]} = \big[1+o(1)\big]\mathrm{e}^{\beta N[f_\beta(z^*)-f_\beta(m_+^*)]}\sqrt{\frac{1-z^{*2}}{1-m_+^{*2}}}. \tag{13.2.7}$$

Inserting this into (13.2.4), we get

$$\mathbb{E}_{m_+^*(N)}[\tau_{m_-^*(N)}] = \mathrm{e}^{\beta N[f_\beta(z^*)-f_\beta(m_+^*)]}\frac{2}{1-z^*}\sqrt{\frac{1-z^{*2}}{1-m_+^{*2}}}\big[1+o(1)\big] \times \sum_{\substack{m,m'\in\Gamma_N \\ |m-z^*|<\varepsilon,\, |m'-m_+^*|<\varepsilon}} \mathrm{e}^{\beta N[f_\beta(m)-f_\beta(z^*)]-\beta N[f_\beta(m')-f_\beta(m_+^*)]}. \tag{13.2.8}$$

To evaluate the sum we write the Taylor expansions

$$\begin{aligned} f_\beta(m) - f_\beta(z^*) &= \tfrac{1}{2}(m-z^*)^2 f_\beta''(z^*) + O\big((m-z^*)^3\big), \\ f_\beta(m') - f_\beta(m_+^*) &= \tfrac{1}{2}(m'-m_+^*)^2 f_\beta''(m_+^*) + O\big((m'-m_+^*)^3\big), \end{aligned} \tag{13.2.9}$$

where we use that $f_\beta'(z^*) = 0$ and $f_\beta'(m_+^*) = 0$. Changing to new variables $u = \sqrt{N}(m-z^*)$ and $u' = \sqrt{N}(m'-m_+^*)$ and recalling (13.1.8), we see that the sum in (13.2.8) equals

$$\big[1+o(1)\big]\frac{N}{4}\int_{\mathbb{R}} du \int_{\mathbb{R}} du' \, \exp\big[\tfrac{1}{2}\beta f_\beta''(z^*)u^2 - \tfrac{1}{2}\beta f_\beta''(m_+^*)u'^2\big]. \tag{13.2.10}$$

Since $f''_\beta(z^*) < 0$ and $f''_\beta(m^*_+) > 0$, the integral converges and equals

$$\frac{2\pi}{\beta\sqrt{[-f''_\beta(z^*)]\,f''_\beta(m^*_+)}}. \tag{13.2.11}$$

Combining (13.2.8) and (13.2.10)–(13.2.11), we end up with (13.2.1). □

The result in Theorem 13.1 fits the classical Arrhenius law with *activation energy* $\beta[f_\beta(z^*) - f_\beta(m^*_+)]$ and *amplitude* given by the prefactor. The former coincides with what was found in (2.1.2) for the Kramers model, with $W = \beta f_\beta$ and $\varepsilon = N^{-1}$, while the latter differs by a factor

$$\frac{N}{2}\,\frac{1}{1-z^*}\sqrt{\frac{1-z^{*2}}{1-m^{*2}_+}}. \tag{13.2.12}$$

This discrepancy comes from the discrete nature of the Curie-Weiss model. In particular, the factor N is due to the fact that time is discrete and only one spin is flipped per time step. In a continuous-time version, we would speed up time by a factor N, after which N would disappear from the last term in the right-hand of (13.2.1).

As a corollary of Theorems 8.43 and 8.45 we get the exponential law of the metastable crossover time.

Theorem 13.2 (Exponential law) *As* $N \to \infty$,

$$\lim_{N\to\infty} \mathbb{P}_{m^*_+(N)}\left(\frac{\tau_{m^*_-(N)}}{\mathbb{E}_{m^*_+(N)}[\tau_{m^*_-(N)}]} > t\right) = \mathrm{e}^{-t} \quad \forall t \geq 0. \tag{13.2.13}$$

13.3 Bibliographical notes

1. The knowledge of the behaviour of $m_N = (m_N(n))_{n\in\mathbb{N}_0}$ does not answer all relevant questions about the dynamics of $\sigma_N = (\sigma(n))_{n\in\mathbb{N}_0}$. This issue was addressed by Levin, Luczak and Peres [164].

2. There are a number of generalised mean-field models that allow for a similar reduction to a *multi-dimensional* diffusive Markov process. See e.g. Bovier, Eckhoff, Gayrard and Klein [33] and Chap. 14 of this book.

3. The calculations in this chapter are not robust against small modifications. Indeed, we are using the *full permutation symmetry* of the Hamiltonian, which is necessary to ensure that $m_N = (m_N(n))_{n\in\mathbb{N}_0}$ is a Markov process. Even when we merely replace the discrete spin variables by continuous spin variables (which leads to the model of mean-field interacting diffusions), the Markov property fails and we are required to consider the empirical measure rather than the empirical magnetisation as the macroscopic variable in order to obtain a Markovian dynamics.

Chapter 14
The Curie-Weiss Model with a Random Magnetic Field: Discrete Distributions

He's a wonderfully clever man, you know. Sometimes he says things that only the Other Professor can understand. Sometimes he says things that nobody *can understand.*

(Lewis Carroll, Sylvie and Bruno)

In Sect. 14.1 we introduce the model. In Sect. 14.2 we define the associated Gibbs measure and the relevant order parameter. In Sect. 14.3 we define the Glauber dynamics and state the main metastability result. Section 14.4 deals with coarse-graining, which works because of the mean-field interaction and because the random fields take finitely many values. We construct the effective Dirichlet form that is obtained after the coarse-graining. Section 14.5 studies the energy landscape near the critical points. Section 14.6 analyses the eigenvalues of the Hessian at the critical points, while Sect. 14.7 looks at the overall topology of the energy landscape, and indicate how the metastability results follow from those in Chap. 10.

14.1 The model

The random-field Curie-Weiss (RFCW) model is one of the simplest examples of a disordered mean-field model. The state space is $S_N = \{-1, +1\}^N$, where N is the number of spins in the system, and the Hamiltonian is given by

$$H_N[\omega](\sigma) = -\frac{1}{2N} \sum_{i,j\in\Lambda} \sigma_i \sigma_j - \sum_{i\in L} h_i[\omega]\sigma_i, \tag{14.1.1}$$

where $\Lambda = \{1, \dots, N\}$ and h_i, $i \in \mathbb{N}$, are i.i.d. random variables on some probability space $(\Omega, \mathscr{F}, \mathbb{P}_h)$.

We will consider two versions of the model that differ substantially in their complexity. If the magnetic field has a *discrete distribution*, then the approach taken for the standard Curie-Weiss model treated in Chap. 13 can be easily extended. This version will be treated in the present chapter. However, if the magnetic field has a *continuous distribution*, then computations become seriously more complex. This version will be treated in Chap. 15.

A. Bovier, F. den Hollander, *Metastability*,
Grundlehren der mathematischen Wissenschaften 351,
DOI 10.1007/978-3-319-24777-9_14

14.2 Gibbs measure and order parameter

We briefly review some key features of the equilibrium behaviour of the RFCW-model, for which we do not need any assumption on the distribution of the random magnetic field.

The Gibbs measure of the RFCW-model is the random probability measure

$$\mu_{\beta,N}[\omega](\sigma) = \frac{2^{-N}\, \mathrm{e}^{-\beta H_N[\omega](\sigma)}}{Z_{\beta,N}[\omega]}, \tag{14.2.1}$$

the partition function is defined as

$$Z_{\beta,N}[\omega] = 2^{-N} \sum_{\sigma\in S_N} \mathrm{e}^{-\beta H_N[\omega](\sigma)}. \tag{14.2.2}$$

As in Chap. 13, the empirical magnetisation

$$m_N(\sigma) = \frac{1}{N}\sum_{i\in\Lambda}\sigma_i \tag{14.2.3}$$

serves as the *order parameter* of the model, and we define its distribution under the Gibbs measure in (14.2.1) as the *induced measure*

$$\mathscr{Q}_{\beta,N} = \mu_{\beta,N} \circ m_N^{-1} \tag{14.2.4}$$

on the set of possible values of m_n, $\Gamma_N = \{-1, -1+2/N, \dots, 1-2/N, 1\}$.

We write

$$Z_{\beta,N}[\omega]\mathscr{Q}_{\beta,N}[\omega](m) = \mathrm{e}^{\frac{1}{2}N\beta m^2}\, Z^1_{\beta,N}[\omega](m), \tag{14.2.5}$$

where

$$Z^1_{\beta,N}[\omega](m) = 2^{-N}\sum_{\sigma\in S_N}\big(\mathrm{e}^{\beta\sum_{i\in\Lambda}h_i[\omega]\sigma_i}\, \mathbb{1}_{\{N^{-1}\sum_{i\in\Lambda}\sigma_i = m\}}\big). \tag{14.2.6}$$

For simplicity, we identify functions f defined on the discrete set Γ_N with functions f defined on the interval $[-1,1]$ by setting $f(m) = f(\lceil 2Nm\rceil/2N)$. Then, by using sharp large deviation estimates (see Chaganty and Sethuraman [55]), $Z^1_N(m)$, $m\in(-1,1)$, can be expressed as

$$Z^1_{\beta,N}[\omega](m) = \frac{\exp[-NI_N[\omega](m)]}{\sqrt{\frac{1}{2}N\pi/I''_N[\omega](m)}}\big[1+o(1)\big], \tag{14.2.7}$$

where $o(1)$ tends to zero as $N \to \infty$, and $I_N[\omega](y)$ is the Legendre-Fenchel transform of the log-moment-generating function

$$U_N[\omega](t) = \frac{1}{N} \ln\left(2^{-N} \sum_{\sigma \in S_N} \left(e^{\beta \sum_{i\in\Lambda} h_i[\omega]\sigma_i} e^{t \sum_{i\in\Lambda} \sigma_i}\right)\right)$$
$$= \frac{1}{N} \sum_{i\in\Lambda} \ln\cosh\bigl(t + \beta h_i[\omega]\bigr). \tag{14.2.8}$$

Hence we can rewrite (14.2.5) as

$$Z_{\beta,N}[\omega]\mathscr{Q}_{\beta,N}[\omega](m) = \sqrt{\frac{2I''_N[\omega](m)}{N\pi}} \exp\bigl[-N\beta F_{\beta,N}[\omega](m)\bigr]\bigl[1+o(1)\bigr], \tag{14.2.9}$$

where

$$F_{\beta,N}[\omega](m) = -\frac{1}{2}m^2 + \frac{1}{\beta} I_N[\omega](m). \tag{14.2.10}$$

We are interested in the behaviour of $Z_{\beta,N}[\omega]\mathscr{Q}_{\beta,N}[\omega]$ near the critical points of $F_{\beta,N}$. These satisfy the equation

$$m^* = \beta^{-1} I'_N\bigl(m^*\bigr) = \beta^{-1} t^*, \tag{14.2.11}$$

or

$$\beta m^* = I'_N\bigl(m^*\bigr) = t^*. \tag{14.2.12}$$

Since I_N is the Legendre-Fenchel transform of U_N, we have $I'_N(x) = [U'_N]^{-1}(x)$, so that

$$m^* = U'_N\bigl(\beta m^*\bigr) = \frac{1}{N} \sum_{i\in\Lambda} \tanh\bigl(\beta\bigl(m^* + h_i[\omega]\bigr)\bigr). \tag{14.2.13}$$

Moreover,

$$a\bigl(m^*\bigr) = F''_{\beta,N}\bigl(m^*\bigr) = -1 + \beta^{-1} I''_N\bigl(m^*\bigr). \tag{14.2.14}$$

Finally, using that $I''_{N,\ell}(m^*) = 1/U''_{N,\ell}(t^*)$, we get the alternative expression

$$a\bigl(m^*\bigr) = -1 + \frac{1}{\beta U''_N(\beta m^*)} = -1 + \frac{1}{\frac{\beta}{N}\sum_{i\in\Lambda}[1 - \tanh^2(\beta(m^* + h_i[\omega]))]}. \tag{14.2.15}$$

Thus, we see that, by the law of large numbers, the set of critical points converges $\mathbb{P}_h$-a.s. to the set of solutions of the equation

$$m^* = \mathbb{E}_h\bigl[\tanh\bigl(\beta\bigl(m^* + h\bigr)\bigr)\bigr], \tag{14.2.16}$$

and the second derivative of $F_{\beta,N}(m^*)$ converges to

$$\lim_{N\to\infty} F''_{\beta,N}\bigl(m^*\bigr) = -1 + \frac{1}{\beta\mathbb{E}_h[1 - \tanh^2(\beta(m^* + h))]}. \tag{14.2.17}$$

Hence m^* is a local minimum when

$$\beta\mathbb{E}_h\big[1-\tanh^2\big(\beta\big(m^*+h\big)\big)\big]<1 \tag{14.2.18}$$

and a local maximum when

$$\beta\mathbb{E}_h\big[1-\tanh^2\big(\beta\big(m^*+h\big)\big)\big]>1. \tag{14.2.19}$$

(The case $\beta\mathbb{E}_h[1-\tanh^2(\beta(m^*+h))]=1$ corresponds to a second-order phase transition and will not be considered here.)

Proposition 14.1 *Let m^* be a critical point of $F_{\beta,N}$. Then $\mathbb{P}_h$-a.s., for all but finitely many values of N,*

$$Z_{\beta,N}\mathscr{Q}_{\beta,N}\big(m^*\big)=\frac{\exp[-\beta N F_{\beta,N}(m^*)][1+o(1)]}{\sqrt{\frac{N\pi}{2}|\mathbb{E}[1-\tanh^2(\beta(m^*+h))]|}} \tag{14.2.20}$$

with

$$F_{\beta,N}\big(m^*\big)=\frac{1}{2}\big(m^*\big)^2-\frac{1}{\beta N}\sum_{i\in\Lambda}\ln\cosh\big(\beta\big(m^*+h_i[\omega]\big)\big). \tag{14.2.21}$$

The above observations provide a detailed picture of the distribution of the order parameter. Note that m^* depends on ω.

14.3 Glauber dynamics

Next we add dynamics. As in Chap. 13, we consider the discrete-time Glauber dynamics with Metropolis transition probabilities (compare with (13.1.4))

$$p_N[\omega]\big(\sigma,\sigma'\big)=\begin{cases}N^{-1}\exp[-\beta[H_N[\omega](\sigma')-H_N[\omega](\sigma)]_+], & \text{if } \|\sigma-\sigma'\|_1=2,\\ 0, & \text{if } \|\sigma-\sigma'\|_1>2,\\ 1-\sum_{\eta\neq\sigma}p(\sigma,\eta), & \text{if } \sigma=\sigma'.\end{cases} \tag{14.3.1}$$

We write $\mathbb{P}_\sigma[\omega]=\mathbb{P}_\sigma$ for the law of this Markov process (for a given realisation of the magnetic fields) starting in σ. Note that this dynamics is ergodic and reversible with respect to the Gibbs measure $\mu_{\beta,N}[\omega]$ for each ω.

A *heuristic picture* for the metastable behaviour of systems like the random-field Curie-Weiss model is based on replacing the *full* Markov process on S_N by an *effective* Markov process for the order parameter, i.e., by a nearest-neighbour random walk on Γ_N with transition probabilities that are reversible with respect to the induced measure $\mathscr{Q}_{\beta,N}$. The ensuing model can be solved exactly. A natural choice for the transition rates of the effective dynamics is

$$r_N[\omega](m,m') = \frac{1}{\mathcal{Q}_{\beta,N}[\omega](m)} \sum_{\sigma\in S_N:\ m_N(\sigma)=m} \mu_{\beta,N}[\omega](\sigma) \times \sum_{\sigma'\in S_N:\ m_N(\sigma')=m'} p_N[\omega](\sigma,\sigma'), \qquad (14.3.2)$$

which are different from zero only when $m' = m - 2/N, m, m + 2/N$. The ensuing Markov process is a one-dimensional nearest-neighbour random walk, for which most quantities of interest can be computed explicitly by elementary means, as in Chap. 13. In particular, it is easy to show that if M is the global minimum of $F_{\beta,N}$ and m^* is a local minimum, then, as Theorem 13.1,

$$\mathbb{E}_{m^*}[\tau_M] = \exp\big[\beta N\big[F_{\beta,N}(z^*) - F_{\beta,N}(m^*)\big]\big] \times \frac{2}{1-z^*}\frac{2\pi N/4}{\beta[-a(z^*)]}\sqrt{\frac{\beta\mathbb{E}_h[1-\tanh^2(\beta(z^*+h))]-1}{1-\beta\mathbb{E}_h[1-\tanh^2(\beta(m^*+h))]}}\,\big[1+o(1)\big], \qquad (14.3.3)$$

where z^* is the saddle point between M and m^*, and $a(z^*)$ is defined in (14.2.15). However, *the prediction of this naive approximation produces the wrong prefactor*, as is shown in our main theorem below.

To obtain precise results, we will need to introduce an exact lumping in the sense of Sect. 9.2.

14.4 Coarse-graining

So far we did not need any assumption on the distribution of the random field. Now we assume that the random field takes values in the finite set $I = \{b_1, \dots, b_n\}$.

Each realisation of the random field $\{h_i[\omega]\}_{i\in\Lambda}$ induces a random partition of the set $\Lambda = \{1, \dots, N\}$ into subsets (see Fig. 14.1)

$$\Lambda_k[\omega] = \big\{i \in \Lambda:\ h_i[\omega] = b_k\big\}, \quad k = 1, \dots, n. \qquad (14.4.1)$$

Accordingly, we introduce n order parameters

$$\mathbf{m}_k[\omega](\sigma) = \frac{1}{N}\sum_{i\in\Lambda_k[\omega]} \sigma_i, \quad k = 1, \dots, n, \qquad (14.4.2)$$

and we denote by $\mathbf{m}[\omega]$ the n-dimensional vector $(\mathbf{m}_1[\omega], \dots, \mathbf{m}_n[\omega])$. In the sequel we will use the convention that boldface symbols denote n-dimensional vectors and their components, while the sum of the components is denoted by the corresponding

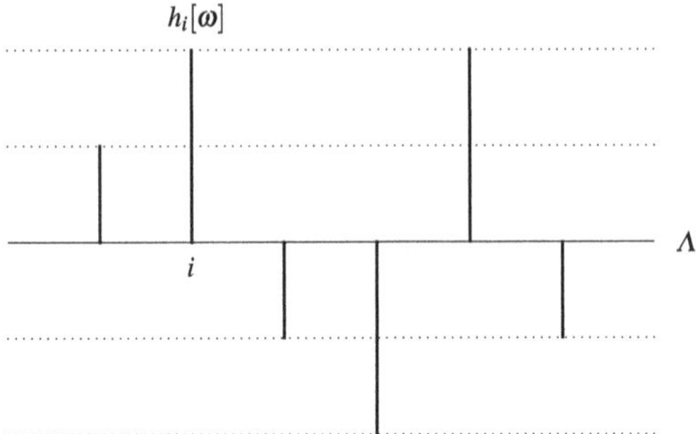

Fig. 14.1 Coarse-graining: Λ is partitioned into sets where the magnetic field takes the same value

plain symbol, e.g. $m[\omega] = \sum_{k=1}^{n} \mathbf{m}_k[\omega]$. The vector $\mathbf{m}$ takes values in the set

$$\Gamma_N^n[\omega] = \bigtimes_{k=1}^{n} \left\{-\rho_{N,k}[\omega], -\rho_{N,k}[\omega] + \tfrac{2}{N}, \dots, \rho_{N,k}[\omega] - \tfrac{2}{N}, \rho_{N,k}[\omega]\right\}, \quad (14.4.3)$$

where

$$\rho_k = \rho_{N,k}[\omega] = \frac{|\Lambda_k[\omega]|}{N}. \quad (14.4.4)$$

We denote by $\mathbf{e}_k$, $k = 1, \dots, n$, the lattice vectors of the set $\Gamma_N^n[\omega]$, i.e., the vectors of length $2/N$ parallel to the unit vectors. Note that the random variables $\rho_{N,k}[\omega]$ concentrate exponentially fast in N around their mean values $\mathbb{E}_h[\rho_{N,k}] = \mathbb{P}_h(h_1 = b_k) = p_k$. In particular, we have the following lemma.

Lemma 14.2 *For all $n \in \mathbb{N}$,*

$$\mathbb{P}\left(\exists_{N_n} \forall_{N \geq N_n} \forall_{1 \leq k \leq n} : |\rho_{N,k} - p_k| \leq \tfrac{1}{2} p_k\right) = 1. \quad (14.4.5)$$

The Hamiltonian takes the form

$$H_N[\omega](\sigma) = -N E\left(\mathbf{m}[\omega](\sigma)\right), \quad (14.4.6)$$

where $E \colon \mathbb{R}^n \to \mathbb{R}$ is the function

$$E(\mathbf{x}) = \frac{1}{2}\left(\sum_{k=1}^{n} \mathbf{x}_k\right)^2 + \sum_{k=1}^{n} b_k \mathbf{x}_k. \quad (14.4.7)$$

The equilibrium distribution of the random variables $\mathbf{m}[\sigma]$ is given by

$$\begin{aligned} \mathscr{Q}_{\beta,N}[\omega](\mathbf{x}) &= \mu_{\beta,N}[\omega]\left(\mathbf{m}[\omega](\sigma) = \mathbf{x}\right) \quad (14.4.8) \\ &= \frac{1}{Z_N[\omega]} e^{\beta N E(\mathbf{x})} 2^{-N} \sum_{\sigma \in S_N} \mathbb{1}_{\{\mathbf{m}[\omega](\sigma) = \mathbf{x}\}}, \quad x \in \Gamma_N^n[\omega], \end{aligned}$$

where $Z_N[\omega]$ is the normalising partition function. We use the same symbols $\mathcal{Q}_{\beta,N}$, $F_{\beta,N}$ for functions defined on the n-dimensional variables $\mathbf{x}$. Since we distinguish vectors from scalars by using boldface type, there should be no confusion possible. Similarly, for a mesoscopic subset $\mathbf{A} \subseteq \Gamma_N^n[\omega]$, we define its microscopic counterpart,

$$A = S_N[\mathbf{A}] = \{\sigma \in S_N : \mathbf{m}(\sigma) \in \mathbf{A}\}. \tag{14.4.9}$$

The vectors $(\mathbf{m}[\omega](\sigma(t)))_{t\in\mathbb{R}_+}$ form a Markov process with transition rates

$$r_N[\omega](\mathbf{x},\mathbf{x}') = \frac{1}{\mathcal{Q}_{\beta,N}[\omega](\mathbf{x})} \sum_{\sigma\in S_N[\mathbf{x}]} \mu_{\beta,N}[\omega](\sigma) \sum_{\sigma'\in S_N[\mathbf{x}']} p[\omega](\sigma,\sigma'). \tag{14.4.10}$$

This can be easily inferred by checking the conditions of Theorem 9.5 in Sect. 9.2.

We can also check that the capacities of these processes are related. Let the sets $A, B \subset S_N$ be defined in terms of the block variables $\mathbf{m}$. This means that, for some $\mathbf{A}, \mathbf{B} \subseteq \Gamma_N^n$, $A = S_N[\mathbf{A}]$ and $B = S_N[\mathbf{B}]$. By symmetry under permutations that leave the partition $\Lambda_k[\omega]$ invariant, we have

$$\begin{aligned}
\operatorname{cap}(A,B) &= \inf_{h\in\mathcal{H}_{A,B}} \frac{1}{2} \sum_{\sigma,\sigma'\in S_N} \mu_{\beta,N}[\omega](\sigma)p(\sigma,\sigma')\big[h(\sigma)-h(\sigma')\big]^2 \\
&= \inf_{u\in\mathcal{G}_{\mathbf{A},\mathbf{B}}} \frac{1}{2} \sum_{\sigma,\sigma'\in S_N} \mu_{\beta,N}[\omega](\sigma)p(\sigma,\sigma')\big[u(\mathbf{m}(\sigma))-u(\mathbf{m}(\sigma'))\big]^2 \\
&= \inf_{u\in\mathcal{G}_{\mathbf{A},\mathbf{B}}} \sum_{\mathbf{x},\mathbf{x}'\in\Gamma_N^n} \big[u(\mathbf{x})-u(\mathbf{x}')\big]^2 \sum_{\sigma\in S_N[\mathbf{x}]} \mu_{\beta,N}[\omega](\sigma) \sum_{\sigma'\in S_N[\mathbf{x}']} p(\sigma,\sigma') \\
&= \inf_{u\in\mathcal{G}_{\mathbf{A},\mathbf{B}}} \sum_{\mathbf{x},\mathbf{x}'\in\Gamma_N^n} \mathcal{Q}_{\beta,N}[\omega](\mathbf{x}) r_N(\mathbf{x},\mathbf{x}')\big[u(\mathbf{x})-u(\mathbf{x}')\big]^2 \\
&= \operatorname{CAP}(\mathbf{A},\mathbf{B}),
\end{aligned} \tag{14.4.11}$$

where

$$\begin{aligned}
\mathcal{H}_{A,B} &= \{h : S_N \to [0,1] : h(\sigma) = 1\ \forall \sigma\in A,\ h(\sigma)=0\ \forall\sigma\in B\}, \\
\mathcal{G}_{\mathbf{A},\mathbf{B}} &= \{u : \Gamma_N^n \to [0,1] : u(\mathbf{x}) = 1\ \forall \mathbf{x}\in \mathbf{A},\ u(\mathbf{x})=0\ \forall\mathbf{x}\in \mathbf{B}\},
\end{aligned} \tag{14.4.12}$$

and we use the symbol CAP for capacity on Γ_N^n.

Most of the interesting issues on the dynamics of the model can now be derived directly from the dynamics on the mesoscopic variables. But for the latter we are now in the setting of Chap. 10 and can harvest the results obtained there. All that is left to do is to analyse the specific energy landscape for the present models.

Theorem 14.3 (Metastable sets) *Let $\mathcal{M}_N$ be the set of (best lattice approximations of) the local minima of the functions $F_{\beta,N}$. Then $\mathcal{M}_N$ is a metastable set in the sense of Definition* 8.2 *for the induced dynamics with transition rates given by r_N.*

Theorem 14.4 (Mean metastable exit times) *Let $\mathbf{x} \in \mathscr{M}_N$. Let $M_{\mathbf{x}}$ be the set of local minima where $F_{\beta,N}$ is smaller than or equal to $F_{\beta,N}(\mathbf{x})$. For every $\sigma \in S[\mathbf{x}]$ and $\mathbf{x} \in \mathscr{M}_N$, $\mathbb{P}_h$-a.s. for all but finitely many values of N,*

$$\mathbb{E}_\sigma[\tau_{S[\mathscr{M}_{\mathbf{x}}]}] = \exp\big[\beta N\big[F_{\beta,N}(z^*) - F_{\beta,N}(x^*)\big]\big] \\ \times \frac{\pi N}{2\beta[-\bar{\gamma}_1]} \sqrt{\frac{\beta \mathbb{E}_h[1-\tanh^2(\beta(z^*+h))]-1}{1-\beta\mathbb{E}_h[1-\tanh^2(\beta(m^*+h))]}}\big[1+o(1)\big], \tag{14.4.13}$$

where $x^* = \sum_{\ell=1}^n \mathbf{x}_\ell$, $z^* = \sum_{\ell=1}^n \mathbf{z}_\ell$, $\mathbf{z}$ is the saddle point between $\mathbf{x}$ and $\mathscr{M}_{\mathbf{x}}$, and $\bar{\gamma}_1$ is the unique negative solution of the equation

$$\mathbb{E}_h\left[\frac{[1-\tanh(\beta(z^*+h))]\exp[-2\beta(z^*+h)_+]}{\frac{\exp[-2\beta(z^*+h)_+]}{\beta[1+\tanh(\beta(z^*+h))]} - 2\gamma}\right] = 1. \tag{14.4.14}$$

Theorem 14.5 (Exponential law) *With the notation of Theorem* 14.4*,*

$$\lim_{N\to\infty} \mathbb{P}_\sigma\big(\tau_{S[\mathscr{M}_{\mathbf{x}}]}/\mathbb{E}_\sigma[\tau_{S[\mathscr{M}_{\mathbf{x}}]}] > t\big) = \mathrm{e}^{-t}, \quad t \geq 0, \quad a.s. \tag{14.4.15}$$

The proofs of these theorems are given in Sect. 14.7.

Note that

$$F_{\beta,N}(z^*) - F_{\beta,N}(m^*) = \exp\bigg[\beta N \frac{(z^*)^2-(m^*)^2}{2} \\ - N^{-1}\sum_{i\in\Lambda}\big[\ln\cosh\big(\beta(z^*+h_i)\big) - \ln\cosh\big(\beta(m^*+h_i)\big)\big]\bigg] \tag{14.4.16}$$

has random fluctuations of order N^{-1}, which lead to strong fluctuations in the metastable crossover time with respect to the disorder variables h_i, $i \in \Lambda$.

14.5 The landscape near critical points

We are very close to the setting of Chap. 10. To complete the connection we need to analyse the measures $\mathscr{Q}_{\beta,N}[\omega](\mathbf{x})$. We henceforth suppress ω from the notation. Note that

$$Z_{\beta,N}\mathscr{Q}_{\beta,N}(\mathbf{x}) = \exp\left[N\beta\left(\frac{1}{2}\left(\sum_{\ell=1}^n \mathbf{x}_\ell\right)^2 + \sum_{\ell=1}^n \mathbf{x}_\ell b_\ell\right)\right]\prod_{\ell=1}^n Z_N^\ell(\mathbf{x}_\ell/\rho_\ell), \tag{14.5.1}$$

where

$$Z_N^\ell(y) = 2^{-|\Lambda_\ell|} \sum_{\sigma \in S_{\Lambda_\ell}} \mathbb{1}_{\{|\Lambda_\ell|^{-1} \sum_{i \in \Lambda_\ell} \sigma_i = y\}}. \tag{14.5.2}$$

For $y \in (-1,1)$, $Z_N^\ell(y)$ can be expressed, via an elementary asymptotics of binomial coefficients, as

$$Z_N^\ell(y) = \frac{\exp[-|\Lambda_\ell| I(y)]}{\sqrt{\frac{\pi}{2}|\Lambda_\ell|/I''(y)}}\big[1+o(1)\big], \tag{14.5.3}$$

where $o(1)$ tends to zero as $|\Lambda_\ell| \to \infty$ and I is Cramèr's rate function (13.1.13) (again we identify functions on Γ_N^n with their natural extensions to $\mathbb{R}^n$). This means that we can express the right-hand side of (14.5.1) as

$$Z_{\beta,N}\mathscr{Q}_{\beta,N}(\mathbf{x}) = \prod_{\ell=1}^n \sqrt{\frac{I''(\mathbf{x}_\ell/\rho_\ell)/\rho_\ell}{N\pi/2}} \exp\big[-N\beta F_{\beta,N}(\mathbf{x})\big]\big[1+o(1)\big], \tag{14.5.4}$$

where

$$F_{\beta,N}(\mathbf{x}) = -\frac{1}{2}\left(\sum_{\ell=1}^n \mathbf{x}_\ell\right)^2 - \sum_{\ell=1}^n \mathbf{x}_\ell b_\ell + \frac{1}{\beta}\sum_{\ell=1}^n \rho_\ell I(\mathbf{x}_\ell/\rho_\ell). \tag{14.5.5}$$

The critical points $\mathbf{z}^*$ of $F_{\beta,N}$ are solutions of the equation

$$\sum_{j=1}^n \mathbf{z}_j^* + b_\ell = \beta^{-1} I'\big(\mathbf{z}_\ell^*/\rho_\ell\big) = \beta^{-1} t_\ell^*, \tag{14.5.6}$$

or, with $z^* = \sum_{j=1}^n \mathbf{z}_\ell^*$,

$$\beta\big(z^* + b_\ell\big) = I'\big(\mathbf{z}_\ell^*/\rho_\ell\big) = t_\ell^*, \tag{14.5.7}$$

which implies

$$\mathbf{z}_\ell^*/\rho_\ell = \tanh\big(\beta\big(\mathbf{z}^* + b_\ell\big)\big). \tag{14.5.8}$$

Summing over ℓ, we see that z^* must satisfy the equation

$$z^* = \frac{1}{N}\sum_{i \in \Lambda} \tanh\big(\beta\big(z^* + h_i\big)\big), \tag{14.5.9}$$

which coincides with (14.2.13) for the one-dimensional order parameter m.

The Hessian matrix $\mathbb{A}(\mathbf{z}^*)$ at a critical point $\mathbf{z}^*$ has elements

$$\big(\mathbb{A}(\mathbf{z}^*)\big)_{k\ell} = \frac{\partial^2 F_{\beta,N}(\mathbf{z}^*)}{\partial \mathbf{z}_k \partial \mathbf{z}_\ell} = -1 + \delta_{k,\ell}\beta^{-1}\rho_\ell^{-1} I''_{N,\ell}\big(\mathbf{z}_\ell^*/\rho_\ell\big) = -1 + \delta_{\ell,k}\hat{\lambda}_\ell, \tag{14.5.10}$$

where the random numbers $\hat{\lambda}_\ell$ are given by

$$\hat{\lambda}_\ell = \frac{1}{\beta\rho_\ell[1-\tanh^2(\beta(z^*+b_\ell))]}. \tag{14.5.11}$$

The determinant of the matrix $\mathbb{A}(\mathbf{z}^*)$ has a simple expression, namely,

$$\begin{aligned}\det\big(\mathbb{A}(\mathbf{z}^*)\big) &= \left(1-\sum_{\ell=1}^{n}\frac{1}{\hat{\lambda}_\ell}\right)\prod_{\ell=1}^{n}\hat{\lambda}_\ell \\ &= \left(1-\frac{\beta}{N}\sum_{i\in\Lambda}[1-\tanh^2\big(\beta\big(z^*+h_i\big)\big)]\right)\prod_{\ell=1}^{n}\hat{\lambda}_\ell \\ &= \big(1-\beta\mathbb{E}_h\big[1-\tanh^2\big(\beta\big(z^*+h\big)\big)\big]\big)\prod_{\ell=1}^{n}\hat{\lambda}_\ell\big[1+o(1)\big].\end{aligned} \tag{14.5.12}$$

Combining these observations, we arrive at the following proposition.

Proposition 14.6 *Let* $\mathbf{z}^*$ *be a critical point of* $F_{\beta,N}$. *Then* $\mathbf{z}^*$ *is given by* (14.5.8) *with* z^* *a solution of* (14.5.9). *Moreover,*

$$\begin{aligned}Z_{\beta,N}\mathscr{Q}_{\beta,N}\big(\mathbf{z}^*\big) &= \frac{\sqrt{-\det(\mathbb{A}(\mathbf{z}^*))}}{\sqrt{(\frac{N\pi}{2\beta})^n(\beta\mathbb{E}_h[1-\tanh^2(\beta(z^*+h))]-1)}} \\ &\quad\times\exp\left[\beta N\left(-\tfrac{1}{2}\big(z^*\big)^2+\frac{1}{\beta N}\sum_{i\in\Lambda}\ln\cosh\big(\beta\big(z^*+h_i\big)\big)\right)\right] \\ &\quad\times\big[1+o(1)\big].\end{aligned} \tag{14.5.13}$$

Proof The proof of the analogous result is given in Chap. 15. □

14.6 Eigenvalues of the Hessian

We next describe the eigenvalues of the Hessian matrix $\mathbb{A}(\mathbf{z}^*)$.

Lemma 14.7 *Let* z^* *be a solution of* (14.5.9). *In addition, assume that all numbers* $\hat{\lambda}_k$ *are distinct. Then* γ *is an eigenvalue of* $\mathbb{A}(\mathbf{z}^*)$ *if and only if it is a solution of the equation*

$$\sum_{\ell=1}^{n}\frac{1}{\frac{1}{\beta\rho_\ell[1-\tanh^2(\beta(z^*+b_\ell))]}-\gamma}=1. \tag{14.6.1}$$

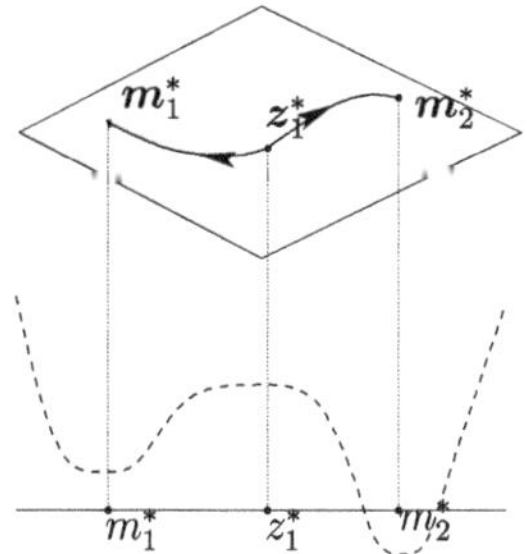

Fig. 14.2 Correspondence of the 1-dimensional and n-dimensional landscapes

Moreover, (14.6.1) *has at most one negative solution, and it has such a solution if and only if*

$$\beta \mathbb{E}_h\big[1 - \tanh^2\big(\beta\big(z^* + h\big)\big)\big] > 1. \tag{14.6.2}$$

Proof To find the eigenvalues of $\mathbb{A}$, simply replace $\hat{\lambda}_k$ by $\hat{\lambda}_k - \gamma$ in the first line of (14.5.12). This gives

$$\det\big(\mathbb{A}\big(\mathbf{z}^*\big) - \gamma\big) = \left(1 - \sum_{\ell=1}^{n} \frac{1}{\hat{\lambda}_\ell - \gamma}\right) \prod_{\ell=1}^{n} (\hat{\lambda}_\ell - \gamma), \tag{14.6.3}$$

provided none of the $\hat{\lambda}_\ell - \gamma$ is zero. Then (14.6.1) is just the requirement that the first factor in the right-hand side of (14.6.3) vanishes. It is easy to see that, under the hypothesis of the lemma, this equation has n solutions, and that exactly one of them is negative under the hypothesis in (14.6.2). □

14.7 Topology of the landscape

From the analysis of the critical points of $F_{\beta,N}$ it follows that the landscape of this function is closely linked to the one-dimensional landscape described in Sect. 11.1 (see Fig. 14.2). We collect the following features:

(i) Let $m_1^* < z_1^* < m_2^* < z_2^* < \cdots < z_k^* < m_{k+1}^*$ be the sequence of minima, respectively, maxima of the one-dimensional function $F_{\beta,N}$ defined in (14.2.10). To each minimum m_i^* corresponds a minimum $\mathbf{m}_i^*$ of $F_{\beta,N}$ such that $\sum_{\ell=1}^n \mathbf{m}_{i,\ell}^* = m_i^*$, and to each maximum z_i^* corresponds a saddle point $\mathbf{z}_i^*$ of $F_{\beta,N}$ such that $\sum_{\ell=1}^n \mathbf{z}_{i,\ell}^* = z_i^*$.

(ii) For any value m of the total magnetisation, the function $F_{\beta,N}(\mathbf{x})$ takes its relative minimum on the set $\{\mathbf{y}\colon \sum \mathbf{y}_\ell = m\}$ at the point $\hat{\mathbf{x}} \in \mathbb{R}^n$ determined (coordinate-wise) by the equation

$$\begin{aligned}\hat{\mathbf{x}}_\ell(m) &= \frac{1}{N}\sum_{i\in\Lambda_\ell}\tanh\bigl(\beta(m+a+h_i)\bigr)\\ &= \rho_\ell\tanh\bigl(\beta(m+a+b_\ell)\bigr),\end{aligned} \tag{14.7.1}$$

where $a = a(m)$ is determined by the equation

$$\begin{aligned}m &= \frac{1}{N}\sum_{i\in\Lambda}\tanh\bigl(\beta(m+a+h_i)\bigr) \\ &= \sum_{\ell=1}^{n}\rho_\ell\tanh\bigl(\beta(m+a+b_\ell)\bigr).\end{aligned} \tag{14.7.2}$$

Moreover,

$$F_{\beta,N}(m) \le F_{\beta,N}(\hat{\mathbf{x}}) \le F_{\beta,N}(m) + O(n\ln N/N). \tag{14.7.3}$$

Remark 14.8 Note that the minimal energy curves $\hat{\mathbf{x}}(\cdot)$ defined by (14.7.1) pass through the minima and the saddle points, but in general are not integral curves of the gradient flow connecting them. Also note that, since we assume that the random fields h_i have bounded support, for every $\delta > 0$ there exist two universal constants $0 < c_1 \le c_2 < \infty$ such that

$$c_1\rho_\ell \le \frac{\mathrm{d}\hat{\mathbf{x}}_\ell(m)}{\mathrm{d}m} \le c_2\rho_\ell, \tag{14.7.4}$$

uniformly in N, $m \in [-1+\delta, 1-\delta]$ and $\ell = 1, \dots, n$.

Finally, in order to apply the results from Chap. 10, we need the form of the transition rates r_ℓ near a saddle point $\mathbf{z}^*$. For $\sigma \in S_N$, put

$$\Lambda_k^{\pm}(\sigma) = \bigl\{i \in \Lambda_k \colon \sigma(i) = \pm 1\bigr\}. \tag{14.7.5}$$

Note that $|\Lambda_k^{\pm}(\sigma)| = \frac{1}{2}(1 \mp \mathbf{x}_k(\sigma))$ is independent of the specific choice of σ. For all $\mathbf{x} \in \Gamma_N^n$, we have

$$\begin{aligned}r_N(\mathbf{x}, \mathbf{x}+\mathbf{e}_\ell) &= \mathscr{Q}_{\beta,N}(\mathbf{x})^{-1}\sum_{\sigma\in S_N[\mathbf{x}]}\mu_{\beta,N}[\omega](\sigma)\sum_{i\in\Lambda_\ell^-(\sigma)}p\bigl(\sigma,\sigma^i\bigr) \\ &= \bigl|\Lambda_\ell^-(\mathbf{x})\bigr|e^{-2\beta[x-\frac{1}{N}+b_\ell]_+}.\end{aligned} \tag{14.7.6}$$

Define, as in Eq. (10.2.8), the sets

$$D_N(\rho) = \bigl\{|\mathbf{z}_\ell - \mathbf{x}_\ell| < \rho,\ \forall \ell = 1, \dots, n\bigr\},$$

with $\rho = C\sqrt{N^{-1}\ln N}$.

It follows easily that, for all $\mathbf{x} \in D_N(\rho)$,

$$\left| \frac{r_N(\mathbf{x}, \mathbf{x}+\mathbf{e}_\ell)}{r_N(\mathbf{z}^*, \mathbf{z}^*+\mathbf{e}_\ell)} - 1 \right| \leq c\beta n\rho, \tag{14.7.7}$$

for some finite constant $c > 0$. Thus, as in Chap. 10, we replace the Dirichlet form near the saddle point by a simplified one, where

$$\widetilde{r}(\mathbf{x}, \mathbf{x}+\mathbf{e}_\ell) = r_N\bigl(\mathbf{z}^*, \mathbf{z}^*+\mathbf{e}_\ell\bigr) = r_\ell, \qquad \widetilde{r}(\mathbf{x}+\mathbf{e}_\ell, \mathbf{x}) = r_\ell \frac{\widetilde{\mathcal{Q}}_{\beta,N}(\mathbf{x})}{\widetilde{\mathcal{Q}}_{\beta,N}(\mathbf{x}+\mathbf{e}_\ell)}, \tag{14.7.8}$$

are the modified rates of a dynamics on $D_N(\rho)$ that is reversible w.r.t. the measure $\widetilde{\mathcal{Q}}_{\beta,N}(\mathbf{x})$. Let $\widetilde{L}_N$ denote the corresponding generator. For $u \in \mathcal{G}_{\mathbf{A},\mathbf{B}}$, we write the corresponding Dirichlet form as

$$\widetilde{\mathcal{E}}_{D_N}(u,u) = \mathcal{Q}_{\beta,N}\bigl(\mathbf{z}^*\bigr) \sum_{\mathbf{x}\in D_N(\rho)} \sum_{\ell=1}^{n} r_\ell \mathrm{e}^{-\beta N((\mathbf{x}-\mathbf{z}^*),\mathbb{A}(\mathbf{z}^*)(\mathbf{x}-\mathbf{z}^*))} \bigl[u(\mathbf{x}) - u(\mathbf{x}+\mathbf{e}_\ell)\bigr]^2. \tag{14.7.9}$$

We now have all the ingredients needed to apply the results of Chap. 10. The only difference is that the free energy functional $F_{\beta,N}$ is random and depends on N. But this presents no obstacle. What is still needed is the computation of the relevant eigenvalues and eigenfunctions of the matrix $\mathbb{B}$ defined in (10.1.6).

Lemma 14.9 *Let z^* be a solution of* (14.5.9) *and assume in addition that*

$$\beta\mathbb{E}_h\bigl[1 - \tanh^2\bigl(\beta\bigl(z^*+h\bigr)\bigr)\bigr] > 1. \tag{14.7.10}$$

Then $\mathbf{z}^*$ defined through (14.5.8) *is a saddle point, and the unique negative eigenvalue of $\mathbb{B}(\mathbf{z}^*)$ is the unique negative solution $\hat{\gamma}_1 = \hat{\gamma}_1(N,n)$ of the equation*

$$\mathbb{E}\left[\frac{[1 - \tanh^2(\beta(z^*+h))]}{1 - 2\gamma \exp(2\beta[z^*+h]_+)\beta[1+\tanh(\beta(z^*+h))]}\right] = 1. \tag{14.7.11}$$

Proof The particular form of the matrix $\mathbb{B}$ allows us to obtain a simple characterisation of all the eigenvalues and eigenvectors. The eigenvalue equations can be written as

$$-\sum_{\ell=1}^{n} \sqrt{r_\ell r_k} u_\ell + (r_k\hat{\lambda}_k - \gamma)u_k = 0 \quad \forall 1 \leq k \leq n. \tag{14.7.12}$$

Assume, for simplicity, that the $r_k\hat{\lambda}_k$ take distinct values. Then there is no non-trivial solution of these equations with $\gamma = r_k\hat{\lambda}_k$, and so $\sum_{\ell=1}^n \sqrt{r_\ell}u_\ell \neq 0$. Thus,

$$u_k = \frac{\sqrt{r_k}\sum_{\ell=1}^n \sqrt{r_\ell}u_\ell}{r_k\hat{\lambda}_k - \gamma}. \tag{14.7.13}$$

Multiplying by $\sqrt{r_k}$ and summing over k, we find that u_k is a solution if and only if γ satisfies the equation

$$\sum_{k=1}^{n} \frac{r_k}{r_k \hat{\lambda}_k - \gamma} = 1. \tag{14.7.14}$$

Inserting the expressions for r_ℓ from (14.7.6), $\mathbf{z}_k^*/\rho_k$ from (14.5.8) and $\hat{\lambda}_k$ from (14.5.11) into (14.7.14), we obtain (14.7.11). Since the left-hand side of (14.7.14) is monotone decreasing in γ as long as $\gamma \geq 0$, it follows that there can be at most one negative solution of this equation, and such a solution exists if and only if the left-hand side is larger than 1 for $\gamma = 0$. □

We can now give the proof of Theorems 14.3–14.5.

Proof The induced dynamics on the block-magnetisations is essentially of the form of the discrete diffusions treated in Chap. 10. To obtain Theorem 14.4, it remains to insert the particular expressions obtained above into the general form of Theorem 10.9. Taking into account that the lattice spacing is $\varepsilon = 2/N$, while the ε in the exponent in the invariant measure is to be replaced by $1/N$, we get

$$\operatorname{cap}(A, B) = \mathscr{Q}_{\beta,N}(\mathbf{z}^*) \frac{\beta|\hat{\gamma}_1|}{2\pi N} \left(\frac{\pi N}{2\beta}\right)^{n/2} \prod_{\ell=1}^{n} \sqrt{\frac{r_\ell}{|\hat{\gamma}_j|}} \left(1 + O\left(\sqrt{(\ln N)^3/N}\right)\right). \tag{14.7.15}$$

Using Proposition 14.6 and the formula for the mean hitting time in Theorem 8.15, we get (14.4.13). Theorem 14.5 follows from Theorem 8.45. □

14.8 Bibliographical notes

1. The random-field Curie-Weiss model was one of the original motivations in Bovier, Eckhoff, Gayrard and Klein [33] (together with the so-called Hopfield model) for the development of the potential-theoretic approach to metastability. It was the main example given in that paper for the case where the distribution of the random field is discrete. Earlier work on the dynamics of this model was done by Matthieu and Picco [180] and Fontes, Matthieu and Picco [114].

2. The equilibrium behaviour of the RFCW-model was analysed first by Salinas and Wereszinski [238], and later in more detail by Amaro de Matos, Baêta Segundo and Perez [5] and Külske [158].

3. For solutions of (14.5.6), see Bovier, Eckhoff, Gayrard and Klein [33] or Bovier, Bianchi and Ioffe [24]. It is straightforward to analyse the case where some of the $\hat{\lambda}_k$'s in Lemma 14.7 coincide.

4. Another model that can be analysed with the methods of this chapter is the Glauber dynamics of the Hopfield model of neural networks (see Bovier and Gayrard [36] for a review) with finitely many stored patterns. This was done in the thesis of an der Heiden [6] under somewhat restrictive conditions.

Chapter 15
The Curie-Weiss Model with Random Magnetic Field: Continuous Distributions

> *"Which contain the greatest amount of Science, do you think, the books, or the minds?" … And I considered a minute before replying: "If you mean living minds, I don't think it's possible to decide. There is so much* written *Science that no living person has ever* read*: and there is so much* thought-out *Science that hasn't yet been written".*
>
> (Lewis Carroll, Sylvie and Bruno)

The random-field Curie-Weiss model with general distributions of the magnetic fields is a key example where non-exact coarse-graining methods can be shown to work efficiently in the context of the potential-theoretic approach. In Sect. 15.1 we state the main results of this chapter. In Sect. 15.2 we set up the coarse-graining and look at the energy landscape near critical points. In Sect. 15.3 we prove the upper bound in Theorem 15.3, which is relatively easy. In Sect. 15.4 we prove the lower bound, which is much harder. In Sect. 15.5 we combine Theorem 15.3 with estates on the harmonic function to complete the proof of Theorem 15.1.

15.1 Main results

We consider the same model with the same dynamics as in Chap. 14, but we drop the assumption made in Sect. 14.4 that the random magnetic fields take on only finitely many values. Instead we will only assume that the common distribution of the random magnetic fields has bounded support. All the results from Sects. 14.1–14.3 remain unchanged. What fails is the exact lumping procedure that allowed us to realise the mesoscopic image of our Markov process as a discrete diffusion process.

Our task is to obtain sharp estimates on metastable exit times. The main result is formulated in the following theorem, whose proof is the content of the present chapter.

Theorem 15.1 (Mean metastable exit times) *Assume that β and the distribution of the magnetic fields are such that there exist more than one local minimum of $F_{\beta,N}$.*

A. Bovier, F. den Hollander, *Metastability*,
Grundlehren der mathematischen Wissenschaften 351,
DOI 10.1007/978-3-319-24777-9_15

Let m^* be a local minimum of $F_{\beta,N}$, $M = M(m^*)$ the set of minima of $F_{\beta,N}$ such that $F_{\beta,N}(m) < F_{\beta,N}(m^*)$, and z^* the minimax between m and M, i.e., the lower of the highest maxima separating m from M to the left, respectively, right. Then, $\mathbb{P}_h$-a.s. and for all but finitely many values of N,

$$\mathbb{E}_{\nu_{S[m^*],S[M]}}[\tau_{S[M]}] = \exp\big[N\big[F_{\beta,N}\big(z^*\big) - F_{\beta,N}\big(m^*\big)\big]\big] \times \frac{\pi N}{2\beta[-\bar{\gamma}_1]}\sqrt{\frac{\beta\mathbb{E}_h(1-\tanh^2(\beta(z^*+h))) - 1}{1-\beta\mathbb{E}_h(1-\tanh^2(\beta(m^*+h)))}}\big[1+o(1)\big], \tag{15.1.1}$$

where $\bar{\gamma}_1$ is the unique negative solution of the equation

$$\mathbb{E}_h\left[\frac{(1-\tanh(\beta(z^*+h)))\exp(-2\beta[z^*+h]_+)}{\frac{\exp(-2\beta[z^*+h]_+)}{\beta(1+\tanh(\beta(z^*+h)))} - 2\gamma}\right] = 1. \tag{15.1.2}$$

Note that

$$F_{\beta,N}\big(z^*\big) - F_{\beta,N}\big(m^*\big) = \exp\left[\beta N\frac{(z^*)^2-(m^*)^2}{2} - \sum_{i\in\Lambda}\big[\ln\cosh\big(\beta\big(z^*+h_i\big)\big) - \ln\cosh\big(\beta\big(m^*+h_i\big)\big)\big]\right]. \tag{15.1.3}$$

Remark 15.2 Theorem 15.1 can be improved with the help of coupling techniques in two ways. First, the starting measure $\nu_{S[m^*],S[M]}$ can be replaced by any configuration σ in a suitably defined subset of $S[m^*]$. Second, the law of the transition time can be shown to be asymptotically exponential. Both these results rely on rather intricate and technical coupling arguments (see Sect. 15.6).

The proof of Theorem 15.1 relies on the following estimate for *capacities*.

Theorem 15.3 (Capacity asymptotics) *With the same notation as in Theorem* 15.1,

$$Z_{\beta,N}\,\mathrm{cap}\big(S\big[m^*\big], S[M]\big) = \frac{\beta|\bar{\gamma}_1|}{2\pi N}\frac{\exp[-\beta N F_{\beta,N}(z^*)][1+o(1)]}{\sqrt{\beta\mathbb{E}_h(1-\tanh^2(\beta(z^*+h))] - 1}}. \tag{15.1.4}$$

15.2 Coarse-graining and the mesoscopic approximation

As in Chap. 14, we want to pass to a coarse-grained description. As exact lumping is not possible, we use a sequence of approximate coarse-grainings.

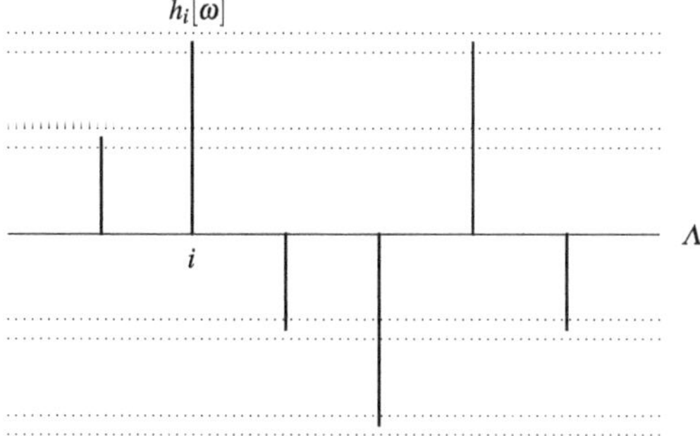

Fig. 15.1 Coarse-graining: Λ is partitioned into sets where the magnetic field takes values in a narrow interval. Compare with Fig. 14.1

15.2.1 Coarse-graining

Let I denote the support of the common distribution of the random fields h_i. Let I_ℓ, $\ell \in \{1,\dots,n\}$, be a partition of I such that $|I_\ell| \le C/n = \varepsilon$ for all ℓ and some $C < \infty$. Each realisation of the random fields $\{h_i[\omega]\}_{i\in\mathbb{N}}$ induces a random partition of the set $\Lambda = \{1,\dots,N\}$ into subsets (see Fig. 15.1)

$$\Lambda_\ell[\omega] = \left\{i \in \Lambda : h_i[\omega] \in I_\ell\right\}, \quad \ell = 1,\dots,n. \tag{15.2.1}$$

In complete analogy with the case of discrete distributions, we introduce n order parameters

$$\mathbf{m}_\ell[\omega](\sigma) = \frac{1}{N} \sum_{i\in\Lambda_\ell[\omega]} \sigma_i, \quad \ell = 1,\dots,n. \tag{15.2.2}$$

All notations from Sect. 14.4 carry over.

Remark 15.4 To simplify the presentation in this chapter, all statements involving random variables on $(\Omega, \mathscr{F}, \mathbb{P}_h)$ are understood to be true with $\mathbb{P}_h$-probability one, for all but finitely many values of N.

We define

$$\bar{h}_\ell = \frac{1}{|\Lambda_\ell|} \sum_{i\in\Lambda_\ell} h_i, \quad \tilde{h}_i = h_i - \bar{h}_\ell. \tag{15.2.3}$$

The Hamiltonian can then be written in the form

$$H_N[\omega](\sigma) = -NE\big(\mathbf{m}[\omega](\sigma)\big) + \sum_{\ell=1}^{n} \sum_{i\in\Lambda_\ell} \sigma_i \tilde{h}_i[\omega], \tag{15.2.4}$$

where $E\colon \mathbb{R}^n \to \mathbb{R}$ is as in (14.4.7), but with $\bar{h}_\ell$ replacing b_ℓ. We define the equilibrium distribution of the variables $\mathbf{m}[\sigma]$ as in (14.4.8), but take a slightly different form:

$$\mathscr{Q}_{\beta,N}[\omega](\mathbf{x}) = \frac{1}{Z_N[\omega]} e^{\beta N E(\mathbf{x})} \mathbb{E}_\sigma \left(\mathbb{1}_{\{\mathbf{m}[\omega](\sigma)=\mathbf{x}\}} e^{\sum_{\ell=1}^n \sum_{i\in\Lambda_\ell} \sigma_i (h_i - \bar{h}_\ell)} \right). \tag{15.2.5}$$

15.2.2 The energy landscape near critical points

We now turn to the precise computation of the measures $\mathscr{Q}_{\beta,N}[\omega](\mathbf{x})$ in the neighbourhood of the critical points of $F_{\beta,N}[\omega](\mathbf{x})$. We will see that this goes very much along the lines of the analysis for discrete distributions. We get the same expression for $Z_{\beta,N}[\omega]\mathscr{Q}_{\beta,N}[\omega](\mathbf{x})$ as in (14.5.1), again with b_ℓ replaced by $\bar{h}_\ell$, and

$$\begin{aligned} Z^\ell_{\beta,N}[\omega](y) &= \mathbb{E}_{\sigma_{\Lambda_\ell}} \left[\exp\left(\beta \sum_{i\in\Lambda_\ell} \tilde{h}_i \sigma_i \right) \mathbb{1}_{\{|\Lambda_\ell|^{-1} \sum_{i\in\Lambda_\ell} \sigma_i = y\}} \right] \\ &= \mathbb{E}^{\tilde{h}}_{\sigma_{\Lambda_\ell}} [\mathbb{1}_{\{|\Lambda_\ell|^{-1} \sum_{i\in\Lambda_\ell} \sigma_i = y\}}]. \end{aligned} \tag{15.2.6}$$

As in Sect. 14.5, we can express $Z_{\beta,N}[\omega]\mathscr{Q}_{\beta,N}[\omega](\mathbf{x})$ in the form of (14.5.4) with $F_{\beta,N}$ given by (14.5.5), but b_ℓ replaced by $\bar{h}_\ell$, where the entropy function $I_{N,\ell}[\omega](y)$ is now defined as the Legendre-Fenchel transform of the log-moment-generating function,

$$\begin{aligned} U_{N,\ell}[\omega](t) &= \frac{1}{|\Lambda_\ell|} \ln \mathbb{E}^{\tilde{h}}_{\sigma_{\Lambda_\ell}} \left[\exp\left(t \sum_{i\in\Lambda_\ell} \sigma_i \right) \right] \\ &= \frac{1}{|\Lambda_\ell|} \sum_{i\in\Lambda_\ell} \ln\cosh(t + \beta\tilde{h}_i). \end{aligned} \tag{15.2.7}$$

The analysis of the free energy functions near critical points $\mathbf{z}^*$ of $F_{\beta,N}$ goes very much as in Sect. 14.5, with the obvious replacements. Using that, by standard properties of Legendre-Fenchel transforms, $I'_{N,\ell}(x) = U'^{-1}_{N,\ell}(x)$, we see that (14.5.8) becomes

$$\mathbf{z}^*_\ell/\rho_\ell = U'_{N,\ell}\big(\beta(z^* + h_\ell)\big) = \frac{1}{|\Lambda_\ell|} \sum_{i\in\Lambda_\ell} \tanh\big(\beta(z^* + h_i)\big), \tag{15.2.8}$$

where z^* again solves (14.5.9), which is independent of n.

Finally, using that $I''_{N,\ell}(\mathbf{z}^*_\ell/\rho_\ell) = 1/U''_{N,\ell}(t^*_\ell)$ at a critical point, we find that the Hessian matrix $\mathbb{A}(\mathbf{z}^*)$ at a critical point $\mathbf{z}^*$ has elements

$$\big(\mathbb{A}(\mathbf{z}^*)\big)_{k\ell} = -1 + \delta_{\ell,k}\hat{\lambda}_\ell, \tag{15.2.9}$$

where

$$\hat{\lambda}_\ell = \frac{I''(\mathbf{z}^*_\ell/\rho_\ell)}{\beta\rho_\ell} = \frac{1}{\beta\rho_\ell U''_{N,\ell}(\beta(z^*+\bar{h}_\ell))} = \frac{1}{\frac{\beta}{N}\sum_{i\in\Lambda_\ell}(1-\tanh^2(\beta(z^*+h_i)))}, \tag{15.2.10}$$

which replaces (14.5.11).

The following is the analogue of Proposition 14.6.

Proposition 15.5 *Let* $\mathbf{z}^*$ *be a critical point of* $F_{\beta,N}$. *Then* $\mathbf{z}^*$ *is given by* (15.2.8), *where* z^* *is a solution of* (14.5.9). *Moreover,*

$$\begin{aligned} &Z_{\beta,N}\mathcal{Q}_{\beta,N}(\mathbf{z}^*) \\ &\quad= \frac{\sqrt{[-\det(\mathbb{A}(\mathbf{z}^*))]}}{\sqrt{(\frac{N\pi}{2\beta})^n|\beta\mathbb{E}_h(1-\tanh^2(\beta(z^*+h)))-1|}} \\ &\qquad\times\exp\left[\beta N\left(-\frac{(z^*)^2}{2}+\frac{1}{\beta N}\sum_{i\in\Lambda}\ln\cosh\big(\beta(z^*+h_i)\big)\right)\right]\big[1+o(1)\big]. \end{aligned} \tag{15.2.11}$$

Proof We start with the representation of $Z_{\beta,N}[\omega]\mathcal{Q}_{\beta,N}[\omega](\mathbf{x})$ given in (14.5.4). Using (15.2.10) and the formula for the determinant of $\mathbb{A}(\mathbf{z}^*)$ given in (14.5.12), we get the prefactor. For the exponential term $F_{\beta,N}$, note that by convex duality

$$I_{N,\ell}(\mathbf{z}^*_\ell/\rho_\ell) = t^*_\ell\mathbf{z}^*_\ell/\rho_\ell - U_{N,\ell}(t^*_\ell) = \beta(z^*+\bar{h}_\ell)\mathbf{z}^*_\ell/\rho_\ell - U_{N,\ell}\big(\beta(z^*+\bar{h}_\ell)\big). \tag{15.2.12}$$

Hence

$$\begin{aligned} &F_{\beta,N}(\mathbf{z}^*) \\ &\quad= \frac{1}{2}(z^*)^2 - \sum_{\ell=1}^n \mathbf{z}^*_\ell\bar{h}_\ell + \frac{1}{\beta}\sum_{\ell=1}^n\big[\rho_\ell\beta(z^*+\bar{h}_\ell)\mathbf{z}^*_\ell/\rho_\ell - \rho_\ell U_{N,\ell}\big(\beta(z^*+\bar{h}_\ell)\big)\big] \\ &\quad= -\frac{1}{2}(z^*)^2 - \sum_{\ell=1}^n\left[\mathbf{z}^*_\ell\bar{h}_\ell - z^*\mathbf{z}^*_\ell - \bar{h}\mathbf{z}^*_\ell + \frac{1}{\beta N}\sum_{i\in\Lambda_\ell}\ln\cosh\big(\beta(z^*+h_i)\big)\right] \\ &\quad= \frac{1}{2}(z^*)^2 - \frac{1}{\beta N}\sum_{i\in\Lambda}\ln\cosh\big(\beta(z^*+h_i)\big), \end{aligned} \tag{15.2.13}$$

which is the desired exponent. □

Remark 15.6 The form given in Proposition 15.5 is highly suitable for our purposes, as the dependence on n appears only in the denominator of the prefactor. We will see that this is just what we need to get a formula for capacities that is independent of the choice of the partition $I=\bigcup_{1\leq\ell\leq n}I_\ell$ and has a limit as $n\to\infty$.

The eigenvalues of the Hessian are characterised in the next lemma, which is the analogue of Lemma 14.7.

Lemma 15.7 *Let* z^* *be a solution of* (14.5.9). *In addition, assume that the distribution of* $(h_i)_{i\in\mathbb{N}}$ *is such that all numbers* $\hat{\lambda}_k$ *are* $\mathbb{P}_h$*-a.s. distinct. Then* γ *is an eigenvalue of* $\mathbb{A}(\mathbf{z}^*)$ *if and only if it is a solution of the equation*

$$\sum_{\ell=1}^{n} \frac{1}{\frac{1}{\frac{\beta}{N}\sum_{i\in\Lambda_\ell}(1-\tanh^2(\beta(z^*+h_i)))} - \gamma} = 1. \tag{15.2.14}$$

Moreover, (15.2.14) *has at most one negative solution, and it has such a negative solution if and only if*

$$\frac{\beta}{N}\sum_{i=1}^{N}\left(1-\tanh^2\left(\beta\left(z^*+h_i\right)\right)\right) > 1. \tag{15.2.15}$$

The proof of Lemma 15.7 is identical to that of Lemma 14.7.

The discussion of the energy landscape carries over unchanged from Chap. 14.

15.3 Upper bounds on capacities

Sections 15.3–15.4 are devoted to the proof of Theorem 15.3. In this section we prove the upper bound. Obtaining upper bounds on capacities just requires guessing a test function. Basically, we may ignore the fact that our coarse-graining is not an exact lumping, since (14.4.11) holds as an upper bound (only the second equality has to be replaced by an inequality).

Let $A = \mathscr{S}_N[\mathbf{A}]$ and $B = \mathscr{S}_N[\mathbf{B}]$, for some $\mathbf{A},\mathbf{B} \subseteq \Gamma_N^n$. Then

$$\begin{aligned}
\operatorname{cap}(A,B) &= \inf_{h\in\mathscr{H}_{A,B}} \frac{1}{2}\sum_{\sigma,\sigma'\in\mathscr{S}_N} \mu_{\beta,N}[\omega](\sigma)p\left(\sigma,\sigma'\right)\left[h(\sigma)-h\left(\sigma'\right)\right]^2 \\
&\le \inf_{u\in\mathscr{G}_{\mathbf{A},\mathbf{B}}} \frac{1}{2}\sum_{\sigma,\sigma'\in\mathscr{S}_N} \mu_{\beta,N}[\omega](\sigma)p\left(\sigma,\sigma'\right)\left[u\left(\mathbf{m}(\sigma)\right)-u\left(\mathbf{m}\left(\sigma'\right)\right)\right]^2 \\
&= \inf_{u\in\mathscr{G}_{\mathbf{A},\mathbf{B}}} \sum_{\mathbf{x},\mathbf{x}'\in\Gamma_N^n} \mathscr{Q}_{\beta,N}[\omega](\mathbf{x})r_N\left(\mathbf{x},\mathbf{x}'\right)\left[u(\mathbf{x})-u\left(\mathbf{x}'\right)\right]^2 \\
&= \mathrm{CAP}_N^n(\mathbf{A},\mathbf{B})
\end{aligned} \tag{15.3.1}$$

with $r_N(\beta,\mathbf{x}')$, $\mathscr{H}_{A,B}$ and $\mathscr{G}_{\mathbf{A},\mathbf{B}}$ defined precisely as in (14.4.10) and (14.4.12).

We proceed from here as in Chap. 10. For this we need the form of the transition rates in the neighbourhood of a critical point. The formulas from Chap. 14 have to be changed only slightly: for all $\mathbf{x} \in \Gamma_N^n$, we now have

$$r_N(\mathbf{x}, \mathbf{x} + \mathbf{e}_\ell) = \mathscr{Q}_{\beta,N}(\mathbf{x})^{-1} \sum_{\sigma \in \mathscr{S}_N[\mathbf{x}]} \mu_{\beta,N}[\omega](\sigma) \sum_{i \in \Lambda_\ell^-(\sigma)} p(\sigma, \sigma^i) \tag{15.3.2}$$

$$= \mathscr{Q}_{\beta,N}(\mathbf{x})^{-1} \sum_{\sigma \in \mathscr{S}_N[\mathbf{x}]} \mu_{\beta,N}[\omega](\sigma) \sum_{i \in \Lambda_\ell^-(\sigma)} \frac{1}{N} \mathrm{e}^{-2\beta[m(\sigma) - N^{-1} + h_i]_+}.$$

Note that, for all $\sigma \in \mathscr{S}_N(\mathbf{x})$, $|\Lambda_\ell^-(\sigma)|$ is a constant depending on $\mathbf{x}$ only. Using that $h_i = \bar{h}_\ell + \widetilde{h}_i$, with $\widetilde{h}_i \in [-\varepsilon, \varepsilon]$, we get the bounds

$$r_N(\mathbf{x}, \mathbf{x} + \mathbf{e}_\ell) = \frac{|\Lambda_\ell^-(\mathbf{x})|}{N} \mathrm{e}^{-2\beta[m(\sigma) + \bar{h}_\ell]_+} \left[1 + O(\varepsilon)\right]. \tag{15.3.3}$$

It follows that, for all $\mathbf{x} \in D_N(\rho)$,

$$\left| \frac{r_N(\mathbf{x}, \mathbf{x} + \mathbf{e}_\ell)}{r_N(\mathbf{z}^*, \mathbf{z}^* + \mathbf{e}_\ell)} - 1 \right| \leq c\beta(\varepsilon + n\rho) \tag{15.3.4}$$

for some finite constant $c > 0$. With these minimal changes we arrive at the same form of the effective Dirichlet form $\widetilde{\mathscr{E}}_{D_N}(u, u)$ as in (14.7.9).

From now on the upper bound follows as in the case of discrete magnetic fields. There is just a slight change in that Lemma 14.9 needs to be replaced by the following.

Lemma 15.8 *Let z^* be a solution of* (14.5.9). *In addition, assume that*

$$\frac{\beta}{N} \sum_{i=1}^{N} \left(1 - \tanh^2\left(\beta\left(z^* + h_i\right)\right)\right) > 1. \tag{15.3.5}$$

Then $\mathbf{z}^*$ defined through (15.2.8) *is a saddle point, and the unique negative eigenvalue of $\mathbb{B}(\mathbf{z}^*)$ is the unique negative solution $\hat{\gamma}_1 = \hat{\gamma}_1(N, n)$ of the equation*

$$\sum_{\ell=1}^{n} \rho_\ell \frac{\frac{1}{|\Lambda_\ell|} \sum_{i \in \Lambda_\ell} (1 - \tanh(\beta(z^* + h_i))) \exp\left(-2\beta[z^* + \bar{h}_\ell]_+\right)}{\frac{\frac{1}{|\Lambda_\ell|} \sum_{i \in \Lambda_\ell} (1 - \tanh(\beta(z^* + h_i))) \exp\left(-2\beta[z^* + \bar{h}_\ell]_+\right)}{\frac{\beta}{|\Lambda_\ell|} \sum_{i \in \Lambda_\ell} (1 - \tanh^2(\beta(z^* + h_i)))} - 2\gamma} = 1. \tag{15.3.6}$$

Moreover,

$$\lim_{n \to \infty} \lim_{N \to \infty} \hat{\gamma}_1(N, n) = \bar{\gamma}_1, \tag{15.3.7}$$

where $\bar{\gamma}_1$ is the unique negative solution of the equation

$$\mathbb{E}_h \left[\frac{(1 - \tanh(\beta(z^* + h))) \exp\left(-2\beta[z^* + h]_+\right)}{\frac{\exp(-2\beta[z^* + h]_+)}{\beta(1 + \tanh(\beta(z^* + h)))} - 2\gamma} \right] = 1. \tag{15.3.8}$$

Proof The proof of (15.3.6) is identical to that of Lemma 14.9. The assertion on the convergence follows from the fact that the size of the small fields tends to zero as $n \to \infty$. □

This result yields the upper bound given in the next proposition.

Proposition 15.9 *With the notation above, for every $n \in \mathbb{N}$,*

$$\mathrm{cap}(A, B) \leq \mathcal{Q}_{\beta,N}(\mathbf{z}^*) \frac{\beta|\hat{\gamma}_1|}{2\pi N} \left(\frac{\pi N}{2\beta}\right)^{n/2} \prod_{\ell=1}^{n} \sqrt{\frac{r_\ell}{|\hat{\gamma}_j|}} \left[1 + O\left(\varepsilon + \sqrt{(\ln N)^3/N}\right)\right]. \tag{15.3.9}$$

Combining Proposition 15.9 with Proposition 14.6, we get (after some computations) the following more explicit representation of the upper bound.

Corollary 15.10 *With the same notation as in Proposition* 15.9,

$$Z_{\beta,N}\, \mathrm{cap}(A, B) \leq \frac{\beta|\bar{\gamma}_1|}{2\pi N} \frac{\exp(-\beta N F_{\beta,N}(z^*))[1 + o(1)]}{\sqrt{\beta N \mathbb{E}_h(1 - \tanh^2(\beta(z^* + h))) - 1}}, \tag{15.3.10}$$

where $\bar{\gamma}_1$ is defined through (15.3.8).

Corollary 15.10 concludes the upper bound in the proof of Theorem 15.3.

15.4 Lower bounds on capacities

To prove the matching lower bound is technically involved, because it requires a more sophisticated use of the Berman-Konsowa principle in Theorem 7.43 that was so successfully used in Sect. 10.3. We have seen in Chap. 14 that we can construct a *defective flow* on mesoscopic variables that produces a good lower bound for capacities (along the lines explained in Sect. 9.1.2). The strategy in the present situation is to try to reproduce this mesoscopic flow on the microscopic level.

15.4.1 Two-scale flows

Let $\mathbf{A}$ and $\mathbf{B}$ be mesoscopic neighbourhoods of two minima of $F_{\beta,N}$, and let $\mathbf{z}^*$ be the corresponding saddle point. Let $A = S_N[\mathbf{A}]$ and $B = S_N[\mathbf{B}]$ be as before. Let $\mathfrak{f}_{\mathbf{A},\mathbf{B}} = \{\mathfrak{f}_{\mathbf{A}}, \mathfrak{f}, \mathfrak{f}_{\mathbf{B}}\}$ be the defective *mesoscopic flow* from $\mathbf{A}$ to $\mathbf{B}$. In this section we are going to construct a subordinate *microscopic flow* $f_{A,B}$ from A to B. In the sequel, given a microscopic bond $b = (\sigma, \sigma')$, we write $\mathbf{e}(b) = (\mathbf{m}(\sigma), \mathbf{m}(\sigma'))$ to denote its mesoscopic pre-image.

We label the realisations of the mesoscopic Markov chain $\mathscr{X}_{\mathbf{A},\mathbf{B}}$ associated with the mesoscopic flow $\mathfrak{f}_{\mathbf{A},\mathbf{B}}$ as $\underline{\mathbf{x}} = (\mathbf{x}_{-\ell_A}, \dots, \mathbf{x}_{\ell_B})$ in such a way that $\mathbf{x}_{-\ell_A} \in \mathbf{A}$, $\mathbf{x}_{\ell_B} \in \mathbf{B}$, and $m(\mathbf{x}_0) = m(\mathbf{z}^*)$. We denote by $\mathbb{P}^{\mathfrak{f}_{\mathbf{A},\mathbf{B}}}$ the corresponding law on the mesoscopic paths. If $\mathbf{e}$ is a mesoscopic bond, then we write $\mathbf{e} \in \underline{\mathbf{x}}$ when $\mathbf{e} = (\mathbf{x}_\ell, \mathbf{x}_{\ell+1})$ for some $\ell = -\ell_A, \dots, \ell_B - 1$. With each path $\underline{\mathbf{x}}$ of positive probability we associate a subordinate microscopic *unit flow* $f^{\underline{\mathbf{x}}}$ such that

$$f^{\underline{\mathbf{x}}}(b) > 0 \quad \text{if and only if} \quad \mathbf{e}(b) \in \underline{\mathbf{x}}, \tag{15.4.1}$$

and

$$\sum_{b:\ \mathbf{e}(b)=\mathbf{e}} f^{\underline{\mathbf{x}}}(b) = 1 \quad \forall\, \underline{\mathbf{x}}, \mathbf{e} \in \underline{\mathbf{x}}. \tag{15.4.2}$$

Finally, since $f^{\underline{\mathbf{x}}}$ is a unit flow on $\underline{\mathbf{x}}$, it defines a Markov chain $\mathbb{P}^{\underline{\mathbf{x}}}$ on the microscopic paths $\mathbb{S}$ whose image is $\underline{\mathbf{x}}$, such that

$$f^{\underline{\mathbf{x}}}(b) = \mathbb{P}^{\underline{\mathbf{x}}}(b \in \mathbb{S}). \tag{15.4.3}$$

We think of $\mathbb{P}^{\underline{\mathbf{x}}}$ as the conditional law, given $\underline{\mathbf{x}}$, of a Markov chain on the microscopic paths, namely,

$$\mathbb{P}^{f_{\mathbf{A},\mathbf{B}}}(\mathbb{S} = \underline{\sigma}) = \sum_{\underline{\mathbf{x}}} \mathbb{P}^{\mathfrak{f}_{\mathbf{A},\mathbf{B}}}(\mathscr{X}_{\mathbf{A},\mathbf{B}} = \underline{\mathbf{x}}) \mathbb{P}^{\underline{\mathbf{x}}}(\mathbb{S} = \underline{\sigma}). \tag{15.4.4}$$

Therefore we have

$$\mathbb{P}^{f_{\mathbf{A},\mathbf{B}}}(b \in \underline{\sigma}) = \sum_{\underline{\mathbf{x}}} \mathbb{P}^{\mathfrak{f}_{\mathbf{A},\mathbf{B}}}(\mathscr{X}_{\mathbf{A},\mathbf{B}} = \underline{\mathbf{x}}) \mathbb{P}^{\underline{\mathbf{x}}}(b \in \underline{\sigma}). \tag{15.4.5}$$

Summing over b giving rise to a mesoscopic bond $\mathbf{e}$, we get

$$\sum_{b:\ \mathbf{e}(b)=\mathbf{e}} \mathbb{P}^{f_{\mathbf{A},\mathbf{B}}}(b \in \underline{\sigma}) = \sum_{\underline{\mathbf{x}}} \mathbb{P}^{\mathfrak{f}_{\mathbf{A},\mathbf{B}}}(\mathscr{X}_{\mathbf{A},\mathbf{B}} = \underline{\mathbf{x}}) \mathbb{1}_{\mathbf{e} \in \underline{\mathbf{x}}} = \mathbb{P}^{\mathfrak{f}_{\mathbf{A},\mathbf{B}}}(\mathbf{e} \in \mathscr{X}_{\mathbf{A},\mathbf{B}}). \tag{15.4.6}$$

This provides the decomposition of unity

$$\mathbb{1}_{\{f_{A,B}(b)>0\}} = \sum_{\underline{\mathbf{x}} \ni \mathbf{e}(b)} \sum_{\underline{\sigma} \ni b} \frac{\mathbb{P}_N^{\mathfrak{f}_{\mathbf{A},\mathbf{B}}}(\mathscr{X}_{\mathbf{A},\mathbf{B}} = \underline{\mathbf{x}}) \mathbb{P}^{\underline{\mathbf{x}}}(\Sigma = \underline{\sigma})}{\mathbb{P}^{\mathfrak{f}_{\mathbf{A},\mathbf{B}}}(\mathbf{e} \in \mathscr{X}_{\mathbf{A},\mathbf{B}}) f^{\underline{\mathbf{x}}}(b)}. \tag{15.4.7}$$

If $\mathfrak{f}_{\mathbf{A},\mathbf{B}}(\mathbf{e}(b))$ would be non-defective, then we could replace $\mathbb{P}^{\mathfrak{f}_{\mathbf{A},\mathbf{B}}}(\mathbf{e} \in \mathscr{X}_{\mathbf{A},\mathbf{B}})$ by $\mathfrak{f}_{\mathbf{A},\mathbf{B}}(\mathbf{e}(b))$. Since we will choose a defective flow, we assume that

$$\mathbb{P}^{\mathfrak{f}_{\mathbf{A},\mathbf{B}}}(\mathbf{e} \in \mathscr{X}_{\mathbf{A},\mathbf{B}}) \leq \mathfrak{f}_{\mathbf{A},\mathbf{B}}\big(\mathbf{e}(b)\big)\big(1 + d(\mathbf{e})\big), \tag{15.4.8}$$

where $d(\mathbf{e})$ depends only on the initial point of the bond $\mathbf{e}$.

As in Lemma 9.4 we get

$$\begin{aligned}&\mathrm{cap}(A,B)\\&\geq \sum_{\underline{\mathbf{x}}} \mathbb{P}_N^{\mathfrak{f}_{\mathbf{A},\mathbf{B}}}(\mathscr{X}_{\mathbf{A},\mathbf{B}}=\underline{\mathbf{x}})\mathbb{E}^{\underline{\mathbf{x}}}\left[\left(\sum_{\ell=-\ell_A}^{\ell_B-1} \frac{\mathfrak{f}_{\mathbf{A},\mathbf{B}}(\mathbf{x}_\ell,\mathbf{x}_{\ell+1})f^{\underline{\mathbf{x}}}(\sigma_\ell,\sigma_{\ell+1})}{(1+d(\mathbf{x}_\ell))\mu_{\beta,N}(\sigma_\ell)p_N(\sigma_\ell,\sigma_{\ell+1})}\right)^{-1}\right]\\&\geq \mathbb{E}_N^{\mathfrak{f}_{\mathbf{A},\mathbf{B}}}\left[\left(\mathbb{E}^{\underline{\mathbf{x}}}\left[\sum_{\ell=-\ell_A}^{\ell_B-1} \frac{\mathfrak{f}_{\mathbf{A},\mathbf{B}}(\mathbf{x}_\ell,\mathbf{x}_{\ell+1})f^{\underline{\mathbf{x}}}(\sigma_\ell,\sigma_{\ell+1})}{(1+d(\mathbf{x}_\ell))\mu_{\beta,N}(\sigma_\ell)p_N(\sigma_\ell,\sigma_{\ell+1})}\right]\right)^{-1}\right], \qquad (15.4.9)\end{aligned}$$

where we use Jensen's inequality to obtain the second inequality. We set

$$\frac{\mathscr{Q}_{\beta,N}(\mathbf{x}_\ell)r_N(\mathbf{x}_\ell,\mathbf{x}_{\ell+1})f^{\underline{\mathbf{x}}}(\sigma_\ell,\sigma_{\ell+1})}{\mu_{\beta,N}(\sigma_\ell)p_N(\sigma_\ell,\sigma_{\ell+1})} = \phi_\ell^{\underline{\mathbf{x}}}(\sigma_\ell,\sigma_{\ell+1}). \qquad (15.4.10)$$

Then (15.4.9) reads

$$\begin{aligned}&\mathrm{cap}(A,B)\\&\geq \mathbb{E}_N^{\mathfrak{f}_{\mathbf{A},\mathbf{B}}}\left[\left(\mathbb{E}^{\underline{\mathbf{x}}}\left[\sum_{\ell=-\ell_A}^{\ell_B-1} \frac{\mathfrak{f}_{\mathbf{A},\mathbf{B}}(\mathbf{x}_\ell,\mathbf{x}_{\ell+1})}{(1+d(\mathbf{x}_\ell))\mathscr{Q}_{\beta,N}(\mathbf{x}_\ell)r_N(\mathbf{x}_\ell,\mathbf{x}_{\ell+1})}\phi_\ell^{\underline{\mathbf{x}}}(\sigma_\ell,\sigma_{\ell+1})\right]\right)^{-1}\right].\end{aligned} \qquad (15.4.11)$$

The point of the above rewrite is that if $\phi_\ell^{\underline{\mathbf{x}}}(\sigma_\ell,\sigma_{\ell+1})$ were equal to one, then we would be in the same situation as in the case of discrete distributions of the magnetic fields, and the left-hand side would be equal to the upper bound in (15.3.9), up to errors of order $N^{-1/2}(\ln N)^{3/2}+\varepsilon$.

Recall from Sect. 10.3 that we can restrict the expectation to a subset of *good* realisations $\underline{\mathbf{x}}$ of the mesoscopic Markov chain $\mathscr{X}_{\mathbf{A},\mathbf{B}}$ whose probability under $\mathbb{P}_N^{\mathfrak{f}_{\mathbf{A},\mathbf{B}}}$ is close to one. It remains to construct $f^{\underline{\mathbf{x}}}$ such that $\phi_\ell^{\underline{\mathbf{x}}}$ in (15.4.10) is close to one (in a weak sense). This requires some additional notation. Given a mesoscopic trajectory $\underline{\mathbf{x}}=(\mathbf{x}_{-\ell_A},\dots,\mathbf{x}_{\ell_B})$, define $k=k(\ell)$ as the direction of the increment of the ℓ-th jump, i.e., $\mathbf{x}_{\ell+1}=\mathbf{x}_\ell+\mathbf{e}_k$. On the microscopic level such a transition corresponds to a flip of a spin from the Λ_k-slot. Thus, recalling the notation $\Lambda_k^{\pm}(\sigma)=\{i\in\Lambda_k\colon \sigma(i)=\pm 1\}$, we have that if $\sigma_\ell\in S_N[\mathbf{x}_\ell]$ and $\sigma_{\ell+1}\in S_N[\mathbf{x}_{\ell+1}]$, then $\sigma_{\ell+1}=\theta_i^+\sigma_\ell$ for some $i\in\Lambda_{k(\ell)}^-(\sigma_\ell)$. By our choice of p_N and r_N,

$$\frac{r_N(\mathbf{x}_\ell,\mathbf{x}_{\ell+1})}{p_N(\sigma_\ell,\sigma_{\ell+1})} = \left|\Lambda_{k(\ell)}^-(\sigma_\ell)\right|\left[1+O(\varepsilon)\right], \qquad (15.4.12)$$

uniformly in ℓ and in all pairs of neighbours $\sigma_\ell,\sigma_{\ell+1}$. Note that the cardinality of $\Lambda_{k(\ell)}^-(\sigma_\ell)$ is the same for all $\sigma_\ell\in S_N[\mathbf{x}_\ell]$.

For $\mathbf{x}\in\Gamma_N^n$, define the measure

$$\mu_{\beta,N}^{\mathbf{x}}(\sigma) = \frac{\mathbb{1}_{\{\sigma\in S_N[\mathbf{x}]\}}\mu_{\beta,N}(\sigma)}{\mathscr{Q}_{\beta,N}(\mathbf{x})} \qquad (15.4.13)$$

$$= \mathbb{1}_{\{\sigma\in S_N[\mathbf{x}]\}} \frac{e^{\beta \sum_{\ell=1}^{n} \sum_{i\in\Lambda_\ell} \sigma_i \tilde{h}_i}}{\sum_{\sigma:\, \mathbf{x}(\sigma)=\mathbf{x}} e^{\beta \sum_{\ell=1}^{n} \sum_{i\in\Lambda_\ell} \sigma_i \tilde{h}_i}}$$

$$= \mathbb{1}_{\{\sigma\in S_N[\mathbf{x}]\}} \prod_{\ell=1}^{n} \frac{e^{\beta \sum_{i\in\Lambda_\ell} \sigma_i \tilde{h}_i}}{\sum_{\sigma_{\Lambda_\ell}:\ \mathbf{x}^\ell(\sigma_{\Lambda(\ell)})=\mathbf{x}^\ell} e^{\beta \sum_{i\in\Lambda_\ell} \sigma_i \tilde{h}_i}} = \prod_{\ell=1}^{n} \mu_{\beta,N}^{\mathbf{x},\ell}(\sigma).$$

Then we can write $\phi_\ell^{\underline{\mathbf{x}}}$ as

$$\phi_\ell^{\underline{\mathbf{x}}}(\sigma_\ell,\sigma_{\ell+1}) = \frac{r_N(\mathbf{x}_\ell,\mathbf{x}_{\ell+1}) f^{\underline{\mathbf{x}}}(\sigma_\ell,\sigma_{\ell+1})}{p_N(\sigma_\ell,\sigma_{\ell+1}) \mu_{\beta,N}^{\underline{\mathbf{x}}}(\sigma_\ell)}. \tag{15.4.14}$$

If the magnetic fields were constant on each set I_k, then we could choose

$$f^{\underline{\mathbf{x}}}(\sigma_\ell,\sigma_{\ell+1}) = \mu_{\beta,N}^{\mathbf{x}_\ell}(\sigma_\ell)/\left|\Lambda_{k(\ell)}^{-}(\sigma_\ell)\right|, \tag{15.4.15}$$

and we would be done. But this is not possible here.

Construction of $f^{\underline{\mathbf{x}}}$

We construct a Markov chain $\mathbb{P}^{\underline{\mathbf{x}}}$ on microscopic trajectories $\Sigma = \{\sigma_0,\dots,\sigma_{\ell_B}\}$ from $S_N[\mathbf{x}_0]$ to B such that $\sigma_\ell \in S_N[\mathbf{x}_\ell]$ for all $\ell = 0,\dots,\ell_B$. The construction of a microscopic flow from A to $S_N[\mathbf{x}_0]$ is just the reversal of the above and we will omit it.

STEP 1. The transition probabilities $q_\ell(\sigma_\ell,\sigma_{\ell+1})$ in (15.4.18) are defined in the following way: all the microscopic jumps are of the form $\sigma_\ell \mapsto \theta_j^+ \sigma_\ell$ for some $j \in \Lambda_{k(\ell)}^{-}(\sigma)$, where θ_j^+ flips the j-th spin from -1 to 1. For such a flip define

$$q_\ell\big(\sigma_\ell,\theta_j^+\sigma_\ell\big) = \frac{e^{2\beta\tilde{h}_j}}{\sum_{i\in\Lambda_k^{-}(\sigma_\ell)} e^{2\beta\tilde{h}_i}}. \tag{15.4.16}$$

Clearly, these ratios sum up to one. Note also that they satisfy

$$q_\ell\big(\sigma_\ell,\theta_j^+\sigma_\ell\big) = \frac{1+O(\varepsilon)}{|\Lambda_{k(\ell)}^{-}|}. \tag{15.4.17}$$

STEP 2. As initial measure $\nu_0^{\underline{\mathbf{x}}}$, choose $\mu_{\beta,N}^{\mathbf{x}_0}$. For $\ell = 0,\dots,\ell_B$, set

$$\nu_{\ell+1}^{\underline{\mathbf{x}}}(\sigma_{\ell+1}) = \sum_{\sigma_\ell\in S_N[\mathbf{x}_\ell]} \nu_\ell^{\underline{\mathbf{x}}}(\sigma) q_\ell(\sigma_\ell,\sigma_{\ell+1}). \tag{15.4.18}$$

Note that these measures are concentrated on $S_N[\mathbf{x}_\ell]$ and are the marginals of $\mathbb{P}^{\underline{\mathbf{x}}}$ at time ℓ.

STEP 3. Define the microscopic flow through an admissible bond $b = (\sigma_\ell, \sigma_{\ell+1})$ as

$$f^{\underline{\mathbf{x}}}(\sigma_\ell, \sigma_{\ell+1}) = \mathbb{P}^{\underline{\mathbf{x}}}(b \in \Sigma) = \nu_\ell^{\underline{\mathbf{x}}}(\sigma_\ell) q_\ell(\sigma_\ell, \sigma_{\ell+1}). \tag{15.4.19}$$

Note that the fact that the q_ℓ are probabilities, together with the definition in (15.4.18), ensures that $f^{\underline{\mathbf{x}}}$ is a unit flow. Consequently,

$$\phi_\ell^{\underline{\mathbf{x}}}(\sigma_\ell, \sigma_{\ell+1}) = \frac{\nu_\ell^{\underline{\mathbf{x}}}(\sigma_\ell)}{\mu_{\beta,N}^{\mathbf{x}_\ell}(\sigma_\ell)} \frac{r_N(\mathbf{x}_\ell, \mathbf{x}_{\ell+1})}{p_N(\sigma_\ell, \sigma_{\ell+1})} q_\ell(\sigma_\ell, \sigma_{\ell+1}). \tag{15.4.20}$$

Using the observations in (15.4.12) and (15.4.17), we see that

$$\phi_\ell^{\underline{\mathbf{x}}}(\sigma_\ell, \sigma_{\ell+1}) = \frac{\nu_\ell^{\underline{\mathbf{x}}}(\sigma_\ell)}{\mu_{\beta,N}^{\mathbf{x}_\ell}(\sigma_\ell)} \big[1 + O(\varepsilon)\big] = \Psi_\ell(\sigma_\ell)\big[1 + O(\varepsilon)\big]. \tag{15.4.21}$$

Note that $\Psi_0(\sigma_0) = 1$. We need to control the evolution of this quantity in time.

Proposition 15.11 *There exists a set* $\mathscr{T}_{\mathbf{A},\mathbf{B}}$ *of* good *mesoscopic trajectories from* **A** *to* **B** *such that*

$$\mathbb{P}_N^{\mathfrak{f}_{\mathbf{A},\mathbf{B}}}(\mathscr{X}_{\mathbf{A},\mathbf{B}} \in \mathscr{T}_{\mathbf{A},\mathbf{B}}) = 1 - o(1), \tag{15.4.22}$$

and, uniformly in $\underline{\mathbf{x}} \in \mathscr{T}_{\mathbf{A},\mathbf{B}}$,

$$\mathbb{E}^{\underline{\mathbf{x}}}\left[\sum_{\ell=-\ell_A}^{\ell_B-1} \Psi_\ell(\sigma_\ell) \frac{\mathfrak{f}_{\mathbf{A},\mathbf{B}}(\mathbf{x}_\ell, \mathbf{x}_{\ell+1})}{(1+d(\mathbf{x}_\ell))\mathscr{Q}_{\beta,N}(\mathbf{x}_\ell) r_N(\mathbf{x}_\ell, \mathbf{x}_\ell + 1)}\right] \le \frac{1}{\mathscr{E}_N(\tilde{g})}\big[1 + O(\varepsilon)\big]. \tag{15.4.23}$$

Proposition 15.11 implies that

$$\mathrm{cap}(A, B) \ge \mathscr{E}_N(\tilde{g})\big[1 - O(\varepsilon)\big], \tag{15.4.24}$$

which is the lower bound necessary to prove Theorem 15.3.

The rest of this section is devoted to the proof of (15.4.23). First of all, we derive recursive estimates on Ψ_ℓ for a given realisation $\underline{\mathbf{x}}$ of the mesoscopic Markov chain. After that it will be obvious how to define $\mathscr{T}_{\mathbf{A},\mathbf{B}}$.

15.4.2 Propagation of errors along microscopic paths

Let $\underline{\mathbf{x}}$ be given. We have seen in (15.4.13) that $\mu_{|b,N}^{\underline{\mathbf{x}}}$ is a product measure. On the other hand, according to (15.4.16), the *large* microscopic Markov chain Σ splits into a direct product of n *small* microscopic Markov chains $\Sigma^{(1)}, \dots, \Sigma^{(n)}$, which independently evolve on $\mathscr{S}_N^{(1)}, \dots, \mathscr{S}_N^{(n)}$. Thus, $k(\ell) = k$ means that the ℓ-th step of the mesoscopic Markov chain induces a step of the k-th small microscopic Markov chain $\Sigma^{(k)}$. Let $\tau_1[\ell],\dots,\tau_n[\ell]$ be the numbers of steps performed by each of the small microscopic Markov chains after ℓ steps of the mesoscopic Markov chain or, equivalently, after ℓ steps of the large microscopic Markov chain Σ. Then the corrector Ψ_ℓ in (15.4.21) also factorises and can be written as

$$\Psi_\ell(\sigma_\ell)=\prod_{j=1}^{n}\psi^{(j)}_{\tau_j[\ell]}\big(\sigma^{(j)}_\ell\big). \tag{15.4.25}$$

Therefore we are left with two separate tasks: On the microscopic level we need to control the propagation of errors along *small* Markov chains, while on the mesoscopic level we need to control the statistics of $\tau_1[\ell],\dots,\tau_n[\ell]$.

Small microscopic Markov chains

To simplify notation we consider the error propagation along the small Markov chains in a more abstract setting. Fix $1\ll M\in\mathbb{N}$ and $0\le\varepsilon\ll 1$. Let $g_1,\dots,g_M\in[-1,1]$. Consider spin configurations $\xi\in\mathscr{S}_M=\{-1,1\}^M$ with product weights

$$w(\xi)=\mathrm{e}^{\varepsilon\sum_i g_i\xi(i)}. \tag{15.4.26}$$

As before, let $\Lambda^\pm(\xi)=\{i\colon \xi(i)=\pm 1\}$. Define layers of fixed magnetisation $\mathscr{S}_M[K]=\{\xi\in\mathscr{S}_M\colon |\Lambda^+(\xi)|=K\}$. Finally, fix $\delta_0,\delta_1\in(0,1)$ such that $\delta_0<\delta_1$.

Set $K_0=\lfloor\delta_0 M\rfloor$ and $r=\lfloor(\delta_1-\delta_0)M\rfloor$. We consider a Markov chain $\Xi=\{\Xi_0,\Xi_1,\dots,\Xi_r\}$ on $\mathscr{S}_M$ such that $\Xi_\tau\in\mathscr{S}_M[K_0+\tau]=\mathscr{S}^\tau_M$ for $\tau=0,1,\dots,r$. Let μ_τ be the probability measure

$$\mu_\tau(\xi)=\frac{w(\xi)\mathbb{1}_{\{\xi\in\mathscr{S}^\tau_M\}}}{Z_\tau}. \tag{15.4.27}$$

We take $\nu_0=\mu_0$ as the initial distribution of Ξ_0 and, following (15.4.16), define transition rates

$$q_\tau\big(\xi_\tau,\theta^+_j\xi_\tau\big)=\frac{\mathrm{e}^{2\varepsilon g_j}}{\sum_{i\in\Lambda^-(\xi_\tau)}\mathrm{e}^{2\varepsilon g_i}}. \tag{15.4.28}$$

We denote by $\mathbb{P}$ the law of this Markov chain and let ν_τ be the distribution of Ξ_τ (which is concentrated on $\mathscr{S}^\tau_M$), i.e., $\nu_\tau(\xi)=\mathbb{P}(\Xi_\tau=\xi)$. The propagation of errors along paths of our Markov chain is then quantified in terms of $\psi_\tau(\cdot)=\nu_\tau(\cdot)/\mu_\tau(\cdot)$.

Proposition 15.12 *For $\tau=1,\dots,r$ and $\xi\in\mathscr{S}^\tau_M$, set*

$$\mathscr{B}_\tau(\xi)=\sum_{i=1}^{M}\mathrm{e}^{2\varepsilon g_i}\mathbb{1}_{\{i\in\Lambda^-(\xi)\}} \tag{15.4.29}$$

$$\mathscr{A}_\tau = \mathbb{E}_{\mu_\tau}\big[\mathscr{B}_\tau(\cdot)\big] = \sum_{i=1}^{M} \mathrm{e}^{2\varepsilon g_i}\, \mu_\tau\big(i \in \Lambda^-(\cdot)\big).$$

Then there exists a $c = c(\delta_0, \delta_1)$ *such that, for any trajectory* $\underline{\xi} = (\xi_0, \dots, \xi_r)$,

$$\psi_\tau(\xi_\tau) \le \left[\frac{\mathscr{A}_0}{\mathscr{B}_0(\xi_0)}\right]^\tau \mathrm{e}^{c\varepsilon\tau^2/M} \tag{15.4.30}$$

for all $\tau = 0, 1, \dots, r$.

Remark 15.13 The second factor in the bound in (15.4.30) will be seen to be what we want, since it grows much slower than $\phi_{\mathbf{A},\mathbf{B}}(\mathbf{x}_\ell, \mathbf{x}_{\ell+1})$ decays. The first factor involves the ratio of $\mathscr{A}_0$ and $\mathscr{B}_0$, which is more delicate. To control it we require a concentration estimate showing that $\mathscr{A}_0/\mathscr{B}_0(\xi_0) \le 1 + O(1/\sqrt{M})$, which will be done later.

Proof By construction, $\psi_0 = 1$. Let $\xi_{\tau+1} \in \mathscr{S}_M^{\tau+1}$. Since ν_τ satisfies the recursion

$$\nu_{\tau+1}(\xi_{\tau+1}) = \sum_{j\in\Lambda^+(\xi_{\tau+1})} \nu_\tau\big(\theta_j^-\xi_{\tau+1}\big) q_\tau\big(\theta_j^-\xi_{\tau+1}, \xi_{\tau+1}\big), \tag{15.4.31}$$

it follows that ψ_τ satisfies

$$\begin{aligned}\psi_{\tau+1}(\xi_{\tau+1}) &= \sum_{j\in\Lambda^+(\xi_{\tau+1})} \frac{\nu_\tau(\theta_j^-\xi_{\tau+1}) q_\tau(\theta_j^-\xi_{\tau+1}, \xi_{\tau+1})}{\mu_{\tau+1}(\xi_{\tau+1})} \qquad (15.4.32)\\ &= \sum_{j\in\Lambda^+(\xi_{\tau+1})} \frac{\mu_\tau(\theta_j^-\xi_{\tau+1}) q_\tau(\theta_j^-\xi_{\tau+1}, \xi_{\tau+1})}{\mu_{\tau+1}(\xi_{\tau+1})} \psi_\tau\big(\theta_j^-\xi_{\tau+1}\big).\end{aligned}$$

By our choice of transition probabilities in (15.4.28),

$$\frac{\mu_\tau(\theta_j^-\xi_{\tau+1}) q_\tau(\theta_j^-\xi_{\tau+1}, \xi_{\tau+1})}{\mu_{\tau+1}(\xi_{\tau+1})} = \frac{Z_{\tau+1}}{Z_\tau}\left\{\sum_{i\in\Lambda^-(\theta_j^-\xi_{\tau+1})} \mathrm{e}^{2\varepsilon g_i}\right\}^{-1}. \tag{15.4.33}$$

Recalling that $|\Lambda^+(\xi_\tau)| = |\Lambda_\tau^+| = K_0 + \tau$ does not depend on the particular value of ξ_τ, we have

$$\begin{aligned}\frac{Z_{\tau+1}}{Z_\tau} &= \frac{1}{Z_\tau}\sum_{\xi\in\mathscr{S}_M^{\tau+1}} w(\xi) = \frac{1}{Z_\tau}\sum_{\xi\in\mathscr{S}_M^{\tau+1}} \frac{1}{|\Lambda^+(\xi)|}\sum_{j\in\Lambda^+(\xi)} w\big(\theta_j^-\xi\big)\mathrm{e}^{2\varepsilon g_j} \qquad (15.4.34)\\ &= \frac{1}{Z_\tau}\sum_{\xi\in\mathscr{S}_M^\tau} w(\xi)\cdot\frac{1}{|\Lambda_{\tau+1}^+|}\sum_{j\in\Lambda^-(\xi)} \mathrm{e}^{2\varepsilon g_j} = \mu_\tau\left(\frac{1}{|\Lambda^+(\xi_{\tau+1})|}\sum_{j\in\Lambda^-(\cdot)} \mathrm{e}^{2\varepsilon g_j}\right).\end{aligned}$$

We conclude that the right-hand side of (15.4.33) equals

$$\frac{1}{|\Lambda^+(\xi_{\tau+1})|}\frac{\mu_\iota\left(\sum_{\iota\in\Lambda^-(\cdot)}e^{2\varepsilon g_i}\right)}{\sum_{i\in\Lambda^-(\theta_j^-\xi_{\tau+1})}e^{2\varepsilon g_i}}=\frac{1}{|\Lambda^+(\xi_{\tau+1})|}\frac{\mathscr{A}_\tau}{\mathscr{B}_\tau(\theta_j^-\xi_{\tau+1})}. \tag{15.4.35}$$

Consequently,

$$\psi_{\tau+1}(\xi_{\tau+1})=\frac{1}{|\Lambda_+(\xi_{\tau+1})|}\sum_{j\in\Lambda_+(\xi_{\tau+1})}\frac{\mathscr{A}_\tau}{\mathscr{B}_\tau(\theta_j^-\xi_{\tau+1})}\psi_\tau\big(\theta_j^-\xi_{\tau+1}\big). \tag{15.4.36}$$

Iterating the above procedure, we arrive at the following conclusion. Consider the set $\mathscr{D}(\xi_{\tau+1})$ of all paths $\underline{\xi}=(\xi_0,\dots,\xi_\tau,\xi_{\tau+1})$ of positive probability from $\mathscr{S}_M^0$ to $\mathscr{S}_M^{\tau+1}$ to $\xi_{\tau+1}$. The number $D_{\tau+1}=|\mathscr{D}(\xi_{\tau+1})|$ of such paths does not depend on $\xi_{\tau+1}$. Therefore, since $\psi_0=1$, we have

$$\psi_{\tau+1}(\xi_{\tau+1})=\frac{1}{D_{\tau+1}}\sum_{\underline{\xi}\in\mathscr{D}(\xi_{\tau+1})}\prod_{s=0}^{\tau}\frac{\mathscr{A}_s}{\mathscr{B}_s(\xi_s)}. \tag{15.4.37}$$

To conclude the proof, we need the following lemma.

Lemma 15.14

$$\frac{\mathscr{A}_s}{\mathscr{B}_s(\xi_s)}=\left(1+\frac{O(\varepsilon)}{M}\right)\frac{\mathscr{A}_{s-1}}{\mathscr{B}_{s-1}(\xi_{s-1})}, \tag{15.4.38}$$

where $O(\varepsilon)$ is uniform in all parameters.

Once (15.4.38) is verified, we have

$$\psi_\tau(\xi_\tau)\le e^{O(\varepsilon)\tau^2/M}\max_{\xi_0\sim\xi_\tau}\left[\frac{\mathscr{A}_0}{\mathscr{B}_0(\xi_0)}\right]^\tau, \tag{15.4.39}$$

where for $\xi_0\in\mathscr{S}_M^0$ the relation $\xi_0\sim\xi_\tau$ means that there is a path of positive probability from ξ_0 to ξ_τ. But all such ξ_0 differ in at most 2τ coordinates. It is straightforward to see that if $\xi_0\sim\xi_\tau$ and $\xi_0'\sim\xi_\tau$, then

$$\frac{\mathscr{B}_0(\xi_0)}{\mathscr{B}_0(\xi_0')}\le e^{O(\varepsilon)\tau/M}, \tag{15.4.40}$$

and (15.4.30) follows. □

It remains to prove Lemma 15.14.

Proof of Lemma 15.14 Let $\xi\in\mathscr{S}_M^s$ and $\xi'=\theta_j^-\xi\in\mathscr{S}_M^{s-1}$. Note that

$$\mathscr{B}_{s-1}\big(\xi'\big)-\mathscr{B}_s(\xi)=e^{2\varepsilon g_j}=1+O(\varepsilon). \tag{15.4.41}$$

Similarly,

$$\begin{aligned}\mathscr{A}_{s-1}-\mathscr{A}_s &= \sum_{i=1}^{M} \mathrm{e}^{2\varepsilon g_i}\left\{\mu_{s-1}\left(i\in\Lambda^-\right)-\mu_s\left(i\in\Lambda^-\right)\right\} \\ &= 1+\sum_{i=1}^{M}\left(\mathrm{e}^{2\varepsilon g_i}-1\right)\left\{\mu_{s-1}\left(i\in\Lambda^-\right)-\mu_s\left(i\in\Lambda^-\right)\right\}.\end{aligned} \tag{15.4.42}$$

By standard local limit results for independent Bernoulli variables,

$$\mu_{s-1}\left(i\in\Lambda^-\right)-\mu_s\left(i\in\Lambda^-\right)=O\left(\frac{1}{M}\right) \tag{15.4.43}$$

uniformly in $s=1,\dots,r-1$ and $i=1,\dots,M$. Hence $\mathscr{A}_{s-1}-\mathscr{A}_s=1+O(\varepsilon)$.

Finally, both $\mathscr{A}_{s-1}$ and $\mathscr{B}_{s-1}(\xi')$ are uniformly $O(M)$, whereas

$$\mathscr{A}_{s-1}-\mathscr{B}_{s-1}\left(\xi'\right)=\sum_{i=1}^{M}\left(\mathrm{e}^{2\varepsilon g_i}-1\right)\left\{\mu_{s-1}\left(i\in\Lambda^-\right)-\mathbb{1}_{\{i\in\Lambda^-(\xi')\}}\right\}=O(\varepsilon)M. \tag{15.4.44}$$

Hence

$$\frac{\mathscr{A}_s}{\mathscr{B}_s(\xi)}=\frac{\mathscr{A}_{s-1}-1+O(\varepsilon)}{\mathscr{B}_{s-1}(\xi')-1+O(\varepsilon)}=\frac{\mathscr{A}_{s-1}}{\mathscr{B}_{s-1}(\xi')}\left[1+\frac{O(\varepsilon)}{M}\right], \tag{15.4.45}$$

which is (15.4.38). □

Lemma 15.15 *Assume that $\tau/M\le C$. Then*

$$\mathbb{E}_{\mu_0}\left[\frac{\mathscr{A}_0}{\mathscr{B}_0(\xi_0)}\right]^{\tau}\le\exp\left[\max\left(O(\varepsilon)\frac{\tau}{\sqrt{M}},O\left(\varepsilon^2\right)\frac{\tau^2}{M}\right)\right]. \tag{15.4.46}$$

Proof First note that

$$\left(\frac{\mathscr{A}_0}{\mathscr{B}_0(\xi_0)}\right)^{\tau}=\left(1+\frac{\mathscr{B}_0(\xi_0)-\mathscr{A}_0}{\mathscr{A}_0}\right)^{-\tau}. \tag{15.4.47}$$

Now define the random variable

$$Y=\frac{\mathscr{B}_0(\xi_0)-\mathscr{A}_0}{\mathscr{A}_0}. \tag{15.4.48}$$

Due to the fact that $|\Lambda^-(\xi)|$ only depends on the magnetisation of ξ, we can rewrite Y as

$$Y=\frac{1}{\mathscr{A}_0}\sum_{i=1}^{M}\left(\mathrm{e}^{2\varepsilon g_i}-1\right)\left[\mathbb{1}_{i\in\Lambda^-(\xi)}-\mu_\tau\left(i\in\Lambda^-\right)\right]. \tag{15.4.49}$$

Since $\mathscr{A}_0 = |\lambda \Lambda^{-1}(\xi(0))|[1 + O(\varepsilon)]$, it follows that

$$|Y| \leq 5\varepsilon, \tag{15.4.50}$$

where 5 is an arbitrary choice for a number larger than 4. This ensures that when we compute $\mathbb{E}_{\mu_0}[(1+Y)^{-\tau}]$, we stay away from the singularity at zero. Note that Y is a centred random variable and is a sum of bounded random variables that are almost independent (their dependence arises only from conditioning on the value of their sum). Moreover, the variance of the summands is of order ε^2/M^2. In this situation, the following lemma holds.

Lemma 15.16 *There exist finite positive constants c, C such that, for any $r > \varepsilon/\sqrt{M}$,*

$$\mu_0\big(|Y| > r\big) \leq C\mathrm{e}^{-cMr^2/\varepsilon^2}. \tag{15.4.51}$$

Next we use that, for $|x| \leq \frac{1}{10}$ say, there is a finite constant $d > 0$ such that $\ln(1+x) \geq x - dx^2$. Hence

$$\begin{aligned}
\mathbb{E}_{\mu_0}[1+Y]^{-\tau} &\leq \mathbb{E}_{\mu_0}\big[\mathrm{e}^{-\tau Y + d\tau Y^2}\big] \qquad (15.4.52)\\
&\leq \mathrm{e}^{\varepsilon\tau/\sqrt{M} + d\varepsilon^2\tau/M} + \tau C \int \mathrm{e}^{-\tau r + d\tau r^2} \mathrm{e}^{-cMr^2/\varepsilon^2} dr\\
&= \mathrm{e}^{\varepsilon\tau/\sqrt{M} + d\varepsilon^2\tau/M} + \frac{\sqrt{2\pi}\tau}{\sqrt{2(cM/\varepsilon^2 - d\tau)}} \mathrm{e}^{\tau^2/(cM/\varepsilon^2 - d\tau)}.
\end{aligned}$$

Since we have assumed that $\tau \leq CM$, and ε is small, the right-hand side of (15.4.52) is as claimed in (15.4.46). □

Back to the large microscopic Markov chain

Going back to (15.4.25), we infer that the corrector of the *large* Markov chain Σ satisfies the following upper bound. Let $\underline{\sigma} = (\sigma_0, \sigma_1, \dots)$ be a trajectory of Σ (as sampled from $\mathbb{P}_{\underline{\mathbf{x}}}$). Then, for every $\ell = 0, 1, \dots, \ell_B - 1$,

$$\Psi_\ell(\sigma_\ell) \leq \exp\left\{c\varepsilon \sum_{j=1}^{n} \frac{\tau_j[\ell]^2}{M_j}\right\} \prod_{j=1}^{n} \left[\frac{\mathscr{A}_0^{(j)}}{\mathscr{B}_0^{(j)}(\sigma_0^{(j)})}\right]^{\tau_j[\ell]}, \tag{15.4.53}$$

where $M_j = |\Lambda_j| = \rho_j N$, and $\mathscr{A}_0^{(j)}, \mathscr{B}_0^{(j)}$ are defined as in (15.4.29) with respect to the corresponding small microscopic Markov chains. We need to check that when this bound is inserted into the left-hand side of (15.4.23), we recover the right-hand side as upper bound.

By the construction of the mesoscopic Markov chain $\mathbb{P}_N^{\mathfrak{f}_{\mathbf{A},\mathbf{B}}}$, and in view of (15.2.8) and (14.5.9), the step frequencies $\tau_j[\ell]/\ell$ are on average proportional to ρ_j. Therefore there exists a constant C_1 such that, up to exponentially negligible $\mathbb{P}_N^{\mathfrak{f}_{\mathbf{A},\mathbf{B}}}$-

probabilities,

$$\max_{1\le j\le n} \frac{\tau_j[\ell_B]}{M_j} \le C_1. \tag{15.4.54}$$

Our mesoscopic trajectories are constructed such that the assumptions of Sect. 15.4.2 hold for each of them. Thus Lemma 15.4.46 together with Proposition 15.12 imply that

$$\begin{aligned} \Psi_\ell(\sigma_\ell) &\le \exp\left(O(\varepsilon) \sum_{j=1}^{n} \max\left(\frac{\tau_j[\ell]}{\sqrt{M_j}}, \frac{\tau_j[\ell]^2}{M_j} \right)\right) \\ &\le \exp\left(\max\left(O(\sqrt{\varepsilon})\frac{\ell}{\sqrt{N}}, O(\varepsilon)\frac{\ell^2}{N}\right)\right), \end{aligned} \tag{15.4.55}$$

uniformly in $\ell = 0, \dots, \ell_B$. Note that to obtain the second line we use the Cauchy-Schwarz inequality and the fact that $\sum_{j=1}^n M_j = N$ and $\varepsilon = O(1/n)$. Inserting this into the bound (15.4.9), we have now proved that

$$\begin{aligned} \operatorname{cap}(A, B) \ge \mathbb{E}_N^{\mathfrak{f}_{\mathbf{A},\mathbf{B}}} \Bigg[\mathbb{1}_{\mathscr{T}_{\mathbf{A},\mathbf{B}}} \\ \times \left(\mathbb{E}^{\underline{\mathbf{x}}} \left[\sum_{\ell=-\ell_A}^{\ell_B-1} \frac{\mathfrak{f}_{\mathbf{A},\mathbf{B}}(\mathbf{x}_\ell, \mathbf{x}_{\ell+1}) \exp(\max(O(\sqrt{\varepsilon})\frac{|\ell|}{\sqrt{N}}, O(\varepsilon)\frac{\ell^2}{N}))}{(1+d(\mathbf{x}_\ell))\mathscr{Q}_{\beta,N}(\mathbf{x}_\ell) r_N(\mathbf{x}_\ell, \mathbf{x}_{\ell+1})} \right]\right)^{-1} \Bigg]. \end{aligned} \tag{15.4.56}$$

Let us now set

$$\phi_{\mathbf{A},\mathbf{B}}(\mathbf{x}_\ell, \mathbf{x}_{\ell+1}) = \frac{\mathfrak{f}_{\mathbf{A},\mathbf{B}}(\mathbf{x}_\ell, \mathbf{x}_{\ell+1})}{(1+d(\mathbf{x}_\ell))\mathscr{Q}_{\beta,N}(\mathbf{x}_\ell) r_N(\mathbf{x}_\ell, \mathbf{x}_{\ell+1})}. \tag{15.4.57}$$

Just as in the proof of the lower bound in Sect. 10.3, from the fact that the free energy is quadratic with a negative eigenvalue in the direction of our paths, we obtain that there exists a $C > 0$ such that, for all $\underline{\mathbf{x}}$ under consideration and for all $\ell = -\ell_A, \dots, \ell_B - 1$,

$$\frac{\mathfrak{f}_{\mathbf{A},\mathbf{B}}(\mathbf{x}_\ell, \mathbf{x}_{\ell+1})}{(1+d(\mathbf{x}_\ell))\mathscr{Q}_{\beta,N}(\mathbf{x}_\ell) r_N(\mathbf{x}_\ell, \mathbf{x}_{\ell+1})} \le e^{-C\ell^2/N} \frac{\mathfrak{f}_{\mathbf{A},\mathbf{B}}(\mathbf{x}_0, \mathbf{x}_1)}{(1+d(\mathbf{x}_0))\mathscr{Q}_{\beta,N}(\mathbf{x}_0) r_N(\mathbf{x}_0, \mathbf{x}_1)}. \tag{15.4.58}$$

From this fact it is elementary to deduce that

$$\begin{aligned} &\sum_{\ell=-\ell_A}^{\ell_B-1} \frac{\mathfrak{f}_{\mathbf{A},\mathbf{B}}(\mathbf{x}_\ell, \mathbf{x}_{\ell+1})}{(1+d(\mathbf{x}_\ell))\mathscr{Q}_{\beta,N}(\mathbf{x}_\ell) r_N(\mathbf{x}_\ell, \mathbf{x}_{\ell+1})} \left| 1 - \exp\left(\max\left(O(\sqrt{\varepsilon})\frac{|\ell|}{\sqrt{N}}, O(\varepsilon)\frac{\ell^2}{N}\right)\right)\right| \\ &\le O(\sqrt{\varepsilon}) \sum_{\ell=-\ell_A}^{\ell_B-1} \frac{\mathfrak{f}_{\mathbf{A},\mathbf{B}}(\mathbf{x}_\ell, \mathbf{x}_{\ell+1})}{(1+d(\mathbf{x}_\ell))\mathscr{Q}_{\beta,N}(\mathbf{x}_\ell) r_N(\mathbf{x}_\ell, \mathbf{x}_{\ell+1})}. \end{aligned} \tag{15.4.59}$$

Thus, we have established that, up to an error of order $\sqrt{\varepsilon}$, the lower bound on the capacity for the coarse-grained model is also a lower bound for the full model. This leads to the inequality in (15.4.24) which, together with the upper bound given in (15.3.10), concludes the proof of Theorem 15.3.

15.5 Estimates on mean hitting times

In this section we conclude the proof of Theorem 15.1. The capacity in the denominator in the right-hand side of (7.1.41) is controlled by Theorem 15.3. It therefore remains to control the equilibrium potential $h_{A,B}(\sigma)$. We are in a situation where the renewal inequality $h_{A,B}(\sigma) \leq \mathrm{cap}(A,\sigma)/\mathrm{cap}(B,\sigma)$ cannot be used because capacities of single configurations are too small. We will need another method to cope with this problem, explained in Sects. 15.5.1–15.5.2.

15.5.1 Mean hitting time and equilibrium potential

Let us start by considering a local minimum m_0^* of the one-dimensional function $F_{\beta,N}$, and denote by M the set of minima m such that $F_{\beta,N}(m) < F_{\beta,N}(m_0^*)$. We consider the disjoint subsets $A = S_N[m_0^*]$ and $B = S_N[M]$, and write (7.1.41) as

$$\sum_{\sigma\in A} \nu_{A,B}(\sigma)\mathbb{E}_\sigma \tau_B = \frac{1}{\mathrm{cap}(A,B)} \sum_{m\in[-1,1]} \sum_{\sigma\in S_N[m]} \mu_{\beta,N}(\sigma) h_{A,B}(\sigma). \tag{15.5.1}$$

We expect the right-hand side of (15.5.1) to be of order $\mathscr{Q}_{\beta,N}(m_0^*)$, so that all terms in the sum over m with $\mathscr{Q}_{\beta,N}(m)$ much smaller than $\mathscr{Q}_{\beta,N}(m_0^*)$ can be ignored. More precisely, we choose $\delta > 0$ in such a way that, for all N large enough, there is no critical point z of $F_{\beta,N}$ with $F_{\beta,N}(z) \in [F_{\beta,N}(m_0^*), F_{\beta,N}(m_0^*) + \delta]$, and we define

$$\mathscr{U}_\delta = \left\{m \in [-1,1]\colon\ F_{\beta,N}(m) \leq F_{\beta,N}\left(m_0^*\right) + \delta\right\}. \tag{15.5.2}$$

Lemma 15.17 *With $\mathscr{U}_\delta^c$ the complement of $\mathscr{U}_\delta$,*

$$\sum_{m\in\mathscr{U}_\delta^c} \sum_{\sigma\in S_N[m]} \mu_{\beta,N}(\sigma) h_{A,B}(\sigma) \leq N\mathrm{e}^{-\beta N\delta} \mathscr{Q}_{\beta,N}\left(m_0^*\right). \tag{15.5.3}$$

The main problem is to control the equilibrium potential $h_{A,B}(\sigma)$ for configurations $\sigma \in S_N[\mathscr{U}_\delta]$. To do so, first note that

$$\mathscr{U}_\delta = \mathscr{U}_\delta\left(m_0^*\right) \bigcup_{m\in M} \mathscr{U}_\delta(m), \tag{15.5.4}$$

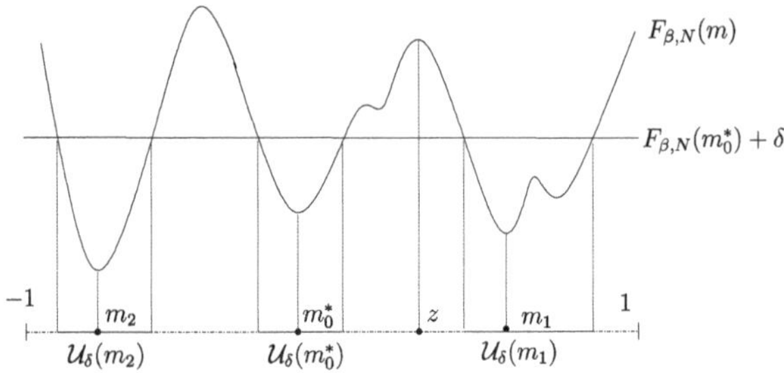

Fig. 15.2 Decomposition of $[-1,1]$: $\mathscr{U}_\delta^c$ is represented by *dotted lines*, $\mathscr{U}_\delta = \mathscr{U}_\delta(m_0^*)\bigcup_{m\in M} \mathscr{U}_\delta(m)$ by *continuous lines*

where $\mathscr{U}_\delta(m)$ is the connected component of $\mathscr{U}_\delta$ containing m (see Fig. 15.2). Note that it may happen that $\mathscr{U}_\delta(m)=\mathscr{U}_\delta(m')$ for two different minima $m, m' \in M$.

With this notation we have the following lemma.

Lemma 15.18 *There exists a constant $c>0$ such that*:

(i)

$$\sum_{\sigma\in S_N[\mathscr{U}_\delta(m)]} \mu_{\beta,N}(\sigma)h_{A,B}(\sigma) \le \mathrm{e}^{-\beta Nc}\mathscr{Q}_{\beta,N}\big(m_0^*\big), \quad m\in M, \tag{15.5.5}$$

(ii)

$$\sum_{\sigma\in S_N[\mathscr{U}_\delta(m_0^*)]} \mu_{\beta,N}(\sigma)\big[1-h_{A,B}(\sigma)\big] \le \mathrm{e}^{-\beta Nc}\mathscr{Q}_{\beta,N}\big(m_0^*\big). \tag{15.5.6}$$

The treatment of (i) and (ii) is completely similar, as both rely on a rough estimate of the probability of leaving the starting valley before visiting its minimum, which will be discussed below.

Assuming Lemma 15.18, we can readily conclude the proof of Theorem 15.1. Indeed, using (15.5.5) together with (15.5.3), we obtain the upper bound

$$\begin{aligned}\sum_{\sigma\in S_N} \mu_{\beta,N}(\sigma)h_{A,B}(\sigma) &\le \sum_{m\in\mathscr{U}_\delta(m_0^*)} \mathscr{Q}_{\beta,N}(m) + O\big(\mathscr{Q}_{\beta,N}\big(m_0^*\big)\mathrm{e}^{-\beta Nc}\big) \\ &= \mathscr{Q}_{\beta,N}\big(m_0^*\big)\sqrt{\frac{\pi N}{2\beta a(m_0^*)}}\big[1+o(1)\big], \end{aligned} \tag{15.5.7}$$

where $a(m_0^*)$ is given in (14.2.17). On the other hand, using (15.5.6) we get the corresponding lower bound

$$\sum_{\sigma\in S_N} \mu_{\beta,N}(\sigma)h_{A,B}(\sigma) \ge \sum_{m\in\mathscr{U}_\delta(m_0^*)} \sum_{\sigma\in S_N[m]} \mu_{\beta,N}(\sigma)\big[1-\big(1-h_{A,B}(\sigma)\big)\big]$$

$$\geq \sum_{m\in\mathscr{U}_\delta(m_0^*)} \mathscr{Q}_{\beta,N}(m) - O\big(\mathscr{Q}_{\beta,N}\big(m_0^*\big)\,\mathrm{e}^{-\beta Nc}\big)$$

$$= \mathscr{Q}_{\beta,N}\big(m_0^*\big)\sqrt{\frac{\pi N}{2\beta a(m_0^*)}}\,\big[1+o(1)\big]. \tag{15.5.8}$$

From (15.2.11) for $\mathscr{Q}_{\beta,N}(m_0^*)$ and (15.1.4) for cap(A,B), we finally obtain

$$\begin{aligned}\mathbb{E}_{\nu_{A,B}}[\tau_B] &= \sum_{\sigma\in S_N}\frac{\mu_{\beta,N}(\sigma)h_{A,B}(\sigma)}{\mathrm{cap}(A,B)}\\ &= \exp\big(\beta N\big[F_{\beta,N}\big(z^*\big)-F_{\beta,N}\big(m_0^*\big)\big]\big)\\ &\quad\times\frac{2\pi N}{\beta|\hat{\gamma}_1|}\sqrt{\frac{\beta\mathbb{E}_h(1-\tanh^2(\beta(z^*+h)))-1}{1-\beta\mathbb{E}_h(1-\tanh^2(\beta(m_0^*+h)))}}\,\big[1+o(1)\big],\end{aligned} \tag{15.5.9}$$

which proves Theorem 15.1.

15.5.2 Upper bounds on harmonic functions

We next prove Lemma 15.18, giving a detailed proof only for (i) because the proof of (ii) is completely analogous. This requires us to get an estimate on the minimiser of the Dirichlet form, the harmonic function $h_{A,B}(\sigma)$.

First note that, since $h_{A,B}(\sigma)=\mathbb{P}_\sigma(\tau_A<\tau_B)$ for all $\sigma\notin A\cup B$, the only non-zero contributions to the sum in (i) come from those sets $\mathscr{U}_\delta(m)$ (at most two) whose corresponding m is such that there are no minima of M between m_0^* and m. By symmetry, we can just as well analyse one of these two sets, denoted by $\mathscr{U}_\delta(m^*)$, assuming for definiteness that $m_0^*<m^*$. Next note that, since $h_{A,B}(\sigma)=0$ for all σ such that $m^*\leq m(\sigma)$, the problem can be reduced further to the set

$$\mathscr{U}_\delta^- - \mathscr{U}_\delta\big(m^*\big)\cap\big\{m\in[0,1]\colon\ m<m^*\big\}. \tag{15.5.10}$$

Define the mesoscopic counterpart of $\mathscr{U}_\delta^-$, namely, for fixed $m^*\in M$ and $n\in\mathbb{N}$, let $\mathbf{m}^*\in\Gamma_N^n$ be the minimum of $F_{\beta,N}(\mathbf{x})$ corresponding to m^*, and define

$$\mathbf{U}_\delta=\mathbf{U}_\delta\big(\mathbf{m}^*\big)=\big\{\mathbf{x}\in\Gamma_N^n: m(\mathbf{x})\in\mathscr{U}_\delta^-\big\}. \tag{15.5.11}$$

We write the boundary of $\mathbf{U}_\delta$ as $\partial\mathbf{U}_\delta=\partial_A\mathbf{U}_\delta\sqcup\partial_B\mathbf{U}_\delta$, where $\partial_B\mathbf{U}_\delta=\partial\mathbf{U}_\delta\cap\mathbf{B}$, and observe that, for all $\sigma\in S_N[\mathbf{U}_\delta]$,

$$h_{A,B}(\sigma)=\mathbb{P}_\sigma[\tau_A<\tau_B]\leq\mathbb{P}_\sigma[\tau_{S[\partial_A\mathbf{U}_\delta]}<\tau_{S[\partial_B\mathbf{U}_\delta]}]. \tag{15.5.12}$$

Let $\max_{1\leq\ell\leq n}\rho_\ell\ll\theta(\varepsilon)\ll 1$, and for $\theta=\theta(\varepsilon)$ define

$$\mathbf{G}_\theta=\left\{\mathbf{m}\in\mathbf{U}_\delta\colon\ \sum_{\ell=1}^n\frac{(\mathbf{m}_\ell-\mathbf{m}_\ell^*)^2}{\rho_\ell}\leq\frac{\varepsilon^2}{\theta}\right\}. \tag{15.5.13}$$

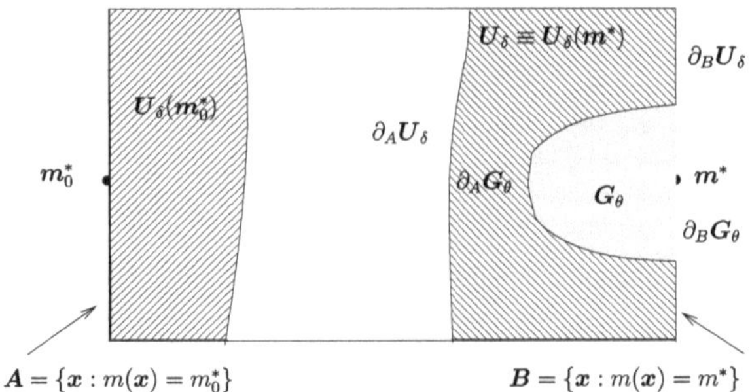

Fig. 15.3 Neighbourhoods of $\mathbf{m}_0^*$ and $\mathbf{m}^*$ in the space Γ_N^n, where $\mathbf{U}_\delta(\mathbf{m}_0^*)$ denotes the mesoscopic counterpart of $\mathscr{U}(m_0^*)$

As before, we denote by $\partial\mathbf{G}_\theta$ the boundary of $\mathbf{G}_\theta$, and write $\partial\mathbf{G}_\theta = \partial_A\mathbf{G}_\theta \cup \partial_B\mathbf{G}_\theta$, where $\partial_B\mathbf{G}_\theta = \partial\mathbf{G}_\theta \cap \mathbf{B}$ (see Fig. 15.3).

The strategy to control the equilibrium potential $\mathbb{P}_\sigma(\tau_A < \tau_B)$ consists in estimating the probabilities $\mathbb{P}_\sigma[\tau_A < \tau_{S_N[\partial_A\mathbf{G}_\theta]\cup B}]$ for $\sigma \in \mathscr{S}[\mathbf{U}_\delta \setminus \mathbf{G}_\theta]$ and $\mathbb{P}_\sigma[\tau_{S_N[\partial_A\mathbf{G}_\theta]} < \tau_B]$ for $\sigma \in \mathbf{G}_\theta$, in order to apply a renewal argument and draw from these estimates a bound on the probability of the original event. Proceeding along this line, we state the following.

Proposition 15.19 *For any $\alpha \in (0,1)$ there exists an $n_0 \in \mathbb{N}$ such that the inequality*

$$\mathbb{P}_\sigma(\tau_A < \tau_{S_N[\partial_A\mathbf{G}_\theta]\cup B}) \leq \mathrm{e}^{-(1-\alpha)\beta N[F_{\beta,N}(m_0^*)+\delta-F_{\beta,N}(\mathbf{m}(\sigma))]} \tag{15.5.14}$$

holds for all $\sigma \in S_N[\mathbf{U}_\delta \setminus \mathbf{G}_\theta]$ and $n \geq n_0$, for all N sufficiently large.

Proof of Proposition 15.19: Super-harmonic barrier functions

Throughout the next computations c, c' and c'' will denote positive constants that are independent of n but may depend on β and on the distribution of h. The value of c and c' may change from line to line.

We first observe that, for all $\sigma \in S_N[\mathbf{U}_\delta \setminus \mathbf{G}_\theta]$,

$$\mathbb{P}_\sigma[\tau_A < \tau_{S_N[\partial_A\mathbf{G}_\theta]\cup B}] \leq \mathbb{P}_\sigma[\tau_{S_N[\partial_A\mathbf{U}_\delta]} < \tau_{S_N[\partial_A\mathbf{G}_\theta]\cup B}]. \tag{15.5.15}$$

The probability in the right-hand side of (15.5.15) is the main object of investigation. The idea behind the proof of bound (15.5.14) is simple. Suppose that ψ is a bounded super-harmonic function defined on $S_N[\mathbf{U}_\delta \setminus \mathbf{G}_\theta]$, with $L = L_N$ the generator of the Markov process defined in Sect. 14.3, i.e.,

$$(L\psi)(\sigma) \leq 0 \quad \forall \sigma \in S_N[\mathbf{U}_\delta \setminus \mathbf{G}_\theta]. \tag{15.5.16}$$

Then $\psi(\sigma_t)$ is a supermartingale, and $T = \tau_{S_N[\partial_A\mathbf{U}_\delta]} \wedge \tau_{S_N[\partial_A\mathbf{G}_\theta]\cup B}$ is an integrable stopping time, so that, by Doob's optional stopping theorem,

$$\mathbb{E}_\sigma \psi(\sigma_T) \leq \psi(\sigma) \quad \forall \sigma \in S_N[\mathbf{U}_\delta \setminus \mathbf{G}_\theta]. \tag{15.5.17}$$

On the other hand,

$$\mathbb{E}_\sigma \psi(\sigma_T) \geq \min_{\sigma' \in S_N[\partial_A \mathbf{U}_\delta]} \psi(\sigma') \mathbb{P}_\sigma(\tau_{S_N[\partial_A \mathbf{U}_\delta]} < \tau_{S_N[\partial_A \mathbf{G}_\theta] \cup B}), \tag{15.5.18}$$

and hence

$$\mathbb{P}_\sigma(\tau_{S_N[\partial_A \mathbf{U}_\delta]} < \tau_{S_N[\partial_A \mathbf{G}_\theta] \cup B}) \leq \max_{\sigma' \in S_N[\partial_A \mathbf{U}_\delta]} \frac{\psi(\sigma)}{\psi(\sigma')}. \tag{15.5.19}$$

The problem is to find a super-harmonic function in order to get a suitable bound in (15.5.19).

Proposition 15.20 *For any $\alpha \in (0, 1)$ there exists $n_0 \in \mathbb{N}$ such that the function $\psi(\sigma) = \phi(\mathbf{m}(\sigma))$ with $\phi \colon \mathbb{R}^n \mapsto \mathbb{R}$ defined by*

$$\phi(\mathbf{x}) = \mathrm{e}^{(1-\alpha)\beta N F_{\beta,N}(\mathbf{x})} \tag{15.5.20}$$

is super-harmonic in $S_N[\mathbf{U}_\delta \setminus \mathbf{G}_\theta]$ for all $n \geq n_0$, for N sufficiently large.

The proof of Proposition 15.20 will involve computations with differences of the function $F_{\beta,N}$. We collect some necessary properties that will be needed along the way. First we need some control on the second derivative of this function. A simple computation shows that

$$\frac{\partial^2 F_{\beta,N}(\mathbf{x})}{\partial \mathbf{x}_\ell^2} = \frac{2}{N}\left(-1 + \frac{1}{\beta \rho_\ell} I''_{N,\ell}(\mathbf{x}_\ell/\rho_\ell)\right). \tag{15.5.21}$$

Thus, we need to estimate the function $I_{N,\ell}$.

Lemma 15.21 *For any $y \in (-1, 1)$,*

$$\tanh^{-1}(y) - \beta\varepsilon \leq I'_{N,\ell}(y) \leq \tanh^{-1}(y) + \beta\varepsilon. \tag{15.5.22}$$

In particular, $\lim_{y \to \pm 1} I'_{N,\ell}(y) = \pm\infty$.

Proof Recall that $I'_{N,\ell}(y) = U'^{-1}_{N,\ell}(y)$. Set $I'_{N,\ell}(y) = t$. Then

$$y = \frac{1}{|\Lambda_\ell|} \sum_{i \in \Lambda_\ell} \tanh(t + \beta \tilde{h}_i), \tag{15.5.23}$$

and hence

$$\tanh(t - \beta\varepsilon) \leq y \leq \tanh(t + \beta\varepsilon), \tag{15.5.24}$$

or, equivalently, (15.5.22). □

Lemma 15.22 *For any* $y \in (-1, 1)$,

$$0 \le I''_{N,\ell}(y) \le \frac{1}{1-[|y|+\varepsilon\beta(1-y^2)]^2}. \tag{15.5.25}$$

In particular, for all $y \in [-1+\nu, 1-\nu]$ *with* $\nu \in (0, 1/2)$,

$$0 \le I''_{N,\ell}(y) \le \frac{1}{2\nu+\nu^2+O(\varepsilon)} \le c, \tag{15.5.26}$$

and, for all $y \in (-1, -1+\nu] \cup [1-\nu, 1)$,

$$0 \le I''_{N,\ell}(y) \le \frac{1}{1-|y|}. \tag{15.5.27}$$

Proof We only consider the case $y \ge 0$, the case $y < 0$ being completely analogous. Using the relation $I''_{N,\ell}(x) = (U''_{N,\ell}(I'_{N,\ell}(x)))^{-1}$, setting $t_\ell = I'_{N,\ell}(y)\,\text{arctanh}(y)$, and using Lemma 15.21, we obtain

$$\begin{aligned} I''_{N,\ell}(y) &= \frac{1}{\frac{1}{|\Lambda_\ell(\mathbf{x})|}\sum_{i\in\Lambda_\ell(\mathbf{x})}[1-\tanh^2(\beta\tilde{h}_i+t_\ell)]} \\ &\le \frac{1}{1-\tanh^2(\varepsilon\beta+t_\ell)} \\ &\le \frac{1}{1-\tanh^2(\tanh^{-1}(y)+2\varepsilon\beta)} \\ &\le \frac{1}{1-[y+2\varepsilon\beta\tanh'(\tanh^{-1}(y))]^2} \\ &= \frac{1}{1-[y+2\varepsilon\beta(1-y^2)]^2}, \end{aligned} \tag{15.5.28}$$

where we use that tanh is monotone increasing. The remainder of the proof is elementary algebra. □

Let us define, for all $\mathbf{m}$ such that $\mathbf{x}_\ell/\rho_\ell \in [-1, 1-2N^{-1}]$,

$$g_\ell(\mathbf{x}) = \frac{N}{2}\big(F_{N,\beta}(\mathbf{x}+\mathbf{e}_\ell) - F_{N,\beta}(\mathbf{x})\big). \tag{15.5.29}$$

Lemma 15.22 has the following corollary.

Corollary 15.23

(i) *If* $\mathbf{x}_\ell/\rho_\ell \in [-1+\nu, 1-\nu]$ *with* $\nu > 0$, *then*

$$g_\ell(\mathbf{x}) = -x - \bar{h}_\ell + \frac{1}{\beta}I'_{N,\ell}(\mathbf{x}_\ell/\rho_\ell) + O(1/N). \tag{15.5.30}$$

(ii) *If* $\mathbf{x}_\ell/\rho_\ell \in [-1,-1+\nu]\cup[1-\nu,1-2N^{-1}]$, *then*

$$g_\ell(\mathbf{x}) = -x - \bar{h}_\ell + \frac{1}{\beta} I'_{N,\ell}(\mathbf{x}_\ell/\rho_\ell) + O(1), \tag{15.5.31}$$

where $O(1)$ is independent of N,n and ν.

(iii) *If* $\mathbf{x}_\ell/\rho_\ell \in [-1+\nu,1-\nu]$ *with* $\nu>0$, *then there exists a* $c<\infty$ *independent of N such that*

$$\left|g_\ell(\mathbf{x}) - g_\ell(\mathbf{x}-\mathbf{e}_\ell)\right| \le \frac{c}{N}. \tag{15.5.32}$$

(iv) *If* $\mathbf{x}_\ell/\rho_\ell \in [-1,-1+\nu]\cup[1-\nu,1-2N^{-1}]$, *then*

$$\left|g_\ell(\mathbf{x}) - g_\ell(\mathbf{x}-\mathbf{e}_\ell)\right| \le C, \tag{15.5.33}$$

where C is a constant independent of N,n and ν.

The proof of this corollary is elementary and will not be detailed. The usefulness of (ii) results from the fact that $|I'_{N,\ell}|$ is large on the relevant domain. More precisely, we have the following lemma.

Lemma 15.24 *There exists a $\nu>0$ independent of N and n such that if $\mathbf{x}_\ell/\rho_\ell > 1-\nu$, then $g_\ell(\mathbf{x})$ is strictly increasing in $\mathbf{x}_\ell$ and tends to ∞ as $\mathbf{x}_\ell/\rho_\ell \uparrow 1$. Similarly, if $\mathbf{x}_\ell/\rho_\ell < -1+\nu$, then $g_\ell(\mathbf{x})$ is strictly decreasing in $\mathbf{x}_\ell$ and tends to $-\infty$ as $\mathbf{x}_\ell/\rho_\ell \downarrow -1$.*

Proof Combine Corollary 15.23(ii) with Lemma 15.21 and note that $\bar{h}_\ell$ is bounded by hypothesis. □

The next step towards the proof of Proposition 15.20 is the following lemma.

Lemma 15.25 *Let $\mathbf{m}\in\mathbf{U}_\delta\setminus\mathbf{G}_\theta$ and put $S(\mathbf{m}) = \{1\le\ell\le n\colon \mathbf{m}_\ell/\rho_\ell \ne 1\}$. Then there exists a constant $c=c(\beta,h)>0$ independent of N and n such that the following holds. If*

$$\sum_{\ell\notin S(\mathbf{m})} \rho_\ell \le \frac{\varepsilon^2}{8\theta}, \tag{15.5.34}$$

then

$$\sum_{\ell\in S(\mathbf{m})} \rho_\ell \big(g_\ell(\mathbf{m})\big)^2 \ge c\frac{\varepsilon^2}{\theta}. \tag{15.5.35}$$

Proof From the relation $I'_{N,\ell}(x) = U'^{-1}_{N,\ell}(x)$ we get that, for all $\ell\in S(\mathbf{m})$,

$$\mathbf{m}_\ell = \frac{1}{N}\sum_{i\in\Lambda_\ell} \tanh\big[\beta\big(g_\ell(\mathbf{m})\big[1+o(1)\big]+m+h_i\big)\big], \tag{15.5.36}$$

where $o(1)$ tends to zero as $N\to\infty$.

We need to be concerned about small $g_\ell(\mathbf{m})$. Subtracting $\frac{1}{N}\sum_{i\in\Lambda_\ell}\tanh(\beta(m+h_i))$ on both sides of (15.5.36) and expanding the right-hand side to first order in $g_\ell(\mathbf{m})$, and afterwards summing over $\ell\in S(\mathbf{m})$, we obtain

$$\left| m-\frac{1}{N}\sum_{i=1}^{N}\tanh\big(\beta(m+h_i)\big)-\sum_{\ell\notin S(\mathbf{m})}\left(\mathbf{m}_\ell-\frac{1}{N}\sum_{i\in\Lambda_\ell}\tanh\big(\beta(m+h_i)\big)\right)\right| \leq c\sum_{\ell\in S(\mathbf{m})}\rho_\ell\big|g_\ell(\mathbf{m})\big|\leq c\left(\sum_{\ell\in S(\mathbf{m})}\rho_\ell g_\ell^2(\mathbf{m})\right)^{1/2}. \tag{15.5.37}$$

Note that the function $m\mapsto m-\frac{1}{N}\sum_{i=1}^{N}\tanh(\beta(m+h_i))$ has, by (14.2.18), a non-zero derivative at m^*. Moreover, by construction, m^* is the only zero of this function in $\mathscr{U}_\delta^-(m^*)$. From this observation, together with (15.5.37), we conclude that

$$\left(\sum_{\ell=1}^{n}\rho_\ell g_\ell^2(\mathbf{m})\right)^{1/2}\geq c\big|m-m^*\big|-2\sum_{\ell\notin S(\mathbf{m})}\rho_\ell \tag{15.5.38}$$

for some constant $c<\infty$, where we use the triangle inequality and the fact that $|\mathbf{m}_\ell-\frac{1}{N}\sum_{i\in\Lambda_\ell}\tanh(\beta(m+h_i))|\leq 2\rho_\ell$. Under the hypothesis of the lemma, this gives the desired bound when $|m-m^*|\geq c''\varepsilon/\sqrt{\theta}$ for some constant $c''<\infty$. On the other hand, we can write for $\ell\in S(\mathbf{m})$,

$$\begin{aligned}\big|\mathbf{m}_\ell-\mathbf{m}_\ell^*\big| &\leq \frac{1}{N}\sum_{i\in\Lambda_\ell}\big|\tanh\big(\beta\big(g_\ell(\mathbf{m})\big[1+\omega(1)\big]+m+h_i\big)\big)-\tanh\big(\beta(m+h_i)\big)\big| \\ &\quad+\frac{1}{N}\sum_{i\in\Lambda_\ell}\big|\tanh\big(\beta(m+h_i)\big)-\tanh\big(\beta\big(m^*+h_i\big)\big)\big| \\ &\leq c\rho_\ell\big|m-m^*\big|+c'\rho_\ell\big|g_\ell(\mathbf{m})\big|.\end{aligned} \tag{15.5.39}$$

Hence we get the bound

$$\begin{aligned}\left(\sum_{\ell\in S(\mathbf{m})}\rho_\ell g_\ell^2(\mathbf{m})\right)^{1/2} &\geq c\left(\sum_{\ell\in S(\mathbf{m})}\frac{(\mathbf{m}_\ell-\mathbf{m}_\ell^*)^2}{\rho_\ell}\right)^{1/2}-c'\big|m-m^*\big| \\ &= c\left(\sum_{\ell=1}^{n}\frac{(\mathbf{m}_\ell-\mathbf{m}_\ell^*)^2}{\rho_\ell}-\sum_{\ell\notin S(\mathbf{m})}\frac{(\mathbf{m}_\ell-\mathbf{m}_\ell^*)^2}{\rho_\ell}\right)^{1/2} \\ &\quad -c'\big|m-m^*\big| \\ &\geq c\left(\varepsilon^2/\theta-4\sum_{\ell\notin S(\mathbf{m})}\rho_\ell\right)^{1/2}-c'\big|m-m^*\big| \\ &\geq c\varepsilon/\sqrt{2\theta}-c'\big|m-m^*\big|,\end{aligned} \tag{15.5.40}$$

where in the last line we use that $\mathbf{m} \notin \mathbf{G}_\theta$. The inequalities in (15.5.38) and (15.5.40) yield (15.5.35). □

Proof of Proposition 15.20 Let $\sigma \in S_N[\mathbf{U}_\delta \setminus \mathbf{G}_\theta]$, and set $\mathbf{x} = \mathbf{m}(\sigma)$ so that, for ψ as in Proposition 15.20, and abbreviate $(L\psi)(\sigma) = (L\phi)(\mathbf{x})$. Let σ^i be the configuration obtained from σ after a spin-flip at i, and introduce the notation

$$(L\phi)(\mathbf{x}) = \sum_{\ell=1}^{n} (L_\ell \phi)(\mathbf{x}), \tag{15.5.41}$$

where

$$\begin{aligned}(L_\ell\phi)(\mathbf{x}) = &\sum_{i\in\Lambda_\ell^-(\mathbf{x})} p_N\big(\sigma,\sigma^i\big)\big[\phi(\mathbf{x}+\mathbf{e}_\ell)-\phi(\mathbf{x})\big]\\ &+\sum_{i\in\Lambda_\ell^+(\mathbf{x})} p_N\big(\sigma,\sigma^i\big)\big[\phi(\mathbf{x}-\mathbf{e}_\ell)-\phi(\mathbf{x})\big].\end{aligned} \tag{15.5.42}$$

Note that if $\mathbf{x}_\ell/\rho_\ell = \pm 1$, then $\Lambda_\ell^{\pm}(\mathbf{x}) = \emptyset$ and the summation over $\Lambda_\ell^{\pm}(\mathbf{x})$ in (15.5.42) disappears.

We define the probabilities

$$\mathbb{P}^\sigma_{\pm,\ell} = \sum_{i\in\Lambda_\ell^{\mp}(\mathbf{x})} p_N\big(\sigma,\sigma^i\big), \tag{15.5.43}$$

and observe that they are uniformly close to the mesoscopic rates r_N, namely,

$$\mathrm{e}^{-c\varepsilon} \le \frac{\mathbb{P}^\sigma_{\pm,\ell}}{r_N(\mathbf{x},\mathbf{x}\pm\mathbf{e}_\ell)} \le \mathrm{e}^{c\varepsilon} \tag{15.5.44}$$

for some $c > 0$ and $\varepsilon = 1/n$. Note also that

$$c\rho_\ell \le \mathbb{P}^\sigma_{+,\ell} + \mathbb{P}^\sigma_{-,\ell} \le c'\rho_\ell. \tag{15.5.45}$$

With the above notation and using the convention $0/0 = 0$, we get

$$\begin{aligned}(L_\ell\phi)(\mathbf{x}) &= \phi(\mathbf{x})\mathbb{P}^\sigma_{+,\ell}\big[\exp\big(2\beta(1-\alpha)g_\ell(\mathbf{x})\big)-1\big]\\ &\quad+\phi(\mathbf{x})\mathbb{P}^\sigma_{-,\ell}\big[\exp\big(-2\beta(1-\alpha)g_\ell(\mathbf{x}-\mathbf{e}_\ell)\big)-1\big]\\ &=\phi(\mathbf{x})\big(\mathbb{1}_{\{\mathbb{P}^\sigma_{+,\ell}\ge\mathbb{P}^\sigma_{-,\ell}\}}\mathbb{P}^\sigma_{+,\ell}G_\ell^+(\mathbf{x})+\mathbb{1}_{\{\mathbb{P}^\sigma_{-,\ell}>\mathbb{P}^\sigma_{+,\ell}\}}\mathbb{P}^\sigma_{-,\ell}G_\ell^-(\mathbf{x})\big),\end{aligned} \tag{15.5.46}$$

where we introduce the functions

$$G_\ell^+(\mathbf{x}) = \exp\big(2\beta(1-\alpha)g_\ell(\mathbf{x})\big)-1+\frac{\mathbb{P}^\sigma_{-,\ell}}{\mathbb{P}^\sigma_{+,\ell}}\big[\exp\big(-2\beta(1-\alpha)g_\ell(\mathbf{x}-\mathbf{e}_\ell)\big)-1\big] \tag{15.5.47}$$

and

$$G_\ell^-(\mathbf{x}) = \exp\big(-2\beta(1-\alpha)g_\ell(\mathbf{x}-\mathbf{e}_\ell)\big) - 1 + \frac{\mathbb{P}^\sigma_{+,\ell}}{\mathbb{P}^\sigma_{-,\ell}}\big[\exp\big(2\beta(1-\alpha)g_\ell(\mathbf{x})\big) - 1\big]. \tag{15.5.48}$$

If $\mathbf{x}_\ell/\rho_\ell = \pm 1$, then the local generator takes the simpler form

$$(L_\ell\phi)(\mathbf{x}) = \begin{cases} \phi(\mathbf{x})\mathbb{P}^\sigma_{-,\ell}[\exp(-2\beta(1-\alpha)g_\ell(\mathbf{x}-\mathbf{e}_\ell)) - 1], & \text{if } \mathbf{x}_\ell/\rho_\ell = 1, \\ \phi(\mathbf{x})\mathbb{P}^\sigma_{+,\ell}[\exp(2\beta(1-\alpha)g_\ell(\mathbf{x})) - 1], & \text{if } \mathbf{x}_\ell/\rho_\ell = -1. \end{cases} \tag{15.5.49}$$

From Lemma 15.24 and inequalities (15.5.45), it follows that, for all ℓ such that $\mathbf{x}_\ell/\rho_\ell = \pm 1$,

$$(L_\ell\phi)(\mathbf{x}) \leq -\big[1+\omega(1)\big]\,\rho_\ell\phi(\mathbf{x}). \tag{15.5.50}$$

Let us now return to the case when $\mathbf{x}$ is not a boundary point. By the reversibility conditions,

$$\begin{aligned} r_N(\mathbf{x},\mathbf{x}+\mathbf{e}_\ell) &= \exp\big(-2\beta g_\ell(\mathbf{x})\big)r_N(\mathbf{x}+\mathbf{e}_\ell,\mathbf{x}), \\ r_N(\mathbf{x},\mathbf{x}-\mathbf{e}_\ell) &= \exp\big(2\beta g_\ell(\mathbf{x}-\mathbf{e}_\ell)\big)r_N(\mathbf{x}-\mathbf{e}_\ell,\mathbf{x}), \end{aligned} \tag{15.5.51}$$

which implies, together with (15.5.44), that

$$\begin{aligned} \exp\big(-2\beta g_\ell(\mathbf{x}) - c\varepsilon\big) &\leq \frac{\mathbb{P}^\sigma_{+,\ell}}{\mathbb{P}^\sigma_{-,\ell}} \leq \exp\big(-2\beta g_\ell(\mathbf{x}) + c\varepsilon\big), \\ \exp\big(2\beta g_\ell(\mathbf{x}-\mathbf{e}_\ell) - c\varepsilon\big) &\leq \frac{\mathbb{P}^\sigma_{-,\ell}}{\mathbb{P}^\sigma_{+,\ell}} \leq \exp\big(2\beta g_\ell(\mathbf{x}-\mathbf{e}_\ell) + c\varepsilon\big). \end{aligned} \tag{15.5.52}$$

Inserting the last bounds into (15.5.47) and (15.5.48), we obtain, after some computations,

$$\begin{aligned} G_\ell^+(\mathbf{x}) \leq{}& \big[\exp\big(2\beta(1-\alpha)g_\ell(\mathbf{x})\big) - 1\big]\big[1 - \exp\big(2\beta\alpha g_\ell(\mathbf{x}-\mathbf{e}_\ell) \mp c\varepsilon\big)\big] \\ &+ \exp\big(2\beta g_\ell(\mathbf{x}-\mathbf{e}_\ell) \mp c\varepsilon\big)\big[\exp 2\beta(1-\alpha)\big(g_\ell(\mathbf{x}) - g_\ell(\mathbf{x}-\mathbf{e}_\ell)\big) - 1\big] \end{aligned} \tag{15.5.53}$$

and

$$\begin{aligned} G_\ell^-(\mathbf{x}) \leq{}& \big[\exp\big(-2\beta(1-\alpha)g_\ell(\mathbf{x}-\mathbf{e}_\ell)\big) - 1\big]\big[1 - \exp\big(-2\beta\alpha g_\ell(\mathbf{x}) \mp c\varepsilon\big)\big] \\ &+ \exp\big(-2\beta g_\ell(\mathbf{x}) \mp c\varepsilon\big)\big[\exp 2\beta(1-\alpha)\big(g_\ell(\mathbf{x}) - g_\ell(\mathbf{x}-\mathbf{e}_\ell)\big) - 1\big], \end{aligned} \tag{15.5.54}$$

where $\mp = -\text{sign}(g_\ell(\mathbf{x})) = -\text{sign}(g_\ell(\mathbf{x}-\mathbf{e}_\ell))$. For all ℓ such that $\mathbf{x}_\ell/\rho_\ell \in [-1+\nu, 1-\nu]$, we can use (15.5.32) to get

$$G_\ell^+(\mathbf{x}) \leq \big[\exp\big(2\beta(1-\alpha)g_\ell(\mathbf{x})\big) - 1\big]\big[1 - \exp\big(2\alpha\beta g_\ell(\mathbf{x}) \mp c\varepsilon\big)\big] + c/N \tag{15.5.55}$$

and

$$G_\ell^-(\mathbf{x}) \leq \left[\exp\left(-2\beta(1-\alpha)g_\ell(\mathbf{x})\right) - 1\right]\left[1 - \exp\left(-2\alpha\beta g_\ell(\mathbf{x}) \mp c\varepsilon\right)\right] + c/N. \tag{15.5.56}$$

The right-hand sides of (15.5.55) and (15.5.56) are negative if and only if $|g_\ell| > \frac{c\varepsilon}{2\alpha\beta}$. Let us define the index sets

$$S^< = \left\{\ell\colon \mathbf{x}_\ell/\rho_\ell \in [-1+\nu, 1-\nu], \left|g_\ell(\mathbf{x})\right| \leq \frac{c\varepsilon}{\alpha\beta}\right\}, \tag{15.5.57}$$

$$S^> = \left\{\ell\colon \mathbf{x}_\ell/\rho_\ell \in [-1+\nu, 1-\nu], \left|g_\ell(\mathbf{x})\right| > \frac{c\varepsilon}{\alpha\beta}\right\}. \tag{15.5.58}$$

If $\ell \in S^<$, then we immediately get that

$$\max\left\{G_\ell^+(\mathbf{x}), G_\ell^-(\mathbf{x})\right\} \leq \frac{c}{\alpha}\varepsilon^2, \tag{15.5.59}$$

and hence from (15.5.45) and (15.5.46) that

$$(L_\ell\phi)(\mathbf{x}) \leq \frac{c'}{\alpha}\varepsilon^2 \rho_\ell \phi(\mathbf{x}). \tag{15.5.60}$$

To control the right-hand side of (15.5.55) and (15.5.56) when $\ell \in S^>$, we set

$$y_\ell = \min\left\{\beta\left|g_\ell(\mathbf{x})\right|, \tfrac{1}{2}\right\} \leq \beta\left|g_\ell(\mathbf{x})\right|. \tag{15.5.61}$$

If $g_\ell(\mathbf{x}) > \frac{c\varepsilon}{\alpha\beta}$, then

$$\exp\left(2\beta(1-\alpha)g_\ell(\mathbf{x})\right) - 1 \geq \exp\left(2(1-\alpha)y_\ell\right) - 1 \geq 2(1-\alpha)y_\ell \tag{15.5.62}$$

and

$$1 - \exp\left(2\beta\alpha g_\ell(\mathbf{x}) - c\varepsilon\right) \leq 1 - \exp\left(\alpha y_\ell\right) \leq -\alpha y_\ell, \tag{15.5.63}$$

so that the product in the right-hand side of (15.5.55) is bounded from above by $-\frac{3}{4}(1-\alpha)\alpha y_\ell^2$. On the other hand, if $g_\ell(\mathbf{x}) < -\frac{c\varepsilon}{\alpha\beta}$, then

$$\exp\left(2\beta(1-\alpha)g_\ell(\mathbf{x})\right) - 1 \leq \exp\left(-2(1-\alpha)y_\ell\right) - 1 \leq -(1-\alpha)y_\ell \tag{15.5.64}$$

and

$$1 - \exp\left(2\beta\alpha g_\ell(\mathbf{x}) + c\varepsilon\right) \geq 1 - \exp\left(-\alpha y_\ell\right) \geq \tfrac{3}{4}\alpha y_\ell, \tag{15.5.65}$$

and the product in the right-hand side of (15.5.55) is bounded from above by $-\frac{3}{4}(1-\alpha)\alpha y_\ell^2$. Altogether this proves that, for all $\ell \in S^>$,

$$G_\ell^+(\mathbf{x}) \leq -\tfrac{3}{4}(1-\alpha)\alpha y_\ell^2, \tag{15.5.66}$$

and with a similar computation that

$$G_\ell^-(\mathbf{x}) \le -\tfrac{3}{4}(1-\alpha)\alpha y_\ell^2. \tag{15.5.67}$$

If $\ell \in S^>$, then we have

$$(L_\ell\phi)(\mathbf{x}) \le -c\alpha\rho_\ell y_\ell^2\phi(\mathbf{x}). \tag{15.5.68}$$

It remains to control the case $\mathbf{x}_\ell/\rho_\ell \in (-1,-1+\nu] \cup [1-\nu,1)$. It follows from Lemma 15.24 that, while the positive contribution to $G_\ell^+(\mathbf{x})$ and $G_\ell^-(\mathbf{x})$ is bounded by a constant, the negative contribution becomes large when ν gets small. More explicitly, for ν small enough we have

$$\begin{aligned} G_\ell^+(\mathbf{x}) &\le -\left(\mathrm{e}^{\pm C'}-1\right)^2 + \mathrm{e}^{\pm C'}\left(\mathrm{e}^{2\beta(1-\alpha)c}-1\right) \le -\left[1+o(1)\right], \\ G_\ell^-(\mathbf{x}) &\le -\left(1-\mathrm{e}^{\mp C'}\right)^2 + \mathrm{e}^{\mp C''}\left(\mathrm{e}^{2\beta(1-\alpha)c}-1\right) \le -\left[1+o(1)\right], \end{aligned} \tag{15.5.69}$$

where C' and C'' are positive constants tending to ∞ as $\nu \downarrow 0$, and the sign $\pm$ is equal to the sign of $\mathbf{x}_\ell$. Together with (15.5.45) and (15.5.46), we finally get

$$(L_\ell\phi)(\mathbf{x}) \le -\left[1+o(1)\right]\rho_\ell\phi(\mathbf{x}). \tag{15.5.70}$$

From (15.5.50), (15.5.60), (15.5.68) and (15.5.70), it turns out that the positive contribution to the generator $(L\phi)(\mathbf{x}) = \sum_{\ell=1}^n (L_\ell\phi)(\mathbf{x})$ comes at most from the indices $\ell \in S^<$, and can be estimated by

$$\frac{c'}{\alpha}\varepsilon^2 \sum_{\ell \in S^<} \rho_\ell \le \frac{c'}{\alpha}\varepsilon^2. \tag{15.5.71}$$

We must distinguish two cases, according to whether the hypothesis of Lemma 15.25 is satisfied or not.

Case 1: $\sum_{\ell \notin S(\mathbf{x})} \rho_\ell > \frac{\varepsilon^2}{8\theta}$. By (15.5.50), we get

$$\begin{aligned} \sum_{\ell=1}^n (L_\ell\phi)(\mathbf{x}) &\le \sum_{\ell \notin S(\mathbf{x})} (L_\ell\phi)(\mathbf{x}) + \sum_{\ell \in S^<} (L_\ell\phi)(\mathbf{x}) \\ &\le -\frac{\varepsilon^2}{8\theta}\left[1+o(1)\right]\phi(\mathbf{x}) + \frac{c'}{\alpha}\varepsilon^2, \end{aligned} \tag{15.5.72}$$

which is as negative as desired when θ is small enough, i.e., when ε is small enough.

Case 2: $\sum_{\ell \notin S(\mathbf{x})} \rho_\ell \le \frac{\varepsilon^2}{8\theta}$. In this case, the assertion of Lemma 15.25 holds. By (15.5.50), (15.5.68), and (15.5.70), we have that, for all $\ell \in S(\mathbf{x}) \setminus L^<$,

$$(L_\ell\phi)(\mathbf{x}) \le -\rho_\ell\phi(\mathbf{x}) \min\left\{c\alpha y_\ell^2, 1\right\} \le -c\alpha\rho_\ell y_\ell^2\phi(\mathbf{x}), \tag{15.5.73}$$

where the last inequality holds for $\alpha < 4/c$. Let us write the generator as

$$(L\psi)(\mathbf{x}) \leq \sum_{\ell \in S(\mathbf{x})\setminus S^<} (L_\ell \psi)(\mathbf{x}) + \sum_{\ell \in S^<} (L_\ell \psi)(\mathbf{x}). \tag{15.5.74}$$

The first sum in (15.5.74) is bounded from above by

$$\begin{aligned} -c\alpha\phi(\mathbf{x}) \sum_{\ell \in S(\mathbf{x})\setminus S^<} \rho_\ell y_\ell^2 &\leq -c\alpha\phi(\mathbf{x}) \sum_{\ell \in s(\mathbf{x})\setminus S^<} \rho_\ell \min\left\{\beta^2 g_\ell^2(\mathbf{x}), \tfrac{1}{4}\right\} \\ &\leq -c\alpha\phi(\mathbf{x}) \min\left\{\beta^2 \sum_{\ell \in S(\mathbf{x})\setminus S^<} \rho_\ell g_\ell^2(\mathbf{x}), \tfrac{1}{4}\right\}. \end{aligned} \tag{15.5.75}$$

But from Lemma 15.25 we know that, for all $\mathbf{x} \in \mathbf{U}_\delta \setminus \mathbf{G}_\theta$,

$$\sum_{\ell \in S(\mathbf{x})\setminus S^<} \rho_\ell g_\ell^2(\mathbf{x}) \geq c\frac{\varepsilon^2}{\theta} - \frac{c'}{\alpha^2}\varepsilon^2 \geq c''\frac{\varepsilon^2}{\theta}, \tag{15.5.76}$$

where c'' is a positive constant, provided that $\alpha \geq c\theta$. Taking n large enough, we find that

$$\min\left\{\beta^2 \sum_{\ell \in s(\mathbf{x})\setminus S^<} \rho_\ell g_\ell^2(\mathbf{x}), \tfrac{1}{4}\right\} \geq \min\left\{c''\frac{\varepsilon^2}{\theta}, \tfrac{1}{4}\right\} = c''\frac{\varepsilon^2}{\theta}, \tag{15.5.77}$$

and then, from (15.5.71) and (15.5.75), we get

$$(L\psi)(\sigma) \leq -\varepsilon^2(1-\alpha)\phi(\mathbf{x})\left(c''\alpha\theta^{-1} - c'\alpha^{-1}\right). \tag{15.5.78}$$

By our choice of θ, taking n large enough we see that the condition $c''\alpha\theta^{-1} - c'\alpha^{-1} > 0$, or $\alpha > c\theta$, is satisfied for any $\alpha \in (0,1)$. Hence, for such n and for N large enough, we get that $(L\psi)(\sigma) \leq 0$, which concludes the proof of Proposition 15.20. □

Substituting the expression for the super-harmonic function in (15.5.20) into (15.5.19), and using (15.5.15), we obtain that, for all $\sigma \in S_N[\mathbf{U}_\delta \setminus \mathbf{G}_\theta]$,

$$\begin{aligned} \mathbb{P}_\sigma[\tau_A < \tau_{S_N[\partial_A \mathbf{G}_\theta] \cup B}] &\leq \max_{\sigma' \in S_N[\partial_A \mathbf{U}_\delta]} \mathrm{e}^{-(1-\alpha)\beta N[F_{\beta,N}(\mathbf{m}(\sigma')) - F_{\beta,N}(\mathbf{m}(\sigma))]} \\ &\leq \mathrm{e}^{-(1-\alpha)\beta N[F_{\beta,N}(m_0^*) + \delta - F_{\beta,N}(\mathbf{m}(\sigma))]}, \end{aligned} \tag{15.5.79}$$

where the last inequality follows from the definition of $\mathbf{U}_\delta$, together with the bounds in (14.7.3). This concludes the proof of Proposition 15.19.

Renewal estimates on escape probabilities

Let us now return to the proof of Lemma 15.18. An easy consequence of (15.5.14) is that, for all $\sigma \in S_N[\partial_A \mathbf{G}_\theta]$,

$$\mathbb{P}_\sigma(\tau_A < \tau_{S_N[\partial_A \mathbf{G}_\theta] \cup B}) \le \mathrm{e}^{-(1-\alpha)\beta N(F_{\beta,N}(m_0^*)+\delta)} \max_{\mathbf{m} \in \partial_A \mathbf{G}_\theta} \mathrm{e}^{(1-\alpha)\beta N F_{\beta,N}(\mathbf{m})}, \tag{15.5.80}$$

while obviously $\mathbb{P}_\sigma(\tau_A < \tau_{S_N[\partial_A \mathbf{G}_\theta] \cup B}) = 0$ for all $\sigma \in S_N[\mathbf{G}_\theta \setminus \partial_A \mathbf{G}_\theta]$. To control the right-hand side of (15.5.80), we need the following lemma.

Lemma 15.26 *There exists a constant $c < \infty$ independent of n such that, for all $\mathbf{m} \in \mathbf{G}_\theta$,*

$$F_{\beta,N}(\mathbf{m}) \le F_{\beta,N}\big(\mathbf{m}^*\big) + c\varepsilon. \tag{15.5.81}$$

Proof Fix $\mathbf{m} \in \mathbf{G}_\theta$ and set $\mathbf{m} - \mathbf{m}^* = \mathbf{v}$. Note that, from the definition of $\mathbf{G}_\theta$, we have

$$\|\mathbf{v}\|_2^2 \le \Big(\max_{1 \le \ell \le n} \rho_\ell\Big) \sum_{\ell=1}^{n} \frac{(\mathbf{m}_\ell - \mathbf{m}_\ell^*)^2}{\rho_\ell} \le \varepsilon^2. \tag{15.5.82}$$

Using Taylor's formula, we have

$$F_{\beta,N}(\mathbf{m}) = F_{\beta,N}\big(\mathbf{m}^*\big) + \tfrac{1}{2}\big(\mathbf{v}, \mathbb{A}\big(\mathbf{m}^*\big)\mathbf{v}\big) + \tfrac{1}{6} D^3 F_{\beta,N}(\mathbf{x})\mathbf{v}^3, \tag{15.5.83}$$

where $\mathbb{A}(\mathbf{m}^*)$ is the positive-definite matrix described in Sect. 15.2.2 (see (15.2.9)) and $\mathbf{x}$ is a suitable element of the ball around $\mathbf{m}^*$. From the explicit representation of the eigenvalues of $\mathbb{A}(\mathbf{m}^*)$, we see that $\|\mathbb{A}(\mathbf{m}^*)\| \le c\varepsilon^{-1}$, and hence

$$\big(\mathbf{v}, \mathbb{A}\big(\mathbf{m}^*\big)\mathbf{v}\big) \le c\varepsilon^{-1} \|\mathbf{v}\|_2^2 \le c\varepsilon. \tag{15.5.84}$$

The remainder is given in explicit form as

$$\begin{aligned} D^3 F_{\beta,N}(\mathbf{x})\mathbf{v}^3 &= \sum_{\ell=1}^{n} \frac{\partial^3 F_{\beta,N}}{\partial \mathbf{x}_\ell^3}(\mathbf{x})\mathbf{v}_\ell^3 = \frac{1}{\beta} \sum_{\ell=1}^{n} \frac{1}{\rho_\ell^2} I_{N,\ell}'''(\mathbf{x}_\ell/\rho_\ell)\mathbf{v}_\ell^3 \\ &= -\frac{1}{\beta} \sum_{\ell=1}^{n} \frac{1}{\rho_\ell^2} \frac{U_{N,\ell}'''(t_\ell)}{[U_{N,\ell}''(t_\ell)]^3} \mathbf{v}_\ell^3 \\ &= -\frac{1}{\beta} \sum_{\ell=1}^{n} \frac{1}{\rho_\ell^2} \frac{|\Lambda_\ell|^{-1} \sum_{i \in \Lambda_\ell} \tanh(t_\ell + \beta \tilde{h}_i)[1 - \tanh^2(t_\ell + \beta \tilde{h}_i)]}{(|\Lambda_\ell|^{-1} \sum_{i \in \Lambda_\ell} [1 - \tanh^2(t_\ell + \beta \tilde{h}_i)])^3} \mathbf{v}_\ell^3, \end{aligned} \tag{15.5.85}$$

where $t_\ell = I_{N,\ell}'(\mathbf{x}_\ell/\rho_\ell)$. Thus,

$$\big|D^3 F_{\beta,N}(\mathbf{x})\mathbf{v}^3\big| \le c \sum_{\ell=1}^{n} \frac{1}{\rho_\ell^2} \mathbf{v}_\ell^3 \le c'\varepsilon^{-1} \|\mathbf{v}\|_2^2 \le c'\varepsilon, \tag{15.5.86}$$

where we use that $|\mathbf{v}_\ell/\rho_\ell| \le 1$. Hence, for some $c < \infty$ independent of n,

$$F_{\beta,N}(\mathbf{m}) < F_{\beta,N}(\mathbf{m}^*) + c\varepsilon. \tag{15.5.87}$$

□

Inserting the result of Lemma 15.26 into (15.5.80), and recalling that $F_{\beta,N}(\mathbf{m}^*) = F_{\beta,N}(m^*)$, we get that, for all $\sigma \in S_N[\partial_A \mathbf{G}_\theta]$,

$$\mathbb{P}_\sigma(\tau_A < \tau_{S_N[\partial_A \mathbf{G}_\theta] \cup B}) \le \mathrm{e}^{-(1-\alpha)\beta N(F_{\beta,N}(m_0^*)+\delta - F_{\beta,N}(m^*) - c\varepsilon)}. \tag{15.5.88}$$

The last ingredient in order to get a suitable estimate on $\mathbb{P}_\sigma(\tau_A < \tau_B)$ is stated in the following lemma.

Lemma 15.27 *For any $\delta_2 > 0$ there exists an $n_0 \in \mathbb{N}$ such that for all $n \ge n_0$, all $\sigma \in S_N[\partial_A \mathbf{G}_\theta]$ and all N large enough,*

$$\mathbb{P}_\sigma(\tau_B < \tau_{S_N[\partial_A \mathbf{G}_\theta]}) \ge \mathrm{e}^{-N\beta\delta_2}. \tag{15.5.89}$$

Proof Fix $\sigma \in S_N[\partial_A \mathbf{G}_\theta]$ and set $\mathbf{m}(0) = \mathbf{m}(\sigma)$. As pointed out in the proof of Lemma 15.26, every $\mathbf{m}(0) \in \partial_A \mathbf{G}_\theta$ can be written in the form $\mathbf{m}(0) = \mathbf{m}^* + \mathbf{v}$ with $\mathbf{v} \in \Gamma_N^n$ such that $\|\mathbf{v}\|_2 \le \varepsilon$. Let $\underline{\mathbf{m}} = (\mathbf{m}(0), \mathbf{m}(1), \dots, \mathbf{m}(\|\mathbf{v}\|_1 N) = \mathbf{m}^*)$ be a nearest-neighbour path in Γ_N^n from $\mathbf{m}(0)$ to $\mathbf{m}^*$ of length $N\|\mathbf{v}\|_1$ with the following property: with ℓ_t the unique index in $\{1, \dots, n\}$ such that $\mathbf{m}_{\ell_t}(t) \ne \mathbf{m}_{\ell_t}(t-1)$,

$$\mathbf{m}_{\ell_t}(t) = \mathbf{m}_{\ell_t}(t-1) + \frac{2}{N} s_t, \quad \forall t \ge 1, \tag{15.5.90}$$

where we define

$$s_t = \mathrm{sign}\big(\mathbf{m}_{\ell_t}^* - \mathbf{m}_{\ell_t}(t-1)\big). \tag{15.5.91}$$

Note that, by property (15.5.90), $\mathbf{m}(t) \in \mathbf{G}_\theta$ for all $t \ge 0$. Thus, all microscopic paths, $(\sigma(t))_{t\ge 0}$ such that $\sigma(0) = \sigma$ and $\mathbf{m}(\sigma(t)) = \mathbf{m}(t)$ for all $t \ge 1$ are contained in the event $\{\tau_B < \tau_{S_N[\partial_A \mathbf{G}_\theta]}\}$. Therefore we get

$$\begin{aligned}
\mathbb{P}_\sigma(\tau_B < \tau_{S_N[\partial_A \mathbf{G}_\theta]}) &\ge \mathbb{P}_\sigma\big(\mathbf{m}(\sigma(t)) = \mathbf{m}(t)\ \forall t = 1, \dots, \|\mathbf{v}\|_1 N\big) \\
&= \prod_{t=1}^{\|\mathbf{v}\|_1 N} \mathbb{P}_\sigma\big(\mathbf{m}(\sigma(t)) = \mathbf{m}(t) \,|\, \mathbf{m}(\sigma(t-1)) = \mathbf{m}(t-1)\big) \\
&= \prod_{t=1}^{\|\mathbf{v}\|_1 N} \sum_{i \in \Lambda_{\ell_t}^{s_t}} p_N\big(\sigma(t-1), \sigma^i(t-1)\big).
\end{aligned} \tag{15.5.92}$$

Note that $\Lambda_{\ell_t}^{s_t}$ is the set of sites in which a spin-flip corresponds to a step from $\mathbf{m}(t-1)$ to $\mathbf{m}(t)$.

The sum of the probabilities in the right-hand side of (15.5.92) corresponds to the quantity $\mathbb{P}^{\sigma(t-1)}_{s_t,\ell_t}$ defined in (15.5.43). From the inequalities in (15.5.44) and (15.3.3) it follows that, for some constant $c > 0$ depending on β and on the distribution of the magnetic field,

$$\mathbb{P}^{\sigma(t-1)}_{s_t,\ell_t} \geq c\big|\Lambda^{s_t}_{\ell_t}(\mathbf{m}(t-1))\big|/N \geq c\big|\Lambda^{s_t}_{\ell_t}(\mathbf{m}^*)\big|/N, \tag{15.5.93}$$

where the second inequality follows by our choice of the path $\underline{\mathbf{m}}$. Now, since $|\Lambda^{\pm}_{\ell}(\mathbf{m}^*)|/N = \frac{1}{2}(\rho_\ell \pm \mathbf{m}^*_\ell)$, we can use the expression in (15.2.8) for $\mathbf{m}^*_{\ell_t}$ and continue from (15.5.93), to obtain

$$\mathbb{P}^{\sigma(t-1)}_{s_t,\ell_t} \geq c'\rho_{\ell_t}. \tag{15.5.94}$$

Inserting the last inequality into (15.5.92) and using that, by the definition of the path $\underline{\mathbf{m}}$, the number of steps corresponding to a spin-flip in Λ_ℓ is equal to $|\mathbf{v}_\ell|N$ for all $\ell = \{1,\dots,n\}$, we get

$$\begin{aligned}\mathbb{P}_\sigma(\tau_B < \tau_{S_N[\partial_A \mathbf{G}_\theta]}) &\geq \prod_{t=1}^{\|\mathbf{v}\|_1 N} c'\rho_{\ell_t} = \mathrm{e}^{\|\mathbf{v}\|_1 N\ln(c')}\prod_{\ell=1}^{n}\rho_\ell^{|\mathbf{v}_\ell|N} \\ &\geq \mathrm{e}^{N\sqrt{\varepsilon}\ln(c')}\mathrm{e}^{-N\sum_{\ell=1}^n \mathbf{v}_\ell\ln(1/\rho_\ell)} \geq \mathrm{e}^{N\sqrt{\varepsilon}\ln(c')}\mathrm{e}^{-N\sum_{\ell=1}^n \mathbf{v}_\ell/\sqrt{\rho_\ell}} \\ &\geq \mathrm{e}^{N\varepsilon\ln(c')}\mathrm{e}^{-N(\sum_{\ell=1}^n \mathbf{v}_\ell^2/\rho_\ell)s^{1/2}\varepsilon^{-1/2}} \geq \mathrm{e}^{-N(\sqrt{\varepsilon/\theta}-\sqrt{\varepsilon}\ln(c'))},\end{aligned} \tag{15.5.95}$$

where in the third line we use the inequality $\|\mathbf{v}\|_1 \leq \varepsilon^{-1/2}\|\mathbf{v}\|_2 \leq \sqrt{\varepsilon}$, and in the last line we use that $\mathbf{m}(0) = \mathbf{m}^* + \mathbf{v} \in \mathbf{G}_\theta$. By our choice of $\theta \gg \varepsilon$, there exists an $n_0 \in \mathbb{N}$ such that, for all $n \geq n_0$, $\sqrt{\varepsilon/\theta} - \sqrt{\varepsilon}\ln(c') \leq \beta\delta_2$. For such n, the inequality in (15.5.95) yields the bound in (15.5.89). □

We finally state the following proposition.

Proposition 15.28 *For all* $\sigma \in \mathscr{S}[\mathbf{U}_\delta]$,

$$\mathbb{P}_\sigma(\tau_A < \tau_B) \leq \mathrm{e}^{-\beta N[(1-\alpha)(F_{\beta,N}(m_0^*)+\delta-F_{\beta,N}(m^*)-c\varepsilon)-\delta_2]}\big[1+o(1)\big]. \tag{15.5.96}$$

Proof We first consider a configuration $\sigma \in S_N[\partial_A \mathbf{G}_\theta]$. Then

$$\begin{aligned}&\mathbb{P}_\sigma(\tau_A < \tau_B) \\ &\quad\leq \mathbb{P}_\sigma(\tau_A < \tau_{S_N[\partial_A\mathbf{G}_\theta]\cup B}) + \sum_{\eta\in S_N[\partial_A\mathbf{G}_\theta]} \mathbb{P}_\sigma(\tau_A < \tau_B,\, \tau_\eta \leq \tau_{S_N[\partial_A\mathbf{G}_\theta]\cup A\cup B}) \\ &\quad\leq \mathbb{P}_\sigma(\tau_A < \tau_{S_N[\partial_A\mathbf{G}_\theta]\cup B}) + \max_{\eta\in S_N[\partial_A\mathbf{G}_\theta]} \mathbb{P}_\eta(\tau_A < \tau_B)\,\mathbb{P}_\sigma(\tau_{S_N[\partial_A\mathbf{G}_\theta]} < \tau_B) \\ &\quad\leq \mathbb{P}_\sigma(\tau_A < \tau_{S_N[\partial_A\mathbf{G}_\theta]\cup B}) + \max_{\eta\in S_N[\partial_A\mathbf{G}_\theta]} \mathbb{P}_\eta(\tau_A < \tau_B)\big(1-\mathrm{e}^{-\beta N\delta_2}\big),\end{aligned} \tag{15.5.97}$$

where in the second line we use the Markov property, and in the last line we insert the result in (15.28). Taking the maximum over $\sigma \in S_N[\partial_A \mathbf{G}_\theta]$ on both sides of (15.5.97), and rearranging the summation, we get

$$\max_{\sigma \in S_N[\partial_A \mathbf{G}_\theta]} \mathbb{P}_\sigma(\tau_A < \tau_B) \leq \max_{\sigma \in S_N[\partial_A \mathbf{G}_\theta \cup B]} \mathbb{P}_\sigma(\tau_A < \tau_{S_N[\partial_A \mathbf{G}_\theta]}) e^{\beta N \delta_2}$$
$$\leq e^{-\beta N((1-\alpha)(F_{\beta,N}(m_0^*)+\delta - F_{\beta,N}(m^*)-c\varepsilon)-\delta_2)}, \quad (15.5.98)$$

where in the last line we use the bound in (15.5.88). This concludes the proof of (15.5.96) for $\sigma \in S_N[\partial_A \mathbf{G}_\theta]$.

Next we consider $\sigma \in S_N[\mathbf{U}_\delta \setminus \partial_A \mathbf{G}_\theta]$. As before,

$$\begin{aligned}
&\mathbb{P}_\sigma(\tau_A < \tau_B)\\
&\leq \mathbb{P}_\sigma(\tau_A < \tau_{S_N[\partial_A \mathbf{G}_\theta] \cup B}) + \sum_{\eta \in S_N[\partial_A \mathbf{G}_\theta]} \mathbb{P}_\sigma(\tau_A < \tau_B,\, \tau_\eta \leq \tau_{S_N[\partial_A \mathbf{G}_\theta] \cup A \cup B})\\
&\leq \mathbb{P}_\sigma(\tau_A < \tau_{S_N[\partial_A \mathbf{G}_\theta] \cup B}) + \max_{\eta \in S_N[\partial_A \mathbf{G}_\theta]} \mathbb{P}_\eta(\tau_A < \tau_B) \mathbb{P}_\sigma(\tau_{S_N[\partial_A \mathbf{G}_\theta]} < \tau_B)\\
&\leq \mathbb{P}_\sigma(\tau_A < \tau_{S_N[\partial_A \mathbf{G}_\theta] \cup B}) + \max_{\eta \in S_N[\partial_A \mathbf{G}_\theta]} \mathbb{P}_\eta(\tau_A < \tau_B), \qquad (15.5.99)
\end{aligned}$$

where $\mathbb{P}_\sigma(\tau_A < \tau_{S_N[\partial_a \mathbf{G}_\theta] \cup B})$ is zero for all $\sigma \in S_N[\mathbf{G}_\theta \setminus \partial_A \mathbf{G}_\theta]$, and is exponentially small in N for all $\sigma \in S_N[\mathbf{U}_\delta \setminus \mathbf{G}_\theta]$ (due to Proposition 15.19). Inserting the bound in (15.5.98) into the last equation, we get (15.5.96) for $\sigma \in S_N[\mathbf{U}_\delta \setminus \partial_A \mathbf{G}]$. □

The proof of (15.5.5) now follows straightforwardly. From (15.5.96) we get

$$\begin{aligned}
&\sum_{\sigma \in S_N[\mathscr{U}_\delta(m^*)]} \mu_{\beta,N}(\sigma) \mathbb{P}_\sigma(\tau_A < \tau_B)\\
&\qquad \leq e^{-\beta N[(1-\alpha)(F_{\beta,N}(m_0^*)+\delta-F_{\beta,N}(m^*)-c\varepsilon)-\delta_2]} \sum_{\mathbf{m} \in \mathbf{U}_\delta} \mathscr{Q}_{\beta,N}(\mathbf{m})\\
&\qquad = \mathscr{Q}_{\beta,N}\left(m_0^*\right) e^{\beta N[\alpha F_{\beta,N}(m_0^*)-(1-\alpha)(\delta - F_{\beta,N}(m^*)-c\varepsilon)+\delta_2]} \sum_{\mathbf{m} \in \mathbf{U}_\delta} e^{-\beta N F_{\beta,N}(\mathbf{m})}\\
&\qquad \leq \mathscr{Q}_{\beta,N}\left(m_0^*\right) N^n e^{\beta N[\alpha(F_{\beta,N}(m_0^*)-F_{\beta,N}(m^*))-(1-\alpha)(\delta-c\varepsilon)+\delta_2]}, \qquad (15.5.100)
\end{aligned}$$

where in the second inequality we use the expression in (14.2.9) for $\mathscr{Q}_{\beta,N}(m_0^*)$, and in the last line we use the bounds $F_{\beta,N}(\mathbf{m}) \leq F_{\beta,N}(\mathbf{m}^*) = F_{\beta,N}(m^*)$ and $|\mathbf{U}_d| \leq N^n$. Finally, choosing α small enough, namely,

$$\alpha < \frac{\delta - c\varepsilon - \delta_2}{F_{\beta,N}(m_0^*) - F_{\beta,N}(m^*) + \delta - c\varepsilon}, \qquad (15.5.101)$$

we can easily ensure that (15.5.100) implies (15.5.5).

In exactly the same way we can prove (15.5.6). This concludes the proof of Lemma 15.18 and hence of Theorem 15.1.

15.6 Bibliographical notes

1. The results of this chapter have been obtained by Bianchi, Bovier and Ioffe in [24]. The proof given here is streamlined, and the use of deficient flows on the mesoscopic level leads to a considerable simplification. Coupling methods have been used by the same authors in [25] to prove that transition times asymptotically exponentially distributed and the independence of the initial law. We have decided that this proof is too technical and too model-specific to be reproduced here. The results have been generalised to the Potts version of the model in the thesis of Slowik [219].

2. The dynamics of the random field Curie-Weiss model has been studied before. Dai Pra and den Hollander [69] studied the short-time dynamics using large deviation results and obtained the analogue of the McKean-Vlasov equations. Mathieu and Picco [180] considered convergence to equilibrium in the particularly simple case where the random field takes only the two values $\pm\varepsilon$ (with further restrictions on the parameters that exclude the presence of more than two minima).

3. The computations in this chapter should be paradigmatic for a wider class of Glauber dynamics for spin systems at low but finite temperatures: through a sequence of approximate coarse-grainings, the problem is mapped to a discrete diffusion in a potential given by a free energy functional, which in turn behaves like a diffusion process. The quality of the approximation is improved by refining the coarse-graining, leading effectively to a limiting SPDE. The random-field Curie-Weiss model is the only case where this program has been carried out in full so far. A natural next candidate would be Glauber dynamics for Ising models with Kac-type interactions (see the monograph by Presutti [202]).

Part VI
Applications: Lattice Systems in Small Volumes at Low Temperatures

In Chaps. 13–15 we studied models with a mean-field interaction at a fixed (subcritical) temperature in the limit of large volume. The parameter controlling the metastable behaviour in these models was the volume. We will now consider situations where the relevant parameter is the temperature.

Part VI looks at lattice models with a short-range interaction in a finite volume in the limit as the temperature tends to zero. These models have been a key success of the large-deviation approach, and they are important targets for the potential-theoretic approach as well. In Chap. 16 we show how to use the theory developed in Part III to establish the *universal metastable behaviour* of these models under a number of *general hypotheses*. These hypotheses will be proved for two specific choices of the dynamics: in Chap. 17 we look at Glauber dynamics for Ising spins, in Chap. 18 at Kawasaki dynamics for lattice gas particles.

In most of Chap. 16 we focus on Metropolis dynamics. At the end we briefly discuss two other dynamics, namely, heat-bath dynamics and probabilistic cellular automata, for which the same universal metastable behaviour can be derived under similar hypotheses.

Chapter 16
Abstract Set-Up and Metastability in the Zero-Temperature Limit

Talking is a wonderful smoother-over of difficulties. When I come upon anything—in Logic or in any other hard subject—that entirely puzzles me, I find it a capital plan to talk it over, aloud, even when I am all alone. One can explain things so clearly to one's self! And then, you know, one is so patient with one's self: one never gets irritated at one's own stupidity!
(Lewis Carroll, A Selection from Symbolic Logic)

This chapter describes the metastable behaviour of lattice systems in small volumes at low temperatures subject to a Metropolis dynamics. These theorems are derived under two hypotheses on the energy landscape, i.e., on the interaction Hamiltonian. These hypotheses, in turn, will be checked for Glauber dynamics in Chap. 17 and for Kawasaki dynamics in Chap. 18. The theorems themselves are *model-independent*, and therefore amplify the *universal nature* of metastability (in the setting considered here). However, they involve a number of quantities that are *model-dependent*. The identification of these quantities will be carried out in Chaps. 17 and 18 as well.

The outline is as follows. In Sect. 16.1 we define Metropolis dynamics on a general configuration space with respect to a general Hamiltonian and for a general set of allowed moves, we state the theorems subject to the hypotheses, and we place the results in their proper context. Section 16.2 explains how this abstract set-up fits into the potential-theoretic framework developed in Part III. The proofs of the theorems are given in Sect. 16.3. In Sect. 16.4 we take a brief look at two other dynamics, namely, heat-bath dynamics and probabilistic cellular automata, and we indicate how these can be included into the same abstract set-up with only minor modifications.

16.1 Hypotheses and universal metastability theorems

In this section we state three metastability theorems under two hypotheses.

A. Bovier, F. den Hollander, *Metastability*,
Grundlehren der mathematischen Wissenschaften 351,
DOI 10.1007/978-3-319-24777-9_16

16.1.1 Metropolis dynamics and geometric definitions

We begin by defining the abstract set-up of lattice models with short-range interaction in finite volume from statistical physics.

Let Λ be a finite set (e.g. $\Lambda \subseteq \mathbb{Z}^d$, $d \geq 1$). We refer to elements of Λ as *sites*. With each site $x \in \Lambda$ we associate a variable $\xi(x) \in \Upsilon$, where Υ is a finite set of spin-values. A configuration $\xi = \{\xi(x)\colon x \in \Lambda\}$ is an element of $S = \Upsilon^\Lambda$. To each configuration ξ we associate an energy given by a *Hamiltonian* $H\colon S \to \mathbb{R}$, which in general depends on one or more parameters. The *Gibbs measure* associated with H is

$$\mu_\beta(\xi) = \frac{1}{Z_\beta} \mathrm{e}^{-\beta H(\xi)}, \quad \xi \in S, \tag{16.1.1}$$

where $\beta \in (0,\infty)$ is the *inverse temperature*, and Z_β is the normalising partition sum.

Equip S with a set of undirected edges E, connecting pairs of elements of S, such that (S,E) is a *connected graph*. Write $\xi \sim \xi'$ when $(\xi,\xi') \in E$. As dynamics we consider the continuous-time Markov process $(\xi_t)_{t\geq 0}$ with state space S whose transition rates are given by

$$c_\beta(\xi,\xi') = \begin{cases} \mathrm{e}^{-\beta[H(\xi')-H(\xi)]_+}, & \xi \sim \xi', \\ 0, & \text{otherwise}, \end{cases} \tag{16.1.2}$$

i.e., transitions occur along edges only. This dynamics is called the *Metropolis dynamics* with respect to H at inverse temperature β. It is ergodic and reversible with respect to μ_β:

$$\mu_\beta(\xi)c_\beta(\xi,\xi') = \mu_\beta(\xi')c_\beta(\xi',\xi) \quad \forall \xi,\xi' \in S, \tag{16.1.3}$$

Choosing a particular model amounts to choosing Λ, S, E, H and β. The generator $\mathscr{L}_\beta$ of the dynamics is

$$(\mathscr{L}_\beta f)(\xi) = \sum_{\xi'\sim\xi} c_\beta(\xi,\xi')[f(\xi')-f(\xi)], \quad f\colon S \to \mathbb{R}. \tag{16.1.4}$$

In order to formulate our metastability theorems, we need some general definitions (compare with Definition 10.2).

Definition 16.1 (Communication heights, communication level sets, stability levels, sets of stable and metastable configurations)

(a) $\Phi(\xi,\xi')$ is the *communication height* between $\xi,\xi' \in S$ defined by

$$\Phi(\xi,\xi') = \min_{\gamma\colon \xi\to\xi'} \max_{\sigma\in\gamma} H(\sigma), \tag{16.1.5}$$

where $\gamma\colon \xi \to \xi'$ is any path of allowed moves from ξ to ξ'. For non-empty sets $A, B \subseteq S$ put

$$\Phi(A,B) = \min_{\xi\in A,\xi'\in B} \Phi(\xi,\xi'). \tag{16.1.6}$$

(b) $\mathscr{S}(\xi,\xi')$ is the *communication level set* between $\xi,\xi' \in S$ defined by

$$\mathscr{S}(\xi,\xi') = \Big\{\zeta \in S\colon\ \exists\gamma\colon \xi \to \xi',\ \gamma \ni \zeta\colon\ \max_{\eta\in\gamma} H(\eta) = H(\zeta) = \Phi(\xi,\xi')\Big\}. \tag{16.1.7}$$

(c) V_ξ is the *stability level* of $\xi \in S$ defined by

$$V_\xi = \Phi(\xi, I_\xi) - H(\xi), \tag{16.1.8}$$

where

$$I_\xi = \{\zeta \in S\colon\ H(\zeta) < H(\xi)\} \tag{16.1.9}$$

is the set of configurations with energy lower than ξ.

(d) S_{stab} is the set of configurations with minimal energy, called *stable configurations*, defined by

$$S_{\text{stab}} = \Big\{\xi \in S\colon\ H(\xi) = \min_{\zeta\in S} H(\zeta)\Big\}. \tag{16.1.10}$$

(e) S_{meta} is the set of non-minimal configurations with maximal stability, called *metastable configurations*, defined by

$$S_{\text{meta}} = \Big\{\xi \in S\colon\ V_\xi = \max_{\zeta\in S\setminus S_{\text{stab}}} V_\zeta\Big\}. \tag{16.1.11}$$

Definition 16.2 (Optimal paths, gates, dead-ends)

(a) $(\xi \to \xi')_{\text{opt}}$ is the set of paths realising the minimax in $\Phi(\xi,\xi')$.
(b) A set $\mathscr{W} \subseteq S$ is called a *gate* for $\xi \to \xi'$ if $\mathscr{W} \subseteq \mathscr{S}(\xi,\xi')$ and $\gamma \cap \mathscr{W} \neq \emptyset$ for all $\gamma \in (\xi \to \xi')_{\text{opt}}$.
(c) A set $\mathscr{W} \subseteq S$ is called a *minimal gate* for $\xi \to \xi'$ if it is a gate for $\xi \to \xi'$ and for any $\mathscr{W}' \subsetneq \mathscr{W}$ there exists a $\gamma' \in (\xi \to \xi')_{\text{opt}}$ such that $\gamma' \cap \mathscr{W}' = \emptyset$.
(d) A priori there may be several (not necessarily disjoint) minimal gates. Their union is denoted by $\mathscr{G}(\xi,\xi')$ and is called the *essential gate* for $(\xi \to \xi')_{\text{opt}}$.
(e) The configurations in $\mathscr{S}(\xi,\xi')\backslash\mathscr{G}(\xi,\xi')$ are called *dead-ends* for $(\xi \to \xi')_{\text{opt}}$.

Armed with these definitions we are ready to state our metastability theorems (see Fig. 16.1).

16.1.2 Metastability theorems and hypotheses

Theorems 16.4–16.6 below involve a pair of configurations $(\mathbf{m},\mathbf{s}) \in S_{\text{meta}} \times S_{\text{stab}}$, which will be referred to as the *metastable configuration*, respectively, the *stable*

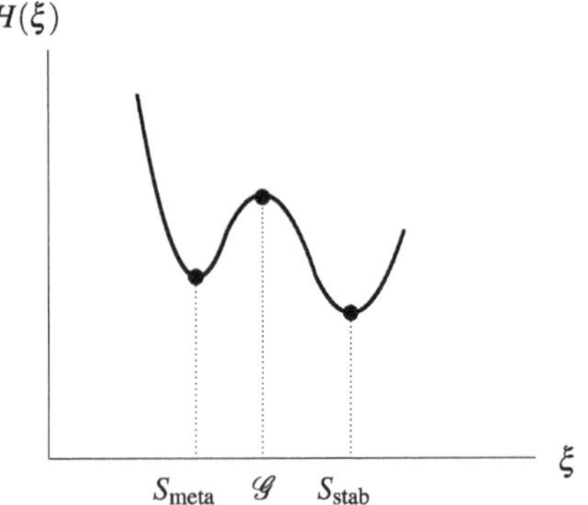

Fig. 16.1 Schematic picture of H, S_{meta}, S_{stab} and $\mathscr{G}$, the essential gate between S_{meta} and S_{stab}

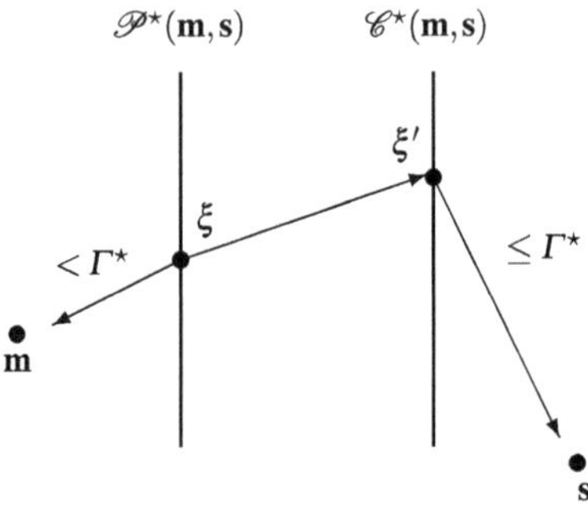

Fig. 16.2 Schematic picture of the protocritical set and the critical set

configuration. Associated with $(\mathbf{m},\mathbf{s})$ is a pair of sets $(\mathscr{P}^\star(\mathbf{m},\mathbf{s}),\mathscr{C}^\star(\mathbf{m},\mathbf{s}))$, which will be referred to as the *protocritical set*, respectively, the *critical set*, defined as follows.

Definition 16.3 (Protocritical and critical sets) (See Fig. 16.2.) Let

$$\Gamma^\star = \Phi(\mathbf{m},\mathbf{s}) - H(\mathbf{m}). \tag{16.1.12}$$

Then $(\mathscr{P}^\star(\mathbf{m},\mathbf{s}),\mathscr{C}^\star(\mathbf{m},\mathbf{s}))$ is the maximal subset of $S\times S$ such that:

(1) $\forall\xi\in\mathscr{P}^\star(\mathbf{m},\mathbf{s})\ \exists\xi'\in\mathscr{C}^\star(\mathbf{m},\mathbf{s})\colon\ \xi\sim\xi'$ and $\forall\xi'\in\mathscr{C}^*(\mathbf{m},\mathbf{s})\ \exists\xi\in\mathscr{P}^*(\mathbf{m},\mathbf{s})$: $\xi'\sim\xi$.
(2) $\forall\xi\in\mathscr{P}^\star(\mathbf{m},\mathbf{s})\colon\ \Phi(\xi,\mathbf{m})<\Phi(\xi,\mathbf{s})$.
(3) $\forall\xi'\in\mathscr{C}^\star(\mathbf{m},\mathbf{s})\ \exists\gamma\colon\ \xi'\to\mathbf{s}\colon\ \max_{\zeta\in\gamma}H(\zeta)-H(\mathbf{m})\leq\Gamma^\star,\ \gamma\cap\{\zeta\in S\colon \Phi(\zeta,\mathbf{m})<\Phi(\zeta,\mathbf{s})\}=\emptyset$.

Think of $\mathscr{P}^\star(\mathbf{m},\mathbf{s})$ as the set of configurations where the dynamics starting from **m** is "almost on top of the hill", and of $\mathscr{C}^\star(\mathbf{m},\mathbf{s})$ as the set of configurations where the dynamics "has reached the top of the hill" and is "capable of crossing over" to **s** without returning to "the valley around $\mathbf{m}$". The latter restriction is put in to remove the dead-ends. Note that

$$H(\xi) - H(\mathbf{m}) = \Gamma^\star \quad \forall\, \xi \in \mathscr{C}^\star(\mathbf{m},\mathbf{s}). \tag{16.1.13}$$

Also note that $\mathscr{C}^\star(\mathbf{m},\mathbf{s}) \subseteq \mathscr{G}(\mathbf{m},\mathbf{s})$, where the inclusion may be strict.

Theorems 16.4–16.6 below will be proved subject to two hypotheses:

(H1) $S_{\mathrm{meta}} = \{\mathbf{m}\}$, $S_{\mathrm{stab}} = \{\mathbf{s}\}$.
(H2) $\xi' \mapsto |\{\xi \in \mathscr{P}^\star(\mathbf{m},\mathbf{s})\colon\, \xi \sim \xi'\}|$ is constant on $\mathscr{C}^\star(\mathbf{m},\mathbf{s})$.

(H1) says that S_{meta} and S_{stab} are singletons, while (H2) says that all configurations in $\mathscr{C}^\star(\mathbf{m},\mathbf{s})$ have the same number of configurations in $\mathscr{P}^\star(\mathbf{m},\mathbf{s})$ from which they can be reached via an allowed move. Any pair of configurations $(\mathbf{m},\mathbf{s})$ satisfying (H1) is referred to as a *metastable pair*. Without loss of generality we may assume that

$$H(\mathbf{m}) = 0. \tag{16.1.14}$$

We write $\mathbb{P}_\xi$ to denote the law of $(\xi_t)_{t\geq 0}$ given $\xi_0 = \xi \in S$, and

$$\tau_A = \inf\{t \geq 0\colon\, \xi_t \in A,\, \exists\, 0 < s < t\colon\, \xi_s \neq \xi_0\} \tag{16.1.15}$$

to denote the *first hitting time of* $A \subseteq S$ *after the starting configuration has been left*. In what follows we abbreviate $\mathscr{P}^\star = \mathscr{P}^\star(\mathbf{m},\mathbf{s})$ and $\mathscr{C}^\star = \mathscr{C}^\star(\mathbf{m},\mathbf{s})$.

Theorem 16.4 (Critical gate and uniform entrance distribution)

(a) $\lim_{\beta\to\infty} \mathbb{P}_{\mathbf{m}}(\tau_{\mathscr{C}^\star} < \tau_{\mathbf{s}} \mid \tau_{\mathbf{s}} < \tau_{\mathbf{m}}) = 1$.
(b) $\lim_{\beta\to\infty} \mathbb{P}_{\mathbf{m}}(\xi_{\tau_{\mathscr{C}^\star}} = \chi) = 1/|\mathscr{C}^\star|$ *for all* $\chi \in \mathscr{C}^\star$.

Theorem 16.5 (Mean crossover time) *There exists a constant* $K \in (0,\infty)$ *such that*

$$\lim_{\beta\to\infty} \mathrm{e}^{-\beta\Gamma^\star}\, \mathbb{E}_{\mathbf{m}}(\tau_{\mathbf{s}}) = K. \tag{16.1.16}$$

Theorem 16.6 (Spectrum and exponential law of crossover time)

(a) $\lim_{\beta\to\infty} \lambda_\beta \mathbb{E}_{\mathbf{m}}(\tau_{\mathbf{s}}) = 1$, *where* λ_β *is the second eigenvalue of* $-\mathscr{L}_\beta$, *with* $\mathscr{L}_\beta$ *the generator of the Metropolis dynamics.*
(b) $\lim_{\beta\to\infty} \mathbb{P}_{\mathbf{m}}(\tau_{\mathbf{s}}/\mathbb{E}_{\mathbf{m}}(\tau_{\mathbf{s}}) > t) = \mathrm{e}^{-t}$ *for all* $t \geq 0$.

It turns out that typically $\Gamma^\star$ is independent of Λ (provided Λ is large enough) and is relatively robust against variations of the dynamics, while K depends on Λ and is rather sensitive to the details of the dynamics.

In Sect. 16.3 we will see that K is given by a *non-trivial variational formula* involving the set of all configurations where the dynamics can enter and exit $\mathscr{S}(\mathbf{m},\mathbf{s})$ (see Lemma 16.17 below). This set includes the border of the "valleys around $\mathbf{m}$ and $\mathbf{s}$", and possibly the border of "wells in $\mathscr{S}(\mathbf{m},\mathbf{s})$", i.e., configurations with energy $< \Gamma^\star$ but communication height $\Gamma^\star$ towards both $\mathbf{m}$ and $\mathbf{s}$. We will see in

Chap. 17 that for Glauber dynamics there are no wells and K can be computed explicitly. We will see in Chap. 18 that for Kawasaki dynamics there are wells, but they are sometimes harmless, e.g. when Λ is a large box in $\mathbb{Z}^2$ whose size tends to infinity (after the limit $\beta \to \infty$ has been taken).

While (H1) plays a central role in the derivation of Theorems 16.4–16.6, (H2) is needed for Theorem 16.4(b) only.

16.1.3 Discussion

1. Theorem 16.4(a) says that $\mathscr{C}^\star$ is a gate for the crossover, i.e., on its way from $\mathbf{m}$ to $\mathbf{s}$ the dynamics passes through $\mathscr{C}^\star$ with a probability tending to one in the limit of low temperature. Theorem 16.4(b) says that, in this limit, all critical configurations are equally likely to be seen upon first entrance in $\mathscr{C}^\star$. Theorem 16.5 says that the average crossover time is asymptotic to $K\mathrm{e}^{\Gamma^\star\beta}$, which is the classical *Arrhenius law* (see Sect. 1.3.1). Theorem 16.6(a) says that the spectral gap $-\mathscr{L}_\beta$ (the first eigenvalue of $-\mathscr{L}_\beta$ is zero) scales like the inverse of the average crossover time, while Theorem 16.6(b) says that asymptotically the crossover time is exponentially distributed on the scale of its average.

2. Theorems 16.4–16.6 are *model-independent*, i.e., they hold in the same form for all stochastic dynamics in a finite volume in the limit of low temperature and for any pair $(\mathbf{m},\mathbf{s})$ satisfying hypotheses (H1–H2). In fact, we will see that (H1–H2) are essentially the minimal hypotheses needed to prove Theorems 16.4–16.6. The *model-dependent* ingredients of Theorems 16.4–16.6 are the pair $(\mathbf{m},\mathbf{s})$ and the triple $(\Gamma^\star,\mathscr{C}^\star,K)$. In Chaps. 17 and 18 we will identify these for Glauber dynamics and Kawasaki dynamics, and prove (H1–H2).

3. There is some flexibility in letting our dynamics start and end at configurations that are different from $\mathbf{m}$ and $\mathbf{s}$. For instance, we will see that the same results apply when the initial configuration is drawn from the "valley around $\mathbf{m}$", and the target configuration is drawn near the bottom of the "valley around $\mathbf{s}$" (see Sect. 16.2.3, Eq. (16.1.17) for precise definitions).

4. Hypothesis (H1) can be relaxed. The Hamiltonian may have valleys that are deeper than $\Gamma^\star$ (the energy barrier between $\mathbf{m}$ and $\mathbf{s}$), but are shielded away from $\mathbf{m}$ by an energy barrier that is higher than $\Gamma^\star$. In that case the dynamics has a negligible probability to enter these valleys, and (H1) is required to hold only on the subset of S obtained by removing all the configurations with energy $> \Gamma^\star + H(\mathbf{m})$. The average crossover time on this subset is the relevant time scale, not the average crossover time on S, which is much longer. See also Item 3 in Sect. 16.5.

16.1.4 Consequences of the hypotheses

Lemmas 16.7–16.10 below are immediate consequences of (H1) and will be needed in Sect. 16.2. Recall that $H(\mathbf{m}) = 0$ by (16.1.14). Recall also Figs. 16.1–16.2.

Lemma 16.7 (H1) *implies that* $V_{\mathbf{m}} = \Gamma^\star$.

Proof By Definition 16.1(c–e), $\mathbf{s} \in I_{\mathbf{m}}$ and hence $V_{\mathbf{m}} \leq \Gamma^\star$. We show that (H1) implies $V_{\mathbf{m}} = \Gamma^\star$. The proof is by contradiction. Suppose that $V_{\mathbf{m}} < \Gamma^\star$. Then there exists a $\xi_0 \in I_{\mathbf{m}}$ such that

$$\Phi(\mathbf{m}, \xi_0) = \Phi(\mathbf{m}, \xi_0) - H(\mathbf{m}) = V_{\mathbf{m}} < \Gamma^\star. \tag{16.1.17}$$

Since (H1) tells us that $\mathbf{m}$ has the largest stability level, we can proceed to reduce the energy further until we hit $\mathbf{s}$. Indeed, the finiteness of S guarantees that there exists an $m \in \mathbb{N}_0$ and a sequence $\xi_1, \dots, \xi_m \in S \backslash \mathbf{m}$ with $\xi_m = \mathbf{s}$ such that $\xi_{i+1} \in I_{\xi_i}$ and $\Phi(\xi_i, \xi_{i+1}) - H(\xi_i) < V_{\mathbf{m}}$ for $i = 0, \dots, m-1$. Therefore we have

$$\begin{aligned} \Phi(\xi_0, \mathbf{s}) &\leq \max_{i=0,\dots,m-1} \Phi(\xi_i, \xi_{i+1}) < \max_{i=0,\dots,m-1} \big[H(\xi_i) + V_{\mathbf{m}}\big] \\ &= H(\xi_0) + V_{\mathbf{m}} < H(\mathbf{m}) + \Gamma^\star = \Gamma^\star, \end{aligned} \tag{16.1.18}$$

where in the first inequality we use the ultrametricity of the communication height,

$$\Phi(\xi, \chi) \leq \max\big\{\Phi(\xi, \zeta), \Phi(\zeta, \chi)\big\} \quad \forall\, \xi, \chi, \zeta \in S \tag{16.1.19}$$

(a property that is closely related to the approximate ultrametricity of capacity encountered in Corollary 8.6), and in the last inequality we use that $V_{\mathbf{m}} \leq \Gamma^\star$ and $H(\xi_0) < H(\mathbf{m})$ because $\xi_0 \in I_{\mathbf{m}}$. It follows from (16.1.17)–(16.1.19) that

$$\Gamma^\star = \Phi(\mathbf{m}, \mathbf{s}) \leq \max\big\{\Phi(\mathbf{m}, \xi_0), \Phi(\xi_0, \mathbf{s})\big\} < \Gamma^\star, \tag{16.1.20}$$

which is a contradiction. □

Lemma 16.8 (H1) *implies that* $H(\xi) > 0$ *for all* $\xi \in S \backslash \mathbf{m}$ *with* $\Phi(\xi, \mathbf{m}) \leq \Phi(\xi, \mathbf{s})$.

Proof The proof is again by contradiction. Fix $\xi_0 \in S \backslash \mathbf{m}$ with $\Phi(\xi_0, \mathbf{m}) \leq \Phi(\xi_0, \mathbf{s})$ and suppose that $H(\xi_0) \leq 0$. Then $\mathbf{m} \notin I_{\xi_0}$. As in the proof of Lemma 16.7, there exist an $m \in \mathbb{N}_0$ and a sequence $\xi_0, \dots, \xi_m \in S$ with $\xi_m = \mathbf{s}$ such that $\xi_{i+1} \in I_{\xi_i}$ and $\Phi(\xi_i, \xi_{i+1}) - H(\xi_i) < V_{\mathbf{m}} = \Gamma^\star$ for $i = 0, \dots, m-1$. Therefore, as in (16.1.18), we get $\Phi(\xi_0, \mathbf{s}) - H(\xi_0) < V_{\mathbf{m}} = \Gamma^\star$. Hence

$$\Gamma^\star = \Phi(\mathbf{m}, \mathbf{s}) \leq \max\big\{\Phi(\mathbf{m}, \xi_0), \Phi(\xi_0, \mathbf{s})\big\} = \Phi(\xi_0, \mathbf{s}) \leq \Phi(\xi_0, \mathbf{s}) - H(\xi_0) < \Gamma^\star, \tag{16.1.21}$$

which is a contradiction. □

Lemma 16.9 (H1) *implies that there exists a* $V^\star < \Gamma^\star$ *such that* $\Phi(\xi,\{\mathbf{m},\mathbf{s}\}) - H(\xi) \le V^\star$ *for all* $\xi \in S\backslash\{\mathbf{m},\mathbf{s}\}$.

Proof In the proof of Lemma 16.8 we have shown that $\Phi(\xi_0,\mathbf{s}) - H(\xi_0) < \Gamma^\star$ for all $\xi_0 \in S\backslash\mathbf{m}$. But

$$\Phi\big(\xi_0,\{\mathbf{m},\mathbf{s}\}\big) = \min\big\{\Phi(\xi_0,\mathbf{m}),\Phi(\xi_0,\mathbf{s})\big\} \le \Phi(\xi_0,\mathbf{s}), \tag{16.1.22}$$

while $\Phi(\mathbf{m},\{\mathbf{m},\mathbf{s}\}) - H(\mathbf{m}) = 0$, and so the claim follows. □

Lemma 16.10 *Let* $\bar{\mathscr{C}}^\star = \{\xi' \in S\backslash[\mathscr{P}^\star \cup \mathscr{C}^\star]\colon H(\xi') \le \Gamma^\star, \exists \xi \in \mathscr{C}^\star\colon \xi \sim \xi'\}$. *Then for every* $\xi' \in \bar{\mathscr{C}}^\star$ *every path in* $(\xi' \to \mathbf{m})_{\mathrm{opt}}$ *passes through* $\mathscr{P}^\star$.

Proof Pick any $\xi' \in \bar{\mathscr{C}}^\star$, any $\gamma \in (\xi' \to \mathbf{m})_{\mathrm{opt}}$ and any $\xi \in \mathscr{C}^\star$ such that $\xi \sim \xi'$. We have $\max_{\zeta\in\gamma} H(\zeta) \le \Gamma^\star$, because $H(\xi') \le \Gamma^\star$ and $\Phi(\mathbf{m},\xi) \le \Gamma^\star$ by Definition 16.3. The reverse of γ can be extended by the single move from ξ' to ξ to obtain a path $\gamma'\colon \mathbf{m} \to \xi$ such that $\max_{\zeta\in\gamma'} H(\zeta) \le \Gamma^\star$. Moreover, by Definition 16.3(3), this path can be further extended by a path $\gamma''\colon \xi \to \mathbf{s}$ such that $\max_{\zeta\in\gamma''} H(\zeta) \le \Gamma^\star$ and $\gamma'' \cap \mathscr{P}^\star = \emptyset$. The concatenation $\gamma' \cup \gamma''$ is an optimal path, i.e., $\gamma' \cup \gamma'' \in (\mathbf{m} \to \mathbf{s})_{\mathrm{opt}}$. However, by the maximality in Definition 16.3, *any* path in $(\mathbf{m} \to \mathbf{s})_{\mathrm{opt}}$ must hit $\mathscr{P}^\star$. Since γ'' does not hit $\mathscr{P}^\star$, it follows that γ' hits $\mathscr{P}^\star$. But $\xi \in \mathscr{C}^\star$ and $\mathscr{P}^\star \cap \mathscr{C}^\star = \emptyset$, and so the piece of γ' between $\mathbf{m}$ and ξ' hits $\mathscr{P}^\star$. □

Lemmas 16.7–16.8 say that $\mathbf{m}$ lies at the bottom of a valley of depth $\Gamma^\star$, Lemma 16.9 says that there are no deeper valleys anywhere else, while Lemma 16.10 says that once an optimal path from $\mathbf{m}$ to $\mathbf{s}$ is over the hill it cannot go back to $\mathbf{m}$ without passing through the protocritical set (see Fig. 16.2).

16.2 Preliminaries

In this section we recall some facts from Part III, adapted to the present context, and use them to derive a few lemmas that are needed in Sect. 16.3 to prove Theorems 16.4–16.6.

16.2.1 Dirichlet form and capacity

As we have seen in Chap. 8, the key object in the potential-theoretic approach to metastability is the *Dirichlet form*

$$\mathscr{E}_\beta(h,h) = \tfrac{1}{2} \sum_{\xi,\xi'\in S} \mu_\beta(\xi) c_\beta(\xi,\xi')\big[h(\xi) - h(\xi')\big]^2, \quad h\colon S \to [0,1], \tag{16.2.1}$$

where μ_β is the Gibbs measure defined in (16.1.1) and c_β is the kernel of transition rates defined in (16.1.2). Given a pair of non-empty disjoint sets $A, B \subseteq S$, the *capacity* of the pair A, B is given by the *Dirichlet principle*,

$$\mathrm{cap}_\beta(A, B) = \min_{\substack{h:\, S\to[0,1]\\ h|_A=1, h|_B=0}} \mathscr{E}_\beta(h, h), \tag{16.2.2}$$

where $h|_A = 1$ means that $h(\xi) = 1$ for all $\xi \in A$ and $h|_B = 0$ means that $h(\xi) = 0$ for all $\xi \in B$. The unique minimizer $h_{A,B}$ of (16.2.2) is called the *equilibrium potential* of the pair A, B, and is the solution of the equation

$$\begin{aligned} (-\mathscr{L}_\beta h)(\xi) &= 0, \quad \xi \in S\backslash(A \cup B),\\ h(\xi) &= 1, \quad \xi \in A,\\ h(\xi) &= 0, \quad \xi \in B, \end{aligned} \tag{16.2.3}$$

which is given by

$$\begin{array}{ll} h_{A,B}(\xi) = \mathbb{P}_\xi(\tau_A < \tau_B), & \xi \in S\backslash(A \cup B),\\ h_{A,B}(\xi) = 1, & \xi \in A,\\ h_{A,B}(\xi) = 0, & \xi \in B. \end{array} \tag{16.2.4}$$

An alternative expression for the capacity is

$$\mathrm{cap}_\beta(A, B) = \sum_{\xi\in A} \mu_\beta(\xi)\, c_\beta(\xi)\, \mathbb{P}_\xi(\tau_B < \tau_A) \tag{16.2.5}$$

with $c_\beta(\xi) = \sum_{\xi'\in S\backslash\xi} c_\beta(\xi, \xi')$ the rate of moving out of ξ (recall (7.1.19)–(7.1.20), (7.1.39) and (16.1.15)).

16.2.2 A priori estimates on the capacity

The following estimates on capacity will be needed later on.

Lemma 16.11 *For every pair of non-empty disjoint sets $A, B \subseteq S$ there exist constants $0 < C_1 \leq C_2 < \infty$ (depending on A, B) such that*

$$C_1 \leq \mathrm{e}^{\beta\Phi(A,B)} Z_\beta\, \mathrm{cap}_\beta(A, B) \leq C_2 \quad \forall \beta \in (0, \infty). \tag{16.2.6}$$

Proof The proof uses basic properties of communication heights.

Upper bound: Suppose that A, B are such that

$$\Phi(\zeta, A) > H(\zeta) \quad \forall \zeta \in B. \tag{16.2.7}$$

Then, picking $h = \mathbb{1}_{K(A,B)}$ in (16.2.2) with

$$K(A,B) = \{\xi \in S\colon\ \Phi(\xi,A) \le \Phi(\xi,B)\}, \tag{16.2.8}$$

we get

$$\mathrm{cap}_\beta(A,B) \le \mathscr{E}_\beta(\mathbb{1}_{K(A,B)}, \mathbb{1}_{K(A,B)}). \tag{16.2.9}$$

Here note that $A \subset K(A,B)$, while (16.2.7) guarantees that $B \subset S\backslash K(A,B)$, so that the boundary conditions on A and B are met.

To estimate $\mathscr{E}_\beta(\mathbb{1}_{K(A,B)}, \mathbb{1}_{K(A,B)})$, the key observation is that if $\xi \sim \xi'$ with $\xi \in K(A,B)$ and $\xi' \in S\backslash K(A,B)$, then

$$\begin{aligned} &(1) \quad H(\xi') < H(\xi),\\ &(2) \quad H(\xi) \ge \Phi(A,B). \end{aligned} \tag{16.2.10}$$

To prove (1), we argue by contradiction. Suppose that $H(\xi') \ge H(\xi)$. Then, because $\xi \sim \xi'$, we have

$$\Phi(\xi',C) = \Phi(\xi,C) \vee H(\xi') \quad \forall C \subseteq S. \tag{16.2.11}$$

But $\xi \in K(A,B)$ tells us that $\Phi(\xi,A) \le \Phi(\xi,B)$, and so (16.2.11) gives

$$\Phi(\xi',A) = \Phi(\xi,A) \vee H(\xi') \le \Phi(\xi,B) \vee H(\xi') = \Phi(\xi',B). \tag{16.2.12}$$

Therefore $\xi' \in K(A,B)$, which is a contradiction. To see (2), note that (1) implies

$$\Phi(\xi,C) = \Phi(\xi',C) \vee H(\xi) \quad \forall C \subseteq S. \tag{16.2.13}$$

Trivially, $H(\xi) \le \Phi(\xi,B)$. We argue by contradiction that equality holds. Suppose that $H(\xi) < \Phi(\xi,B)$. Then (16.2.13) gives

$$\begin{aligned} H(\xi) < \Phi(\xi,B) &= \Phi(\xi',B) \vee H(\xi) = \Phi(\xi',B)\\ &< \Phi(\xi',A) = \Phi(\xi',A) \vee H(\xi) = \Phi(\xi,A), \end{aligned} \tag{16.2.14}$$

where the second inequality uses that $\xi' \in S\backslash K(A,B)$. Thus, we have $\Phi(\xi,A) > \Phi(\xi,B)$, which contradicts $\xi \in K(A,B)$. From the equality $H(\xi) = \Phi(\xi,B)$ and (16.1.19) we obtain $\Phi(A,B) \le \Phi(A,\xi) \vee \Phi(\xi,B) = \Phi(\xi,B) = H(\xi)$, which proves (2).

Combining (16.2.10) with (16.1.1)–(16.1.3), we find that

$$\begin{aligned} \mu_\beta(\xi)c_\beta(\xi,\xi') &= \frac{1}{Z_\beta}\mathrm{e}^{-\beta[H(\xi)\vee H(\xi')]}\\ &\le \frac{1}{Z_\beta}\mathrm{e}^{-\beta\Phi(A,B)} \quad \forall \xi \in K(A,B),\ \xi' \in S\backslash K(A,B). \end{aligned} \tag{16.2.15}$$

Hence

$$\mathscr{E}_\beta(\mathbb{1}_{K(A,B)}, \mathbb{1}_{K(A,B)}) \le C_2\frac{1}{Z_\beta}\mathrm{e}^{-\beta\Phi(A,B)} \tag{16.2.16}$$

with $C_2 = |\{(\xi,\xi') \in K(A,B) \times S\backslash K(A,B)\colon \xi \sim \xi'\}|$. Together with (16.2.9) this completes the proof subject to (16.2.7).

Reversing the roles of A and B, we see that the same bound holds when

$$\Phi(\zeta', B) > H(\zeta') \quad \forall \zeta' \in A. \tag{16.2.17}$$

Thus it remains to consider A, B such that

$$\begin{aligned} &\exists \zeta \in B\colon \ \Phi(\zeta, A) = H(\zeta), \\ &\exists \zeta' \in A\colon \ \Phi(\zeta', B) = H(\zeta'). \end{aligned} \tag{16.2.18}$$

Estimating

$$\begin{aligned} \mathrm{cap}_\beta(A,B) &\le \mathscr{E}_\beta(\mathbb{1}_A, \mathbb{1}_A) = \sum_{\xi \in A, \xi' \in S\backslash A} \mu_\beta(\xi) c_\beta(\xi, \xi') \\ &= \sum_{\substack{\xi \in A, \xi' \in S\backslash A \\ \xi \sim \xi'}} \frac{1}{Z_\beta} \mathrm{e}^{-\beta[H(\xi) \vee H(\xi')]} \le C_2 \frac{1}{Z_\beta} \mathrm{e}^{-\beta \Phi(A, S\backslash A)} \end{aligned} \tag{16.2.19}$$

with $C_2 = |\{(\xi,\xi')\colon \xi \sim \xi', \xi \in A, \xi' \in S\backslash A\}|$, and using that $\Phi(A, S\backslash A) = \Phi(A,B)$ by (16.2.18), we get the claim.

Lower bound: The lower bound is obtained by picking any self-avoiding path

$$\gamma = (\gamma_0, \gamma_1, \dots, \gamma_L) \tag{16.2.20}$$

that realizes the minimax in $\Phi(A,B)$ and ignore all the transitions that are not in this path, i.e.,

$$\mathrm{cap}_\beta(A,B) \ge \min_{\substack{h\colon \gamma \to [0,1] \\ h(\gamma_0)=1, h(\gamma_L)=0}} \mathscr{E}_\beta^\gamma(h,h), \tag{16.2.21}$$

where the Dirichlet form $\mathscr{E}_\beta^\gamma$ is defined as $\mathscr{E}_\beta$ in (16.2.1) but with S replaced by γ. Due to the one-dimensional nature of the set γ, the variational problem in the right-hand side can be solved explicitly by elementary computations (recall Sect. 7.1.4). We find that the minimum equals

$$M = \left[\sum_{l=0}^{L-1} \frac{1}{\mu_\beta(\gamma_l) c_\beta(\gamma_l, \gamma_{l+1})} \right]^{-1}, \tag{16.2.22}$$

and is uniquely attained at h given by

$$h(\gamma_l) = M \sum_{k=0}^{l-1} \frac{1}{\mu_\beta(\gamma_k) c_\beta(\gamma_k, \gamma_{k+1})}, \quad l = 0, 1, \dots, L. \tag{16.2.23}$$

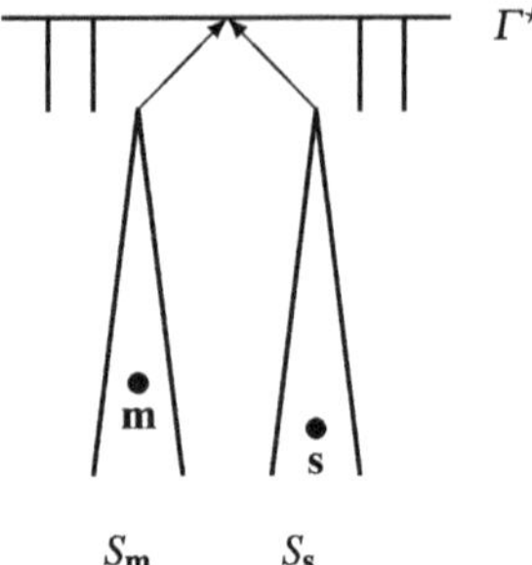

Fig. 16.3 Schematic picture of the subgraphs $S^\star$ (on or below the top line) and $S^{\star\star}$ (below the top line) and of the connected components $S_{\mathbf{m}}$ and $S_{\mathbf{s}}$. The four vertical lines represent dead-ends

We thus have

$$\begin{aligned} \mathrm{cap}_\beta(A, B) &\geq M \geq \frac{1}{L} \min_{l=0,1,\ldots,L-1} \mu_\beta(\gamma_l) c_\beta(\gamma_l, \gamma_{l+1}) \\ &= \frac{1}{L} \frac{1}{Z_\beta} \min_{l=0,1,\ldots,L-1} \mathrm{e}^{-\beta[H(\gamma_l) \vee H(\gamma_{l+1})]} = C_1 \frac{1}{Z_\beta} \mathrm{e}^{-\beta \Phi(A,B)} \end{aligned} \tag{16.2.24}$$

with $C_1 = 1/L$. □

16.2.3 Graph structure of the energy landscape

In this section we have a closer look at the geometric structure of the set S.

Theorem 16.12 (Graph structure of the energy landscape) *View S as a graph whose vertices are the configurations and whose edges connect configurations that can be obtained from each other via an allowed move, i.e., (ξ, ξ') is an edge if and only if $\xi \sim \xi'$. Define (see Fig. 16.3)*

- *$S^\star$ is the subgraph of S obtained by removing all vertices ξ with $H(\xi) > \Gamma^\star$ and all edges incident to these vertices;*
- *$S^{\star\star}$ is the subgraph of $S^\star$ obtained by removing all vertices ξ with $H(\xi) = \Gamma^\star$ and all edges incident to these vertices;*
- *$S_{\mathbf{m}}$ and $S_{\mathbf{s}}$ are the connected components of $S^{\star\star}$ containing $\mathbf{m}$ and $\mathbf{s}$, respectively.*

Then

$$\begin{aligned} S_{\mathbf{m}} &= \left\{\xi \in S\colon\ \Phi(\xi, \mathbf{m}) < \Phi(\xi, \mathbf{s}) = \Gamma^\star\right\}, \\ S_{\mathbf{s}} &= \left\{\xi \in S\colon\ \Phi(\xi, \mathbf{s}) < \Phi(\xi, \mathbf{m}) = \Gamma^\star\right\}. \end{aligned} \tag{16.2.25}$$

Moreover, $S_{\mathbf{m}}$ and $S_{\mathbf{s}}$ are disconnected in $S^{\star\star}$, and

$$\begin{aligned} &\mathscr{P}^\star \subseteq S_{\mathbf{m}}, \qquad \mathscr{C}^\star \subseteq S^\star \backslash S_{\mathbf{m}}, \\ &\forall \xi \in \mathscr{C}^\star \ \exists \gamma: \ \xi \to S_{\mathbf{s}} \quad \textit{such that } \gamma \backslash \xi \subseteq S^\star \backslash S_{\mathbf{m}}. \end{aligned} \tag{16.2.26}$$

Proof All paths connecting **m** and **s** reach energy level $\geq \Gamma^\star$ (recall that $H(\mathbf{m}) = 0$ by (16.1.14)). Therefore $S_{\mathbf{m}}$ and $S_{\mathbf{s}}$ are disconnected in $S^{\star\star}$ (because $S^{\star\star}$ does not contain vertices with energy $\geq \Gamma^\star$). The claims in (16.2.25) are immediate from the definition of $S_{\mathbf{m}}$ and $S_{\mathbf{s}}$. The claims in (16.2.26) are immediate consequences of Definition 16.3. □

16.2.4 Metastable pair

An important consequence of (H1) and Lemma 16.11 is the following.

Lemma 16.13 (Metastable pair) *The pair* $\{\mathbf{m}, \mathbf{s}\}$ *is a* metastable set *in the sense of Definition* 8.2:

$$\lim_{\beta\to\infty} \frac{\max_{\xi\notin\{\mathbf{m},\mathbf{s}\}} \mu_\beta(\xi)/\mathrm{cap}_\beta(\xi, \{\mathbf{m}, \mathbf{s}\})}{\min_{\xi\in\{\mathbf{m},\mathbf{s}\}} \mu_\beta(\xi)/\mathrm{cap}_\beta(\xi, \{\mathbf{m}, \mathbf{s}\}\backslash\xi)} = 0. \tag{16.2.27}$$

Proof Note that (16.1.1), Lemma 16.9 and the lower bound in (16.2.6) give that the numerator is bounded from above by $e^{\beta(V^\star - H(\mathbf{m}))}/C_1 = e^{\beta(\Gamma^\star - \delta)}/C_1$ for some $\delta > 0$, while (16.1.1), the definition of $\Gamma^\star$ and the upper bound in (16.2.6) give that the denominator is bounded from below by $e^{\Gamma^\star \beta}/C_2$ (the minimum being attained at **m**). □

The property in (16.2.27) has an important consequence.

Lemma 16.14 (Mean crossover time asymptotics) $\mathbb{E}_{\mathbf{m}}(\tau_{\mathbf{s}}) = [Z_\beta \, \mathrm{cap}_\beta(\mathbf{m}, \mathbf{s})]^{-1}[1 + o(1)]$ *as* $\beta \to \infty$.

Proof According to (8.2.10) and Theorem 8.15,

$$\mathbb{E}_{\mathbf{m}}(\tau_{\mathbf{s}}) = \frac{\mu_\beta(A(\mathbf{m}))}{\mathrm{cap}_\beta(\mathbf{m}, \mathbf{s})} \big[1 + o(1)\big], \quad \beta \to \infty, \tag{16.2.28}$$

where

$$\begin{aligned} A(\mathbf{m}) &= \big\{\xi \in S: \ \mathbb{P}_\xi(\tau_{\mathbf{m}} < \tau_{\mathbf{s}}) \geq \mathbb{P}_\xi(\tau_{\mathbf{s}} < \tau_{\mathbf{m}})\big\} \\ &= \big\{\xi \in S: \ h_{\mathbf{m},\mathbf{s}}(\xi) \geq \tfrac{1}{2}\big\}. \end{aligned} \tag{16.2.29}$$

It follows from Lemma 16.15 below that

$$\lim_{\beta\to\infty} \min_{\xi\in S_{\mathbf{m}}} h_{\mathbf{m},\mathbf{s}}(\xi) = 1, \qquad \lim_{\beta\to\infty} \max_{\xi\in S_{\mathbf{s}}} h_{\mathbf{m},\mathbf{s}}(\xi) = 0. \tag{16.2.30}$$

Hence, for large enough β,

$$S_{\mathbf{m}} \subseteq A(\mathbf{m}) \subseteq S \backslash S_{\mathbf{s}}. \tag{16.2.31}$$

By Lemma 16.8, we have $H(\xi) > 0 = H(\mathbf{m})$ for all $\xi \neq \mathbf{m}$ such that $\Phi(\xi, \mathbf{m}) \leq \Phi(\xi, \mathbf{s})$. Therefore, by the second inclusion in (16.2.31),

$$\min_{\xi \in A(\mathbf{m}) \backslash \mathbf{m}} H(\xi) > 0. \tag{16.2.32}$$

The latter in turn implies that $\mu_\beta(A(\mathbf{m}))/\mu_\beta(\mathbf{m}) = 1 + o(1)$. Since $\mu_\beta(\mathbf{m}) = 1/Z_\beta$, we get the claim. □

What Lemma 16.14 shows is that the proof of Theorem 16.5 revolves around getting sharp bounds on $Z_\beta \operatorname{cap}_\beta(\mathbf{m}, \mathbf{s})$. The a priori estimates in (16.2.6) serve as a jump board, because together with Lemma 16.14 they already yield the estimate

$$\frac{1}{C_2} \leq \mathrm{e}^{-\beta \Gamma^\star} \mathbb{E}_{\mathbf{m}}(\tau_{\mathbf{s}}) \leq C_1. \tag{16.2.33}$$

Thus, our task is to narrow down the constants leading to the identification of the prefactor K. The strategy in Sect. 16.3 to do so is the following:

- Note that all terms in the Dirichlet form in (16.2.1) involving configurations ξ with $H(\xi) > \Gamma^\star$, i.e., $\xi \in S \backslash S^\star$, contribute at most $C\mathrm{e}^{-\beta(\Gamma^\star + \delta)}$ for some $\delta > 0$ and can be neglected. Thus, effectively we can replace S by $S^\star$.
- Show that $h_{\mathbf{m},\mathbf{s}} = 1 - O(\mathrm{e}^{-\beta\delta})$ on $S_{\mathbf{m}}$ and $h_{\mathbf{m},\mathbf{s}} = O(\mathrm{e}^{-\beta\delta})$ on $S_{\mathbf{s}}$ for some $\delta > 0$. Thus, effectively we can replace $h_{\mathbf{m},\mathbf{s}}$ by 1 on $S_{\mathbf{m}}$ and by 0 on $S_{\mathbf{s}}$.
- Derive sharp estimates for $h_{\mathbf{m},\mathbf{s}}$ on $S^\star \backslash (S_{\mathbf{m}} \cup S_{\mathbf{s}})$ in terms of a variational formula involving only the vertices and the edges that are on or incident to $S^\star \backslash (S_{\mathbf{m}} \cup S_{\mathbf{s}})$. Use this variational formula to identify K.

16.3 Proof of the metastability theorems

With the preparations done in Sect. 16.2, we are now ready to prove Theorems 16.4–16.6. This will be done in Sects. 16.3.1–16.3.3, in reverse order.

16.3.1 Exponential distribution of the crossover time

Proof Theorem 16.6 follows from the general theory in Sect. 8.4. The intuition behind the exponential distribution of the crossover time is simple: each time the dynamics reaches $\mathscr{C}^\star(\mathbf{m}, \mathbf{s})$ but fails to enter $S_{\mathbf{s}}$ and instead falls back into $S_{\mathbf{m}}$, it has a probability exponentially close to 1 to return to $\mathbf{m}$ because $\mathbf{m}$ lies at the bottom of $S_{\mathbf{m}}$ (recall Lemma 16.8). Each time the dynamics returns to $\mathbf{m}$, it starts from scratch.

Thus, the dynamics manages to reach a critical configuration and go over the hill only after a number of unsuccessful attempts that tends to infinity as $\beta \to \infty$, each having a small probability that tends to zero as $\beta \to \infty$. Consequently, the time to go over the hill is exponentially distributed on the scale of its average. □

16.3.2 Average crossover time

In this section we prove Theorem 16.5.

Proof Our starting point is Lemma 16.14. Recalling (16.2.1)–(16.2.4), our task is to show that

$$\begin{aligned} Z_\beta \operatorname{cap}_\beta(\mathbf{m},\mathbf{s}) &= \tfrac{1}{2} \sum_{\xi,\xi' \in S} Z_\beta \mu_\beta(\xi) c_\beta(\xi,\xi') \left[h_{\mathbf{m},\mathbf{s}}(\xi) - h_{\mathbf{m},\mathbf{s}}(\xi')\right]^2 \\ &= \left[1+o(1)\right] \Theta\, \mathrm{e}^{-\beta\Gamma^\star}, \quad \beta \to \infty, \end{aligned} \tag{16.3.1}$$

and to identify the constant Θ, since (16.3.1) will imply (16.1.16) with $\Theta = 1/K$. This is done in three steps: in the first two steps we derive sharp estimates on $h^\star_{\mathbf{m},\mathbf{s}}$, in the third step we use these estimates to derive a variational formula for Θ.

1. For all $\xi \in S \backslash S^\star$ we have $H(\xi) > \Gamma^\star$, and so there exists a $\delta > 0$ such that $Z_\beta \mu_\beta(\xi) \leq \mathrm{e}^{-\beta(\Gamma^\star + \delta)}$. Since $c_\beta(\xi,\xi') \leq 1$ for all $\xi,\xi' \in S$, we can therefore replace S by $S^\star$ in the sum in (16.3.1) at the cost of a prefactor $1 + O(\mathrm{e}^{-\beta\delta})$ (for details, see the proof of Lemma 16.17 below).

Lemma 16.15 *There exist $C < \infty$ and $\delta > 0$ such that*

$$\min_{\xi \in S_{\mathbf{m}}} h_{\mathbf{m},\mathbf{s}}(\xi) \geq 1 - C\mathrm{e}^{-\beta\delta}, \qquad \max_{\xi \in S_{\mathbf{s}}} h_{\mathbf{m},\mathbf{s}}(\xi) \leq C\mathrm{e}^{-\beta\delta}, \quad \forall \beta \in (0,\infty). \tag{16.3.2}$$

Proof Combine Lemma 8.4 with Lemma 16.11. □

2. Because of Lemma 16.15, on the set $S_{\mathbf{m}} \cup S_{\mathbf{s}}$, $h_{\mathbf{m},\mathbf{s}}$ is trivial and its contribution to the sum in (16.3.1) can be put into the prefactor $1 + o(1)$ (for details, see the proof of Lemma 16.17 below). Consequently, all that is needed is to understand what $h_{\mathbf{m},\mathbf{s}}$ looks like on the set

$$S^\star \backslash (S_{\mathbf{m}} \cup S_{\mathbf{s}}) = \left\{\xi \in S^\star \colon\ \Phi(\xi,\mathbf{m}) = \Phi(\xi,\mathbf{s}) = \Gamma^\star\right\}. \tag{16.3.3}$$

However, Lemma 16.16 below shows that $h_{\mathbf{m},\mathbf{s}}$ is also trivial on the set

$$S^{\star\star} \backslash (S_{\mathbf{m}} \cup S_{\mathbf{s}}) = \bigcup_{i=1}^{I} S_i, \tag{16.3.4}$$

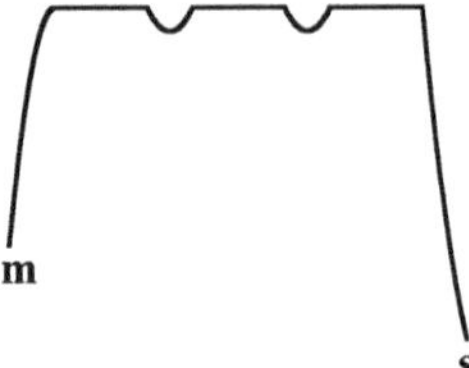

Fig. 16.4 Schematic picture of the wells S_i

which is a union of wells S_i, $i = 1, \dots, I$, in $\mathscr{S}(\mathbf{m}, \mathbf{s})$ for some $I \in \mathbb{N}$. Each S_i is a maximal set of communicating configurations with energy $< \Gamma^\star$ and with communication height $\Gamma^\star$ towards both **m** and **s** (recall Fig. 16.3 and see Fig. 16.4).

Lemma 16.16 *There exist $C < \infty$ and $\delta > 0$ such that*

$$\max_{\xi,\xi' \in S_i} \left| h_{\mathbf{m},\mathbf{s}}(\xi) - h_{\mathbf{m},\mathbf{s}}(\xi') \right| \leq C e^{-\beta\delta}, \quad \forall i = 1, \dots, I,\ \beta \in (0, \infty). \tag{16.3.5}$$

Proof Fix $i \in \{1, \dots, I\}$ and $\xi, \xi' \in S_i$. Estimate

$$h_{\mathbf{m},\mathbf{s}}(\xi) = \mathbb{P}_\xi(\tau_\mathbf{m} < \tau_\mathbf{s}) \leq \mathbb{P}_\xi(\tau_\mathbf{m} < \tau_{\xi'}) + \mathbb{P}_\xi(\tau_{\xi'} < \tau_\mathbf{m} < \tau_\mathbf{s}). \tag{16.3.6}$$

Combining Lemma 8.4 with Lemma 16.11, we have

$$\mathbb{P}_\xi(\tau_\mathbf{m} < \tau_{\xi'}) \leq \frac{\mathrm{cap}_\beta(\xi, \mathbf{m})}{\mathrm{cap}_\beta(\xi, \xi')} \leq C\, e^{-\beta[\Phi(\xi,\mathbf{m}) - \Phi(\xi,\xi')]} \leq C\, e^{-\beta\delta}, \tag{16.3.7}$$

where we use that $\Phi(\xi, \mathbf{m}) = \Gamma^\star$ and $\Phi(\xi, \xi') < \Gamma^\star$. But

$$\mathbb{P}_\xi(\tau_{\xi'} < \tau_\mathbf{m} < \tau_\mathbf{s}) = \mathbb{P}_\xi(\tau_{\xi'} < \tau_{\mathbf{m}\cup\mathbf{s}})\, \mathbb{P}_{\xi'}(\tau_\mathbf{m} < \tau_\mathbf{s}) \leq \mathbb{P}_{\xi'}(\tau_\mathbf{m} < \tau_\mathbf{s}) = h_{\mathbf{m},\mathbf{s}}(\xi'). \tag{16.3.8}$$

Combining (16.3.6)–(16.3.8), we therefore get

$$h_{\mathbf{m},\mathbf{s}}(\xi) \leq C\, e^{-\beta\delta} + h_{\mathbf{m},\mathbf{s}}(\xi'). \tag{16.3.9}$$

Interchange ξ and ξ' to get the claim. □

Lemma 16.16 shows that the contribution to the sum in (16.3.1) of the transitions inside a well can also be put into the prefactor $1 + o(1)$ (for details, see the proof of Lemma 16.17 below). Thus, only the transitions *in and out of wells* contribute.

3. In view of the above observations, the estimation of $Z_\beta\, \mathrm{cap}_\beta(\mathbf{m}, \mathbf{s})$ reduces to the study of a simpler variational problem.

Lemma 16.17 (Variational formula for the prefactor) *As $\beta \to \infty$,*

$$Z_\beta\, \mathrm{cap}_\beta(\mathbf{m}, \mathbf{s}) = \left[1 + o(1)\right] \Theta\, e^{-\beta\Gamma^\star} \tag{16.3.10}$$

with

$$\Theta = \min_{C_1,\dots,C_I} \min_{\substack{h:\, S^\star\to[0,1] \\ h|_{S_\mathbf{m}}=1,\, h|_{S_\mathbf{s}}=0,\, h|_{S_i}=C_i\,\forall i=1,\dots,I}} \tfrac{1}{2} \sum_{\xi,\xi'\in S^\star} \mathbb{1}_{\{\xi\sim\xi'\}} \left[h(\xi)-h(\xi')\right]^2 \tag{16.3.11}$$

Proof First, recalling (16.1.1)–(16.1.2) and (16.2.1)–(16.2.2), we have

$$\begin{aligned} &Z_\beta \operatorname{cap}_\beta(\mathbf{m},\mathbf{s}) \\ &\quad = Z_\beta \min_{\substack{h:\, S\to[0,1] \\ h(\mathbf{m})=1,\, h(\mathbf{s})=0}} \tfrac{1}{2} \sum_{\xi,\xi'\in S} \mu_\beta(\xi) c_\beta(\xi,\xi') \left[h(\xi)-h(\xi')\right]^2 \\ &\quad = O\left(\mathrm{e}^{-(\Gamma^\star+\delta)\beta}\right) \\ &\qquad + Z_\beta \min_{\substack{h:\, S^\star\to[0,1] \\ h(\mathbf{m})=1,\, h(\mathbf{s})=0}} \tfrac{1}{2} \sum_{\xi,\xi'\in S^\star} \mu_\beta(\xi) c_\beta(\xi,\xi') \left[h(\xi)-h(\xi')\right]^2. \end{aligned} \tag{16.3.12}$$

Next, with the help of Lemmas 16.11–16.15, we get

$$\begin{aligned} &\min_{\substack{h:\, S^\star\to[0,1] \\ h(\mathbf{m})=1,\, h(\mathbf{s})=0}} \tfrac{1}{2} \sum_{\xi,\xi'\in S^\star} \mu_\beta(\xi) c_\beta(\xi,\xi') \left[h(\xi)-h(\xi')\right]^2 \\ &\quad = \min_{\substack{h:\, S^\star\to[0,1] \\ h=h^\star_{\mathbf{m},\mathbf{s}} \text{ on } S_\mathbf{m}\cup S_\mathbf{s}\cup(S_1,\dots,S_I)}} \tfrac{1}{2} \sum_{\xi,\xi'\in S^\star} \mu_\beta(\xi) c_\beta(\xi,\xi') \left[h(\xi)-h(\xi')\right]^2 \\ &\quad = \min_{C_1,\dots,C_I} \min_{\substack{h:\, S^\star\to[0,1] \\ h|_{S_\mathbf{m}}=1-O(\mathrm{e}^{-\beta\delta}),\, h|_{S_\mathbf{s}}=O(\mathrm{e}^{-\beta\delta}),\, h|_{S_i}=C_i+O(\mathrm{e}^{-\beta\delta})\,\forall i=1,\dots,I}} \\ &\qquad \tfrac{1}{2} \sum_{\xi,\xi'\in S^\star} \mu_\beta(\xi) c_\beta(\xi,\xi') \left[h(\xi)-h(\xi')\right]^2 \\ &\quad = \left[1-O\left(\mathrm{e}^{-\delta\beta}\right)\right] \min_{C_1,\dots,C_I} \min_{\substack{h:\, S^\star\to[0,1] \\ h|_{S_\mathbf{m}}=1,\, h|_{S_\mathbf{s}}=0,\, h|_{S_i}=C_i\,\forall i=1,\dots,I}} \\ &\qquad \tfrac{1}{2} \sum_{\xi,\xi'\in S^\star} \mu_\beta(\xi) c_\beta(\xi,\xi') \left[h(\xi)-h(\xi')\right]^2, \end{aligned} \tag{16.3.13}$$

where the error term $O(\mathrm{e}^{-\delta\beta})$ arises after we replace the *approximate* boundary conditions

$$h = \begin{cases} 1-O(\mathrm{e}^{-\beta\delta}) & \text{on } S_\mathbf{m}, \\ O(\mathrm{e}^{-\beta\delta}) & \text{on } S_\mathbf{s}, \\ C_i+O(\mathrm{e}^{-\beta\delta}) & \text{on } S_i,\ i=1,\dots,I, \end{cases} \tag{16.3.14}$$

coming from Lemmas 16.15–16.16 by the *sharp* boundary conditions

$$h = \begin{cases} 1 & \text{on } S_\mathbf{m}, \\ 0 & \text{on } S_\mathbf{s}, \\ C_i & \text{on } S_i,\ i=1,\dots,I. \end{cases} \tag{16.3.15}$$

The minimum with the sharp boundary conditions is an upper bound for the minimum with the approximate boundary conditions. Conversely, removal from the minimum with the approximate boundary conditions of all the transitions that stay inside $S_{\mathbf{m}}$, $S_{\mathbf{s}}$ or S_i for some $i = 1, \dots, I$ yields a lower bound that is within a factor $1 - O(\mathrm{e}^{-\beta\delta})$ of the minimum with the sharp boundary conditions.

Finally, by (16.1.1)–16.1.3) we have

$$Z_\beta \mu_\beta(\xi) c_\beta(\xi, \xi') = \mathbb{1}_{\{\xi \sim \xi'\}} \, \mathrm{e}^{-\beta\Gamma^\star} \tag{16.3.16}$$

for all $\xi, \xi' \in S^\star$ that are not both in $S_{\mathbf{m}}$ or both in $S_{\mathbf{s}}$ or both in S_i for some $i = 1, \dots, I$. Indeed, by Theorem 16.12 and the decomposition in (16.3.4), in each of these cases either $H(\xi) = \Gamma^\star > H(\xi')$ or $H(\xi) < \Gamma^\star = H(\xi')$, because there are no allowed moves between $S_{\mathbf{m}}$, $S_{\mathbf{s}}$ and S_i, $i = 1, \dots, I$. Combining (16.3.12)–(16.3.13) and (16.3.16), we arrive at the claim. □

Combining Lemma 16.14 with (16.3.10)–(16.3.11), we see that we have completed the proof of (16.1.16) with $K = 1/\Theta$. □

The variational formula for Θ is *non-trivial* because it depends on the geometry of the wells S_i, $i = 1, \dots, I$. In Chaps. 17 and 18 we will see how to compute K for Glauber dynamics and Kawasaki dynamics.

16.3.3 Gate for the crossover and uniform entrance distribution

In this section we prove Theorem 16.4.

Proof (a) We will show that there exist $\delta > 0$ and $C < \infty$ such that for all β,

$$\mathbb{P}_{\mathbf{m}}(\tau_{\mathscr{C}^\star} < \tau_{\mathbf{s}} \mid \tau_{\mathbf{s}} < \tau_{\mathbf{m}}) \geq 1 - C\mathrm{e}^{-\beta\delta}, \tag{16.3.17}$$

which implies the claim. The proof goes as follows.

By (16.2.5), $\mathrm{cap}_\beta(\mathbf{m}, \mathbf{s}) = \mu_\beta(\mathbf{m})\, c_\beta(\mathbf{m}) \mathbb{P}_{\mathbf{m}}(\tau_{\mathbf{s}} < \tau_{\mathbf{m}})$ with $\mu_\beta(\mathbf{m}) = 1/Z_\beta$. From the lower bound in Lemma 16.11 it therefore follows that

$$\mathbb{P}_{\mathbf{m}}(\tau_{\mathbf{s}} < \tau_{\mathbf{m}}) \geq C_1 \, \mathrm{e}^{-\beta\Gamma^\star} \frac{1}{c_\beta(\mathbf{m})}. \tag{16.3.18}$$

We will show that

$$\mathbb{P}_{\mathbf{m}}\big(\{\tau_{\mathscr{C}^\star} < \tau_{\mathbf{s}}\}^c, \, \tau_{\mathbf{s}} < \tau_{\mathbf{m}}\big) \leq C_2 \, \mathrm{e}^{-\beta(\Gamma^\star + \delta)} \frac{1}{c_\beta(\mathbf{m})}. \tag{16.3.19}$$

Combining (16.3.18)–(16.3.19), we get (16.3.17) with $C = C_2/C_1$.

Because $\mathscr{C}^\star \subseteq \mathscr{G}(\mathbf{m},\mathbf{s})$, any path from $\mathbf{m}$ to $\mathbf{s}$ that does not pass through $\mathscr{C}^\star$ must hit a configuration ξ with $H(\xi) > \Gamma^\star$. Therefore there exists a set U, with $H(\xi) \geq \Gamma^\star + \delta$ for all $\xi \in U$ and some $\delta > 0$, such that

$$\mathbb{P}_{\mathbf{m}}\big(\{\tau_{\mathscr{C}^\star} < \tau_{\mathbf{s}}\}^c,\ \tau_{\mathbf{s}} < \tau_{\mathbf{m}}\big) \leq \mathbb{P}_{\mathbf{m}}(\tau_U < \tau_{\mathbf{m}}). \tag{16.3.20}$$

Now estimate, with the help of reversibility,

$$\begin{aligned}
\mathbb{P}_{\mathbf{m}}(\tau_U < \tau_{\mathbf{m}}) &\leq \sum_{\xi \in U} \mathbb{P}_{\mathbf{m}}(\tau_\xi < \tau_{\mathbf{m}}) \\
&= \sum_{\xi \in U} \frac{\mu_\beta(\xi) c_\beta(\xi)}{\mu_\beta(\mathbf{m}) c_\beta(\mathbf{m})} \mathbb{P}_\xi(\tau_{\mathbf{m}} < \tau_\xi) \\
&\leq \frac{1}{c_\beta(\mathbf{m})} \sum_{\xi \in U} \big|\{\xi' \in S\backslash\xi \colon \xi \sim \xi'\}\big| \,\mathrm{e}^{-\beta H(\xi)} \\
&\leq \frac{1}{c_\beta(\mathbf{m})} C_2\, \mathrm{e}^{-\beta(\Gamma^\star+\delta)}
\end{aligned} \tag{16.3.21}$$

with $C_2 = |\{(\xi,\xi') \in U \times S\backslash\xi \colon \xi \sim \xi'\}|$, where we use that $H(\mathbf{m}) = 0$ and $c_\beta(\xi,\xi') \leq 1$. Combine (16.3.20)–(16.3.21) to get the claim in (16.3.19).

(b) Write

$$\mathbb{P}_{\mathbf{m}}(\xi_{\tau_{\mathscr{C}^\star}} = \xi \mid \tau_{\mathscr{C}^\star} < \tau_{\mathbf{m}}) = \frac{\mathbb{P}_{\mathbf{m}}(\xi_{\tau_{\mathscr{C}^\star}} = \xi,\ \tau_{\mathscr{C}^\star} < \tau_{\mathbf{m}})}{\mathbb{P}_{\mathbf{m}}(\tau_{\mathscr{C}^\star} < \tau_{\mathbf{m}})}, \quad \xi \in \mathscr{C}^\star. \tag{16.3.22}$$

By reversibility,

$$\begin{aligned}
\mathbb{P}_{\mathbf{m}}(\xi_{\tau_{\mathscr{C}^\star}} = \xi,\ \tau_{\mathscr{C}^\star} < \tau_{\mathbf{m}}) &= \frac{\mu_\beta(\xi) c_\beta(\xi)}{\mu_\beta(\mathbf{m}) c_\beta(\mathbf{m})} \mathbb{P}_\xi(\tau_{\mathbf{m}} < \tau_{\mathscr{C}^\star}) \\
&= \mathrm{e}^{-\Gamma^\star \beta} \frac{c_\beta(\xi)}{c_\beta(\mathbf{m})} \mathbb{P}_\xi(\tau_{\mathbf{m}} < \tau_{\mathscr{C}^\star}), \quad \xi \in \mathscr{C}^\star.
\end{aligned} \tag{16.3.23}$$

Moreover (recall (16.2.4)–(16.2.3)),

$$\mathbb{P}_\xi(\tau_{\mathbf{m}} < \tau_{\mathscr{C}^\star}) = \sum_{\substack{\xi' \in S\backslash\mathscr{C}^\star \\ \xi \sim \xi'}} \frac{c_\beta(\xi,\xi')}{c_\beta(\xi)} h_{\mathbf{m},\mathscr{C}^\star}(\xi'), \quad \xi \in \mathscr{C}^\star, \tag{16.3.24}$$

where

$$h_{\mathbf{m},\mathscr{C}^\star}(\xi') = \begin{cases} 0 & \text{if } \xi' \in \mathscr{C}^\star, \\ 1 & \text{if } \xi' = \mathbf{m}, \\ \mathbb{P}_{\xi'}(\tau_{\mathbf{m}} < \tau_{\mathscr{C}^\star}) & \text{otherwise.} \end{cases} \tag{16.3.25}$$

Because $\mathscr{P}^\star \subseteq S_{\mathbf{m}}$ by Theorem 16.12 and $\mathscr{C}^\star \subseteq \mathscr{G}(\mathbf{m},\mathbf{s})$ by Definition 16.3, we have $\Phi(\xi',\mathscr{C}^\star) - \Phi(\xi',\mathbf{m}) = \Gamma^\star - \Phi(\xi',\mathbf{m}) \geq \delta$ for all $\xi' \in \mathscr{P}^\star$ and some $\delta > 0$. Therefore, as in the proof of Lemma 8.4, it follows that

$$\min_{\xi' \in \mathscr{P}^\star} h_{\mathbf{m},\mathscr{C}^\star}(\xi') \geq 1 - C\mathrm{e}^{-\beta\delta}. \tag{16.3.26}$$

Moreover, let

$$\bar{\mathscr{C}}^\star = \{\xi' \in S\backslash[\mathscr{P}^\star \cup \mathscr{C}^\star]\colon\ H(\xi') \leq \Gamma^\star,\ \exists \xi \in \mathscr{C}^\star\colon\ \xi \sim \xi'\}. \tag{16.3.27}$$

By Lemma 16.10, any path from $\bar{\mathscr{C}}^\star$ to $\mathbf{m}$ that avoids $\mathscr{C}^\star$ must reach an energy level above $\Gamma^\star$, and so $h_{\mathbf{m},\mathscr{C}^\star}(\xi') \leq h_{S\backslash S^\star,\mathscr{C}^\star}(\xi')$ for all $\xi' \in \bar{\mathscr{C}}^\star$. But $\Phi(\xi', S\backslash S^\star) - \Phi(\xi',\mathscr{C}^\star) = \Phi(\xi', S\backslash S^\star) - \Gamma^\star \geq \delta$ for all $\xi' \in \bar{\mathscr{C}}^\star \cap S^\star$ and some $\delta > 0$. Therefore, again as in the proof of Lemma 8.4, it follows that

$$\max_{\xi' \in \bar{\mathscr{C}}^\star \cap S^\star} h_{\mathbf{m},\mathscr{C}^\star}(\xi') \leq C\mathrm{e}^{-\beta\delta}. \tag{16.3.28}$$

The estimates in (16.3.26)–(16.3.28) can be used as follows. By restricting the sum in (16.3.24) to $\xi' \in \mathscr{P}^\star$ and using (16.3.26), we get the lower bound

$$\mathbb{P}_\xi(\tau_{\mathbf{m}} < \tau_{\mathscr{C}^\star}) \geq \left(1 - C\mathrm{e}^{-\beta\delta}\right) \frac{c_\beta(\xi,\mathscr{P}^\star)}{c_\beta(\xi)}, \quad \xi \in \mathscr{C}^\star. \tag{16.3.29}$$

On the other hand, by using (16.3.28) in combination with the fact that $c_\beta(\xi, S\backslash[\mathscr{C}^\star \cup \bar{\mathscr{C}}^\star]) = c_\beta(\xi,\mathscr{P}^\star)$ for all $\xi \in \mathscr{C}^\star$ (recall Fig. 16.2) and $c_\beta(\xi,\xi') \leq \mathrm{e}^{-\beta\delta}$ for all $\xi \in \mathscr{C}^\star$ and $\xi' \in S\backslash S^\star$, we get the upper bound

$$\mathbb{P}_\xi(\tau_{\mathbf{m}} < \tau_{\mathscr{C}^\star}) \leq \frac{c_\beta(\xi,\mathscr{P}^\star)}{c_\beta(\xi)} + C\mathrm{e}^{-\beta\delta}|\bar{\mathscr{C}}^\star| + \mathrm{e}^{-\beta\delta}|S\backslash S^\star|, \quad \xi \in \mathscr{C}^\star. \tag{16.3.30}$$

Because $H(\xi') < H(\xi) = \Gamma^\star$ for all $\xi \in \mathscr{C}^\star$ and $\xi' \in \mathscr{P}^\star$, we have

$$c_\beta(\xi,\mathscr{P}^\star) = \sum_{\xi' \in \mathscr{P}^\star} c_\beta(\xi,\xi') = |\{\xi' \in \mathscr{P}^\star\colon\ \xi \sim \xi'\}|, \quad \xi \in \mathscr{C}^\star, \tag{16.3.31}$$

and, since $c_\beta(\xi) \leq |S|$, it follows that $\xi \mapsto c_\beta(\xi,\mathscr{P}^\star)/c_\beta(\xi) \geq C > 0$. Combine this observation with (16.3.29)–(16.3.30), to get

$$\mathbb{P}_\xi(\tau_{\mathbf{m}} < \tau_{\mathscr{C}^\star}) = [1 + O(\mathrm{e}^{-\beta\delta})] \frac{c_\beta(\xi,\mathscr{P}^\star)}{c_\beta(\xi)}, \quad \xi \in \mathscr{C}^\star. \tag{16.3.32}$$

Combine this in turn with (16.3.22)–(16.3.23), to arrive at

$$\begin{aligned}\mathbb{P}_{\mathbf{m}}(\xi_{\tau_{\mathscr{C}^\star}} = \xi \mid \tau_{\mathscr{C}^\star} < \tau_{\mathbf{m}}) &= \frac{c_\beta(\xi)\,\mathbb{P}_\xi(\tau_{\mathbf{m}} < \tau_{\mathscr{C}^\star})}{\sum_{\xi' \in \mathscr{C}^\star} c_\beta(\xi')\,\mathbb{P}_{\xi'}(\tau_{\mathbf{m}} < \tau_{\mathscr{C}^\star})} \\ &= [1 + O(\mathrm{e}^{-\beta\delta})] \frac{c_\beta(\xi,\mathscr{P}^\star)}{\sum_{\xi' \in \mathscr{C}^\star} c_\beta(\xi',\mathscr{P}^\star)}, \quad \xi \in \mathscr{C}^\star.\end{aligned} \tag{16.3.33}$$

Finally, by (H2) and (16.3.31), $\xi \mapsto c_\beta(\xi, \mathscr{P}^\star)$ is constant on $\mathscr{C}^\star$. Together with (16.3.33) this proves the claim. □

16.4 Beyond Metropolis dynamics

There is nothing that prevents us from choosing a dynamics that is different from the Metropolis dynamics in (16.1.2). We take a brief look at two examples, namely, *heat-bath dynamics* (Sect. 16.4.1) and *probabilistic cellular automata* (Sect. 16.4.2). We show that Theorems 16.4–16.6 in Sect. 16.1 carry over provided we modify hypothesis (H2).

16.4.1 Heat-bath dynamics

Return to the setting of Sect. 16.1.1. The *heat-bath dynamics* is the continuous-time Markov process with state space $S = \Upsilon^\Lambda$ and transition rates

$$c_\beta(\xi, \xi') = \begin{cases} [1 + \mathrm{e}^{\beta[H(\xi') - H(\xi)]}]^{-1} & \xi \sim \xi', \\ 0 & \text{otherwise.} \end{cases} \tag{16.4.1}$$

This Markov process is reversible with respect to μ, the Gibbs measure associated with H. Note that for large β the transition rates of the heat-bath dynamics and the Metropolis dynamics are close to each other, except when $H(\xi') = H(\xi)$, in which case the former gives $c_\beta(\xi, \xi') = \frac{1}{2}$ while the latter gives $c_\beta(\xi, \xi') = 1$.

Theorem 16.18 (Metastability for heat-bath dynamics) *Theorems* 16.4–16.6 *are valid for heat-bath dynamics subject to* (H1) *and*

(H2′) (H2) *holds and* $\xi' \mapsto H(\xi')$ *is constant on* $\mathscr{P}^\star$.

Proof The same proofs as in Sects. 16.2–16.3 apply, except for minor modifications in a few spots:

1. In Sect. 16.2.2, the only modification is that, because

$$\mu_\beta(\xi) c_\beta(\xi, \xi') = \left[\mathrm{e}^{\beta H(\xi)} + \mathrm{e}^{\beta H(\xi')}\right]^{-1} \geq \tfrac{1}{2}\,\mathrm{e}^{-\beta[H(\xi) \vee H(\xi')]}, \tag{16.4.2}$$

the lower bound in (16.2.24) holds with $C_1 = 1/2L$ instead if $C_1 = 1/L$. This does not affect the a priori estimates in Lemma 16.11.

2. In Sect. 16.3.2, the only modification is that instead of (16.3.16) we have

$$Z_\beta \mu_\beta(\xi) c_\beta(\xi, \xi') = \mathbb{1}_{\{\xi \sim \xi'\}}\,\mathrm{e}^{-\beta\Gamma^\star}\left[1 + O(\mathrm{e}^{-\beta\delta})\right] \tag{16.4.3}$$

for all $\xi, \xi' \in S^\star$ that are not both in $S_{\mathbf{m}}$ or both in $S_{\mathbf{s}}$ or both in S_i for some $i = 1, \dots, I$. This does not affect Lemma 16.17, and so the same variational formula for $\Theta = 1/K$ as in (16.3.11) holds.

3. In Sect. 16.3.3, no modification is needed all the way up to and including (16.3.30) (the last term in the right-hand side of (16.3.30) comes with a factor $[1 + e^{\beta\delta}]^{-1} \leq e^{-\beta\delta}$). Also, instead of (16.3.31), we can estimate $c_\beta(\xi, \mathscr{P}^\star) \geq \frac{1}{2}|\{\xi' \in \mathscr{P}^\star \colon \xi \sim \xi'\}|$, so that also (16.3.32) and (16.3.33) carry over. Finally, $\xi \mapsto c_\beta(\xi, \mathscr{P}^\star)$ is not constant on $\mathscr{C}^\star$. However,

$$\begin{aligned} c_\beta(\xi, \xi') = \left[1 + e^{\beta[H(\xi') - H(\xi)]}\right]^{-1} = e^{-\beta[H(\xi') - H(\xi)]}\left[1 + O\left(e^{-\beta\delta}\right)\right], \\ \xi \in \mathscr{C}^\star \, \xi' \in \mathscr{P}^\star. \end{aligned} \tag{16.4.4}$$

Consequently, if we strengthen (H2) to (H2′), then (16.3.33) again gives the uniform entrance distribution (use that $H(\xi) = \Gamma^\star$ for all $\xi \in \mathscr{C}^\star$). □

Since the triple $(\Gamma^\star, \mathscr{C}^\star, K)$ only depends on H, $\sim$, $\mathbf{m}$ and $\mathbf{s}$, even this is not affected by the choice of (16.4.1) over Metropolis.

16.4.2 Probabilistic cellular automata

Again return to the setting of Sect. 16.1.1. A *probabilistic cellular automaton* (PCA) is a discrete-time Markov chain with state space $S = \Upsilon^\Lambda$ and transition matrix

$$p(\xi, \xi') = \prod_{x \in \Lambda} p_{x,\xi}\left(\xi'(x)\right), \quad \xi, \xi' \in S, \tag{16.4.5}$$

where, for each $x \in \Lambda$ and $\xi \in S$, $p_{x,\xi}(\cdot)$ is a probability measure on S with full support. This transition matrix corresponds to independent updates of all the spins simultaneously at each unit of time ("parallel dynamics"), according to local updating rules that take into account both the location of the spin and the values of the spins in its surroundings. Typically, $p_{x,\xi}(\cdot)$ is assumed to depend on ξ only through the spins $\xi(y)$, $y \in N(x)$, in some small neighbourhood $N(x)$ of x. If Λ has a lattice structure (e.g. a torus in $\mathbb{Z}^d$, $d \geq 1$), then typically $N(x) = x + N$, $x \in \Lambda$, for some small $N \subseteq \Lambda$.

What makes PCA's into challenging objects is that they evolve via *global moves* rather than local moves: all transitions—between any pair of configurations in S—have positive probability, and therefore all transitions are allowed. This means that $\sim$ loses the role it played for Metropolis dynamics,

For $\beta > 0$, the PCA in (16.4.5) is reversible with respect to the Gibbs measure $\mu_\beta(\xi) = e^{-\beta H(\xi)}/Z_\beta$, $\xi \in S$, associated with the Hamiltonian $H \colon S \to \mathbb{R}$ if

$$\mu_\beta(\xi) p(\xi, \xi') = \mu_\beta(\xi') p(\xi', \xi) \quad \forall \xi, \xi' \in S, \tag{16.4.6}$$

i.e., if the "dynamic Hamiltonian" defined by

$$\mathscr{H}(\xi,\xi') = H(\xi) - \frac{1}{\beta}\ln p(\xi,\xi') \tag{16.4.7}$$

is a symmetric function on $S \times S$. For a given choice of H and β, this condition puts a constraint on the choice of PCA in (16.4.5).

The communication height between two configurations $\xi,\xi' \in S$ with $\xi \neq \xi'$ is defined to be

$$\Phi(\xi,\xi') = \min_{\gamma:\,\xi\to\xi'} \max_{e\in\gamma} \mathscr{H}(e), \tag{16.4.8}$$

where the maximum runs over all edges e in γ, i.e., over all pairs of successive configurations visited by the path. This is different from Definition 16.1(a), where the maximum runs over the single configurations in γ, and H was used instead of $\mathscr{H}$ (note that $\Phi(\xi,\xi) = H(\xi)$ by convention). The definition of the communication level set $\mathscr{S}(\xi,\xi')$ in Definition 16.1(b) must be adapted accordingly: this becomes a set of pairs of configurations rather than single configurations ($\mathscr{S}(\xi,\xi) = \xi$ by convention). A similar change applies to the definition of gates and dead-ends in Definition 16.2. What makes (16.4.8) non-trivial to compute is the fact that the Hamiltonian and the transition probabilities compete with each other: to make $\mathscr{H}(\cdot,\cdot)$ small we must make $H(\cdot)$ small and $p(\cdot,\cdot)$ large simultaneously.

Definition 16.3 must be changed into the following.

Definition 16.19 (Protocritical and critical sets for PCA dynamics) Let

$$\Gamma^\star = \Phi(\mathbf{m},\mathbf{s}) - H(\mathbf{m}). \tag{16.4.9}$$

Then $(\mathscr{P}^\star(\mathbf{m},\mathbf{s}), \mathscr{C}^\star(\mathbf{m},\mathbf{s}))$ is the maximal subset of $S \times S$ such that:

(1) $\forall \xi \in \mathscr{P}^\star(\mathbf{m},\mathbf{s})\ \exists \gamma\colon \xi \to \mathbf{m}\colon\ \max_{e\in\gamma} \mathscr{H}(e) - H(\mathbf{m}) < \Gamma^\star$.
(2) $\forall \xi' \in \mathscr{C}^\star(\mathbf{m},\mathbf{s})\ \exists \gamma\colon \xi' \to \mathbf{s}\colon\ \max_{e\in\gamma} \mathscr{H}(e) - H(\mathbf{m}) \le \Gamma^\star,\ \gamma \cap \{\zeta \in S\colon \Phi(\zeta,\mathbf{m}) < \Phi(\zeta,\mathbf{s})\} = \emptyset$.

With this change we can now state the following.

Theorem 16.20 (Metastability for PCA dynamics) *Theorems* 16.4–16.6 *are valid for PCA dynamics subject to* (H1) *and*

(H2″) $(\xi,\xi') \mapsto p(\xi,\xi')$ *is constant on* $\mathscr{C}^\star \times \mathscr{P}^\star$.

Proof We will again go through the proofs in Sects. 16.2–16.3 to see what needs to be modified.

1. In Sect. 16.2.1, the definition of the Dirichlet form in (16.2.1) becomes

$$\mathscr{E}(h,h) = \tfrac{1}{2} \sum_{\xi,\xi'\in S} \mu_\beta(\xi) p(\xi,\xi') \big[h(\xi) - h(\xi')\big]^2, \quad h\colon S \to [0,1], \tag{16.4.10}$$

while the definition of the capacity in (16.2.2) remains the same. Note that

$$\mu_\beta(\xi)p(\xi,\xi') = \mathrm{e}^{-\beta\mathscr{H}(\xi,\xi')}/Z_\beta. \tag{16.4.11}$$

Replace (16.2.5) by

$$\mathrm{cap}_\beta(A,B) = \sum_{\xi\in A}\mu_\beta(\xi)p(\xi,\xi')\mathbb{P}_\xi(\tau_B < \tau_A), \tag{16.4.12}$$

which is simpler than (16.2.5) because the PCA dynamics evolves is discrete rather than continuous time.

2. Throughout Sect. 16.2.2, the new definition of communication height in (16.4.8) must be used, and $\mu_\beta(\xi)c_\beta(\xi,\xi')$ must be replaced by $\mu_\beta(\xi)p(\xi,\xi')$. Otherwise there are no changes.

3. In Sect. 16.2.3, Theorem 16.12 needs to be adapted as follows: S is the graph consisting of all vertices and all edges; $S^\star$ is the subgraph of S obtained by removing all edges e with $\mathscr{H}(e) > \Gamma^\star$; $S^{\star\star}$ is the subgraph of $S^\star$ obtained by removing all edges e with $\mathscr{H}(e) = \Gamma^\star$; $S_{\mathbf{m}}$ and $S_{\mathbf{s}}$ are the connected components of $S^{\star\star}$ containing $\mathbf{m}$ and $\mathbf{s}$, respectively. With these modifications the claim in Theorem 16.12 stays the same.

4. There are no changes in Sects. 16.2.4 and 16.3.1. Throughout Sect. 16.3.2, $\mu_\beta(\xi)c_\beta(\xi,\xi')$ must be replaced by $\mu_\beta(\xi)p(\xi,\xi')$, while the indicator $\mathbb{1}\{\xi\sim\xi'\}$ must be removed from (16.3.11) and (16.3.16). The definition of S_i, $i=1,\dots,I$ in (16.3.4) stays the same, and so does the variational formula for $\Theta = 1/K$ in (16.3.11).

5. Throughout Sect. 16.3.3, remove the terms $c_\beta(\xi)$ and $c_\beta(\mathbf{m})$ (in view of (16.4.12)) and replace $c_\beta(\xi,\xi')$ by $p(\xi,\xi')$. Finally, if we replace (H2) by (H2$''$), then (16.3.33) again gives the uniform entrance distribution. □

16.5 Bibliographical notes

1. The line of reasoning pursued in Sects. 16.2–16.3 was put forward in Bovier and Manzo [39] for Glauber dynamics and in Bovier, Nardi and den Hollander [31] for Kawasaki dynamics, both in their model-specific context (see Chaps. 17 and 18). The two hypotheses in Sect. 16.1.2 have been stripped of this context in order to capitalise as much as possible on the general theory developed in Part III: they are the *essentially minimal hypotheses* that are needed to obtain the universal metastable behaviour expressed in Theorems 16.4–16.6. Part of this stripping was already done in den Hollander, Nardi and, Troiani [87], where Kawasaki dynamics with two types of particles was considered (see Sect. 18.7).

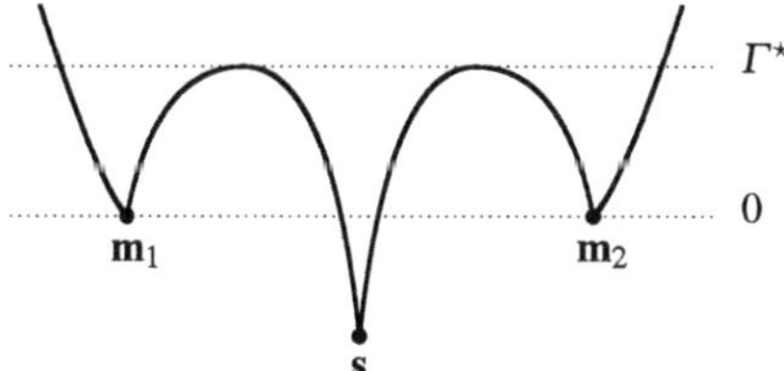

Fig. 16.5 Example where $S_{\mathrm{meta}} = \{\mathbf{m}_1, \mathbf{m}_2\}$, yet for both these metastable configurations the same results apply as in Theorems 16.4–16.6

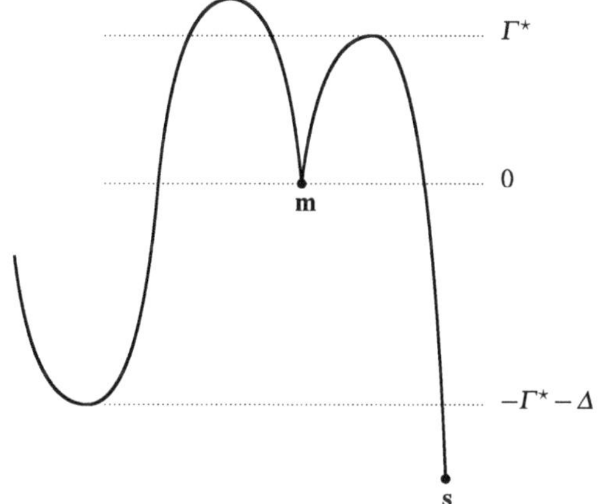

Fig. 16.6 Example where there is a well of depth $> \Gamma^\star$. The presence of this well does not influence the typical crossover time from $\mathbf{m}$ to $\mathbf{s}$, but enlarges its average by a factor $e^{\beta\Delta}$

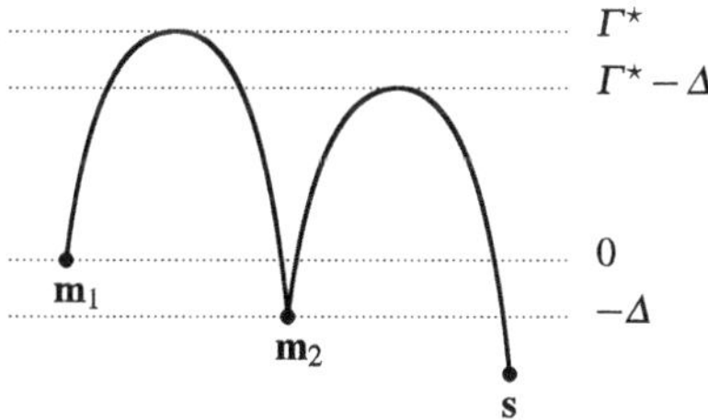

Fig. 16.7 Example where $S_{\mathrm{meta}} = \{\mathbf{m}_1, \mathbf{m}_2\}$, with $\mathbf{s}$ separated from $\mathbf{m}_1$ by $\mathbf{m}_2$. In this case the distribution of the crossover time from $\mathbf{m}_1$ to $\mathbf{s}$ divided by its average is one half times the convolution of two unit exponentials

2. An "axiomatisation" of the essential features of metastability in the context of the pathwise approach to metastability can be found in Manzo, Nardi, Olivieri and Scoppola [171]. Here, hypotheses similar to (H1) are formulated, and are used to derive the results in Theorem 16.4(a) and Theorem 16.5 without the prefactor (i.e., the average crossover time is identified up to a multiplicative factor $e^{o(\beta)}$). This paper also contains a careful analysis of the role of minimal gates and essential gates for the metastable pair.

3. The non-degeneracy assumption in (H1) can be relaxed. For instance, if $S_{\rm meta}$ is not a singleton, then the same results as in Theorems 16.4–16.6 apply for each choice of $\mathbf{m} \in S_{\rm meta}$ as long as $(\mathbf{m} \to \mathbf{s})_{\rm opt}$ does not need to cross $S_{\rm meta} \backslash \mathbf{m}$. An example is given in Fig. 16.5. See Cirillo and Nardi [65] for an analysis of what may happen in degenerate situations.

Figures 16.6–16.7 exhibit two examples where (H1) fails and the metastable behaviour is different.

Chapter 17
Glauber Dynamics

"You have no right to grow here," *said the Dormouse. "Don't talk nonsense," said Alice more boldly: "you know you're growing too." "Yes, but I grow at a reasonable pace," said the Dormouse: "not in a ridiculous fashion... "*
(Lewis Carroll, Alice's Adventures in Wonderland)

In this chapter we apply the results obtained in Chap. 16 to *Ising spins* in two and three dimensions subject to *Glauber dynamics*. Spins live in a finite box, flip up and down, want to align when they sit next to each other, and want to align with an external magnetic field. We are interested in how the system *magnetises*, i.e., how the dynamics aligns the spins with the magnetic field when initially all the spins are pointing in the opposite direction. Our goal will be to prove hypotheses (H1–H2) in Sect. 16.1.2, implying that Theorems 16.4–16.6 are valid. In two dimensions we will identify $(\Gamma^\star, \mathscr{C}^\star, K)$. In three dimensions we will also identify $\Gamma^\star$, but we will obtain only partial information on $\mathscr{C}^\star$ and K.

17.1 Introduction and main results

17.1.1 Model

Let $\Lambda \subset \mathbb{Z}^2$ be a large square torus, centred at the origin. With each site $x \in \Lambda$ we associate a spin variable $\sigma(x)$ assuming the values -1 or $+1$, indicating whether the spin at x is pointing *down* or *up* (see Fig. 17.1). A configuration is denoted by $\sigma \in S = \{-1, +1\}^\Lambda$. Each configuration $\sigma \in S$ has an energy given by the Hamiltonian

$$H(\sigma) = -\frac{J}{2} \sum_{\{x,y\} \in \Lambda^*} \sigma(x)\sigma(y) - \frac{h}{2} \sum_{x \in \Lambda} \sigma(x), \tag{17.1.1}$$

where

$$\Lambda^* = \big\{\{x, y\}\colon\ x, y \in \Lambda,\ \|x - y\| = 1\big\} \tag{17.1.2}$$

A. Bovier, F. den Hollander, *Metastability*,
Grundlehren der mathematischen Wissenschaften 351,
DOI 10.1007/978-3-319-24777-9_17

$$\begin{matrix} - & - & + & - & - \\ + & - & - & - & + \\ + & - & - & + & - \\ - & + & + & + & - \\ + & - & - & + & - \end{matrix}$$

Fig. 17.1 An Ising-spin configuration

is the set of non-oriented nearest-neighbour bonds in Λ. The interaction consists of a *ferromagnetic pair potential* $J > 0$ for each pair of neighbouring spins in Λ and a *magnetic field* $h > 0$ for each spin in Λ.

The Hamiltonian in (17.1.1) models Ising spins in Λ that want to align with neighbouring spins and with an external magnetic field that is pointing upwards. We are interested in *Glauber dynamics* on Λ. This is the Metropolis dynamics with respect to H at inverse temperature β defined in (16.1.2) with *single-spin flips* as allowed moves, i.e., -1 changes to $+1$ or $+1$ changes to -1 at single sites in Λ. Clearly, this dynamics is a *finite-state* Markov process, and hence fits into the general theory described in Chap. 16.

The Gibbs measure μ_β defined in (16.1.1) is the equilibrium of the dynamics with transition rates c_β defined in (16.1.2) and satisfies the reversibility property in (16.1.3).

17.1.2 Metastable regime and critical droplet size

Throughout the sequel we assume that

$$h \in (0, 2J). \tag{17.1.3}$$

This parameter range will be seen to correspond to metastable behaviour in the limit as $\beta \to \infty$ (see Fig. 17.4 below). A key role will be played by what we call the *critical droplet size*:

$$\ell_c = \left\lceil \frac{2J}{h} \right\rceil \tag{17.1.4}$$

($\lceil \cdot \rceil$ denotes the upper integer part). For reasons that will become clear later on, we will assume that

$$\frac{2J}{h} \notin \mathbb{N}. \tag{17.1.5}$$

Thus, an $(\ell_c - 1) \times (\ell_c - 1)$ droplet will be "subcritical" while an $\ell_c \times \ell_c$ droplet will be "supercritical". Moreover, we will assume that Λ is large enough so that it contains an $2\ell_c \times 2\ell_c$ square, which is necessary for (H1) and will also prevent the critical droplet to be a ring that wraps around Λ.

Analogous assumptions are needed in three dimensions (see Sect. 17.6).

Fig. 17.2 Configurations in $\mathcal{Q}$, $\mathcal{Q}^{1\text{pr}}$ and $\mathcal{Q}^{2\text{pr}}$. Inside the contours sit the up-spins, outside the contours sit the down-spins

17.1.3 Main theorems

Each configuration can be decomposed into maximally connected components, called *clusters*.

Definition 17.1

(a) Let

$$\begin{aligned} \boxminus &= \{\sigma \in S \colon \sigma(x) = -1 \ \forall x \in \Lambda\}, \\ \boxplus &= \{\sigma \in S \colon \sigma(x) = +1 \ \forall x \in \Lambda\}, \end{aligned} \tag{17.1.6}$$

denote the configurations where all spins in Λ are down, respectively, up.

(b) Let $\mathcal{Q}$ be the set of configurations where the up-spins form a single $(\ell_c - 1) \times \ell_c$ *quasi-square* anywhere in Λ.

(c) Let $\mathcal{Q}^{1\text{pr}}$ be the set of configurations where the up-spins form a single quasi-square $(\ell_c - 1) \times \ell_c$ anywhere in Λ with a *single protuberance* attached anywhere to one of its longest sides.

(d) Let $\mathcal{Q}^{2\text{pr}}$ be the set of configurations where the up-spins form a single quasi-square $(\ell_c - 1) \times \ell_c$ anywhere in Λ with a *double protuberance* attached anywhere to one of its longest sides.

See Fig. 17.2 for a picture of the configurations in $\mathcal{Q}$, $\mathcal{Q}^{1\text{pr}}$ and $\mathcal{Q}^{2\text{pr}}$.

The main metastability theorems for Glauber dynamics are the following. Recall Definition 16.3.

Theorem 17.2 *The pair* $(\boxminus, \boxplus)$ *satisfies hypotheses* (H1–H2) *in Section* 16.1.2 *and hence Theorems* 16.4–16.6 *hold.*

Theorem 17.3 *The pair* $(\boxminus, \boxplus)$ *has protocritical set* $\mathcal{P}^\star(\boxminus, \boxplus) = \mathcal{Q}$, *critical set* $\mathcal{C}^\star(\boxminus, \boxplus) = \mathcal{Q}^{1\text{pr}}$, *and communication height*

$$\Gamma^\star = \Gamma^\star(\boxminus, \boxplus) = H(\mathcal{Q}^{1\text{pr}}) - H(\boxminus) = J[4\ell_c] - h[\ell_c(\ell_c - 1) + 1]. \tag{17.1.7}$$

Theorem 17.4 *The prefactor* $K = K(\Lambda)$ *equals* $K(\Lambda) = \frac{3}{4(2\ell_c - 1)} \frac{1}{|\Lambda|}$.

Fig. 17.3 Transitions over the hill: $\mathscr{Q} \to \mathscr{Q}^{1\text{pr}} \to \mathscr{Q}^{2\text{pr}}$

In addition, we have the following geometric description of the configurations in the valleys $S_\boxminus$, $S_\boxplus$ around $\boxminus$, $\boxplus$ defined in (16.2.25). Let

$$\begin{aligned} V_{\leq \mathscr{Q}} &= \{\sigma \in S\colon\ \sigma \leq \sigma' \text{ for some } \sigma' \in \mathscr{Q}\}, \\ V_{\geq \mathscr{Q}^{2\text{pr}}} &= \{\sigma \in S\colon\ \sigma \geq \sigma' \text{ for some } \sigma' \in \mathscr{Q}^{2\text{pr}}\}, \end{aligned} \tag{17.1.8}$$

where we write $\sigma \leq \sigma'$ when $\sigma(x) \leq \sigma'(x)$ for all $x \in \Lambda$, and vice versa.

Theorem 17.5 $S_\boxminus \supseteq V_{\leq \mathscr{Q}}$, $S_\boxplus \supseteq V_{\geq \mathscr{Q}^{2\text{pr}}}$.

17.1.4 Discussion

1. The proof of Theorem 17.2 is given in Sect. 17.3. (H2) is easy to check, (H1) is more involved and relies on certain *isoperimetric inequalities*.

2. The heuristics behind Theorem 17.3 is as follows. In Sect. 17.4 we will see that $\mathscr{Q}^{1\text{pr}} \subseteq \mathscr{S}(\boxminus, \boxplus)$, the communication level set of the pair $(\boxminus, \boxplus)$. We will see that on its way from $\boxminus$ to $\boxplus$ the dynamics passes through $\mathscr{S}(\boxminus, \boxplus)$ in three steps: (1) first it creates a quasi-square of up-spins; (2) next it attaches a single protuberance; (3) finally it turns this single protuberance into a double protuberance (see Fig. 17.3). After these three steps are completed, the dynamics is "over the hill" and proceeds downwards in energy to fill the box with up-spins. This also explains where Theorem 17.5 comes from.

3. The heuristics behind Theorem 17.4 is as follows. The average time it takes for the dynamics to enter $\mathscr{C}^\star(\boxminus, \boxplus) = \mathscr{Q}^{1\text{pr}}$ when starting from $\boxminus$ is

$$\frac{1}{|\mathscr{Q}^{1\text{pr}}|}\, e^{\beta \Gamma^\star} \big[1 + o(1)\big], \quad \beta \to \infty, \tag{17.1.9}$$

where $|\mathscr{Q}^{1\text{pr}}|$ counts the number of critical droplets. Let $\pi(\ell_c)$ be the probability that the single protuberance is turned into a double protuberance rather than is being removed. Then

$$\frac{1}{\pi(\ell_c)} \big[1 + o(1)\big], \quad \beta \to \infty, \tag{17.1.10}$$

is the average number of times a critical droplet just created attempts to move over the hill before it finally manages to do so. The average nucleation time is the product of (17.1.9) and (17.1.10), and so we conclude that

$$K = \frac{1}{|\mathscr{Q}^{1\mathrm{pr}}|\,\pi(\ell_c)}. \tag{17.1.11}$$

To compute $|\mathscr{Q}^{1\mathrm{pr}}|$, note that

$$\left|\mathscr{Q}^{1\mathrm{pr}}\right| = |\Lambda|\,N(\ell_c) \quad \text{with } N(\ell_c) = 4\ell_c. \tag{17.1.12}$$

Indeed, the $(\ell_c - 1) \times \ell_c$ quasi-square can be located anywhere in Λ (which is a torus) in two possible orientations, while the single protuberance can be attached in any of the $2\ell_c$ possible locations on one of the sides of length ℓ_c (see Fig. 17.2). To compute $\pi(\ell_c)$, note that if the protuberance sits at one of the two extreme ends of the side it is attached to, then the probability is $\frac{1}{2}$ that its *one* neighbouring spin flips up before the spin itself flips down. On the other hand, if the protuberance sits somewhere else, then the probability is $\frac{2}{3}$ that one of its *two* neighbouring spins on the same side flip up before the spin itself flips down. Since the location of the protuberance in uniform (because of the uniform exit distribution stated in Theorem 16.4(b)), we therefore get

$$\pi(\ell_c) = \frac{1}{\ell_c}\left\{2\,\frac{1}{2} + (\ell_c - 2)\,\frac{2}{3}\right\} = \frac{2\ell_c - 1}{3\ell_c}. \tag{17.1.13}$$

Combine (17.1.11)–(17.1.13) to get the formula for K in Theorem 17.4.

Outline The outline of the remainder of this chapter is as follows. In Sect. 17.2 we introduce some geometric definitions that are needed for the proof of Theorems 17.2–17.5. These theorems are proved in Sects. 17.3–17.5. Section 17.6 looks at the extension from two to three dimensions.

17.2 Geometric definitions

In order to prove Theorems 17.2–17.5, we need some further definitions.

1. Throughout the sequel, we identify a configuration $\sigma \in S$ with the set of locations of its up-spins $\mathrm{supp}(\sigma) = \{x \in \Lambda\colon\ \sigma(x) = +1\}$, and write $x \in \sigma$ to indicate that σ has an up-spin at x.

2. Given a configuration $\sigma \in S$, consider the set $C(\sigma) \subseteq \mathbb{R}^2$ defined as the union of the closed unit squares centred at the sites of $\mathrm{supp}(\sigma)$. The maximal connected components $C_1, \ldots, C_m$, $m \in \mathbb{N}$, of $C(\sigma)$ are called *clusters* of σ (two unit squares touching only at the corners are not connected). There is a one-to-one correspondence between configurations $\sigma \in S$ and sets $C(\sigma)$.

3. For $\sigma \in S$, let $|\sigma|$ be the volume of $C(\sigma)$, $\partial(\sigma)$ the Euclidean boundary of $C(\sigma)$, called the *contour* of σ, and $|\partial(\sigma)|$ the length of $\partial(\sigma)$. Then the energy associated with σ is given by

$$H(\sigma) = J\big|\partial(\sigma)\big| - h|\sigma| + H(\boxminus). \tag{17.2.1}$$

4. To describe the shape of clusters, we need the following:

- An $\ell_1 \times \ell_2$ *rectangle* is a union of closed unit squares centered at the sites in Λ with side lengths $\ell_1, \ell_2 \geq 1$. We use the convention $\ell_1 \leq \ell_2$ and collect rectangles in *equivalence classes modulo translations and rotations*.
- A *quasi-square* is an $\ell \times (\ell + \delta)$ rectangle with $\ell \geq 1$ and $\delta \in \{0, 1\}$. A square is a quasi-square with $\delta = 0$.
- A *bar* is a $1 \times k$ rectangle with $k \geq 1$. A bar is called a *row* or a *column* if it fills a side of a rectangle.
- A *corner* of a rectangle is an intersection of two bars attached to the rectangle.
- A *1-protuberance* is a 1×1 bar attached to one side of a rectangle.
- A *2-protuberance* is a 1×2 bar attached to one side of a rectangle.

5. The configuration space S can be partitioned as

$$S = \bigcup_{n=0}^{|\Lambda|} \mathscr{V}_n, \tag{17.2.2}$$

where

$$\mathscr{V}_n = \{\sigma \in S\colon\ |\sigma| = n\} \tag{17.2.3}$$

is the set of configurations with n up-spins.

17.3 Verification of the two hypotheses

In this section we verify (H1) and (H2) for Glauber dynamics and thereby prove Theorem 17.2.

17.3.1 *First hypothesis*

Proof Let D_ℓ denote the set of configurations where the up-spins form a single $\ell \times \ell$ square anywhere in Λ. The energy of the configurations in D_ℓ equals (recall (17.1.1) and see Fig. 17.4)

$$E(\ell) = H(D_\ell) - H(\boxminus) = J[4\ell] - h[\ell]^2, \tag{17.3.1}$$

which is maximal at $\ell = 2J/h$ and is negative for $l > 4J/h$. Since Λ is chosen large enough so that it contains an $2\ell_c \times 2\ell_c$ square, it follows that $H(\boxminus) = H(0 \times 0) > H(\boxplus)$. It is obvious from (17.2.1) that $\boxplus$ is the global minimum of H, while $\boxminus$ is a

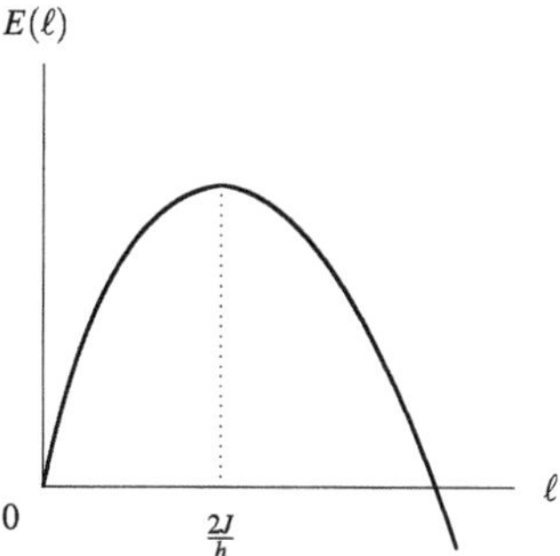

Fig. 17.4 $\ell \mapsto E(\ell)$ (compare with Fig. 1.1)

local minimum of H. Thus, to settle (H1) it remains to show that $\boxminus$ has the unique maximal stability level on $S\backslash\boxplus$.

Let $\gamma^\star = (\gamma_0^\star, \dots, \gamma_{|\Lambda|}^\star)\colon \boxminus \to \boxplus$ be any path that grows a droplet of up-spins by successively adding rows and bars to a quasi-square or square. We refer to this as the *reference path*. In Sect. 17.4 we will show that:

$$\begin{aligned} &\gamma^\star \in (\boxminus \to \boxplus)_{\mathrm{opt}}, \\ &H\big(\gamma_k^\star\big) = \min_{\sigma \in \mathscr{V}_k} H(\sigma). \end{aligned} \tag{17.3.2}$$

Let

$$k^\star = \min\big\{k \in \mathbb{N}\colon\; H\big(\gamma_k^\star\big) \le H(\boxminus)\big\} \ge 2 \tag{17.3.3}$$

be the first time the reference path after it has left $\boxminus$ hits an energy not exceeding that of $\boxminus$.

For $\sigma, \sigma' \in S$, let $\sigma \vee \sigma'$ and $\sigma \wedge \sigma'$ denote the componentwise maximum, respectively, minimum of σ and σ'. An easy computation shows that, for all $\sigma, \sigma' \in S$,

$$\begin{aligned} &\big|\partial\big(\sigma \vee \sigma'\big)\big| + \big|\partial\big(\sigma \wedge \sigma'\big)\big| < \big|\partial(\sigma)\big| + \big|\partial\big(\sigma'\big)\big|, \\ &\big|\sigma \vee \sigma'\big| + \big|\sigma \wedge \sigma'\big| = |\sigma| + \big|\sigma'\big|. \end{aligned} \tag{17.3.4}$$

Pick any $\sigma \in S \setminus [\boxminus \cup \boxplus]$. Then there exists at least one pair of neighbouring sites x and y in Λ such that $\sigma(x) = -1$ and $\sigma(y) = +1$. By translation invariance we may assume without loss of generality that the first two spins that are flipped up in $\gamma^\star$ are located at x and y, respectively. Then

$$\begin{aligned} &\sigma \wedge \gamma_1^\star = \boxminus, \\ &1 \le \big|\sigma \wedge \gamma_k^\star\big| < k \quad \forall k \ge 2. \end{aligned} \tag{17.3.5}$$

In what follows we will consider the path $\sigma \vee \gamma_k^\star$ for $0 \le k \le k^\star$. We have

$$H\big(\sigma \vee \gamma_1^\star\big) - H(\sigma) < H\big(\boxminus \vee \gamma_1^\star\big) - H(\boxminus) = H\big(\gamma_1^\star\big) - H(\boxminus), \tag{17.3.6}$$

where the inequality comes from the interaction between the up-spins at x and y.

Moreover, for $2 \le k \le k^\star$ we can estimate

$$\begin{aligned} H(\sigma \vee \gamma_k^\star) - H(\sigma) &= J\big[|\partial(\sigma \vee \gamma_k^\star)| - |\partial(\sigma)|\big] - h\big[|\sigma \vee \gamma_k^\star| - |\sigma|\big] \\ &\le J\big[|\partial(\gamma_k^\star)| - |\partial(\sigma \wedge \gamma_k^\star)|\big] - h\big[|\gamma_k^\star| - |\sigma \wedge \gamma_k^\star|\big] \\ &= H(\gamma_k^\star) - H(\sigma \wedge \gamma_k^\star) \\ &< H(\gamma_k^\star) - H(\boxminus), \end{aligned} \tag{17.3.7}$$

where we use (17.2.1) and (17.3.2)–(17.3.5). (Note that the second lines of (17.3.2) and (17.3.5) imply that $H(\sigma \wedge \gamma_k^\star) > H(\boxminus)$ for $2 \le k \le k^\star$.) By picking $k = k^\star$ in (17.3.7), we get

$$H(\sigma \vee \gamma_{k^\star}^\star) - H(\sigma) < H(\gamma_{k^\star}^\star) - H(\boxminus) \le 0. \tag{17.3.8}$$

Combining (17.3.6)–(17.3.8), we find that

$$H(\sigma \vee \gamma_{k^\star}^\star) < H(\sigma), \quad \Phi(\sigma, \sigma \vee \gamma_{k^\star}^\star) - H(\sigma) < 0 \vee \max_{1 \le k \le k^\star} H(\gamma_k^\star) - H(\boxminus) \le \Gamma^\star, \tag{17.3.9}$$

where the second inequality uses that $\sigma = \sigma \vee \gamma_0^\star$ because $\gamma_0^\star = \boxminus$ and the third inequality uses that $\gamma^\star \in (\boxminus \to \boxplus)_{\mathrm{opt}}$. Because of Definition 16.1(c), what (17.3.9) says is that the stability level of σ is $< \Gamma^\star$.

Since $\Phi(\boxminus, \boxplus) - H(\boxminus) = \Gamma^\star$, it follows that $\boxminus$ has the unique maximal stability level on $S \backslash \boxplus$ (recall Lemmas 16.7). □

17.3.2 *Second hypothesis*

Proof It is obvious from Definitions 17.1(b–c) and Theorem 17.3 that (H3) is satisfied. Indeed, each configuration in $\mathscr{C}^\star(\boxminus, \boxplus) = \mathscr{Q}^{1\mathrm{pr}}$ has exactly one configuration in $\mathscr{P}^\star(\boxminus, \boxplus) = \mathscr{Q}$ from which it can be reached via an allowed move, namely, the configuration that is obtained from it by removing the single protuberance. □

17.4 Structure of the communication level set

In this section we prove Theorems 17.3 and 17.5.

Proposition 17.6

(i) $\Phi(\boxminus, \boxplus) = \Gamma^\star$.
(ii) $\mathscr{S}(\boxminus, \boxplus) \supseteq \mathscr{Q}^{1\mathrm{pr}}$.

Proof The proof is based on four lemmas (Lemmas 17.7–17.10 below).

(i) We prove that $\Phi(\boxminus, \boxplus) \le \Gamma^\star$ and $\Phi(\boxminus, \boxplus) \ge \Gamma^\star$.

• $\Phi(\boxminus, \boxplus) \le \Gamma^\star$: All we need to do is to construct a path that connects $\boxminus$ and $\boxplus$ without exceeding energy $\Gamma^\star$. The proof comes in three steps.

1. We first show that the configurations in $\mathscr{Q}$ are connected to $\boxminus$ by a path that stays below $\Gamma^\star$.

Lemma 17.7 *For any $\sigma \in \mathscr{Q}$ there exists an $\gamma\colon \sigma \to \boxminus$ such that* $\max_{\xi\in\omega} H(\xi) < \Gamma^\star$.

Proof Fix $\sigma \in \mathscr{Q}$. Note that, by (17.1.7), we have

$$H(\sigma) = \Gamma^\star - (2J - h). \tag{17.4.1}$$

First, we flip down a spin at a corner of the quasi-square, which increases the energy by h. Next, we repeat this operation another $\ell_c - 3$ times, each time picking a spin from a corner on the same shortest side. To guarantee that we never reach energy $\Gamma^\star$, we must have that

$$h(\ell_c - 2) < 2J - h, \tag{17.4.2}$$

or

$$\ell_c < \frac{2J}{h} + 1. \tag{17.4.3}$$

But this inequality holds by the definition of ℓ_c in (17.1.4) and the non-degeneracy hypothesis in (17.1.5). Finally, we flip down the last spin, which lowers the energy by $2J - h$, so that we arrive at energy

$$\Gamma^\star - (2J - h) - \bigl[2J - h(\ell_c - 1)\bigr], \tag{17.4.4}$$

which is strictly smaller than (17.4.1) by (17.4.3). Thus, *the removal of a row of length $\ell_c - 1$ from the $(\ell_c - 1) \times \ell_c$ quasi-square in σ lowers the energy* (see Fig. 17.5). We now have a square of side length $\ell_c - 1$. It is obvious that we can remove further rows without encountering new conditions, until we reach $\boxminus$. □

2. We next show that the configurations in $\mathscr{Q}^{2\mathrm{pr}}$ are connected to $\boxplus$ by a path that stays below $\Gamma^\star$.

Lemma 17.8 *For any $\sigma \in \mathscr{Q}^{2\mathrm{pr}}$ there exists an $\gamma\colon \sigma \to \boxplus$ such that* $\max_{\xi\in\omega} H(\xi) < \Gamma^\star$.

Proof Fix $\sigma \in \mathscr{Q}^{2\mathrm{pr}}$. Note that

$$H(\sigma) = \Gamma^\star - h. \tag{17.4.5}$$

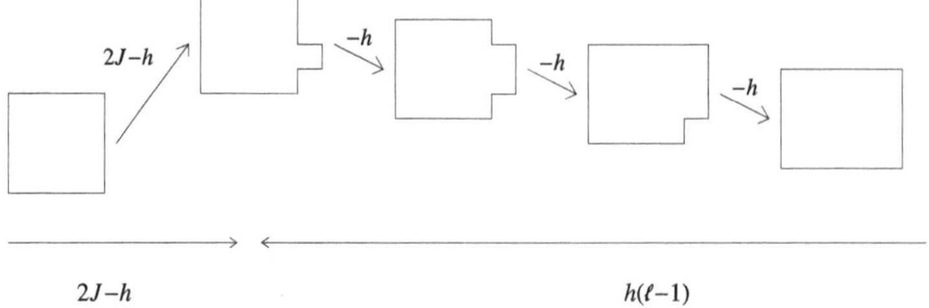

Fig. 17.5 Cost of adding or removing a row of length ℓ

First, we flip up a spin next to the 2-protuberance. This lowers the energy by h. We can repeat this operation another $\ell_c - 3$ times until the row is filled. By that time we have a square of side length ℓ_c and energy

$$\Gamma^\star - h(\ell_c - 1). \tag{17.4.6}$$

Next, we flip up a spin to form a new 1-protuberance. This raises the energy by $2J - h$. To make sure that we do not reach energy $\Gamma^\star$, we must have

$$h(\ell_c - 1) > 2J - h, \tag{17.4.7}$$

or

$$\ell_c > \frac{2J}{h}, \tag{17.4.8}$$

which holds by the definition of ℓ_c and the non-degeneracy hypothesis in (17.1.5). We now have a square of side length ℓ_c with a 1-protuberance. By flipping up a spin next to this 1-protuberance, we get a 2-protuberance and reach energy

$$\Gamma^\star - h(\ell_c - 1) + (2J - h) - h, \tag{17.4.9}$$

which is strictly smaller than (17.4.5) by (17.4.8). Thus, *the completion of a row of length ℓ_c with a 2-protuberance and the creation of a new 2-protuberance lowers the energy* (see Fig. 17.5). It is obvious that we can complete further rows and create further 2-protuberances without encountering new conditions, until we reach ⊞. □

3. We can now conclude the proof of $\Phi(⊟, ⊞) \leq \Gamma^\star$ as follows. The desired path γ: ⊟ → ⊞ is realized by tracing the path in Lemma 17.7 in the reverse direction, from ⊟ to $\sigma \in \mathscr{Q}$, then going from σ to $\sigma' \in \mathscr{Q}^{1\mathrm{pr}}$ by adding a 1-protuberance and from σ' to $\sigma'' \in \mathscr{Q}^{2\mathrm{pr}}$ by extending this 1-protuberance to a 2-protuberance, and finally following the path in Lemma 17.8 from σ'' to ⊞. This γ will be called the *reference path for the magnetisation*.

• $\Phi(⊟, ⊞) \geq \Gamma^\star$: The proof comes in two more steps.

4. The first crucial ingredient in the proof is the following observation:

Lemma 17.9 *Any $\omega \in (\boxminus \to \boxplus)_{\mathrm{opt}}$ must pass through $\mathscr{Q}$.*

Proof Any path $\gamma: \boxminus \to \boxplus$ must cross the set $\mathscr{V}_{\ell_c(\ell_c-1)}$. As shown in Alonso and Cerf [4], Theorem 2.6, the following *isoperimetric inequality* holds as a consequence of (17.2.1): in $\mathscr{V}_{\ell_c(\ell_c-1)}$ the unique (modulo translations and rotations) configuration of minimal energy is the $(\ell_c - 1) \times \ell_c$ quasi-square, which has energy $H(\sigma) = \Gamma^\star - (2J - h)$. All other configurations in $\mathscr{V}_{\ell_c(\ell_c-1)}$ have energy at least $\Gamma^\star + h$, and thus any path not hitting $\mathscr{Q}$ exceeds energy $\Gamma^\star$. □

5. The second crucial ingredient in the proof is the following observation:

Lemma 17.10 *Any $\gamma \in (\boxminus \to \boxplus)_{\mathrm{opt}}$ must pass through $\mathscr{Q}^{\mathrm{1pr}}$.*

Proof Follow the path until it hits the set $\mathscr{V}_{\ell_c(\ell_c-1)}$. According to Lemma 17.9, the configuration in this set must be an $(\ell_c - 1) \times \ell_c$ quasi-square. Since we need not consider any paths that return to the set $\mathscr{V}_{\ell_c(\ell_c-1)}$ afterwards, a first step beyond the quasi-square must be the creation of a 1-protuberance. This brings us to energy $\Gamma^\star$. If the 1-protuberance is created on the side of length ℓ_c, then we have a configuration in $\mathscr{Q}^{\mathrm{1pr}}$. If, on the other hand, it is created on the side of length $\ell_c - 1$, then completion of the row leads an $(\ell_c - 1) \times (\ell_c + 1)$ rectangle with energy $\Gamma^\star - h(\ell_c - 2)$. After that the creation of a 1-protuberance brings us to energy $\Gamma^\star - h(\ell_c - 2) + (2J - h)$, which exceeds energy $\Gamma^\star$ because of (17.4.2). Since $(\ell_c - 1) \times (\ell_c + 1) + 1 = \ell_c \times \ell_c$, any other path that proceeds from the $(\ell_c - 1) \times \ell_c$ quasi-square with a 1-protuberance on the side of length $\ell_c - 1$ to the set $\mathscr{V}_{\ell_c^2}$ without returning to the set $\mathscr{V}_{\ell_c(\ell_c-1)}$ also exceeds energy $\Gamma^\star$. Indeed, according to Alonso and Cerf [4], Theorem 2.6, the unique configuration with minimal energy in the set $\mathscr{V}_{\ell_c^2}$ is the $\ell_c \times \ell_c$ square (modulo rotations and translations). □

Lemmas 17.9–17.10 imply that $\Phi(\boxminus, \boxplus) \geq \Gamma^\star$, and together with Steps 1–3 complete the proof of Proposition 17.6(i).

(ii) Proposition 17.6(ii) follows from Lemma 17.10 because $H(\mathscr{Q}^{\mathrm{1pr}}) = \Gamma^\star$. □

The relations $\mathscr{P}^\star(\boxminus, \boxplus) = \mathscr{Q}$ and $\mathscr{C}^\star(\boxminus, \boxplus) = \mathscr{Q}^{\mathrm{1pr}}$ and the formula for $\Gamma^\star$ claimed in Theorem 17.3 are an immediate consequence of Definition 16.3 and Lemmas 17.7–17.10.

The claim in Theorem 17.5 is immediate from Lemmas 17.7–17.8 in combination with Proposition 16.12, Lemma 16.15, (16.2.4) and (17.3.6)–(17.3.7).

17.5 Computation of the prefactor

In this section we prove Theorem 17.4.

Proof Our starting point is the variational formula for $\Theta = 1/K$ in Lemma 16.17. This variational problem simplifies considerably because of the following two facts that are specific to our Glauber dynamics (abbreviate $\mathscr{C}^\star = \mathscr{C}^\star(\boxminus, \boxplus)$):

- $S^\star \setminus [S_\boxminus \cup S_\boxplus] = \mathscr{C}^\star$, i.e., there are no wells inside $\mathscr{C}^\star$.
- There are no allowed moves within $\mathscr{C}^\star$, i.e., critical droplets cannot transform into each other via single spin-flips.

Consequently, (16.3.11) reduces to

$$\Theta = \min_{h:\ \mathscr{Q}^{1\mathrm{pr}} \to [0,1]} \sum_{\sigma \in \mathscr{Q}^{1\mathrm{pr}}} \left\{ [1 - h(\sigma)]^2 N^-(\sigma) + [0 - h(\sigma)]^2 N^+(\sigma) \right\},$$
$$= \sum_{\sigma \in \mathscr{Q}^{1\mathrm{pr}}} \frac{N^-(\sigma) N^+(\sigma)}{N^-(\sigma) + N^+(\sigma)}, \tag{17.5.1}$$

where

$$N^-(\sigma) = \left| \{\sigma' \in \mathscr{Q}\colon \sigma \sim \sigma'\} \right|,$$
$$N^+(\sigma) = \left| \{\sigma' \in \mathscr{Q}^{2\mathrm{pr}}\colon \sigma \sim \sigma'\} \right|, \tag{17.5.2}$$

is the number of configurations in $\mathscr{Q}$, respectively, $\mathscr{Q}^{2\mathrm{pr}}$ that can reached from $\sigma \in \mathscr{Q}^{1\mathrm{pr}}$ by a single spin-flip (use that $\mathscr{Q} \subseteq S_\boxminus$ and $\mathscr{Q}^{2\mathrm{pr}} \subseteq S_\boxplus$). For all $\sigma \in \mathscr{Q}^{1\mathrm{pr}}$ we have $N^-(\sigma) = 1$, $N^+(\sigma) = 1$ when the 1-protuberance in σ sits at a corner, and $N^+(\sigma) = 2$ when it does not. Hence

$$\Theta = 2|\Lambda| \left[2(\ell_c - 2)\tfrac{2}{3} + 4\tfrac{1}{2} \right] = |\Lambda|\,\tfrac{4}{3}(2\ell_c - 1), \tag{17.5.3}$$

where $2|\Lambda|$ counts the number of locations and rotations of the protocritical droplet. Since $K = 1/\Theta$, this completes the proof of Theorem 17.4. □

17.6 Extension to three dimensions

In this section we briefly indicate how to extend the main definitions and results from two to three dimensions. No proofs are given. See Sect. 17.7 for references.

Let $\Lambda \subset \mathbb{Z}^3$ be a large cubic box, centred at the origin. The metastable parameter range replacing (17.1.3) is

$$h \in (0, 3J), \tag{17.6.1}$$

and, similarly as in (17.1.5), we assume that

$$\frac{2J}{h} \notin \mathbb{N}, \qquad \frac{4J}{h} \notin \mathbb{N}. \tag{17.6.2}$$

The analogue of Definitions 17.1(b–c) reads:

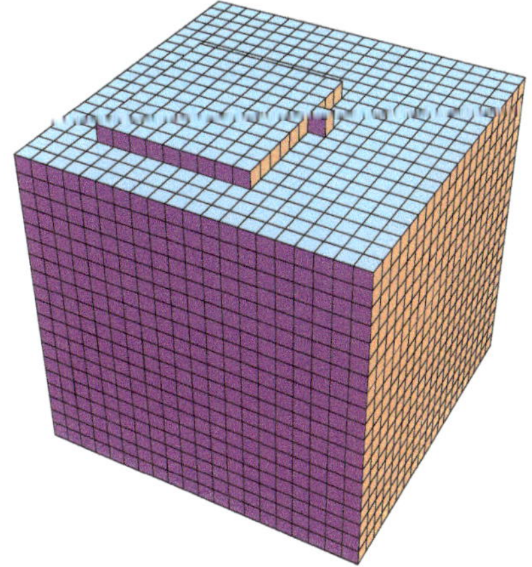

Fig. 17.6 An element of $\mathscr{Q}^{1\mathrm{pr}}$ for $\ell_c = 10$, $m_c = 20$ and $\delta_c = 0$

Definition 17.11

(a) Let $\mathscr{Q}$ be the set of configurations where the up-spins form an $(m_c - 1) \times (m_c - \delta_c) \times m_c$ quasi-cube with, attached to one of its faces, an $(\ell_c - 1) \times \ell_c$ quasi-square, anywhere in Λ. Here, $\delta_c \in \{0, 1\}$ depends on the arithmetic properties of J and h, while

$$\ell_c = \left\lceil \frac{2J}{h} \right\rceil, \qquad m_c = \left\lceil \frac{4J}{h} \right\rceil, \tag{17.6.3}$$

are the two-dimensional critical droplet size on a face, respectively, the three-dimensional critical droplet size, replacing (17.1.4). Note that $m_c \in \{2\ell_c - 1, 2\ell_c\}$.

(b) Let $\mathscr{Q}^{1\mathrm{pr}}$ be the set of configurations obtained from $\mathscr{Q}$ by adding a single protuberance anywhere to one of the longest sides of the quasi-square (see Fig. 17.6).

(c) Let

$$\begin{aligned} \Gamma^\star = \Gamma^\star(\boxminus, \boxplus) &= H\big(\mathscr{Q}^{1\mathrm{pr}}\big) - H(\boxminus) \\ &= J\big[2m_c(m_c - \delta_c) + 2m_c(m_c - 1) + 2(m_c - \delta_c)(m_c - 1) + 4\ell_c\big] \\ &\quad - h\big[m_c(m_c - \delta_c)(m_c - 1) + \ell_c(\ell_c - 1) + 1\big]. \end{aligned} \tag{17.6.4}$$

Theorem 17.3 carries over: $\mathscr{P}^\star(\boxminus, \boxplus) = \mathscr{Q}$ and $\mathscr{C}^\star(\boxminus, \boxplus) = \mathscr{Q}^{1\mathrm{pr}}$. Also Theorem 17.2 carries over: the proof of (H1–H2) is the same as in Sects. 17.3.1–17.3.2. As to Theorem 17.4, the prefactor K can be computed explicitly, namely,

$$K = K_{d=3} = \frac{K_{d=2}}{M_{d=3}} \tag{17.6.5}$$

with $K_{d=2}$ the prefactor in two dimensions and $M_{d=3}$ the number of quasi-cubes in three dimensions that are contained in a three-dimensional critical droplet. The rationale behind (17.6.5) is that a three-dimensional critical droplet is obtained by first growing a quasi-cube with the appropriate side lenghts and then growing a two-dimensional critical droplet on one side of this quasi-cube.

17.7 Bibliographical notes

1. The results in this chapter are taken from Bovier and Manzo [39]. Cruder versions of the main results in Chap. 16 for Glauber dynamics, derived with the help of the pathwise approach to metastability, were obtained by Neves and Schonmann [193] in two dimensions and by Ben Arous and Cerf [19] in three dimensions.

2. The formula for K claimed in [39] contains a small error. This is corrected in Theorem 17.4. The argument in Sect. 17.3.1 first appeared in den Hollander, Nardi, Olivieri and Scoppola [84].

3. It is possible to extend the analysis in Sect. 17.6 to arbitrary dimension. As shown in Neves [192], $\Gamma^\star$ can be computed in a recursive manner, based on the observation that a critical droplet in dimension d can be obtained by attaching a critical droplet in dimension $d-1$ to the appropriate side of a quasi-hypercube with the appropriate side lengths. The main difficulty is to show that all the configurations in $\mathscr{C}^\star$ can be obtained in this way, which remains open. For the computation of K the simple structure of $\mathscr{C}^\star$, exploited in Sect. 17.5 for the case of two dimensions, prevails in higher dimensions, provided we assume that $2J/h, \dots, (d-1)J/h \notin \mathbb{N}$. In particular, we have $K = K_d = K_{d-1}/M_d$ with M_d the number of quasi-cubes in d dimensions that are contained in a d-dimensional critical droplet. For details and relevant formulas, see [39].

4. Detailed results are known about the *tube of typical trajectories*, i.e., the set of paths within which the crossover from ⊟ to ⊞ takes place, also referred to as the *nucleation pattern*. The identification of this tube requires an analysis of the dynamics on *shorter time scales*, in particular, the typical times scales on which rows and columns are grown. This is the realm of the pathwise approach to metastability. Such a refined analysis is also necessary to improve on the result in Theorem 17.5. See Olivieri and Vares [198, Sects. 7.3–7.4], and Gaudillière, Olivieri and Scoppola [122].

5. An anisotropic version of Glauber dynamics, in which the Hamiltonian in (18.1.2) is modified by allowing for different pair potentials $J_h > J_v > 0$ in the horizontal and the vertical direction, was studied in Kotecký and Olivieri [155]. Surprisingly, despite the anisotropy the critical droplet still is a quasi-square with a single protuberance with side length $\ell_c = \lceil 2J_v/h \rceil$ under the assumption that $0 < h < 2J_v \wedge 2(J_h - J_v)$. Only after this critical droplet has been grown does the nucleation proceed in an anisotropic manner, by first fully expanding in the horizontal direction and afterwards fully expanding in the vertical direction, after which the box is filled with plus spins. For details, see [198, Sect. 7.7].

6. A version where a next-to-nearest-neighbour interaction with pair potential $\bar{J} > 0$ is added to the Hamiltonian in (18.1.2) was considered by Kotecký and Olivieri [156]. Here the critical droplet is expected to have an octagonal shape, with side lengths that depend on the values of $J, \bar{J}$. However, the situation turns out to be different. Under the assumption that $h < \frac{1}{7}\bar{J} < \frac{1}{70}J$, initially the nucleation pattern follows a sequence of regular octogons whose sides are equal, up to length $\bar{\ell}_c = \lceil 2\bar{J}/h \rceil$. After that the oblique sides remain of fixed length $\bar{\ell}_c$ while the horizontal and the vertical sides continue to grow longer, until they reach length $\ell_c - 2(\bar{\ell}_c - 1)$ with $\ell_c = \lceil 2J/h \rceil$, at which stage the critical droplet is reached. After that the horizontal and the vertical sides continue to grow longer until the box is filled with plus-spins. The analysis is rather delicate, because standard isoperimetric inequalities can no longer be used. For details, see [198, Sect. 7.9].

7. A staggered version of Glauber dynamics, in which the Hamiltonian in (17.1.1) is modified by allowing for opposite magnetic fields $h_{\text{even}} > 0 > h_{\text{odd}}$ at the even and the odd numbered sites, was studied in Nardi and Olivieri [188]. Once again the nucleation pattern is unusual. There are three regimes, corresponding to the three equilibrium phases of the model (plus-phase, minus-phase and staggered phase). See [198, Sect. 7.10].

8. A version of the Glauber dynamics with three spin-values called the Blume-Capel model, namely, $\Upsilon = \{-1, 0, +1\}$ and allowing for single-site changes of the spins, was considered by Manzo and Olivieri [172, 173]. There are three regimes, corresponding to the three equilibrium phases of the model (plus-phase, zero-phase and minus-phase). The nucleation pattern is fairly complex. See [198, Sect. 7.11].

9. Metastability for Ising spins subject to a PCA spin-flip dynamics (defined in Sect. 16.4.2) was considered in a series of papers by Bigilis, Cirillo, Lebowitz and Speer [27], Cirillo, Nardi and Polosa [61], Cirillo [59], Cirillo and Nardi [60], Cirillo, Nardi and Spitoni [62–64], and Nardi and Spitoni [190]. The model studied most closely is the one where Λ is a torus in $\mathbb{Z}^2$ and $\Upsilon = \{-1, +1\}$ is the spin space, like in the present chapter, but the Hamiltonian is

$$H(\sigma) = -h \sum_{x \in \Lambda} \sigma(x) - \frac{1}{\beta} \sum_{x \in \Lambda} \ln\cosh\bigl(\beta\bigl[U_\sigma(x) + h\bigr]\bigr), \quad \sigma \in S = \Upsilon^\Lambda, \qquad (17.7.1)$$

where $\beta, h > 0$ and $U_\sigma(x) = \sum_{y \in N} \sigma(x+y)$ with $N = \{z \in \Lambda \colon \|z\| \leq 1\}$ and $\|\cdot\|$ the lattice norm on $\mathbb{Z}^2$, and the single-spin transition probabilities are

$$p_{x,\sigma}(s) = \frac{1}{2}\bigl[1 + s\tanh\bigl(\beta\bigl[U_\sigma(x) + h\bigr]\bigr)\bigr], \quad x \in \Lambda, \sigma \in S, s \in \Upsilon. \qquad (17.7.2)$$

The choice in (17.7.1)–(17.7.2) matches the reversibility condition in (16.4.6). Note that H depends on β, but

$$\lim_{\beta\to\infty} H(\sigma) = \bar{H}(\sigma) = -h\sum_{x\in\Lambda}\sigma(x) - \sum_{x\in\Lambda}\left|U_\sigma(x)+h\right|. \tag{17.7.3}$$

As shown in [190] (recall (16.4.7)),

$$\mathscr{H}\left(\sigma,\sigma'\right) = \bar{\mathscr{H}}\left(\sigma,\sigma'\right) + \frac{|\Lambda|\ln 2}{\beta} \tag{17.7.4}$$

with

$$\bar{\mathscr{H}}\left(\sigma,\sigma'\right) \geq \bar{H}(\sigma)\vee\bar{H}\left(\sigma'\right). \tag{17.7.5}$$

Therefore, in the limit as $\beta\to\infty$, the PCA behaves like a discrete-time Markov chain driven by the Hamiltonian $\bar{H}$, similar to the Metropolis dynamics. A full identification of the triple $(\Gamma^\star, \mathscr{C}^\star, K)$ was achieved for the choice where $\mathbf{m} = ⊟$ and $\mathbf{s} = ⊞$ in the metastable regime where $0 < h < 1$ and $\beta\to\infty$. The results are similar to what we found in Sect. 17.1. However, the proofs are more difficult, because there are many configurations in which the dynamics stays trapped for a long time, e.g. when the plus-spins form a rectangle. There are also many pairs of configurations between which the dynamics oscillates for a long time, e.g. when the plus-spins form two alternate checkerboards in a rectangle. This complexity hampers the geometric analysis of $\mathscr{H}$ in (16.4.7) that is needed to identify $(\Gamma^\star, \mathscr{C}^\star, K)$. The model in which 0 is removed from N ("no self-interaction") is even harder to analyse. It turns out that on its way from ⊟ to ⊞ the PCA visits the two configurations where the spins form a checkerboard in Λ (provided Λ has even side length). This does not happen for the model where 0 is included in N.

Chapter 18
Kawasaki Dynamics

"All right," said the Cat; and this time it vanished quite slowly, beginning with the end of the tail, and ending with the grin, which remained some time after the rest of it had gone.
(Lewis Carroll, Alice's Adventures in Wonderland)

In this chapter we apply the results obtained in Chap. 16 to the *lattice gas* in two and three dimensions subject to *Kawasaki dynamics*. Particles live in a finite box, hop between nearest-neighbour sites, feel an attractive interaction when they sit next to each other, and are created, respectively, annihilated at the boundary of the box in a way that reflects the presence of an infinite gas reservoir. We are interested in how the system *nucleates*, i.e., how the box fills up when it is initially empty. Our goal will be to prove hypotheses (H1–H2) in Sect. 16.1.2, implying that Theorems 16.4–16.6 are valid. In two dimensions we will further identify $(\Gamma^\star, \mathscr{C}^\star)$ and obtain the asymptotics of K in the limit as the size of the box tends to infinity. In three dimensions we will also identify $\Gamma^\star$, but we will obtain only partial information on $\mathscr{C}^\star$ and K.

Kawasaki differs from Glauber, treated in Chap. 17, in that it is a *conservative dynamics*: particles are conserved in the interior of the box. Consequently, during the growing and the shrinking of droplets, particles must travel between the droplet and the boundary of the box, which causes several complications. Moreover, it turns out that in the metastable regime *particles move along the border of a droplet more rapidly than they arrive from the boundary of the box*. This leads to a shape of the critical droplet that is more complicated than the one for Glauber dynamics. This complexity needs to be handled in order to obtain information on $\mathscr{C}^\star$ and K.

18.1 Introduction and main results

18.1.1 Model

Let $\Lambda \subset \mathbb{Z}^2$ be a large square box, centered at the origin. Let

$$\partial^- \Lambda = \{x \in \Lambda:\ \exists\, y \notin \Lambda:\ \|y - x\| = 1\} \tag{18.1.1}$$

A. Bovier, F. den Hollander, *Metastability*,
Grundlehren der mathematischen Wissenschaften 351,
DOI 10.1007/978-3-319-24777-9_18

0	0	1	0	0
0	0	0	1	0
0	1	1	0	0
0	1	1	0	0
0	0	0	0	0

Fig. 18.1 A lattice-gas configuration

be the internal boundary of Λ, and put $\Lambda^- = \Lambda \setminus \partial^-\Lambda$. With each site $x \in \Lambda$ we associate an occupation variable $\eta(x)$ assuming the values 0 or 1, indicating the *absence* or *presence* of a particle at x (see Fig. 18.1). A configuration is denoted by $\eta \in S = \{0, 1\}^\Lambda$. Each configuration $\eta \in S$ has an energy given by the Hamiltonian

$$H(\eta) = -U \sum_{\{x,y\}\in(\Lambda^-)^*} \eta(x)\eta(y) + \Delta \sum_{x\in\Lambda} \eta(x), \tag{18.1.2}$$

where

$$(\Lambda^-)^* = \big\{\{x, y\}:\ x, y \in \Lambda^-,\ \|x - y\| = 1\big\} \tag{18.1.3}$$

is the set of non-oriented nearest-neighbour bonds in Λ^-. The interaction consists of a *binding energy* $-U < 0$ for each neighbouring pair of particles in Λ^- and an *activation energy* $\Delta > 0$ for each particle in Λ. Note that particles in $\partial^-\Lambda$ do not interact with particles anywhere in Λ.

The Hamiltonian in (18.1.2) models a lattice gas in Λ. We are interested in *Kawasaki dynamics* on Λ with an *open boundary*. This is the Metropolis dynamics with respect to H at inverse temperature β defined in (16.1.2) with two types of allowed moves: (1) *particle hop*: 0 and 1 interchange at a pair of neighbouring sites in Λ; (2) *particle creation or annihilation*: 0 changes to 1 or 1 changes to 0 at a single site in $\partial^-\Lambda$. Clearly, this dynamics is a finite-state Markov process, and hence fits into the general theory described in Chap. 16.

Kawasaki dynamics models the behaviour inside Λ of a lattice gas in $\mathbb{Z}^2$, consisting of particles subject to random hopping with hard core repulsion inside Λ, neighbouring attraction inside Λ^-, and creation and annihilation in $\partial^-\Lambda$. We may think of $\mathbb{Z}^2 \setminus \Lambda$ as an *infinite reservoir* that keeps the particle density inside Λ fixed at $e^{-\beta\Delta}$. In our model this reservoir is replaced by an *open boundary* $\partial^-\Lambda$. Note that a move of particles inside $\partial^-\Lambda$ does not involve a change of energy because the interaction acts only inside Λ^-.

The Gibbs measure μ_β defined in (16.1.1) is the equilibrium of the dynamics with transition rates c_β defined in (16.1.2) and satisfies the reversibility property in (16.1.3).

18.1.2 Metastable regime and critical droplet size

Throughout the sequel we assume that

$$\Delta \in (U, 2U). \tag{18.1.4}$$

We will see that this parameter range corresponds to metastable behaviour in the limit as $\beta \to \infty$ (see Fig. 18.3 below). A key role will be played by what we call the *critical droplet size*:

$$\ell_c = \left\lceil \frac{U}{2U - \Delta} \right\rceil. \tag{18.1.5}$$

For reasons that will become clear later on, we will assume that

$$\frac{U}{2U - \Delta} \notin \mathbb{N}. \tag{18.1.6}$$

Thus, an $(\ell_c - 1) \times (\ell_c - 1)$ droplet will be "subcritical" while an $\ell_c \times \ell_c$ droplet will be "supercritical". Moreover, we will assume that Λ is large enough so that Λ^- contains an $2\ell_c \times 2\ell_c$ square.

Analogous assumptions are needed in three dimensions (see Sect. 18.6).

18.1.3 Main theorems

Each configuration can be decomposed into maximally connected components, called *clusters*. A *free particle* is a particle in Λ not interacting with other particles.

Definition 18.1

(a) Let

$$\begin{aligned} \square &= \{\eta \in S\colon\ \eta(x) = 0\ \forall x \in \Lambda\}, \\ \blacksquare &= \{\eta \in S\colon\ \eta(x) = 1\ \forall x \in \Lambda^-,\ \eta(x) = 0\ \forall x \in \partial^-\Lambda\}, \end{aligned} \tag{18.1.7}$$

denote the configurations where Λ is empty, respectively, Λ^- is full and $\partial^-\Lambda$ is empty.

(b) Let

$$\mathscr{D} = \bar{\mathscr{D}} \cup \tilde{\mathscr{D}}, \tag{18.1.8}$$

where

- $\bar{\mathscr{D}}$ is the set of configurations with a single cluster anywhere in Λ^- consisting of an $(\ell_c - 2) \times (\ell_c - 2)$ *square* with four *bars* of lengths $\bar{k}_i$, $i = 1, 2, 3, 4$,

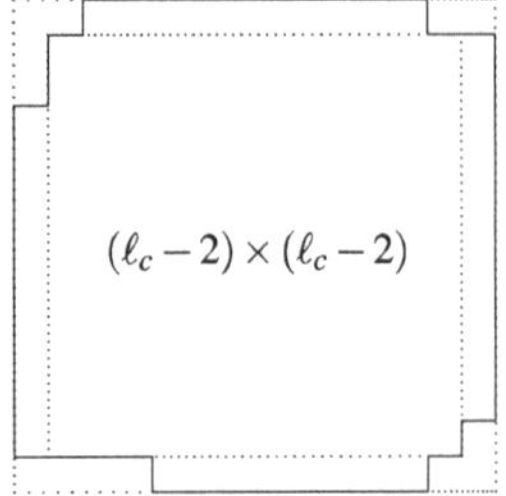

Fig. 18.2 A configuration in $\bar{\mathscr{D}}$ with an $(\ell_c-2)\times(\ell_c-2)$ square in the center and four bars attached to it. A similar picture applies for $\widetilde{\mathscr{D}}$ with an $(\ell_c-3)\times(\ell_c-1)$ rectangle in the center

attached to its four sides satisfying

$$1 \le \bar{k}_i \le \ell_c - 1, \quad \sum_i \bar{k}_i = 3\ell_c - 3. \tag{18.1.9}$$

– $\widetilde{\mathscr{D}}$ is the set of configurations with a single cluster anywhere in Λ^- consisting of an $(\ell_c-3)\times(\ell_c-1)$ *rectangle* with four *bars* of lengths $\widetilde{k}_i$, $i=1,2,3,4$, attached to its four sides satisfying

$$1 \le \widetilde{k}_i \le \ell_c - 1, \quad \sum_i \widetilde{k}_i = 3\ell_c - 2. \tag{18.1.10}$$

(c) Let $\mathscr{D}^{\mathrm{fp}}$ denote the set of configurations obtained from $\mathscr{D}$ by *adding a free particle* anywhere in $\partial^-\Lambda$.

In the definition of $\bar{\mathscr{D}}$, the four bars may be placed anywhere in the ring around the square, i.e., anywhere in the union of the two rows and the two columns forming the outer layer of the square (see Fig. 18.2). A total of $3\ell_c-3$ particles must be accommodated in this ring in such a way that each side of the ring, i.e., each row or column, contains precisely one bar. A bar may include a corner of the ring provided the neighbouring bar also includes this corner. Similarly for $\widetilde{\mathscr{D}}$.

In Sect. 18.1.4, item 2, we will see that the configurations in $\mathscr{D}$ arise from each other via *motion of particles along the border of the droplet*, a phenomenon that is specific to Kawasaki dynamics.

The main metastability theorems for Kawasaki dynamics are the following. Recall Definition 16.3.

Theorem 18.2 *The pair* $(\square,\blacksquare)$ *satisfies hypotheses* (H1–H2) *in Sect.* 16.1.2, *and hence Theorems* 16.4–16.6 *hold.*

Theorem 18.3 *The pair* $(\square, \blacksquare)$ *has protocritical set* $\mathscr{P}^\star(\square, \blacksquare) = \mathscr{D}$, *critical set* $\mathscr{C}^\star(\square, \blacksquare) = \mathscr{D}^{\mathrm{fp}}$, *and communication height*

$$\begin{aligned}\Gamma^\star &= \Gamma^\star(\square, \blacksquare) = H\big(\mathscr{D}^{\mathrm{fp}}\big) - H(\square) = H(\mathscr{D}) + \Delta \\ &= -U\big[(\ell_c - 1)^2 + \ell_c(\ell_c - 2) + 1\big] + \Delta\big[\ell_c(\ell_c - 1) + 2\big] \\ &= 2U[\ell_c + 1] - (2U - \Delta)\big[\ell_c(\ell_c - 1) + 2\big].\end{aligned} \tag{18.1.11}$$

Theorem 18.4 *For large* Λ *the prefactor* $K = K(\Lambda)$ *scales like*

$$\lim_{\Lambda \to \mathbb{Z}^2} \frac{|\Lambda|}{\ln |\Lambda|} K(\Lambda) = \frac{1}{4\pi N(\ell_c)} \tag{18.1.12}$$

with

$$N(\ell_c) = \sum_{k=1,2,3,4} \binom{4}{k} \left[\binom{\ell_c + k - 2}{2k - 1} + 2\binom{\ell_c + k - 3}{2k - 1}\right] \tag{18.1.13}$$

the cardinality of $\mathscr{D}$ *modulo shifts.*

Remark The asymptotics in (18.1.12) does not depend on the shape of Λ, e.g. it would be the same if Λ were a large circle rather than a large square.

In addition, we have the following geometric description of the configurations in the valleys $S_\square$, $S_\blacksquare$ around $\square$, $\blacksquare$ defined in (16.2.25). Let

$$\begin{aligned}V_{\leq \mathscr{D}} &= \big\{\eta \in S\colon\ \eta \leq \eta' \text{ for some } \eta' \in \mathscr{D}\big\}, \\ V_{\geq \mathscr{C}^G} &= \big\{\eta \in S\colon\ \eta \geq \eta' \text{ for some } \eta' \in \mathscr{C}^G\big\},\end{aligned} \tag{18.1.14}$$

where $\mathscr{C}^G$ is the set of configurations obtained from $\mathscr{C}^\star = \mathscr{D}^{\mathrm{fp}}$ by moving the free particle from $\partial^-\Lambda$ to the cluster and attaching it at a "good" site in the outer layer of the cluster (i.e., next to two other particles; see Fig. 18.10 below).

Theorem 18.5 $S_\square \supseteq V_{\leq \mathscr{D}}$, $S_\blacksquare \supseteq V_{\geq \mathscr{C}^G}$.

18.1.4 Discussion

1. The proof of Theorem 18.2 is given in Sect. 18.3. (H2) is easy to check, (H1) is more involved and relies on certain *isoperimetric inequalities*.

2. The heuristics behind Theorem 18.3 is as follows. In Sect. 18.4 we will see that $\mathscr{D}^{\mathrm{fp}} \subseteq \mathscr{S}(\square, \blacksquare)$, the communication level set of the pair $(\square, \blacksquare)$. We will see that the dynamics passes through $\mathscr{S}(\square, \blacksquare)$ in four steps: (1) first it creates a "canonical protocritical droplet", namely, a configuration in $\mathscr{D}$ with the property that three bars have full length and one bar consists of a single protuberance; (2) next it allows particles to "move along the border of the droplet", thereby forming all the other "protocritical droplets" in $\mathscr{D}$; (3) after that it brings in a free particle, thereby forming a "critical droplet"; (4) finally it attaches this free particle to the boundary of the protocritical droplet. After these four steps are completed, the dynamics is "over the hill" and proceeds downwards in energy to fill up the box. This also explains where Theorem 18.5 comes from.

Note: If the free particle attaches itself at a "bad site" in the outer layer of the protocritical droplet (i.e., next to one other particle), then either it may again detach itself or it may cause a *motion of particles along the border of the droplet*, after which another particle may detach itself, possibly leaving behind a *different* protocritical droplet. However, since for large Λ a free particle has a small probability to escape from the protocritical droplet and return to $\partial^-\Lambda$, it must eventually attach itself at a "good site". See Sect. 18.4.4 for more details.

3. The heuristics behind Theorem 18.4 is as follows. The average time it takes for the dynamics to enter $\mathscr{C}^\star(\square, \blacksquare) = \mathscr{D}^{\mathrm{fp}}$ when starting from $\square$ is

$$\frac{1}{|\mathscr{D}|}\frac{1}{|\partial^-\Lambda|}\mathrm{e}^{\beta\Gamma^\star}\big[1+o(1)\big], \quad \beta\to\infty, \tag{18.1.15}$$

where $|\mathscr{D}|$ counts the number of protocritical droplets and $|\partial^-\Lambda|$ counts the number of locations where the free particle can be created. Let $\pi(\Lambda,\ell_c)$ be the probability that the free particle moves from $\partial^-\Lambda$ to the protocritical droplet and attaches itself at a good site, i.e., the probability that the dynamics after it enters $\mathscr{C}^\star(\square, \blacksquare)$ moves onwards to $\blacksquare$ rather than returns to $\square$. Then

$$\frac{1}{\pi(\Lambda,\ell_c)}\big[1+o(1)\big], \quad \beta\to\infty, \tag{18.1.16}$$

is the average number of times a free particle just created in $\partial^-\Lambda$ attempts to move to the protocritical droplet and attach itself at a good site before it finally manages to do so. The average nucleation time is the product of (18.1.15) and (18.1.16), and so we conclude that

$$K = \frac{1}{|\mathscr{D}|\,|\partial^-\Lambda|\,\pi(\Lambda,\ell_c)}. \tag{18.1.17}$$

To compute $|\mathscr{D}|$, note that

$$|\mathscr{D}| = \big[1+o(1)\big]|\Lambda|\,N(\ell_c), \quad \Lambda\to\mathbb{Z}^2. \tag{18.1.18}$$

To compute $\pi(\Lambda,\ell_c)$, note that

$$|\partial^-\Lambda|\,\pi(\Lambda,\ell_c) = \big[1+o(1)\big]\frac{4\pi}{\ln|\Lambda|}, \quad \Lambda\to\mathbb{Z}^2. \tag{18.1.19}$$

Indeed, as we will see in Sect. 18.5, the right-hand side of (18.1.19) is the probability for large Λ that a particle detaching itself from the protocritical droplet reaches $\partial^-\Lambda$ before re-attaching itself. Due to the recurrence of simple random walk in two dimensions, for large Λ this probability is independent of the shape and the location of the protocritical droplet, as long as it is far from $\partial^-\Lambda$. By reversibility, the reverse motion has the same probability, which explains (18.1.19). Combine (18.1.17)–(18.1.19) to get (18.1.12).

4. In the limit of weak supersaturation, $\Delta \uparrow 2U$, we have $\ell_c \to \infty$. In this limit, the formula in (18.1.13) gives $N(\ell_c) \sim \ell_c^7/2520$.

5. In Sect. 18.5 we will derive a representation for the prefactor K in terms of certain capacities associated with two-dimensional simple random walk on Λ in the presence of a protocritical droplet. We will see that this representation is non-trivial because of the presence of so-called "good sites" and "bad sites" on the border of the protocritical droplet. Consequently, no easily computable expression is available for K for finite Λ, only bounds. Theorem 18.4 shows that these bounds merge in the limit as $\Lambda \to \mathbb{Z}^2$.

Outline The outline of the remainder of this chapter is as follows. In Sect. 18.2 we introduce some key geometric definitions that are needed for the proof of Theorems 18.2–18.5. These theorems are proved in Sects. 18.3–18.5. Section 18.6 looks at the extension from two to three dimensions.

Throughout the sequel we assume that $\ell_c \geq 3$. The case $\ell_c = 2$ is trivial: $\mathscr{P}^\star(\square, \blacksquare) = \mathscr{D}$ is the set of configurations consisting of three particles forming a cluster anywhere in Λ^-, $\mathscr{C}^\star(\square, \blacksquare) = \mathscr{D}^{\mathrm{fp}}$ is the set of configurations obtained from these by adding a free particle anywhere in $\partial^-\Lambda$, and $\Gamma^\star(\square, \blacksquare) = -2U + 4\Delta$.

18.2 Geometric definitions

In order to prove Theorems 18.2–18.4, we need some further definitions.

1. Throughout the sequel, we identify a configuration $\eta \in S$ with its support $\mathrm{supp}(\eta) = \{x \in \Lambda \colon \eta(x) = 1\}$, and write $x \in \eta$ to indicate that η has a particle at x. Free particles, 1-protuberances and corners are defined as follows:

- For $x \in \Lambda^-$, let $\mathscr{N}(x) = \{y \in \Lambda^- \colon |y - x| = 1\}$ be the set of nearest-neighbour sites of x in Λ^-.
- A *free particle* in $\eta \in S$ is a site $x \in \eta \cap \partial^-\Lambda$ or a site $x \in \eta \cap \Lambda^-$ such that $\sum_{y \in \mathscr{N}(x)} \eta(y) = 0$, i.e., a particle not in interaction with any other particle (remember from (18.1.2) that particles in the interior boundary $\partial^-\Lambda$ have no interaction with particles in the interior Λ^-).
- A *1-protuberance* in $\eta \in S$ is a site $x \in \eta \cap \Lambda^-$ such that $\sum_{y \in \mathscr{N}(x)} \eta(y) = 1$.

- A *corner* in $\eta \in S$ is a site $x \in \Lambda^-$ such that $\sum_{y\in\mathcal{N}(x)} \eta(y) \ge 2$. A corner in η can be either occupied or vacant.

2. Given a configuration $\eta \in S$, consider the set $C(\eta) \subseteq \mathbb{R}^2$ defined as the union of the closed unit squares centred at the sites inside Λ^- where η has a particle. The maximal connected components $C_1, \ldots, C_m$, $m \in \mathbb{N}$, of $C(\eta)$ are called *clusters* of η (two unit squares touching only at the corners are not connected). There is a one-to-one correspondence between configurations $\eta \subseteq \Lambda^-$ and sets $C(\eta)$. A configuration $\eta \in S$ is characterised by a set $C(\eta)$, depending only on $\eta \cap \Lambda^-$, plus possibly a set of particles in $\partial^-\Lambda$, namely, $\eta \cap \partial^-\Lambda$. Thus, we are actually identifying two different objects: a configuration $\eta \in S$ and the pair $(C(\eta), \eta \cap \partial^-\Lambda)$.

3. For $\eta \in S$, let $|\eta|$ be the number of particles in η, $\partial(\eta)$ the Euclidean boundary of $C(\eta)$, called the *contour* of η, and $|\partial(\eta)|$ the length of $\partial(\eta)$. Then the energy associated with η is given by

$$H(\eta) = \frac{U}{2}\big|\partial(\eta)\big| - (2U - \Delta)\big|\eta \cap \Lambda^-\big| + \Delta\big|\eta \cap \partial^-\Lambda\big|. \tag{18.2.1}$$

4. To describe the shape of clusters, we need the following:

- An $\ell_1 \times \ell_2$ *rectangle* is a union of closed unit squares centred at the sites inside Λ^- with side lengths $\ell_1, \ell_2 \ge 1$. We use the convention $\ell_1 \le \ell_2$ and collect rectangles in *equivalence classes modulo translations and rotations*.
- A *bar* is a $1 \times k$ rectangle with $k \ge 1$. A bar is called a *row* or a *column* if it fills a side of a rectangle.
- A *corner* of a rectangle is an intersection of two bars attached to the rectangle.
- A *quasi-square* is an $\ell \times (\ell + \delta)$ rectangle with $\ell \ge 1$ and $\delta \in \{0, 1\}$. A square is a quasi-square with $\delta = 0$.
- If η is a configuration with a single contour, then we denote by $\mathrm{CR}(\eta)$ the rectangle *circumscribing* η, i.e., the smallest rectangle containing η. We write

$$\begin{aligned} \partial^-\mathrm{CR}(\eta) &= \big\{x \in \mathrm{CR}(\eta)\colon\ \exists\, y \notin \mathrm{CR}(\eta)\colon\ \|y - x\| = 1\big\},\\ \partial^+\mathrm{CR}(\eta) &= \big\{x \notin \mathrm{CR}(\eta)\colon\ \exists\, y \in \mathrm{CR}(\eta)\colon\ \|y - x\| = 1\big\}, \end{aligned} \tag{18.2.2}$$

to denote the interior, respectively, external boundary of $\mathrm{CR}(\eta)$, and put

$$\begin{aligned} \mathrm{CR}^-(\eta) &= \mathrm{CR}(\eta) \setminus \partial^-\mathrm{CR}(\eta),\\ \mathrm{CR}^+(\eta) &= \mathrm{CR}(\eta) \cup \partial^+\mathrm{CR}(\eta). \end{aligned} \tag{18.2.3}$$

Note that here we identify particles with unit squares.
- Given η such that $\eta \supseteq \mathrm{CR}^-(\eta)$, we say that it is possible to move a particle from *row* $r_\alpha(\eta) \subseteq \partial^-\mathrm{CR}(\eta)$ to *row* $r_{\alpha'}(\eta) \subseteq \partial^-\mathrm{CR}(\eta)$ via *corner* $c_{\alpha,\alpha'}(\eta) \in \partial^-\mathrm{CR}(\eta)$ if (see Figs. 18.5–18.6 in Sect. 18.4.1)

$$\big|c_{\alpha\alpha'}(\eta) \cap \eta\big| = 0, \quad \big|r_\alpha(\eta) \cap \eta\big| \ge 1, \quad 1 \le \big|r_{\alpha'}(\eta) \cap \eta\big| \le \big|r_{\alpha'}(\eta)\big|, \tag{18.2.4}$$

where $\alpha\alpha' \in \{ne, nw, se, sw\}$ with $n =$ north, $s =$ south, etc. By convention, corners are not part of rows. If equality holds in the last inequality, then we need to place the bar in the row opposite to $r_\alpha(\eta)$, say $r_{\alpha''}(\eta)$, a distance 1 away from $c_{\alpha'\alpha''}(\eta)$ in order to be able to accommodate the shift of a bar in $r_{\alpha'}(\eta)$ that is necessary to accommodate the particle that moves around the corner.

5. For $\eta, \eta' \in S$, a path $\gamma : \eta \to \eta'$ of allowed moves is called a *U-path* if

$$\begin{aligned} &\text{(i)} \quad H(\eta) = H\big(\eta'\big), \\ &\text{(ii)} \quad \max_i H(\gamma_i) \le H(\eta) + U, \\ &\text{(iii)} \quad |\gamma_i| = |\eta| \text{ for all } i. \end{aligned} \tag{18.2.5}$$

6. The configuration space S can be partitioned as

$$S = \bigcup_{n=0}^{|\Lambda|} \mathscr{V}_n, \tag{18.2.6}$$

where

$$\mathscr{V}_n = \big\{\eta \in S\colon\ |\eta| = n\big\} \tag{18.2.7}$$

is the set of configurations with n particles.

18.3 Verification of the two hypotheses

18.3.1 First hypothesis

Proof Let D_ℓ denote the set of configurations where the particles form a single $\ell \times \ell$ square anywhere inside Λ^-. The energy $E(\ell)$ of the configurations in D_ℓ equals (recall (18.1.2) and see Fig. 18.3)

$$E(\ell) = H(D_\ell) - H(\square) = -U\big[2\ell(\ell-1)\big] + \Delta\ell^2 = 2U\ell - (2U - \Delta)\ell^2, \tag{18.3.1}$$

which is maximal at $\ell = U/(2U - \Delta)$ and is negative for $l > 2U/(2U - \Delta)$. Since Λ is chosen large enough so that Λ^- contains an $2\ell_c \times 2\ell_c$ square, it follows that $H(\square) = H(0 \times 0) > H(\blacksquare)$. It is obvious from (18.2.1) that $\blacksquare$ is the global minimum of H, while $\square$ is a local minimum of H. Thus, to settle (H1) it remains to show that $\square$ has the unique maximal stability level on $S\backslash\blacksquare$.

We can repeat the argument for Glauber dynamics in Sect. 17.3.1 by thinking of up-spins as particles and down-spins as vacancies. The additional obstacle under Kawasaki dynamics is that, when we are growing the configuration by considering the union of η with the droplets in the reference path, *particles cannot be created where needed but have to arrive from* $\partial^-\Lambda$. We have to make sure that at any stage the configuration is such that a particle coming from $\partial^-\Lambda$ can be moved to where it is needed. This requires a technical construction with "pistons enclosing η", for which we refer to the literature (see the reference in Sect. 18.7). $\square$

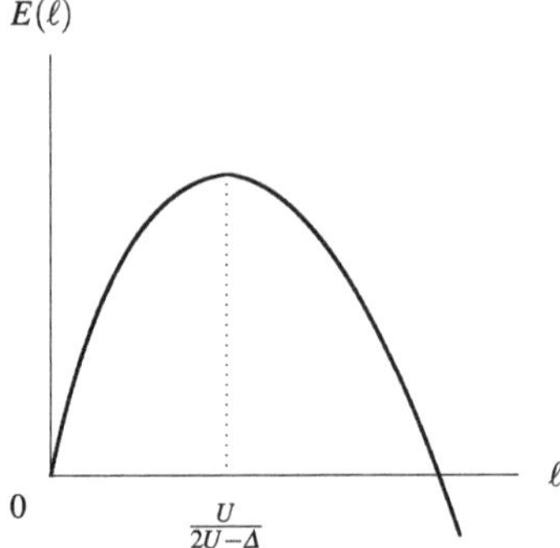

Fig. 18.3 $\ell \mapsto E(\ell)$ (compare with Fig. 1.1)

18.3.2 Second hypothesis

Proof It is obvious from Definitions 18.1(b–c) and Theorem 18.3 that (H2) is satisfied. Indeed, each configuration in $\mathscr{C}^\star(\square, \blacksquare) = \mathscr{D}^{\mathrm{fp}}$ has exactly one configuration in $\mathscr{P}^\star(\square, \blacksquare) = \mathscr{D}$ from which it can be reached via an allowed move, namely, the configuration that is obtained from it by removing the free particle in $\partial^- \Lambda$. $\square$

18.4 Structure of the communication level set

In this section we prove Theorems 18.3 and 18.5. In Sect. 18.4.1 we consider the set $\mathscr{Q}$ consisting of those configurations in $\mathscr{D}$ where the single cluster is an $(\ell_c - 1) \times \ell_c$ quasi-square with a protuberance attached to one of its sides. We show that $\mathscr{D}$, our target protocritical set, coincides with $\mathscr{Q}^U$, *the set all configurations that can be obtained from* $\mathscr{Q}$ *via a* U*-path*. In Sect. 18.4.2 we use the identity $\mathscr{D} = \mathscr{Q}^U$ to show that $\Phi(\square, \blacksquare) = \Gamma^\star$ with $\Gamma^\star$ given by (18.1.11) and $\mathscr{S}(\square, \blacksquare) \supseteq \mathscr{D}^{\mathrm{fp}}$. In Sect. 18.4.3 we combine the results obtained in Sect. 18.4.2 to show that $\mathscr{P}^\star(\square, \blacksquare) = \mathscr{D}$ and $\mathscr{C}^\star(\square, \blacksquare) = \mathscr{D}^{\mathrm{fp}}$, thereby completing the proof of Theorem 18.3. In Sect. 18.4.4 we take a closer look at what happens when the free particle in $\mathscr{D}^{\mathrm{fp}}$ attaches itself to the single cluster, where we distinguish between "good sites" and "bad sites" on the border of the single cluster. The latter distinction will be needed in Sect. 18.5 for the proof of Theorem 18.4. In Sect. 18.4.5, finally, we compute the cardinality of $\mathscr{D}$ modulo shifts, which will also be needed in Sect. 18.5 for the proof of Theorem 18.4.

18.4.1 Canonical protocritical droplets

The following definition formalises the notion of canonical protocritical droplet and protocritical droplet mentioned in Item 2 of Sect. 18.1.4.

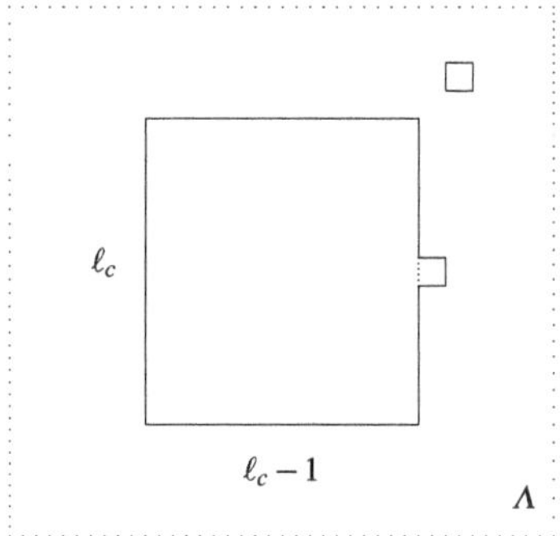

Fig. 18.4 A canonical critical droplet: an element of $\mathscr{Q}^{\text{fp}} \subseteq \mathscr{D}^{\text{fp}}$

Definition 18.6

(a) Let $\mathscr{Q} \subseteq \mathscr{D}$ be the set of configurations consisting of an $(\ell_c - 1) \times \ell_c$ quasi-square anywhere in Λ^- with a protuberance attached to one of its sides (see Fig. 18.4). These configurations are called *canonical protocritical droplets.*

(b) Let $\mathscr{Q}^U$ be the set of configurations that can be reached from some configuration in $\mathscr{Q}$ via a U-path, i.e.,

$$\mathscr{Q}^U = \{\eta' \in \mathscr{V}_{n_c} \colon \exists \eta \in \mathscr{Q} \colon H(\eta) = H(\eta'),\ \Phi_{\mathscr{V}_{n_c}}(\eta, \eta') \leq H(\eta) + U\}, \tag{18.4.1}$$

where $n_c = \ell_c(\ell_c - 1) + 1$ is the volume of the clusters in $\mathscr{Q}$ and $\Phi_{\mathscr{V}_{n_c}}$ is the communication height within $\mathscr{V}_{n_c}$. These configurations are called *protocritical droplets*.

Note that $\mathscr{Q} = \bar{\mathscr{Q}} \cup \tilde{\mathscr{Q}}$, where

- $\bar{\mathscr{Q}}$ are those configurations where the single particle is attached to one of the *longest* sides of the $(\ell_c - 1) \times \ell_c$ quasi-square.
- $\tilde{\mathscr{Q}}$ are those configurations where the single particle is attached to one of the *shortest* sides of the $(\ell_c - 1) \times \ell_c$ quasi-square.

Thus, $\bar{\mathscr{Q}}$ consists of precisely those configurations in $\bar{\mathscr{D}}$ where in (18.1.9) one $\bar{k}_i$ equals 1 and the others are maximal. Similarly, $\tilde{\mathscr{Q}}$ consists of precisely those configurations in $\tilde{\mathscr{D}}$ where in (18.1.10) one $\tilde{k}_i$ equals 1 and the others are maximal. We will see in Sect. 18.4 that the configurations in $\bar{\mathscr{D}}, \tilde{\mathscr{D}}$ arise from those in $\bar{\mathscr{Q}}, \tilde{\mathscr{Q}}$ via a *motion of particles along the border of the droplet* (see Figs. 18.5–18.6). This property is special for Kawasaki dynamics.

Our main result in this section is the following relation, which will be needed in Sect. 18.4.2.

Proposition 18.7 $\mathscr{D} = \mathscr{Q}^U$.

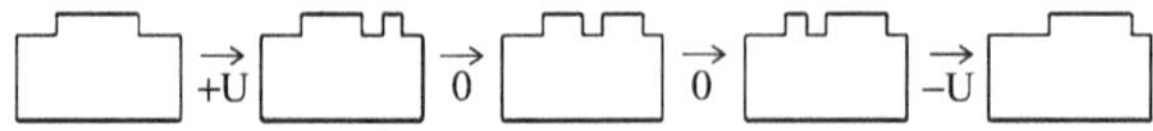

Fig. 18.5 Translation of a bar on a side of a rectangle at cost U

Fig. 18.6 Motion of a particle around a corner of a rectangle at cost U

Proof The proof is split into two parts:

$$\begin{aligned} &\text{(i)} \quad \mathscr{D} \subseteq \mathscr{Q}^U, \\ &\text{(ii)} \quad \mathscr{D} \supseteq \mathscr{Q}^U. \end{aligned} \tag{18.4.2}$$

• *Proof of* (i): Recall the definition of U-path in (18.2.5) and of the protocritical set $\mathscr{D} = \bar{\mathscr{D}} \cup \tilde{\mathscr{D}}$ in Definition 18.1(b). To prove (i) we must show that for all $\eta \in \mathscr{D}$,

$$\begin{aligned} &\text{(i1)} \quad H(\eta) = H(\mathscr{Q}), \\ &\text{(i2)} \quad \exists\, \gamma\colon \mathscr{Q} \to \eta\colon\ \max_i H(\gamma_i) \leq H(\mathscr{Q}) + U,\ |\gamma_i| = n_c \text{ for all } i. \end{aligned} \tag{18.4.3}$$

• *Proof of* (i1): Any $\eta \in \mathscr{D}$ has a single contour $\partial(\sigma)$ inside Λ^- of length $|\partial(\sigma)| = 4\ell_c$ and volume $|\eta \cap \Lambda^-| = \ell_c(\ell_c - 1) + 1 = n_c$, while $|\eta \cap \partial^- \Lambda| = 0$ (see Fig. 18.2). Thus, by (18.2.1), H is constant on $\mathscr{D}$. Since $\mathscr{Q} \subseteq \mathscr{D}$, this completes the proof of (i1).

• *Proof of* (i2): Note that, because $\bar{\mathscr{Q}}$ and $\tilde{\mathscr{Q}}$ are connected via a U-path (disconnect the 1-protuberance and re-attach it to one of the neighboring sides of the $(\ell_c - 1) \times \ell_c$ quasi-square), we have

$$\mathscr{Q}^U = \{\eta \in S\colon \exists\, U\text{-path from } \bar{\mathscr{Q}} \text{ to } \eta\} = \{\eta \in S\colon \exists\, U\text{-path from } \tilde{\mathscr{Q}} \text{ to } \eta\}. \tag{18.4.4}$$

First we prove that for any $\eta \in \bar{\mathscr{D}}$ there exists a $\gamma\colon \bar{\mathscr{Q}} \to \eta$ such that $\max_i H(\gamma_i) \leq H(\bar{\mathscr{Q}}) + U$ and $|\gamma_i| = n_c$ for all i. We start the path from some $\zeta \in \bar{\mathscr{Q}}$. Then, recalling the labelling in Definition 18.1(b), we have

- $\bar{k}_1(\zeta) = 1$ contained in $r_e(\zeta)$;
- $\bar{k}_2(\zeta) = \ell_c - 2$ contained in $r_n(\zeta)$;
- $\bar{k}_3(\zeta) = \bar{k}_4(\zeta) = \ell_c - 1$ contained in $r_w(\zeta) \cup c_{nw}(\zeta)$ and $r_s(\zeta) \cup c_{sw}(\zeta)$, respectively.

Here, without loss of generality, we assume that the 1-protuberance is attached to $r_e(\zeta)$ and proceed anti-clockwise. Using the mechanism described in Figs. 18.5–18.6, we move $\bar{k}_2(\zeta) - \bar{k}_2(\eta)$ particles from $r_n(\zeta)$ to $r_e(\zeta)$, one by one. After that we move $\bar{k}_3(\zeta) - \bar{k}_3(\eta) + \bar{k}_4(\zeta) - \bar{k}_4(\eta)$ particles from $r_s(\zeta) \cup c_{sw}(\zeta)$ to $r_e(\zeta)$. Finally, we move $\bar{k}_3(\zeta) - \bar{k}_3(\eta)$ particles from $r_w(\zeta) \cup c_{nw}(\zeta)$ to $r_s(\zeta) \cup c_{sw}(\zeta)$. The result is a configuration $\eta \in \bar{\mathscr{D}}$.

Next we prove that for any $\eta \in \widetilde{\mathscr{D}}$ there exists a $\gamma\colon \widetilde{\mathscr{Q}} \to \eta$ such that $\max_i H(\gamma_i) \le H(\widetilde{\mathscr{Q}}) + U$ and $|\gamma_i \cap \Lambda| = n_c$ for all i. We start the path from some $\zeta \in \widetilde{\mathscr{Q}}$. We have

- $\widetilde{k}_1(\zeta) = 1$ contained in $r_e(\zeta)$;
- $\widetilde{k}_2(\zeta) = \widetilde{k}_4(\zeta) = \ell_c - 1$ contained in $r_n(\zeta)$ and $r_s(\zeta)$;
- $\widetilde{k}_3(\zeta) = \ell_c - 1$ contained in $r_w(\zeta) \cup c_{nw}(\zeta) \cup c_{sw}(\zeta)$.

We move $\widetilde{k}_2(\zeta) - \widetilde{k}_2(\eta)$ particles from $r_n(\zeta)$ to $r_e(\zeta)$. After that we move $\widetilde{k}_3(\zeta) - \widetilde{k}_3(\eta) + \widetilde{k}_4(\zeta) - \widetilde{k}_4(\eta)$ particles from $r_s(\zeta) \cup c_{sw}(\zeta)$ to $r_e(\zeta)$. Finally, we move $\widetilde{k}_3(\zeta) - \widetilde{k}_3(\eta)$ particles from $r_w(\zeta) \cup c_{nw}(\zeta)$ to $r_s(\zeta) \cup c_{sw}(\zeta)$. The result is a configuration $\eta \in \widetilde{\mathscr{D}}$. This completes the proof of (i2).

• *Proof of* (ii): By (18.4.3), all configurations in $\mathscr{D}$ are connected via a U-path. Since $\mathscr{Q} \subseteq \mathscr{Q}^U \cap \mathscr{D}$, in order to prove (ii) it suffices to show that $\mathscr{D}$ cannot be exited via a U-path (recall (18.4.4)).

Call a path *clustering* if all the configurations in the path consist of a single cluster and no free particles. Below we will prove that for any $\eta \in \mathscr{D}$ and any η' connected to η by a clustering U-path,

$$\begin{aligned} &\text{(a)} \quad \mathrm{CR}(\eta') = \mathrm{CR}(\eta), \\ &\text{(b)} \quad \eta' \supseteq \mathrm{CR}^-(\eta). \end{aligned} \tag{18.4.5}$$

What (18.4.5) says is that neither $\bar{\mathscr{D}}$ nor $\widetilde{\mathscr{D}}$ can be exited via a clustering U-path. From this in turn we deduce that for any $\eta \in \mathscr{D}$ and any η' connected to η by a U-path we must have that $\eta' \in \mathscr{D}$, which is what we want to prove. The argument for the latter goes as follows. Detaching a particle costs $2U$ unless the particle is a 1-protuberance, in which case the cost is U. The only configurations in $\mathscr{D}$ having a 1-protuberance are those in $\mathscr{Q}$. If we detach the 1-protuberance from a configuration in $\mathscr{Q}$, at cost U, then we obtain an $(\ell_c - 1) \times \ell_c$ quasi-square plus a free particle. Since now only moves at zero cost are allowed, only the free particle can move. Since in a U-path the particle number is conserved, the only way to regain U and complete the U-path is to re-attach the free particle to the quasi-square, in which case we return to $\mathscr{Q}$.

Remark Note that the motion of particles along the border of a droplet may shift the droplet. Indeed, from any configuration in $\mathscr{Q}$ the 1-protuberance may detach itself and re-attach itself to a different side of the quasi-square or rectangle. Thus, the U-path may shift the protocritical droplet to anywhere in Λ^-.

• *Proof of* (a): Starting from any $\eta \in S$, it is geometrically impossible to modify $\mathrm{CR}(\eta)$ without detaching a particle.

• *Proof of* (b): Fix $\eta \in \mathscr{D}$. The proof is done in two steps.

1. Let us first consider clustering U-paths along which we do not move a particle from $\mathrm{CR}^-(\eta)$. Along such paths we only encounter configurations in $\mathscr{D}$ or configu-

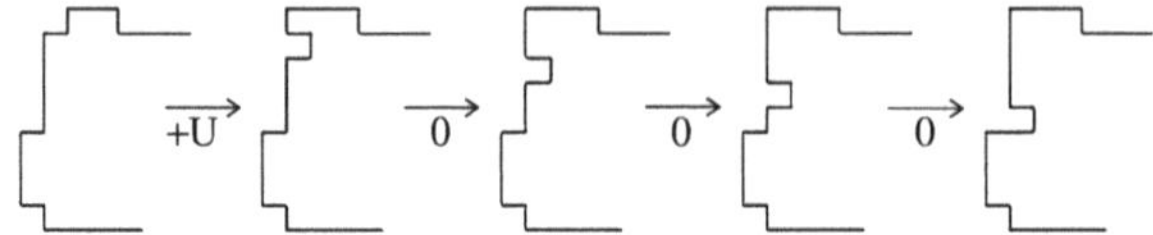

Fig. 18.7 Creation and motion of the hole at cost 0

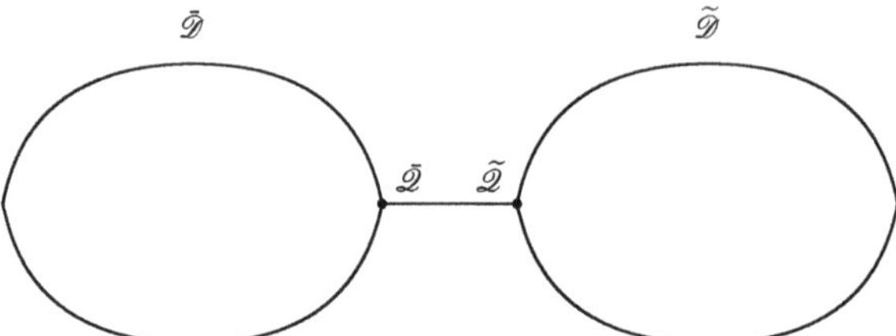

Fig. 18.8 Dumb-bell shape of $\mathscr{D} = \bar{\mathscr{D}} \cup \widetilde{\mathscr{D}}$ for U-paths: the canonical protocritical droplets $\bar{\mathscr{Q}}$ and $\widetilde{\mathscr{Q}}$ are the gateways between the sets of protocritical droplets $\bar{\mathscr{D}}$ and $\widetilde{\mathscr{D}}$

rations obtained from $\mathscr{D}$ by breaking one of the bars in $\partial^- \text{CR}(\eta)$ into two pieces, at cost U (because there is no particle outside $\text{CR}(\eta)$ that can help to lower the cost). From the latter only moves at zero cost are possible, so no particle can be detached, and the only way to regain U and complete the U-path is to restore a bar.

2. Let us next consider clustering U-paths along which we move a particle from a corner of $\text{CR}^-(\eta)$. This move costs $2U$, which exceeds U. The overshoot U must be regained by letting the particle slide next to a bar that is attached to a side of $\text{CR}^-(\eta)$ (see Fig. 18.7). Since there are never two bars attached to the same side, we can at most gain U. This is why it is not possible to move a particle from $\text{CR}^-(\eta)$ other than from a corner.

From here only moves at zero cost are allowed. There are no 1-protuberances present anymore, because only the configurations in $\mathscr{Q}$ have a 1-protuberance. Thus, no particle outside $\text{CR}^-(\eta)$ can move, except the one that just detached itself from $\text{CR}^-(\eta)$. This particle can move back, in which case we return to the same configuration η. In fact, all possible moves at zero cost consist in moving the "hole" just created in $\text{CR}^-(\eta)$ along the side of $\text{CR}^-(\eta)$, until it reaches the height of the top of the bar attached to this side of $\text{CR}^-(\eta)$, after which it cannot advance anymore at zero cost (see Fig. 18.7). All these moves do not change the energy, except the one that returns the particle to its original position and regains U.

This proves our claim in (18.4.5), completes the proof of (ii) in (18.4.2), and hence of Proposition 18.7. □

We saw above that U-paths cannot exit $\mathscr{D} = \bar{\mathscr{D}} \cup \widetilde{\mathscr{D}}$, but can make a crossover between $\bar{\mathscr{D}}$ and $\widetilde{\mathscr{D}}$. This crossover can, however, only occur between $\bar{\mathscr{Q}}$ and $\widetilde{\mathscr{Q}}$. A schematic picture of $\mathscr{D}$ therefore is as in Fig. 18.8.

18.4.2 Protocritical and critical droplets

Most of this section revolves around getting a precise description of $(\square \to \blacksquare)_{\text{opt}}$, the set of *optimal paths for the nucleation* (recall Definition 16.2(a)).

Proposition 18.8

(i) $\Phi(\square, \blacksquare) = \Gamma^\star$.
(ii) $\mathscr{S}(\square, \blacksquare) \supseteq \mathscr{D}^{\text{fp}}$.

Proof The proof is based on five lemmas (Lemmas 18.9–18.13 below).

(i) We prove that $\Phi(\square, \blacksquare) \le \Gamma^\star$ and $\Phi(\square, \blacksquare) \ge \Gamma^\star$.

• $\Phi(\square, \blacksquare) \le \Gamma^\star$: All we need to do is to construct a path that connects $\square$ and $\blacksquare$ without exceeding energy $\Gamma^\star$. The proof comes in three steps.

1. We first show that the configurations in $\mathscr{Q}$ are connected to $\square$ by a path that stays below $\Gamma^\star$.

Lemma 18.9 *For any* $\eta^{1\text{pr}} \in \mathscr{Q}$ *there exists a* $\gamma: \eta^{1\text{pr}} \to \square$ *such that* $\max_{\xi \in \gamma} H(\xi) < \Gamma^\star$.

Proof Fix $\eta^{1\text{pr}} \in \mathscr{Q}$. Note that, by (18.1.11), we have $H(\eta^{1\text{pr}}) = \Gamma^\star - \Delta$. First, we detach the 1-protuberance from the $(\ell_c - 1) \times \ell_c$ quasi-square, which costs U and raises the energy to $\Gamma^\star - \Delta + U (< \Gamma^\star)$, move the particle to the boundary of the box, which costs nothing, and move it out of the box, which pays Δ. We are then left with a quasi-square of energy

$$\Gamma^\star - (2\Delta - U). \tag{18.4.6}$$

Second, we detach a particle from a corner of the quasi-square, which costs $2U$, and move it out of the box, which pays Δ. Thus, the energy increases by $2U - \Delta$ when detaching and removing a particle from a corner of the quasi-square. We repeat this operation another $\ell_c - 3$ times, each time picking particles from the bar on the same shortest side. To guarantee that we never reach energy $\Gamma^\star$, we have the condition that

$$(2U - \Delta)k + 2U < 2\Delta - U \quad \text{for } 0 \le k \le \ell_c - 3, \tag{18.4.7}$$

or

$$3 \le \ell_c < \frac{U}{2U - \Delta} + 1. \tag{18.4.8}$$

The second inequality holds by the definition of ℓ_c in (18.1.5) and the non-degeneracy assumption in (18.1.6), the first inequality by our exclusion of $\ell_c = 2$ (recall the statement made at the end of Sect. 18.3). Third, detaching the last par-

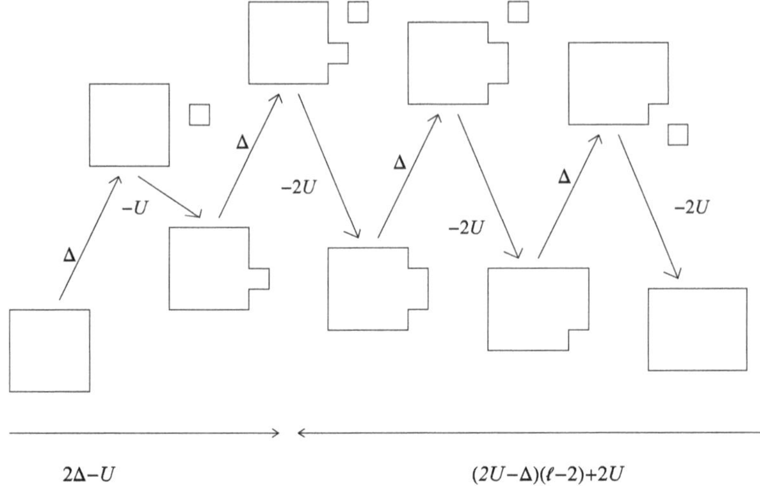

Fig. 18.9 Cost of adding or removing a row of length ℓ

ticle costs U instead of $2U$. To guarantee that we still do not reach energy $\Gamma^\star$, we have the condition that

$$(2U - \Delta)(\ell_c - 2) + U < 2\Delta - U, \tag{18.4.9}$$

which is weaker than (18.4.7) because $2U - \Delta < U$. Removal of the last particle pays Δ, so that we arrive at energy

$$\big(\Gamma^\star - (2\Delta - U)\big) + (2U - \Delta)(\ell_c - 2) + U - \Delta = \Gamma^\star - 2\Delta + (2U - \Delta)(\ell_c - 1), \tag{18.4.10}$$

which is strictly smaller than (18.4.6) by the second inequality in (18.4.8). Thus, removal of a row of length $\ell_c - 1$ from the $(\ell_c - 1) \times \ell_c$ quasi-square in $\eta^{1\mathrm{pr}} \in \mathscr{Q}$ lowers the energy (see Fig. 18.9). We now have a square of side length $\ell_c - 1$. It is obvious that we can remove further rows without encountering new conditions, until we reach $\square$. $\square$

2. For $\eta^{1\mathrm{pr}} \in \mathscr{Q}$, let $\eta^{2\mathrm{pr}}$ be the configuration obtained from $\eta^{1\mathrm{pr}}$ by attaching an extra particle next to the 1-protuberance, thereby forming a 2-protuberance. We next show that $\eta^{2\mathrm{pr}}$ is connected to $\blacksquare$ by a path that stays below $\Gamma^\star$.

Lemma 18.10 *For any* $\eta^{1\mathrm{pr}} \in \mathscr{Q}$ *there exists a* $\gamma\colon \eta^{2\mathrm{pr}} \to \blacksquare$ *such that* $\max_{\xi\in\gamma} H(\xi) < \Gamma^\star$.

Proof Without loss of generality we may assume that $\eta^{1\mathrm{pr}} \in \bar{\mathscr{D}}$ because of Proposition 18.7. Fix $\eta^{1\mathrm{pr}} \in \mathscr{Q}$. Note that $H(\eta^{2\mathrm{pr}}) = \Gamma^\star - 2U$. First, we create a particle, which costs Δ and raises the energy to $\Gamma^\star - (2U - \Delta)(< \Gamma^\star)$, move it to the droplet, which costs nothing, and attach it next to the 2-protuberance, which pays $2U$, thereby forming a bar of length 3. This operation pays $2U - \Delta$. We can repeat this operation another $\ell_c - 3$ times until the row is filled. By that time we have a square of side length ℓ_c and energy

$$\Gamma^\star - 2U - (2U - \Delta)(\ell_c - 2). \tag{18.4.11}$$

Second, we create another particle and attach it anywhere to the square to form a new 1-protuberance. This operation costs $\Delta - U$. We must make sure that we can still create a particle without reaching energy $\Gamma^\star$, which gives us the condition

$$\Delta - U + \Delta < 2U + (2U - \Delta)(\ell_c - 2), \tag{18.4.12}$$

or

$$\ell_c > \frac{U}{2U - \Delta}, \tag{18.4.13}$$

which holds by the definition of ℓ_c and the non-degeneracy assumption in (18.1.6). Third, we create another particle and attach it next to the new 1-protuberance. This brings us to energy

$$\Gamma^\star - U - (2U - \Delta)\ell_c, \tag{18.4.14}$$

which is below the energy of $\eta^{2\mathrm{pr}}$ by (18.4.13). It is obvious that we can add further rows without encountering new conditions, until we reach $\blacksquare$. $\square$

3. We can now conclude the proof of $\Phi(\square, \blacksquare) \leq \Gamma^\star$ by constructing a bridge between $\eta^{1\mathrm{pr}}$ and $\eta^{2\mathrm{pr}}$ that does not exceed $\Gamma^\star$. Namely, create a particle at the boundary, which costs Δ and raises the energy to $\Gamma^\star$, move it to the droplet, which costs nothing, and place it next to the 1-protuberance, which pays $2U$. The desired path $\gamma : \square \to \blacksquare$ is realized by tracing the path in Lemma 18.9 in the reverse direction, back from $\square$ to $\eta^{1\mathrm{pr}}$, going over the bridge from $\eta^{1\mathrm{pr}}$ to $\eta^{2\mathrm{pr}}$, and then following the path in Lemma 18.10 from $\eta^{2\mathrm{pr}}$ to $\blacksquare$. This γ will be called the *reference path through* η *for the nucleation.*

- $\Phi(\square, \blacksquare) \geq \Gamma^\star$: The proof comes in three more steps.

4. The first crucial ingredient in the proof is the following observation:

Lemma 18.11 *Any* $\gamma \in (\square \to \blacksquare)_{\mathrm{opt}}$ *must pass through a configuration consisting of a single* $(\ell_c - 1) \times \ell_c$ *quasi-square somewhere in* Λ^-.

Proof Any path $\gamma : \square \to \blacksquare$ must cross the set $\mathcal{V}_{\ell_c(\ell_c-1)}$. As shown in Alonso and Cerf [4], Theorem 2.6, in $\mathcal{V}_{\ell_c(\ell_c-1)}$ the unique (modulo translations and rotations) configuration of minimal energy is the $(\ell_c - 1) \times \ell_c$ quasi-square, which we denote by η and which has energy

$$H(\eta) = \Gamma^\star - (2\Delta - U). \tag{18.4.15}$$

All other configurations in $\mathcal{V}_{\ell_c(\ell_c-1)}$ have energy at least $\Gamma^\star - 2\Delta + 2U$. To increase the particle number starting from any such configuration, we must create a particle at cost Δ. But the resulting configuration would have energy $\Gamma^\star - \Delta + 2U (> \Gamma^\star)$ and thus would lead to a path exceeding energy $\Gamma^\star$. □

5. The second crucial ingredient in the proof is the following observation:

Lemma 18.12 *Any* $\gamma \in (\square \to \blacksquare)_{\mathrm{opt}}$ *must pass through* $\mathscr{Q}$.

Proof Follow the path until it hits the set $\mathscr{V}_{\ell_c(\ell_c-1)}$. According to Lemma 18.11, the configuration in this set must be an $(\ell_c - 1) \times \ell_c$ quasi-square. Since we need not consider any paths that return to the set $\mathscr{V}_{\ell_c(\ell_c-1)}$ afterwards, a first step beyond the quasi-square must be the creation of a new particle. This brings us to energy

$$\Gamma^\star - \Delta + U. \tag{18.4.16}$$

Before any new particle is created, we must lower the energy by at least U. The obviously only possible way to do this is to move the particle to the quasi-square and attach it to one of its sides, which reduces the energy to

$$\Gamma^\star - \Delta \tag{18.4.17}$$

and gives us a configuration in $\mathscr{Q}$. □

6. It now suffices to show that to reach $\blacksquare$ from $\mathscr{Q}$ we must reach energy $\Gamma^\star$. This goes as follows. Starting from $\mathscr{Q}$, it is impossible to reduce the energy without lowering the particle number. Indeed, this follows from Alonso and Cerf [4], Theorem 2.6, which asserts that the minimal energy in $\mathscr{V}_{\ell_c(\ell_c-1)+1}$ is realised (although not uniquely) by the configurations in $\mathscr{Q}$. Since any further move to increase the particle number involves the creation of a new particle, the energy must reach $\Gamma^\star$.

Lemmas 18.11–18.12 imply that $\Phi(\square, \blacksquare) = \Gamma^\star$, and together with Steps 1–3 completes the proof of Proposition 18.8(i).

(ii) Our final observation is the following:

Lemma 18.13 *The set of configurations in* $\mathscr{V}_{\ell_c(\ell_c-1)+1}$ *that can be reached from* $\square$ *by a path that stays below* $\Gamma^\star$ *and for which it is possible to add a particle without exceeding* $\Gamma^\star$ *coincides with the set* $\mathscr{Q}^U$ *defined in Definition* 18.6(*b*).

Proof From step 2 above it is clear that the definition of $\mathscr{Q}^U$ precisely assures that the assertion holds true. Indeed, by Lemma 18.12, any $\gamma \in (\square \to \blacksquare)_{\mathrm{opt}}$ crosses $\mathscr{V}_{\ell_c(\ell_c-1)+1}$ in $\mathscr{Q}$. Once it is in $\mathscr{Q}$, before the arrival of the next particle, which costs Δ, it can reach all configurations that have the same energy, the same particle number, and can be reached at cost $\leq U < \Delta$. □

We know from Proposition 18.7 that $\mathscr{Q}^U = \mathscr{D}$. By adding a free particle in $\partial^- \Lambda$ to a configuration in $\mathscr{D}$ we obtain a configuration in $\mathscr{D}^{\mathrm{fp}}$. Hence Lemma 18.13 implies that any optimal path passes through $\mathscr{D}^{\mathrm{fp}}$. This completes the proof of Proposition 18.8(ii). □

18.4.3 Identification of the protocritical and the critical set

The relations $\mathscr{P}^\star(\square \to \blacksquare) = \mathscr{D}$ and $\mathscr{C}^\star(\square \to \blacksquare) = \mathscr{D}^{\mathrm{fp}}$ and the formula for $\Gamma^\star$ claimed in Theorem 18.3 are an immediate consequence of Definition 16.3, Lemmas 18.9–18.10 and the following proposition:

Proposition 18.14 *Any $\gamma \in (\square \to \blacksquare)_{\mathrm{opt}}$ passes first through $\mathscr{Q}$, then (possibly) through $\mathscr{D} \setminus \mathscr{Q}$, and finally through $\mathscr{D}^{\mathrm{fp}}$.*

Proof Combine Lemmas 18.12–18.13 and Proposition 18.8(i). □

The claim in Theorem 18.5 is immediate from Lemmas 18.9–18.10 in combination with Proposition 16.12, Lemma 16.15, (16.2.4) and (17.3.6)–(17.3.7). As argued in Sect. 18.3.1, the latter two equations (which were derived for Glauber dynamics) continue to be valid for Kawasaki dynamics as well. In Sect. 18.4.4 below we will see why it is important to attach the free particle at a "good site", i.e., next to two other particles in the protocritical droplet.

Think of $\mathscr{Q}$ as the set of *canonical protocritical droplets*: $\mathscr{D}$, the set of *protocritical droplets*, is the set of all configurations the dynamics can reach *after* hitting $\mathscr{Q}$ *before* the creation of the next free particle in $\partial^- \Lambda$. This particle completes the formation of a critical droplet (= a protocritical droplet + a free particle at the boundary) that triggers the nucleation. If subsequently the free particle moves to the protocritical droplet and attaches itself at a "good site", then the dynamics has "moved over the hill" and proceeds to fill up Λ^-.

18.4.4 Motion on the plateau

The following observations, which constitute a refinement of what the dynamics does when it is close to forming a critical droplet, will be needed in Sect. 18.5.

(1) Starting from $\mathscr{D}^{\mathrm{fp}} \setminus \mathscr{Q}^{\mathrm{fp}}$, the only transitions that do not raise the energy are motions of the free particle, as long as the free particle is at lattice distance ≥ 3 from the protocritical droplet.
(2) Starting from $\mathscr{Q}^{\mathrm{fp}}$, the only transitions that do not raise the energy are motions of the free particle and motions of the 1-protuberance along the side of the quasi-square where it is attached, as long as the free particle is at lattice distance ≥ 3 from the protocritical droplet. When the lattice distance is 2, either the free particle can be attached to the protocritical droplet or the 1-protuberance can be detached from the protocritical droplet and attached to the free particle, to form a quasi-square plus a dimer. From the latter configuration the only transition that does not raise the energy is the reverse move.
(3) Starting from $\mathscr{D}^{\mathrm{fp}}$, the only configurations that can be reached by a path that lowers the energy and does not decrease the particle number are those where the free particle is attached to the protocritical droplet.

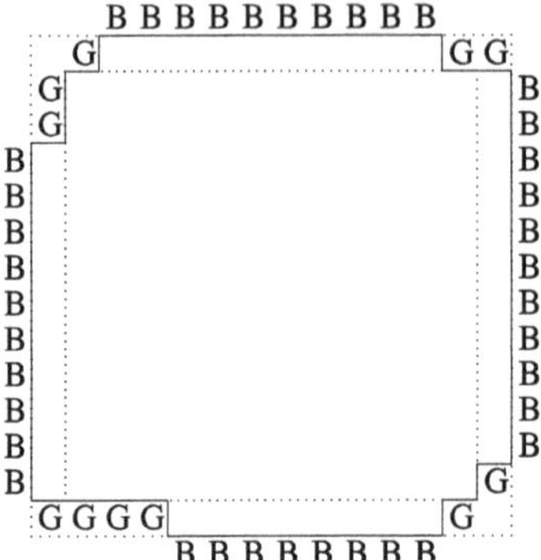

Fig. 18.10 Good sites (G) and bad sites (B)

The restriction in observation (1) that the free particle must be at lattice distance ≥ 3 from the protocritical droplet is needed for the following reason: If the protocritical droplet is a configuration in $\mathscr{D} \setminus \mathscr{Q}$ and the free particle sits at lattice distance 2 from a corner of a bar, diagonally opposite the particle that sits in the corner of the bar, then at zero cost this particle may detach itself from the bar and slide inbetween the quasi-square and the free particle. For observation (3) note the following: if we start from the configuration described above and slide the remaining particles in the bar one by one, all at zero cost except the last one, which pays U, then we reach a configuration where the free particle is attached to the protocritical droplet with the bar shifted.

The following definition introduces the notion of *good sites* (G) and *bad sites* (B) on the border of protocritical droplets (see Fig. 18.10).

Definition 18.15

(a) For $\eta \in \mathscr{D}^{\mathrm{fp}}$, write $\eta = (\hat{\eta}, x)$ with $\hat{\eta} \in \mathscr{D}$ the protocritical droplet and $x \in \partial^- \Lambda$ the position of the free particle.
(b) Let the configurations that can be reached from $\eta = (\hat{\eta}, x) \in \mathscr{D}^{\mathrm{fp}}$ according to observation (3) be denoted by

$$\begin{aligned} &\mathscr{C}^G(\hat{\eta}) \text{ if the particle is attached in } \partial^- \mathrm{CR}(\hat{\eta}), \\ &\mathscr{C}^B(\hat{\eta}) \text{ if the particle is attached in } \partial^+ \mathrm{CR}(\hat{\eta}). \end{aligned} \tag{18.4.18}$$

(c) Let

$$\mathscr{C}^G = \bigcup_{\hat{\eta} \in \mathscr{D}} \mathscr{C}^G(\hat{\eta}), \qquad \mathscr{C}^B = \bigcup_{\hat{\eta} \in \mathscr{D}} \mathscr{C}^B(\hat{\eta}). \tag{18.4.19}$$

The next proposition, which is the main result of this section, shows that when the dynamics reaches $\mathscr{C}^G$ it has gone "over the hill", while when it reaches $\mathscr{C}^B$ it has not.

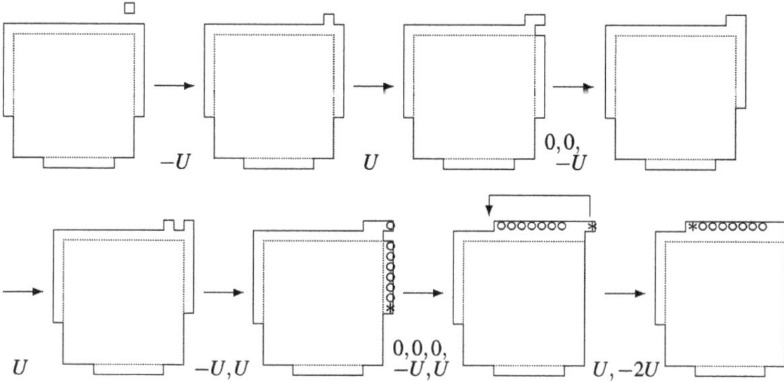

Fig. 18.11 An example of a path from $\mathscr{C}^B$ to ■

Proposition 18.16

(i) *If* $\eta \in \mathscr{C}^G$, *then there exists a* $\gamma : \eta \to$ ■ *such that* $\max_{\xi \in \gamma} H(\xi) < \Gamma^*$.
(ii) *If* $\eta \in \mathscr{C}^B$, *then there are no* $\gamma : \eta \to \square$ *or* $\gamma : \eta \to$ ■ *such that* $\max_{\xi \in \gamma} H(\xi) < \Gamma^*$.

Proof (i) If $\eta \in \mathscr{C}^G$, then its energy is either $\Gamma^\star - 2U$ or $\Gamma^\star - U$, depending on whether the particle was attached in a corner or as a 1-protuberance. In the latter case we can move the particle at no cost into a corner and gain an extra $-U$. After that it is possible to create a new particle and re-attach it, which leads to energy $\Gamma^\star - 2U - (2U - \Delta)$. We can continue in this way, filling up all rows in $\partial^- \mathrm{CR}(\eta)$, until we reach either an $\ell_c \times \ell_c$ square or an $(\ell_c - 1) \times (\ell_c + 1)$ rectangle, depending on whether η arose from $\bar{\mathscr{D}}$ or $\tilde{\mathscr{D}}$ (recall Definition 18.1(b)). In the first case we can proceed along the reference path for the nucleation constructed in the proof of Proposition 18.8. In the latter case, however, we can connect to this reference path as follows. The energy of the $(\ell_c - 1) \times (\ell_c + 1)$ rectangle is $\Gamma^\star - 2U - (2U - \Delta)(\ell_c - 3)$. This is lower than $\Gamma^\star - \Delta$, because $\ell_c \geq 3$. Create a particle, which costs Δ, and attach it to one of the longest sides of the rectangle, which pays U. Now slide particles along the corner of the rectangle, following the mechanism described in Figs. 18.5–18.6, until an $\ell_c \times \ell_c$ square is reached. This costs U and keeps the energy below $\Gamma^\star$. From there again proceed along the reference path for the nucleation.

(ii) If $\eta \in \mathscr{C}^B$, then $H(\eta) = \Gamma^\star - U$, so as long as the energy stays below $\Gamma^\star$ it is impossible to create a new particle before further lowering the energy. But there are no moves available to lower the energy. The only moves available are those where the particle that was last attached is moving along the side or is detached again, which brings us back to $\mathscr{D}^{\mathrm{fp}}$, or those that start a motion of particles along the border of the droplet (as in Fig. 18.6), which may or may not bring us back to $\mathscr{D}^{\mathrm{fp}}$. In both cases the cost is U and the energy returns to $\Gamma^\star$.

An example of a path from $\mathscr{C}^B$ to ■ that does not return to a protocritical droplet plus a free particle is obtained as follows (see Fig. 18.11). Suppose that $\hat{\eta} \in \mathscr{D}$ is such that one bar completes one side of $\partial^- \mathrm{CR}(\hat{\eta})$, and suppose that the free particle attaches itself on top of that bar, forming a 1-protuberance. Then the energy is $\Gamma^\star - U$. Slide this bar to the end of the side it is attached to (at cost and gain U) and slide the two bars on the neighboring sides to the end as well (at cost and gain U). Then the energy is again $\Gamma^\star - U$. Next move the shorter bar on top of the longer bar via a motion as in Fig. 18.6. When the last particle of the bar is moved, it can be detached (at cost U) and re-attached (at gain $2U$). Then the energy is $\Gamma^\star - 2U$. Now create a free particle (at cost Δ), move it to the droplet (at cost 0), and attach it in a corner of the droplet (at gain $2U$). Continue "downhill" in this way, adding on successive rows as in the reference path that was used above, until ■ is reached. □

Proposition 18.16(ii) shows that the configurations in $\mathscr{C}^B$ are *wells*, i.e., their energy is $< \Gamma^\star$, but to move to either □ or ■ the energy must return to $\Gamma^\star$. The configurations of the form "quasi-square plus dimer" described in observation (2) are elements of $\mathscr{S}(\square, \blacksquare)$ but not of $\mathscr{C}^\star(\square, \blacksquare)$. Indeed, the only possible move at zero cost is the one where the free particle jumps back to the quasi-square.

Summarizing the above, we have the following:

- The set of configurations through which all optimal paths must pass is a union of *plateaus*, indexed by $\hat{\eta} \in \mathscr{D}$.
- Each plateau consists of a protocritical droplet $\hat{\eta}$ and a collection of positions of the free particle, indexed by $\Lambda \setminus (\hat{\eta} \cup \partial^+ \hat{\eta})$.
- Each plateau has *wells* and *dead-ends* when the free particle is close to the protocritical droplet.

This geometric structure is special for Kawasaki dynamics. We will not attempt to describe the wells and the dead-ends in detail. For the proof of Theorem 18.4 in Sect. 18.5 this will not be needed.

18.4.5 Cardinality of the set of protocritical droplets

In this section we show that the cardinality of $\mathscr{D}$ modulo shifts of the protocritical droplet equals the formula given in (18.1.13).

Proof First we consider $\bar{\mathscr{D}}$. We have to count the number of different shapes of the clusters in $\bar{\mathscr{D}}$ (recall Fig. 18.2). We do this by counting in how many ways $\ell_c - 1$ particles can be removed from the four bars of an $\ell_c \times \ell_c$ square starting from the four corners (recall Definition 18.1(b)). We split the counting according to the number $k = 1, 2, 3, 4$ of corners from which particles are removed. The number of ways in which we can choose k corners is $\binom{4}{k}$. After we have removed the particles at these corners, we need to remove $\ell_c - 1 - k$ more particles from either side of each corner. The number of ways in which this can be done is

$$
\begin{aligned}
&\left|\left\{(m_1,\dots,m_{2k})\in\mathbb{N}_0^{2k}:\ m_1+\cdots+m_{2k}=\ell_c-1-k\right\}\right|\\
&\quad=\left|\left\{(m_1,\dots,m_{2k})\in\mathbb{N}^{2k}:\ m_1+\cdots+m_{2k}=\ell_c-1+k\right\}\right|\\
&\quad=\binom{\ell_c-2+k}{2k-1}.
\end{aligned}
\tag{18.4.20}
$$

The counting for $\widetilde{\mathscr{D}}$ is the same, except that we start from an $(\ell_c-1)\times(\ell_c+1)$ rectangle and count in how many ways ℓ_c-2 particles can be removed from the four bars. The answer is the same as in (18.4.20) with ℓ_c-1 replaced by ℓ_c-2, except for an extra factor 2 that counts the two orientations of the rectangle. □

18.5 Asymptotics of the prefactor for large volumes

In this section we prove Theorem 18.4. Our starting point is the variational formula for $\Theta=1/K$ given in Lemma 16.17. In Sect. 18.5.1 we define certain objects that capture the geometry of critical droplets and wells. In Sect. 18.5.2 we derive upper and lower bounds for Θ in terms of certain capacities of simple random walk on Λ^+ restricted not to enter the support of a protocritical droplet. In Sect. 18.5.3 we compute the asymptotics of these capacities in the limit as $\Lambda\to\mathbb{Z}^2$, and show that the upper and lower bounds merge because of the recurrence of simple random walk on $\mathbb{Z}^2$.

18.5.1 Geometry of critical droplets and wells

In the proof we need one more definition, which relies on the geometric structure outlined in Sect. 18.4.4. Recall the definition of $S_\square$, $S_\blacksquare$ and S_i, $i=1,\dots,I$, from (16.2.25) and (16.3.3)–(16.3.4). Abbreviate $\mathrm{supp}^+(\hat\eta)=\mathrm{supp}(\hat\eta)\cup\partial^+\mathrm{supp}(\hat\eta)$.

Definition 18.17

(a) Let $\mathscr{D}_\Lambda^{\mathrm{fp}}=\{\eta=(\hat\eta,x):\ \hat\eta\in\mathscr{D},\ x\in\Lambda\setminus\mathrm{supp}^+(\hat\eta)\}$.
(b) For $\hat\eta\in\mathscr{D}$, let

$$
\begin{aligned}
G(\hat\eta)&=\left\{x\in\partial^+\mathrm{supp}(\hat\eta):\ (\hat\eta,x)\in S_\blacksquare\right\},\\
B(\hat\eta)&=\left\{x\in\partial^+\mathrm{supp}(\hat\eta):\ \exists\, i=1,\dots,I:\ (\hat\eta,x)\in S_i\right\},
\end{aligned}
\tag{18.5.1}
$$

be the set of good sites, respectively, bad sites for $\hat\eta$. Note that $(\hat\eta,x)$ may be in the same S_i for different $x\in B(\hat\eta)$.

(c) For $\hat{\eta} \in \mathscr{D}$, let

$$I(\hat{\eta}) = \left\{ i \in 1, \dots, I \colon \exists x \in B(\hat{\eta}) \colon (\hat{\eta}, x) \in S_i \right\}. \tag{18.5.2}$$

Note that $B(\hat{\eta})$ can be partitioned into disjoint sets $B_1(\hat{\eta}), \dots, B_{|I(\hat{\eta})|}(\hat{\eta})$ according to which S_i the configuration $(\hat{\eta}, x)$ belongs to.

(d) Write

$$\begin{aligned} &\mathrm{CS}(\hat{\eta}) = \mathrm{supp}(\hat{\eta}) \cup G(\hat{\eta}), \quad \mathrm{CS}^+(\hat{\eta}) = \mathrm{CS}(\hat{\eta}) \cup \partial^+ \mathrm{CS}(\hat{\eta}), \\ &\mathrm{CS}^{++}(\hat{\eta}) = \mathrm{CS}^+(\hat{\eta}) \cup \partial^+ \mathrm{CS}^+(\hat{\eta}). \end{aligned} \tag{18.5.3}$$

By Proposition 18.16, the link between the sets in Definitions 18.15(b) and 18.17(b) is

$$\begin{aligned} \mathscr{C}^G(\hat{\eta}) &= \bigcup_{x \in G(\hat{\eta})} (\hat{\eta}, x), \\ \mathscr{C}^B(\hat{\eta}) &= \bigcup_{x \in B(\hat{\eta})} (\hat{\eta}, x). \end{aligned} \tag{18.5.4}$$

For the argument below it is important that $G(\hat{\eta}) \neq \emptyset$ for all $\hat{\eta} \in \mathscr{D}$. On the other hand, the sets $B(\hat{\eta})$, $\hat{\eta} \in \mathscr{D}$, will turn out to play no role for the asymptotics of K as $\Lambda \to \mathbb{Z}^2$.

18.5.2 Capacity bounds on the prefactor

We have the following bounds on $\Theta = 1/K$.

Lemma 18.18 *$\Theta \in [\Theta_1, \Theta_2]$ with*

$$\begin{aligned} \Theta_1 &= \left[1 + o(1)\right] \sum_{\hat{\eta} \in \mathscr{D}} \mathrm{cap}^{\Lambda^+}\left(\partial^+ \Lambda, \mathrm{CS}(\hat{\eta})\right), \\ \Theta_2 &= \sum_{\hat{\eta} \in \mathscr{D}} \mathrm{cap}^{\Lambda^+}\left(\partial^+ \Lambda, \mathrm{CS}^{++}(\hat{\eta})\right), \end{aligned} \tag{18.5.5}$$

where

$$\mathrm{cap}^{\Lambda^+}\left(\partial^+ \Lambda, F\right) = \min_{\substack{g \colon \Lambda^+ \to [0,1] \\ g|_{\partial^+ \Lambda} = 1,\, g|_F = 0}} \frac{1}{2} \sum_{(x,x') \in (\Lambda^+)^\star} \left[g(x) - g(x')\right]^2, \quad F \subset \Lambda, \tag{18.5.6}$$

with $(\Lambda^+)^\star = \{(x, y) \colon x, y \in \Lambda^+, \|x - y\| = 1\}$ is the capacity of simple random walk on Λ modulo normalisation, and $o(1)$ is an error term that tends to zero as $\Lambda \to \mathbb{Z}^2$.

Proof The variational problem in (16.3.11) decomposes into disjoint variational problems for the maximally connected components of $S^\star$. Only those components that contain $S_\sqcap$ or $S_\blacksquare$ contribute, since for the other components the minimum is achieved by picking h constant.

• $\Theta \geq \Theta_1$: A lower bound is obtained from (16.3.11) by removing all transitions that do not involve a fixed protocritical droplet and a move of the free/attached particle. This removal gives

$$\Theta \geq \sum_{\hat\eta \in \mathscr{D}} \min_{C_i(\hat\eta),\, i \in I(\hat\eta)} \min_{\substack{g:\, \Lambda^+ \to [0,1] \\ g|_{G(\hat\eta)}=0,\, g|_{B_i(\hat\eta)}=C_i(\hat\eta),\, i\in I(\hat\eta),\, g|_{\partial^+\Lambda}=1}} \tfrac12 \sum_{(x,x')\in[\Lambda^+ \backslash \mathrm{supp}(\hat\eta)]^\star} \left[g(x) - g(x')\right]^2. \tag{18.5.7}$$

To see how this bound arises from (16.3.11), pick h in (16.3.11) and g in (18.5.7) such that

$$h(\eta) = h(\hat\eta, x) = g(x), \qquad \hat\eta \in \mathscr{D},\ x \in \Lambda^+ \backslash \mathrm{supp}(\hat\eta), \tag{18.5.8}$$

and use that, by Definitions 18.17(b–c),

$$\begin{array}{ll} (\hat\eta, x) \in S_\blacksquare, & x \in G(\hat\eta), \\ (\hat\eta, x) \in S_i, & x \in B_i(\hat\eta),\ i \in I(\hat\eta), \\ (\hat\eta, x) \in \mathscr{D} \subset S_\square, & x \in \partial^+\Lambda. \end{array} \tag{18.5.9}$$

A further lower bound is obtained by removing from the right-hand side of (18.5.9) the boundary condition on the sets $B_i(\hat\eta)$, $i \in I(\hat\eta)$. This gives

$$\begin{aligned} \Theta &\geq \sum_{\hat\eta \in \mathscr{D}} \min_{\substack{g:\, \Lambda^+ \to [0,1] \\ g|_{G(\hat\eta)}=0,\, g|_{\partial^+\Lambda}=1}} \tfrac12 \sum_{(x,x')\in[\Lambda^+ \backslash \mathrm{supp}(\hat\eta)]^\star} \left[g(x) - g(x')\right]^2 \\ &= \sum_{\hat\eta \in \mathscr{D}} \mathrm{cap}^{\Lambda^+ \backslash \mathrm{supp}(\hat\eta)}\big(\partial^+\Lambda, G(\hat\eta)\big), \end{aligned} \tag{18.5.10}$$

where the upper index $\Lambda^+ \backslash \mathrm{supp}(\hat\eta)$ refers to the fact that no moves in and out of $\mathrm{supp}(\hat\eta)$ are allowed (i.e., this set acts as an obstacle for the free particle). To complete the proof we show that, in the limit as $\Lambda \to \mathbb{Z}^2$,

$$\begin{aligned} \mathrm{cap}^{\Lambda^+}\big(\partial^+\Lambda, \mathrm{supp}(\hat\eta) \cup G(\hat\eta)\big) &\geq \mathrm{cap}^{\Lambda^+ \backslash \mathrm{supp}(\hat\eta)}\big(\partial^+\Lambda, G(\hat\eta)\big) \\ &\geq \mathrm{cap}^{\Lambda^+}\big(\partial^+\Lambda, \mathrm{supp}(\hat\eta) \cup G(\hat\eta)\big) - O\big([1/\ln|\Lambda|]^2\big). \end{aligned} \tag{18.5.11}$$

We will show in Sect. 18.5.2 that $\mathrm{cap}^{\Lambda^+}(\partial^+\Lambda, \mathrm{CS}(\hat\eta))$ decays like $1/\ln|\Lambda|$. Since $\mathrm{CS}(\hat\eta) = \mathrm{supp}(\hat\eta) \cup G(\hat\eta)$ by Definition 18.17(d), the lower bound $\Theta \geq \Theta_1$ follows.

Remark 18.19 Before we prove (18.5.11), note that the capacity in the right-hand side of (18.5.11) includes more transitions than the capacity in the left-hand side, namely, all transitions from $\mathrm{supp}(\hat{\eta})$ to $B(\hat{\eta})$. Let

$$g_{\partial^+\Lambda,G(\hat{\eta})}^{\Lambda^+\backslash\mathrm{supp}(\hat{\eta})}(x)=\text{equilibrium potential for }\mathrm{cap}^{\Lambda^+\backslash\mathrm{supp}(\hat{\eta})}\big(\partial^+\Lambda,G(\hat{\eta})\big)\text{ at }x. \tag{18.5.12}$$

Below we will show that

$$g_{\partial^+\Lambda,G(\hat{\eta})}^{\Lambda^+\backslash\mathrm{supp}(\hat{\eta})}(x)\le C/\ln|\Lambda| \quad \forall x\in B(\hat{\eta}) \text{ for some } C<\infty. \tag{18.5.13}$$

Since in the Dirichlet form in (18.5.6) the equilibrium potential appears squared, the error made by adding to the capacity in the left-hand side of (18.5.11) the transitions from $\mathrm{supp}(\hat{\eta})$ to $B(\hat{\eta})$ is of order $[1/\ln|\Lambda|]^2$ times $|B(\hat{\eta})|$, which explains how (18.5.11) arises.

Formally, let $\mathbb{P}_x^{\hat{\eta}}$ be the law of the simple random walk that starts at $x\in B(\hat{\eta})$ and is forbidden to visit the sites in $\mathrm{supp}(\hat{\eta})$. Let $y\in G(\hat{\eta})$. As in the proof of Lemma 8.4, we have

$$\begin{aligned} g_{\partial^+\Lambda,G(\hat{\eta})}^{\Lambda^+\backslash\mathrm{supp}(\hat{\eta})}(x)&=\mathbb{P}_x^{\hat{\eta}}(\tau_{\partial^+\Lambda}<\tau_{G(\hat{\eta})})=\frac{\mathbb{P}_x^{\hat{\eta}}(\tau_{\partial^+\Lambda}<\tau_{G(\hat{\eta})\cup x})}{\mathbb{P}_x^{\hat{\eta}}(\tau_{G(\hat{\eta})\cup\partial^+\Lambda}<\tau_x)}\\ &\le\frac{\mathbb{P}_x^{\hat{\eta}}(\tau_{\partial^+\Lambda}<\tau_x)}{\mathbb{P}_x^{\hat{\eta}}(\tau_y<\tau_x)}\le\frac{\mathrm{cap}^{\Lambda^+\backslash\mathrm{supp}(\hat{\eta})}(x,\partial^+\Lambda)}{\mathrm{cap}^{\Lambda^+\backslash\mathrm{supp}(\hat{\eta})}(x,y)}. \end{aligned} \tag{18.5.14}$$

The denominator of (18.5.14) can be bounded from below by some $C'>0$ that is independent of x, y and $\mathrm{supp}(\hat{\eta})$. To see why, pick a path from x to y that avoids $\mathrm{supp}(\hat{\eta})$ but stays inside a layer around $\mathrm{supp}(\hat{\eta})$, and argue as in the proof of the lower bound of Lemma 6.11. On the other hand, the numerator is bounded from above by $\mathrm{cap}^{\Lambda^+}(x,\partial^+\Lambda)$, i.e., by the capacity of the same pair of sets for a random walk that is not forbidden to visit $\mathrm{supp}(\hat{\eta})$, since the Dirichlet problem associated to the latter has the same boundary conditions but includes more transitions. In the proof of Lemma 18.20 below, we will see that $\mathrm{cap}^{\Lambda^+}(x,\partial^+\Lambda)$ decays like $C''/\ln|\Lambda|$ for some $C''<\infty$ (see (18.5.21)–(18.5.22) below). We therefore conclude that indeed (18.5.13) holds with $C=C''/C'$.

• $\Theta\le\Theta_2$: The upper bound is obtained from (16.3.11) by picking $C_i=0$, $i=1,\dots,I$, and

$$h(\eta)=\begin{cases}1 & \text{for } \eta\in S_\square,\\ g(x) & \text{for } \eta=(\hat{\eta},x)\in\mathscr{C}^{++},\\ 0 & \text{for } \eta\in S^\star\backslash[S_\square\cup\mathscr{C}^{++}],\end{cases} \tag{18.5.15}$$

where

$$\mathscr{C}^{++}=\big\{\eta=(\hat{\eta},x)\colon\ \hat{\eta}\in\mathscr{D},\ x\in\Lambda\backslash\mathrm{cs}^{++}(\hat{\eta})\big\} \tag{18.5.16}$$

consists of those configurations in $\mathscr{D}_\Lambda^{\mathrm{fp}}$ for which the free particle is at distance ≥ 2 of the protocritical droplet and the set of good sites. The choice in (18.5.15) gives

$$\Theta \leq \sum_{\hat{\eta}\in\mathscr{D}} \mathrm{cap}^{\Lambda^+}\big(\partial^+\Lambda, \mathrm{CS}^{++}(\hat{\eta})\big). \tag{18.5.17}$$

To see how this upper bound arises, note that:

- The choice in (18.5.15) satisfies the boundary conditions in (16.3.11) because (recall (16.3.3)–(16.3.4))

$$\begin{aligned}
&\mathscr{C}^{++} \subseteq \mathscr{D}_\Lambda^{\mathrm{fp}},\ [S_\square \cup \mathscr{D}_\Lambda^{\mathrm{fp}}] \cap \left[S_\blacksquare \cup \left(\bigcup_{i=1}^{I} S_i\right)\right] = \emptyset \\
&\implies \quad S^\star\backslash[S_\square \cup \mathscr{C}^{++}] \supset \left[S_\blacksquare \cup \left(\bigcup_{i=1}^{I} S_i\right)\right].
\end{aligned} \tag{18.5.18}$$

- Since $\mathscr{D} \subset S_\square$, the first line of (18.5.15) implies that $h(\eta) = 1$ for $\eta = (\hat{\eta}, x)$ with $\hat{\eta} \in \mathscr{D}$ and $x \in \partial^+\Lambda$, which is consistent with the boundary condition $g|_{\partial^+\Lambda} = 1$ in (18.5.6).
- The third line of (18.5.15) implies that $h(\eta) = 0$ for $\eta = (\hat{\eta}, x)$ with $\hat{\eta} \in \mathscr{D}$ and $x \in \mathrm{CS}^{++}(\hat{\eta})$, which is consistent with the boundary condition $g|_F = 0$ in (18.5.6) for $F = \mathrm{CS}^{++}(\hat{\eta})$.

Note further that:

- The only transitions in $S^\star$ between $S_\square$ and $\mathscr{C}^{++}$ are those where a free particle enters $\partial^-\Lambda$.
- The only transitions in $S^\star$ between $\mathscr{C}^{++}$ and $S^\star\backslash[S_\square \cup \mathscr{C}^{++}]$ are those where the free particle moves from distance 2 to distance 1 of the protocritical droplet. All other transitions either involve a detachment of a particle from the protocritical droplet (which raises the number of droplets) or an increase in the number of particles in Λ. Such transitions lead to energy $> \Gamma^\star$, which is not possible in $S^\star$.
- There are no transitions between $S_\square$ and $S^\star\backslash[S_\square \cup \mathscr{C}^{++}]$.

The latter arguments show that (18.5.6) includes all the transitions in (16.3.11). □

18.5.3 Capacity asymptotics

With Lemma 18.18 we have obtained upper and lower bounds on Θ in terms of capacities for simple random walk on $\mathbb{Z}^2$ of the pairs of sets $\partial^+\Lambda$ and $\mathrm{CS}(\hat{\eta})$, respectively, $\mathrm{CS}^{++}(\hat{\eta})$, with $\hat{\eta}$ summed over $\mathscr{D}$. We use these bounds to prove Theorem 18.4. The transition rates of the simple random walk are 1 between neighbouring pairs of sites.

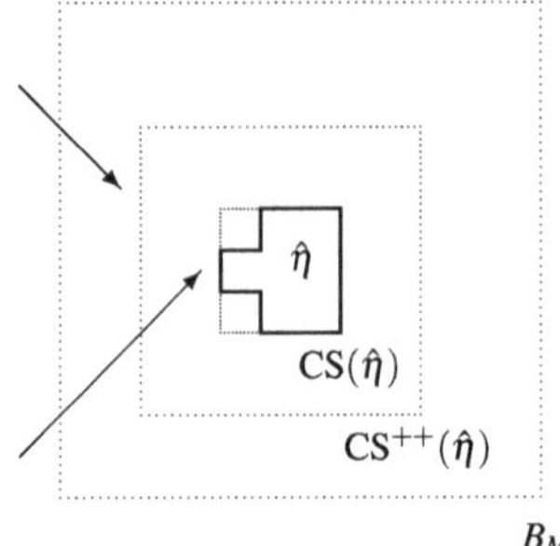

Fig. 18.12 Simple random walk of a free particle moving from $\partial^+ B_M$ to $\mathrm{CS}(\hat{\eta})$, respectively, $\mathrm{CS}^{++}(\hat{\eta})$

Proof Lemma 18.20 below shows that, in the limit as $\Lambda \to \mathbb{Z}^2$, each of the capacities in the upper and lower bound on Θ has the same asymptotic behaviour, namely, $[1+o(1)]\,4\pi/\ln|\Lambda|$, *irrespective* of the location and shape of the protocritical droplet (provided it is not too close to $\partial^+\Lambda$, which is a negligible fraction of the possible locations). In what follows we take $\Lambda = B_M = [-M,+M]^2 \cap \mathbb{Z}^2$ for some $M \in \mathbb{N}$ large enough ($M > 2\ell_c$).

Lemma 18.20 *For any* $\varepsilon > 0$ *(see Fig.* 18.12*),*

$$\begin{aligned} &\lim_{M\to\infty} \max_{\substack{\hat{\eta}\in\mathscr{D}\\ d(\partial^+ B_M,\mathrm{supp}(\hat{\eta}))\geq \varepsilon M}} \left|\frac{\ln M}{2\pi}\,\mathrm{cap}^{B_M^+}\big(\partial^+ B_M, \mathrm{CS}(\hat{\eta})\big) - 1\right| = 0,\\ &\lim_{M\to\infty} \max_{\substack{\hat{\eta}\in\mathscr{D}\\ d(\partial^+ B_M,\mathrm{supp}(\hat{\eta}))\geq \varepsilon M}} \left|\frac{\ln M}{2\pi}\,\mathrm{cap}^{B_M^+}\big(\partial^+ B_M, \mathrm{CS}^{++}(\hat{\eta})\big) - 1\right| = 0, \end{aligned} \tag{18.5.19}$$

where $d(\partial^+ B_M, \mathrm{supp}(\hat{\eta})) = \min\{\|x-y\| : x \in \partial^+ B_M,\ y \in \mathrm{supp}(\hat{\eta})\}$.

Proof We only prove the first line of (18.5.19). The proof of the second line is similar.

• *Lower bound*: For $\hat{\eta} \in \mathscr{D}$, let $y \in \mathrm{CS}(\hat{\eta}) \subset B_M$ denote the site closest to the center of $\mathrm{CS}(\hat{\eta})$. The capacity decreases when we enlarge the set over which the Dirichlet form is minimised. Therefore we have

$$\begin{aligned} \mathrm{cap}^{B_M^+}\big(\partial^+ B_M, \mathrm{CS}(\hat{\eta})\big) &\geq \mathrm{cap}^{B_M^+}\big(\partial^+ B_M, y\big)\\ &= \mathrm{cap}^{(B_M-y)^+}\big(\partial^+(B_M-y), 0\big) \geq \mathrm{cap}^{B_{2M}^+}\big(\partial^+ B_{2M}, 0\big), \end{aligned} \tag{18.5.20}$$

where the last equality uses that $(B_M - y)^+ \subset B_{2M}^+$ because $y \in B_M$. By the analogue of (16.2.5) for simple random walk, we have (compare (18.5.6) with (16.2.1)–(16.2.2))

$$\mathrm{cap}^{B_{2M}^+}\big(\partial^+ B_{2M}, 0\big) = \mathrm{cap}^{B_{2M}^+}\big(0, \partial^+ B_{2M}\big) = 4\,\mathbb{P}_0(\tau_{\partial^+ B_{2M}} < \tau_0), \tag{18.5.21}$$

where $\mathbb{P}_0$ is the law on path space of the *discrete-time* simple random walk on $\mathbb{Z}^2$ starting at 0. It is a standard fact (see e.g. Révész [205], Lemma 22.1) that

$$\mathbb{P}_0(\tau_{\partial^+ B_{2M}} < \tau_0) = \big[1+o(1)\big]\frac{\pi}{2\ln(2M)}, \qquad M \to \infty. \tag{18.5.22}$$

Combining (18.5.20)–(18.5.22), we get the desired lower bound.

- *Upper bound*: As in (18.5.20), we have

$$\begin{aligned} &\operatorname{cap}^{B_M^+}\big(\partial^+ B_M, \operatorname{CS}(\hat\eta)\big) \le \operatorname{cap}^{B_M^+}\big(\partial^+ B_M, S_y(\hat\eta)\big) \\ &\quad = \operatorname{cap}^{(B_M-y)^+}\big(\partial^+(B_M-y), S_y(\hat\eta)-y\big) \le \operatorname{cap}^{B_{\varepsilon M}^+}\big(\partial^+ B_{\varepsilon M}, S_y(\hat\eta)-y\big), \end{aligned} \tag{18.5.23}$$

where $S_y(\hat\eta)$ is the smallest square centered at y containing $\operatorname{CS}(\hat\eta)$, and the last inequality uses that $(B_M-y)^+ \supset B_{\varepsilon M}^+$ when $d(\partial^+ B_M, \operatorname{supp}(\hat\eta)) \ge \varepsilon M$. By the recurrence of simple random walk, we have

$$\operatorname{cap}^{B_{\varepsilon M}^+}\big(\partial^+ B_{\varepsilon M}, S_y(\hat\eta)-y\big) = \big[1+o(1)\big]\operatorname{cap}^{B_{\varepsilon M}^+}\big(\partial^+ B_{\varepsilon M}, 0\big), \qquad M \to \infty. \tag{18.5.24}$$

Combining (18.5.22)–(18.5.24), we get the desired upper bound. □

We are now ready to complete the proof of Theorem 18.4. Combining Lemmas 18.18–18.20, we find that $\Theta \in [\Theta_1, \Theta_2]$ with

$$\begin{aligned} \Theta_1 &= O(\varepsilon M) + \sum_{\substack{\hat\eta \in \mathscr{D} \\ d(\partial^+ B_M, \operatorname{supp}(\hat\eta)) \ge \varepsilon M}} \operatorname{cap}^{B_M^+}\big(\partial^+ B_M, \operatorname{CS}(\hat\eta)\big) \\ &= O(\varepsilon M) + \frac{2\pi}{\ln M}\big|\{\hat\eta \in \mathscr{D} \colon d\big(\partial^+ B_M, \operatorname{supp}(\hat\eta)\big) \ge \varepsilon M\}\big|\,\big[1+o(1)\big] \\ &= O(\varepsilon M) + \frac{2\pi}{\ln M} N(\ell_c)\big[2(1-\varepsilon)M\big]^2\big[1+o(1)\big], \qquad M \to \infty, \end{aligned} \tag{18.5.25}$$

for any $\varepsilon > 0$ and the same expression for Θ_2, where we use that

$$\operatorname{cap}^{B_M^+}\big(\partial^+ B_M, \operatorname{CS}(\hat\eta)\big) \le \operatorname{cap}^{B_M^+}\big(B_M^+\backslash \operatorname{CS}(\hat\eta), \operatorname{CS}(\hat\eta)\big) = \tfrac{1}{2}\big|\operatorname{CS}^+(\hat\eta)\big| \le \tfrac{1}{2}(\ell_c+2)^2, \tag{18.5.26}$$

and we recall that $N(\ell_c)$ is the cardinality of $\mathscr{D}$ modulo shifts of the protocritical droplets. Let $M \to \infty$ followed by $\varepsilon \downarrow 0$, to conclude that $\Theta \sim 2\pi N(\ell_c)(2M)^2/\ln M$. Since $|\Lambda| = (2M+1)^2$ and $K = 1/\Theta$, this proves (18.1.12) in Theorem 18.4. □

18.6 Extension to three dimensions

In this section we briefly indicate how to extend the main definitions and results from two to three dimensions.

Let $\Lambda \subset \mathbb{Z}^3$ be a large cubic box, centred at the origin. The metastable parameter range replacing (18.1.4) is

$$\Delta \in (U, 3U), \tag{18.6.1}$$

and, similarly as in (18.1.6), we assume that

$$\frac{U}{3U-\Delta} \notin \mathbb{N}, \qquad \frac{2U}{3U-\Delta} \notin \mathbb{N}. \tag{18.6.2}$$

The analogue of Definitions 18.1(b–c) and 18.6 reads:

Definition 18.21

(a) Let $\mathscr{Q}$ denote the set of configurations having one cluster anywhere in Λ^- consisting of an $(m_c-1)\times(m_c-\delta_c)\times m_c$ quasi-cube with, attached to one of its faces, an $(\ell_c-1)\times\ell_c$ quasi-square with, attached to one of its sides, a single particle. Here, $\delta_c \in \{0,1\}$ depends on the arithmetic properties of U and Δ, while

$$\ell_c = \left\lceil \frac{U}{3U-\Delta} \right\rceil, \qquad m_c = \left\lceil \frac{2U}{3U-\Delta} \right\rceil, \tag{18.6.3}$$

are the two-dimensional critical droplet size on a face, respectively, the three-dimensional critical droplet size, replacing (18.1.5). Note that $m_c \in \{2\ell_c-1, 2\ell_c\}$.

(b) For $\Delta \in (2U, 3U)$, let $\mathscr{Q}^{2U}$ denote the set of configurations that can be reached from some configuration in $\mathscr{Q}$ via a $2U$-path, i.e.,

$$\mathscr{Q}^{2U} = \left\{\eta' \in \mathscr{V}_{n_c}\colon \exists \eta \in \mathscr{Q}\colon H(\eta) = H(\eta'),\ \Phi_{\mathscr{V}_{n_c}}(\eta,\eta') \le H(\eta) + 2U\right\}, \tag{18.6.4}$$

where $n_c = m_c(m_c-\delta_c)(m_c-1) + \ell_c(\ell_c-1)+1$ is the volume of the clusters in $\mathscr{Q}$. For $\Delta \in (U, 2U)$, use U instead of $2U$ in (18.6.4).

(c) Let $[\mathscr{Q}^{2U}]^{\mathrm{fp}}$ denote the set of configurations obtained from $\mathscr{Q}^{2U}$ by adding a free particle anywhere in $\partial^-\Lambda$ (see Fig. 18.13).

(d) Let

$$\begin{aligned}\Gamma^\star = \Gamma^\star(\square,\blacksquare) &= H\big([\mathscr{Q}^{2U}]^{\mathrm{fp}}\big) = H\big(\mathscr{Q}^{2U}\big) + \Delta = H(\mathscr{Q}) + \Delta \\ &= U\big[m_c(m_c-\delta_c) + m_c(m_c-1) + (m_c-\delta_c)(m_c-1) + 2\ell_c + 3\big] \\ &\quad - (3U-\Delta)\big[m_c(m_c-\delta_c)(m_c-1) + \ell_c(\ell_c-1) + 2\big].\end{aligned} \tag{18.6.5}$$

Theorem 18.3 carries over: $\mathscr{P}^\star(\square,\blacksquare) = \mathscr{Q}^{2U}$ and $\mathscr{C}^\star(\square,\blacksquare) = [\mathscr{Q}^{2U}]^{\mathrm{fp}}$. Unfortunately, we are not able to fully identify the geometry of $\mathscr{Q}^{2U}$, i.e., the analogue of Fig. 18.2 is missing. This is due to the fact that *the motion of particles along the border of the droplet is much more complex in three than in two dimensions* (see Fig. 18.14 for an example).

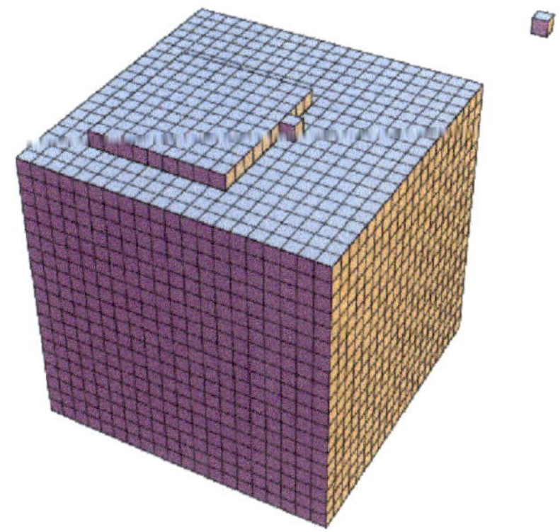

Fig. 18.13 An element of $\mathscr{Q}^{\mathrm{fp}} \subseteq \mathscr{D}^{\mathrm{fp}}$ for $\ell_c = 10$, $m_c = 20$ and $\delta_c = 0$

Fig. 18.14 An example of motion of particles along the border of the droplet

Also Theorem 18.2 carries over: the proof of (H1–H2) is the same as in Sects. 18.3.1–18.3.2, except that for (H1) a little extra care is needed to handle the geometry in three dimensions.

As in two dimensions, no easily computable formula for K is available. Similarly as in Sect. 18.5, however, the prefactor K can be estimated in terms of capacities associated with three-dimensional simple random walk. Since the latter is *transient*, the large volume scaling of these capacities is no longer independent of the shape and the location of the protocritical droplet. Therefore Theorem 18.4 carries over in a somewhat weaker form.

Theorem 18.22 *For large Λ,*

$$\lim_{\Lambda \to \mathbb{Z}^3} |\Lambda| \, K(\Lambda, \ell_c, m_c, \delta_c) = \frac{1}{M(\ell_c, m_c, \delta_c) N(\ell_c, m_c, \delta_c)}, \tag{18.6.6}$$

where $N(\ell_c, m_c, \delta_c)$ *is the cardinality of* $\mathscr{D}$ *modulo shifts, and* $M(\ell_c, m_c, \delta_c)$ *satisfies the bounds*

$$\kappa\big(m_c - \lceil\sqrt{m_c}\,\rceil\big) \le M(\ell_c, m_c, \delta_c) \le \kappa(m_c + 3) \tag{18.6.7}$$

with $\kappa(m)$ *the capacity of the* $m \times m \times m$ *cube for simple random walk on* $\mathbb{Z}^3$.

Proof The extension of the proof in Sect. 18.5 from two to three dimensions is in principle straightforward and involves no new ideas. The geometry of the communication level set is less explicit, but no detail is needed for the proof.

By the transience of simple random walk in three dimensions, we know that

$$\lim_{\Lambda\to\mathbb{Z}^3} \operatorname{cap}^{\Lambda^+}\big(\partial^+\Lambda, F\big) = \operatorname{cap}^{\mathbb{Z}^3}(F) \tag{18.6.8}$$

exists for any finite non-empty $F \subset \mathbb{Z}^3$. The limit, which is positive and finite, is the capacity of F. Let $\kappa(m) = \operatorname{cap}^{\mathbb{Z}^3}(m \times m \times m)$ be the capacity of the $m \times m \times m$ cube for simple random walk on $\mathbb{Z}^3$. Then we know that

$$\lim_{m\to\infty} \kappa(m)/m = \kappa \tag{18.6.9}$$

with κ the capacity of the unit cube for standard Brownian motion on $\mathbb{R}^3$. Since $2\pi R$ is the capacity of the ball with radius R for standard Brownian motion on $\mathbb{R}^3$, we have that $\kappa \in (2\pi, 2\pi\sqrt{3})$.

The lower bound in (18.6.7) comes from the fact that all protocritical droplets contain a cube of side length $m_c - \sqrt{m_c}$. The upper bound comes from the fact that all protocritical droplets are contained in a cube of side length $m_c + 1$, and that as long as the free particle is at distance ≥ 2 from the protocritical droplet no border motion is possible. Both these facts are easy to establish. □

With the help of (18.6.7) and (18.6.9), we have good control over $M(\ell_c, m_c, \delta_c)$ for m_c large, i.e., for Δ close to $3U$. We have no formula for $N(\ell_c, m_c, \delta_c)$ analogous to (18.1.13). It would be nice to know its asymptotics for m_c large.

18.7 Bibliographical notes

1. The results in this chapter are taken from Bovier, den Hollander and Nardi [31], with geometric input from den Hollander, Nardi, Olivieri and Scoppola [84]. Cruder versions of the main results in Chap. 16 for Kawasaki dynamics, derived with the help of the pathwise approach to metastability, were obtained by den Hollander, Olivieri and Scoppola [88–90] in two dimensions and by den Hollander, Nardi, Olivieri and Scoppola [84] in three dimensions. The latter paper contains the "piston construction" mentioned in Sect. 18.3.1.

2. The formula for the number of protocritical droplets modulo shifts claimed in [31] is wrong. The correct formula is (18.1.13), as shown in Sect. 18.4.5. The authors are grateful to Markus Mayer for pointing out the error.

3. For details of the argument needed in Sect. 18.3.1 to extend the proof of (H1) from Glauber dynamics to Kawasaki dynamics, see [84]. For a comparison of Glauber dynamics and Kawasaki dynamics, see den Hollander [81].

4. The results in this chapter extend to arbitrary shapes of Λ (instead of a square or a cube), provided $|\partial\Lambda|/|\Lambda|$ tends to zero as $\Lambda \to \mathbb{Z}^2$, respectively, $\Lambda \to \mathbb{Z}^3$. For the relevant capacity asymptotics, needed in Sects. 18.5.2 and 18.6, see van den Berg [227].

5. For more information on the tube of typical trajectories, or nucleation pattern, see Olivieri and Vares [198], Sect. 7.13.

6. It would appear that the analysis in Sect. 18.6 could be extended to arbitrary dimension, like for Glauber dynamics (recall Sect. 17.6, Item 3). However, this extension has never been written out in detail. The set of critical droplets is quite complex due to the motion of particles along the border of droplets. In two dimensions we have a full understanding of this motion, in three dimensions a partial understanding (see [84]), while in higher dimensions we know very little. It is clear that the critical droplets for Glauber dynamics all are protocritical droplets for Kawasaki dynamics. But the border motion can create many additional shapes, all via V-paths with $V < \Delta$.

7. An anisotropic version of Kawasaki dynamics, in which the Hamiltonian in (18.1.2) is modified by allowing for different binding energies $U_h < 0$ and $U_v < 0$ in the horizontal and the vertical direction, was studied in Nardi, Olivieri and Scoppola [189]. Different nucleation patterns occur for weak and strong anisotropy. In both cases the critical droplets are different from what is naively expected, similarly as for the anisotropic Glauber dynamics described in Item 5 of Chap. 17.

8. Kawasaki dynamics with two types of particles, with binding energy $-U < 0$ between particles of different types (and no binding energy between particles of the same type) and with different activation energies $\Delta_1 > 0$ and $\Delta_2 > 0$, was studied in den Hollander, Nardi and Troiani [85–87]. There are several regimes, with critical droplets being either square-shaped or rhombus-shaped. The proof of (H1)–(H2) is quite involved, and is hampered by the fact that droplets with fixed volume and minimal surface change shape when they come close to $\partial\Lambda$.

Part VII
Applications: Lattice Systems in Large Volumes at Low Temperatures

Part VII looks at nucleation in lattice systems that grow to infinity as the temperature tends to zero. *Spatial entropy* comes into play: in large volumes, even at low temperatures, entropy is competing with energy because the metastable state and the states that evolve from it under the dynamics have a non-trivial spatial structure. Chapter 19 looks at Glauber dynamics, Chap. 20 at Kawasaki dynamics.

The transition from the metastable state (with only subcritical droplets) to the stable state (with one or more supercritical droplets) is triggered by the appearance of a single critical droplet *somewhere* in the system. The main property driving the results in Chaps. 19–20 is that the average time until this appearance is inversely proportional to the volume. This property is referred to as *homogeneous nucleation*, because it says that the critical droplet for the transition appears essentially independently in small volumes that partition the large volume.

No information will be obtained about what happens to the system after the critical droplet has appeared. This belongs to the *post-nucleation* regime, which is much harder than the *pre-nucleation* regime considered here, and which will be briefly addressed in Chap. 23. Our results are further limited in the sense that we need to draw the initial configuration according to a specific distribution on the set of subcritical configurations, namely, the *last-exit biased distribution* introduced in Chap. 8. To show that the same results hold for more general initial distributions we would need to establish strong recurrence properties of the dynamics within the metastable state. Another limitation is that there will be no proof that the nucleation time divided by its average converges to the exponential distribution.

Contrary to Chap. 16, where for small volumes we were able to deal with a general dynamics under a general set of hypotheses, the situation for large volumes is significantly more difficult. This is why we can so far offer results only for Glauber and Kawasaki. It remains a challenge to develop a more abstract set-up.

Chapter 19
Glauber Dynamics

La complexion qui fait le talent pour les petites choses est contraire à celle qu'il faut pour le talent des grandes.
(François de La Rochefoucauld, Réflexions)

The goal of this chapter is to extend the analysis of Chap. 17 to volumes that grow moderately fast as the temperature decreases. Let $\Lambda_\beta \subset \mathbb{Z}^2$ be a square box with periodic boundary conditions such that $\lim_{\beta\to\infty} |\Lambda_\beta| = \infty$. We run the Glauber dynamics on Λ_β starting from a *random initial configuration* where all the droplets (= clusters of plus-spins) are small. For large β, and in the parameter range corresponding to the metastable regime (recall Sect. 17.1.2), the transition from the metastable state (with only subcritical droplets) to the stable state (with one or more supercritical droplets) is triggered by the appearance of a single critical droplet somewhere in Λ_β. We will show that the average time until this happens scales like $e^{\Gamma^\star\beta}/N(\ell_c)|\Lambda|$, where $\Gamma^\star$ and $N(\ell_c)$ are the quantities as for small volumes (recall Sect. 17.1.3). This scaling is valid as long as the average nucleation time tends to infinity.

19.1 Introduction and main results

19.1.1 Glauber dynamics in large volumes

We retain the setting of Sect. 17.1.1, expect that we replace the torus $\Lambda \subset \mathbb{Z}^2$ by a β-dependent torus $\Lambda_\beta \subset \mathbb{Z}^2$. Accordingly, we write S_β, H_β instead of S, H to indicate that the configuration space and the Hamiltonian also depend on β.

Subcritical, protocritical and critical configurations We want to start our Glauber dynamics on Λ_β from an initial configuration in which all droplets are sufficiently small. To make this notion precise, we need the following definitions.

Definition 19.1

(a) Let $C_B(\sigma)$, $\sigma \in S_\beta$, be the configuration that is obtained from σ by a "bootstrap percolation map", i.e., by circumscribing all the droplets in σ with rectangles,

A. Bovier, F. den Hollander, *Metastability*,
Grundlehren der mathematischen Wissenschaften 351,
DOI 10.1007/978-3-319-24777-9_19

and continuing to do so in an iterative manner until a union of disjoint rectangles is obtained.

(b) Call $C_B(\sigma)$ *subcritical* if all its rectangles fit inside the protocritical droplets for Glauber dynamics in Chap. 17, and are at distance ≥ 2 from each other (i.e., are non-interacting).

Definition 19.2

(a) $\mathscr{S} = \{\sigma \in S_\beta \colon\ C_B(\sigma) \text{ is subcritical}\}$.
(b) $\mathscr{P} = \{\sigma \in \mathscr{S} \colon c_\beta(\sigma, \sigma') > 0 \text{ for some } \sigma' \in \mathscr{S}^c\}$.
(c) $\mathscr{C} = \{\sigma' \in \mathscr{S}^c \colon c_\beta(\sigma, \sigma') > 0 \text{ for some } \sigma \in \mathscr{S}\}$.

We refer to $\mathscr{S}$, $\mathscr{P}$ and $\mathscr{C}$ as the set of *subcritical*, *protocritical*, respectively, *critical* configurations. Note that, for every $\sigma \in S_\beta$, each step in the bootstrap percolation map $\sigma \to C_B(\sigma)$ decreases the energy, and therefore the Glauber dynamics moves from σ to $C_B(\sigma)$ in a time of order one. This is why $C_B(\sigma)$ appears in the definition of $\mathscr{S}$. The subcritical configurations therefore are the analogues of the subcritical droplets we encountered in Sect. 17.1.

Remark 19.3 The sets $\mathscr{P}, \mathscr{C}$ will play a similar rôle as, but are not directly comparable with, the sets $\mathscr{P}^\star, \mathscr{C}^\star$ in Chap. 17.

Sets of starting configurations For $\ell_1, \ell_2 \in \mathbb{N}$, let $R_{\ell_1,\ell_2}(x) \subset \Lambda_\beta$ be the $\ell_1 \times \ell_2$ rectangle whose lower-left corner is x. (We always take $\ell_1 \leq \ell_2$ and allow for both orientations of the rectangle, i.e., $R_{\ell_1,\ell_2}(x)$ actually represents two rectangles.) For $L = 1, \ldots, 2\ell_c - 3$, let $Q_L(x)$ denote the L-th element in the *canonical* sequence of growing squares and quasi-squares

$$R_{1,2}(x),\ R_{2,2}(x),\ R_{2,3}(x),\ R_{3,3}(x), \ldots,\ R_{\ell_c-1,\ell_c-1}(x),\ R_{\ell_c-1,\ell_c}(x). \tag{19.1.1}$$

Our starting configurations will be drawn from one of the sets $\mathscr{S}_L \subset \mathscr{S}$ defined by

$$\mathscr{S}_L = \big\{\sigma \in \mathscr{S} \colon\ \text{each rectangle in } C_B(\sigma) \text{ fits inside } Q_L(x) \text{ for some } x \in \Lambda_\beta\big\}, \tag{19.1.2}$$

for any $L \in \mathbb{N}$ that satisfies $L^* \leq L \leq 2\ell_c - 3$ with

$$L^* = \min\Big\{1 \leq L \leq 2\ell_c - 3\colon\ \lim_{\beta\to\infty} \mu_\beta(\mathscr{S}_L)/\mu_\beta(\mathscr{S}) = 1\Big\}. \tag{19.1.3}$$

In words, $\mathscr{S}_L$ is the subset of those subcritical configurations whose droplets fit inside a square or quasi-square labelled by L, with L chosen large enough so that $\mathscr{S}_L$ is typical within $\mathscr{S}$ under the Gibbs measure μ_β associated with H_β in the limit as $\beta \to \infty$. (Our main theorem in Sect. 19.1.2 turns out not to depend on the choice of L subject to these restrictions.)

Note that $\mathscr{S}_{2\ell_c-3} = \mathscr{S}$. The value of L^* depends on how fast Λ_β grows with β. In Sect. 19.6 we show that, for every $1 \leq L \leq 2\ell_c - 4$,

$$\lim_{\beta\to\infty} \mu_\beta(\mathscr{S}_L)/\mu_\beta(\mathscr{S}) = 1 \quad \text{if and only if} \quad \lim_{\beta\to\infty} |\Lambda_\beta| \mathrm{e}^{-\beta\Gamma^\star_{L+1}} = 0 \tag{19.1.4}$$

with $\Gamma^\star_{L+1}$ the energy needed to create a droplet $Q_{L+1}(0)$ at the origin. Thus, if $|\Lambda_\beta| = e^{\theta\beta}$, then $L^* = L^*(\theta) = (2\ell_c - 3) \wedge \min\{L \in \mathbb{N}\colon \Gamma^\star_{L+1} > \theta\}$, which increases stepwise from 1 to $2\ell_c - 3$ as θ increases from 0 to $\Gamma^\star$, with $\Gamma^\star$ the communication height in Chap. 17.

Initial distribution For non-empty disjoint sets $A, B \subset S_\beta$, we recall that $\nu_{A,B}$ denotes the *last-exit biased distribution on A for the crossover to B*, defined in (7.1.38) as

$$\nu_{A,B}(\sigma) = \frac{\mu_\beta(\sigma) e_{A,B}(\sigma)}{\mathrm{cap}(A,B)}, \qquad \sigma \in A, \tag{19.1.5}$$

where $e_{A,B}$ is the equilibrium measure defined in (7.1.21).

We choose the initial distribution to be biased according to the last exit of $\mathscr{S}_L$ for the transition from $\mathscr{S}_L$ to a target set in $\mathscr{S}^c$. Three choices for this target set are made in Sect. 19.1.2, namely, $\mathscr{S}^c$, $\mathscr{S}^c \backslash \mathscr{C}$ and $\mathscr{D}_M$, $M \in \mathbb{N}$, $M \geq \ell_c$, defined by

$$\mathscr{D}_M = \big\{\sigma \in S_\beta\colon \exists x \in \Lambda_\beta \text{ such that } \mathrm{supp}\big[C_B(\sigma)\big] \supset R_{M,M}(x)\big\}, \tag{19.1.6}$$

which is the set of configurations containing a supercritical droplet of size M.

19.1.2 Main theorem

Throughout this chapter we assume that we are in the metastable regime where $h \in (0, 2J)$ (recall Sect. 17.1.2). We further assume that

$$\lim_{\beta\to\infty} |\Lambda_\beta| = \infty, \qquad \lim_{\beta\to\infty} |\Lambda_\beta| \, e^{-\beta\Gamma^\star} = 0. \tag{19.1.7}$$

The second condition ensures that the existence of a critical droplet anywhere in the box is still a rare event and does only occur after a large time. If this condition were violated, then the metastable transition would no longer be dominated by the time of nucleation, but by the growth of supercritical droplets that exist somewhere far away.

For $\sigma \in S_\beta$, let $\mathbb{P}_\sigma$ denote the law of the Glauber dynamics starting from σ. For ν a probability distribution on S_β, write

$$\mathbb{P}_\nu(\cdot) = \sum_{\sigma \in S_\beta} \nu(\sigma) \, \mathbb{P}_\sigma(\cdot). \tag{19.1.8}$$

Abbreviate

$$N_1 = N_1(\ell_c) = 4\ell_c, \qquad N_2 = N_2(\ell_c) = \tfrac{4}{3}(2\ell_c - 1). \tag{19.1.9}$$

Theorem 19.4 (Mean crossover time) *Subject to* (19.1.3) *and* (19.1.7), *the following hold*:

(a)

$$\lim_{\beta\to\infty} |\Lambda_\beta|\, \mathrm{e}^{-\beta\Gamma^\star}\, \mathbb{E}_{\nu_{\mathscr{S}_L,\mathscr{S}^c}}(\tau_{\mathscr{S}^c}) = \frac{1}{N_1}. \tag{19.1.10}$$

(b)

$$\lim_{\beta\to\infty} |\Lambda_\beta|\, \mathrm{e}^{-\beta\Gamma^\star}\, \mathbb{E}_{\nu_{\mathscr{S}_L,\mathscr{S}^c\backslash\mathscr{C}}}(\tau_{\mathscr{S}^c\backslash\mathscr{C}}) = \frac{1}{N_2}. \tag{19.1.11}$$

(c)

$$\lim_{\beta\to\infty} |\Lambda_\beta|\, \mathrm{e}^{-\beta\Gamma^\star}\, \mathbb{E}_{\nu_{\mathscr{S}_L,\mathscr{D}_M}}(\tau_{\mathscr{D}_M}) = \frac{1}{N_2}, \quad \forall \ell_c \le M \le 2\ell_c - 1. \tag{19.1.12}$$

19.1.3 Discussion

1. Theorem 19.4(a) says that the average time to create a critical droplet is $[1 + o(1)]\mathrm{e}^{\beta\Gamma^\star}/N_1|\Lambda_\beta|$. Theorems 19.4(b–c) say that the average time to go beyond this critical droplet and to grow a droplet that is twice as large is $[1+o(1)]\mathrm{e}^{\beta\Gamma^\star}/N_2|\Lambda_\beta|$. The factor N_1 counts the number of *shapes* of the critical droplet, while $|\Lambda_\beta|$ counts the number of *locations*. The average times to create a critical, respectively, a supercritical droplet differ by a factor $N_2/N_1 < 1$. This is because, as we saw in Sect. 17.1.4, item 3, once the dynamics is "on top of the hill" $\mathscr{C}$ it has a positive probability to "fall back" to $\mathscr{S}$. On average the dynamics makes $N_1/N_2 > 1$ attempts to reach the top $\mathscr{C}$ before it finally "falls over" to $\mathscr{S}^c\backslash\mathscr{C}$. After that, it rapidly grows a large droplet.

2. If the second condition in (19.1.7) fails, then there is a positive probability to see a protocritical droplet in Λ_β under the starting measure $\nu_{\mathscr{S}_L,\mathscr{S}^c}$, and nucleation sets in immediately. In that situation different questions about the system become relevant, which are no longer nucleation-driven but are growth-driven (see Chap. 23). Theorem 19.4(a) continues to be true, but it no longer describes metastable behaviour.

3. The *average* probability under the Gibbs measure μ_β of destroying a supercritical droplet and returning to a configuration in $\mathscr{S}_L$ is exponentially small in β. We defer the proof of this fact to Chap. 20, where we consider Kawasaki dynamics. The proof for Glauber is easily read off from the one for Kawasaki. Thus, the crossover from $\mathscr{S}_L$ to $\mathscr{S}^c\backslash\mathscr{C}$ truly represents the threshold for nucleation, and Theorem 19.4(b) truly represents the nucleation time.

Outline Theorem 19.4 is proved in Sects. 19.2–19.4. Along the way we need two technical facts whose proofs are deferred to Sects. 19.5–19.6. These deal with sparseness of subcritical droplets and typicality of starting configurations, respectively.

19.2 Average time to create a critical droplet

To estimate the average crossover time from $\mathscr{S}_L \subset \mathscr{S}$ to $\mathscr{S}^c$ in Theorem 19.4(a), we will use (7.1.41) in Corollary 7.11. With $\mathscr{A} = \mathscr{S}_L$ and $\mathscr{B} = \mathscr{S}^c$, this relation reads

$$\sum_{\sigma \in \mathscr{S}_L} \nu_{\mathscr{S}_L,\mathscr{S}^c}(\sigma)\, \mathbb{E}_\sigma(\tau_{\mathscr{S}^c}) = \frac{1}{\text{CAP}(\mathscr{S}_L, \mathscr{S}^c)} \sum_{\sigma \in \mathscr{S}} \mu_\beta(\sigma)\, h_{\mathscr{S}_L,\mathscr{S}^c}(\sigma). \tag{19.2.1}$$

The left-hand side is the quantity of interest in (19.1.10). In Sects. 19.2.1–19.2.2 we estimate $\sum_{\sigma \in \mathscr{S}} \mu_\beta(\sigma) h_{\mathscr{S}_L,\mathscr{S}^c}(\sigma)$ and $\text{CAP}(\mathscr{S}_L, \mathscr{S}^c)$. The estimates will show that

$$\text{r.h.s. (19.2.1)} = \frac{1}{N_1 |\Lambda_\beta|} e^{\beta \Gamma^\star} \big[1 + o(1)\big], \quad \beta \to \infty. \tag{19.2.2}$$

19.2.1 Estimate of the equilibrium potential

Lemma 19.5 *$\sum_{\sigma \in \mathscr{S}} \mu_\beta(\sigma) h_{\mathscr{S}_L,\mathscr{S}^c}(\sigma) = \mu_\beta(\mathscr{S})[1 + o(1)]$ as $\beta \to \infty$.*

Proof Write, using (7.1.16),

$$\begin{aligned} \sum_{\sigma \in \mathscr{S}} \mu_\beta(\sigma) h_{\mathscr{S}_L,\mathscr{S}^c}(\sigma) &= \sum_{\sigma \in \mathscr{S}_L} \mu_\beta(\sigma) h_{\mathscr{S}_L,\mathscr{S}^c}(\sigma) + \sum_{\sigma \in \mathscr{S} \backslash \mathscr{S}_L} \mu_\beta(\sigma) h_{\mathscr{S}_L,\mathscr{S}^c}(\sigma) \\ &= \mu_\beta(\mathscr{S}_L) + \sum_{\sigma \in \mathscr{S} \backslash \mathscr{S}_L} \mu_\beta(\sigma) \mathbb{P}_\sigma(\tau_{\mathscr{S}_L} < \tau_{\mathscr{S}^c}). \end{aligned} \tag{19.2.3}$$

The last sum is bounded above by $\mu_\beta(\mathscr{S} \backslash \mathscr{S}_L)$. But $\mu_\beta(\mathscr{S} \backslash \mathscr{S}_L) = o(\mu_\beta(\mathscr{S}))$ as $\beta \to \infty$ by our choice of L in (19.1.3). □

19.2.2 Estimate of the capacity

Lemma 19.6 *$\text{CAP}(\mathscr{S}_L, \mathscr{S}^c) = N_1\, |\Lambda_\beta| e^{-\beta \Gamma^\star} \mu_\beta(\mathscr{S})[1 + o(1)]$ as $\beta \to \infty$ with $N_1 = 4\ell_c$.*

Proof The proof proceeds via upper and lower bounds, which are written out below. □

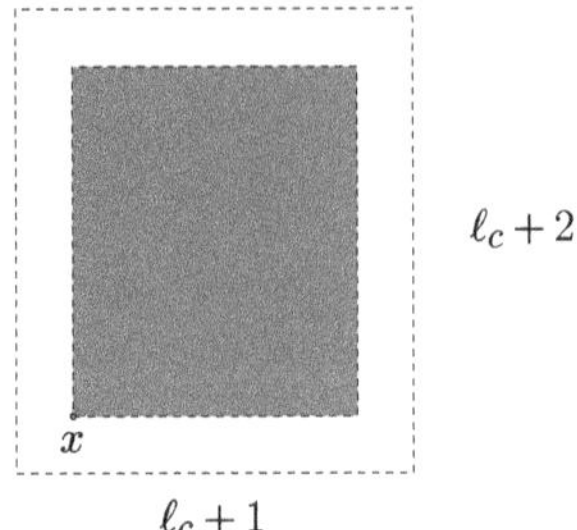

Fig. 19.1 $R_{\ell_c-1,\ell_c}(x)$ (*shaded box*) and $[R_{\ell_c+1,\ell_c+2}(x-(1,1))]^c$ (complement of *dotted box*)

Upper bound

Proof We use the Dirichlet principle and a test function that is equal to 1 on $\mathscr{S}$ to get the upper bound

$$\begin{aligned}\mathrm{CAP}\big(\mathscr{S}_L,\mathscr{S}^c\big)\le\mathrm{CAP}\big(\mathscr{S},\mathscr{S}^c\big)&=\sum_{\substack{\sigma\in\mathscr{S},\sigma'\in\mathscr{S}^c\\ c_\beta(\sigma,\sigma')>0}}\mu_\beta(\sigma)c_\beta\big(\sigma,\sigma'\big)\\&=\sum_{\substack{\sigma\in\mathscr{S},\sigma'\in\mathscr{S}^c\\ c_\beta(\sigma,\sigma')>0}}\big[\mu_\beta(\sigma)\wedge\mu_\beta\big(\sigma'\big)\big]\le\mu_\beta(\mathscr{C}),\end{aligned}\tag{19.2.4}$$

where the second equality uses reversibility in combination with the fact that $c_\beta(\sigma,\sigma')\vee c_\beta(\sigma',\sigma)=1$. Thus, it suffices to show that

$$\mu_\beta(\mathscr{C})\le N_1\,|\Lambda_\beta|\,\mathrm{e}^{-\beta\Gamma^\star}\,\mu_\beta(\mathscr{S})\big[1+o(1)\big]\quad\text{as }\beta\to\infty.\tag{19.2.5}$$

For every $\sigma\in\mathscr{P}$ there are one or more rectangles $R_{\ell_c-1,\ell_c}(x)$, $x=x(\sigma)\in S_\beta$, that are filled by $(+1)$-spins in $C_B(\sigma)$. If $\sigma'\in\mathscr{C}$ is such that $\sigma'=\sigma^y$ for some $y\in\Lambda_\beta$, then σ' has a $(+1)$-spin at y situated on the boundary of one of these rectangles (recall Definition 19.2). Let

$$\begin{aligned}\hat{\mathscr{S}}(x)&=\big\{\sigma\in\mathscr{S}:\ \mathrm{supp}[\sigma]\subseteq R_{\ell_c-1,\ell_c}(x)\big\},\\ \check{\mathscr{S}}(x)&=\big\{\sigma\in\mathscr{S}:\ \mathrm{supp}[\sigma]\subseteq\big[R_{\ell_c+1,\ell_c+2}\big(x-(1,1)\big)\big]^c\big\}.\end{aligned}\tag{19.2.6}$$

For every $\sigma\in\mathscr{P}$, we have $\sigma=\hat\sigma\vee\check\sigma$ for some $\hat\sigma\in\hat{\mathscr{S}}(x)$ and $\check\sigma\in\check{\mathscr{S}}(x)$ with $x=x(\sigma)$, uniquely decomposing the configuration into two non-interacting parts inside $R_{\ell_c-1,\ell_c}(x)$ and $[R_{\ell_c+1,\ell_c+2}(x-(1,1))]^c$ (see Fig. 19.1). We have

$$H_\beta(\sigma)-H_\beta(\boxminus)=\big[H_\beta(\hat\sigma)-H_\beta(\boxminus)\big]+\big[H_\beta(\check\sigma)-H_\beta(\boxminus)\big].\tag{19.2.7}$$

Moreover, for any $y\notin\mathrm{supp}[C_B(\sigma)]$, we have

$$H_\beta\big(\sigma^y\big)\ge H_\beta(\sigma)+2J-h.\tag{19.2.8}$$

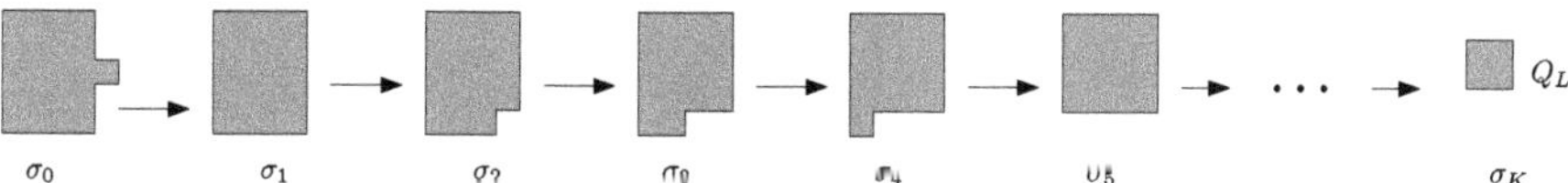

Fig. 19.2 Canonical order to break down a critical droplet

Hence

$$
\begin{aligned}
\mu_\beta(\mathscr{C}) &= \frac{1}{Z_\beta} \sum_{\sigma\in\mathscr{P}} \sum_{\substack{x\in\Lambda_\beta\\ \sigma^x\in\mathscr{C}}} \mathrm{e}^{-\beta H_\beta(\sigma^x)} \\
&\le \frac{1}{Z_\beta} N_1\,\mathrm{e}^{-\beta[2J-h-H_\beta(\boxminus)]} \sum_{x\in\Lambda_\beta} \sum_{\check\sigma\in\check{\mathscr{S}}(x)} \mathrm{e}^{-\beta H_\beta(\check\sigma)} \sum_{\substack{\hat\sigma\in\hat{\mathscr{S}}(x)\\ \hat\sigma\vee\check\sigma\in\mathscr{P}}} \mathrm{e}^{-\beta H_\beta(\hat\sigma)} \\
&\le \big[1+o(1)\big] \frac{1}{Z_\beta} N_1\,|\Lambda_\beta|\,\mathrm{e}^{-\beta\Gamma^\star} \sum_{\check\sigma\in\check{\mathscr{S}}(0)} \mathrm{e}^{-\beta H_\beta(\check\sigma)} \\
&= \big[1+o(1)\big] N_1\,|\Lambda_\beta|\,\mathrm{e}^{-\beta\Gamma^\star}\,\mu_\beta\big(\check{\mathscr{S}}(0)\big),
\end{aligned}
\tag{19.2.9}
$$

where the first inequality uses (19.2.7)–(19.2.8), with $N_1 = 2\times 2\ell_c = 4\ell_c$ counting the number of critical droplets that can arise from a protocritical droplet via a spin flip, and the second inequality uses that

$$
\hat\sigma\in\hat{\mathscr{S}}(0),\ \hat\sigma\vee\check\sigma\in\mathscr{P} \Longrightarrow H_\beta(\hat\sigma) \ge H_\beta\big(R_{\ell_c-1,\ell_c}(0)\big) = \Gamma^\star - (2J-h) + H_\beta(\boxminus)
\tag{19.2.10}
$$

with equality in the right-hand side if and only if $\mathrm{supp}[\hat\sigma] = R_{\ell_c-1,\ell_c}(0)$. Combining (19.2.4) and (19.2.9) with the inclusion $\check{\mathscr{S}}(0)\subset\mathscr{S}$, we get the upper bound in (19.2.5). □

Lower bound

Proof We exploit Theorem 7.43 by making a judicious choice for the flow f. In fact, for Glauber dynamics this choice will be simple: with each configuration $\sigma\in\mathscr{S}_L$ we associate a configuration in $\mathscr{C}\subset\mathscr{S}^c$ containing a unique critical droplet and a flow that, from each such configuration, follows a unique *deterministic path* along which this droplet is broken down in the *canonical order* (see Fig. 19.2) until the set $\mathscr{S}_L$ is reached, i.e., a square or quasi-square droplet with label L is left over (recall (19.1.1)–(19.1.2)).

The proof comes in 5 steps.

1. Let $w(\beta)$ be such that

$$
\lim_{\beta\to\infty} w(\beta) = \infty, \qquad \lim_{\beta\to\infty} \frac{1}{\beta}\ln w(\beta) = 0, \qquad \lim_{\beta\to\infty} |\Lambda_\beta|/w(\beta) = \infty,
\tag{19.2.11}
$$

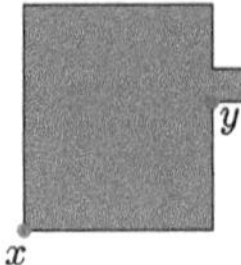

Fig. 19.3 The critical droplet $P_{(y)}(x)$

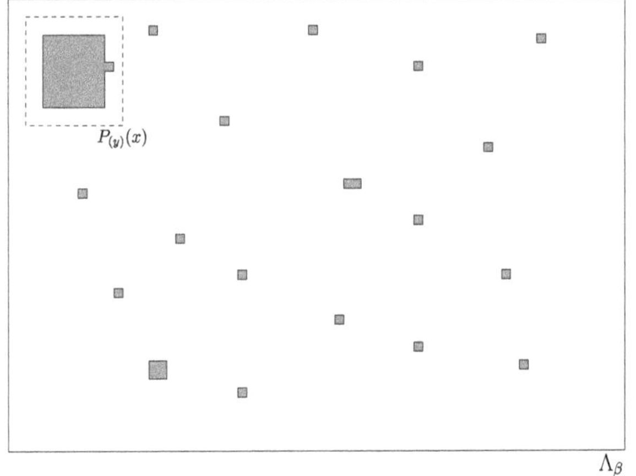

Fig. 19.4 Going from $\mathscr{S}_L$ to $\mathscr{C}_L$ by adding a critical droplet $P_{(y)}(x)$ somewhere in Λ_β

and define

$$\mathscr{W} = \{\sigma \in \mathscr{S} \colon |\mathrm{supp}[\sigma]| \le |\Lambda_\beta|/w(\beta)\}. \tag{19.2.12}$$

Let $\mathscr{C}_L \subset \mathscr{C} \subset \mathscr{S}^c$ be the set of configurations obtained by picking any $\sigma \in \mathscr{S}_L \cap \mathscr{W}$ and adding somewhere in Λ_β a critical droplet at distance ≥ 2 from supp$[\sigma]$. Note that the density restriction imposed on $\mathscr{W}$ guarantees that adding such a droplet is possible almost everywhere in Λ_β for β large enough. Denoting by $P_{(y)}(x)$ the critical droplet obtained by adding a protuberance at y along the longest side of the rectangle $R_{\ell_c-1,\ell_c}(x)$, we may write

$$\mathscr{C}_L = \{\sigma \cup P_{(y)}(x) \colon \sigma \in \mathscr{S} \cap \mathscr{W},\ x, y \in \Lambda_\beta,\ (x,y) \perp \sigma\}, \tag{19.2.13}$$

where $(x,y)\perp\sigma$ stands for the restriction that the critical droplet $P_{(y)}(x)$ is not interacting with supp$[\sigma]$, which implies that $H_\beta(\sigma \cup P_{(y)}(x)) = H_\beta(\sigma) + \Gamma^\star$ (see Figs. 19.3 and 19.4).

2. For each $\sigma \in \mathscr{C}_L$, we let $\gamma_\sigma = (\gamma_\sigma(0), \gamma_\sigma(1), \ldots, \gamma_\sigma(K))$ be the canonical path from $\sigma = \gamma_\sigma(0)$ to $\mathscr{S}_L$ along which the critical droplet is broken down ($\gamma_\sigma(k) = \sigma_k$ in Fig. 19.2), where $K = v(2\ell_c - 3) - v(L)$ with

$$v(L) = |Q_L(0)| \tag{19.2.14}$$

(recall (19.1.1)). We will choose our flow such that

$$f(\sigma',\sigma'') = \begin{cases} \nu_0(\sigma), & \text{if } \sigma'=\sigma,\ \sigma''=\gamma_\sigma(1) \text{ for some } \sigma\in\mathscr{C}_L, \\ \sum_{\tilde\sigma\in\mathscr{C}_L} f(\gamma_{\tilde\sigma}(k-1),\gamma_\sigma(k)), & \text{if } \sigma'=\gamma_\sigma(k),\ \sigma''=\gamma_\sigma(k+1) \\ & \text{for some } k\ge 1,\ \sigma\in\mathscr{C}_L, \\ 0, & \text{otherwise.} \end{cases} \tag{19.2.15}$$

Here, ν_0 is some initial distribution on $\mathscr{C}_L$ that will turn out to be arbitrary as long as its support is all of $\mathscr{C}_L$.

3. We see from (19.2.15) that the flow increases whenever paths merge. In our case this happens only after the first step, when the protuberance at y is removed. Therefore we get the explicit form

$$f(\sigma',\sigma'') = \begin{cases} \nu_0(\sigma), & \text{if } \sigma'=\sigma,\ \sigma''=\gamma_\sigma(1) \text{ for some } \sigma\in\mathscr{C}_L, \\ C\nu_0(\sigma), & \text{if } \sigma'=\gamma_\sigma(k),\ \sigma''=\gamma_\sigma(k+1) \text{ for some } k\ge 1,\ \sigma\in\mathscr{C}_L, \\ 0, & \text{otherwise,} \end{cases} \tag{19.2.16}$$

where $C=2\ell_c$ is the number of possible positions of the protuberance on the protocritical droplet (see Fig. 19.2). Using Theorem 7.43, we therefore have

$$\begin{aligned} &\mathrm{CAP}(\mathscr{S}_L,\mathscr{S}^c) \\ &\quad= \mathrm{CAP}(\mathscr{S}^c,\mathscr{S}_L) \ge \mathrm{CAP}(\mathscr{C}_L,\mathscr{S}_L) \\ &\quad\ge \sum_{\sigma\in\mathscr{C}_L}\nu_0(\sigma)\left[\sum_{k=0}^{K-1}\frac{f(\gamma_\sigma(k),\gamma_\sigma(k+1))}{\mu_\beta(\gamma_\sigma(k))c_\beta(\gamma_\sigma(k),\gamma_\sigma(k+1))}\right]^{-1} \\ &\quad= \sum_{\sigma\in\mathscr{C}_L}\left[\frac{1}{\mu_\beta(\sigma)c_\beta(\gamma_\sigma(0),\gamma_\sigma(1))}+\sum_{k=1}^{K-1}\frac{C}{\mu_\beta(\gamma_\sigma(k))c_\beta(\gamma_\sigma(k),\gamma_\sigma(k+1))}\right]^{-1}. \end{aligned} \tag{19.2.17}$$

Thus, all we have to do is to control the sum between square brackets.

4. Because $c_\beta(\gamma_\sigma(0),\gamma_\sigma(1))=1$ (removing the protuberance lowers the energy), the term with $k=0$ equals $1/\mu_\beta(\sigma)$. To show that the terms with $k\ge 1$ are of higher order, we argue as follows. Abbreviate $\Xi=h(\ell_c-2)$. For every $k\ge 1$ and $\sigma(0)\in\mathscr{C}_L$, we have (see Fig. 19.5)

$$\begin{aligned} \mu_\beta\big(\gamma_\sigma(k)\big)c_\beta\big(\gamma_\sigma(k),\gamma_\sigma(k+1)\big) &= \frac{1}{Z_\beta}\,\mathrm{e}^{-\beta[H_\beta(\gamma_\sigma(k))\vee H_\beta(\gamma_\sigma(k+1))]} \\ &\ge \mu_\beta(\sigma_0)\,\mathrm{e}^{\beta[2J-h-\Xi]} = \mu_\beta(\sigma)\mathrm{e}^{\delta\beta}, \end{aligned} \tag{19.2.18}$$

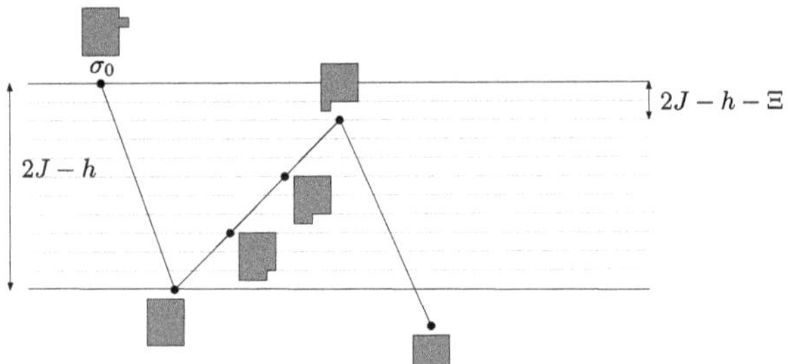

Fig. 19.5 Visualization of (19.2.18)

where $\delta = 2J - h - \Xi = 2J - h(\ell_c - 1) > 0$. Therefore

$$\sum_{k=1}^{K-1} \frac{C}{\mu_\beta(\gamma_\sigma(k))c_\beta(\gamma_\sigma(k), \gamma_\sigma(k+1))} \leq \frac{1}{\mu_\beta(\sigma)} C K \mathrm{e}^{-\delta\beta}, \tag{19.2.19}$$

and so from (19.2.17) we get

$$\mathrm{CAP}\big(\mathscr{S}_L, \mathscr{S}^c\big) \geq \sum_{\sigma\in\mathscr{C}_L} \frac{\mu_\beta(\sigma)}{1+CK\mathrm{e}^{-\beta\delta}} = \frac{\mu_\beta(\mathscr{C}_L)}{1+CK\mathrm{e}^{-\beta\delta}} = \big[1+o(1)\big]\mu_\beta(\mathscr{C}_L). \tag{19.2.20}$$

5. Finally, we estimate, with the help of (19.2.13),

$$\begin{aligned}
\mu_\beta(\mathscr{C}_L) &= \frac{1}{Z_\beta}\sum_{\sigma\in\mathscr{C}_L} \mathrm{e}^{-\beta H_\beta(\sigma)} = \frac{1}{Z_\beta}\sum_{\sigma\in\mathscr{S}_L\cap\mathscr{W}} \sum_{\substack{x,y\in\Lambda_\beta\\(x,y)\perp\sigma}} \mathrm{e}^{-\beta H_\beta(\sigma\cup P_{(y)}(x))} \\
&= \mathrm{e}^{-\beta\Gamma^\star}\frac{1}{Z_\beta}\sum_{\sigma\in\mathscr{S}_L\cap\mathscr{W}} \mathrm{e}^{-\beta H_\beta(\sigma)} \sum_{\substack{x,y\in\Lambda_\beta\\(x,y)\perp\sigma}} 1 \\
&\geq \mathrm{e}^{-\beta\Gamma^\star}\mu_\beta(\mathscr{S}_L\cap\mathscr{W})\, N_1\, |\Lambda_\beta| \big[1-(\ell_c+1)^2/w(\beta)\big].
\end{aligned} \tag{19.2.21}$$

The last inequality uses that $|\Lambda_\beta|(\ell_c+1)^2/w(\beta)$ is the maximal number of sites in Λ_β where it is not possible to insert a non-interacting critical droplet (recall (19.2.12) and note that a critical droplet fits inside an $\ell_c \times \ell_c$ square). Finally, according to Lemma 19.9 in Sect. 19.5, we have

$$\mu_\beta(\mathscr{S}_L\cap\mathscr{W}) = \mu_\beta(\mathscr{S}_L)\big[1+o(1)\big], \tag{19.2.22}$$

while conditions (19.1.2)–(19.1.3) imply that $\mu_\beta(\mathscr{S}_L) = \mu_\beta(\mathscr{S})[1+o(1)]$. Combining the latter with (19.2.20)–(19.2.21), we obtain the desired lower bound. □

19.3 Average time to go beyond the critical droplet

To prove Theorem 19.4(b) we use the same technique as in Sect. 19.2. Therefore we only give a sketch of the proof.

To estimate the average crossover time from $\mathscr{S}_L \subset \mathscr{S}$ to $\mathscr{S}^c\backslash\mathscr{C}$, we again use Corollary 7.11, this time with $\mathscr{A} = \mathscr{S}_L$ and $\mathscr{B} = \mathscr{S}^c\backslash\mathscr{C}$:

$$\sum_{\sigma\in\mathscr{S}_L} \nu_{\mathscr{S}_L,\mathscr{S}^c\backslash\mathscr{C}}(\sigma)\,\mathbb{E}_\sigma(\tau_{\mathscr{S}^c\backslash\mathscr{C}}) = \frac{1}{\mathrm{CAP}(\mathscr{S}_L,\mathscr{S}^c\backslash\mathscr{C})} \sum_{\sigma\in\mathscr{S}\cup\mathscr{C}} \mu_\beta(\sigma)\,h_{\mathscr{S}_L,\mathscr{S}^c\backslash\mathscr{C}}(\sigma). \tag{19.3.1}$$

The left-hand side is the quantity of interest in (19.1.11). In Sects. 19.3.1–19.3.2 we estimate both $\sum_{\sigma\in\mathscr{S}\cup\mathscr{C}} \mu_\beta(\sigma) h_{\mathscr{S}_L,\mathscr{S}^c\backslash\mathscr{C}}(\sigma)$ and $\mathrm{CAP}(\mathscr{S}_L,\mathscr{S}^c\backslash\mathscr{C})$. The estimates will show that

$$\text{r.h.s. } (19.3.1) = \frac{1}{N_2|\Lambda_\beta|}\,e^{\beta\Gamma^\star}\left[1+o(1)\right], \quad \beta\to\infty. \tag{19.3.2}$$

19.3.1 Estimate of the equilibrium potential

Lemma 19.7 $\sum_{\sigma\in\mathscr{S}\cup\mathscr{C}} \mu_\beta(\sigma) h_{\mathscr{S}_L,\mathscr{S}^c\backslash\mathscr{C}}(\sigma) = \mu_\beta(\mathscr{S})[1+o(1)]$ *as* $\beta\to\infty$.

Proof Write, using (7.1.16),

$$\sum_{\sigma\in\mathscr{S}\cup\mathscr{C}} \mu_\beta(\sigma) h_{\mathscr{S}_L,\mathscr{S}^c\backslash\mathscr{C}}(\sigma) = \mu_\beta(\mathscr{S}_L) + \sum_{\sigma\in(\mathscr{S}\backslash\mathscr{S}_L)\cup\mathscr{C}} \mu_\beta(\sigma)\mathbb{P}_\sigma(\tau_{\mathscr{S}_L} < \tau_{\mathscr{S}^c\backslash\mathscr{C}}). \tag{19.3.3}$$

The last sum is bounded above by $\mu_\beta(\mathscr{S}\backslash\mathscr{S}_L)+\mu_\beta(\mathscr{C})$. As before, $\mu_\beta(\mathscr{S}\backslash\mathscr{S}_L) = o(\mu_\beta(\mathscr{S}))$ as $\beta\to\infty$. But (19.1.7) and (19.2.9) imply that $\mu_\beta(\mathscr{C}) = o(\mu_\beta(\mathscr{S}))$ as $\beta\to\infty$. □

19.3.2 Estimate of the capacity

Lemma 19.8 $\mathrm{CAP}(\mathscr{S},\mathscr{S}^c\backslash\mathscr{C}) = N_2\,|\Lambda_\beta| e^{-\beta\Gamma^\star}\mu_\beta(\mathscr{S})[1+o(1)]$ *as* $\beta\to\infty$ *with* $N_2 = \frac{4}{3}(2\ell_c - 1)$.

Proof The proof is similar as that of Lemma 19.6, except that it takes care of the transition probabilities away from the critical droplet (see Fig. 19.6, where σ' is the configuration that is reached through these transitions). The proof again proceeds via upper and lower bounds, which are written out below. □

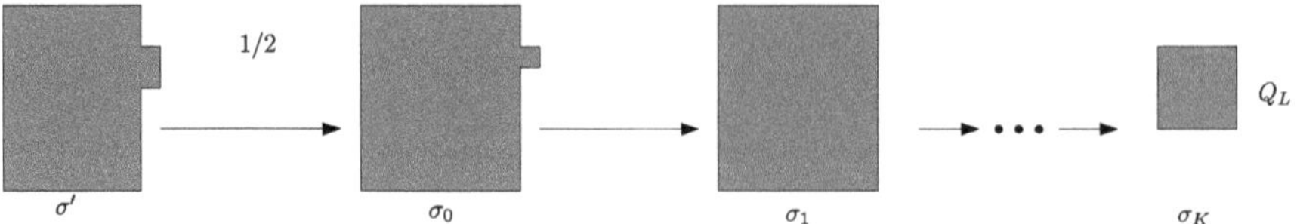

Fig. 19.6 Canonical order to break down a proto-critical droplet plus a double protuberance. In the first step, the double protuberance has probability $\frac{1}{2}$ to be broken down in either of the two possible ways. The subsequent steps are deterministic as in Fig. 19.2

Upper bound

Proof Recalling (7.1.35) and Lemma 7.12, and noting that Glauber dynamics does not allow transitions within $\mathscr{C}$, we have, for all $h\colon \mathscr{C} \to [0,1]$,

$$\begin{aligned} \text{CAP}\big(\mathscr{S}_L, \mathscr{S}^c\backslash\mathscr{C}\big) &\le \text{CAP}\big(\mathscr{S}, \mathscr{S}^c\backslash\mathscr{C}\big) \\ &\le \sum_{\sigma\in\mathscr{C}} \mu_\beta(\sigma)\big[\hat{c}_\sigma\big(h(\sigma)-1\big)^2 + \check{c}_\sigma\big(h(\sigma)-0\big)^2\big], \end{aligned} \tag{19.3.4}$$

where $\hat{c}_\sigma = \sum_{\eta\in\mathscr{S}} c_\beta(\sigma,\eta)$ and $\check{c}_\sigma = \sum_{\eta\in\mathscr{S}^c\backslash\mathscr{C}} c_\beta(\sigma,\eta)$. The quadratic form in the right-hand side of (19.3.4) achieves its minimum for $h(\sigma) = \hat{c}_\sigma/(\hat{c}_\sigma + \check{c}_\sigma)$, so

$$\text{CAP}\big(\mathscr{S}_L, \mathscr{S}^c\backslash\mathscr{C}\big) \le \sum_{\sigma\in\mathscr{C}} C_\sigma\, \mu_\beta(\sigma) \tag{19.3.5}$$

with $C_\sigma = \hat{c}_\sigma\check{c}_\sigma/(\hat{c}_\sigma + \check{c}_\sigma)$. We have

$$\begin{aligned} \sum_{\sigma\in\mathscr{C}} C_\sigma\, \mu_\beta(\sigma) &= \frac{1}{Z_\beta} \sum_{\sigma\in\mathscr{P}} \sum_{\substack{x\in\Lambda_\beta \\ \sigma^x\in\mathscr{C}}} C_{\sigma^x}\, \mathrm{e}^{-\beta H_\beta(\sigma^x)} \\ &= \mathrm{e}^{-\beta(2J-h)} \frac{1}{Z_\beta} \sum_{\sigma\in\mathscr{P}} \mathrm{e}^{-\beta H_\beta(\sigma)}\, 2\big(\tfrac{1}{2}4 + \tfrac{2}{3}(2\ell_c - 4)\big) \\ &= \mathrm{e}^{-\beta(2J-h)}\, \mu_\beta(\mathscr{P})\, N_2 = \frac{1}{N_1}\, \mu_\beta(\mathscr{C})\, N_2, \end{aligned} \tag{19.3.6}$$

where in the second line we use that $C_\sigma = \frac{1}{2}$ if σ has a protuberance in a corner (2×4 choices) and $C_\sigma = \frac{2}{3}$ otherwise ($2 \times (2\ell_c - 4)$ choices). □

Lower bound

Proof In analogy with (19.2.13), denoting by $P^2_{(y)}(x)$ the droplet obtained by adding a double protuberance at y along the longest side of the rectangle $R_{\ell_c-1,\ell_c}(x)$, we define the set $\mathscr{D}_L \subset \mathscr{S}^c\backslash\mathscr{C}$ by

$$\mathscr{D}_L = \big\{\sigma \cup P^2_{(y)}(x)\colon\ \sigma \in \mathscr{S}_L \cap \mathscr{W},\ x, y \in \Lambda_\beta,\ (x,y)\bot\sigma\big\}. \tag{19.3.7}$$

As in (19.2.15), we may choose any starting measure ν_0 on $\mathscr{D}_L$. We choose the flow as follows. For the first step we choose

$$f(\sigma',\sigma) = \tfrac{1}{2}\,\nu_0(\sigma), \qquad \sigma' \in \mathscr{D}_L,\ \sigma \in \mathscr{C}_L, \tag{19.3.8}$$

which reduces the double protuberance to a single protuberance (compare (19.2.13) and (19.3.7)). For all subsequent steps we follow the deterministic paths γ_σ used in Sect. 19.2.2, which start from $\gamma_\sigma(0) = \sigma$. Note, however, that we get different values for the flows $f(\gamma_\sigma(0), \gamma_\sigma(1))$ depending on whether the protuberance sits in a corner or not. In the former case, it has only one possible antecedent, and so

$$f\big(\gamma_\sigma(0), \gamma_\sigma(1)\big) = \tfrac{1}{2}\,\nu_0(\sigma), \tag{19.3.9}$$

while in the latter case it has two antecedents, and so

$$f\big(\gamma_\sigma(0), \gamma_\sigma(1)\big) = \nu_0(\sigma). \tag{19.3.10}$$

This time the terms $k = 0$ and $k = 1$ are of the same order while, as in (19.2.19), all the subsequent terms give a contribution that is a factor $O(\mathrm{e}^{-\delta\beta})$ smaller. Indeed, in analogy with (19.2.17) we obtain, writing $\sigma \sim \sigma'$ when $c_\beta(\sigma',\sigma) > 0$,

$$\begin{aligned}
\mathrm{CAP}\big(\mathscr{S}_L, \mathscr{S}^c\backslash\mathscr{C}\big) &= \mathrm{CAP}\big(\mathscr{S}^c\backslash\mathscr{C}, \mathscr{S}_L\big) \geq \mathrm{CAP}(\mathscr{D}_L, \mathscr{S}_L) \\
&\geq \sum_{\sigma'\in\mathscr{D}_L} \frac{1}{2} \sum_{\substack{\sigma\in\mathscr{C}_L \\ \sigma\sim\sigma'}} \Bigg[\frac{f(\sigma',\sigma)}{\mu_\beta(\sigma)} + \frac{f(\sigma,\gamma_\sigma(1))}{\mu_\beta(\sigma)} \\
&\qquad + \sum_{k=1}^{K-1} \frac{f(\gamma_\sigma(k),\gamma_\sigma(k+1))}{\mu_\beta(\gamma_\sigma(k))c_\beta(\gamma_\sigma(k),\gamma_\sigma(k+1))}\Bigg]^{-1} \\
&\geq \sum_{\sigma'\in\mathscr{D}_L} \frac{1}{2} \sum_{\substack{\sigma\in\mathscr{C}_L \\ \sigma\sim\sigma'}} \mu_\beta(\sigma)\big[f(\sigma',\sigma) + f(\sigma,\gamma_\sigma(1)) + CK\mathrm{e}^{-\beta\delta}\big]^{-1} \\
&= \big[1+o(1)\big]\,\mu_\beta(\mathscr{C}_L)\left(\frac{2\ell_c - 4}{2\ell_c}\,\frac{1}{1+\frac{1}{2}} + \frac{1}{2}\,\frac{4}{2\ell_c}\,\frac{1}{\frac{1}{2}+\frac{1}{2}}\right) \\
&= \big[1+o(1)\big]\,\mu_\beta(\mathscr{C}_L)\,\frac{N_2}{N_1}.
\end{aligned} \tag{19.3.11}$$

Using (19.2.21) and the remarks following it, we get the desired lower bound. □

Figure 19.6 depicts the sequence of steps taken to break a protocritical droplet down.

19.4 Average time to grow a droplet twice the critical size

The proof of Theorem 19.4(c) follows along the same lines as that of Theorem 19.4(a–b) in Sects. 19.2–19.3. The starting point is the analogue of (19.3.1) with $\mathscr{S}^c\backslash\mathscr{C}$ replaced by $\mathscr{D}_M$ and $\mathscr{S}\cup\mathscr{C}$ by $\mathscr{D}_M^c$.

19.4.1 Estimate of the equilibrium potential

Proof Write

$$\sum_{\sigma\in\mathscr{D}_M^c}\mu_\beta(\sigma)h_{\mathscr{S}_L,\mathscr{D}_M}(\sigma)=\sum_{\sigma\in\mathscr{S}_L}\mu_\beta(\sigma)h_{\mathscr{S}_L,\mathscr{D}_M}(\sigma)+\sum_{\sigma\in\mathscr{D}_M^c\backslash\mathscr{S}_L}\mu_\beta(\sigma)h_{\mathscr{S}_L,\mathscr{D}_M}(\sigma)$$
$$=\mu_\beta(\mathscr{S}_L)+\sum_{\sigma\in\mathscr{D}_M^c\backslash\mathscr{S}_L}\mu_\beta(\sigma)\mathbb{P}_\sigma(\tau_{\mathscr{S}_L}<\tau_{\mathscr{D}_M}). \tag{19.4.1}$$

The last sum is bounded above by $\mu_\beta(\mathscr{S}\backslash\mathscr{S}_L)+\mu_\beta(\mathscr{D}_M^c\backslash\mathscr{S})$. But $\mu_\beta(\mathscr{S}\backslash\mathscr{S}_L)=o(\mu_\beta(\mathscr{S}))$ as $\beta\to\infty$ by our choice of L in (19.1.3), while $\mu_\beta(\mathscr{D}_M^c\backslash\mathscr{S})=o(\mu_\beta(\mathscr{S}))$ as $\beta\to\infty$ because of the restriction $\ell_c\le M\le 2\ell_c-1$. Indeed, under that restriction the energy of a square droplet of size M is strictly larger than the energy of a critical droplet. □

19.4.2 Estimate of the capacity

Proof The main point is to prove that $\mathrm{CAP}(\mathscr{S}_L,\mathscr{D}_M)=[1+o(1)]\mathrm{CAP}(\mathscr{S}_L,\mathscr{S}^c\backslash\mathscr{C})$. But $\mathrm{CAP}(\mathscr{S}_L,\mathscr{D}_M)\le\mathrm{CAP}(\mathscr{S}_L,\mathscr{S}^c\backslash\mathscr{C})$. The latter was estimated in Sect. 19.3, and so we need only prove a lower bound on $\mathrm{CAP}(\mathscr{S}_L,\mathscr{D}_M)$. This is done by using a flow that breaks down an $M\times M$ droplet to a square or quasi-square droplet Q_L in the canonical way, which takes $M^2-v(L)$ steps (recall Fig. 19.2 and (19.2.14)). The leading terms are still the protocritical droplet with a single and a double protuberance. To each $M\times M$ droplet is associated a unique critical droplet, so that the prefactor in the lower bound is the same as in the proof of Theorem 19.4(b).

Note that we can even allow M to grow with β as $M=\mathrm{e}^{o(\beta)}$. Indeed, (19.2.11)–(19.2.12) imply that there is room enough to add a droplet of size $\mathrm{e}^{o(\beta)}$ almost everywhere in Λ_β, and the factor $M^2\mathrm{e}^{-\delta\beta}$ replacing $K\mathrm{e}^{-\delta\beta}$ in (19.2.20) still is $o(1)$. □

19.5 Sparseness of subcritical droplets

Recall Definition 19.2(a) and (19.2.11)–(19.2.12). In this section we prove the claim made in (19.2.22).

Lemma 19.9 $\lim_{\beta\to\infty}\frac{1}{\beta}\ln\frac{\mu_\beta(\mathscr{S}\backslash\mathscr{W})}{\mu_\beta(\mathscr{S})}=-\infty$.

Proof We will prove that $\lim_{\beta\to\infty}\frac{1}{\beta}\ln\mu_\beta(\mathscr{S}\backslash\mathscr{W})/\mu_\beta(\boxminus)=-\infty$. Since $\boxminus\in\mathscr{S}$, this will prove the claim.

Let $w(\beta)$ be the function satisfying (19.2.11). We begin by noting that

$$\mu_\beta(\mathscr{S}\backslash\mathscr{W}) \le \mu_\beta(\mathscr{I}) \quad \text{with } \mathscr{I} = \big\{\sigma \in \mathscr{S} \colon \big|\mathrm{supp}\big[C_B(\sigma)\big]\big| > |\Lambda_\beta|/w(\beta)\big\}, \tag{19.5.1}$$

because the bootstrap percolation map increases the number of (+1)-spins. Let $\mathscr{D}(k)$ denote the set of configurations whose support consists on k non-interacting subcritical rectangles. Put $C_1 = (\ell_c+2)(\ell_c+1)$. Since the union of a subcritical rectangle and its exterior boundary has at most C_1 sites, it follows that in $\mathscr{I}$ there are at least $|\Lambda_\beta|/C_1 w(\beta)$ non-interacting rectangles. Thus, we have

$$\mu_\beta(\mathscr{I}) \le \sum_{k=\frac{|\Lambda_\beta|}{C_1 w(\beta)}}^{K_{\max}} F(k) \quad \text{with } F(k) = \frac{1}{Z_\beta} \sum_{\substack{\sigma\in S_\beta\colon\\ C(\sigma)\in\mathscr{D}(k)}} \mathrm{e}^{-\beta H_\beta(\sigma)}, \tag{19.5.2}$$

where $K_{\max} \le |\Lambda_\beta|$.

Next, note that

$$F(k) \le \big(2^{C_1}\big)^k \frac{1}{Z_\beta} \sum_{\sigma\in\mathscr{D}(k)} \mathrm{e}^{-\beta H_\beta(\sigma)}. \tag{19.5.3}$$

Since the bootstrap percolation map is downhill, the energy of a subcritical rectangle is bounded below by $C_2 = 2J-h$ (recall Fig. 19.5), and the number of ways to place k rectangles in Λ_β is at most $\binom{|\Lambda_\beta|}{k}$, it follows that for k large enough

$$\begin{aligned} F(k) &\le 2^{C_1 k} \binom{|\Lambda_\beta|}{k} \mu_\beta(\boxminus)\,\mathrm{e}^{-C_2\beta k} \\ &\le 2^{C_1 k} \big(C_1 e w(\beta)\big)^k \mu_\beta(\boxminus)\,\mathrm{e}^{-C_2\beta k} \le \mu_\beta(\boxminus)\,\exp\big[-\tfrac{1}{2}C_2\,\beta k\big], \end{aligned} \tag{19.5.4}$$

where the second inequality uses that $k! \ge k^k \mathrm{e}^{-k}$, $k\in\mathbb{N}$, and the third inequality uses that $w(\beta) = \mathrm{e}^{o(\beta)}$. We thus have

$$\sum_{k=\frac{|\Lambda_\beta|}{C_1 w(\beta)}}^{K_{\max}} F(k) \le 2\mu_\beta(\boxminus)\, w(\beta)\, \frac{|\Lambda_\beta|}{w(\beta)} \exp\left[-\tfrac{1}{2}\frac{C_2}{C_1}\,\beta\, \frac{|\Lambda_\beta|}{w(\beta)}\right], \tag{19.5.5}$$

which is the desired estimate because $|\Lambda_\beta|/w(\beta)$ tends to infinity as $\beta\to\infty$. □

19.6 Typicality of starting configurations

In this section we prove the claim made in (19.1.4).

Proof Split

$$\mathscr{S} = \mathscr{S}_L \cup (\mathscr{S}\setminus\mathscr{S}_L) = \mathscr{S}_L \cup \mathscr{U}_{>L}, \tag{19.6.1}$$

where $\mathscr{U}_{>L} \subset \mathscr{S}$ are those configurations σ for which $C_B(\sigma)$ has at least one rectangle that is larger than $Q_L(0)$. We have

$$C_B(\sigma) = \bigcup_{x \in X(\sigma)} R_{\ell_1(x),\ell_2(x)}(x), \tag{19.6.2}$$

where $X(\sigma)$ is the set of lower-left corners of the rectangles in $C_B(\sigma)$, which in turn can be split as

$$X(\sigma) = X^{>L}(\sigma) \cup X^{\le L}(\sigma), \tag{19.6.3}$$

where $X^{>L}(\sigma)$ labels the rectangles that are larger than $Q_L(0)$ and $X^{\le L}(\sigma)$ labels the rest.

Let $\sigma|_A$ denote the restriction of σ to the set $A \subset \mathbb{Z}^2$. Then, for any $x \in X(\sigma)$, we have

$$H(\sigma) = H(\sigma|_{R_{\ell_1(x),\ell_2(x)}(x)}) + H(\sigma|_{R^c_{\ell_2(x),\ell_2(x)}(x)}), \tag{19.6.4}$$

because the rectangles in $C_B(\sigma)$ are non-interacting. Since for $\sigma \in \mathscr{U}_{>L}$ there is at least one rectangle with lower-left corner in $X^{>L}(\sigma)$, we have

$$\begin{aligned} &\mu_\beta(\mathscr{U}_{>L}) \\ &\quad \le \sum_{x \in \Lambda_\beta} \sum_{\sigma \in \mathscr{S}} \mathbb{1}_{\{x \in X^{>L}(\sigma)\}} \, \mu_\beta(\sigma) \\ &\quad = \sum_{x \in \Lambda_\beta} \sum_{\sigma \in \mathscr{S}} \mathbb{1}_{\{x \in X^{>L}(\sigma)\}} \, \frac{1}{Z_\beta} \exp\{-\beta[H(\sigma|_{R_{\ell_1(x),\ell_2(x)}(x)}) + H(\sigma|_{R^c_{\ell_1(x),\ell_2(x)}(x)})]\} \\ &\quad \le e^{-\beta\Gamma_{L+1}} \sum_{x \in \Lambda_\beta} \sum_{\sigma \in \mathscr{S}} \mathbb{1}_{\{x \in X^{>L}(\sigma)\}} \, \frac{1}{Z_\beta} e^{-\beta H(\sigma|_{R^c_{\ell_1(x),\ell_2(x)}(x)})}, \end{aligned} \tag{19.6.5}$$

where Γ_{L+1} is the energy of $Q_{L+1}(0)$. In the last step we use the fact that the bootstrap map is downhill and that the energy of $Q_L(0)$ is increasing with L. Since the energy of a subcritical rectangle is non-negative, we get

$$\mu_\beta(\mathscr{U}_{>L}) \le N_{L+1}\, e^{-\beta\Gamma_{L+1}}\, |\Lambda_\beta|\, \mu_\beta(\mathscr{S}) \tag{19.6.6}$$

with N_{L+1} counting the number of configurations with support in $Q_{L+1}(0)$.

On the other hand, by considering only those configurations in $\mathscr{U}_{>L}$ that have a $Q_{L+1}(0)$ droplet, we get

$$\mu_\beta(\mathscr{U}_{>L}) \ge N_{L+1}\, e^{-\beta\Gamma_{L+1}}\, |\Lambda_\beta|\, \mu_\beta^{[Q_{L+1}(0)]^c}(\mathscr{S}), \tag{19.6.7}$$

where the last factor is the Gibbs weight of the configurations in $\mathscr{S}$ with support outside $[Q_{L+1}(0)]^c$. It easy to show that $\mu_\beta^{[Q_{L+1}(0)]^c}(\mathscr{S}) = \mu_\beta(\mathscr{S})[1+o(1)]$ as $\beta \to \infty$ and so

$$\mu_\beta(\mathscr{U}_{>L}) \ge N_{L+1}\, e^{-\beta\Gamma_{L+1}}\, |\Lambda_\beta|\, \mu_\beta(\mathscr{S})\,[1+o(1)], \quad \beta \to \infty. \tag{19.6.8}$$

Combining (19.6.6) and (19.6.8), we conclude that $\lim_{\beta\to\infty} \mu_\beta(\mathscr{U}_{>L})/\mu_\beta(\mathscr{S}) = 0$ if and only if

$$\lim_{\beta\to\infty} |\Lambda_\beta|\, \mathrm{e}^{-\Gamma_{L+1}} = 0, \tag{19.6.9}$$

which proves the claim. □

19.7 Bibliographical notes

1. The results in this chapter are taken from Bovier, den Hollander and Spitoni [32]. The "bootstrap percolation map" in Definition 19.1 is taken from Kotecký and Olivieri [155].

2. If we draw the starting configuration from some subset of $\mathscr{S}$ that has a strong recurrence property under the dynamics, then the choice of initial distribution on this subset should not matter. This issue remains to be resolved. Gaudilliere, den Hollander, Nardi, Olivieri and Scoppola [118–120] provide a partial answer within the pathwise approach to metastability, i.e., up to exponential order in β.

3. We expect Theorem 19.4(c) to hold for values of M that grow with β as $M = \mathrm{e}^{o(\beta)}$. As we saw in Sect. 19.4, the necessary capacity estimates carry over, but the necessary equilibrium potential estimates do not. Also this issue remains to be resolved.

4. The extension of the main theorem in Sect. 19.1.2 from two to three (and higher) dimensions is straightforward. See also Sect. 17.6.

5. Theorem 19.4 identifies the first time when a critical droplet appears *somewhere* in Λ_β. It is a different issue to compute the first time when the plus-phase appears near the origin. Two regimes have been studied: (1) $|\Lambda| = \infty$, $h \in (0, 2J)$, $\beta \to \infty$; (2) $|\Lambda| = \infty$, $J > 0$, $\beta > 0$ large enough, $h \downarrow 0$. Regime (1) was considered in two dimensions by Dehghanpour and Schonmann [77, 78], and in three and higher dimensions by Cerf and Manzo [54]. Regime (2) was considered in two dimensions by Schonmann [211–214], and Shlosman and Schonmann [215]. The invasion time is identified up to errors that are subexponential in β, respectively, $1/h$. Proofs are hard because the invasion time depends on *where* critical droplets appear for the first time, how they grow and diffuse, how they meet other droplets along the way and possibly merge with them, and how they eventually invade the origin. We will return to this problem in Chap. 23.

6. The analogue of regime (1) in item 5 for the Blume-Capel model (recall Sect. 17.7, item 8), was studied in Manzo and Olivieri [173].

Chapter 20
Kawasaki Dynamics

Tout le monde trouve à redire en autrui ce qu'on trouve à redire en lui. (François de La Rochefoucauld, Réflexions)

The goal of this chapter is to extend the analysis in Chap. 19 to Kawasaki dynamics. We will see that, again, the average time until the appearance of a critical droplet somewhere is inversely proportional to the volume, and is driven by the same quantities $\Gamma^\star$ and K as for small volumes. However, in the proof we encounter several difficult issues, all coming from the fact that Kawasaki dynamics is *conservative*. The first is to understand why $\Gamma^\star$, representing the energetic cost to create a critical droplet in a small box with an *open boundary*, i.e., in a *grand-canonical* setting, reappears even though we choose our box to have a *closed boundary*, i.e., we work in a *canonical* setting. This "mystery" will be resolved by the observation that the formation of a critical droplet reduces the entropy of the system: the precise computation of this entropy loss yields $\Gamma^\star$ via *dynamical equivalence of ensembles*. The second problem is to control the probability of a particle moving from the gas to the protocritical droplet at the last stage of the nucleation, which plays a key role in understanding how K comes up. This non-locality issue will be dealt with via upper and lower estimates. As we will see, the latter in fact causes the scaling to be slightly different than for small volumes.

20.1 Introduction and main results

20.1.1 Kawasaki dynamics in large volumes

We retain the setting of Sect. 18.1.1, and again let Λ_β, S_β and H_β depend on β. The main difference with the small volume situation described in Chap. 18 is that we consider the dynamics on a torus rather than on a box with an open boundary, and do not allow particles to be created or annihilated. Indeed, as Hamiltonian we choose

$$H_\beta(\sigma) = -U \sum_{\{x,y\}\in(\Lambda_\beta)^*} \sigma(x)\sigma(y), \quad \sigma \in S_\beta, \tag{20.1.1}$$

A. Bovier, F. den Hollander, *Metastability*,
Grundlehren der mathematischen Wissenschaften 351,
DOI 10.1007/978-3-319-24777-9_20

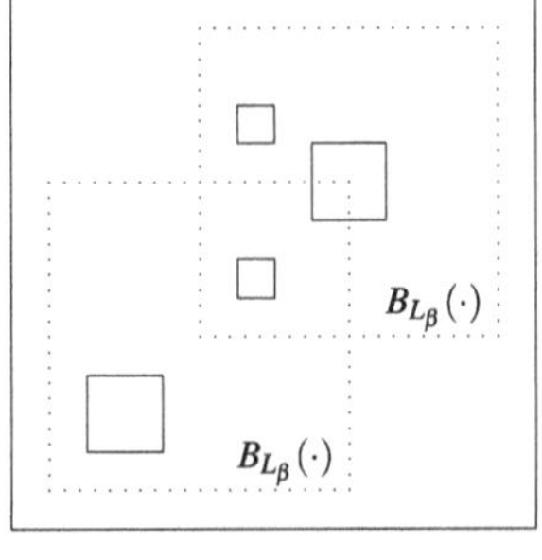

Fig. 20.1 An example of a configuration in $\mathscr{S}$: no box $B_{L_\beta}(\cdot)$ of size L_β contains more than a protocritical number of particles

and we work in the *canonical ensemble*, i.e., the second term in (18.1.2) is removed. The number of particles in Λ_β is taken to be

$$n_\beta = \lceil \rho_\beta |\Lambda_\beta| \rceil, \tag{20.1.2}$$

where ρ_β is the particle density, which is chosen to be

$$\rho_\beta = \mathrm{e}^{-\beta\Delta}, \quad \Delta > 0. \tag{20.1.3}$$

Here, the activity parameter Δ that was removed from the Hamiltonian resurfaces via the density in Λ_β, i.e., we view Λ_β as a gas reservoir surrounding local volumes. Because of *particle conservation*, the state space of our dynamics is the set

$$S_\beta^{(n_\beta)} = \{\sigma \in S_\beta \colon |\mathrm{supp}[\sigma]| = n_\beta\}, \tag{20.1.4}$$

where $\mathrm{supp}[\sigma] = \{x \in \Lambda_\beta \colon \sigma(x) = 1\}$.

Subcritical, protocritical and critical configurations Let L_β be a reference distance, defined as

$$L_\beta^2 = \mathrm{e}^{\beta(\Delta-\delta_\beta)} = \frac{1}{\rho_\beta}\,\mathrm{e}^{-\beta\delta_\beta} \tag{20.1.5}$$

with δ_β chosen such that

$$\lim_{\beta\to\infty} \delta_\beta = 0, \qquad \lim_{\beta\to\infty} \beta\delta_\beta = \infty, \tag{20.1.6}$$

and such that L_β is odd. What this says is that L_β is marginally below the *typical interparticle distance*.

Definition 20.1 Let $B_{L_\beta}(x)$, $x \in \Lambda_\beta$, be the square box with side length L_β centred at x (see Fig. 20.1).

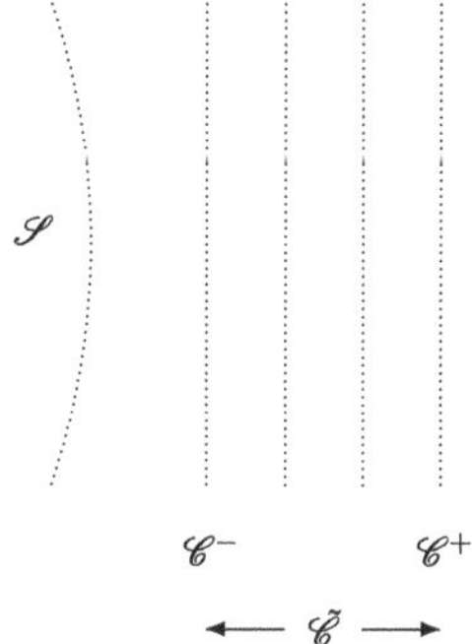

Fig. 20.2 Schematic picture of the sets $\mathscr{S}$, $\mathscr{C}^-$, $\mathscr{C}^+$ defined in Definition 20.1 and the set $\tilde{\mathscr{C}}$ interpolating between $\mathscr{C}^-$ and $\mathscr{C}^+$

(a) $\mathscr{S} = \{\sigma \in S_\beta^{(n_\beta)} \colon |\mathrm{supp}[\sigma] \cap B_{L_\beta}(x)| \le \ell_c(\ell_c - 1) + 1 \ \forall x \in \Lambda_\beta\}$.
(b) $\mathscr{P} = \{\sigma \in \mathscr{S} \colon c_\beta(\sigma, \sigma') > 0$ for some $\sigma' \in \mathscr{S}^c\}$.
(c) $\mathscr{C} = \{\sigma' \in \mathscr{S}^c \colon c_\beta(\sigma, \sigma') > 0$ for some $\sigma \in \mathscr{S}\}$.
(d) $\mathscr{C}^- = \{\sigma \in \mathscr{C} \colon \exists x \in \Lambda_\beta$ such that $B_{L_\beta}(x)$ contains a protocritical droplet whose lower-left corner is at x plus a free particle$\}$.
(e) $\mathscr{C}^+ =$ the set of configurations obtained from $\mathscr{C}^-$ by moving the free particle to a site at distance 2 from the protocritical droplet, i.e., next to its boundary.
(f) $\tilde{\mathscr{C}} =$ the set of configurations "interpolating" between $\mathscr{C}^-$ and $\mathscr{C}^+$, i.e., the free particle is somewhere between the boundary of the protocritical droplet and the boundary of the box of size L_β around it (see Fig. 20.2).

As in Chap. 19, we refer to $\mathscr{S}$, $\mathscr{P}$ and $\mathscr{C}$ as the set of *subcritical*, *protocritical*, respectively, *critical* configurations. Note that, for every $\sigma \in \mathscr{S}$, the number of particles in a box of size L_β does not exceed the number of particles in a protocritical droplet. These particles do not have to form a cluster or to be near to each other, because the Kawasaki dynamics brings them together in a time of order $L_\beta^2 = o(1/\rho_\beta)$.

Remark 20.2 The sets $\mathscr{P}, \mathscr{C}$ will play a similar rôle as, but are not directly comparable with, the sets $\mathscr{P}^\star, \mathscr{C}^\star$ in Chap. 18.

Sets of starting configurations The initial distribution will again be concentrated on sets $\mathscr{S}_L \subset \mathscr{S}$, this time defined by

$$\mathscr{S}_L = \left\{\sigma \in S_\beta^{(n_\beta)} \colon \left|\mathrm{supp}[\sigma] \cap B_{L_\beta}(x)\right| \le L \ \forall x \in \Lambda_\beta\right\}, \tag{20.1.7}$$

for any $L \in \mathbb{N}$ that satisfies $L^* \le L \le \ell_c(\ell_c - 1) + 1$ with

$$L^* = \min\left\{1 \le L \le \ell_c(\ell_c - 1) + 1 \colon \lim_{\beta \to \infty} \frac{\mu_\beta(\mathscr{S}_L)}{\mu_\beta(\mathscr{S})} = 1\right\}, \tag{20.1.8}$$

where μ_β is the canonical Gibbs measure associated with H_β living on $S_\beta^{(n_\beta)}$. In words, $\mathscr{S}_L$ is the subset of those subcritical configurations for which no box of size L_β carries more than L particles, with L chosen such that $\mathscr{S}_L$ is typical within $\mathscr{S}$ under the Gibbs measure μ_β as $\beta \to \infty$.

Note that $\mathscr{S}_{\ell_c(\ell_c-1)+1} = \mathscr{S}$. As for Glauber, the value of L^* depends on how fast Λ_β grows with β. In Sect. 20.4.4 we will show that, for every $1 \leq L \leq \ell_c(\ell_c-1)$,

$$\lim_{\beta\to\infty} \mu_\beta(\mathscr{S}_L)/\mu_\beta(\mathscr{S}) = 1 \quad \text{if and only if} \quad \lim_{\beta\to\infty} |\Lambda_\beta| \mathrm{e}^{-\beta(\Gamma_{L+1}-(\Delta-\delta_\beta))} = 0 \tag{20.1.9}$$

with Γ_{L+1} the energy needed to create a droplet of $L+1$ particles (closest in shape to a square or quasi-square) in $B_{L_\beta}(0)$ under the grand-canonical Hamiltonian on this box. Thus, if $|\Lambda_\beta| = \mathrm{e}^{\theta\beta}$, then $L^* = L^*(\theta) = [\ell_c(\ell_c-1)+1] \wedge \min\{L \in \mathbb{N}\colon \Gamma_{L+1} - \Delta > \theta\}$, which increases stepwise from 1 to $\ell_c(\ell_c-1)+1$ as θ increases from Δ to $\Gamma^\star$, the communication height in Chap. 18.

Initial distribution We choose the initial distribution to be the last-exit biased distribution on $\mathscr{S}$ for the crossover to $\mathscr{S}^c\backslash\tilde{\mathscr{C}}$, respectively, $\mathscr{D}_M$, $M \in \mathbb{N}$, $M \geq \ell_c$, defined by

$$\mathscr{D}_M = \big\{\sigma \in S_\beta\colon \exists x \in \Lambda_\beta \text{ such that } \mathrm{supp}[\sigma] \supset R_{M,M}(x)\big\}, \tag{20.1.10}$$

i.e., the set of configurations containing a supercritical droplet of size M.

20.1.2 Main theorem

Throughout this chapter we assume that we are in the metastabe regime where $\Delta \in (U, 2U)$ (recall Sect. 18.1.2). We further assume that

$$\lim_{\beta\to\infty} |\Lambda_\beta|\, \rho_\beta = \infty, \qquad \lim_{\beta\to\infty} |\Lambda_\beta|\, L_\beta^2\, \mathrm{e}^{-\beta\Gamma^\star} = 0. \tag{20.1.11}$$

This first condition says that the number of particles tends to infinity, and ensures that the formation of a critical droplet somewhere does not globally deplete the surrounding gas. The second condition ensures that the set of configurations with a protocritical droplet and a free particle within distance L_β is atypical compared to $\mathscr{S}$.

Write $N = N(\ell_c)$ to denote the number of protocritical droplets modulo shifts for Kawasaki dynamics in small volumes, which was identified in (18.1.13).

Theorem 20.3 (Mean crossover time) *Subject to* (20.1.8) *and* (20.1.11), *the following hold*:

(a)

$$\lim_{\beta\to\infty} |\Lambda_\beta|\, \frac{4\pi}{\beta\Delta}\, \mathrm{e}^{-\beta\Gamma^\star}\, \mathbb{E}_{\nu_{\mathscr{S}_L,(\mathscr{S}^c\backslash\tilde{\mathscr{C}})\cup\mathscr{C}^+}}(\tau_{(\mathscr{S}^c\backslash\tilde{\mathscr{C}})\cup\mathscr{C}^+}) = \frac{1}{N}. \tag{20.1.12}$$

(b)

$$\lim_{\beta\to\infty} |\Lambda_\beta| \frac{4\pi}{\beta\Delta} \mathrm{e}^{-\beta\Gamma^\star} \mathbb{E}_{\nu_{\mathscr{S}_L,\mathscr{D}_M}}(\tau_{\mathscr{D}_M}) = \frac{1}{N}, \quad \forall \ell_c \le M \le 2\ell_c - 1. \quad (20.1.13)$$

20.1.3 Discussion

1. Theorem 20.3(a) says that the average time to create a critical droplet is $[1+o(1)](\beta\Delta/4\pi)\mathrm{e}^{\beta\Gamma^\star}/N|\Lambda_\beta|$. The factor $\beta\Delta/4\pi$ comes from the simple random walk that is performed by the free particle "from the gas to the protocritical droplet" (i.e., as the dynamics goes from $\mathscr{C}^-$ to $\mathscr{C}^+$), while the factor N counts the number of shapes of the protocritical droplet. Theorem 20.3(b) says that, once the critical droplet is created, it rapidly grows to a droplet that has twice the size.

2. In Sect. 20.5 we will show that the *average* probability under the Gibbs measure μ_β of destroying a supercritical droplet and returning to a configuration in $\mathscr{S}_L$ is exponentially small in β. Hence, the crossover from $\mathscr{S}_L$ to $\mathscr{S}^c\backslash\tilde{\mathscr{C}}\cup\mathscr{C}^+$ represents the threshold for nucleation, and Theorem 20.3(a) represents the nucleation time.

3. The Λ_β-dependence in Theorem 20.3(a) matches the Λ-dependence in Theorem 18.4, with the logarithmic factor in (18.1.12) being linked to the extra factor $\beta\Delta$ in (20.1.12). Note that this factor is particularly interesting, since it says that the *effective box size* responsible for the formation of a critical droplet is L_β.

Outline Theorem 20.3 is proved in Sects. 20.2–20.3. Along the way we need several technical facts whose proofs are deferred to Sects. 20.4–20.5. These are all related to the difficult issues mentioned in the opening of this chapter.

20.2 Average time to create a critical droplet

In this section we prove Theorem 20.3(a). Our starting point is the analogue of (19.3.1) with $\mathscr{S}\cup\mathscr{C}$ and $\mathscr{S}^c\backslash\mathscr{C}$ replaced by $\mathscr{S}\cup(\tilde{\mathscr{C}}\backslash\mathscr{C}^+)$ and $(\mathscr{S}^c\backslash\tilde{\mathscr{C}})\cup\mathscr{C}^+$.

20.2.1 Estimate of the equilibrium potential

Lemma 20.4 $\sum_{\sigma\in\mathscr{S}\cup(\tilde{\mathscr{C}}\backslash\mathscr{C}^+)} \mu_\beta(\sigma) h_{\mathscr{S}_L,(\mathscr{S}^c\backslash\tilde{\mathscr{C}})\cup\mathscr{C}^+}(\sigma) = \mu_\beta(\mathscr{S})[1+o(1)]$ *as* $\beta\to\infty$.

Proof Write, using (7.1.16),

$$\sum_{\sigma\in\mathscr{S}\cup(\tilde{\mathscr{C}}\backslash\mathscr{C}^+)} \mu_\beta(\sigma)h_{\mathscr{S}_L,(\mathscr{S}^c\backslash\tilde{\mathscr{C}})\cup\mathscr{C}^+}(\sigma)$$

$$= \mu_\beta(\mathscr{S}_L) + \sum_{\sigma\in(\mathscr{S}\backslash\mathscr{S}_L)\cup(\tilde{\mathscr{C}}\backslash\mathscr{C}^+)} \mu_\beta(\sigma)\mathbb{P}_\sigma(\tau_{\mathscr{S}_L} < \tau_{(\mathscr{S}^c\backslash\tilde{\mathscr{C}})\cup\mathscr{C}^+}). \tag{20.2.1}$$

The last sum is bounded above by $\mu_\beta(\mathscr{S}\backslash\mathscr{S}_L) + \mu_\beta(\tilde{\mathscr{C}}\backslash\mathscr{C}^+)$. But $\mu_\beta(\mathscr{S}\backslash\mathscr{S}_L) = o(\mu_\beta(\mathscr{S}))$ as $\beta\to\infty$ by our choice of L in (20.1.8). In Lemma 20.11 in Sect. 20.4.3 we will show that $\mu_\beta(\tilde{\mathscr{C}}\backslash\mathscr{C}^+) = o(\mu_\beta(\mathscr{S}))$ as $\beta\to\infty$. □

20.2.2 Estimate of the capacity

Lemma 20.5 $\mathrm{cap}(\mathscr{S}_L, (\mathscr{S}^c\backslash\tilde{\mathscr{C}}) \cup \mathscr{C}^+) = N|\Lambda_\beta|\,\frac{4\pi}{\beta\Delta}\,\mathrm{e}^{-\beta\Gamma^\star}\mu_\beta(\mathscr{S})[1+o(1)]$ *as* $\beta\to\infty$.

Proof The argument is in the same spirit as that in Sect. 19.2.2. However, a number of additional hurdles need to be taken that come from the conservative nature of Kawasaki dynamics. The proof proceeds via upper and lower bounds, written out below. Both take up quite a bit of space. □

Upper bound

Proof The proof comes in 7 steps.

1. Protocritical droplet and free particle. We have

$$\mathrm{cap}\big(\mathscr{S}_L, (\mathscr{S}^c\backslash\tilde{\mathscr{C}})\cup\mathscr{C}^+\big) \le \mathrm{cap}\big(\mathscr{S}\cup\mathscr{C}^-, (\mathscr{S}^c\backslash\tilde{\mathscr{C}})\cup\mathscr{C}^+\big)$$

$$= \min_{\substack{h:\, S_\beta^{(n_\beta)}\to[0,1]\\ h|_{\mathscr{S}\cup\mathscr{C}^-}=1,\, h|_{(\mathscr{S}^c\backslash\tilde{\mathscr{C}})\cup\mathscr{C}^+}=0}} \tfrac{1}{2}\sum_{\sigma,\sigma'\in S_\beta^{(n_\beta)}} \mu_\beta(\sigma)c_\beta(\sigma,\sigma')\left[h(\sigma)-h(\sigma')\right]^2. \tag{20.2.2}$$

Split the right-hand side into a contribution coming from $\sigma,\sigma'\in\tilde{\mathscr{C}}$ and the rest, i.e.,

$$\text{r.h.s. } (20.2.2) = I + \gamma_1(\beta), \tag{20.2.3}$$

where

$$I = \min_{\substack{h:\, \tilde{\mathscr{C}}\to[0,1]\\ h|_{\mathscr{C}^-}=1,\, h|_{\mathscr{C}^+}=0}} \tfrac{1}{2}\sum_{\sigma,\sigma'\in\tilde{\mathscr{C}}} \mu_\beta(\sigma)c_\beta(\sigma,\sigma')\left[h(\sigma)-h(\sigma')\right]^2 \tag{20.2.4}$$

and $\gamma_1(\beta)$ is an error term that will be estimated in Step 7. This term will turn out to be small because $\mu_\beta(\sigma)c_\beta(\sigma,\sigma')$ is small when either $\sigma \in S_\beta^{(n_\beta)}\backslash\tilde{\mathscr{C}}$ or $\sigma' \in S_\beta^{(n_\beta)}\backslash\mathscr{C}$. Next, partition $\tilde{\mathscr{C}}$, $\mathscr{C}^-$, $\mathscr{C}^+$ into sets $\tilde{\mathscr{C}}(x)$, $\mathscr{C}^-(x)$, $\mathscr{C}^+(x)$, $x \in \Lambda_\beta$, by requiring that the lower-left corner of the protocritical droplet is in the center of the box $B_{L_\beta}(x)$. Then, because $c_\beta(\sigma,\sigma')=0$ when $\sigma \in \tilde{\mathscr{C}}(x)$ and $\sigma' \in \tilde{\mathscr{C}}(x')$ for some $x \neq x'$, we may write

$$I = |\Lambda_\beta| \min_{\substack{h:\,\tilde{\mathscr{C}}(0)\to[0,1]\\ h|_{\mathscr{C}^-(0)}=1,\, h|_{\mathscr{C}^+(0)}=0}} \tfrac{1}{2} \sum_{\sigma,\sigma'\in\tilde{\mathscr{C}}(0)} \mu_\beta(\sigma)c_\beta(\sigma,\sigma')\big[h(\sigma)-h(\sigma')\big]^2. \tag{20.2.5}$$

2. Decomposition of configurations. Define (compare with (19.2.6))

$$\begin{aligned}\hat{\mathscr{C}}(0) &= \big\{\sigma\mathbb{1}_{B_{L_\beta}(0)}\colon\ \sigma\in\tilde{\mathscr{C}}(0)\big\},\\ \check{\mathscr{C}}(0) &= \big\{\sigma\mathbb{1}_{[B_{L_\beta}(0)]^c}\colon\ \sigma\in\tilde{\mathscr{C}}(0)\big\}.\end{aligned} \tag{20.2.6}$$

Then every $\sigma \in \tilde{\mathscr{C}}(0)$ can be uniquely decomposed as $\sigma = \hat{\sigma}\vee\check{\sigma}$ for some $\hat{\sigma}\in\hat{\mathscr{C}}(0)$ and $\check{\sigma}\in\check{\mathscr{C}}(0)$. Note that $\hat{\mathscr{C}}(0)$ has $K=\ell_c(\ell_c-1)+2$ particles and $\check{\mathscr{C}}(0)$ has $n_\beta - K$ particles (and recall that, by the first half of (20.1.11), $n_\beta\to\infty$ as $\beta\to\infty$). Define

$$\mathscr{C}^{\mathrm{fp}}(0) = \big\{\sigma\in\tilde{\mathscr{C}}(0)\colon\ H_\beta(\sigma)=H_\beta(\hat{\sigma})+H_\beta(\check{\sigma})\big\}, \tag{20.2.7}$$

i.e., the set of configurations consisting of a protocritical droplet and a free particle inside $B_{L_\beta}(0)$ not interacting with the particles outside $B_{L_\beta}(0)$. Write $\mathscr{C}^{\mathrm{fp},-}(0)$ and $\mathscr{C}^{\mathrm{fp},+}(0)$ to denote the subsets of $\mathscr{C}^{\mathrm{fp}}(0)$ where the free particle is at distance L_β, respectively, 2 from the protocritical droplet. Split the right-hand side of (20.2.5) into a contribution coming from $\sigma,\sigma'\in\mathscr{C}^{\mathrm{fp}}(0)$ and the rest, i.e.,

$$\text{r.h.s. (20.2.5)} = |\Lambda_\beta|\big[II+\gamma_2(\beta)\big], \tag{20.2.8}$$

where

$$II = \min_{\substack{h:\,\mathscr{C}^{\mathrm{fp}}(0)\to[0,1]\\ h|_{\mathscr{C}^{\mathrm{fp},-}(0)}=1,\, h|_{\mathscr{C}^{\mathrm{fp},+}(0)}=0}} \tfrac{1}{2} \sum_{\sigma,\sigma'\in\mathscr{C}^{\mathrm{fp}}(0)} \mu_\beta(\sigma)c_\beta(\sigma,\sigma')\big[h(\sigma)-h(\sigma')\big]^2 \tag{20.2.9}$$

and $\gamma_2(\beta)$ is an error term that will be estimated in Step 6. This term will turn out to be small because of loss of entropy when the particle is at the boundary.

3. Reduction to capacity of simple random walk. Estimate

$$
\begin{aligned}
II = & \min_{\substack{h:\,\mathscr{C}^{\mathrm{fp}}(0)\to[0,1]\\ h|_{\mathscr{C}^{\mathrm{fp},-}(0)}=1,\,h|_{\mathscr{C}^{\mathrm{fp},+}(0)}=0}} \frac{1}{2}\sum_{\check\sigma,\check\sigma'\in\check{\mathscr{C}}(0)} \sum_{\substack{\hat\sigma,\hat\sigma'\in\hat{\mathscr{C}}(0):\\ \hat\sigma\vee\check\sigma,\hat\sigma'\vee\check\sigma'\in\mathscr{C}^{\mathrm{fp}}(0)}} \\
& \mu_\beta(\hat\sigma\vee\check\sigma)\,c_\beta\big(\hat\sigma\vee\check\sigma,\hat\sigma'\vee\check\sigma'\big)\big[h(\hat\sigma\vee\check\sigma)-h\big(\hat\sigma'\vee\check\sigma'\big)\big]^2 \\
\le & \min_{\substack{g:\,\hat{\mathscr{C}}(0)\to[0,1]\\ g|_{\hat{\mathscr{C}}^-(0)}=1,\,g|_{\hat{\mathscr{C}}^+(0)}=0}} \frac{1}{2}\sum_{\check\sigma\in\check{\mathscr{C}}(0)} \sum_{\substack{\hat\sigma,\hat\sigma'\in\hat{\mathscr{C}}(0):\\ \hat\sigma\vee\check\sigma,\hat\sigma'\vee\check\sigma\in\mathscr{C}^{\mathrm{fp}}(0)}} \\
& \mu_\beta(\hat\sigma\vee\check\sigma)\,c_\beta\big(\hat\sigma\vee\check\sigma,\hat\sigma'\vee\check\sigma\big)\big[g(\hat\sigma)-g\big(\hat\sigma'\big)\big]^2,
\end{aligned}
\tag{20.2.10}
$$

where $\hat{\mathscr{C}}^-(0)$, $\hat{\mathscr{C}}(0)^+$ denote the subsets of $\hat{\mathscr{C}}(0)$ where the free particle is at distance L_β, respectively, 2 from the protocritical droplet, and the inequality comes from substituting

$$
h(\hat\sigma\vee\check\sigma)=g(\hat\sigma),\quad \hat\sigma\in\hat{\mathscr{C}}(0),\,\check\sigma\in\check{\mathscr{C}}(0),
\tag{20.2.11}
$$

and afterwards replacing the double sum over $\check\sigma,\check\sigma'\in\check{\mathscr{C}}(0)$ by the single sum over $\check\sigma\in\check{\mathscr{C}}(0)$ because $c_\beta(\hat\sigma\vee\check\sigma,\hat\sigma'\vee\check\sigma')>0$ only if either $\hat\sigma=\hat\sigma'$ or $\check\sigma=\check\sigma'$ (the dynamics updates one pair of neighbouring sites at a time). Next, estimate

$$
\begin{aligned}
&\text{r.h.s. (20.2.10)}\\
&\le \sum_{\check\sigma\in\check{\mathscr{C}}(0)} \frac{1}{Z_\beta^{(n_\beta)}}\mathrm{e}^{-\beta H_\beta(\check\sigma)} \min_{\substack{g:\,\hat{\mathscr{C}}(0)\to[0,1]\\ g|_{\hat{\mathscr{C}}^-(0)}=1,\,g|_{\hat{\mathscr{C}}^+(0)}=0}} \frac{1}{2}\sum_{\substack{\hat\sigma,\hat\sigma'\in\hat{\mathscr{C}}(0)\\ \hat\sigma\vee\check\sigma,\hat\sigma'\vee\check\sigma\in\mathscr{C}^{\mathrm{fp}}(0)}} \\
&\qquad \mathrm{e}^{-\beta H_\beta(\hat\sigma)}\,c_\beta\big(\hat\sigma,\hat\sigma'\big)\big[g(\hat\sigma)-g\big(\hat\sigma'\big)\big]^2,
\end{aligned}
\tag{20.2.12}
$$

where we used $H_\beta(\sigma)=H_\beta(\hat\sigma)+H_\beta(\check\sigma)$ from (20.2.7) and write $c_\beta(\hat\sigma,\hat\sigma')$ to denote the transition rate associated with the Kawasaki dynamics restricted to $B_{L_\beta}(0)$, which clearly equals $c_\beta(\hat\sigma\vee\check\sigma,\hat\sigma'\vee\check\sigma)$ for every $\check\sigma\in\check{\mathscr{C}}(0)$ such that $\hat\sigma\vee\check\sigma,\hat\sigma'\vee\check\sigma\in\mathscr{C}^{\mathrm{fp}}(0)$ because there is no interaction between the particles inside and outside $B_{L_\beta}(0)$. The minimum in the r.h.s. of (20.2.12) can be estimated from above by

$$
\text{minimum in (20.2.12)}\le\sum_{\sigma\in\mathscr{P}(0)}\mathscr{V}_\beta(\sigma)
\tag{20.2.13}
$$

with $\mathscr{P}(0)$ the set of protocritical droplets with lower-left corner at 0, and

$$
\mathscr{V}_\beta(\sigma)=\min_{\substack{f:\,\mathbb{Z}^2\to[0,1]\\ f|_{P_\sigma(0)}=1,\,f|_{[B_{L_\beta}(0)]^c}=0}} \frac{1}{2}\sum_{\substack{x,x'\in\mathbb{Z}^2\\ x\sim x'}}\big[f(x)-f\big(x'\big)\big]^2,
\tag{20.2.14}
$$

where $P_\sigma(0)$ is the support of the protocritical droplet in σ, and $x \sim x'$ means that x and x' are neighbouring sites. Indeed, (20.2.13) is obtained from the expression in (20.2.12) by dropping the restriction $\hat{\sigma} \vee \check{\sigma}, \hat{\sigma}' \vee \check{\sigma} \in \mathscr{C}^{\mathrm{fp}}(0)$, substituting

$$g\big(P_\sigma(0) \cup \{x\}\big) = f(x), \quad \sigma \in \mathscr{P}(0),\ x \in B_{L_\beta}(0)\backslash P_\sigma(0), \tag{20.2.15}$$

and noting that $c_\beta(P_\sigma(0) \cup \{x\}, P_\sigma(0) \cup \{x'\}) = 1$ when $x \sim x'$ and zero otherwise. What (20.2.14) says is that

$$\mathscr{V}_\beta(\sigma) = \mathrm{cap}\big(P_\sigma(0), \big[B_{L_\beta}(0)\big]^c\big) \tag{20.2.16}$$

is the capacity of simple random walk between the protocritical droplet $P_\sigma(0)$ in σ and the exterior of $B_{L_\beta}(0)$. Now, define

$$\check{Z}_\beta^{(n_\beta - K)}(0) = \sum_{\check{\sigma} \in \check{\mathscr{C}}(0)} \mathrm{e}^{-\beta H_\beta(\check{\sigma})}. \tag{20.2.17}$$

Then we obtain from (20.2.12)–(20.2.13) that

$$\text{r.h.s. } (20.2.12) \le \mathrm{e}^{-\beta \bar{\Gamma}^\star} \frac{\check{Z}_\beta^{(n_\beta - K)}(0)}{Z_\beta^{(n_\beta)}} \sum_{\sigma \in \mathscr{P}(0)} \mathscr{V}_\beta(\sigma), \tag{20.2.18}$$

where $\bar{\Gamma}^\star = -U[(\ell_c - 1)^2 + \ell_c(\ell_c - 2) + 1]$ is the binding energy of the protocritical droplet.

4. Capacity estimate. For future reference we state the following estimate on capacities for simple random walk.

Lemma 20.6 *Let $U \subset \mathbb{Z}^2$ be any set such that $\{0\} \subset U \subset B_k(0)$, with $k \in \mathbb{N}_0$ independent of β. Let $V \subset \mathbb{Z}^2$ be any set such that $[B_{KL_\beta}(0)]^c \subset V \subset [B_{L_\beta}(0)]^c$, with $K \in \mathbb{N}$ independent of β. Then*

$$\mathrm{cap}\big(\{0\}, \big[B_{KL_\beta}(0)\big]^c\big) \le \mathrm{cap}(U, V) \le \mathrm{cap}\big(B_k(0), \big[B_{L_\beta}(0)\big]^c\big). \tag{20.2.19}$$

Moreover, via (20.1.5)–(20.1.6),

$$\begin{aligned}\mathrm{cap}\big(B_k(0), \big[B_{KL_\beta}(0)\big]^c\big) &= \big[1 + o(1)\big] \frac{2\pi}{\ln(KL_\beta) - \ln k} \\ &= \big[1 + o(1)\big] \frac{4\pi}{\beta\Delta}, \quad \beta \to \infty.\end{aligned} \tag{20.2.20}$$

Proof The inequalities in (20.2.19) follow from standard monotonicity properties of capacities. The asymptotic estimate in (20.2.20) for capacities of concentric boxes are standard (see e.g. Lawler [160], Sect. 2.3), and also follow by comparison to Brownian motion. □

We can apply Lemma 20.6 to estimate $\mathscr{V}_\beta(\sigma)$ in (20.2.16), since the protocritical droplet with lower-left corner in 0 fits inside the box $B_{2\ell_c}(0)$. This gives

$$\mathscr{V}_\beta(\sigma) = \frac{4\pi}{\beta\Delta}\left[1+o(1)\right], \quad \forall \sigma \in \mathscr{P}(0),\ \beta \to \infty. \tag{20.2.21}$$

Moreover, from Theorem 18.4 we know that $|\mathscr{P}(0)|$, the number of shapes of the protocritical droplet, equals N.

5. Equivalence of ensembles. According to Lemma 20.8 in Sect. 20.4.1, we have

$$\frac{\check{Z}_\beta^{(n_\beta-K)}(0)}{Z_\beta^{(n_\beta)}} = (\rho_\beta)^K\, \mu_\beta(\mathscr{S})\left[1+o(1)\right], \quad \beta \to \infty. \tag{20.2.22}$$

This is an "equivalence of ensembles" property relating the probabilities to find $n_\beta - K$, respectively, n_β particles inside $[B_{L_\beta}(0)]^c$ (recall (20.2.6)). Combining (20.2.2)–(20.2.3), (20.2.5), (20.2.8), (20.2.10), (20.2.12), (20.2.18) and (20.2.21)–(20.2.22), we get

$$\begin{aligned}\mathrm{cap}\big(\mathscr{S},\mathscr{C}^+\big) &\le \gamma_1(\beta) + |\Lambda_\beta|\gamma_2(\beta) + N\,|\Lambda_\beta|\,\frac{4\pi}{\beta\Delta}\,\mathrm{e}^{-\beta\Gamma^\star}\,\mu_\beta(\mathscr{S})\left[1+o(1)\right],\\ &\beta \to \infty,\end{aligned} \tag{20.2.23}$$

where we use that $\bar{\Gamma}^\star + \Delta K = \Gamma^\star$. This completes the proof of the upper bound, provided that the error terms $\gamma_1(\beta)$ and $\gamma_2(\beta)$ are negligible.

6. Second error term. To estimate the error term $\gamma_2(\beta)$, note that the configurations in $\tilde{\mathscr{C}}(0)\backslash\mathscr{C}^{\mathrm{fp}}(0)$ are those for which inside $B_{L_\beta}(0)$ there is a protocritical droplet whose lower-left corner is at 0, and at the boundary of $\beta_{L_\beta}(0)$ there is a particle that is attached to some cluster outside $\beta_{L_\beta}(0)$. Recalling (20.2.5)–(20.2.9), we therefore have

$$\begin{aligned}\gamma_2(\beta) &\le \sum_{\sigma\in\tilde{\mathscr{C}}(0)\backslash\mathscr{C}^{\mathrm{fp}}(0)}\ \sum_{\sigma'\in\tilde{\mathscr{C}}(0)} \mu_\beta(\sigma)c_\beta\big(\sigma,\sigma'\big)\left[h(\sigma)-h\big(\sigma'\big)\right]^2\\ &\le 6\mu_\beta\big(\tilde{\mathscr{C}}(0)\backslash\mathscr{C}^{\mathrm{fp}}(0)\big),\end{aligned} \tag{20.2.24}$$

where we use that $h\colon \tilde{\mathscr{C}}(0) \to [0,1]$, $\mu_\beta(\sigma)c_\beta(\sigma,\sigma') = \mu_\beta(\sigma)\wedge\mu_\beta(\sigma')$, and there are at most 6 possible transitions from $\tilde{\mathscr{C}}(0)\backslash\mathscr{C}^{\mathrm{fp}}(0)$ to $\tilde{\mathscr{C}}(0)$: 3 through a move by the particle at the boundary of $B_{L_\beta}(0)$ and 3 through a move by a particle in the cluster outside $B_{L_\beta}(0)$. Since

$$H_\beta(\sigma) \ge H_\beta(\hat{\sigma}) + H_\beta(\check{\sigma}) - U, \quad \sigma \in \tilde{\mathscr{C}}(0)\backslash\mathscr{C}^{\mathrm{fp}}(0), \tag{20.2.25}$$

it follows from the same argument as in Steps 3 and 5 that

$$\mu_\beta\big(\tilde{\mathscr{C}}(0)\backslash\mathscr{C}^{\mathrm{fp}}(0)\big) < N\,\mathrm{e}^{-\beta\bar{\Gamma}^\star}\,(\rho_\beta)^{K+1}\,\mu_\beta(\mathscr{S})\,\mathrm{e}^{\beta U}\,4(K-1)\,\big[1+o(1)\big], \tag{20.2.26}$$

where $(\rho_\beta)^{K+1}$ comes from the fact that there are $n_\beta-(K+1)$ particles outside $B_{L_\beta+1}(0)$ (once more use Lemma 20.8 in Sect. 20.4.1), $\mathrm{e}^{\beta U}$ comes from the gap in (20.2.25), and $4(K-1)$ counts the maximal number of places at the boundary of $B_{L_\beta}(0)$ where the particle can interact with particles outside $B_{L_\beta}(0)$ due to the constraint that defines $\mathscr{S}$ (recall Definition 20.1)(a)). Since $\rho_\beta\mathrm{e}^{\beta U}=o(1)$, we therefore see that $\gamma_2(\beta)$ indeed is small compared to the main term of (20.2.23).

7. First error term. To estimate the error term $\gamma_1(\beta)$, we define the sets of pairs of configurations

$$\begin{aligned}\mathscr{I}_1&=\big\{(\sigma,\eta)\in\big[S_\beta^{(n_\beta)}\big]^2\colon\ \sigma\in\mathscr{S},\eta\in\mathscr{S}^c\backslash\tilde{\mathscr{C}}\big\},\\ \mathscr{I}_2&=\big\{(\sigma,\eta)\in\big[S_\beta^{(n_\beta)}\big]^2\colon\ \sigma\in\tilde{\mathscr{C}},\eta\in\mathscr{S}^c\backslash\tilde{\mathscr{C}}\big\},\end{aligned} \tag{20.2.27}$$

and estimate

$$\gamma_1(\beta)\le\sum_{i=1}^{2}\sum_{(\sigma,\eta)\in\mathscr{I}_i}\mu_\beta(\sigma)\,c_\beta(\sigma,\eta)=\tfrac12\Sigma(\mathscr{I}_1)+\tfrac12\Sigma(\mathscr{I}_2). \tag{20.2.28}$$

The sum $\Sigma(\mathscr{I}_1)$ can be written as

$$\Sigma(\mathscr{I}_1)=|\Lambda_\beta|\sum_{\sigma\in\mathscr{P}}\sum_{\eta\in\mathscr{S}^c\backslash\tilde{\mathscr{C}}}c_\beta(\eta,\sigma)\,1\big\{\big|\mathrm{supp}[\eta]\cap B_{L_\beta}(0)\big|=K\big\}\,\frac{1}{Z_\beta^{(n_\beta)}}\,\mathrm{e}^{-\beta H_\beta(\eta)}, \tag{20.2.29}$$

where we use that $\mu_\beta(\sigma)c_\beta(\sigma,\eta)=\mu_\beta(\eta)c_\beta(\eta,\sigma)$, $\sigma,\eta\in S_\beta^{(n_\beta)}$, and $c_\beta(\eta,\sigma)=0$, $\eta\in\mathscr{S}^c\backslash\tilde{\mathscr{C}}$, $\sigma\notin\mathscr{P}$ (recall Definition 20.1(b)). We have

$$H_\beta(\eta)\ge H_\beta(\hat{\eta})+H_\beta(\check{\eta})-kU,\qquad \eta\in\mathscr{S}^c\backslash\tilde{\mathscr{C}}, \tag{20.2.30}$$

where k counts the number of pairs of particles interacting across the boundary of $B_{L_\beta}(0)$. Moreover, since $\eta\notin\tilde{\mathscr{C}}$, we have

$$H_\beta(\hat{\eta})\ge\bar{\Gamma}^\star+U. \tag{20.2.31}$$

Inserting (20.2.30)–(20.2.31) into (20.2.29), we obtain

$$\begin{aligned}\Sigma(\mathscr{I}_1)&\le|\Lambda_\beta|\,\mathrm{e}^{-\beta\bar{\Gamma}^\star}\,\mu_\beta(\mathscr{S})\big[1+o(1)\big]\sum_{k=0}^{K}(\rho_\beta)^{K+k}\big[4(K-1)\big]^k\,\mathrm{e}^{\beta(k-1)U}\\ &=|\Lambda_\beta|\,\mathrm{e}^{-\beta\bar{\Gamma}^\star}\,\mu_\beta(\mathscr{S})\big[1+o(1)\big]\,\mathrm{e}^{-\beta U},\end{aligned} \tag{20.2.32}$$

where $(\rho_\beta)^{K+k}$ comes from the fact that there are $n_\beta - (K+k)$ particles outside $B_{L_\beta+1}(0)$ (once more use Lemma 20.8 in Sect. 20.4.1), and the inequality again uses an argument similar as in Steps 3 and 5. Therefore $\Sigma(\mathscr{I}_1)$ is small compared to the main term of (20.2.23). The sum $\Sigma(\mathscr{I}_2)$ can be estimated as

$$\begin{aligned}\Sigma(\mathscr{I}_2) &= \sum_{\sigma\in\tilde{\mathscr{C}}}\sum_{\eta\in\mathscr{S}^c\backslash\tilde{\mathscr{C}}} \mu_\beta(\sigma)\,c_\beta(\sigma,\eta)\\ &= |\Lambda_\beta| \sum_{\sigma\in\tilde{\mathscr{C}}(0)} \mu_\beta(\sigma) \sum_{\eta\in\mathscr{S}^c\backslash\tilde{\mathscr{C}}(0)} c_\beta(\sigma,\eta)\\ &\le |\Lambda_\beta|\,\mu_\beta\big(\tilde{\mathscr{C}}(0)\big)\left\{\mathrm{e}^{-\beta U} + (4L_\beta)\,\rho_\beta\big[1+o(1)\big]\right\},\end{aligned} \tag{20.2.33}$$

where the first term comes from detaching a particle from the critical droplet and the second term from a extra particle entering $B_{L_\beta}(0)$. The term between braces is $o(1)$. Moreover, $\mu_\beta(\tilde{\mathscr{C}}(0)) = \mu_\beta(\mathscr{C}^{\mathrm{fp}}(0)) + \mu_\beta(\tilde{\mathscr{C}}(0)\backslash\mathscr{C}^{\mathrm{fp}}(0))$. The second term was estimated in (20.2.26), the first term can again be estimated as in Steps 3 and 5:

$$\begin{aligned}\mu_\beta\big(\mathscr{C}^{\mathrm{fp}}(0)\big) &= \sum_{\hat\sigma\in\hat{\mathscr{C}}(0)} \sum_{\substack{\check\sigma\in\check{\mathscr{C}}(0)\\ \hat\sigma\vee\check\sigma\in\mathscr{C}^{\mathrm{fp}}(0)}} \mu_\beta(\hat\sigma\vee\check\sigma)\\ &= N\,\mathrm{e}^{-\beta\bar\Gamma^\star}\,\frac{\check Z_\beta^{(n_\beta-K)}(0)}{Z_\beta^{(n_\beta)}} = N\,\mathrm{e}^{-\beta\Gamma^\star}\,\mu_\beta(\mathscr{S})\big[1+o(1)\big].\end{aligned} \tag{20.2.34}$$

Therefore also $\Sigma(\mathscr{I}_2)$ is small compared to the main term of (20.2.23). □

Having completed the proof of the upper bound in Lemma 20.5, we next turn to the proof of the lower bound.

Lower bound

For future reference we state the following property of the harmonic function for simple random walk on $\mathbb{Z}^2$.

Lemma 20.7 *Let g be the harmonic function of simple random walk on $B_{2L_\beta}(0)$ (which is equal to 1 on $\{0\}$ and 0 on $[B_{2L_\beta}(0)]^c$). Then there exists a constant $C<\infty$ such that*

$$\sum_e \big[g(z) - g(z+e)\big]_+ \le C/L_\beta \quad \forall z \in \big[B_{L_\beta}(0)\big]^c. \tag{20.2.35}$$

Proof See e.g. Lawler, Schramm and Werner [161], Lemma 5.1. The proof can be given via the estimates in Lawler [160], Sect. 1.7, or via a coupling argument. □

The proof of the lower bound follows the same line of argument as for Glauber dynamics in that it relies on the construction of a suitable unit flow. This flow will, however, be considerably more difficult. In particular, we will no longer be able to get away with choosing a deterministic flow, and the full power of the Berman-Konsowa variational principle has to be brought to bear.

Proof The proof comes in 5 steps.

1. Starting configurations. We start our flow on a subset of the configurations in $\mathscr{C}^+$ that is sufficiently large and sufficiently convenient. Let $\mathscr{C}_2^+ \subset \mathscr{C}^+$ denote the set of configurations having a protocritical droplet with lower-left corner at some site $x \in \Lambda_\beta$, a free particle at distance 2 from this protocritical droplet, no other particles in the box $B_{2L_\beta}(x)$, and satisfying the constraints in $\mathscr{S}_L$, i.e., all other boxes of size $2L_\beta$ carry no more particles than there are in a protocritical droplet. This is the same as $\mathscr{C}^+$, except that the box around the protocritical droplet has size $2L_\beta$ rather than L_β.

Let $K = \ell_c(\ell_c - 1) + 2$ be the volume of the critical droplet, and let $\mathscr{S}_2^{(n_\beta - K)}$ be the analogue of $\mathscr{S}$ when the total number of particles is $n_\beta - K$ and the boxes in which we count particles have size $2L_\beta$ (compare with Definition 20.1). Similarly as in (19.2.17), our task is to derive a lower bound for $\mathrm{cap}(\mathscr{S}_L, (\mathscr{S}^c \backslash \tilde{\mathscr{C}}) \cup \mathscr{C}^+) = \mathrm{cap}((\mathscr{S}^c \backslash \tilde{\mathscr{C}}) \cup \mathscr{C}^+, \mathscr{S}_L) \geq \mathrm{cap}(\mathscr{C}_L, \mathscr{S}_L)$, where $\mathscr{C}_L \subset \mathscr{C}_2^+ \subset \mathscr{C}^+$ defined by

$$\mathscr{C}_L = \left\{\sigma \cup P_{(y)}(x,z) \colon \sigma \in \mathscr{S}_2^{(n_\beta - K)},\ x, y \in \Lambda_\beta,\ (x,y,z) \perp \sigma\right\} \tag{20.2.36}$$

is the analogue of (19.2.13), namely, the set of configurations obtained from $\mathscr{S}_2^{(n_\beta - K)}$ by adding a critical droplet somewhere in Λ_β (lower-left corner at x, protuberance at y, free particle at z) such that it does not interact with the particles in σ and has an empty box of size $2L_\beta$ around it. Note that the $n_\beta - K$ particles can block at most $n_\beta (2L_\beta)^2 = o(|\Lambda_\beta|)$ sites from being the center of an empty box of size $2L_\beta$, and so the critical particle can be added at $|\Lambda_\beta| - o(|\Lambda_\beta|)$ locations.

We partition $\mathscr{C}_L$ into sets $\mathscr{C}_L(x)$, $x \in \Lambda_\beta$, according to the location of the protocritical droplet. It suffices to consider the case where the critical droplet is added at $x = 0$, because the union over x trivially produces a factor $|\Lambda_\beta|$.

2. Overall strategy. Starting from a configuration in $\mathscr{C}_L(0)$, we will successively pick $K - L$ particles from the critical droplet (starting with the free particle at z at distance 2) and move them out of the box $B_{L_\beta}(0)$, placing them essentially uniformly in the annulus $B_{2L_\beta}(0) \backslash B_{L_\beta}(0)$. Once this has been achieved, the configuration is in $\mathscr{S}_L$. Each such move will produce an entropy of order L_β^2, which will be enough to compensate for the loss of energy in tearing down the droplet. The order in which the particles are removed follows the canonical order employed in the lower bound for Glauber dynamics (recall Fig. 19.2). As for Glauber, we will use Theorem 7.43 to estimate

$$\mathrm{cap}(\mathscr{C}_L,\mathscr{S}_L) \geq |\Lambda_\beta| \sum_{\sigma\in\mathscr{C}_L(0)} \sum_{\gamma:\ \gamma_0=\sigma} \mathbb{P}^f(\gamma) \sum_{k=0}^{\tau(\gamma)} \left[\frac{f(\gamma_k,\gamma_{k+1})}{\mu_\beta(\gamma_k)c_\beta(\gamma_k,\gamma_{k+1})}\right]^{-1} \tag{20.2.37}$$

for a suitably constructed flow f and associated path measure $\mathbb{P}^f$, starting from some initial distribution on $\mathscr{C}_L(0)$ (which as for Glauber will be irrelevant), and $\tau(\gamma)$ the time at which the last of the $K-L$ particles exits the box $B_{L_\beta}(0)$.

The difference between Glauber and Kawasaki is that, while in Glauber the droplet can be torn down via single spin-flips, in Kawasaki after we have detached a particle from the droplet we need to move it out of the box $B_{L_\beta}(0)$, which takes a large number of steps. Thus, $\tau(\gamma)$ is the sum of $K-L$ stopping times, each of which, except the first, is a sum of two stopping times itself, one to detach the particle and one to move it out of the box $B_{L_\beta}(0)$. With each motion of a single particle we need to gain an entropy factor of order close to $1/\rho_\beta$. This will be done by constructing a flow that involves only the motion of this single particle, based on the harmonic function of the simple random walk in the box $B_{2L_\beta}(0)$ up to the boundary of the box $B_{L_\beta}(0)$. Outside $B_{L_\beta}(0)$ the flow becomes more complex: we modify it in such a way that a small fraction of the flow, of order $L_\beta^{-1+\varepsilon}$ for some $\varepsilon>0$ small enough, is going into the direction of removing the next particle from the droplet. The reason for this choice is that we want to make sure that the flow becomes sufficiently small, of order $L_\beta^{-2+\varepsilon}$, so that this can compensate for the fact that the Gibbs weight in the denominator of the lower bound in Theorem 7.43 is reduced by a factor $\mathrm{e}^{-\beta U}$ when the protuberance is detached. The reason for the extra ε is that we want to make sure that, along most of the paths, the protuberance is detached before the first particle leaves the box $B_{2L_\beta}(0)$.

Once the protuberance detaches itself from the protocritical droplet, the first particle stops and the second particle moves in the same way as the first particle did when it moved away from the protocritical droplet, and so on. This is repeated until no more than L particles remain in $B_{L_\beta}(0)$, by which time we have reached $\mathscr{S}_L$. As we will see, the only significant contribution to the lower bound comes from the motion of the first particle (as for Glauber), and this coincides with the upper bound established earlier. The details of the construction are to some extent arbitrary and there are many other choices imaginable.

3. First particle. We first construct the flow that moves the particle at distance 2 from the protocritical droplet to the boundary of the box $B_{L_\beta}(0)$. This flow will consist of independent flows for each fixed shape and location of the critical droplet, and will be seen to produce the essential contribution to the lower bound.

We label the configurations in $\mathscr{C}_L(0)$ by σ, describing the shape of the critical droplet, as well as the configuration outside the box $B_{2L_\beta}(0)$, and we label the position of the free particle in σ by $z_1(\sigma)$.

Let g be the harmonic function for simple random walk with boundary conditions 0 on $[B_{2L_\beta}(0)]^c$ and 1 on the critical droplet. Then we choose our flow to be

$$f\big(\sigma(z),\sigma(z')\big)=\begin{cases}C_1\,[g(z)-g(z+e)]_+, & \text{if } z'=z+e,\ \|e\|=1,\\ 0, & \text{otherwise},\end{cases} \tag{20.2.38}$$

where $\sigma(z)$ is the configuration obtained from σ by placing the first particle at site z. The constant C_1 is chosen to ensure that f defines a unit flow, i.e.,

$$\begin{aligned}&\sum_{\sigma\in\mathscr{C}_L(0)} C_1 \sum_{z_1(\sigma),e}\big[g\big(z_1(\sigma)\big)-g\big(z_1(\sigma)+e\big)\big]\\ &\quad = C_1\sum_{\sigma\in\mathscr{C}_L(0)} \operatorname{cap}\big(P_\sigma(0),\big[B_{2L_\beta}(0)\big]^c\big)=1,\end{aligned} \tag{20.2.39}$$

where $P_\sigma(0)$ denotes the support of the protocritical droplet in σ, and the capacity refers to the simple random walk.

Now, let $z^1(k)$ be the location of the first particle at time k, and

$$\tau^1=\inf\big\{k\in\mathbb{N}\colon\ z^1(k)\in\big[B_{L_\beta}(0)\big]^c\big\} \tag{20.2.40}$$

be the first time when, under the Markov chain associated to the flow f, it exits $B_{L_\beta}(0)$. Let γ be a path of this Markov chain. Then, by (20.2.38)–(20.2.39), we have

$$\sum_{k=0}^{\tau^1}\frac{f(\gamma_k,\gamma_{k+1})}{\mu_\beta(\gamma_k)c_\beta(\gamma_k,\gamma_{k+1})}=\frac{C_1[g(z^1(0))-g(z^1(\tau^1))]}{\mu_\beta(\gamma_0)} \tag{20.2.41}$$

where the sum over the g's is telescoping because only paths along which the g-function decreases carry positive probability, and $c_\beta(\gamma_k,\gamma_{k+1})=1$ for all $0\le k\le\tau^1$ because the first particle is free. We have $g(z^1(0))=1$, while, by Lemma 20.7, there exists a $C<\infty$ such that

$$g(x)\le C/\ln L_\beta,\quad x\in\big[B_{L_\beta}(0)\big]^c. \tag{20.2.42}$$

Therefore

$$\sum_{k=0}^{\tau^1}\frac{f(\gamma_k,\gamma_{k+1})}{\mu_\beta(\gamma_k)c_\beta(\gamma_k,\gamma_{k+1})}=\frac{C_1}{\mu_\beta(\gamma_0)}\big[1+o(1)\big]. \tag{20.2.43}$$

Next, by Lemma 20.6, we have

$$\operatorname{cap}\big(P_\sigma(0),\big[B_{2L_\beta}(0)\big]^c\big)=\frac{4\pi}{\beta\Delta}\big[1+o(1)\big],\quad \sigma\in\mathscr{C}_L(0),\ \beta\to\infty, \tag{20.2.44}$$

(because $\{0\}\subset P_\sigma(0)\subset B_{2\ell_c}(0)$ for all $\sigma\in\mathscr{C}_L(0)$). Since $N=|\mathscr{C}_L(0)|$, it follows from (20.2.39) that

$$\frac{1}{C_1}=N\,\frac{4\pi}{\beta\Delta}\big[1+o(1)\big], \tag{20.2.45}$$

and so (20.2.43) becomes

$$\left[\sum_{k=0}^{\tau^1}\frac{f(\gamma_k,\gamma_{k+1})}{\mu_\beta(\gamma_k)c_\beta(\gamma_k,\gamma_{k+1})}\right]^{-1}=\mu_\beta(\gamma_0)\,N\,\frac{4\pi}{\beta\Delta}\big[1+o(1)\big],\quad \beta\to\infty. \tag{20.2.46}$$

This is the contribution we want, because when we sum (20.2.46) over $\gamma_0=\sigma\in\mathscr{C}_L(0)$ (recall (20.2.37)), we get a factor

$$\mu_\beta\big(\mathscr{C}_L(0)\big)=\mathrm{e}^{-\beta\Gamma}\,\mu_\beta(\mathscr{S})\big[1+o(1)\big]. \tag{20.2.47}$$

To see why (20.2.47) is true, recall from (20.2.36) that $\mathscr{C}_L(0)$ is obtained from $\mathscr{S}_2^{(n_\beta-K)}$ by adding a critical droplet with lower-left corner at the origin that does not interact with the $n_\beta-K$ particles elsewhere in Λ_β. Hence

$$\mu_\beta\big(\mathscr{C}_L(0)\big)=\mathrm{e}^{-\beta\bar{\Gamma}^\star}\,\frac{\tilde{Z}_\beta^{(n_\beta-K)}(0)}{Z_\beta^{(n_\beta)}}, \tag{20.2.48}$$

where $\tilde{Z}_\beta^{(n_\beta-K)}(0)$ is the analogue of $\check{Z}_\beta^{(n_\beta-K)}(0)$ (defined in (20.2.17)) obtained by requiring that the $n_\beta-K$ particles are in $[R_{\ell_c,\ell_c}(0)]^c$ instead of $[B_{L_\beta}(0)]^c$. However, it will follow from the proofs of Lemmas 20.8–20.10 in Sect. 20.4 that, similarly as in (20.2.22),

$$\frac{\tilde{Z}_\beta^{(n_\beta-K)}(0)}{Z_\beta^{(n_\beta)}}=(\rho_\beta)^K\,\mu_\beta(\mathscr{S})\big[1+o(1)\big],\quad \beta\to\infty, \tag{20.2.49}$$

which yields (20.2.47) because $\Gamma^\star=\bar{\Gamma}^\star+K\Delta$. For the remaining part of the construction of the flow it therefore suffices to ensure that the sum beyond τ^1 gives a smaller contribution.

4. Second particle. Once the first particle (i.e., the free particle) has left the box $B_{L_\beta}(0)$, we need to allow the second particle (i.e., the protuberance) to detach itself from the protocritical droplet and to move out of $B_{L_\beta}(0)$ as well. The problem is that detaching the second particle reduces the Gibbs weight appearing in the denominator by $\mathrm{e}^{-U\beta}$, while the increments of the flow are reduced only to about $1/L_\beta$. Thus, we cannot immediately detach the second particle. Instead, we do this with probability $L_\beta^{-1+\varepsilon}$ only. The idea is that, once the first particle is outside $B_{L_\beta}(0)$, we *leak* some of the flow that drives the motion of the first particle into a flow that detaches the second particle. To do this, we have to first construct a *leaky flow* in $B_{2L_\beta}(0)\backslash B_{L_\beta}(0)$ for simple random walk. This goes as follows.

Let $p(z,z+e)$ denote the transition probabilities of simple random walk driven by the harmonic function g on $B_{2L_\beta}(0)$. Put

$$\tilde{p}(z,z+e)=\begin{cases}p(z,z+e), & \text{if } z\in B_{L_\beta}(0),\\ (1-L_\beta^{-1+\varepsilon})\,p(z,z+e), & \text{if } z\in B_{2L_\beta}(0)\backslash B_{L_\beta}(0).\end{cases} \tag{20.2.50}$$

Use the transition probabilities $\tilde{p}(z,z+e)$ to define a path measure $\tilde{P}$. This path measure describes simple random walk driven by g, but with a killing probability $L_\beta^{-1+\varepsilon}$ inside the annulus $B_{2L_\beta}(0)\backslash B_{L_\beta}(0)$. Put

$$k(z,z+e)=\sum_\gamma \tilde{P}(\gamma)\mathbb{1}_{(z,z+e)\in\gamma},\quad z\in B_{2L_\beta}(0). \tag{20.2.51}$$

This edge function satisfies the following equations:

$$\begin{aligned}
&\bullet\, k(z,z+e)=\big[g(z)-g(z+e)\big]_+,\\
&\quad \text{if } z\in B_{L_\beta}(0),\\
&\bullet\, k(z,z+e)=0,\\
&\quad \text{if } z\in B_{2L_\beta}(0)\backslash B_{L_\beta}(0) \text{ and } \big[g(z)-g(z+e)\big]_+=0,\\
&\bullet\, \big(1-L_\beta^{-1+\varepsilon}\big)\sum_e k(z+e,z)\mathbb{1}_{g(z+e)-g(z)>0}=\sum_e k(z,z+e)\mathbb{1}_{g(z)-g(z+e)>0}\\
&\quad \text{if } z\in B_{2L_\beta}(0)\backslash B_{L_\beta}(0).
\end{aligned} \tag{20.2.52}$$

Note that inside the annulus $B_{2L_\beta}(0)\backslash B_{L_\beta}(0)$ at each site the flow out is less than the flow in by a leaking factor $1-L_\beta^{-1+\varepsilon}$. We pick $\varepsilon>0$ so small that

$$\mathrm{e}^{\beta U} \text{ is exponentially smaller in } \beta \text{ than } L_\beta^{2-\varepsilon} \tag{20.2.53}$$

(which is possible by (20.1.5)–(20.1.6)). The important fact for us is that this leaky flow is dominated by the harmonic flow associated with g, in particular, the flow in satisfies

$$\sum_e k(z+e,z)\le\sum_e\big[g(z+e)-g(z)\big]_+\quad \forall z\in B_{2L_\beta}(0) \tag{20.2.54}$$

(and the same applies for the flow out). This inequality holds because g satisfies the same equations as in (20.2.50)–(20.2.51) but without the leaking factor $1-L_\beta^{-1+\varepsilon}$.

Using this leaky flow, we can now construct a flow involving the first two particles, as follows:

$$\begin{aligned}
&\bullet\, f\big(\sigma(z_1,a),\sigma(z_1+e,a)\big)=C_1k(z_1,z_1+e), \qquad\qquad (20.2.55)\\
&\quad \text{if } z_1\in B_{2L_\beta}(0),\\
&\bullet\, f\big(\sigma(z_1,a),\sigma(z_1,b)\big)=C_1L_\beta^{-1+\varepsilon}\sum_e k(z_1,z_1+e),\\
&\quad \text{if } z_1\in B_{2L_\beta}(0)\backslash B_{L_\beta}(0),\\
&\bullet\, f\big(\sigma(z_1,z_2),\sigma(z_1,z_2+e)\big)=\Big\{C_1L_\beta^{-1+\varepsilon}\sum_e k(z_1,z_1+e)\Big\}\big[g(z_2)-g(z_2+e)\big]_+,\\
&\quad \text{if } z_1\in B_{2L_\beta}(0)\backslash B_{L_\beta}(0),\ z_2\in B_{L_\beta}(0)\backslash P_\sigma(0).
\end{aligned}$$

Here, we write a and b for the locations of the second particle prior and after it detaches itself from the protocritical droplet, and $\sigma(z_1, z_2)$ for the configuration obtained from σ by placing the first particle (that was at distance 2 from the protocritical droplet) at site z_1 and the second particle (that was the protuberance) at site z_2. The flow for other motions is zero, and the constant C_1 is the *same* as in (20.2.38)–(20.2.39).

We next define two further stopping times, namely,

$$\zeta^2 = \inf\{k \in \mathbb{N}\colon\ z^2(\gamma_k) = b\}, \tag{20.2.56}$$

i.e., the first time the second particle (the protuberance) detaches itself from the protocritical droplet, and

$$\tau^2 = \inf\{k \in \mathbb{N}\colon\ z^2(\gamma_k) \in \left[B_{L_\beta}(0)\right]^c\}, \tag{20.2.57}$$

i.e., the first time the second particle exits the box $B_{L_\beta}(0)$. Note that, since we choose the leaking probability to be $L_\beta^{-1+\varepsilon}$, the probability that ζ^2 is larger than the first time the first particle exits $B_{2L_\beta}(0)$ is of order $\exp[-L_\beta^\varepsilon]$ and hence is negligible. We will disregard the contributions of such paths in the lower bound. These paths will be called *good*.

We will next show that (20.2.41) also holds if we extend the sum along any path of positive probability up to ζ^2. The reason for this lies in Lemma 20.7. Let γ be a path that has a positive probability under the path measure $\mathbb{P}^f$ associated with f stopped at τ^2. We will assume that this path is good in the sense described above. To that end we decompose

$$\begin{aligned}
&\sum_{k=0}^{\tau^2} \frac{f(\gamma_k, \gamma_{k+1})}{\mu_\beta(\gamma_k)c_\beta(\gamma_k, \gamma_{k+1})} \\
&\quad = \sum_{k=0}^{\tau^1} \frac{f(\gamma_k, \gamma_{k+1})}{\mu_\beta(\gamma_k)c_\beta(\gamma_k, \gamma_{k+1})} + \sum_{k=\tau^1+1}^{\zeta^2-2} \frac{f(\gamma_k, \gamma_{k+1})}{\mu_\beta(\gamma_k)c_\beta(\gamma_k, \gamma_{k+1})} \\
&\qquad + \sum_{k=\zeta^2-1}^{\tau^2} \frac{f(\gamma_k, \gamma_{k+1})}{\mu_\beta(\gamma_k)c_\beta(\gamma_k, \gamma_{k+1})} = I + II + III.
\end{aligned} \tag{20.2.58}$$

The term I was already estimated in (20.2.41)–(20.2.47). To estimate II, we use (20.2.42) and (20.2.54)–(20.2.55) to bound (compare with (20.2.41))

$$II \le C_1 \frac{g(z^1(\zeta^2)) - g(z^1(\tau^1))}{\mu_\beta(\gamma_0)} \le C_1 \frac{[C/\ln L_\beta]}{\mu_\beta(\gamma_0)}, \tag{20.2.59}$$

which is negligible compared to I due to the factor $C/\ln L_\beta$. It remains to estimate III. Note that

$$III = \frac{f(\gamma_{\zeta^2-1},\gamma_{\zeta^2})}{\mu_\beta(\gamma_{\zeta^2-1})c_\beta(\gamma_{\zeta^2-1},\gamma_{\zeta^2})} + \sum_{k=\zeta^2}^{\tau^2} \frac{f(\gamma_k,\gamma_{k+1})}{\mu_\beta(\gamma_k)c_\beta(\gamma_k,\gamma_{k+1})}. \tag{20.2.60}$$

The first term corresponds to the move when the protuberance detaches itself from the protocritical droplet. Its numerator is given by $f(\sigma(z_1,a),\sigma(z_1,b))$ (for some $z_1 \in [B_{L_\beta}(0)]^c$) which, by Lemma 20.7 and (20.2.54)–(20.2.55), is smaller than $C_1 L_\beta^{-1+\varepsilon} C L_\beta^{-1} = C_1 C L_\beta^{-2+\varepsilon}$. On the other hand, its denominator is given by

$$\mu(\gamma_{\zeta^2-1})c_\beta(\gamma_{\zeta^2-1},\gamma_{\zeta^2}) = \mu_\beta(\gamma_0)\mathrm{e}^{-U\beta}. \tag{20.2.61}$$

The same holds for the denominators in all the other terms in III, while the numerators in these terms satisfy the bound

$$f(\gamma_k,\gamma_{k+1}) \leq C_1\, C\, L_\beta^{-2+\varepsilon}\big[g\big(z^2(\gamma_k)\big) - g\big(z^2(\gamma_{k+1})\big)\big]. \tag{20.2.62}$$

Adding up the various terms, we get that

$$III \leq \frac{C_1}{\mu_\beta(\gamma_0)} L_\beta^{-2+\varepsilon}\mathrm{e}^{\beta U}\big(1+\big[g\big(z^2(\zeta^2)\big) - g\big(z^2(\tau^2)\big)\big]\big) \leq \frac{2C_1}{\mu_\beta(\gamma_0)} L_\beta^{-2+\varepsilon}\mathrm{e}^{\beta U}. \tag{20.2.63}$$

The right-hand side is smaller than I by a factor $L_\beta^{-2+\varepsilon}\mathrm{e}^{\beta U}$, which, by (20.2.53), is exponentially small in β.

5. Remaining particles. The lesson from the previous steps is that we can construct a flow with the property that each time we remove a particle from the droplet we gain a factor $L_\beta^{-2+\varepsilon}$, i.e., almost $\mathrm{e}^{-\Delta\beta}$. (This entropy gain corresponds to the gain from the magnetic field in Glauber dynamics, or from the activity in Kawasaki dynamics on a finite open box.) We can continue our flow by tearing down the critical droplet in the same order as we did for Glauber dynamics. Each removal corresponds to a flow that is built in the same way as described in Step 4 for the second particle. There will be some minor modifications involving a negligible fraction of paths where a particle hits a particle that was moved out earlier, but this is of no consequence. As a result of the construction, the sums along the remainders of these paths will give only negligible contributions.

Thus, we have shown that the lower bound coincides, up to a factor $1+o(1)$, with the upper bound and the lemma is proven. □

Combining the upper bound obtained in Sect. 20.2.2 with the lower bound obtained in Sect. 20.2.2, we have finally completed the proof of Lemma 20.5, and therefore of Theorem 20.3(a).

20.3 Average time to grow a droplet twice the critical size

In this section we prove Theorem 20.3(b). The starting point is again the analogue of (19.3.1) with $\mathscr{S}^c\backslash\mathscr{C}$ replaced by $\mathscr{D}_M$ and $\mathscr{S}\cup\mathscr{C}$ by $\mathscr{D}_M^c$.

Proof The same observation holds as in (19.4.1). Therefore the proof follows along the same lines as that of Theorem 20.3(a). The main point is to prove $\mathrm{cap}(\mathscr{D}_M,\mathscr{S}_L)=[1+o(1)]\mathrm{cap}(\mathscr{C}^+,\mathscr{S}_L)$. Since $\mathrm{cap}(\mathscr{S}_L,\mathscr{D}_M)\leq\mathrm{cap}(\mathscr{S}_L,\mathscr{C}^+)$, all we need to do is prove a lower bound on $\mathrm{cap}(\mathscr{D}_M,\mathscr{S}_L)$. This is done in almost exactly the same way as for Glauber, by using the construction given there and substituting each Glauber move by a flow involving the motion of just two particles.

Note that, as long as $M=\mathrm{e}^{o(\beta)}$, an $M\times M$ droplet can be added at $|\Lambda_\beta|-o(|\Lambda_\beta|)$ locations to a configuration $\sigma\in\mathscr{S}$ (compare with (20.2.36)). The only novelty is that we have to eventually remove the cloud of particles that is produced in the annulus $B_{2L_\beta}(0)\backslash B_{L_\beta}(0)$. This is done in much the same way as before. As long as only $\mathrm{e}^{o(\beta)}$ particles have to be removed, potential collisions between particles can be ignored as they are sufficiently unlikely. □

20.4 Equivalence of ensembles

Recall that $K=\ell_c(\ell_c-1)+2$ is the number of particles in a critical droplet. For $m\in\mathbb{N}_0$, let

$$\mathscr{S}^{(n_\beta-m)}=\left\{\sigma\in S_\beta^{(n_\beta-m)}\colon\ \left|\mathrm{supp}[\sigma]\cap B_{L_\beta}(x)\right|<K\ \forall x\in\Lambda_\beta\right\} \tag{20.4.1}$$

and

$$\begin{aligned} Z_\beta^{(n_\beta-m)} &= \sum_{\sigma\in\mathscr{S}^{(n_\beta-m)}} \mathrm{e}^{-\beta H_\beta(\sigma)}, \\ \check{Z}_\beta^{(n_\beta-m)} &= \sum_{\sigma\in\mathscr{S}^{(n_\beta-m)}} \mathrm{e}^{-\beta H_\beta(\sigma)}\,\mathbb{1}_{\{\mathrm{supp}[\sigma]\subset\Lambda_\beta\backslash B_{L_\beta}(0)\}}. \end{aligned} \tag{20.4.2}$$

The first is the partition function with $n_\beta-m$ particles restricted such that no box of size L_β has $\geq K$ particles. The second is the same partition function but with the additional restriction that no particle falls in $B_{L_\beta}(0)$.

The following lemma was used in (20.2.22), (20.2.26), (20.2.32) and (20.2.49).

Lemma 20.8 $\check{Z}_\beta^{(n_\beta-m)}/Z_\beta^{(n_\beta)}=(\rho_\beta)^m\,\mu_\beta(\mathscr{S}^{(n_\beta)})\,[1+o(1)]$ *as* $\beta\to\infty$ *for all* $m\in\mathbb{N}_0$.

In Sects. 20.4.1–20.4.2 two lemmas are proved that combine to yield Lemma 20.8. Sections 20.4.3–20.4.4 prove atypicality of critical droplets and typicality of starting configurations.

20.4.1 Partition functions for different numbers of particles

Lemma 20.9 $Z_\beta^{(n_\beta-m)}/Z_\beta^{(n_\beta)} = (\rho_\beta)^m\,[1+o(1)]$ *as* $\beta\to\infty$ *for all* $m\in\mathbb{N}$.

Proof The proof comes in 5 steps.

1. It suffices to give the proof for $m=1$. The same proof works for $m\geq 2$ after we replace n_β by $n_\beta-m+1$. Write

$$\begin{aligned} Z_\beta^{(n_\beta)} &= \sum_{\substack{\mathrm{supp}[\sigma]\subset\Lambda_\beta\\ |\sigma|=n_\beta-1}} \frac{1}{n_\beta} \sum_{x\in\Lambda_\beta\backslash\mathrm{supp}[\sigma]} \mathrm{e}^{-\beta H_\beta(\sigma\vee\mathbb{1}_x)}\,\mathbb{1}_{\{\sigma\vee\mathbb{1}_x\in\mathscr{S}^{(n_\beta)}\}}\\ &= \sum_{\substack{\mathrm{supp}[\sigma]\subset\Lambda_\beta\\ |\sigma|=n_\beta-1}} \mathrm{e}^{-\beta H_\beta(\sigma)}\big[I(\sigma)+II(\sigma)\big] = I+II, \end{aligned} \tag{20.4.3}$$

where

$$\begin{aligned} I(\sigma) &= \frac{1}{n_\beta} \sum_{\substack{x\in\Lambda_\beta\\ \mathrm{dist}(x,\mathrm{supp}[\sigma])>1}} \mathbb{1}_{\{\sigma\vee\mathbb{1}_x\in\mathscr{S}^{(n_\beta)}\}},\\ II(\sigma) &= \frac{1}{n_\beta} \sum_{\substack{x\in\Lambda_\beta\\ \mathrm{dist}(x,\mathrm{supp}[\sigma])=1}} \mathrm{e}^{-\beta[H_\beta(\sigma\vee\mathbb{1}_x)-H_\beta(\sigma)]}\,\mathbb{1}_{\{\sigma\vee\mathbb{1}_x\in\mathscr{S}^{(n_\beta)}\}}. \end{aligned} \tag{20.4.4}$$

In the first sum the particle at x is free and $H_\beta(\sigma\vee\mathbb{1}_x)=H_\beta(\sigma)$, while in the second sum it is not free and $H_\beta(\sigma\vee\mathbb{1}_x)<H_\beta(\sigma)$. For every $\sigma\in\mathscr{S}^{(n_\beta-1)}$, we have (recall (20.4.1))

$$|\Lambda_\beta|-(2L_\beta+1)^2(n_\beta-1)\leq \sum_{\substack{x\in\Lambda_\beta\\ \mathrm{dist}(x,\mathrm{supp}[\sigma])>1}} \mathbb{1}_{\{\sigma\vee\mathbb{1}_x\in\mathscr{S}^{(n_\beta)}\}}\leq|\Lambda_\beta|. \tag{20.4.5}$$

Moreover, by (20.1.2)–(20.1.3) and (20.1.5)–(20.1.6), we have $L_\beta^2 n_\beta = o(|\Lambda_\beta|)$, and so it follows that

$$\begin{aligned} I &= \frac{|\Lambda_\beta|}{n_\beta}\big[1+o(1)\big]\sum_{\sigma\in\mathscr{S}^{(n_\beta-1)}} \mathrm{e}^{-\beta H_\beta(\sigma)} = \frac{|\Lambda_\beta|}{n_\beta}\big[1+o(1)\big]\,Z_\beta^{(n_\beta-1)}\\ &= \frac{1}{\rho_\beta}\big[1+o(1)\big]\,Z_\beta^{(n_\beta-1)}. \end{aligned} \tag{20.4.6}$$

We will show that II is exponentially smaller than I, which will prove the claim.

2. Let us define a 1-cluster as a maximal set of particles such that for each particle in the cluster there is another particle in the cluster at distance ≤ 2. Write

$$II = \frac{1}{n_\beta} \sum_{N=1}^{n_\beta - 1} \sum_{\substack{C_1,\dots,C_N \\ \sum_{m=1}^N |C_m| = n_\beta - 1}} \sum_{a=1}^{N} \sum_{x \in \partial C_a} \mathrm{e}^{-\beta \sum_{b \neq a} H_\beta(C_b) - \beta H_\beta(C_a \vee \mathbb{1}_x)} \mathbb{1}_{\{x \cup [\cup_{a=1}^N C_a] \in \mathscr{S}^{(n_\beta)}\}}, \tag{20.4.7}$$

where N counts the number of 1-clusters, labelled $C_1, \dots, C_N$. Order the 1-clusters according to the number of particles they contain, by writing

$$\begin{aligned} II = \frac{1}{n_\beta} &\sum_{N_1,\dots,N_{K-1}=0}^{n_\beta - 1} \mathbb{1}_{\{\sum_{k=1}^{K-1} k N_k = n_\beta - 1\}} \left(\prod_{k=1}^{K-1} \frac{1}{N_k!} \sum_{\substack{C_1^k,\dots,C_{N_k}^k \\ |C_1^k| = \dots = |C_{N_k}^k| = k}} \right) \\ &\sum_{k=1}^{K-2} \sum_{l=1}^{N_k} \sum_{x \in \partial C_l^k} \exp\left[-\beta \sum_{k'=1}^{K-1} \sum_{l'=1}^{N_{k'}} H_\beta(C_{l'}^{k'}) \mathbb{1}_{\{(k',l') \neq (k,l)\}} - \beta H_\beta(C_l^k \vee \mathbb{1}_x) \right] \\ &\times \mathbb{1}_{\{x \cup [\cup_{k=1}^{K-1} \cup_{l=1}^{N_k} C_l^k] \in \mathscr{S}^{(n_\beta)}\}}, \end{aligned} \tag{20.4.8}$$

where N_k counts the number of 1-clusters of size k labelled $C_1^k, \dots, C_{N_k}^k$, and the sum over k in the second line does not include the term with $k = K - 1$ because $C_l^{K-1} \vee \mathbb{1}_x$ contains K particles, making it a supercritical cluster that is excluded by the indicator in the third line (recall (20.4.1)).

3. By a standard isoperimetric inequality, we have that

$$H_\beta(C_l^k \vee \mathbb{1}_x) \geq H_{k+1} \quad \forall x \in \partial C_l^k \tag{20.4.9}$$

with H_k denoting the energy of a droplet of k particles that is closest to a square or quasi-square. Therefore we may estimate

$$\begin{aligned} &\sum_{x \in \partial C_l^k} \exp\left[-\beta \sum_{k'=1}^{K-1} \sum_{l'=1}^{N_{k'}} H_\beta(C_{l'}^{k'}) \mathbb{1}_{\{(k',l') \neq (k,l)\}} - \beta H_\beta(C_l^k \vee \mathbb{1}_x) \right] \\ &\quad \leq 4k\, \mathrm{e}^{-\beta H_{k+1}} \exp\left[-\beta \sum_{k'=1}^{K-1} \sum_{l'=1}^{N_{k'}} H_\beta(C_{l'}^{k'}) \mathbb{1}_{\{(k',l') \neq (k,l)\}} \right]. \end{aligned} \tag{20.4.10}$$

The last sum no longer contains the 1-cluster C_l^k that x is attached to. Since $|C_l^k| = k$, the other 1-clusters contain a total of $n_\beta - (k+1)$ particles. Hence, inserting (20.4.10) into (20.4.8), we arrive at the estimate

$$II \le \frac{|\Lambda_\beta|}{n_\beta} \sum_{k=1}^{K-2} 4k\, \mathrm{e}^{-\beta H_{k+1}}\, Z_\beta^{(n_\beta-k-1)}, \tag{20.4.11}$$

where $|\Lambda_\beta|$ counts the possible locations of the 1-cluster that has been removed, we trace back the decomposition in (20.4.7)–(20.4.8), and we use that if $x \in \partial C_l^k$, then

$$\left\{ x \cup \left[\bigcup_{k'=1}^{K-1} \bigcup_{l'=1}^{N_{k'}} C_{l'}^{k'} \right] \in \mathscr{S}^{(n_\beta)} \right\} \subset \left\{ \left[\bigcup_{k'=1}^{K-1} \bigcup_{l'=1}^{N_{k'}} C_{l'}^{k'} \right] \backslash C_l^k \in \mathscr{S}^{(n_\beta-k-1)} \right\}. \tag{20.4.12}$$

However, the same argument as in Step 1 yields

$$Z_\beta^{(n_\beta-k-1)} \le (\rho_\beta)^k\, Z_\beta^{(n_\beta-1)} \left[1+o(1)\right]. \tag{20.4.13}$$

Combining (20.4.6), (20.4.11) and (20.4.13), we get

$$II \le \left[\sum_{k=1}^{K-2} 4k\, \mathrm{e}^{-\beta H_{k+1}}\, (\rho_\beta)^k \right] I \left[1+o(1)\right]. \tag{20.4.14}$$

4. In Step 5 below we will prove that, for $\ell_c \ge 3$,

$$H_k + (k-1)\Delta > 0 \quad \forall\, 2 \le k \le (2\ell_c - 3)^2. \tag{20.4.15}$$

Inserting this bound into (20.4.14) and using (20.1.3), we see that the sum in (20.4.14) is $O(\mathrm{e}^{-\beta\varepsilon})$ for some $\varepsilon > 0$, and so II indeed is exponentially smaller than I. Here, note that if $\ell_c \ge 3$, then $K \le (2\ell_c - 3)^2$, which means that (20.4.15) covers the range of k-values needed in (20.4.14). For $\ell_c = 2$ we have $K = 4$, but $H_2 + \Delta = -U + \Delta > 0$ and $H_3 + 2\Delta = -2U + 2\Delta > 0$ because $\Delta > U$, and so we are done as well.

5. It remains to prove (20.4.15). Let $|\sigma|$ denote the volume of σ (the number of particles) and $\gamma(\sigma)$ the perimeter of σ (the number of holes next to a particle). Then the energy of σ equals (recall (18.2.1))

$$H_\beta(\sigma) = -U\left[2|\sigma| - \tfrac{1}{2}\gamma(\sigma)\right], \quad \sigma \in S_\beta. \tag{20.4.16}$$

By the standard isoperimetric inequality, we have

$$|\sigma| \le \left[\tfrac{1}{4}\gamma(\sigma)\right]^2 \quad \forall\, \sigma \in S_\beta. \tag{20.4.17}$$

Hence

$$H_k + (k-1)\Delta \ge -2U[k - \sqrt{k}] + (k-1)\Delta = -(2U - \Delta)k + 2U\sqrt{k} - \Delta. \tag{20.4.18}$$

Let $\ell^* = U/(2U - \Delta)$. Then the right-hand side equals $(2U - \Delta)[-k + 2\ell^*\sqrt{k} - (2\ell^* - 1)]$, which is $= 0$ for $k = 1$ and > 0 for $2 \le k < (2\ell^* - 1)^2$. Since $\ell_c = \lceil \ell^* \rceil$

and $\ell^* \notin \mathbb{N}$ (recall (18.1.5)–(18.1.6)), we have $2\ell^* - 1 > 2\ell_c - 3$, which proves the claim. □

20.4.2 Partition functions for different volumes

Lemma 20.10 $\lim_{\beta\to\infty} \check{Z}_\beta^{(n_\beta - m)} / Z_\beta^{(n_\beta - m)} = 1$ *for all* $m \in \mathbb{N}_0$.

Proof It suffices to give the proof for $m = 0$. The same proof works for $m \geq 1$ after we replace n_β by $n_\beta - m$. Since $\check{Z}_\beta^{(n_\beta)} \leq Z_\beta^{(n_\beta)}$, it suffices to prove the lower bound. Write

$$
\begin{aligned}
Z_\beta^{(n_\beta)} &= \check{Z}_\beta^{(n_\beta)} \\
&\quad + \sum_{m=1}^{K} \sum_{\eta \in S_\beta^{(m)}} \sum_{\substack{\zeta \in S_\beta^{(n_\beta - m)} \\ \eta\vee\zeta \in \mathscr{S}^{(n_\beta)}}} \mathrm{e}^{-\beta H_\beta(\eta\vee\zeta)} \mathbb{1}_{\{\mathrm{supp}[\eta]\subset B_{L_\beta}(0)\}} \mathbb{1}_{\{\mathrm{supp}[\zeta]\subset [B_{L_\beta}(0)]^c\}} \\
&\leq \check{Z}_\beta^{(n_\beta)} + \gamma_1(\beta) + \gamma_2(\beta),
\end{aligned}
\tag{20.4.19}
$$

where

$$
\gamma_1(\beta) = \sum_{m=1}^{K} \sum_{\eta \in S_\beta^{(m)}} \sum_{\substack{\zeta \in S_\beta^{(n_\beta - m)} \\ \eta\vee\zeta \in \mathscr{S}^{(n_\beta)}}} \mathrm{e}^{-\beta[H_\beta(\eta)+H_\beta(\zeta)]} \mathbb{1}_{\{\mathrm{supp}[\eta]\subset B_{L_\beta}(0)\}} \mathbb{1}_{\{\mathrm{supp}[\zeta]\subset [B_{L_\beta}(0)]^c\}}
\tag{20.4.20}
$$

and $\gamma_2(\beta)$ is a term that arises from particles interacting across the boundary of $B_{L_\beta}(0)$. We will show that both $\gamma_1(\beta)$ and $\gamma_2(\beta)$ are negligible.

Estimate

$$
\begin{aligned}
&\gamma_1(\beta) \\
&\quad \leq \sum_{m=1}^{K} \check{Z}_\beta^{(n_\beta - m)} \sum_{\eta\in\mathscr{S}^{(m)}} \mathrm{e}^{-\beta H_\beta(\eta)} \mathbb{1}_{\{\mathrm{supp}[\eta]\subset B_{L_\beta}(0)\}} \\
&\quad = [1+o(1)]\, \check{Z}_\beta^{(n_\beta)} \sum_{m=1}^{K} (\rho_\beta)^m \sum_{\eta\in\mathscr{S}^{(m)}} \mathrm{e}^{-\beta H_\beta(\eta)} \mathbb{1}_{\{\mathrm{supp}[\eta]\subset B_{L_\beta}(0)\}} \\
&\quad = [1+o(1)]\, \check{Z}_\beta^{(n_\beta)} \sum_{m=1}^{K} (\rho_\beta)^m \sum_{j=1}^{m} \sum_{\substack{2\leq k_1,\dots,k_j\leq K \\ \sum_{i=1}^{j} k_i = m}} \sum_{\substack{C=\cup_{i=1}^{j} C_i \subset B_{L_\beta}(0) \\ |C_i|=k_i\,\forall i}} \mathrm{e}^{-\beta \sum_{i=1}^{j} H_\beta(C_i)},
\end{aligned}
\tag{20.4.21}
$$

where the first equality uses Lemma 20.9 with Λ_β replaced by $\Lambda_\beta\backslash B_{L_\beta}(0)$, while the second equality is an expansion in terms of clusters. Using once more the isoperimetric inequality in (20.4.15), we get (recall (20.1.5))

$$\begin{aligned}
\gamma_1(\beta) &\leq \big[1+o(1)\big]\check{Z}_\beta^{(n_\beta)} \sum_{m=1}^{K} (\rho_\beta)^m \sum_{j=1}^{m} \sum_{\substack{2\leq k_1,\dots,k_j\leq K\\ \sum_{i=1}^j k_i=m}} \mathrm{e}^{-\beta\sum_{i=1}^j H_{k_i}} \Bigg(\sum_{\substack{C=\cup_{i=1}^j C_i\\ |C_i|=k_i\,\forall i}} 1\Bigg)\\
&\leq A\,\check{Z}_\beta^{(n_\beta)} \sum_{m=1}^{K} (\rho_\beta)^m \sum_{j=1}^{m} (L_\beta^2)^j \sum_{\substack{2\leq k_1,\dots,k_j\leq K\\ \sum_{i=1}^j k_i=m}} \mathrm{e}^{-\beta\sum_{i=1}^j H_{k_i}}\\
&= A\,\check{Z}_\beta^{(n_\beta)} \sum_{m=1}^{K}\sum_{j=1}^{m} \sum_{\substack{2\leq k_1,\dots,k_j\leq K\\ \sum_{i=1}^j k_i=m}} \mathrm{e}^{-\beta\sum_{i=1}^j [H_{k_i}+k_i\Delta-(\Delta-\delta_\beta)]}\\
&\leq B\,\check{Z}_\beta^{(n_\beta)}\,\mathrm{e}^{-\beta\varepsilon}
\end{aligned} \tag{20.4.22}$$

for some $\varepsilon>0$ and some constants $A, B<\infty$ that are independent of β, i.e., $\gamma_1(\beta)$ is negligible. Estimate

$$\begin{aligned}
&\gamma_2(\beta) \qquad\qquad (20.4.23)\\
&\leq \sum_{m-1}^{K} \sum_{\eta\in\mathscr{S}(m)} \mathrm{e}^{-\beta H_\beta(\eta)} \sum_{k=1}^{m} \mathrm{e}^{\beta kU}\, \mathbb{1}_{\{\mathrm{supp}[\eta]\subset B_{L_\beta}(0)\}}\, \check{Z}_\beta^{(n_\beta-m-k)}\\
&\leq \sum_{m=1}^{K} \sum_{\eta\in\mathscr{S}(m)} \mathrm{e}^{-\beta H_\beta(\eta)} \sum_{k=1}^{m} \mathrm{e}^{\beta kU}\, \mathbb{1}_{\{\mathrm{supp}[\eta]\subset B_{L_\beta}(0)\}}\, (\rho_\beta)^{m+k}\, \check{Z}_\beta^{(n_\beta)}\big[1+o(1)\big]\\
&\leq \big[1+o(1)\big]\check{Z}_\beta^{(n_\beta)} \sum_{m=1}^{K} (\rho_\beta)^m \sum_{\eta\in\mathscr{S}(m)} \mathrm{e}^{-\beta H_\beta(\eta)} \sum_{k=1}^{m} \mathrm{e}^{-\beta k(\Delta-U)}\, \mathbb{1}_{\{\mathrm{supp}[\eta]\subset B_{L_\beta}(0)\}},
\end{aligned}$$

and we can proceed as in (20.4.21)–(20.4.22) to show that

$$\gamma_2(\beta) \leq C\,\check{Z}_\beta^{(n_\beta)}\,\mathrm{e}^{-\beta\varepsilon'} \tag{20.4.24}$$

for some $\varepsilon>0$ and some constant $C<\infty$ that is independent of β, i.e., $\gamma_2(\beta)$ is negligible. □

20.4.3 Atypicality of critical droplets

The following lemma was used in Sect. 20.2.1. Recall Definition 20.1, and note that $\mathscr{S}=\mathscr{S}^{(n_\beta)}$.

Lemma 20.11 $\lim_{\beta\to\infty}\mu_\beta(\tilde{\mathscr{C}}\backslash\mathscr{C}^+)/\mu_\beta(\mathscr{S})=0$.

Proof Similarly as in (20.4.19), we first write

$$\mu_\beta\big(\tilde{\mathscr{C}}\backslash\mathscr{C}^+\big)\le\mu_\beta(\tilde{\mathscr{C}})=\gamma_3(\beta)+|\Lambda_\beta|\big[1+o(1)\big]\sum^{*}_{\eta\in S_\beta^{(K)}}\sum_{\zeta\in\mathscr{S}_\beta^{(n_\beta-K)}}\\ \times\frac{e^{-\beta[H_\beta(\eta)+H_\beta(\zeta)]}}{Z_\beta^{(n_\beta)}}\mathbb{1}_{\{\mathrm{supp}[\eta]\subset B_{L_\beta}(0)\}}\mathbb{1}_{\{\mathrm{supp}[\zeta]\subset[B_{L_\beta}(0)]^c\}},\tag{20.4.25}$$

where the first sum runs over all configurations in $B_{L_\beta}(0)$ consisting of a protocritical droplet centred at the origin and a free particle elsewhere, and $\gamma_3(\beta)$ is a negligible term that arises from particles interacting across the boundary of $B_{L_\beta}(0)$, similar as the term $\gamma_2(\beta)$ in Sect. 20.4.2. The double sum in the right-hand side of (20.4.25) equals

$$\frac{\check{Z}_\beta^{(n_\beta-K)}}{Z_\beta^{(n_\beta)}}\big|B_{L_\beta}(0)\big|\big[1+o(1)\big]N\,e^{-\beta\bar{\Gamma}^\star}\big[1+o(1)\big],\tag{20.4.26}$$

where $\bar{\Gamma}^\star(=\Gamma^\star-K\Delta)$ is the energy of a critical droplet and N is the number of shapes of a critical droplet. By Lemma 20.8, $\check{Z}_\beta^{(n_\beta-K)}/Z_\beta^{(n_\beta)}=\mu_\beta(\mathscr{S}^{(n_\beta)})(\rho_\beta)^K[1+o(1)]$. Hence

$$\begin{aligned}\text{r.h.s. (20.4.25)}&=N\,|\Lambda_\beta|\,L_\beta^2\,e^{-\beta\bar{\Gamma}^\star}\,(\rho_\beta)^K\,\mu_\beta\big(\mathscr{S}^{(n_\beta)}\big)\big[1+o(1)\big]\\&=N\,|\Lambda_\beta|\,L_\beta^2\,e^{-\beta\Gamma^\star}\,\mu_\beta\big(\mathscr{S}^{(n_\beta)}\big)\big[1+o(1)\big],\quad\beta\to\infty,\end{aligned}\tag{20.4.27}$$

which is $o(\mu_\beta(\mathscr{S}^{(n_\beta)}))$ by (20.1.11). □

20.4.4 Typicality of starting configurations

In this section we prove the claim made in (20.1.9).

Proof Split

$$\mathscr{S}^{(n_\beta)}=\mathscr{S}=\mathscr{S}_L\cup(\mathscr{S}\setminus\mathscr{S}_L)=\mathscr{S}_L\cup\mathscr{U}_{>L},\tag{20.4.28}$$

where $\mathscr{U}_{>L}\subset\mathscr{S}$ are those configurations σ for which there exists an x such that $|\mathrm{supp}[\sigma]\cap B_{L_\beta}(x)|>L$. Then

$$\mu_\beta(\mathscr{U}_{>L})=\sum_{x\in\Lambda_\beta}\sum_{\sigma\in\mathscr{S}^{(n_\beta)}}\sum_{m=L+1}^{K}\mu_\beta(\sigma)\,\mathbb{1}_{\{|\mathrm{supp}[\sigma]\cap B_{L_\beta}(x)|=m\}}=|\Lambda_\beta|\big[\varphi(\beta)+\gamma(\beta)\big],\tag{20.4.29}$$

where

$$\varphi(\beta)=\sum_{m=L+1}^{K}\sum_{\eta\in S_\beta^{(m)}}\sum_{\substack{\zeta\in S_\beta^{(n_\beta-m)}\\ \eta\vee\zeta\in\mathscr{S}^{(n_\beta)}}}\frac{\mathrm{e}^{-\beta[H_\beta(\eta)+H_\beta(\zeta)]}}{Z_\beta^{(n_\beta)}}$$

$$\times\, \mathbb{1}_{\{\mathrm{supp}[\eta]\subset B_{L_\beta}(0)\}}\,\mathbb{1}_{\{\mathrm{supp}[\zeta]\subset[B_{L_\beta}(0)]^c\}} \qquad (20.4.30)$$

and $\gamma(\beta)$ is an error term arising from particles interacting across the boundary of $B_{L_\beta}(0)$. By the same argument as in (20.4.23), this term is negligible. Moreover,

$$\varphi(\beta)\leq\sum_{m=L+1}^{K}\frac{\check{Z}_\beta^{(n_\beta-m)}}{Z_\beta^{(n_\beta)}}\left(\sum_{\eta\in\mathscr{S}^{(m)}}\mathrm{e}^{-\beta\,H_\beta(\eta)}\,\mathbb{1}_{\{\mathrm{supp}[\eta]\subset B_{L_\beta}(0)\}}\right) \qquad (20.4.31)$$

$$\leq\big[1+o(1)\big]\,\mu_\beta\big(\mathscr{S}^{(n_\beta)}\big)\sum_{m=L+1}^{K}(\rho_\beta)^m\left(\sum_{\eta\in\mathscr{S}^{(m)}}\mathrm{e}^{-\beta H_\beta(\eta)}\,\mathbb{1}_{\{\mathrm{supp}[\eta]\subset B_{L_\beta}(0)\}}\right),$$

where in the last inequality we use Lemmas 20.8–20.10. Now proceed as in (20.4.21)–(20.4.22), via the cluster expansion, to get

$$\varphi(\beta)\leq\big[1+o(1)\big]\,A\,\mu_\beta\big(\mathscr{S}^{(n_\beta)}\big)\sum_{m=L+1}^{K}\sum_{j=1}^{m}\sum_{\substack{2\leq k_1,\dots,k_j\leq K\\ \sum_{i=1}^j k_i=m}}\mathrm{e}^{-\beta\sum_{i=1}^j[H_{k_i}+k_i\Delta-(\Delta-\delta_\beta)]}$$

$$\leq\big[1+o(1)\big]\,B\,\mu_\beta\big(\mathscr{S}^{(n_\beta)}\big)\,\mathrm{e}^{-\beta[\Gamma_{L+1}-(\Delta-\delta_\beta)]}, \qquad (20.4.32)$$

where H_k is the energy of a droplet with k particles that is closest to a square or quasi-square, $\Gamma_{L+1}=H_{L+1}+(L+1)\Delta$, and the second inequality uses the isoperimetric inequality together with the fact that $H_k+k\Delta$ is increasing in k for subcritical droplets.

On the other hand, by considering only those configurations in $\mathscr{U}_{>L}$ that have a droplet with $L+1$ particles, we get

$$\varphi(\beta)\geq\big[1+o(1)\big]\,A\,\mu_\beta\big(\mathscr{S}^{(n_\beta)}\big)\,\mathrm{e}^{-\beta[\Gamma_{L+1}-(\Delta-\delta_\beta)]}. \qquad (20.4.33)$$

Combining (20.4.29) and (20.4.32)–(20.4.33), we conclude that

$$\lim_{\beta\to\infty}\mu_\beta(\mathscr{U}_{>L})/\mu_\beta\big(\mathscr{S}^{(n_\beta)}\big)=0 \text{ if and only if } \lim_{\beta\to\infty}|\Lambda_\beta|\,\mathrm{e}^{-\beta[\Gamma_{L+1}-(\Delta-\delta_\beta)]}=0. \qquad (20.4.34)$$

□

20.5 The critical droplet is the threshold

We show that our estimates on capacities imply that the *average* probability under the Gibbs measure μ_β of destroying a supercritical droplet and returning to a configuration in $\mathscr{S}_L$ is exponentially small in β. The proof for Glauber dynamics can be read off immediately, and fullfills the promise made in Sect. 19.1.3, Item 3.

Proof Pick $M \geq \ell_c$. Recall (7.1.19)–(7.1.20) that (7.1.39). Summing over $\sigma \in \partial\mathscr{D}_M$, the internal boundary of $\mathscr{D}_M$, we get that

$$\frac{\sum_{\sigma\in\partial\mathscr{D}_M} \mu_\beta(\sigma)c_\beta(\sigma)\mathbb{P}_\sigma(\tau_{\mathscr{S}_L} < \tau_{\mathscr{D}_M})}{\sum_{\sigma\in\partial\mathscr{D}_M} \mu_\beta(\sigma)c_\beta(\sigma)} = \frac{\mathrm{cap}(\mathscr{S}_L, \mathscr{D}_M)}{\sum_{\sigma\in\partial\mathscr{D}_M} \mu_\beta(\sigma)c_\beta(\sigma)}. \tag{20.5.1}$$

Clearly, the left-hand side of (20.5.1) is the escape probability to $\mathscr{S}_L$ from $\partial\mathscr{D}_M$ *averaged* with respect to the canonical Gibbs measure μ_β conditioned on $\partial\mathscr{D}_M$ and weighted by the outgoing rate c_β. To show that this quantity is exponentially small in β, which is our goal, it suffices to show that in the right-hand side of (20.5.1) the denominator is large compared to the numerator.

By Lemma 20.5,

$$\mathrm{cap}(\mathscr{S}_L, \mathscr{D}_M) \leq \mathrm{cap}\big(\mathscr{S}_L, (\mathscr{S}^c \setminus \tilde{\mathscr{C}}) \cup \mathscr{C}^+\big) = N\,|\Lambda_\beta|\,\frac{4\pi}{\Delta\beta}\,\mathrm{e}^{-\beta\Gamma^\star}\,\mu_\beta(\mathscr{S})\big[1+o(1)\big]. \tag{20.5.2}$$

On the other hand, note that $\partial\mathscr{D}_M$ contains all configurations σ for which there is an $M \times M$ droplet somewhere in Λ_β, all L_β-boxes not containing this droplet carry at most K particles, and there is a free particle somewhere in Λ_β. The last condition ensures that $c_\beta(\sigma) \geq 1$. Therefore we can use Lemma 20.8 to estimate

$$\begin{aligned}\sum_{\sigma\in\partial\mathscr{D}_M} \mu_\beta(\sigma)c_\beta(\sigma) &\geq |\Lambda_\beta|\,\mathrm{e}^{-\beta H_{M^2}}\,\frac{\check{Z}_\beta^{(n_\beta - M^2)}}{Z_\beta^{(n_\beta)}} \\ &= |\Lambda_\beta|\,\mathrm{e}^{-\beta H_{M^2}}\,(\rho_\beta)^{M^2}\,\mu_\beta(\mathscr{S})\big[1+o(1)\big],\end{aligned} \tag{20.5.3}$$

where H_{M^2} is the energy of an $M \times M$ droplet. Combining (20.5.2)–(20.5.3) we find that the right-hand side of (20.5.1) is bounded from above by

$$\left(N\frac{4\pi}{\Delta\beta}\right)\frac{\exp[-\beta\Gamma^\star]}{\exp[-\beta(H_{M^2} + \Delta M^2)]}\big[1+o(1)\big], \tag{20.5.4}$$

which is exponentially small in β because $\Gamma^\star > H_{M^2} + \Delta M^2$ for all $M \geq \ell_c$. $\square$

20.6 Bibliographical notes

1. The results in this chapter are taken from Bovier, den Hollander and Spitoni [32]. Section 20.4 corrects flaws in the original proofs of Lemmas 20.8 and 20.11.

2. In Gaudillière, den Hollander, Nardi, Olivieri, and Scoppola [118–120] the same nucleation problem as in Sect. 20.1 is studied with the help of the pathwise approach to metastability. Only the exponential asymptotics of the nucleation time is obtained, but for a much wider class of initial distributions than we can presently handle with the potential-theoretic approach. The techniques developed in these papers center around the idea of *approximating* the low-temperature and low-density Kawasaki lattice gas by an *ideal gas* (without interaction) and showing that this ideal gas stays *close to equilibrium* while exchanging particles with droplets that are growing and shrinking. In this way, the large system is shown to behave essentially like the union of many small independent systems, leading to homogeneous nucleation. The proofs are long and complicated, but they provide considerable detail about the *typical trajectory* of the system prior to and shortly after the onset of nucleation, something the potential-theoretic approach cannot offer.

3. If we worked in the *grand-canonical ensemble*, then we would have to consider Kawasaki dynamics on Λ_β with an open boundary and with Hamiltonian

$$H^{gc}(\sigma) = -U \sum_{\substack{(x,y)\in\Lambda_\beta \\ x\sim y}} \sigma(x)\sigma(y) + \Delta \sum_{x\in\Lambda_\beta} \sigma(x), \qquad \sigma \in S_\beta, \tag{20.6.1}$$

where $\Delta > 0$ is the usual activity parameter mimicking the presence of an infinite gas reservoir around Λ_β. This was the setting of Chap. 17 for small volumes. For large volumes, however, even with this Hamiltonian we still have to face all the difficult issues of non-locality we struggled with in Sects. 20.2–20.5.

4. As for Theorem 19.4(c), we expect Theorem 20.3(b) to hold for values of M that grow with β as $M = e^{o(\beta)}$.

5. Gois and Landim [128] consider Kawasaki dynamics at inverse temperature β on a two-dimensional torus of size $L(\beta)$ in the limit as $\beta \downarrow 0$ when $L(\beta) \to \infty$. Initially the particles form a square of size n. It is shown that, under certain growth restrictions for $L(\beta)$, most of the time the particles form a square of size n, and that there is a time scale $L^2(\beta)\theta(\beta)$ on which the centre of the square performs a Brownian motion. It is shown that $C_1(n/L^2(\beta))\,e^{2\beta} \leq \theta(\beta) \leq C_2 n^2\,e^{2\beta}$ for some $C_1, C_2 \in (0,\infty)$.

6. The extension of our results to higher dimensions is limited only by the combinatorial problems involved in the computation of the number of critical droplets (which is hard in the case of Kawasaki dynamics) and of the probability for simple random walk to hit a critical droplet of a given shape when coming from far. Recall Sect. 18.6.

7. There appears to be no work that deals with metastability of Kawasaki dynamics in infinite volume, contrary to Glauber dynamics (recall Sect. 19.7, Item 5).

Part VIII
Applications: Lattice Systems in Small Volumes at High Densities

Part VIII describes lattice systems in small volumes at high densities. The focus is on the zero-range process, which consists of a collection of continuous-time simple random walks with on-site attraction and no on-site repulsion. We consider the limit where the particle density is high, show that the process spends most of its time in a "condensed state", i.e., a configuration where most of the particles pile up on a single site, and prove that the process evolves via a "metastable hopping" of this pile from one site to another. Both the hopping time and the hopping distribution are computed.

Chapter 21
The Zero-Range Process

The shop seemed to be full of all manner of curious things—but the oddest part of it all was that, whenever she looked at any shelf, to make out exactly what it had on it, that particular shelf was always quite empty, though the others round it were crowded as full as they could hold.

(Lewis Carroll, Through the Looking-Glass, and what Alice found there)

The zero-range process offers yet another example of a system for which potential-theoretic methods can be used to describe metastable behaviour. The free energy landscape is of a different nature than what we encountered in the models treated so far. In particular, there is no temperature parameter, and the key quantity to control is entropy. This necessitates a different approach to the choice of test functions to estimates capacities, which is worthwhile to expose.

21.1 Model and basic properties

Let $N \in \mathbb{N}$ and $S = \{1, \dots, L\}$, $L \in \mathbb{N}$. The zero-range process $Y = (\eta(t))_{t\geq 0}$ on S with N particles is the continuous time Markov process with state space

$$E_{N,S} = \left\{ \eta = (\eta_x)_{x\in S} \in \mathbb{N}_0^S \colon \sum_{x\in S} \eta_x = N \right\} \tag{21.1.1}$$

such that in configuration η a particle jumps from site x to site y at rate $g(\eta_x) r(x,y)$. Here, $\eta_x \in \mathbb{N}_0$ represents the number of particles at site $x \in S$, $r(\cdot,\cdot)$ is an irreducible probability transition kernel associated with a reversible random walk $X = (X(t))_{t\geq 0}$ on S, and g is chosen as

$$g(0) = 0, \qquad g(1) = 1, \qquad g(n) = \frac{a(n)}{a(n-1)}, \qquad n \in \mathbb{N}\backslash\{1\}, \tag{21.1.2}$$

with

$$a(0) = 1, \qquad a(n) = n^\alpha \quad \text{for some } \alpha \in (1,\infty). \tag{21.1.3}$$

A. Bovier, F. den Hollander, *Metastability*,
Grundlehren der mathematischen Wissenschaften 351,
DOI 10.1007/978-3-319-24777-9_21

Formally, Y is defined through its generator $L_{N,S}$ acting on functions $F \in C(E_{N,S}, \mathbb{R})$ as

$$(L_{N,S}F)(\eta) = \sum_{x,y \in S} g(\eta_x) r(x,y) \big[F\big(\eta^{x,y}\big) - F(\eta) \big], \tag{21.1.4}$$

where $\eta^{x,y}$ is the configuration obtained from η by moving a particle from site x to site y. Note that the zero-range dynamics preserve particles.

Lemmas 21.1–21.2 and Theorem 21.3 below are well-known results for the zero-range process in equilibrium. For references, see the bibliography in Sect. 21.6.

Lemma 21.1 *Y is irreducible, and is reversible with respect to the unique invariant probability measure μ_N given by*

$$\mu_{N,S}(\eta) = \frac{N^\alpha}{Z_{N,S}} \frac{m_*^\eta}{a(\eta)}, \qquad \eta \in E_{N,S}, \tag{21.1.5}$$

with

$$m_*^\eta = \prod_{x \in S} m_*(x)^{\eta_x}, \qquad a(\eta) = \prod_{x \in S} a(\eta_x), \tag{21.1.6}$$

where

$$m_*(x) = \frac{m(x)}{M_*}, \qquad M_* = \max_{x \in S} m(x), \tag{21.1.7}$$

with m the invariant measure of the random walk X, and $Z_{N,S}$ denotes the normalising partition function

$$Z_{N,S} = N^\alpha \sum_{\zeta \in E_{N,S}} \frac{m_*^\zeta}{a(\zeta)}. \tag{21.1.8}$$

Note that $m_*(x) = 1$ for all $x \in S_*$ with

$$S_* = \big\{ y \in S \colon m(y) = M_* \big\}. \tag{21.1.9}$$

Lemma 21.2 *For L fixed,*

$$\lim_{N\to\infty} Z_{N,S} = Z_S = \frac{|S_*|}{\Gamma(\alpha)} \prod_{x \in S} \Gamma_x = |S_*| \, \Gamma(\alpha)^{|S_*|-1} \prod_{y \notin S_*} \Gamma_y, \tag{21.1.10}$$

where

$$\Gamma_x = \sum_{j \in \mathbb{N}_0} \frac{m_*(x)^j}{a(j)}, \qquad \Gamma(\alpha) = \sum_{j \in \mathbb{N}_0} \frac{1}{a(j)}. \tag{21.1.11}$$

The interesting feature of the zero-range process with a g-function given by (21.1.2) is that it exhibits a *condensation phenomenon*. Namely, for large N the invariant measure μ_N concentrates on disjoint sets of configurations $\mathscr{E}_N^x$, $x \in S_*$, which are defined as follows. Given a sequence $(\ell_N)_{N\in\mathbb{N}}$ such that

$$\lim_{N\to\infty} \ell_N = \infty, \qquad \lim_{N\to\infty} \frac{\ell_N}{N} = 0, \tag{21.1.12}$$

we say that a configuration has a *condensate* at site $x \in S_*$ when it belongs to the set

$$\mathscr{E}_N^x = \{\eta \in E_{N,S} \colon \eta_x \geq N - \ell_N\}. \tag{21.1.13}$$

Theorem 21.3 *Suppose that* $L = L(N)$ *is such that* $\lim_{N\to\infty} L(N)/N = 0$. *Then there exists a sequence* $(\ell_N)_{N\in\mathbb{N}}$ *satisfying* (21.1.12) *such that*

$$\lim_{N\to\infty} \mu_{N,S}\left(\bigcup_{x\in S_*} \mathscr{E}_N^x\right) = 1. \tag{21.1.14}$$

Note that the configurations η^x, $x \in S_*$, given by

$$\eta_x^x = N, \qquad \eta_y^x = 0, \quad y \in S\backslash\{x\}, \tag{21.1.15}$$

have maximal measure. When L is independent of N, this maximal measure is bounded away from zero uniformly in N.

The condensation phenomenon in Theorem 21.3 already occurs when the particle density $\rho = N/L$ exceeds a certain critical particle density $\rho_c \in (0,\infty)$, which depends on the parameters of the model.

21.2 Metastable behaviour

The question that will be addressed in this section is how Y moves between the different condensate configurations.

21.2.1 Finite system size

First we state our results when L is kept fixed and $N \to \infty$. Recall Definition 8.1.4.

Theorem 21.4 Y *is metastable with respect to the set* $\mathscr{M} = \bigcup_{x\in S_*} \eta^x$ *with* $\rho = O(L(L^2+N)/N^{\alpha+1}r')$, *where* $r' = \inf_{u\in S} r(u, u\pm 1)$.

The proof of Theorem 21.4 will be given in Sect. 21.5 and is based on a computation of capacities $\text{cap}_{N,S}(\eta,\zeta)$ between configurations $\eta,\zeta\in E_{N,S}$, where $\text{cap}_{N,S}$ refers to capacity associated with Y.

Define

$$I_\alpha(s)=\int_s^{1-s}u^\alpha(1-u)^\alpha\,du,\quad 0\le s\le\tfrac{1}{2},\tag{21.2.1}$$

and

$$\eta^U=\bigcup_{x\in U}\eta^x,\quad \emptyset\neq U\subseteq S_*.\tag{21.2.2}$$

Theorem 21.5 (Sharp asymptotics of capacities) *Let $S_*^1,S_*^2\subsetneq S_*$ be non-empty disjoint sets. Then*

$$\begin{aligned}\text{cap}_{N,S}\big(\eta^{S_*^1},\eta^{S_*^2}\big)&=\big[1+o(1)\big]\frac{1}{2N^{\alpha+1}\,M_*\,|S_*|\,I_\alpha(0)\,\Gamma(\alpha)}\\&\quad\times\inf_{W\in\mathscr{W}(S_*^1,S_*^2)}\sum_{x,y\in S_*}\text{cap}_S(x,y)[W_y-W_x]^2,\end{aligned}\tag{21.2.3}$$

where $\text{cap}_{N,S}$ denotes capacity for Y, cap_S denotes capacity for X, and

$$\mathscr{W}\big(S_*^1,S_*^2\big)=\big\{W=(W_z)_{z\in S}\in[0,1]^{S_*}\colon\ W|_{S_*^1}=1,\ W|_{S_*^2}=0\big\}.\tag{21.2.4}$$

Remark 21.6 Note that the second line in (21.2.3) is the conductance between S_*^1 and S_*^2 of a resistor network on S_* with conductances $\text{cap}_S(x,y)$ between sites $x,y\in S_*$.

Theorem 21.5 allows us to use Corollary 7.11 and Theorem 8.45 to obtain the following result for the metastable exit times $\tau_{\mathscr{M}\backslash\eta^x}$, $x\in S_*$.

Corollary 21.7 (Mean and exponential law of metastable exit times) *For every $x\in S_*$ the metastable exit time $\tau_{\mathscr{M}\backslash\eta^x}$*

(i) *has asymptotic mean*

$$\mathbb{E}_{\eta^x}[\tau_{\mathscr{M}\backslash\eta^x}]=\big[1+o(1)\big]\frac{N^{\alpha+1}\,M_*\,I_\alpha(0)\,\Gamma(\alpha)}{\sum_{y\in S_*\backslash\{x\}}\text{cap}_S(x,y)},\quad N\to\infty,\tag{21.2.5}$$

(ii) *on the scale of its mean has asymptotic exponential distribution*

$$\mathbb{P}_{\eta^x}\big[\tau_{\mathscr{M}\backslash\eta^x}>t\,\mathbb{E}_{\eta^x}[\tau_{\mathscr{M}\backslash\eta^x}]\big]=\big[1+o(1)\big]\,\mathrm{e}^{-t[1+o(1)]},\quad N\to\infty.\tag{21.2.6}$$

Remark 21.8 Combining the previous remark with Corollary 21.7 we see that, in the limit as $N\to\infty$, on the time scale $N^{\alpha+1}$ the zero-range process observed when it hits the set $\mathscr{M}=\cup_{x\in S_*}\eta^x$ behaves like a continuous-time random walk with transition rates $\bar r(x,y)$ given by $\bar r(x,y)=M_*I_\alpha(0)\Gamma(\alpha)\text{cap}_S(x,y)/\sum_{z\in S_*\backslash\{x\}}\text{cap}_S(x,z)$.

21.2.2 Diverging system size

Next we state our results when $L = L(N)$ and $N \to \infty$ with

$$\lim_{N\to\infty} L(N) = \infty, \qquad \lim_{N\to\infty} \frac{L(N)}{N} = 0. \tag{21.2.7}$$

In this case the transitions rates $r(x, y)$ and the set S_* will typically depend on N. We suppress this dependence to lighten the notation.

Define

$$\mathscr{E}_N = \mathscr{E}_N(S_*) = \bigcup_{x\in S_*} \mathscr{E}_N^x. \tag{21.2.8}$$

For general disjoint non-empty sets S_*^1, S_*^2 we can only derive a lower bound and an upper bound for $\mathrm{cap}_{N,S}(\mathscr{E}_N(S_*^1), \mathscr{E}_N(S_*^2))$ that coincide up to a constant. But for partitions of S_* we can get more.

Theorem 21.9 (Sharp asymptotics of capacities) *Suppose that $L(N)$ satisfies* (21.2.7). *Let S_*^1, S_*^2 be a partition of S_*. Then*

$$\begin{aligned}&\mathrm{cap}_{N,S}\big(\mathscr{E}_N(S_*^1), \mathscr{E}_N(S_*^2)\big)\\ &\quad = \big[1+o(1)\big]\,\frac{1}{N^{\alpha+1}\,M_*\,|S_*|\,I_\alpha(0)\,\Gamma(\alpha)} \sum_{x\in S_*^1, y\in S_*^2} \mathrm{cap}_S(x,y).\end{aligned} \tag{21.2.9}$$

As before, Theorem 21.9 allows us to use Corollary 7.11 to obtain the following result for the metastable exit times $\tau_{\mathscr{E}_N\backslash\mathscr{E}_N^x}$, $x \in S_*$, where we recall that $\nu_{A,B}$ denotes the last-exit biased distribution on A for the transition from A to B.

Corollary 21.10 (Mean metastable exit times) *Suppose that $L(N)$ satisfies condition* (21.2.7). *For every $x \in S_*$ the metastable exit time $\tau_{\mathscr{E}_N\backslash\mathscr{E}_N^x}$ has asymptotic mean*

$$\mathbb{E}_{\nu_{\mathscr{E}_N^x,\mathscr{E}_N\backslash\mathscr{E}_N^x}}[\tau_{\mathscr{E}_N\backslash\mathscr{E}_N^x}] = \big[1+o(1)\big]\,\frac{N^{\alpha+1}\,M_* I_\alpha(0)\,\Gamma(\alpha)}{\sum_{y\in S_*\backslash\{x\}} \mathrm{cap}_S(x,y)}, \qquad N\to\infty. \tag{21.2.10}$$

Remark 21.11 We would like to show that the assertion in Corollary 21.10 also holds for the process starting in a single configuration $\eta^x \in \mathscr{M}$, and that the law of the exit time is exponential. In Bovier, Bianchi and Ioffe [25] such results were obtained for the Curie-Weiss model with random magnetic field described in Chap. 15, through the use of coupling techniques. Such techniques, however, seem difficult to implement for the zero-range model.

21.3 Capacity estimates

In this section we derive lower bounds and upper bounds on capacities that coincide in the limit as $N \to \infty$ with L fixed. These bounds will be used in Sect. 21.4 to prove Theorem 21.5.

21.3.1 Lower bound

We begin by proving an a priori bound showing that the equilibrium potential is almost constant on the sets $\mathscr{E}_N^x$, $x \in S_*$.

Lemma 21.12 *Let $S_*^1, S_*^2 \subsetneq S_*$ be non-empty disjoint sets, and let W denote the equilibrium potential for the capacitor $(\eta^{S_*^1}, \eta^{S_*^2})$. Then there is a constant K_α such that*

$$\left|W(\xi) - W(\xi')\right| \le K_\alpha \frac{L(L^2+N)}{N^{\alpha+1} r'}, \qquad \xi, \xi' \in \mathscr{E}_N^z,\ z \in S_*, \tag{21.3.1}$$

where $r' = \inf_{u \in S} r(u, u \pm 1)$.

Proof Clearly,

$$\begin{aligned} W(\xi) &= \mathbb{P}_\xi[\tau_{\eta^{S_*^1}} < \tau_{\eta^{S_*^2}}] \\ &= \mathbb{P}_\xi[\tau_{\xi'} < \tau_{\eta^{S_*^1}}, \tau_{\eta^{S_*^1}} < \tau_{\eta^{S_*^2}}] + \mathbb{P}_\xi[\tau_{\eta^{S_*^1}} < \tau_{\eta^{S_*^2}}, \tau_{\xi'} > \tau_{\eta^{S_*^1}}] \\ &= \mathbb{P}_\xi[\tau_{\xi'} < \tau_{\eta^{S_*^1}}]\mathbb{P}_{\xi'}[\tau_{\eta^{S_*^1}} < \tau_{\eta^{S_*^2}}] + \mathbb{P}_\xi[\tau_{\eta^{S_*^1}} < \tau_{\xi' \cup \eta^{S_*^2}}] \\ &= \left(1 - \mathbb{P}_\xi[\tau_{\eta^{S_*^1}} < \tau_{\xi'}]\right) W(\xi') + \mathbb{P}_\xi[\tau_{\eta^{S_*^1}} < \tau_{\xi' \cup \eta^{S_*^2}}]. \end{aligned} \tag{21.3.2}$$

With the help of the renewal equation in (8.2.2), this yields

$$\left(1 - \frac{\mathbb{P}_\xi[\tau_{\eta^{S_*^1}} < \tau_\xi]}{\mathbb{P}_\xi[\tau_{\xi'} < \tau_\xi]}\right) W(\xi') \le W(\xi) \le W(\xi') + \frac{\mathbb{P}_\xi[\tau_{\eta^{S_*^1}} < \tau_\xi]}{\mathbb{P}_\xi[\tau_{\xi'} < \tau_\xi]}. \tag{21.3.3}$$

In Proposition 21.24 below we show that

$$\frac{\mathbb{P}_\xi[\tau_{\eta^{S_*^1}} < \tau_\xi]}{\mathbb{P}_\xi[\tau_{\xi'} < \tau_\xi]} \le K_\alpha \frac{L(L^2+N)}{N^{\alpha+1} r'}. \tag{21.3.4}$$

Combining (21.3.3)–(21.3.4), we get the claim. □

Remark 21.13 Lemma 21.12 is most useful when L is fixed and $N \to \infty$, but it also allows us to include cases with slowly growing $L = L(N)$.

We first prove a lower bound on capacities for arbitrary L and N.

Proposition 21.14 *Let $S_*^1, S_*^2 \subsetneq S_*$ be non-empty disjoint sets. Put*

$$\delta = \max\left\{\frac{L(L^2+N)}{N^{\alpha+1}r'}, \frac{\ell_N}{N}\right\}, \tag{21.3.5}$$

and assume that $0 < \delta \ll 1$. Then

$$\begin{aligned}\mathrm{cap}_{N,S}\big(\mathscr{E}_N(S_*^1), \mathscr{E}_N(S_*^2)\big) \geq \big[1+O(\delta)\big] \inf_{W\in\mathscr{W}(S_*^1,S_*^2)} \sum_{x,y\in S_*} \mathrm{cap}_S(x,y)[W_y - W_x]^2 \\ \times \frac{1}{2N^{\alpha+1}\, M_* I_\alpha(0)\, Z_{N,S}} \sum_{k=0}^{\ell_N} \sum_{\xi\in E_{k,S_0}} \frac{m_*^\xi}{a(\xi)},\end{aligned} \tag{21.3.6}$$

where $S_0 = S\backslash\{u,v\}$ for any $u, v \in S_*$.

Proof The proof comes in 4 Steps.

1. As usual, a lower bound is obtained by using the monotonicity of the Dirichlet form

$$\begin{aligned}&\mathscr{E}_N(h) = \tfrac{1}{2} \sum_{z,w\in S} \sum_{\eta\in E_{N,S}} \mu_{N,S}(\eta) g(\eta_z) r(z,w)\big[h\big(\eta^{z,w}\big) - h(\eta)\big]^2,\\ &h \in \mathscr{H}_N\big(\mathscr{E}_N\big(S_*^1\big), \mathscr{E}_N\big(S_*^2\big)\big),\end{aligned} \tag{21.3.7}$$

where

$$\mathscr{H}_N(A,B) = \{h\colon\ E_{N,S} \to \mathbb{R}_+\colon\ h|_A = 1,\ h|_B = 0\}. \tag{21.3.8}$$

The strategy is to set rates to zero so that disjoint one-dimensional paths are obtained. Afterwards we can use that the sum of the Dirichlet forms over the one-dimensional paths, which are computable, yields a lower bound for the Dirichlet form in (21.3.7), and hence for the capacity.

The construction goes as follows (see Figs. 21.1, 21.2, 21.3):

- For each $\xi \in E_{k,S}$, $k \in \{0,\dots,\ell_N\}$, we obtain a one-dimensional path as follows. Let $\{\xi, p_{x,y}\} \in E_{N-1,S}$ be the configuration given by $\{\xi, p_{x,y}\}_z = \xi_z$ for $z \in S\backslash\{x,y\}$, and $\{\xi, p_{x,y}\}_x = \xi_x + p$ and $\{\xi, p_{x,y}\}_y = \xi_y + N - k - p - 1$ for $p \in \{0,\dots,N-k-1\}$. For each pair $x, y \in S_*$, the one-dimensional path consists of the path-segments $\{\xi, p_{x,y}\}$, where the excess $N-k$ particles on site x jump one by one until they reach site y (only one particle is jumping at any time).
- The path-segments are disjoint for the following reason. Let $\{\xi, p_{x,y}\}, \{\xi', p_{x,y}\} \in E_{N-1,S}$ be two different path-segments. Suppose that at some time t these paths-segments coincide in a single configuration due to a jump of the jumping particle. However, since $\{\xi, p_{x,y}\}$ and $\{\xi', p_{x,y}\}$ are different, the sites at which this particle is at time t in these segments must differ. In the next step particles from

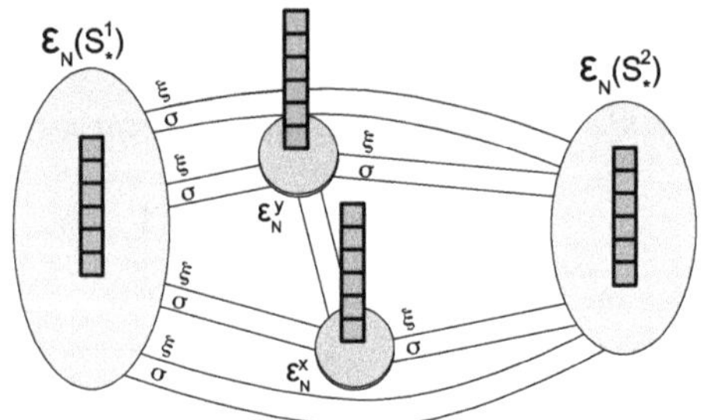

Fig. 21.1 Restriction of the transition rates

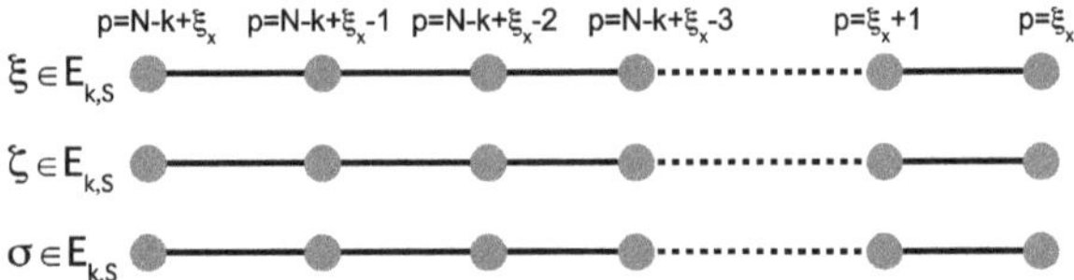

Fig. 21.2 Jump of a particle in a path-segment

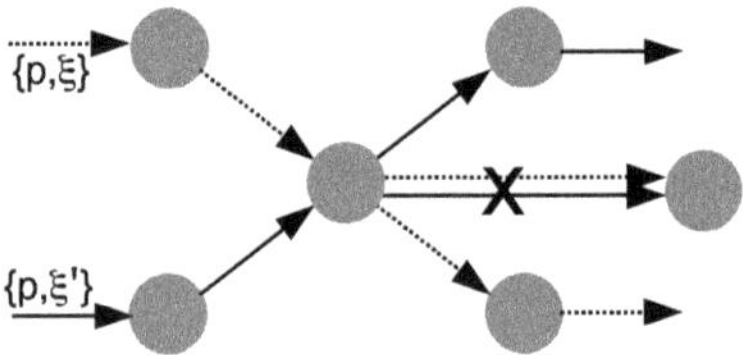

Fig. 21.3 Disjoint paths

different sites jump in such a way that the resulting configurations are different, and hence the paths-segments cannot merge.

With the construction described above, we obtain one-dimensional paths that consist of a Dirichlet form of a zero-range process on two sites multiplied by a term that we can estimate by the capacity of the underlying random walk.

2. Let $\mathfrak{d}_z \in E_{1,S}$ be the configuration with exactly one particle at site $z \in S$ (the jumping particle). Let $h^* \in \mathscr{H}_N(\mathscr{E}_N(S_*^1), \mathscr{E}_N(S_*^2))$ be the minimiser of (21.3.7). Then

$$\begin{aligned}\operatorname{cap}_{N,S}\big(\mathscr{E}_N(S_*^1), \mathscr{E}_N(S_*^2)\big) &\geq \mathscr{E}_N^{\mathrm{lb}}(h^*)\\ &= \tfrac{1}{4}\sum_{k=0}^{\ell_N}\sum_{\xi\in E_{k,S}}\sum_{x,y\in S_*}\sum_{p=0}^{N-k-1}\sum_{z,w\in S}\mu_{N,S}\big(\{\xi,p_{x,y}\}+\mathfrak{d}_z\big)g\big(\{\xi,p_{x,y}\}_z+1\big)r(z,w)\\ &\quad\times\big[h^*\big(\{\xi,p_{x,y}\}+\mathfrak{d}_w\big)-h^*\big(\{\xi,p_{x,y}\}+\mathfrak{d}_z\big)\big]^2. \end{aligned} \tag{21.3.9}$$

Here an extra factor $\frac{1}{2}$ arises because the sum over x, y counts the configurations $\{\xi, p_{x,y}\}_{p=0}^{N-k-1}$, $\xi \in E_{k,S}$, $0 \le k \le \ell_N$, twice. Inserting the definition of g in (21.1.2) and μ_N in (21.1.5) into (21.3.9), we get

$$\begin{aligned}
&\mathscr{E}_N^{\mathrm{lb}}(h^*)\\
&\quad = \frac{N^\alpha}{4Z_{N,S}M_*}\sum_{k=0}^{\ell_N}\sum_{\xi\in E_{k,S}}\sum_{x,y\in S_*}\frac{m_*^\xi}{a(\xi\backslash\{x,y\})}\\
&\qquad \times \sum_{p=0}^{N-k-1}\frac{1}{a(\xi_x+p)a(\xi_y+N-k-p-1)}\\
&\qquad \times \sum_{z,w\in S} m(z)r(z,w)\big[h^*\big(\{\xi,p_{x,y}\}+\mathfrak{d}_w\big)-h^*\big(\{\xi,p_{x,y}\}+\mathfrak{d}_z\big)\big]^2, \qquad (21.3.10)
\end{aligned}$$

where $\xi\backslash\{x,y\}$ is the configuration without the sites x, y. Next, fix $x, y \in S_*$ and $\xi \in E_{k,S}$, and let $f_{x,y}\colon S \to \mathbb{R}$ be given by

$$f_{x,y}(v) = \frac{h^*(\{\xi,p_{x,y}\}+\mathfrak{d}_v)-h^*(\{\xi,p_{x,y}\}+\mathfrak{d}_y)}{h^*(\{\xi,p_{x,y}\}+\mathfrak{d}_x)-h^*(\{\xi,p_{x,y}\}+\mathfrak{d}_y)}. \qquad (21.3.11)$$

Note that

$$f_{x,y} \in \mathscr{B}(x,y) = \big\{f\colon S \to \mathbb{R}_+\colon\ f(x)=1,\ f(y)=0\big\}. \qquad (21.3.12)$$

Inserting $f_{x,y}$ into (21.3.10), we see that the sum over $z, w \in S$ equals

$$2\mathscr{E}_S(f_{x,y})\big[h^*\big(\{\xi,p_{x,y}\}+\mathfrak{d}_x\big)-h^*\big(\{\xi,p_{x,y}\}+\mathfrak{d}_y\big)\big]^2, \qquad (21.3.13)$$

where $\mathscr{E}_S$ is the Dirichlet form associated with the random walk on S. Since $f_{x,y} \in \mathscr{B}(x,y)$ and $\mathscr{E}_S(f_{x,y}) \ge \mathrm{cap}_S(x,y)$, we get from (21.3.10) that

$$\begin{aligned}
&\mathscr{E}_N^{\mathrm{lb}}(h^*)\\
&\quad \ge \frac{N^\alpha}{2Z_{N,S}M_*}\sum_{k=0}^{\ell_N}\sum_{\xi\in E_{k,S}}\sum_{x,y\in S_*}\mathrm{cap}_S(x,y)\frac{m_*^\xi}{a(\xi\backslash\{x,y\})}\\
&\qquad \times \sum_{p=0}^{N-k-1}\frac{1}{a(\xi_x+p)a(\xi_y+N-k-p-1)}\\
&\qquad \times \big[h^*\big(\{\xi,p_{x,y}\}+\mathfrak{d}_x\big)-h^*\big(\{\xi,p_{x,y}\}+\mathfrak{d}_y\big)\big]^2\\
&\quad \ge \frac{N^\alpha}{2Z_{N,S}M_*}\sum_{k=0}^{\ell_N}\sum_{\xi\in E_{k,S}}\inf_{W(\xi)\in\mathscr{W}(S_*^1,S_*^2)}\sum_{x,y\in S_*}\mathrm{cap}_S(x,y)\frac{m_*^\xi}{a(\xi\backslash\{x,y\})}
\end{aligned}$$

$$\times \inf_{\substack{h_{W(\xi)}\in\mathscr{H}_N(\mathscr{E}_N(S_*^1),\mathscr{E}_N(S_*^2))\\ h_{W(\xi)}(\eta)=W_x(\xi)\ \forall\eta\in\mathscr{E}_N^x\\ h_{W(\xi)}(\eta)=W_y(\xi)\ \forall\eta\in\mathscr{E}_N^y}} \sum_{p=0}^{N-k-1} \frac{[h_{W(\xi)}(\{\xi,p_{x,y}\}+\mathfrak{d}_x)-h_{W(\xi)}(\{\xi,p_{x,y}\}+\mathfrak{d}_y)]^2}{a(\xi_x+p)a(\xi_y+N-k-p-1)}$$

$$= \frac{N^\alpha}{2Z_{N,S}M_*}\sum_{k=0}^{\ell_N}\sum_{\xi\in E_{k,S}} \inf_{W(\xi)\in\mathscr{W}(S_*^1,S_*^2)} \sum_{x,y\in S_*} \mathrm{cap}_S(x,y)\frac{m_*^\xi}{a(\xi\backslash\{x,y\})}$$
$$\times\left[W_x(\xi)-W_y(\xi)\right]^2$$
$$\times \inf_{h_\xi\in\mathscr{H}_N(\mathscr{E}_N(S_*^1\cup x),\mathscr{E}_N(S_*^2\cup y))} \sum_{p=0}^{N-k-1}\frac{[h_\xi(\xi_x+p+1)-h_\xi(\xi_x+p)]^2}{a(\xi_x+p)a(\xi_y+N-k-p-1)}. \tag{21.3.14}$$

3. Due to the boundary conditions on the function h_ξ (recall (21.3.8)), the last factor in (21.3.14) reduces to

$$\inf_{h_\xi\in\mathscr{H}_N(\mathscr{E}_N(S_*^1\cup x),\mathscr{E}_N(S_*^2\cup y))} \sum_{p=\ell_N-k+\xi_y}^{N-\ell_N-\xi_x-1}\frac{[h_\xi(\xi_x+p+1)-h_\xi(\xi_x+p)]^2}{a(\xi_x+p)a(\xi_y+N-k-p-1)}. \tag{21.3.15}$$

This is just the Dirichlet form of a zero-range process living on the two sites x and y only, which is minimized by the function

$$H(x)=\frac{\sum_{q=\ell_N-k+\xi_x+\xi_y+1}^{x} a(q-1)a(\xi_x+\xi_y+N-k-q)}{\sum_{q=\ell_N-k+\xi_x+\xi_y+1}^{N-\ell_N} a(q-1)a(\xi_x+\xi_y+N-k-q)}. \tag{21.3.16}$$

For $p\in[\ell_N-k+\xi_y, N-\ell_N-\xi_x-1]$ we have

$$H(\xi_x+p+1)-H(\xi_x+p)=\frac{a(\xi_x+p)a(\xi_y+N-k-p-1)}{\sum_{q=\ell_N-k+\xi_x+\xi_y}^{N-\ell_N-1} a(q)a(\xi_x+\xi_y+N-k-q-1)}. \tag{21.3.17}$$

Inserting (21.3.17) into (21.3.15), we obtain

$$\frac{1}{\sum_{q=\ell_N-k+\xi_x+\xi_y}^{N-\ell_N-1} a(q)a(\xi_x+\xi_y+N-k-q-1)}. \tag{21.3.18}$$

Since this expression depends on the configuration ξ only through the number of particles k, for fixed k it is bounded from below by $[1-O(\ell_N/N)]/N^{2\alpha+1}I_\alpha(0)$ (recall (21.2.1)).

4. Combining (21.3.14) with Lemma 21.12, we arrive at

$$\mathscr{E}_N^{\mathrm{lb}}(h^*)=\frac{1}{2Z_{N,S}\,N^{\alpha+1}\,M_*\,I_\alpha(0)}\sum_{k=0}^{\ell_N}\sum_{\xi\in E_{k,S}}$$

$$
\begin{aligned}
&\inf_{W\in\mathscr{W}(S_*^1,S_*^2)} \sum_{x,y\in S_*} \frac{m_*^\xi}{a(\xi\backslash\{x,y\})}\mathrm{cap}_S(x,y)\big[W_x-W_y+O(\delta)\big]^2 \\
&\times\big[1-O(\delta)\big]^2 \\
&= \frac{1}{2Z_{N,S}\,N^{\alpha+1}\,M_*\,I_\alpha(0)} \\
&\times\big[1+O(\delta)\big]\inf_{W\in\mathscr{W}(S_*^1,S_*^2)} \sum_{x,y\in S_*}\mathrm{cap}_S(x,y)[W_x-W_y]^2\sum_{k=0}^{\ell_N}\sum_{\xi\in E_{k,S_0}}\frac{m_*^\xi}{a(\xi)},
\end{aligned}
\tag{21.3.19}
$$

where we recall (21.3.5). By (21.3.9), this gives the claim. □

Remark 21.15

(1) Note that if $S_*^2 = S_*\backslash S_*^1$, then Lemma 21.12 is not needed in the proof of Proposition 21.14 and the bound in (21.3.6) holds with $\delta = \ell_N/N$. This fact establishes the lower bound in Theorem 21.9.
(2) If $\delta \downarrow 0$, which occurs when L is independent of N, then the same bound as in (21.3.6) holds for $\mathrm{cap}_{N,S}(\eta^{S_*^1},\eta^{S_*^2})$. This is because Lemma 21.12 implies that on the sets $\mathscr{E}_N^x$, $x\in S_*^1\cup S_*^2$, the equilibrium potential W is close to 1 or to 0. This fact establishes the lower bound in Theorem 21.5.

21.3.2 Upper bound

Let $\rho_c \in (0,\infty)$ denote the critical particle density mentioned below Theorem 21.3.

Proposition 21.16 *Let $S_*^1, S_*^2 \subset S_*$ be non-empty disjoint sets. Then, for $\varepsilon > 0$,*

$$
\begin{aligned}
\mathrm{cap}_{N,S}\big(\eta^{S_*^1},\eta^{S_*^2}\big) \le \mathrm{cap}_{N,S}\big(\mathscr{E}_N(S_*^1),\mathscr{E}_N(S_*^2)\big) \le \frac{LC_\varepsilon N^{-\alpha-1}}{(\ell_N-L\rho_c)^\alpha} \\
+\inf_{W\in\mathscr{W}(S_*^1,S_*^2)}\sum_{x,y\in S_*}\mathrm{cap}_S(x,y)[W_y-W_x]^2 \\
\times\frac{N^{-\alpha-1}}{Z_{N,S}M_*I_\alpha(3\varepsilon)}\sum_{m=0}^{\ell_N}\sum_{\xi\in E_{m,S_0}}\frac{m_*^\xi}{a(\xi)}\left(1+O\left(\frac{\ell_N}{\varepsilon N}\right)\right),
\end{aligned}
\tag{21.3.20}
$$

where C_ε is a constant that depends on ε.

Proof The first inequality in (21.3.20) is obvious. The proof of the second inequality comes in 8 Steps.

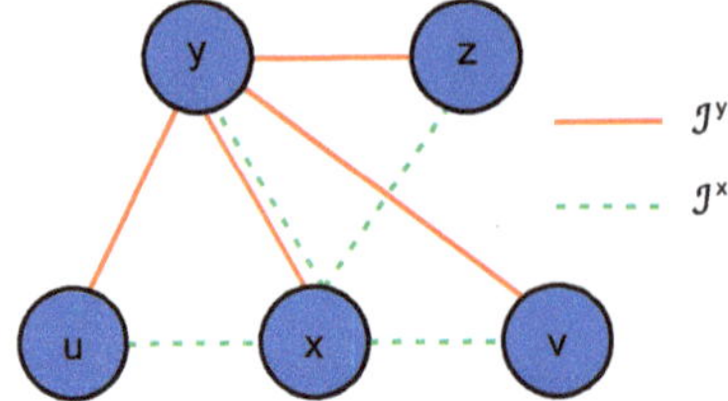

Fig. 21.4 Sharing configurations of $\mathscr{I}^x$, $\mathscr{I}^y$

1. Let $\mathscr{U} = \{u \in \mathbb{R}_+^S : \sum_{x \in S} u_x = 1\}$. Define the sets

- $\mathscr{F}^{x,y} = \{u \in \mathscr{U} : u_x + u_y \geq 1 - \varepsilon\}$,
- $\mathscr{L}^x = \{u \in \mathscr{U} : u_x > 1 - 3\varepsilon\}$,
- $\mathscr{D}^{x,y} = \mathscr{F}^{x,y} \backslash \mathscr{L}^x$,
- $\mathscr{I}^{x,y} = \mathscr{F}^{x,y} \backslash \{\mathscr{L}^x \cup \mathscr{L}^y\}$ and
- $\mathscr{I}^x = \bigcup_{y \in S_*} \mathscr{I}^{x,y}$, for $x, y \in S_*$.

We need the following facts.

Lemma 21.17 *The sets $\mathscr{D}^{x,y}$ and $\mathscr{I}^{x,y}$, $x \neq y \in S_*$ are mutually disjoint.*

Proof Since $\mathscr{I}^{x,y} \cup \mathscr{L}^y = \mathscr{D}^{x,y}$, it is enough to prove the assertion for the sets $\mathscr{D}^{x,y}$, $x \neq y \in S_*$. Let $x \neq y \neq z \in S_*$. Assume that $\frac{\eta}{N} \in \mathscr{D}^{x,y} \cap \mathscr{D}^{x,z}$. Then $\frac{\eta_y}{N}, \frac{\eta_z}{N} \geq 2\varepsilon$, and we get a contradiction because $1 \geq \frac{\eta_x + \eta_y + \eta_z}{N} \geq 1 - \varepsilon + \frac{\eta_z}{N} \geq 1 - \varepsilon + 2\varepsilon = 1 + \varepsilon$. □

Note that the sets $\mathscr{I}^{x,y}$ are symmetric for $x \neq y \in S_*$. From Lemma 21.17 we know that if $\frac{\eta}{N} \in \mathscr{I}^x$, then there exists a unique $y \in S_* \backslash \{x\}$ and therefore $\frac{\eta}{N} \in \mathscr{I}^y$ (see Fig. 21.4).

2. For the construction of the test function we define a smooth function $h^{x,y} : \mathscr{U} \to [0, 1]$, $y \in S_* \backslash \{x\}$, such that

$$\sum_{y \in S_* \backslash \{x\}} h^{x,y}(u) = 1, \quad u \in \mathscr{U}, \qquad h^{x,y}(u) = 1, \quad u \in \mathscr{D}^{x,y}, \tag{21.3.21}$$

and a smooth function $g^x : \mathscr{U} \to [0, 1]$, for $x \in S_*$ such that

$$\sum_{x \in S_*} g^x(u) = 1, \quad u \in \mathscr{U}, \qquad g^x(u) = \tfrac{1}{2}, \quad u \in \mathscr{I}^x, \qquad g^x(u) = 1, \quad u \in \mathscr{L}^x. \tag{21.3.22}$$

Let $\varepsilon > 0$ such that $\ell_N < \varepsilon N$. We start by choosing a test function for the capacity between two sites $x, y \in S_*$ out of the set $\mathscr{H}_N(\mathscr{E}_N^x, \mathscr{E}_N^y)$. This test function depends on the function that solves the variational problem for the capacity of the underlying random walk and on the harmonic function of the zero-range process on two sites,

$$G^{x,y}(\eta) = \sum_{k=1}^{L-1} \big[f_{xy}(z_k) - f_{xy}(z_{k+1})\big]\, H\left(\frac{\eta_x}{N} + \min\left\{\frac{1}{N}\sum_{l=2}^{k} \eta_{z_l}, \varepsilon\right\}\right), \tag{21.3.23}$$

where $x = z_1, z_2, \ldots, z_L = y$ is an enumeration of S such that $f_{xy}(z_i) \geq f_{xy}(z_j)$ for all $1 \leq i < j \leq L$, and f_{xy} is the harmonic function in $\mathscr{B}(x, y)$. The function $H\colon \{0, \ldots, \frac{N-m}{N}\} \to \mathbb{R}_+$ is the harmonic function of the zero-range process on two sites,

$$H(z) = \frac{\sum_{q=\lfloor 3\varepsilon N\rfloor}^{\lfloor zN\rfloor} a(q-1)a(N-m-q)}{\sum_{q=\lfloor 3\varepsilon N\rfloor}^{N-\lfloor 3\varepsilon N\rfloor-1} a(q)a(N-m-q-1)}, \tag{21.3.24}$$

with boundary conditions

$$\begin{aligned} H(z) &= 0, \quad z \in \big\{0, \ldots, \lfloor 3\varepsilon N\rfloor\big\},\\ H(z) &= 1, \quad z \in \big\{N - \lfloor 3\varepsilon N\rfloor, \ldots, N\big\}. \end{aligned} \tag{21.3.25}$$

Lemma 21.18 *$G^{x,y}$ belongs to the set $\mathscr{H}_N(\mathscr{E}_N^x, \mathscr{E}_N^y)$.*

Proof Let $\eta \in \mathscr{E}_N^x$. Then, for N large enough, $\frac{\eta_x}{N} > 1 - 3\varepsilon$. Due to the boundary condition in (21.3.25) the harmonic function H in (21.3.23) takes the value 1 for each k. Hence

$$G^{x,y}(\eta) = \sum_{k=1}^{L-1} \big[f_{xy}(z_k) - f_{xy}(z_{k+1})\big] = 1. \tag{21.3.26}$$

Let $\eta \in \mathscr{E}_N^y$. Then, for N large enough, $\sum_{z\in S\backslash\{y\}} \frac{\eta_z}{N} < 2\varepsilon$, and again, through the boundary condition (21.3.25), the harmonic function H is always 0, which implies $G^{x,y}(\eta) = 0$. □

3. We are now ready to choose the test function on $\eta \in E_{N,S}$, namely,

$$G_W^S(\eta) = \sum_{x\in S_*} g^x(\eta/N) G_W^x(\eta), \tag{21.3.27}$$

where, for $x, y \in S_*$ and $\eta \in E_{N,S}$,

$$G_W^{x,y}(\eta) = W_y + G^{x,y}(\eta)(W_x - W_y), \qquad G_W^x(\eta) = \sum_{y\in S_*\backslash\{x\}} h^{x,y}(\eta/N) G_W^{x,y}(\eta), \tag{21.3.28}$$

with $W \in \mathscr{W}(S_*^1, S_*^2)$.

Lemma 21.19 $G_W^S \in \mathscr{H}_N(\mathscr{E}_N(S_*^1), \mathscr{E}_N(S_*^2))$.

Proof Let $\eta \in \mathscr{E}_N^x$, $x \in S_*^1$. Then $\frac{\eta}{N} \in \mathscr{L}^x$ and $g^{x'}(\eta/N) = \mathbb{1}_{\{x'=x\}}$. Moreover, $G_W^{x,y}(\eta) = W_y + G^{x,y}(\eta)(W_x - W_y) = W_y + 1 - W_y = 1$ because $\eta \in \mathscr{E}_N^x$. Therefore $G^{x,y}(\eta) = 1$ for all $y \in S_* \backslash \{x\}$, and $W_x = 1$ because $x \in S_*^1$. Hence

$$\begin{aligned} G_W^S(\eta) &= \sum_{x' \in S_*} g^{x'}(\eta/N) G_W^{x'}(\eta) = \sum_{y \in S_* \backslash \{x\}} h^{x,y}(\eta/N) G_W^{x,y}(\eta) \\ &= \sum_{y \in S_* \backslash \{x\}} h^{x,y}(\eta/N) = 1. \end{aligned} \tag{21.3.29}$$

Let $\eta \in \mathscr{E}_N^x$, $x \in S_*^2$. Then $\frac{\eta}{N} \in \mathscr{L}^x$ and $g^{x'}(\eta/N) = \mathbb{1}_{\{x'=x\}}$. Moreover, $W_x = 0$ because $x \in S_*^2$. Therefore $G^{x,y} = 1$ because $\eta \in \mathscr{E}_N^x$, and $G_W^{x,y}(\eta) = W_y + 0 - W_y = 0$ for all $y \in S_* \backslash \{x\}$. Hence

$$G_W^S(\eta) = \sum_{x' \in S_*} g^{x'}(\eta/N) G_W^{x'}(\eta) = \sum_{y \in S_* \backslash \{x\}} h^{x,y}(\eta/N) G_W^{x,y}(\eta) = 0, \tag{21.3.30}$$

which settles the claim. □

Lemma 21.20 *Let* $\eta \in F_N^{x,y} = \{\eta \in E_N : \eta_x + \eta_y \geq N - \ell_N\}$, $x, y \in S_*$. *Then*

$$\begin{aligned} &G_W^S(\eta) = W_x, && \forall \eta \in \mathscr{E}_N^x, \\ &G_W^S(\eta) = W_y, && \forall \eta \in \mathscr{E}_N^y, \\ &G_W^S(\eta) = G_W^{x,y}(\eta) = G_W^{y,x}(\eta), && \forall \eta \in F_N^{x,y} \backslash \{\mathscr{E}_N^x \cup \mathscr{E}_N^y\}. \end{aligned} \tag{21.3.31}$$

Proof We start with the last equation. For this we show that $G^{x,y}(\eta) + G^{y,x}(\eta) = 1$ for $\frac{\eta}{N} \in \mathscr{F}^{x,y}$. If $\frac{\eta}{N} \in \mathscr{L}^x$, then this equality holds because $G^{x,y}(\eta) = 1$ and $G^{y,x}(\eta) = 0$. The same holds for $\frac{\eta}{N} \in \mathscr{L}^y$. For $\frac{\eta}{N} \in \mathscr{I}^{x,y}$, let $z_1, \ldots, z_L$ be the enumeration obtained from $f_{x,y}$ and $w_1, \ldots, w_L$ the enumeration obtained from $f_{y,x}$. Since $f_{x,y} + f_{y,x} = 1$, we can choose $w_{k+1} = z_{L-k}$ and get

$$\begin{aligned} G^{x,y}(\eta) + G^{y,x}(\eta) &= \sum_{k=1}^{L-1} \big[f_{x,y}(z_k) - f_{x,y}(z_{k+1}) \big] H\left(\sum_{n=1}^{k} \frac{\eta_{z_n}}{N} \right) \\ &\quad + \sum_{k=1}^{L-1} \big[f_{y,x}(w_k) - f_{y,x}(w_{k+1}) \big] H\left(\sum_{n=1}^{k} \frac{\eta_{w_{z_n}}}{N} \right) \\ &= \sum_{k=1}^{L-1} \big[f_{x,y}(z_k) - f_{x,y}(z_{k+1}) \big] H\left(\sum_{n=1}^{k} \frac{\eta_{z_n}}{N} \right) \\ &\quad + \sum_{k=1}^{L-1} \big[f_{y,x}(z_{L-k+1}) - f_{y,x}(z_{L-k}) \big] H\left(\sum_{n=1}^{k} \frac{\eta_{z_{L-n+1}}}{N} \right) \end{aligned}$$

$$= \sum_{k=1}^{L-1} [f_{x,y}(z_k) - f_{x,y}(z_{k+1})] H\left(\sum_{n=1}^{k} \frac{\eta_{z_n}}{N}\right)$$
$$+ \sum_{k=1}^{L-1} [f_{x,y}(z_k) - f_{x,y}(z_{k+1})] H\left(\sum_{n=k+1}^{L} \frac{\eta_{z_n}}{N}\right)$$
$$= \sum_{k=1}^{L-1} [f_{x,y}(z_k) - f_{x,y}(z_{k+1})] \left(H\left(\sum_{n=1}^{k} \frac{\eta_{z_n}}{N}\right) + H\left(\sum_{n=k+1}^{L} \frac{\eta_{z_n}}{N}\right)\right)$$
$$= \sum_{k=1}^{L-1} [f_{x,y}(z_k) - f_{x,y}(z_{k+1})] H(N - \lfloor 3\varepsilon N \rfloor) = 1. \quad (21.3.32)$$

With this equation, we get

$$G_W^{x,y}(\eta) = W_y + G^{x,y}(\eta)(W_x - W_y) = W_y + \big(1 - G^{y,x}(\eta)\big)(W_x - W_y)$$
$$= W_y + W_x - W_y + G^{y,x}(\eta)(W_y - W_x) = G_W^{y,x}(\eta). \quad (21.3.33)$$

Now observe that, for N large enough, $\frac{F_N^{x,y}}{N} = \mathscr{I}^{x,y} \cup \mathscr{L}^x \cup \mathscr{L}^y$. For $\eta/N \in \frac{1}{N}\mathscr{E}_N^x \subseteq \mathscr{L}^x$ and $\eta/N \in \frac{1}{N}\mathscr{E}_N^y \subseteq \mathscr{L}^y$ the assertion follows from Lemma 21.19. Let $\eta/N \in \mathscr{I}^{x,y}$. Then $\eta/N \in \mathscr{I}^x$ and $\eta/N \in \mathscr{I}^y$. Hence $g^{x'}(\eta/N) = \frac{\mathbb{1}_{\{x' \in \{x,y\}\}}}{2}$, and we get

$$G_W^S(\eta) = \sum_{x' \in S_*} g^{x'}(\eta/N) G_W^{x'}(\eta) = \tfrac{1}{2} G_W^x(\eta) + \tfrac{1}{2} G_W^y(\eta). \quad (21.3.34)$$

Moreover, $\eta/N \in \mathscr{D}^{x,y}$ and $\eta/N \in \mathscr{D}^{y,x}$. Hence $h^{x,y'}(\eta/N) = \mathbb{1}_{\{y'=y\}}$ and $h^{y,y'}(\eta/N) = \mathbb{1}_{\{y'=x\}}$. Thus, (21.3.34) equals

$$\tfrac{1}{2} \sum_{y' \in S_* \backslash \{x\}} h^{x,y'}(\eta/N) G_W^{x,y'}(\eta) + \tfrac{1}{2} \sum_{y' \in S_* \backslash \{x\}} h^{y,y'}(\eta/N) G_W^{y,y'}(\eta)$$
$$= \tfrac{1}{2} G_W^{x,y}(\eta) + \tfrac{1}{2} G_W^{y,x}(\eta) = \tfrac{1}{2} G_W^{x,y}(\eta) + \tfrac{1}{2} G_W^{x,y}(\eta) = G_W^{x,y}(\eta), \quad (21.3.35)$$

which settles the claim. □

4. Using the test function G_W^S, we can now derive an upper bound for the desired capacity. For $x \in S_*$, let $F_N^x = \bigcup_{y \in S_* \backslash \{x\}} F_N^{x,y}$ and $F_N = \bigcup_{x \in S_*} F_N^x$. The Dirichlet form $\mathscr{E}_N(G_W^S)$ of G_W^S (not to be confused with the set $\mathscr{E}_N$ in (21.2.8)) is

$$\mathscr{E}_N\big(G_W^S\big) = \tfrac{1}{2} \sum_{\eta \in E_{N,S}} \sum_{z,w \in S} \mu_{N,S}(\eta) g(\eta_z) r(z,w) \big[G_W^S(\eta^{z,w}) - G_W^S(\eta)\big]^2$$

$$
\begin{aligned}
&= \tfrac{1}{2} \sum_{x,y\in S_*} \sum_{\eta\in F_N^{x,y}} \sum_{z,w\in S} \mu_{N,S}(\eta) g(\eta_z) r(z,w) \big[G_W^S(\eta^{z,w}) - G_W^S(\eta) \big]^2 \\
&\quad + \tfrac{1}{2} \sum_{\eta\in F_N^c} \sum_{z,w\in S} \mu_{N,S}(\eta) g(\eta_z) r(z,w) \big[G_W^S(\eta^{z,w}) - G_W^S(\eta) \big]^2 \\
&= \sum_{x,y\in S_*} \mathscr{E}_N\big(G_W^S \mid F_N^{x,y}\big) + \mathscr{E}_N\big(G_W^S \mid F_N^c\big).
\end{aligned} \tag{21.3.36}
$$

Thus, we have to estimate the Dirichlet form on the set of configurations F_N^c and the set of configurations $F_N^{x,y}$, $x, y \in S_*$.

5. We start with the set of configurations $F_N^{x,y}$ with fixed $x \neq y \in S_*$. It follows from Lemma 21.20 that

$$
\begin{aligned}
&\mathscr{E}_N\big(G_W^S \mid F_N^{x,y}\big) \\
&\quad = \tfrac{1}{2} \sum_{1\le i,j\le L} \sum_{\eta\in F_N^{x,y}} \mu_{N,S}(\eta) g(\eta_{z_i}) r(z_i, z_j) \big[G_W^S(\eta^{z_i,z_j}) - G_W^S(\eta) \big]^2 \\
&\quad = \tfrac{1}{2} \sum_{1\le i,j\le L} \sum_{\eta\in F_N^{x,y}} \mu_{N,S}(\eta) g(\eta_{z_i}) r(z_i, z_j) \big[\big(G_W^{x,y}(\eta^{z_i,z_j}) - G_W^{x,y}(\eta) \big) \big]^2 \\
&\quad = \tfrac{1}{2} \sum_{1\le i,j\le L} \sum_{\eta\in F_N^{x,y}} \mu_{N,S}(\eta) g(\eta_{z_i}) r(z_i, z_j) \\
&\qquad \times \big[(W_x - W_y) \big(G^{x,y}(\eta^{z_i,z_j}) - G^{x,y}(\eta) \big) \big]^2 \\
&\quad = (W_x - W_y)^2 \mathscr{E}_N\big(G^{x,y} \mid F_N^{x,y}\big).
\end{aligned} \tag{21.3.37}
$$

Furthermore,

$$
\begin{aligned}
&\mathscr{E}_N\big(G^{x,y} \mid F_N^{x,y}\big) \\
&\quad = \tfrac{1}{2} \sum_{1\le i,j\le L} \sum_{\eta\in F_N^{x,y}} \mu_{N,S}(\eta) g(\eta_{z_i}) r(z_i, z_j) \big[G^{x,y}(\eta^{z_i,z_j}) - G^{x,y}(\eta) \big]^2 \\
&\quad = \frac{N^\alpha}{2Z_{N,S}} \sum_{1\le i,j\le L} \sum_{\eta\in F_N^{x,y}} \frac{m_*^\eta}{a(\eta)} \frac{a(\eta_{z_i})}{a(\eta_{z_i}-1)} r(z_i, z_j) \big[G^{x,y}(\eta^{z_i,z_j}) - G^{x,y}(\eta) \big]^2 \\
&\quad \le \frac{N^\alpha}{2Z_{N,S} M_*} \sum_{1\le i,j\le L} m(z_i) r(z_i, z_j) \\
&\qquad \times \sum_{\substack{\xi\in E_{N-1,S} \\ \xi_x+\xi_y\ge N-\ell_N-1}} \frac{m_*^\xi}{a(\xi)} \big[G^{x,y}(\xi + \mathfrak{d}_{z_j}) - G^{x,y}(\xi + \mathfrak{d}_{z_i}) \big]^2,
\end{aligned} \tag{21.3.38}
$$

where we set $\eta = \xi + \partial_{z_i}$. By the definition of $G^{x,y}$, we obtain

$$\mathscr{E}_N\big(G^{x,y} \mid F_N^{x,y}\big)$$
$$\le \frac{N^\alpha}{2Z_{N,S}M_*} \sum_{1\le i,j\le L} m(z_i)r(z_i,z_j) \sum_{\substack{\xi\in E_{N-1,S}\\ \xi_x+\xi_y\ge N-\ell_N-1}} \frac{m_*^\xi}{a(\xi)} \Bigg[\sum_{k=1}^{L-1}[f_{xy}(z_k)-f_{xy}(z_{k+1})]$$
$$\times \left(H\left(\frac{\xi_x}{N}+\sum_{l=2}^{k}\frac{\xi_{z_l}}{N}+\frac{1\!\!1_{\{k\ge i\}}}{N}\right)-H\left(\frac{\xi_x}{N}+\sum_{l=2}^{k}\frac{\xi_{z_l}}{N}+\frac{1\!\!1_{\{k\ge j\}}}{N}\right)\right)\Bigg]^2. \tag{21.3.39}$$

Fix two sites $z_i \ne z_j \in S$ with $i<j$. Since $m_*(x)=m_*(y)=1$, by setting $m_k = \sum_{l=2}^k \xi_{z_l}$ we get for the sum over the configurations ξ in (21.3.39) the upper bound

$$\sum_{m=0}^{\ell_N} \sum_{\zeta\in E_{m,S\backslash\{x,y\}}} \frac{m_*^\zeta}{a(\zeta)} \sum_{p=\lfloor 2\varepsilon N\rfloor}^{N-\lfloor 3\varepsilon N\rfloor-1} \frac{1}{a(p)a(N-m-p-1)}$$
$$\times \left[\sum_{k=i}^{j-1}[f_{xy}(z_k)-f_{xy}(z_{k+1})]\left(H\left(\frac{p+m_k+1}{N}\right)-H\left(\frac{p+m_k}{N}\right)\right)\right]^2. \tag{21.3.40}$$

The sum over p only runs from $\lfloor 2\varepsilon N\rfloor$ to $N-\lfloor 3\varepsilon N\rfloor-1$, due to the boundary conditions in (21.3.25). Inserting the explicit form (21.3.24), we get

$$\sum_{m=0}^{\ell_N} \sum_{\zeta\in E_{m,S\backslash\{x,y\}}} \frac{m_*^\zeta}{a(\zeta)} \sum_{p=\lfloor 2\varepsilon N\rfloor}^{N-\lfloor 3\varepsilon N\rfloor-1} \frac{1}{a(p)a(N-m-p-1)}$$
$$\times \left[\sum_{k=i}^{j-1}[f_{xy}(z_k)-f_{xy}(z_{k+1})]\frac{a(p+m_k)a(N-m-m_k-p-1)}{\sum_{q=\lfloor 3\varepsilon N\rfloor}^{N-\lfloor 3\varepsilon N\rfloor-1} a(q)a(N-m-q-1)}\right]^2. \tag{21.3.41}$$

Since $m_k \le \ell_N$ and $p \ge \lfloor 2\varepsilon N\rfloor$, we can estimate

$$a(p+m_k) = a(p)\left(1+\frac{m_k}{p}\right)^\alpha \le a(p)\left(1+\frac{\ell_N}{\lfloor 2\varepsilon N\rfloor}\right)^\alpha \le a(p)\left[1+O\left(\frac{\ell_N}{\varepsilon N}\right)\right], \tag{21.3.42}$$

and $a(N-m-m_k-p-1) \le a(N-m-p-1)$. Inserting these estimates into (21.3.41) we get the upper bound

$$\sum_{m=0}^{\ell_N} \sum_{\zeta\in E_{m,S\backslash\{x,y\}}} \frac{m_*^\zeta}{a(\zeta)} \sum_{p=\lfloor 2\varepsilon N\rfloor}^{N-\lfloor 3\varepsilon N\rfloor-1} \frac{1}{a(p)a(N-m-p-1)}$$

$$\times \left[\sum_{k=i}^{j-1} [f_{xy}(z_k) - f_{xy}(z_{k+1})] \frac{a(p)a(N-m-p-1)}{\sum_{q=\lfloor 3\varepsilon N \rfloor}^{N-\lfloor 3\varepsilon N \rfloor -1} a(q)a(N-m-q-1)} \right.$$

$$\left. \times \left[1 + O\left(\frac{\ell_N}{\varepsilon N} \right) \right] \right]^2$$

$$= \sum_{m=0}^{\ell_N} \sum_{\zeta \in E_{m,S \setminus \{x,y\}}} \frac{m_*^\zeta}{a(\zeta)} \left[\sum_{k=i}^{j-1} [f_{xy}(z_i) - f_{xy}(z_j)] \right]^2$$

$$\times \frac{\sum_{p=\lfloor 2\varepsilon N \rfloor}^{N-\lfloor 3\varepsilon N \rfloor -1} a(p)a(N-m-p-1)}{(\sum_{q=\lfloor 3\varepsilon N \rfloor}^{N-\lfloor 3\varepsilon N \rfloor -1} a(q)a(N-m-q-1))^2} \left[1 + O\left(\frac{\ell_N}{\varepsilon N} \right) \right]. \tag{21.3.43}$$

For $j < i$ we get the same bound. Now note that

$$\frac{\sum_{p=\lfloor 2\varepsilon N \rfloor}^{N-\lfloor 3\varepsilon N \rfloor -1} a(p)a(N-m-p-1)}{(\sum_{q=\lfloor 3\varepsilon N \rfloor}^{N-\lfloor 3\varepsilon N \rfloor -1} a(q)a(N-m-q-1))^2} = \frac{1+\mathscr{R}}{\sum_{p=\lfloor 3\varepsilon N \rfloor}^{N-\lfloor 3\varepsilon N \rfloor -1} a(p)a(N-m-p-1)} \tag{21.3.44}$$

with

$$\mathscr{R} = \frac{\sum_{p=\lfloor 2\varepsilon N \rfloor}^{\lfloor 3\varepsilon N \rfloor -1} a(p)a(N-m-p-1)}{\sum_{p=\lfloor 3\varepsilon N \rfloor}^{N-\lfloor 3\varepsilon N \rfloor -1} a(p)a(N-m-p-1)}. \tag{21.3.45}$$

It is easy to verify that (recall (21.2.1))

$$\mathscr{R} = O\left(\frac{\lfloor \varepsilon N \rfloor}{N - \lfloor \varepsilon N \rfloor} \right). \tag{21.3.46}$$

Thus, we obtain for (21.3.39) the upper bound

$$\frac{N^\alpha}{2Z_{N,S}M_*} \sum_{1 \le i,j \le L} m(z_i)r(z_i,z_j)[f_{xy}(z_i) - f_{xy}(z_j)]^2$$

$$\times \sum_{m=0}^{\ell_N} \sum_{\zeta \in E_{m,S \setminus \{x,y\}}} \frac{m_*^\zeta}{a(\zeta)} \frac{1}{\sum_{p=\lfloor 3\varepsilon N \rfloor}^{N-\lfloor 3\varepsilon N \rfloor -1} a(p)a(N-m-p-1)} \left[1 + O\left(\frac{\ell_N}{\varepsilon N} \right) \right]. \tag{21.3.47}$$

The sum over i, j in (21.3.47) is just the capacity of the underlying random walk between the two sites x and y. Since

$$\sum_{p=\lfloor 3\varepsilon N \rfloor}^{N-\lfloor 3\varepsilon N \rfloor -1} a(p)a(N-m-p-1) \ge N^{2\alpha+1} I_\alpha(3\varepsilon) \left[1 - O\left(\frac{\ell_N}{\varepsilon N} \right) \right], \tag{21.3.48}$$

we get for (21.3.47)

$$\mathscr{E}_N\big(G^{x,y} \mid F_N^{x,y}\big) \leq \frac{\mathrm{cap}_S(x,y)}{N^{\alpha+1} Z_{N,S} M_* I_\alpha(3\varepsilon)} \sum_{m=0}^{\ell_N} \sum_{\zeta\in E_{m,S_0}} \frac{m_*^\zeta}{a(\zeta)}\left[1+O\left(\frac{\ell_N}{\varepsilon N}\right)\right]. \tag{21.3.49}$$

6. Next we do the computation of $\mathscr{E}_N(G_W^S \mid F_N^c)$.

Lemma 21.21 *There exists an ε-dependent constant C_ε such that*

$$\max_{\eta\in E_{N,S}\setminus F_N} \big|G_W^S(\eta^{z,w}) - G_W^S(\eta)\big| \leq \frac{C_\varepsilon}{N}. \tag{21.3.50}$$

Proof Write

$$\big|G_W^S(\eta^{z,w}) - G_W^S(\eta)\big| = \left|\sum_{x\in S_*} g^x(\eta^{z,w}/N) G_W^x(\eta^{z,w}) - g^x(\eta/N) G_W^x(\eta)\right|. \tag{21.3.51}$$

Since g^x is a smooth function, there exists a constant C such that $|g^x(\eta^{z,w}/N) - g^x(\eta/N)| \leq \frac{C}{N}$, and hence (21.3.51) can be bounded from above by

$$\begin{aligned}
&\left|\sum_{x\in S_*} g^x(\eta/N)\left(G_W^x(\eta^{z,w})\left(1+\frac{C}{N}\right) - G_W^x(\eta)\right)\right| \\
&\quad = \left|\sum_{x\in S_*} g^x(\eta/N)\left(\sum_{y\in S_*\setminus\{x\}} h^{x,y}(\eta^{z,w}/N) G_W^{x,y}(\eta^{z,w})\left(1+\frac{C}{N}\right)\right.\right. \\
&\qquad \left.\left. - h^{x,y}(\eta/N) G_W^{x,y}(\eta)\right)\right|.
\end{aligned} \tag{21.3.52}$$

Since also $h^{x,y}$ is a smooth function, there exist a constant C such that $|h^{x,y}(\eta^{z,w}/N) - h^{x,y}(\eta/N)| \leq \frac{C}{N}$. Hence (21.3.52) is at most

$$\begin{aligned}
&\leq \left|\sum_{x\in S_*} g^x(\eta/N)\left(\sum_{y\in S_*\setminus\{x\}} h^{x,y}(\eta/N)\left(G_W^{x,y}(\eta^{z,w})\left(1+\frac{C}{N}\right)^2 - G_W^{x,y}(\eta)\right)\right)\right| \\
&= \left|\sum_{x\in S_*} g^x(\eta/N)\left(\sum_{y\in S_*\setminus\{x\}} h^{x,y}(\eta/N)\right.\right. \\
&\quad \left.\left.\times\left(G_W^{x,y}(\eta^{z,w}) - G_W^{x,y}(\eta) + \left(\frac{2C}{N}+\frac{C^2}{N^2}\right) G_W^{x,y}(\eta^{z,w})\right)\right)\right|
\end{aligned}$$

$$
\begin{aligned}
&\leq \Bigg| \sum_{x\in S_*} g^x(\eta/N) \\
&\quad \times \Bigg(\sum_{y\in S_*\setminus\{x\}} h^{x,y}(\eta/N)\Big((W_x - W_y)\big(G^{x,y}(\eta^{z,w}) - G^{x,y}(\eta)\big) + \frac{C}{N}\Big)\Bigg)\Bigg| \\
&\leq \sum_{x\in S_*} |g^x(\eta/N)| \sum_{y\in S_*\setminus\{x\}} |h^{x,y}(\eta/N)||W_x - W_y||G^{x,y}(\eta^{z,w}) - G^{x,y}(\eta)| + \left|\frac{C}{N}\right|,
\end{aligned}
\tag{21.3.53}
$$

where we use that $G_W^{x,y}(\eta) \leq 1$ for all $\eta \in E_{N,S}$. It remains to estimate

$$
\max_{\eta\in F_N^c} \big|G^{x,y}(\eta^{z,w}) - G^{x,y}(\eta)\big|. \tag{21.3.54}
$$

We may assume that $z = z_j$ and $w = z_i$ with $i < j$. Then

$$
\begin{aligned}
&\max_{\eta\in F_N^c} \big|G^{x,y}(\eta^{z_i,z_j}) - G^{x,y}(\eta)\big| \\
&\quad = \max_{\eta\in F_N^c} \left| \sum_{k=i}^{j-1} (f_{x,y}(z_k) - f_{x,y}(z_{k+1}))\left(H\left(\frac{\eta_x}{N} + \frac{m_k + 1}{N}\right) - H\left(\frac{\eta_x}{N} + \frac{m_k}{N}\right)\right)\right| \\
&\quad = \max_{\eta\in F_N^c} \left| \sum_{k=i}^{j-1} (f_{x,y}(z_k) - f_{x,y}(z_{k+1}))\frac{a(\eta_x + m_k)a(N - \eta_x - m_k - 1)}{\sum_{p=\lfloor 3\varepsilon N\rfloor+1}^{N-\lfloor 3\varepsilon N\rfloor} a(p-1)a(N-p)}\right| \\
&\quad \leq \max_{\eta\in F_N^c} \left| \sum_{k=i}^{j-1} (f_{x,y}(z_k) - f_{x,y}(z_{k+1}))\frac{a(N/2)a(N/2)}{\sum_{p=\lfloor 3\varepsilon N\rfloor+1}^{N-\lfloor 3\varepsilon N\rfloor} a(p-1)a(N-p)}\right| \\
&\quad \leq \frac{(N/2)^{2\alpha}}{\sum_{p=\lfloor 3\varepsilon N\rfloor+1}^{N-\lfloor 3\varepsilon N\rfloor} a(p-1)a(N-p)}.
\end{aligned}
\tag{21.3.55}
$$

Since $\sum_{p=\lfloor 3\varepsilon N\rfloor+1}^{N-\lfloor 3\varepsilon N\rfloor} a(p-1)a(N-p) \geq N^{2\alpha+1} I_\alpha(3\varepsilon)[1 - O(\frac{\ell_N}{\varepsilon N})]$, we get that there exists a constant C_ε such that

$$
\text{r.h.s. (21.3.55)} \leq \frac{(N/2)^{2\alpha}}{N^{2\alpha+1} I_\alpha(3\varepsilon)}\left[1 + O\left(\frac{\ell_N}{\varepsilon N}\right)\right] = \frac{C_\varepsilon}{N}. \tag{21.3.56}
$$

Thus

$$
\text{r.h.s. (21.3.53)} \leq \frac{C_\varepsilon}{N} \sum_{x\in S_*} g^x(\eta/N) \sum_{y\in S_*\setminus\{x\}} h^{x,y}(\eta/N) = \frac{C_\varepsilon}{N}, \tag{21.3.57}
$$

because $g^x, h^{x,y}$, $x \neq y \in S_*$ are positive functions and $|W_x - W_y| \leq 1$. □

7. In order to proceed with the computation, we need a technical result that follows from Großkinsky and Spohn [132]. The first statement says that all excess particles accumulate on a single site in S_*. The second statement says that if there is a constraint on the maximal occupation number of a single site, then as many excess particles as possible accumulate on a single site.

Proposition 21.22 *Let $Z_{N,S}(k)$ be the constrained partition function with the condition $\eta_z < k$ for all $\eta \in E_{N,S}$ and $z \in S$. Then:*

(i) $Z_{N,S}(k) = \sum_{x \in S_*} \frac{N^\alpha}{(N-\rho_c L)^\alpha (\rho_c L)^\alpha} Z_{\rho_c L, S\backslash\{x\}}[1+o(1)]$ *for* $k \geq N - \rho_c L$.
(ii) $Z_{N,S}(k) = \sum_{x \in S_*} \frac{N^\alpha}{(N-k)^\alpha k^\alpha} Z_{N-k, S\backslash\{x\}}(k)[1+o(1)]$ *for* $k < N - \rho_c L$.

Proposition 21.22 is needed for the following lemma.

Lemma 21.23 *Let $A^x = \{\eta \in E_{N,S}\colon \lfloor \varepsilon N \rfloor \leq \eta_x \leq N - \lfloor \varepsilon N \rfloor,\ \eta_x + \eta_y < N - \ell_N, \forall y \in S\backslash\{x\}\}$, $x \in S_*$. Then there is a constant C_* (depending on α and $|S_*|$) such that*

$$\sum_{\eta \in A^x} \mu_{N,S}(\eta) \leq \frac{C_*}{(\ell_N - \rho_c L)^\alpha} \frac{1}{N^{\alpha-1}}. \tag{21.3.58}$$

Proof Write

$$\begin{aligned} \sum_{\eta \in A^x} \mu_{N,S}(\eta) &= \frac{N^\alpha}{Z_{N,S}} \sum_{k=\lfloor \varepsilon N \rfloor}^{N-\lfloor \varepsilon N \rfloor} \frac{1}{k^\alpha} \sum_{\substack{\xi \in E_{N-k,S\backslash\{x\}} \\ \xi_y < N-k-\ell_N\ \forall y \in S\backslash\{x\}}} \frac{m_*^\xi}{a(\xi)} \\ &= \frac{N^\alpha}{Z_{N,S}} \sum_{k=\lfloor \varepsilon N \rfloor}^{N-\lfloor \varepsilon N \rfloor} \frac{1}{k^\alpha} \frac{1}{(N-k)^\alpha} Z_{N-k,S\backslash\{x\}}(N-k-\ell_N). \end{aligned} \tag{21.3.59}$$

Now we apply Proposition 21.22(ii) to $Z_{N-k,S\backslash\{x\}}(N-k-\ell_N)$. Since $N-k-\ell_N < N - \rho_c(L-1)$ for all $k \in \{3\varepsilon N, \ldots, N-3\varepsilon N\}$, we find that (21.3.59) equals

$$\begin{aligned} &= N^\alpha \sum_{\lfloor \varepsilon N \rfloor}^{N-\lfloor \varepsilon N \rfloor} \frac{1}{k^\alpha (N-k)^\alpha} \sum_{y \in S_*\backslash\{x\}} \frac{(N-k)^\alpha}{\ell_N^\alpha (N-k-\ell_N)^\alpha} \frac{Z_{\ell_N, S\backslash\{x,y\}}(N-k-\ell_N)}{Z_{N,S}} \\ &\quad \times \left[1+o(1)\right]. \end{aligned} \tag{21.3.60}$$

Since $N-k-\ell_N \geq \ell_N - \rho_c(L-2)$ for all $k \in \{3\varepsilon N, \ldots, N-3\varepsilon N\}$, we can apply Proposition 21.22(i) to $Z_{\ell_N, S\backslash\{x,y,\}}$. Hence (21.3.60) equals

$$N^\alpha \sum_{k=\lfloor \varepsilon N \rfloor}^{N-\lfloor \varepsilon N \rfloor} \frac{1}{k^\alpha} \sum_{y \in S_* \setminus \{x\}} \frac{1}{\ell_N^\alpha (N-k-\ell_N)^\alpha}$$
$$\times \sum_{z \in S_* \setminus \{x,y\}} \frac{\ell_N^\alpha [1+o(1)]}{(\ell_N - (L-2)\rho_c)^\alpha (\rho_c(L-2))^\alpha} \frac{Z_{(L-2)\rho_c, S\setminus\{x,y,z\}}}{Z_{N,S}}. \tag{21.3.61}$$

Note that $Z_{N,S} \geq Z_{N,S\setminus\{x,y\}}$. We apply Proposition 21.22(i) with $k = N$ for $Z_{N,S\setminus\{x,y\}}$,

$$Z_{N,S\setminus\{x,y\}} = \sum_{z \in S_* \setminus \{x,y\}} \frac{N^\alpha Z_{(L-2)\rho_c, S\setminus\{x,y,z\}}}{(N-\rho_c(L-2))^\alpha (\rho_c(L-2))^\alpha} \big[1+o(1)\big]$$
$$= \frac{N^\alpha |S_* - 2| (1+o(1))}{(N-(L-2)\rho_c)^\alpha (\rho_c(L-2))^\alpha} Z_{(L-2)\rho_c, S\setminus\{x_*,y_*,z_*\}}, \tag{21.3.62}$$

where $x_* \neq y_* \neq z_* \in S_*$ are arbitrary condensate sites. Inserting (21.3.62), we see that

r.h.s. (21.3.61)

$$\leq \sum_{k=\lfloor \varepsilon N \rfloor}^{N-\lfloor \varepsilon N \rfloor} \frac{1}{k^\alpha} \frac{|S_*-1|}{(N-k-\ell_N)^\alpha} \frac{|S_*-2|}{(\ell_N-(L-2)\rho_c)^\alpha} \frac{(N-(L-2)\rho_c)^\alpha}{|S_*-2|} \big[1+o(1)\big]$$
$$= |S_*-1| \left(\frac{N-(L-2)\rho_c}{\ell_N-(L-2)\rho_c}\right)^\alpha \sum_{k=\lfloor \varepsilon N \rfloor}^{N-\lfloor \varepsilon N \rfloor} \frac{1}{k^\alpha} \frac{1}{(N-k-\ell_N)^\alpha} \big[1+o(1)\big]$$
$$\leq 2^{\alpha+1} |S_*-1| \left(\frac{N-(L-2)\rho_c}{N-\ell_N}\right)^\alpha \frac{1}{(\ell_N-(L-2)\rho_c)^\alpha} \sum_{k=\lfloor \varepsilon N \rfloor}^{(N-\ell_N)/2} \frac{1}{k^\alpha} \big[1+o(1)\big]$$
$$\leq \frac{2^{\alpha+1}|S_*-1|}{\alpha-1} \left(1+\frac{\ell_N-(L-2)\rho_c}{N-\ell_N}\right)^\alpha \frac{1}{(\ell_N-(L-2)\rho_c)^\alpha} \frac{1}{\varepsilon^{\alpha-1} N^{\alpha-1}}$$
$$\times \big[1+o(1)\big]$$
$$\leq \frac{C_*}{(\ell_N - L\rho_c)^\alpha} \frac{1}{\varepsilon^{\alpha-1} N^{\alpha-1}}. \tag{21.3.63}$$

This settles the claim. □

8. Observe that configurations out of the set $\bigcup_{z \in S\setminus S_*} \{\eta \in E_{N,S} \colon \eta_z \geq N - \ell_N\}$ do not contribute to the Dirichlet form, because for these configurations $G^{x,y} = 0$ for all x, $y \in S_*$. Combining Lemmas 21.21–21.23, we get

$$\mathscr{E}_N\big(G_W^S \mid F_N^c\big) = \frac{1}{2} \sum_{\eta \in F_N^c} \sum_{z,w \in S} \mu_{N,S}(\eta) g(\eta_z) r(z,w) \big[G_W^S(\eta^{z,w}) - G_W^S(\eta)\big]^2$$

$$\le \frac{1}{2}\sum_{x\in S_*}\sum_{\eta\in E_N\backslash F_N^x}\sum_{z,w\in S}\mu_{N,S}(\eta)r(z,w)\big[G_W^S(\eta^{z,w})-G_W^S(\eta)\big]^2$$

$$\le \frac{C_\varepsilon^2}{2N^2}\sum_{x\in S_*}\sum_{\eta\in A^x}\mu_{N,S}(\eta)\sum_{z,w\in S}r(z,w)$$

$$\le \frac{LC_*C_\varepsilon^2}{(\ell_N-L\rho_c)^\alpha}\frac{1}{N^{\alpha+1}}. \tag{21.3.64}$$

Combine (21.3.36), (21.3.37), (21.3.49) and (21.3.64) to arrive at

$$\begin{aligned}
&\mathrm{cap}_{N,S}\big(\mathscr{E}_N(S_*^1),\mathscr{E}_N(S_*^2)\big)\\
&\quad= \inf_{G\in\mathscr{H}_N(\mathscr{E}_N(S_*^1),\mathscr{E}_N(S_*^2))}\mathscr{E}_N(G)\\
&\quad\le \inf_{W\in\mathscr{W}(S_*^1,S_*^2)}\mathscr{E}_N\big(G_W^S\big)\\
&\quad= \inf_{W\in\mathscr{W}(S_*^1,S_*^2)}\sum_{x,y\in S_*}\mathscr{E}_N\big(G^S\mid F_N^{x,y}\big)[W_y-W_x]^2+\mathscr{E}_N\big(G_W^S\mid F_N^c\big)\\
&\quad\le \inf_{W\in\mathscr{W}(S_*^1,S_*^2)}\sum_{x,y\in S_*}\mathrm{cap}_S(x,y)[W_y-W_x]^2\\
&\qquad\times\frac{N^{-\alpha-1}}{2Z_{N,S}M_*I_\alpha(3\varepsilon)}\sum_{m=0}^{\ell_N}\sum_{\xi\in E_{m,S_0}}\frac{m_*^\xi}{a(\xi)}\left[1+O\left(\frac{\ell_N}{\varepsilon N}\right)\right]\\
&\qquad+\frac{LC_*C_\varepsilon^2}{(\ell_N-L\rho_c)^\alpha}\frac{1}{N^{\alpha+1}}.
\end{aligned}\tag{21.3.65}$$

This completes the proof. □

21.4 Proof of the main theorems

21.4.1 Finite system size

Proof To prove Theorem 21.5, recall the remark made after the proof of Proposition 21.14. Using Lemma 21.2, we get for L fixed and $N\to\infty$ the lower bound

$$\begin{aligned}
\mathrm{cap}_{N,S}\big(\eta^{S_*^1},\eta^{S_*^2}\big)\ge{}&\frac{N^{-\alpha-1}[1+o(1)]}{2Z_SM_*I_\alpha(0)}\\
&\times\inf_{W\in\mathscr{W}(S_*^1,S_*^2)}\sum_{x,y\in S_*}\mathrm{cap}_S(x,y)[W_x-W_y]^2\sum_{k=0}^{\ell_N}\sum_{\xi\in E_{k,S_0}}\frac{m_*^\xi}{a(\xi)}.
\end{aligned}\tag{21.4.1}$$

With the definitions in (21.1.10)–(21.1.11), the sum over k in (21.4.1) converges as $N \to \infty$ to

$$\sum_{k\in\mathbb{N}_0}\sum_{\zeta\in E_{k,S_0}} \frac{m_*^\zeta}{a(\zeta)} = \prod_{z\in S_0}\sum_{j\in\mathbb{N}_0} \frac{m_*(z)^j}{a(j)} = \prod_{z\in S_0} \Gamma_z = \frac{Z_S}{|S_*|\Gamma(\alpha)}. \tag{21.4.2}$$

This gives the lower bound for (21.2.3).

Insert the bound

$$\sum_{m=0}^{\ell_N}\sum_{\xi\in E_{m,S_0}} \frac{m_*^\xi}{a(\xi)} \le \frac{Z_S}{|S_*|\Gamma(\alpha)} \tag{21.4.3}$$

into (21.3.20) and use Lemma 21.2. This gives the upper bound for (21.2.3). □

21.4.2 Diverging system size

Proof To prove Theorem 21.9, suppose that $L(N)$ satisfies (21.2.7).

1. In the case where S_*^1 and S_*^2 are a partition of S_*, the lower bound in Proposition 21.14 takes the form

$$\begin{aligned}\mathrm{cap}_{N,S}\big(\mathscr{E}_N(S_*^1),\mathscr{E}_N(S_*^2)\big) \ge \frac{1}{N^{\alpha+1}\,M_*\,I_\alpha(0)} \sum_{x\in S_*^1,\,y\in S_*^2} \mathrm{cap}_S(x,y)\\ \times \frac{1}{Z_{N,S}} \sum_{k=0}^{\ell_N}\sum_{\xi\in E_{k,S_0}} \frac{m_*^\xi}{a(\xi)}\big[1-O(\delta)\big].\end{aligned} \tag{21.4.4}$$

From Proposition 21.22 we know that

$$\begin{aligned} Z_{N,S} &= N^\alpha \sum_{x\in S_*}\sum_{k=0}^{\ell_N} \frac{1}{(N-k)^\alpha} \sum_{\xi\in E_{k,S\setminus\{x\}}} \frac{m_*^\xi}{a(\xi)}\big[1+o(1)\big]\\ &\le \left(1+\frac{\ell_N}{N-\ell_N}\right)^\alpha \sum_{x\in S_*}\sum_{k=0}^{\ell_N}\sum_{\xi\in E_{k,S\setminus\{x\}}} \frac{m_*^\xi}{a(\xi)}\big[1+o(1)\big]\\ &= |S_*|\sum_{k=0}^{\ell_N}\sum_{\xi\in E_{k,S\setminus\{x_*\}}} \frac{m_*^\xi}{a(\xi)}\big[1+o(1)\big]\\ &\le |S_*| \prod_{z\in S\setminus\{x_*\}}\sum_{k\ge0} \frac{m_*(z)^k}{a(k)}\big[1+o(1)\big]\end{aligned}$$

$$= |S_*| \prod_{z\in S\backslash\{x_*\}} \Gamma_z[1+o(1)]$$
$$= |S_*|\Gamma(\alpha) \prod_{z\in S_0} \Gamma_z[1+o(1)]. \tag{21.4.5}$$

Hence we get

$$\begin{aligned}&\mathrm{cap}_{N,S}(\mathscr{E}_N(S_*^1), \mathscr{E}_N(S_*^2))\\ &\quad\geq [1+o(1)]\,\frac{1}{N^{\alpha+1}\,M_*\,|S_*|\,I_\alpha(0)\,\Gamma(\alpha)} \sum_{x\in S_*^1, y\in S_*^2} \mathrm{cap}_S(x,y),\end{aligned} \tag{21.4.6}$$

which is the desired lower bound.

2. In the case where S_*^1 and S_*^2 are a partition of S_*, we get from (21.3.20), together with the estimate in (21.4.3), the upper bound

$$\begin{aligned}\mathrm{cap}_{N,S}(\mathscr{E}_N(S_*^1), \mathscr{E}_N(S_*^2)) &\leq \frac{Z_S}{Z_{N,S}}\,\frac{1}{N^{\alpha+1}\,M_*\,|S_*|\,I_\alpha(0)\,\Gamma(\alpha)}\\ &\quad\times \sum_{x\in S_*^1, y\in S_*^2} \mathrm{cap}_S(x,y)\left[1+O\left(\frac{\ell_N C_N}{N}\right)\right].\end{aligned} \tag{21.4.7}$$

Since $\frac{Z_S}{Z_{N,S}} = 1+o(1)$, we obtain

$$\begin{aligned}\mathrm{cap}_{N,S}(\mathscr{E}_N(S_*^1), \mathscr{E}_N(S_*^2)) &\leq [1+o(1)]\,\frac{1}{N^{\alpha+1}\,M_*\,|S_*|\,I_\alpha(0)\,\Gamma(\alpha)}\\ &\quad\times \sum_{x\in S_*^1, y\in S_*^2} \mathrm{cap}_S(x,y),\end{aligned} \tag{21.4.8}$$

which is the desired upper bound.

3. Combining (21.4.6) and (21.4.8), we get Theorem 21.9. □

21.5 Proof that the condensate configurations form a metastable set

We now prove Proposition 21.4 for L fixed and $N \to \infty$.

Proposition 21.24

$$\frac{\sup_{\eta\in\mathscr{M}}[\mathrm{cap}_{N,S}(\eta,\mathscr{M}\backslash\eta)/\mu_{N,S}(\eta)]}{\inf_{\xi\in\mathscr{M}^c}[\mathrm{cap}_{N,S}(\xi,\mathscr{M})/\mu_{N,S}(\xi)]} \leq O\big(L(L^2+N)/N^{\alpha+1}r'\big). \tag{21.5.1}$$

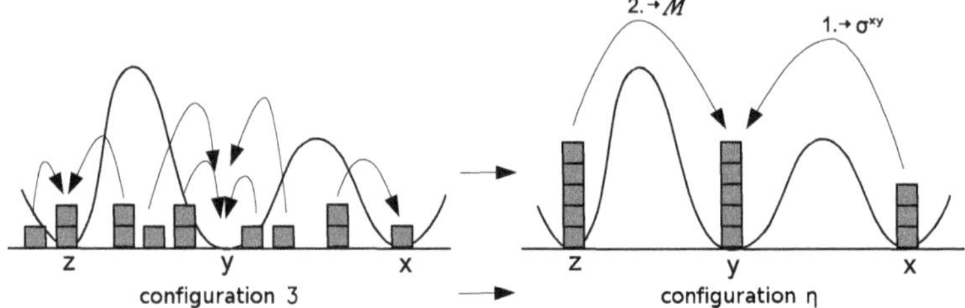

Fig. 21.5 Motion of the particles in the chosen path

Proof The following lemma bounds the denominator in (21.5.1) from below.

Lemma 21.25 *Let* $r' = \min_{u \in S} r(u, u \pm 1) > 0$. *Then*

$$\frac{\mathrm{cap}_{N,S}(\xi, \mathscr{M})}{\mu_{N,S}(\xi)} \geq \frac{r'}{L(L^2 + N)|S_*|K(\alpha)}, \qquad \xi \in \mathscr{M}^c, \tag{21.5.2}$$

where $K(\alpha)$ *is a constant that depends on* α.

Proof First we give the proof for $|S_*| = 3$. After that we extend the proof to $|S_*| > 3$ with the help of an algorithm.

• $|S_*| = 3$. Let $S_* = \{x, y, z\}$ and $r(x, y) \geq r(x, z) \geq r(y, z)$. To estimate the capacity from below we only need one path from ξ to $\mathscr{M}$. First, we let all particles of a valley (recall (8.2.10) for the definition of valleys) jump onto its attractor. The resulting configuration, where only the three attractors x, y, z in the three valleys $A(x), A(y), A(z)$ are occupied, is called η. Assume that on y there are more particles than on x (see Fig. 21.5). Next, since $r(x, y) \geq r(x, z), r(y, z)$, and since y is occupied by more particles than x, we let all particles from x jump to y. The resulting configuration with only the sites y and z occupied, is called $\sigma^{x,y}$ (see Fig. 21.5). Finally, we pick the site y or z where σ^{xy} has the highest occupation number and move all particles to that site.

Since

$$\mathrm{cap}_{N,S}(\xi, \mathscr{M}) \geq \frac{1}{\mathrm{cap}_{N,S}^{-1}(\xi, \eta) + \mathrm{cap}_{N,S}^{-1}(\eta, \sigma^{xy}) + \mathrm{cap}_{N,S}^{-1}(\sigma^{xy}, \mathscr{M})}, \tag{21.5.3}$$

we must calculate a lower bound for each of the three capacities in the right-hand side of (21.5.3).

For $w \in S_*$, let $(w)_n$ be a distance-to-w-increasing enumeration of $A(w)\backslash\{w\}$, and let $\xi_w^i = \xi_w + \sum_{j=1}^{i-1} \xi_{w_j}$ be the number of particles on site w before the particles on site w_i jump onto w. Let $|w| = \max\{\mathrm{dist}(w, w_i)\colon\ w \in S_*, w_i \in (w)_n\}$, where $\mathrm{dist}(w, w_i)$ is the Euclidean distance. For each transition of the particles from site w_i to site w via nearest-neighbour jumps, we use the explicit formula for the capacity of the one-dimensional chain. Doing so, we get for the capacity between ξ and η the lower bound

$$\mathrm{cap}_{N,S}^{-1}(\xi,\eta) \le \sum_{w\in S_*} \sum_{i=1}^{|(w)_n|} \sum_{k=0}^{\xi_{w_i}-1} \frac{|w|(N-1)^\alpha}{\mu_{N,S}(\{\xi_{w_i}-k,\xi_w^i+k\})r'N^\alpha}, \tag{21.5.4}$$

where $\{\xi_{w_i}-k,\xi_w^i+k\}$ is the configuration

$$\{\xi_{w_i}-k,\xi_w^i+k\}_v = \begin{cases} \xi_v, & v\in S\backslash A(w) \text{ and } v\in\{w_n:\ n>i\},\\ 0, & v\in\{w_n:\ n<i\},\\ \xi_{w_i}-k, & v=w_i,\\ \xi_w^i+k, & v=w. \end{cases} \tag{21.5.5}$$

Note that the μ_N-measure of the configuration increases after each transition of a particle, because there are more particles on condensate-sites. Therefore (21.5.4) equals

$$\begin{aligned} &\sum_{w\in S_*} \sum_{i=1}^{|(w)_n|} \sum_{k=0}^{\xi_{w_i}-1} \frac{Z_{N,S}|w|}{N^\alpha} \frac{a(\{\xi_{w_i}-k,\xi_w^i+k\}\backslash\{w_i,w\})}{m_*^{\{\xi_{w_i}-k,\xi_w^i+k\}\backslash\{w_i\}}} \\ &\quad\times \frac{a(\xi_{w_i}-k)a(\xi_w^i+k)(N-1)^\alpha}{m_*(w_i)^{\xi_{w_i}-k}N^\alpha r'} \\ &= \sum_{w\in S_*} \sum_{i=1}^{|(w)_n|} \frac{Z_{N,S}|w|}{N^\alpha} \frac{a(\{\xi_{w_i}-k,\xi_w^i+k\}\backslash\{w_i,w\})a(\xi_w^i)a(\xi_{w_i})}{m_*^{\{\xi_{w_i}-k,\xi_w^i+k\}\backslash\{w_i\}} m_*(w_i)^{\xi_{w_i}} r'} \\ &\quad\times \sum_{k=0}^{\xi_{w_i}-1} \left(1-\frac{k}{\xi_{w_i}}\right)^\alpha \left(1+\frac{k}{\xi_w^i}\right)^\alpha \left(1-\frac{1}{N}\right)^\alpha m_*(w_i)^k. \end{aligned} \tag{21.5.6}$$

The sum over k in (21.5.6) is bounded by L times an α-dependent constant $K'(\alpha)$. Observe that $a(\xi_w^1)=1$ when $\xi_w^1=0$, and we can use the same estimation. We therefore obtain

$$\mathrm{cap}_{N,S}^{-1}(\xi,\eta) \le \sum_{w\in S_*} \sum_{i=1}^{|(w)_n|} \frac{|w|LK'(\alpha)}{\mu_{N,S}(\{\xi_{w_i},\xi_w^i\})r'} \le \sum_{w\in S_*} \sum_{i=1}^{|(w)_n|} \frac{L^2K'(\alpha)}{\mu_{N,S}(\{\xi_{w_i},\xi_w^i\})r'}. \tag{21.5.7}$$

If $\eta_x = N$ or $\eta_y = N$, then we can stop here, because we already have a configuration in $\mathscr{M}$. Otherwise we continue the estimation of the capacities.

Without loss of generality, let $\eta_x \le \eta_y$. A similar estimation of the formula for the one-dimensional chain yields

$$
\begin{aligned}
&\operatorname{cap}_{N,S}^{-1}(\eta, \sigma^{xy}) \\
&\quad \le \sum_{k=0}^{\eta_x - 1} \frac{\operatorname{dist}(x, y) K''(\alpha)(N-1)^\alpha}{\mu_{N,S}(\{\eta_x - k, \eta_y + k\}) r' N^\alpha} \\
&\quad = \frac{Z_{N,S} \operatorname{dist}(x, y) K''(\alpha)}{N^\alpha} \frac{a(\eta \backslash \{x, y\})}{r' m_*^\eta} \sum_{k=0}^{\eta_x - 1} \frac{a(\eta_x - k) a(\eta_y + k)(N-1)^\alpha}{N^\alpha} \\
&\quad = \frac{\operatorname{dist}(x, y) K''(\alpha)}{\mu_{N,S}(\eta) r'} \sum_{k=0}^{\eta_x - 1} \left(1 - \frac{k}{\eta_x}\right)^\alpha \left(1 + \frac{k}{\eta_y}\right)^\alpha \left(1 - \frac{1}{N}\right)^\alpha,
\end{aligned} \tag{21.5.8}
$$

where $\{\eta_x - k, \eta_y + k\} \in E_{N,S}$ is the configuration where all sites are empty except for site x with $\eta_x - k$ particles, site y with $\eta_y + k$ particles and site z with η_z particles. Observe that $\eta_x / \eta_y \le 1$. For all $k \in \{1, \ldots, \eta_x - 1\}$, we have

$$
\left(1 - \frac{1}{N}\right)^\alpha \le 1, \qquad \left(1 - \frac{k}{\eta_x}\right)^\alpha \le 1, \qquad \left(1 + \frac{k}{\eta_y}\right)^\alpha \le 2^\alpha. \tag{21.5.9}
$$

Since $\eta_x, \eta_y < N$, we can estimate the sum in (21.5.8) from above by $2^\alpha N$, and get

$$
\operatorname{cap}_{N,S}^{-1}(\eta, \sigma^{xy}) \le \frac{2^\alpha K''(\alpha) N \operatorname{dist}(x, y)}{\mu_{N,S}(\eta) r'} \le \frac{2^\alpha K''(\alpha) N L}{\mu_{N,S}(\eta) r'}. \tag{21.5.10}
$$

If $\sigma_y^{xy} = N$, then we can stop here. Otherwise we continue the estimation of the capacities.

Without loss of generality, let $\sigma_z^{xy} \le \sigma_y^{xy}$. A similar estimation yields

$$
\operatorname{cap}_{N,S}^{-1}(\sigma^{xy}, \mathscr{M}) \le \frac{2^\alpha K''(\alpha) N \operatorname{dist}(y, z)}{\mu_N(\sigma^{xy}) r'} \le \frac{2^\alpha K''(\alpha) N L}{\mu_{N,S}(\sigma^{xy}) r'}. \tag{21.5.11}
$$

Combining (21.5.7), (21.5.10) and (21.5.11), we get the lower bound

$$
\begin{aligned}
\frac{\operatorname{cap}_{N,S}(\xi, \mathscr{M})}{\mu_{N,S}(\xi)} \ge \Bigg[& \frac{L^2 K'(\alpha)}{r'} \sum_{w \in S_*} \sum_{i=1}^{|(w)_n|} \frac{\mu_{N,S}(\xi)}{\mu_{N,S}(\{\xi_{w_i}, \xi_w^i\})} \\
& + \frac{2^\alpha K''(\alpha) N L}{r'} \left(\frac{\mu_{N,S}(\xi)}{\mu_{N,S}(\eta)} + \frac{\mu_{N,S}(\xi)}{\mu_{N,S}(\sigma^{xy})} \right) \Bigg]^{-1}.
\end{aligned} \tag{21.5.12}
$$

Since there are more particles on the sites of S_* in the configurations $\{\xi_{w_i}, \xi_w^i\}$, η and σ^{xy} than in the configuration ξ, we have

$$
\frac{\mu_{N,S}(\xi)}{\mu_{N,S}(\{\xi_{w_i}, \xi_w^i\})}, \frac{\mu_{N,S}(\xi)}{\mu_{N,S}(\eta)}, \frac{\mu_{N,S}(\xi)}{\mu_{N,S}(\sigma^{xy})} \le 1. \tag{21.5.13}
$$

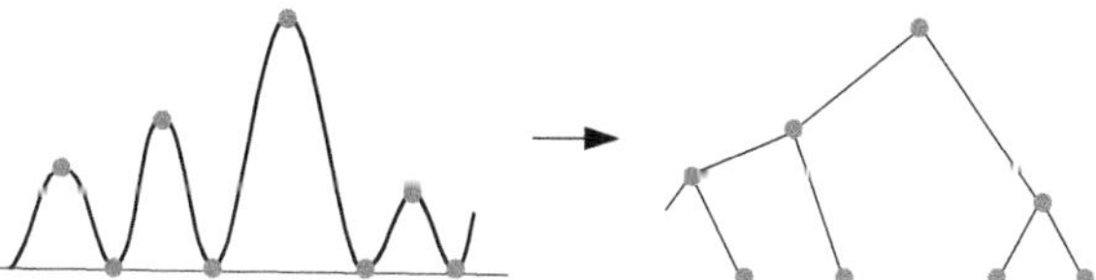

Fig. 21.6 Construction of the tree

Thus, we can continue by estimating

$$\frac{\mathrm{cap}_{N,S}(\xi,\mathscr{M})}{\mu_{N,S}(\xi)} \geq \left[\frac{L^3K'(\alpha)|S_*|}{r'} + \frac{2^\alpha K''(\alpha)(|S_*|-1)NL}{r'}\right]^{-1}$$
$$\geq \frac{r'}{L(L^2+N)|S_*|K(\alpha)}, \tag{21.5.14}$$

and we are done.

• $|S_*|>3$. The following algorithm generalizes the argument to $|S_*|>3$. Fix a configuration $\xi\notin\mathscr{M}$. First, we let all particles in a valley jump onto its attractor. From the resulting configuration we construct a labeled tree. Each attractor corresponds to a leaf of the tree that is labeled with its occupation number. The local maxima of the potential of the random walk on S are the vertices of the tree, where the largest local maximum is the root of the tree. The root is connected with the vertices of the next two largest local maxima, and so on, as illustrated in Fig. 21.6.

The algorithm works as follows. For each pair of leaves we calculate the length of the shortest path between them and choose the pair of leaves with the shortest path. If there are multiple shortest paths, then we choose the one with the lowest-labeled leave. Next, we increase the label of the highest-labeled leaf in the pair by the value of the label of the lowest-labeled leaf, and delete the latter. We continue until we obtain a tree with only one leaf. This algorithm describes a path from the configuration $\xi\notin\mathscr{M}$ to a configuration in $\mathscr{M}$ (because the final tree corresponds to a configuration in $\mathscr{M}$). Thus, for the general case we have to calculate at most $|S_*|-1$ transitions between condensate-sites, i.e., we have to estimate at most $|S_*|-1$ capacities. Hence (21.5.14) also holds in the general case. Figure 21.7 illustrates the algorithm for the case $|S_*|=3$. □

Having concluded the proof of Lemma 21.25, we can now conclude the proof of Proposition 21.24. Using Theorem 21.5 and Lemma 21.2, we obtain

$$\sup_{w\in S_*}\frac{\mathrm{cap}_{N,S}(\eta^w,\mathscr{M}\backslash\eta^w)}{\mu_{N,S}(\eta^w)}$$
$$= \sup_{w\in S_*}\frac{Z_{N,S}[1+o(1)]}{N^{\alpha+1}\,|S_*|\,M_*\,I_\alpha(0)\,\Gamma(\alpha)}\sum_{v\in S_*\backslash\{w\}}\mathrm{cap}_S(w,v)$$

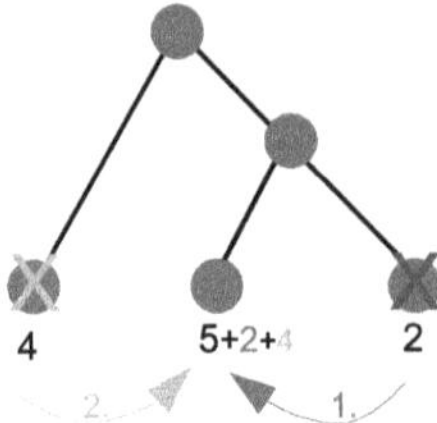

Fig. 21.7 Algorithm for $|S_*| = 3$

$$= \big[1 + o(1)\big] \frac{Z_S}{N^{\alpha+1}\, |S_*|\, M_*\, I_\alpha(0)\, \Gamma(\alpha)} \sup_{w \in S_*} \sum_{v \in S_* \setminus \{w\}} \mathrm{cap}_S(w, v), \qquad (21.5.15)$$

where $\mu_{N,S}(\eta^w) = 1/Z_{N,S}$ is the equilibrium weight of the configuration η^w where all N particles are at site w. The expression in (21.5.15) gives us control of the numerator in (21.5.1) (recall (21.1.15) and the fact that $\mathscr{M} = \bigcup_{w \in S_*} \eta^w$). For the denominator in (21.5.1) we use Lemma 21.25. For the quotient we therefore get

$$\begin{aligned}
&\frac{\sup_{\eta \in \mathscr{M}} [\mathrm{cap}_{N,S}(\eta, \mathscr{M} \setminus \eta)/\mu_{N,S}(\eta)]}{\inf_{\xi \in \mathscr{M}^c} [\mathrm{cap}_{N,S}(\xi, \mathscr{M})/\mu_{N,S}(\xi)]} \\
&\quad \leq \big[1 + o(1)\big] \frac{Z_S L(L^2 + N) K(\alpha)}{N^{\alpha+1}\, M_*\, I_\alpha(0)\, \Gamma(\alpha) r'} \sup_{w \in S_*} \sum_{v \in S_* \setminus \{w\}} \mathrm{cap}_S(w, v) \\
&\quad = O\big(L(L^2 + N)/N^{\alpha+1} r'\big),
\end{aligned} \qquad (21.5.16)$$

which is the desired result. □

21.6 Bibliographical notes

1. Metastability for the zero-range process in the case where the random walk is reversible was first investigated by Beltrán and Landim [16]. Convergence of the time-scaled process, observed when it visits the metastable set, to a continuous-time random walk was proved with the help of martingale techniques (recall Remark 21.8). The results were recovered in the framework of the potential-theoretic approach to metastability in the Diploma Thesis of Rebecca Neukirch [191], and were published in Bovier and Neukirch [40]. The presentation in this chapter is based on this work.

2. Lemma 21.1 was proved, in increasing generality, in Andjel [7], Evans [105], Großkinsky and Spohn [132]. Lemma 21.2 was proved in Großkinsky and Spohn [132], Beltrán and Landim [16]. The condensation phenomenon in Theorem 21.3

was proved in Großkinsky and Spohn [132], Großkinsky, Schütz and Spohn [131], Großkinsky and Schütz [130].

3. Landim [159] deals with the zero-range process in the case where the random walk is totally asymmetric (so that reversibility fails).

Part IX
Challenges

Part IX describes a number of models whose metastable behaviour is in principle understood but still presents major technical challenges. Some of these challenges may eventually become tractable with the potential-theoretic methods outlined in this monograph.

In Chap. 22 we list some challenges *within* metastability. In particular, we look at Glauber dynamics in *large volumes* of Ising spins with a nearestneighbour interaction in *small magnetic fields*, respectively, at birth-death dynamics of point particles with a spatial interaction in *small volumes* at *low temperatures*. The former is a variant of the model treated in Chaps. 17 and 19 where the critical droplets are so large that they take on a macroscopic shape, called the Wulff shape. The latter is a continuum variant of the model treated in Chapts. 18 and 20, and is linked to the phenomenon of crystallisation. For both we state a few theorems and point to a few open problems.

In Chap. 23 we list some challenges *outside* metastability. In particular, we look at Glauber dynamics in *infinite volume* and study what happens *after* the system has created a critical droplet. The focus is on how long it takes for the origin to be invaded by the plus phase when the system starts in the minus-phase. It turns out that this crossover is caused by a critical droplet that is created far away from the origin and subsequently grows until it reaches the origin.

Chapter 22
Challenges Within Metastability

Under this modification, the process of division was repeated, but with the old negative result: the attempt was therefore abandoned, though not without hope that future mathematicians, by introducing a number of hitherto undetermined constants, raised to the second degree, might succeed in obtaining a positive result.
(Lewis Carroll, A New Method of Evaluation)

There are several challenges *within* metastability that as yet remain unsolved, but are potentially within reach of the conceptual and technical machinery described in the present monograph. This chapter is devoted to two models representing some of these challenges. We state a few theorems—without proofs—and point to a few open problems.

Section 22.1 looks at Ising spins in a small magnetic field subject to Glauber dynamics. This is the same model as treated in Chaps. 17 and 19, but in a different metastable regime, namely, where the critical droplet is very large. In Sect. 22.2 we discuss a model of an interacting particle system in the continuum, to which the methods described in the present monograph apply after appropriate modifications.

22.1 Glauber dynamics in large volumes at small magnetic fields

In Part VII we already dealt with large volumes for Glauber dynamics and Kawasaki dynamics at low temperatures. In this section we again look at Glauber dynamics in large volumes, but now at positive temperatures and small magnetic fields. The temperature is chosen strictly below the *critical temperature* of the Ising model on the infinite lattice $\mathbb{Z}^2$ in zero magnetic field. In the limit as the magnetic field tends to zero the size of the critical droplet tends to infinity. The main idea is that its asymptotic shape is the *Wulff shape* from equilibrium statistical physics, i.e., the shape that minimises the integrated surface tension between the minus-phase outside the droplet and the plus-phase inside the droplet. In what follows we consider volumes that are comparable to the volume of the critical droplet. In Chap. 23 we will see what happens in larger volumes.

A. Bovier, F. den Hollander, *Metastability*,
Grundlehren der mathematischen Wissenschaften 351,
DOI 10.1007/978-3-319-24777-9_22

Return to the setting of Sect. 17.1.1. The Ising-spin Hamiltonian on a finite square box $\Lambda \subset \mathbb{Z}^2$ reads (recall (17.1.1)–(17.1.2))

$$H(\sigma) = -\frac{J}{2} \sum_{\{x,y\} \in \Lambda^*} \sigma(x)\sigma(y) - \frac{h}{2} \sum_{x \in \Lambda} \sigma(x), \quad \sigma \in S = \{-1, +1\}^\Lambda, \quad (22.1.1)$$

with $J, h > 0$, where we use periodic boundary conditions. The system follows a Metropolis dynamics $(\sigma_t)_{t \geq 0}$ with spin-flip rates

$$c_\beta(\sigma, \sigma^x) = \begin{cases} e^{-\beta[H(\sigma^x) - H(\sigma)]_+}, & \sigma \in S, \ x \in \Lambda, \\ 0, & \text{otherwise,} \end{cases} \quad (22.1.2)$$

where σ^x is the configuration obtained from σ by flipping the spin at site x. We write $\mathbb{E}_\sigma$ to denote expectation w.r.t. the law of this dynamics starting from $\sigma_0 = \sigma$.

We are interested in the metastable behaviour of the system in the limit as $h \downarrow 0$ for fixed $\beta \in (\beta_c, \infty)$, where $\beta_c = \frac{1}{2J} \ln(1 + \sqrt{2})$ is the *critical inverse temperature* of the model on $\mathbb{Z}^2$ at $h = 0$ (see e.g. Georgii [125], p. 451). In this limit the critical droplet will be large, namely, it has a linear size of order $1/h$ (recall (17.1.4)). In order to accommodate this droplet, we pick $\Lambda = \Lambda_h$ with

$$\Lambda_h = \left[-\frac{C}{h}, \frac{C}{h}\right]^2 \cap \mathbb{Z}^2 \quad \text{with } C \in (0, \infty) \text{ large enough.} \quad (22.1.3)$$

As initial configuration we take $\sigma_0 = \boxminus_h$, i.e., all spins in Λ_h are pointing downwards.

In Sect. 22.1.1 we state the main theorem for the crossover time, in Sect. 22.1.2 we explain the Wulff construction, while in Sect. 22.1.3 we give the heuristics behind the theorem.

22.1.1 Metastable crossover time

Intuitively, we expect the system to quickly relax to a distribution that is close to the *minus-phase of the Ising model on the infinite lattice* $\mathbb{Z}^2$ *in zero magnetic field* $h = 0$. Indeed, since h is small the barrier for tunnelling to a distribution that is close to the plus-phase is very high. We are interested in the crossover time. To that end, let $f : S \to \mathbb{R}$ be called local if it depends on finitely many spin variables only. Let $\langle f \rangle_-$ and $\langle f \rangle_+$ be the average of f in the minus-phase, respectively, the plus-phase.

Theorem 22.1 (Mean crossover time from minus-phase to plus-phase) *Fix* $\beta \in (\beta_c, \infty)$. *If f is a local function, then*

$$\lim_{h \downarrow 0} \mathbb{E}_{\boxminus_h}\big(f(\sigma_{\tau(h;\kappa)})\big) = \begin{cases} \langle f \rangle_-, & \text{if } \kappa < \kappa_\beta, \\ \langle f \rangle_+, & \text{if } \kappa > \kappa_\beta, \end{cases} \quad (22.1.4)$$

where $\tau(h;\kappa) = \exp[\kappa/h]$ *and*

$$\kappa_\beta = \frac{\beta w^*(\beta)^2}{4m^*(\beta)}, \tag{22.1.5}$$

with $m^*(\beta)$ *the spontaneous magnetisation of the plus-phase and* $w^*(\beta)$ *the integrated surface tension of the Wulff droplet of unit volume.*

Theorem 22.1 says that the crossover from the minus-phase to the plus-phase occurs at time $\exp[(\kappa_\beta/h)[1+o(1)]]$. What is remarkable about (22.1.4) is that it relates the crossover, which is a *non-equilibrium* quantity, to the spontaneous magnetisation and the integrated surface tension, which are *equilibrium* quantities. *A priori* there is no reason why the critical droplet should have an equilibrium shape (= Wulff shape). In fact, in Sect. 17.7, Items 5–8, we saw examples where (sub)critical droplets do *not* take on an equilibrium shape.

22.1.2 Wulff construction

Here is a brief description of the construction of the Wulff droplet (see Fig. 22.2). Let $\mathscr{S}^1 = \{x \in \mathbb{R}^2\colon \|x\|_2 = 1\}$ be the surface of the Euclidean ball of radius 1. The surface tension in the Ising model on $\mathbb{Z}^2$ at $h = 0$ in the direction perpendicular to $n \in \mathscr{S}^1$ is defined as

$$T_\beta(n) = -\lim_{\ell\to\infty} \frac{1}{2\beta\|y(\ell)\|_2} \ln\left(\frac{Z_{\ell,\sigma(n)}}{Z_{\ell,+}}\right). \tag{22.1.6}$$

Here, $y(\ell)$ and $-y(\ell)$ are the points where the straight line $\{x \in \mathbb{R}^2\colon (x,n) = 0\}$ intersects the boundary of the box $\Lambda^\ell = [-\ell,\ell]^2$, $Z_{\ell,\sigma(n)}$ is the partition sum on $\Lambda^\ell \cap \mathbb{Z}^2$ with the boundary condition $\sigma(n)$ given by

$$\sigma(n)(x) = \begin{cases} +1 & \text{if } (x,n) \geq 0, \\ -1 & \text{if } (x,n) < 0, \end{cases} \qquad x \in \partial\Lambda^\ell, \tag{22.1.7}$$

and $Z_{\ell,+}$ is the partition sum with the plus boundary condition.

Let $\mathscr{D}$ denote the set of closed self-avoiding rectifiable curves in $\mathbb{R}^2$ that are the boundary of a bounded region in $\mathbb{R}^2$. For $\gamma \in \mathscr{D}$, define the surface tension along γ as (see Fig. 22.1)

$$I_\beta(\gamma) = \int_\gamma T_\beta(n_s)\, dn_s, \tag{22.1.8}$$

where s parametrises γ according to the Euclidean length measure, and n_s is the unit outward normal vector at the point $s \in \gamma$ (which exists for almost every $s \in \gamma$). For $n \in \mathscr{S}^1$ and $\lambda \in (0,\infty)$, define the region

$$\mathscr{W}_\beta^\lambda(n) = \left\{x \in \mathbb{R}^2\colon (x,n) \leq \lambda T_\beta(n)\right\}. \tag{22.1.9}$$

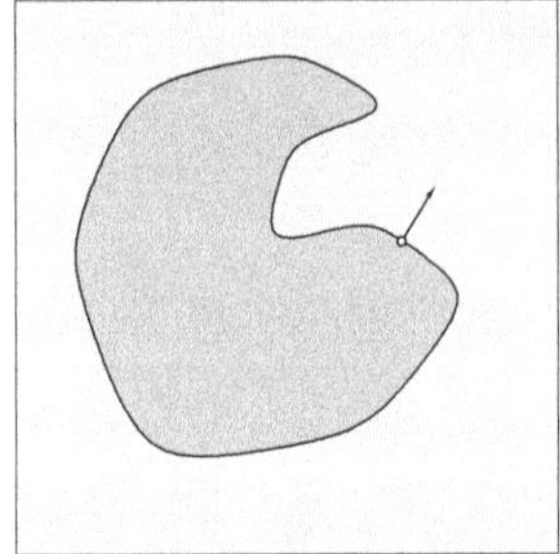

Fig. 22.1 The surface tension of a droplet equals the integral of the local surface tension over the boundary of the droplet. The local surface tension depends on the direction perpendicular to the boundary

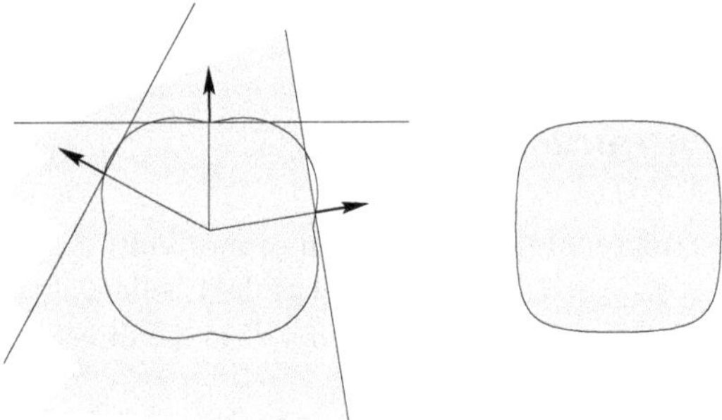

Fig. 22.2 Wulff construction. *Left:* Polar plot of the function $n \mapsto T_\beta(n)$, with three outward directions and three orthogonal tangent lines demarking three inward half-spaces (of which only one has been shaded). *Right:* The intersection of all the half-spaces (= the inner envelope of the tangent lines) gives rise to the *Wulff shape*. The *Wulff droplet* is the scaling of the Wulff shape that has unit volume

For $\lambda \in (0, \infty)$, define the intersection

$$\mathscr{W}_\beta^\lambda = \bigcap_{n \in \mathscr{S}^1} \mathscr{W}_\beta^\lambda(n). \tag{22.1.10}$$

The latter region satisfies the scaling relation $\mathscr{W}_\beta^\lambda = \lambda \mathscr{W}_\beta^1$, i.e., its shape stays the same as λ is varied. The *Wulff droplet* is defined as the region

$$\mathscr{W}_\beta = \mathscr{W}_\beta^{\lambda(\beta)}, \tag{22.1.11}$$

where $\lambda(\beta)$ is chosen such that $\mathscr{W}_\beta$ has volume 1 (see Fig. 22.2). Clearly, $\mathscr{W}_\beta$ is convex and hence $\partial\mathscr{W}_\beta \in \mathscr{D}$. The integrated surface tension of the Wulff droplet, the quantity that appears in (22.1.5), reads

$$w^*(\beta) = I_\beta(\partial\mathscr{W}_\beta). \tag{22.1.12}$$

It is known that the Wulff droplet satisfies the variational principle

$$w^*(\beta) \leq I_\beta(\gamma') \quad \forall \gamma' \subset \mathscr{B} \text{ with interiour volume } 1, \tag{22.1.13}$$

with equality if and only if γ is a translation of $\partial \mathscr{W}_\beta$.

22.1.3 Heuristics

The heuristics behind Theorem 22.1 is as follows. Consider a droplet of the plus-phase inside the minus-phase. Let S be the shape of this droplet and ℓ^2 its volume (i.e., the number of sites inside). For large ℓ, the free energy of this droplet is roughly

$$\Phi_S(\ell) = -m^*(\beta)h\ell^2 + w_S(\beta)\ell. \tag{22.1.14}$$

Here, $-m^*(\beta)h\ell^2$ is the change of the free energy due to the fact that inside the droplet the minus-phase is replaced by the plus-phase, and $w_S(\beta)\ell$ is the change of the free energy due to the surface tension along the border of the droplet. The two terms are of the same order of magnitude when ℓ is of order $1/h$. Therefore, putting $\ell = b/h$ and $\Phi_S(\ell) = \phi_S(b)/h$, we get

$$\phi_S(b) = -m^*(\beta)b^2 + w_S(\beta)b. \tag{22.1.15}$$

This function takes its maximal value at $b_c = w_S(\beta)/2m^*(\beta)$, reaching the value $\phi_S(b_c) = w_S(\beta)^2/4m^*(\beta)$. The height of this barrier is minimised by the Wulff shape, i.e., for S with $w_S(\beta) = w^*(\beta)$.

22.2 Crystallisation in small volumes at low temperatures

When we want to analyse the metastable behaviour of interacting particle systems in the continuum rather than on a lattice, a number of technical difficulties arise that need to be handled. In this section we describe a system of point particles living on a finite torus in $\mathbb{R}^d$, $d \geq 2$, and interacting with each other through a pair potential that has a hard-core repulsion, a finite range, and a unique attractive minimum. Particles are randomly created and annihilated according to a Metropolis dynamics, with the outside of the torus acting as an infinite particle reservoir with a fixed chemical potential. Once on the torus, particles cannot move.

Our goal will be to derive the metastable behaviour of this system in a way that is analogous to what was done in Chaps. 16–18. The chemical potential is chosen such that the system is metastable: starting from the vacuum configuration where the torus is empty, the system wants to nucleate (i.e., fill up the torus with particles in close packing). However, in order to do so it has to create a *critical droplet* that is large enough to trigger the nucleation. We are again interested in the nucleation time and in the size and shape of the critical droplets in the limit as the temperature tends to zero. We will show that this can be achieved subject to a number of hypotheses on the energy landscape, which replace the hypotheses (H1) and (H2) in Chap. 16. It will turn out that the *prefactor* depends in a delicate way on the temperature, the chemical potential, and the shape of the pair potential near its minimum, which is different from what we found in Chaps. 16–18 for lattice systems.

The problem with working in the continuum is that it is hard to control the energy landscape, especially in the vicinity of the set of critical droplets. We rely on properties derived in the literature for minimal-energy configurations at fixed particle numbers. Our assumptions on the energy landscape are expected to be true for a large class of pair potentials, but as yet can be proven only for a particular pair potential in $d=2$, called the *soft disk potential.*

22.2.1 Static model

Let

$$\Omega_* = \big\{\omega \subset \mathbb{R}^d\colon\ \mathrm{card}(\omega) \in \mathbb{N}_0\big\} \tag{22.2.1}$$

with $\mathrm{card}(\omega)$ the cardinality of ω. The set Ω_* represents all the finite-particle configurations in $\mathbb{R}^d$ (particles have locations, do not overlap, and are indistinguishable), and is endowed with the Hausdorff metric $d_H\colon\ \Omega_* \times \Omega_* \to [0,\infty)$.

Particles interact with each other through a *pair potential* $v\colon\ [0,\infty) \to \mathbb{R}\cup\{\infty\}$. Let $U\colon\ \Omega_* \to \mathbb{R}\cup\{\infty\}$ be the energy function defined by

$$U(\omega) = U\big(\{x_1,\dots,x_N\}\big) = \sum_{1\le i<j\le N} v\big(\|x_i - x_j\|\big), \tag{22.2.2}$$

i.e., $U(\omega)$ is the energy of the configuration $\omega = \{x_1,\dots,x_N\} \in \Omega_*$ with $N \in \mathbb{N}_0$ particles. Note that $U(\emptyset)=0$ and $U(\{x\})=0$, $x \in \mathbb{R}^d$.

We will be interested in particle configurations restricted to a *finite torus*. To that end, fix $L \in (0,\infty)$ large, let $\Lambda = [-L,L]^d$, give Λ periodic boundary conditions, and put

$$\Omega = \Omega_\Lambda = \big\{\omega \subset \Lambda\colon\ \mathrm{card}(\omega) \in \mathbb{N}_0\big\} \subset \Omega_*. \tag{22.2.3}$$

For $A \subset \Lambda$ and $\omega \in \Omega$, let $n_A(\omega) = \mathrm{card}(\omega \cap A) \in \mathbb{N}_0$ denote the number of particles in A. Let $\mathscr{F}$ be the smallest σ-algebra on Ω such that for every Borel measurable $A \subset \Lambda$ the map $n_A\colon\ \Omega \to \mathbb{N}_0$ is measurable with respect to $\mathscr{F}$. Write $\mathbb{Q} = \mathbb{Q}_\Lambda$ for the Poisson point process on Λ with intensity 1, which is a probability measure on $(\Omega,\mathscr{F})$. For later purpose, note the formula

$$\int_\Omega \mathbb{Q}(d\omega) \int_\Lambda F(\omega, \omega\cup x)\,dx = \int_\Omega \mathbb{Q}(d\omega) \sum_{x\in\omega} F(\omega\backslash x, \omega) \quad \forall F \in C_b(\Omega), \tag{22.2.4}$$

with the usual convention that the sum over $\emptyset$ is zero.

Let $\beta \in (0, \infty)$ denote the inverse temperature and $\mu \in \mathbb{R}$ the chemical potential. The *grand-canonical Hamiltonian* $H = H_{\mu,\Lambda}$ on Λ with chemical potential $\mu \in \mathbb{R}$ is the function on Ω defined by

$$H(\omega) = U(\omega) - \mu\, n(\omega), \tag{22.2.5}$$

where $n(\omega) = n_\Lambda(\omega)$. The *grand-canonical Gibbs measure* $\mathbb{P}_\beta = \mathbb{P}_{\beta,\mu,\Lambda}$ is the probability measure on $(\Omega, \mathscr{F})$ defined by

$$\frac{d\mathbb{P}_\beta}{d\mathbb{Q}}(\omega) = \frac{1}{\Xi}\,\mathrm{e}^{-\beta H(\omega)}, \tag{22.2.6}$$

where $\Xi_\beta = \Xi_{\beta,\mu,\Lambda}$ is the grand-canonical partition function

$$\Xi_\beta = \int_\Omega \mathrm{e}^{-\beta H(\omega)}\,\mathbb{Q}(d\omega). \tag{22.2.7}$$

Throughout the paper the labels μ, Λ will be suppressed from the notation.

22.2.2 Dynamic model

The particle configuration evolves in time according to a Markov process

$$X = \big(X(t)\big)_{t\geq 0} \tag{22.2.8}$$

on Ω with càdlàg paths. Particles are randomly created and annihilated inside Λ as if the outside of Λ were an infinite gas reservoir with chemical potential μ. The dynamics is Metropolis with grand-canonical Hamiltonian H in (22.2.5). Once inside Λ, particles cannot move.

Our dynamics is the Markov process with generator $\mathscr{L}_\beta$ given by

$$\begin{aligned}(\mathscr{L}_\beta f)(\omega) = &\int_\Lambda b_\beta(x,\omega)\big[f(\omega\cup x) - f(\omega)\big]\,\mathrm{d}x \\ &+ \sum_{x\in\omega} d_\beta(x,\omega)\big[f(\omega\setminus x) - f(\omega)\big], \quad f \in C_b(\Omega),\ \omega\in\Omega,\end{aligned} \tag{22.2.9}$$

with Metropolis rates

$$b_\beta(x,\omega) = e^{-\beta[H(\omega\cup x) - H(\omega)]_+}, \qquad d_\beta(x,\omega) = e^{\beta[H(x,\omega\setminus x) - H(\omega)]_-}. \tag{22.2.10}$$

Under this dynamics, in configuration ω a particle is added to an infinitesimal neighbourhood dx of a site $x \notin \omega$ with rate $b_\beta(x,\omega)dx$, while a particle is removed at site $x \in \omega$ with rate $d_\beta(x,\omega)$. Apart from this creation (= birth) and annihilation (= death), no particle motion takes place. We write $\mathbb{P}_\omega$ to denote the law of $(X_t)_{t\geq 0}$ given $X_0 = \omega$.

The grand-canonical Gibbs measure in (22.2.6) is reversible w.r.t. this dynamics because b_β, d_β satisfy the relation

$$\mathrm{e}^{-\beta H(\omega)} b_\beta(x,\omega) = \mathrm{e}^{-\beta H(\omega\cup x)}\, d_\beta(x,\omega\cup x), \quad x\notin\omega\in\Omega. \tag{22.2.11}$$

22.2.3 Metastability theorems for the soft disk potential

The set of ground states will be denoted by

$$\blacksquare = \operatorname{argmin} H, \tag{22.2.12}$$

the vacuum by $\emptyset$. In what follows we will be interested in the metastable crossover from $\emptyset$ to $\mathscr{N}$, with $\mathscr{N} \supsetneq \blacksquare$ representing a state where the system has *sufficiently nucleated*. This set must satisfy $\max_{\omega\in\mathscr{N}} H(\omega) < H(\emptyset)$ and hypothesis (H4) in Sect. 22.2.4, but is otherwise general. An example is the set of all configurations with a sufficiently large number of particles. The set of protocritical and critical configurations for the transition from $\emptyset$ to $\mathscr{N}$ will be denoted by $\mathscr{P}^\star$ and $\mathscr{C}^\star$.

Pick any $\beta\to\varepsilon(\beta)$ such that $\lim_{\beta\to\infty}\varepsilon(\beta)=0$ and $\lim_{\beta\to\infty}\beta\varepsilon(\beta)=\infty$, and define

$$\mathscr{C}^\star(\beta) = \big\{\omega\in\Omega\colon\, d_H\big(\omega,\mathscr{C}^\star\big)\le\varepsilon(\beta)\big\}. \tag{22.2.13}$$

This set of *near-critical configurations* turns out to play an important role in our analysis of the nucleation time.

Theorems 22.4–22.5 below are valid for the special case where the pair potential is the *soft disk potential* given by (see Fig. 22.3)

$$v(r) = \begin{cases} \infty, & 0\le r<1,\\ 24r-25, & 1\le r<\frac{25}{24},\\ 0, & \frac{25}{24}\le r<\infty. \end{cases} \tag{22.2.14}$$

Put $h=\mu+3$, and assume that $h\in(0,1)$ with $h^{-1}\notin\frac{1}{2}\mathbb{N}$. Let

$$\ell_c = \lfloor h^{-1}\rfloor, \tag{22.2.15}$$

note that $(\ell_c+1)^{-1}<h<\ell_c^{-1}$, and define

$$k_c = \begin{cases} 3\ell_c^2+4\ell_c+2, & h\in((\ell_c+1)^{-1},(\ell_c+\frac{1}{2})^{-1}),\\ 3\ell_c^2+2\ell_c+1, & h\in((\ell_c+\frac{1}{2})^{-1},\ell_c^{-1}). \end{cases} \tag{22.2.16}$$

Theorem 22.2 (Geometry of critical droplets)

(i) *If* $h\in((\ell_c+1)^{-1},(\ell_c+\frac{1}{2})^{-1})$, *then* $\mathscr{C}^\star$ *is the set of all configurations consisting of a single hexagon of side length* ℓ_c *located anywhere in* Λ, *with a bar of* ℓ_c *particles added on any side.*

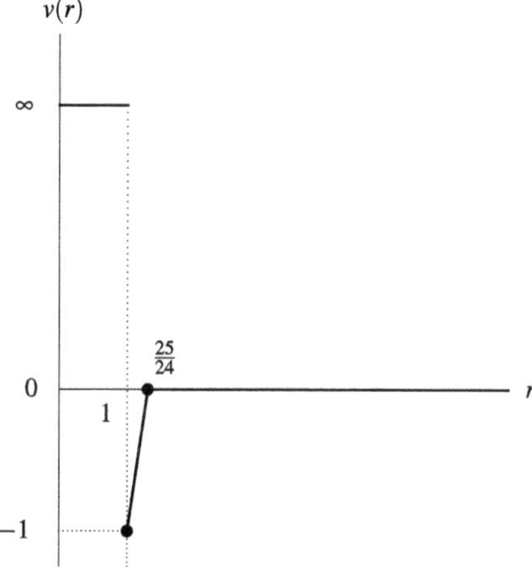

Fig. 22.3 Soft disk potential

(ii) *If* $h \in ((\ell_c + \frac{1}{2})^{-1}, \ell_c^{-1})$, *then* $\mathscr{C}^\star$ *is the set of all configurations consisting of a single hexagon of side length* ℓ_c *located anywhere in* Λ, *with a bar of* $\ell_c + 1$ *particles removed from any side.*

Theorem 22.3 (Critical gate) $\lim_{\beta\to\infty} \mathbb{P}_\emptyset(\tau_{\mathscr{C}^\star(\beta)} < \tau_{\mathscr{N}}) = 1$.

Theorem 22.4 (Mean crossover time)

(i) *There exists a* $K_{\mathscr{N}}(\mu) \in (0, \infty)$ *such that*

$$\lim_{\beta\to\infty} (24\beta)^{-(2k_c-3)}\, \mathrm{e}^{-\beta\Gamma}\, \mathbb{E}_\emptyset(\tau_{\mathscr{N}}) = \frac{K_{\mathscr{N}}(\mu)}{2\pi|\Lambda|}, \tag{22.2.17}$$

where

$$\Gamma = E_{k_c} - k_c\mu \tag{22.2.18}$$

with

$$E_{k_c} = -3k_c + \lceil \sqrt{12k_c - 3}\,\rceil. \tag{22.2.19}$$

(ii) *There exists a* $\chi \in \mathbb{R}$ *such that*

$$\lim_{\mu\downarrow -3} (\mu+3)^2 \ln K_{\mathscr{N}}(\mu) = \chi. \tag{22.2.20}$$

Theorem 22.5 (Exponential distribution of crossover time)

$$\lim_{\beta\to\infty} \mathbb{P}_\emptyset\big(\tau_{\mathscr{N}}/\mathbb{E}_\emptyset(\tau_{\mathscr{N}}) > t\big) = e^{-t} \quad \textit{for all } t \geq 0.$$

Theorem 22.2 identifies the critical droplets as *quasi-hexagons* with k_c particles. These take over the rôle of the quasi-squares for the lattice dynamics studied in Chap. 17.

Theorem 22.3 shows that nucleation takes place through the set $C^\star(\beta)$ of near-critical configurations, where the particles are located in regions with a linear size of order $1/\beta$ around the set of locations corresponding to the critical droplets. The volume of order $\beta^{2(k_c-2)}$ of these regions in $(\mathbb{R}^2)^{k_c-2}$ is large enough so that the dynamics is sufficiently likely to create particles inside (two particles drop out because of the arbitrary location and orientation of the critical droplet captured by the factor $2\pi|\Lambda|$ in the right-hand side of (22.2.17)).

Theorem 22.4(i) identifies the average nucleation time. The factor $(24\beta)^{2k_c-3}$ in the left-hand side of (22.2.17) is proportional to the inverse of the volume just described, with the factor 24 coming from the slope of the soft disk potential at its minimum. The factor $K_N(\mu)$ turns out to be a complicated object, since it incorporates all the possible shapes of the critical droplet, and depends in a delicate manner on the fine details of the soft disk potential near its minimum.

Theorem 22.4(ii) shows that scaling of $K_N(\mu)$ for $\mu \downarrow e_\infty$, which corresponds to the regime of weak metastability where the size of the critical droplet tends to infinity, leads to a limit χ that has the interpretation of the free energy per site in a triangular crystal of particles interacting through the soft disk potential.

22.2.4 Extension to other pair potentials

In this section we state *four hypotheses* on the energy landscape under which the results in Sect. 22.2.3 carry over to other pair potentials, modulo minor modifications. The class of pair potentials we are interested in is the following.

Definition 22.6 (Class of pair potentials) The pair potential $v\colon [0,\infty) \to (-\infty,\infty]$ is assumed to be lower-semicontinuous, to be non-positive, continuous and Lipschitz where finite, to have a hard-core repulsion, a finite range, and a unique strictly negative minimum, i.e., there exist $0 < r_1 \le r_2 < r_3 < \infty$ such that (see Fig. 22.4)

$$v(r)\begin{cases} =\infty, & \text{for } 0 \le r < r_1,\\ \le 0, & \text{for } r_1 \le r < \infty,\\ > v(r_2), & \text{for } r \ne r_2,\\ =0, & \text{for } r \ge r_3. \end{cases} \tag{22.2.21}$$

The conditions in (22.2.21) imply that v is stable, i.e., there exists a $C \in (0,\infty)$ such that $U(\omega) \ge -C\,\mathrm{card}(\omega)$ for all $\omega \in \Omega_*$ (see Ruelle [209, Sect. 3.2]).

For $k \in \mathbb{N}$, define

$$E_k = \inf_{\substack{\omega\in\Omega_*\\ \mathrm{card}(\omega)=k}} U(\omega). \tag{22.2.22}$$

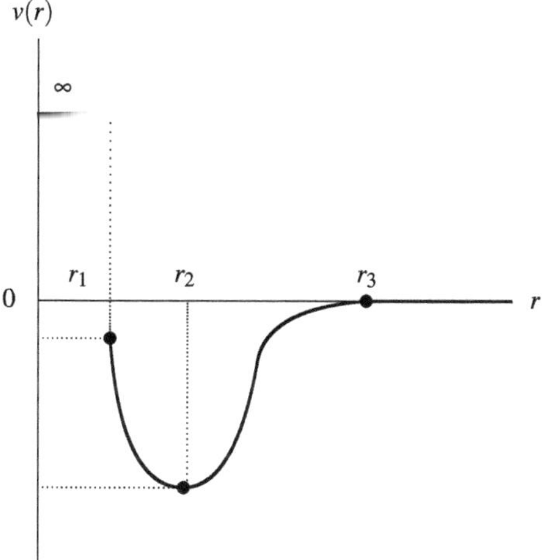

Fig. 22.4 Shape of the pair potential $r \mapsto v(r)$. The soft disk potential has $r_1 = r_2 = 1$, $r_3 = \frac{25}{24}$ and $v(r_1) = v(r_2) = -1$ (see Fig. 22.3)

Any configuration achieving the infimum is called a k-particle ground state in $\mathbb{R}^d$. By subadditivity, we have the existence of

$$e_\infty = \lim_{k\to\infty} E_k/k = \inf_{k\in\mathbb{N}} E_k/k. \tag{22.2.23}$$

The analogue of (22.2.22) on Λ reads

$$E_{k,\Lambda} = \inf_{\substack{\omega\in\Omega\\ \mathrm{card}(\omega)=k}} U(\omega). \tag{22.2.24}$$

Our first three hypotheses read:

Assumption 22.7 (Ground state properties)

(H1) $k \mapsto E_k - k\mu$ has a unique maximizer k_c on $\mathbb{N}_0$ that satisfies $2 \le k_c < \infty$.
(H2) $\lim_{k\to\infty}(E_k - ke_\infty) = \infty$.
(H3) There exists an $\varepsilon > 0$ such that $E_{k,\Lambda} = E_k$ for all $0 \le k \le \varepsilon|\Lambda|$.

Hypothesis (H1) precludes certain special values of the chemical potential, and guarantees that our system is in the *metastable regime*: the threshold for nucleation is neither one particle nor infinitely many particles (Λ needs to be chosen large enough so that the k_c-particle ground states fit inside). The restrictions on the pair potential in Definition 22.6 imply that $e_\infty \in (-\infty, 0)$. Since $E_0 = 0$ and $E_k > ke_\infty$ for all $k \in \mathbb{N}$, it follows that $k_c = \infty$ for all $\mu < e_\infty$. Hypothesis (H2) implies that $k_c = \infty$ also for $\mu = e_\infty$. On the other hand, $k_c < \infty$ for all $\mu > e_\infty$. In order to have $k_c \ge 2$, as required in Hypothesis (H1), we need to constrain μ from above. Under Hypothesis (H2) there exists an $h_c \in (0, \infty)$ (depending on the fine details of the pair potential) such that

$$2 \le k_c < \infty \quad \text{if and only if} \quad \mu \in (e_\infty, e_\infty + h_c). \tag{22.2.25}$$

Hypothesis (H3), finally, says that the k-particle ground states in Λ coincide with those in $\mathbb{R}^d$ when k is not too large. (Λ needs to be chosen large enough so that no interaction around the torus occurs.)

Our fourth hypothesis is analogous to what we had in Lemma 16.9.

Assumption 22.8 (No-deep-well property)

(H4) There exists a $V^* < \Gamma$ such that $\max_{\substack{\omega \in \Omega \backslash \{\emptyset, \mathcal{N}\} \\ H(\omega) \le \Gamma}} [\Phi(\omega, \{\emptyset, \mathcal{N}\}) - H(\omega)] \le V^*$.

Hypothesis (H4) says that the deepest wells in the energy landscape are those containing $\emptyset$ and $\mathcal{N}$, which makes $(\emptyset, \mathcal{N})$ into a *metastable pair* in the sense of Definition 8.2.

The following theorem states that the results in Sect. 22.2.3 indeed carry over.

Theorem 22.9 (Metastability for other pair potentials) *Theorems* 22.3–22.5 *hold under Hypotheses* (H1)–(H4), *with k_c and E_{k_c} as defined above, and with an appropriate modification of the scaling factors in* (22.2.17) *and* (22.2.20).

Different pair potentials will have different scaling in β. E.g. when the pair potential is twice differentiable near its minimum, the scaling factor in the left-hand side of (22.2.17) becomes $[\sqrt{2\pi v''(1)\beta}]^{-(2k_c-3)}$ and the formula for $K_{\mathcal{N}}$ in the right-hand side of (22.2.17) involves the determinant of a certain $(2k_c-3)$-dimensional quadratic form. In the near-critical configurations the regions where the particles are located have a linear size of order $1/\sqrt{\beta}$.

Hypotheses (H1)–(H4) are satisfied for the soft disk potential in $d=2$, with $e_\infty = -3$ and $h_c = 1$. We expect that they are in fact satisfied for a large class of pair potentials. For instance, (H1)–(H3) should hold under the conditions on the pair potential stated in Definition 22.6, with $E_k - ke_\infty \asymp k^{(d-1)/d}$ as $k \to \infty$. Moreover, if the well of the pair potential is *narrow and deep enough* (see Fig. 22.4), then Theorem 22.2 should carry over as well, with the spacing in the triangular lattice equal to r_2 and with E_k multiplied by $-v(r_2) > 0$. Hypothesis (H4) is more delicate, and will no doubt require stronger conditions on the pair potential, e.g. unimodality.

Settling (H1)–(H4) beyond the soft disk potential in $d=2$ represents *a hard open problem in the analytic theory of crystallisation*. In $d=3$, for instance, it is believed that the ground states consist of stacked layers of triangular lattices, but the relative position of these layers is a matter of debate.

22.3 Bibliographical notes

1. Theorem 22.1 was proved by Shlosman and Schonmann [215]. The proof is long and technical. To obtain control on the growing and shrinking of large droplets, delicate coupling and coarse-graining techniques are needed in which the microscopic regions where the system changes from the minus-phase to the plus-phase are approximated on a mesoscopic scale by local pieces of a continuum interface, and the cost of these pieces is related to the direction-dependent surface tension, as explained in Sects. 22.1.2–22.1.3.

2. Vanheuverzwijn [231] proved the existence of metastable states for the Ising model on $\mathbb{Z}^2$. Numerical studies by Rikvold, Tomita, Miyashita and Sides [206] confirm Theorem 22.1.

3. The extension of Theorem 22.1 to $d \geq 3$ was obtained in Bodineau, Graham and Wouts [29] (see Item 7 below). A weaker version of this extension was conjectured in Aizenman and Lebowitz [1] and was proved in Schonmann [212].

4. The proof of Theorem 22.1 given in [215] in fact applies to spin-flip rates that are more general than (22.1.1)–(22.1.2): translation invariant, finite range, attractive, monotone in the magnetic field, uniformly bounded away from zero and infinity. Also the initial condition can be more general: any starting distribution that is stochastically below the minus-phase.

5. It is shown in [215] that the metastable state, i.e., the state at time $\tau(h;\kappa)$ with $\kappa < \kappa_\beta$, is "infinitesimally larger" than the minus-phase. An asymptotic expansion in powers of h is derived for the difference of a local average under the metastable state and the minus-phase, which can be interpreted as describing the C^∞-continuation in h of the family of Gibbs distributions with negative h into the region of positive h. This continuation is expected not to be analytic, a situation that should be typical for metastable states. It is known that there is no analytic continuation of the minus-phase across $h = 0$.

6. The extension of Theorems 22.1 to Kawasaki dynamics, where the limit of small magnetic field $h \downarrow 0$ is taken over by the limit of weak supersaturation $\Delta \uparrow 2U$, is also still open. The main difficulty is that Kawasaki dynamics is *conservative*: the growing of large droplets is hampered because the gas around the droplet gets depleted. None of the techniques developed for Glauber dynamics seems easily transportable.

7. Bodineau, Graham and Wouts [29] look at a *diluted* version of the model studied in Sect. 22.1, where the pair interaction is switched off on a random set of sites with density $p \in (0,1)$. The relaxation time is shown to be $\exp[\kappa_\beta(p)/h^{d-1}]$, $h \downarrow 0$, with $\kappa_\beta(p) = w^*(p,\beta)^2/m^*(p,\beta)$, where $m^*(p,\beta)$ and $w^*(p,\beta)$ are the analogues of $m^*(\beta)$ and $w^*(\beta)$ in the non-diluted model, i.e., the spontaneous magnetisation of the plus-phase and the integrated surface tension of the Wulff droplet of unit volume. Intuitively, dilution enhances relaxation because there is no surface tension in diluted areas, and so it is expected that $p \mapsto \kappa_\beta(p)$ is non-increasing. However, no proof is available, even though both $p \mapsto m^*(p, \beta)$ and $p \mapsto w^*(p, \beta)$ are known to be non-increasing. It is shown that $\lim_{\beta \to \infty} \kappa_\beta(p)/\kappa_\beta = 0$ for all $p \in (0, 1)$.

8. The results in Sect. 22.2 are taken from den Hollander and Jansen [83].

9. A substantial body of literature is concerned with the speed of convergence to equilibrium in both Glauber dynamics and Kawasaki dynamics at finite temperature, both for high temperature and for low temperature. See e.g. Cancrini and Martinelli [47], Cancrini, Martinelli and Roberto [48, 49], Lubetzky and Sly [168], Lubetzky, Martinelli, Sly and Toninelli [167]. The slow mixing at low temperatures found in Cancrini, Cesi and Martinelli [46], Martinelli and Toninelli [178] is certainly a signature of metastable behaviour.

10. A major challenge consists in understanding metastable behaviour of stochastic dynamics on *large random graphs*. The theory developed in Chap. 16 applies to finite graphs, but assumption (H1) needs verification and may fail for certain realisations of the graph. (Assumption (H2) typically fails, which means we lose Theorem 16.4(b).) The key quantities in Theorems 16.4–16.6, i.e., the critical set, the communication height and the prefactor, all depend on the realisation of the graph. Understanding the scaling of these quantities for large graphs is interesting. Dommers [93] contains rough results for Glauber dynamics on random regular graphs.

Chapter 23
Challenges Beyond Metastability

Learning, undigested by thought, is labor lost;
thought, unassisted by learning, is perilous.
(Kong-tze, Lun Yu)

Parts VI and VII dealt with nucleation in small and large volumes at low temperatures. In the former, where the volume was kept fixed, we were able to arrive at a full description of metastability, with detailed computations for both Glauber dynamics and Kawasaki dynamics. In the latter, where the volume grew exponentially with the inverse temperature, we restricted ourselves to the computation of the time of first appearance of a critical droplet somewhere in the large volume. We were unable to follow the subsequent growth of this droplet beyond twice its initial size. In particular, it remained open how this droplet eventually invaded the large volume. Especially for Kawasaki dynamics this is a formidable challenge, because a large droplet tends to deplete the surrounding gas.

What happens in *infinite* volume? In that case a new mechanism of nucleation becomes possible: the critical droplet is created somewhere far from the origin and *invades* the neighbourhood of the origin by growing. The key question reads: Is this mechanism more efficient than nucleation close to the origin? It turns out that the answer is yes.

In this chapter we look at Glauber dynamics in infinite volume in *two metastable regimes*. In Sect. 23.1 we consider the limit of low temperature at positive magnetic field, while in Sect. 23.2 we turn to the limit of small magnetic field at positive temperature. We restrict ourselves to presenting the main ideas only, omitting proofs. For references we refer the reader to the bibliographical notes.

Post-nuclear growth is not part of metastability theory, the latter being concerned with pre-nucleation and nucleation phenomena only. Key features, such as the renewal structure created by repeated unsuccessful trials to form a critical droplet, are lost. In fact, so far potential theory has rather little to say about post-nuclear growth. Consequently, sharp results are hard to get, and fully rely on ad hoc methods.

A. Bovier, F. den Hollander, *Metastability*,
Grundlehren der mathematischen Wissenschaften 351,
DOI 10.1007/978-3-319-24777-9_23

23.1 Low temperatures

Return to the setting of Sect. 17.1.1. Replace Λ by $\mathbb{Z}^d$, $d \geq 2$. The infinite-volume Ising-spin Hamiltonian reads

$$H(\sigma) = -\frac{J}{2} \sum_{\{x,y\} \in (\mathbb{Z}^d)^*} \sigma(x)\sigma(y) - \frac{h}{2} \sum_{x \in \mathbb{Z}^d} \sigma(x), \quad \sigma \in S = \{-1,+1\}^{\mathbb{Z}^d}, \tag{23.1.1}$$

with $J, h > 0$. The system follows a Metropolis dynamics $(\sigma_t)_{t \geq 0}$ with spin-flip rates given by

$$c(\sigma, \sigma^x) = \begin{cases} \mathrm{e}^{-\beta[\Delta_x H(\sigma)]_+}, & \sigma \in S,\ x \in \mathbb{Z}^d, \\ 0, & \text{otherwise}, \end{cases} \tag{23.1.2}$$

where

$$\Delta_x H(\sigma) = \sigma(x) \Bigg[\sum_{\substack{y \in \mathbb{Z}^d \\ (x,y) \in (\mathbb{Z}^d)^*}} J\sigma(y) + h \Bigg]. \tag{23.1.3}$$

The reason for writing $\Delta_x H(\sigma)$ instead of $H(\sigma^x) - H(\sigma)$ is that H is infinite (see Liggett [166], Chapter III, for a precise definition). We again write $\mathbb{E}_\sigma$ to denote expectation w.r.t. the law of this dynamics starting from $\sigma_0 = \sigma$.

As in Sects. 17.1.2 and 17.6, we assume that $h \in (0, dJ)$ with dJ/h non-integer. As in Sect. 17.1.3, we let $\boxminus$ and $\boxplus$ denote the configurations where all the spins are down, respectively, up. We say that a function $f\colon S \to \mathbb{R}$ is local if it depends on finitely many spin variables only. The main result of this section is the following.

Theorem 23.1 (Mean crossover time from minus to plus) *If f is a local function, then*

$$\lim_{\beta \to \infty} \mathbb{E}_{\boxminus}\big(f(\sigma_{\tau(\beta;\kappa)})\big) = \begin{cases} f(\boxminus), & \text{if } \kappa < \kappa_d, \\ f(\boxplus), & \text{if } \kappa > \kappa_d, \end{cases} \tag{23.1.4}$$

where $\tau(\beta;\kappa) = \exp(\beta\kappa)$ and

$$\kappa_d = \frac{1}{d+1} \sum_{k=1}^{d} \Gamma_k^\star, \tag{23.1.5}$$

with $\Gamma_k^\star$ the energy of the critical droplet in k dimensions.

Recall from Sect. 17.7, Item 3, that an explicit formula is available for $\Gamma_k^\star$.

The heuristics behind Theorem 23.1 is as follows. The most efficient mechanism for relaxation from minus to plus near the origin is for the system to create a critical droplet of plus-spins somewhere far away from the origin and let this droplet grow and invade the origin. Suppose that nucleation in a finite box occurs at rate

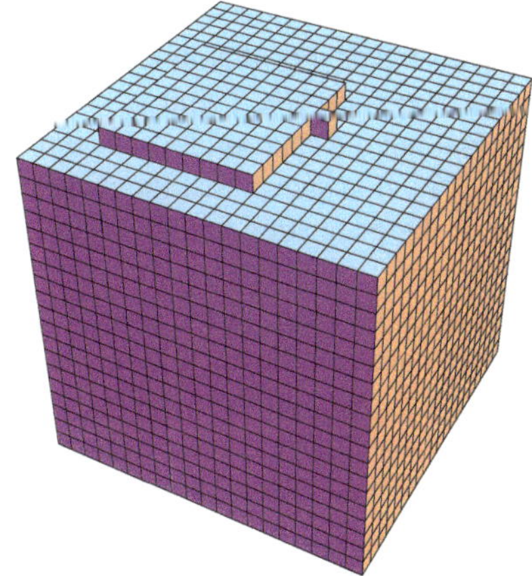

Fig. 23.1 Recursion in the dimension: A three-dimensional critical droplet, consisting of a quasi-cube with a two-dimensional critical droplet attached to one of its faces

$\exp(-\beta\Gamma_d^\star)$ (which we know from Chap. 19 is true if we ignore terms of order 1). Suppose further that the speed of growth of a large supercritical droplet is v_d (i.e., the speed at which the faces move outwards). Then, to invade the origin at time t, the droplet must be born inside the *space-time cone* whose basis is a d-dimensional hypercube with side length $v_d t$ and whose height is t. The critical space-time cone is such that the nucleation rate is of order 1. Therefore, writing τ_d for the time when the origin is invaded, we have

$$\tau_d\,(v_d\tau_d)^d\,\exp\bigl(-\beta\Gamma_d^\star\bigr)=1, \tag{23.1.6}$$

where we ignore terms of order $\exp(o(\beta))$. Since large droplets are approximately parallelepipeds, the dynamics on a face behaves like a $d-1$-dimensional Glauber dynamics, and so the time needed to fill a face is τ_{d-1}. Hence

$$v_d=1/\tau_{d-1}. \tag{23.1.7}$$

Combining (23.1.6)–(23.1.7), and putting $\tau_d=\exp(\beta\kappa_d)$, we obtain the recursion relation

$$(d+1)\kappa_d=\Gamma_d^\star+d\kappa_{d-1}. \tag{23.1.8}$$

Since $\kappa_0=0$, this yields the formula for κ_d in (23.1.5). Figure 23.1 (which is the same as Fig. 17.6 in Chap. 17) illustrates (23.1.8) for $d=3$.

In order to turn the above heuristics into the statement in (23.1.4), three major obstacles have to be overcome:

(1) **Speed of growth:** To control the speed of growth v_d, a comparison is made with a simpler *nucleation-and-growth model* in the spirit of bootstrap percolation, namely, all sites are initially unoccupied, a site becomes occupied at rate $e^{-\beta\Gamma_d^\star}$ if it has no occupied neighbours, at rate $\varepsilon=e^{-\beta\gamma}$ if it has one occupied

neighbour, and at rate 1 if it has two or more occupied neighbours, and occupied sites stay occupied forever. For this model it is shown that the speed is $\varepsilon^{1/d}$ and the nucleation time is $\exp[\beta\kappa_c]$ with $\kappa_c = \max\{\gamma, (\Gamma_d^\star + \gamma)/(d+1)\}$, provided $\Gamma_d^\star \geq \gamma$. By making the appropriate choice for γ as a function of J, h (e.g. $\gamma = 2J - h$ in $d = 2$), it is shown that the nucleation time in the original model is close to that of the nucleation-and-growth model. The proof requires delicate coupling techniques.

(2) **Energy landscape:** A detailed study of the energy landscape is necessary in order to show that the dynamics does not get caught in a deep well. For $d = 2, 3$ this can be done with the help of the combinatorial techniques mentioned and used in Chap. 17, but for higher dimensions no analogous techniques are available. The necessary estimates are obtained via rougher arguments.

(3) **Space-time clusters:** Some control on the size of space-time clusters is needed, e.g. to show that it is very unlikely for large space-time clusters to be formed prior to nucleation or for subcritical clusters to move over long distances. This requires estimates on recurrence times as well as an analysis of "cycle compounds".

The factor $\frac{1}{d+1}$ in (23.1.5) shows that the mechanism of far-away nucleation followed by invasion is faster than the mechanism of close-by nucleation alone. Thus, *space-time entropy* places a crucial role in infinite volume.

23.2 Small magnetic fields

Return to the model in Sect. 22.1 with Λ replaced by $\mathbb{Z}^2$ and again consider the metastable regime $\beta \in (0, \beta_c)$ and $h \downarrow 0$.

Theorem 23.2 (Mean crossover time from minus to plus) *The same result as in Theorem* 22.1 *holds with κ_β replaced by $\frac{1}{3}\kappa_\beta$.*

The heuristics behind this theorem is the same as for the model in Sect. 23.1. The extra factor $\frac{1}{d+1} = \frac{1}{3}$ in $d = 2$ again comes from space-time entropy: invasion of a growing droplet that is created somewhere in a space-time cone of the appropriate size. The same three obstacles have to be overcome to build a proof.

23.3 Bibliographical notes

1. Aizenman and Lebowitz [1] argued that a nucleation center in bootstrap percolation plays a similar rôle as a growing Wulff droplet in the Ising model subject to Glauber dynamics, and that therefore bootstrap percolation can serve as a paradigm for the description of metastable behaviour. They sketched a program for Glauber dynamics, which was subsequently carried out in Schonmann and Shlosman [215].

2. Theorem 23.1 was proved for $d = 2$ in Dehghanpour and Schonmann [77, 78], and for $d \geq 3$ in Cerf and Manzo [54]. Theorem 23.2 was proved for $d = 2$ in Schonmann and Shlosman [215]. The extension to $d \geq 3$ is still open. It is conjectured in Schonmann [213] that $v_2 \sim Ch$, $h \downarrow 0$, which is much stronger than the scaling property $v_2 = \exp[o(1/h)]$, $h \downarrow 0$, that is needed for the proof of Theorem 23.2.

3. It remains a challenge to obtain a sharper estimate of the crossover time, i.e., to find the function $\beta \mapsto T_d(\beta)$ such that $f(\sigma_t) \approx f(\boxminus)$ for $t \ll T_d(\beta)\exp(\beta\kappa_d)$ and $f(\sigma_t) \approx f(\boxplus)$ for $t \gg T_d(\beta)\exp(\beta\kappa_d)$ as $\beta \to \infty$. This function takes over the role of the prefactor K in Chaps. 16–18.

4. No analogues of Theorems 23.1–23.2 for Kawasaki dynamics have been proved yet. This is a formidable challenge because Kawasaki dynamics is conservative. In the post-nuclear phase there are droplets of varying sizes. Small droplets tend to shrink and be absorbed by large droplets, a phenomenon referred to as *Ostwald ripening*. Becker-Döring theory [15] is a phenomenological theory that tries to capture the size distribution of the droplets as a function of time. It turns out that, at low densities, to a good approximation the average radius of droplets grows like a fractional power of time (exponent $\frac{1}{3}$ in $d = 3$), which is supported by simulations. For mathematical background, see Ball, Carr and Penrose [10].

Glossary

$(A \to B)_{\text{opt}}$ optimal path between sets A and B
β inverse temperature
$(B_t)_{t \in \mathbb{R}_+}$ Brownian motion
$\mathscr{B}$ Borel σ-algebra
$B_\rho(x)$ ball of radius ρ centred at site x
$\text{cap}(A, B)$ capacity between sets A and B
$\mathscr{C}^*$ critical set
$e(x, t)$ heat kernel at site x at time t
$\mathbb{E}$ expectation with respect to a probability measure $\mathbb{P}$
$\mathbb{E}_\mu$ expectation with respect to a Markov process with initial law μ
$\mathbb{E}_x$ expectation with respect to a Markov process starting at site x
$\mathscr{E}(f, g)$ Dirichlet form for functions f, g
$e_{A,B}$ equilibrium measure on A for a pair of sets A, B
F potential function
$\mathscr{F}, \mathscr{G}$ σ-algebra's
$(\mathscr{F}_n)_{n \in \mathbb{N}}, (\mathscr{F}_t)_{t \in \mathbb{R}_+}$ filtration in discrete, respectively, continuous time
$\Phi(A, B)$ communication height between sets A and B
$\mathscr{G}(A, B)$ essential gate between sets A and B
$G_{D^c}(x, y)$ Green function at $x, y \in D$ with Dirichlet boundary conditions
$H(\xi)$ Hamiltonian at configuration ξ
$H_{D^c}(x, y)$ Poisson kernel at $x, y \in D$ with Dirichlet boundary conditions
$h_{A,B}$ equilibrium potential between sets A and B
$I(x), I(\gamma)$ large deviation rate functions at site x, respectively, path γ
L Markov generator, discrete time
$\mathscr{L}$ Markov generator, continuous time
$\mathscr{L}^p(\mu)$ space of functions whose p-th power is integrable w.r.t. μ
$L^p(\mu)$ space of equivalence classes of functions whose p-th power is integrable w.r.t. μ
$\mathscr{L}(\gamma, \dot{\gamma}, s)$ Lagrangian in large deviation theory at time s for path γ
λ_x, λ_k eigenvalue of generator associated with $x \in \mathscr{M}$, respectively, $x_k \in \mathscr{M}$
$\lambda_0^{\mathscr{N}}$ principal eigenvalue of generator with Dirichlet boundary conditions in $\mathscr{N}$

A. Bovier, F. den Hollander, *Metastability*,
Grundlehren der mathematischen Wissenschaften 351,
DOI 10.1007/978-3-319-24777-9

$(M_n)_{n\in\mathbb{N}}, (M_t)_{t\in\mathbb{R}_+}$ martingales in discrete, respectively, continuous time
$\mathscr{M}$ set of metastable points
$\mathscr{M}_x$ set of metastable points that are more stable than metastable point x
μ measure, often invariant measure or Gibbs measure
$\nu_{A,B}$ last-exit biased distribution on A for Markov process killed in B
$(\Omega, \mathscr{F}, \mathbb{P})$ probability space, consisting of a set, a sigma-algebra and a probability measure
$(\Omega, \mathscr{F}, \mathbb{P}, (\mathscr{F}_t)_{t\in I})$ filtered probability space, consisting of a set, a sigma-algebra, a probability measure and a filtration
$\mathscr{P}^*$ protocritical set
$P(x, A)$ one-step transition kernel from site x to set A
$\mathscr{P}_t(x, A)$ Markov semigroup at time t from site x to set A
$\mathbb{P}^h, \mathbb{E}^h$ Doob transformed law and expectation with harmonic function h
R_λ resolvent of a semigroup
S often state space of a stochastic process
$\mathscr{S}(A, B)$ communication level set between sets A and B
σ, η configurations (spin variables, occupation variables, etc.)
τ stopping time
$W(x, t)$ space-time white noise
$(X_t)_{t\in\mathbb{R}_+}, (Y_t)_{t\in\mathbb{R}_+}$ stochastic processes
Z_N partition function for system size N

References

1. Aizenman, M., Lebowitz, J.: Metastability effects in bootstrap percolation. J. Phys. A **21**, 3801–3813 (1988)
2. Aldous, D., Fill, J.A.: Reversible Markov Chains and Random Walks on Graphs. https://www.stat.berkeley.edu/~aldous/RWG/book.pdf (2002/2014)
3. Allen, S., Cahn, J.: Ground state structures in ordered binary alloys with second neighbor interactions. Acta Metall. **20**, 423–433 (1972)
4. Alonso, L., Cerf, R.: The three-dimensional polyominoes of minimal area. Electron. J. Comb. **3**, 1–39 (1996)
5. Amaro de Matos, J.M.G., Baêta Segundo, J.A., Perez, J.F.: Fluctuations in dilute antiferromagnets: Curie-Weiss models. J. Phys. A **25**, 2819–2830 (1992)
6. an der Heiden, M.: Metastability of Markov chains and in the Hopfield model. Ph.D. thesis, Technische Universtität Berlin (2007)
7. Andjel, E.D.: Invariant measures for the zero range processes. Ann. Probab. **10**, 525–547 (1982)
8. Arrhenius, S.: On the reaction rate of the inversion of non-refined sugar upon souring. Z. Phys. Chem. **4**, 226–248 (1889)
9. Baake, E., Baake, M., Bovier, A., Klein, M.: An asymptotic maximum principle for essentially linear evolution models. J. Math. Biol. **50**, 83–114 (2005)
10. Ball, J.M., Carr, J., Penrose, O.: The Becker-Döring cluster equations: basic properties and asymptotic behaviour of solutions. Commun. Math. Phys. **104**, 657–692 (1986)
11. Barret, F.: Sharp asymptotics of metastable transition times for one-dimensional SPDEs. Ann. Inst. Henri Poincaré Probab. Stat. **51**, 129–166 (2015)
12. Barret, F., Bovier, A., Méléard, S.: Uniform estimates for metastable transition times in a coupled bistable system. Electron. J. Probab. **15**, 323–345 (2010)
13. Bauer, H.: Probability Theory and Elements of Measure Theory. Academic Press Inc. [Harcourt Brace Jovanovich Publishers], London (1981)
14. Baur, E.: Metastabilität von reversiblen Diffusionsprozessen. Diploma thesis, Bonn University (2011)
15. Becker, R., Döring, W.: Kinetische Behandlung der Keimbildung in übersättigten Dämpfen. Ann. Phys. (Leipz.) **24**, 719–752 (1935)
16. Beltrán, J., Landim, C.: Tunneling and metastability of continuous time Markov chains II, the nonreversible case. J. Stat. Phys. **149**, 598–618 (2012)
17. Beltrán, J., Landim, C.: A martingale approach to metastability. Probab. Theory Relat. Fields **161**, 267–307 (2015)
18. Ben Arous, G., Bovier, A., Gayrard, V.: Glauber dynamics of the random energy model. I. Metastable motion on the extreme states. Commun. Math. Phys. **235**, 379–425 (2003)

A. Bovier, F. den Hollander, *Metastability*,
Grundlehren der mathematischen Wissenschaften 351,
DOI 10.1007/978-3-319-24777-9

19. Ben Arous, G., Cerf, R.: Metastability of the three-dimensional Ising model on a torus at very low temperatures. Electron. J. Probab. **1**, 1–55 (1996)
20. Berglund, N., Dutercq, S.: The Eyring–Kramers law for Markovian jump processes with symmetries. J. Theor. Probab. (2015). doi:10.1007/s10959-015-0617-9
21. Berglund, N., Gentz, B.: The Eyring-Kramers law for potentials with nonquadratic saddles. Markov Process. Relat. Fields **16**, 549–598 (2010)
22. Berglund, N., Gentz, B.: Sharp estimates for metastable lifetimes in parabolic SPDEs: Kramers' law and beyond. Electron. J. Probab. **18**, 1–58 (2013)
23. Berman, K.A., Konsowa, M.H.: Random paths and cuts, electrical networks, and reversible Markov chains. SIAM J. Discrete Math. **3**, 311–319 (1990)
24. Bianchi, A., Bovier, A., Ioffe, D.: Sharp asymptotics for metastability in the random field Curie-Weiss model. Electron. J. Probab. **14**, 1541–1603 (2009)
25. Bianchi, A., Bovier, A., Ioffe, D.: Pointwise estimates and exponential laws in metastable systems via coupling methods. Ann. Probab. **40**, 339–379 (2012)
26. Bianchi, A., Gaudillière, A.: Metastable states, quasi-stationary and soft measures, mixing time asymptotics via variational principles. arXiv:1103.1143v1, to appear in Stoch. Proc. Appl. (2011)
27. Bigelis, S., Cirillo, E., Lebowitz, J., Speer, E.: Critical droplets in metastable probabilistic cellular automata. Phys. Rev. E **59**, 3935–3941 (1999)
28. Billingsley, P.: Probability and Measure. Wiley Series in Probability and Mathematical Statistics. Wiley, New York (1995)
29. Bodineau, T., Graham, B., Wouts, M.: Metastability in the dilute Ising model. Probab. Theory Relat. Fields **157**, 955–1009 (2013)
30. Bouchaud, J.-P., Cugliandolo, L., Kurchan, J., Mézard, M.: Out of equilibrium dynamics in spin-glasses and other glassy systems. In: Young, A.P. (ed.) Spin Glasses and Random Fields. World Scientific, Singapore (1998)
31. Bovier, A., den Hollander, F., Nardi, F.R.: Sharp asymptotics for Kawasaki dynamics on a finite box with open boundary. Probab. Theory Relat. Fields **135**, 265–310 (2006)
32. Bovier, A., den Hollander, F., Spitoni, C.: Homogeneous nucleation for Glauber and Kawasaki dynamics in large volumes and low temperature. Ann. Probab. **38**, 661–713 (2010)
33. Bovier, A., Eckhoff, M., Gayrard, V., Klein, M.: Metastability in stochastic dynamics of disordered mean-field models. Probab. Theory Relat. Fields **119**, 99–161 (2001)
34. Bovier, A., Eckhoff, M., Gayrard, V., Klein, M.: Metastability and low lying spectra in reversible Markov chains. Commun. Math. Phys. **228**, 219–255 (2002)
35. Bovier, A., Eckhoff, M., Gayrard, V., Klein, M.: Metastability in reversible diffusion processes. I. Sharp asymptotics for capacities and exit times. J. Eur. Math. Soc. (JEMS) **6**, 399–424 (2004)
36. Bovier, A., Gayrard, V.: Hopfield models as generalized random mean field models. In: Mathematical Aspects of Spin Glasses and Neural Networks. Progr. Probab., vol. 41, pp. 3–89. Birkhäuser, Boston (1998)
37. Bovier, A., Gayrard, V.: Sample path large deviations for a class of Markov chains related to disordered mean field models. Preprint available at arXiv:math/9905022, 487, WIAS Berlin (1999)
38. Bovier, A., Gayrard, V., Klein, M.: Metastability in reversible diffusion processes. II. Precise asymptotics for small eigenvalues. J. Eur. Math. Soc. (JEMS) **7**, 69–99 (2005)
39. Bovier, A., Manzo, F.: Metastability in Glauber dynamics in the low-temperature limit: beyond exponential asymptotics. J. Stat. Phys. **107**, 757–779 (2002)
40. Bovier, A., Neukirch, R.: A note on metastable behaviour in the zero-range process. In: Griebel, M. (ed.) Singular Phenomena and Scaling in Mathematical Models, pp. 365–376. Springer, Berlin (2013)
41. Brassesco, S.: Some results on small random perturbations of an infinite-dimensional dynamical system. Stoch. Process. Appl. **38**, 33–53 (1991)
42. Brassesco, S., Buttà, P.: Interface fluctuations for the $D = 1$ stochastic Ginzburg-Landau equation with nonsymmetric reaction term. J. Stat. Phys. **93**, 1111–1142 (1998)

43. Burke, C.J., Rosenblatt, M.: A Markovian function of a Markov chain. Ann. Math. Stat. **29**, 1112–1122 (1958)
44. Buslov, V.A., Makarov, K.A.: A time-scale hierarchy with small diffusion. Teor. Mat. Fiz. **76**, 219–230 (1988)
45. Buslov, V.A., Makarov, K.A.: Life spans and least eigenvalues of an operator of small diffusion. Mat. Zametki **51**, 160 (1992)
46. Cancrini, N., Cesi, F., Martinelli, F.: The spectral gap for the Kawasaki dynamics at low temperature. J. Stat. Phys. **95**, 215–271 (1999)
47. Cancrini, N., Martinelli, F.: On the spectral gap of Kawasaki dynamics under a mixing condition revisited. J. Math. Phys. **41**, 1391–1423 (2000)
48. Cancrini, N., Martinelli, F., Roberto, C.: The logarithmic Sobolev constant of Kawasaki dynamics under a mixing condition revisited. Ann. Inst. Henri Poincaré Probab. Stat. **38**, 385–436 (2002)
49. Cancrini, N., Martinelli, F., Roberto, C.: Spectral gap and logarithmic Sobolev constant of Kawasaki dynamics under a mixing condition revisited. In: In and out of Equilibrium (Mambucaba, 2000). Progr. Probab., vol. 51, pp. 259–271. Birkhäuser, Boston (2002)
50. Caputo, P., Lacoin, H., Martinelli, F., Simenhaus, F., Toninelli, F.L.: Polymer dynamics in the depinned phase: metastability with logarithmic barriers. Probab. Theory Relat. Fields **153**, 587–641 (2012)
51. Cassandro, M., Galves, A., Olivieri, E., Vares, M.E.: Metastable behavior of stochastic dynamics: a pathwise approach. J. Stat. Phys. **35**, 603–634 (1984)
52. Cassandro, M., Olivieri, E., Picco, P.: Small random perturbations of infinite-dimensional dynamical systems and nucleation theory. Ann. Inst. Henri Poincaré, Phys. Théor. **44**, 343–396 (1986)
53. Catoni, O., Cerf, R.: The exit path of a Markov chain with rare transitions. ESAIM Probab. Stat. **1**, 95–144 (1995/97)
54. Cerf, R., Manzo, F.: Nucleation and growth for the Ising model in d dimensions at very low temperatures. Ann. Probab. **41**, 3697–3785 (2013)
55. Chaganty, N.R., Sethuraman, J.: Strong large deviation and local limit theorems. Ann. Probab. **21**, 1671–1690 (1993)
56. Chatterjee, S., Durrett, R.: Contact processes on random graphs with power law degree distributions have critical value 0. Ann. Probab. **37**, 2332–2356 (2009)
57. Chenal, F., Millet, A.: Uniform large deviations for parabolic SPDEs and applications. Stoch. Process. Appl. **72**, 161–186 (1997)
58. Chow, Y.S., Teicher, H.: Probability Theory, 3rd edn. Springer Texts in Statistics. Springer, New York (1997)
59. Cirillo, E.: A note on the metastability of the Ising model: the alternate updating case. J. Stat. Phys. **106**, 335–390 (2002)
60. Cirillo, E., Nardi, F.: Metastability for the Ising model with a parallel dynamics. J. Stat. Phys. **110**, 183–217 (2003)
61. Cirillo, E., Nardi, F., Polosa, A.: Magnetic order in the Ising model with parallel dynamics. Phys. Rev. E **64**, 57103 (2001)
62. Cirillo, E., Nardi, F., Spitoni, C.: Competitive nucleation in reversible probabilistic cellular automata. Phys. Rev. E **78**, 040601 (2008)
63. Cirillo, E., Nardi, F., Spitoni, C.: Metastability for reversible probabilistic cellular automata with self-interaction. J. Stat. Phys. **132**, 431–471 (2008)
64. Cirillo, E., Nardi, F., Spitoni, C.: Competitive nucleation in metastable systems. Commun. SIMAI Congr. **3**, 040601(R) (2009)
65. Cirillo, E.N.M., Nardi, F.R.: Relaxation height in energy landscapes: an application to multiple metastable states. J. Stat. Phys. **150**, 1080–1114 (2013)
66. Coddington, E., Levinson, N.: Theory of Ordinary Differential Equations. McGraw-Hill, New York-Toronto-London (1955)
67. Da Prato, G., Debussche, A.: Strong solutions to the stochastic quantization equations. Ann. Probab. **31**, 1900–1916 (2003)

68. Da Prato, G., Zabczyk, J.: Stochastic Equations in Infinite Dimensions. Encyclopedia of Mathematics and Its Applications, vol. 44. Cambridge University Press, Cambridge (1992)
69. Dai Pra, P., den Hollander, F.: McKean-Vlasov limit for interacting random processes in random media. J. Stat. Phys. **84**, 735–772 (1996)
70. Davies, E.: Dynamical stability of metastable states. J. Funct. Anal. **46**, 373–386 (1982)
71. Davies, E.: Spectral properties of metastable Markov semigroups. J. Funct. Anal. **52**, 315–329 (1983)
72. Davies, E.B.: Metastability and the Ising model. J. Stat. Phys. **27**, 657–675 (1982)
73. Davies, E.B.: Metastable states of symmetric Markov semigroups. I. Proc. Lond. Math. Soc. **45**, 133–150 (1982)
74. Davies, E.B.: Metastable states of symmetric Markov semigroups. II. J. Lond. Math. Soc. **26**, 541–556 (1982)
75. Dawson, D., Greven, A.: Spatial Fleming-Viot Models with Selection and Mutation. Lecture Notes in Mathematics, vol. 2092. Springer, Berlin (2014)
76. de Hoog, F.R., Anderssen, R.S.: Asymptotic formulas for discrete eigenvalue problems in Liouville normal form. Math. Models Methods Appl. Sci. **11**, 43–56 (2001)
77. Dehghanpour, P., Schonmann, R.H.: Metropolis dynamics relaxation via nucleation and growth. Commun. Math. Phys. **188**, 89–119 (1997)
78. Dehghanpour, P., Schonmann, R.H.: A nucleation-and-growth model. Probab. Theory Relat. Fields **107**, 123–135 (1997)
79. Dembo, A., Zeitouni, O.: Large Deviations Techniques and Applications, 2nd edn. Applications of Mathematics (New York), vol. 38. Springer, New York (1998)
80. den Hollander, F.: Large Deviations. Fields Institute Monographs, vol. 14. American Mathematical Society, Providence (2000)
81. den Hollander, F.: Metastability under stochastic dynamics. Stoch. Process. Appl. **114**, 1–26 (2004)
82. den Hollander, F., Jansen, S.: Berman-Konsowa principle for reversible Markov jump processes. arXiv:1309.1305, to appear in Markov Process. Relat. Fields (2015)
83. den Hollander, F., Jansen, S.: Metastability at low temperature for continuum interacting particle systems (2015, in preparation)
84. den Hollander, F., Nardi, F., Olivieri, E., Scoppola, E.: Droplet growth for three-dimensional Kawasaki dynamics. Probab. Theory Relat. Fields **125**, 153–194 (2003)
85. den Hollander, F., Nardi, F.R., Troiani, A.: Kawasaki dynamics with two types of particles: stable/metastable configurations and communication heights. J. Stat. Phys. **145**, 1423–1457 (2011)
86. den Hollander, F., Nardi, F.R., Troiani, A.: Kawasaki dynamics with two types of particles: critical droplets. J. Stat. Phys. **149**, 1013–1057 (2012)
87. den Hollander, F., Nardi, F.R., Troiani, A.: Metastability for Kawasaki dynamics at low temperature with two types of particles. Electron. J. Probab. **17**, 26 (2012)
88. den Hollander, F., Olivieri, E., Scoppola, E.: Metastability and nucleation for conservative dynamics. J. Math. Phys. **41**, 1424–1498 (2000)
89. den Hollander, F., Olivieri, E., Scoppola, E.: Nucleation in fluids: some rigorous results. Physica A **279**, 110–122 (2000)
90. den Hollander, F., Olivieri, E., Scoppola, E.: Metastability and nucleation for conservative dynamics. Markov Process. Relat. Fields **7**, 51–53 (2001)
91. Deuschel, J.-D., Stroock, D.: Large Deviations. Pure and Applied Mathematics, vol. 137. Academic Press, Boston (1989)
92. Dobrushin, R., Shlosman, S.: "Non-Gibbsian" states and their Gibbs description. Commun. Math. Phys. **200**, 125–179 (1999)
93. Dommers, S.: Metastability of the Ising model on random regular graphs at zero temperature. arXiv:1411.6802 (2014)
94. Donsker, M., Varadhan, S.: A law of the iterated logarithm for total occupation times of transient Brownian motion. Commun. Pure Appl. Math. **33**, 365–393 (1980)

95. Doob, J.L.: Classical Potential Theory and Its Probabilistic Counterpart. Grundlehren der Mathematischen Wissenschaften [Fundamental Principles of Mathematical Sciences], vol. 262. Springer, New York (1984)
96. Doyle, P., Snell, J.: Random Walks and Electric Networks. Carus Mathematical Monographs, vol. 22. Mathematical Association of America, Washington (1984)
97. Doyle, P.G.: Energy for Markov chains. Preprint available at http://math.dartmouth.edu/~doyle/docs/energy/energy.pdf (1994)
98. Dupuis, P., Ellis, R.: A Weak Convergence Approach to the Theory of Large Deviations. Wiley Series in Probability and Statistics: Probability and Statistics. Wiley, New York (1997)
99. E, W., Ren, W., Vanden-Eijnden, E.: Energy landscapes and rare events. In: Proceedings of the International Congress of Mathematicians, Vol. I (Beijing, 2002), pp. 621–630. Higher Ed. Press, Beijing (2002)
100. E, W., Vanden-Eijnden, E.: Towards a theory of transition paths. J. Stat. Phys. **123**, 503–523 (2006)
101. Eckhoff, M.: Capacity and the Low Lying Spectrum in Attractive Markov Chains. Ph.D. thesis, Universität Potsdam (2000)
102. Eckhoff, M.: The low lying spectrum of irreversible, infinite state Markov chains in the metastable regime. Technical report. Preprint, available at http://www.math.uzh.ch/fileadmin/user/eckhoff/publikation/specirrev.pdf (2002)
103. Ellis, R.: Entropy, Large Deviations, and Statistical Mechanics. Grundlehren der Mathematischen Wissenschaften, vol. 271. Springer, New York (1985)
104. Ethier, S., Kurtz, T.: Markov Processes. Characterization and Convergence. Wiley Series in Probability and Mathematical Statistics: Probability and Mathematical Statistics. Wiley, New York (1986)
105. Evans, M.: Phase transitions in one-dimensional nonequilibrium systems. Braz. J. Phys. **30**, 42–57 (2000)
106. Eyring, H.: The activated complex in chemical reactions. J. Chem. Phys. **3**, 107–115 (1935)
107. Faris, W., Jona-Lasinio, G.: Large fluctuations for a nonlinear heat equation with noise. J. Phys. A **15**, 3025–3055 (1982)
108. Feller, W.: An Introduction to Probability Theory and Its Applications. Vol. I, 3rd edn. Wiley, New York (1968)
109. Feller, W.: An Introduction to Probability Theory and Its Applications. Vol. II, 2nd edn. Wiley, New York (1971)
110. Feng, J., Kurtz, T.: Large Deviations for Stochastic Processes. Mathematical Surveys and Monographs, vol. 131. American Mathematical Society, Providence (2006)
111. Fernandez, R., Manzo, F., Nardi, F., Scoppola, E., Sohier, J.: Conditioned, quasi-stationary, restricted measures and escape from metastable states. ArXiv e-prints (Oct. 2014)
112. Fernandez, R., Manzo, F., Nardi, F.R., Scoppola, E.: Asymptotically exponential hitting times and metastability. arXiv:1406.2637 (2014)
113. Fiedler, B., Rocha, C.: Heteroclinic orbits of semilinear parabolic equations. J. Differ. Equ. **125**, 239–281 (1996)
114. Fontes, L.R., Mathieu, P., Picco, P.: On the averaged dynamics of the random field Curie-Weiss model. Ann. Appl. Probab. **10**, 1212–1245 (2000)
115. Freidlin, M.I., Wentzell, A.D.: Random Perturbations of Dynamical Systems. Grundlehren der Mathematischen Wissenschaften, vol. 260. Springer, New York (1984)
116. Fukushima, M., Oshima, Y., Takeda, M.: Dirichlet Forms and Symmetric Markov Processes. de Gruyter Studies in Mathematics, vol. 19. de Gruyter, Berlin (2011), extended edition
117. Funaki, T.: Random motion of strings and related stochastic evolution equations. Nagoya Math. J. **89**, 129–193 (1983)
118. Gaudillière, A., den Hollander, F., Nardi, F.R., Olivieri, E., Scoppola, E.: Ideal gas approximation for a two-dimensional rarefied gas under Kawasaki dynamics. Stoch. Process. Appl. **119**, 737–774 (2009)
119. Gaudillière, A., den Hollander, F., Nardi, F.R., Olivieri, E., Scoppola, E.: Droplet dynamics in a two-dimensional rarified gas under Kawasaki dynamics (2015, in preparation)

120. Gaudillière, A., den Hollander, F., Nardi, F.R., Olivieri, E., Scoppola, E.: Homogeneous nucleation for two-dimensional Kawasaki dynamics (2015, in preparation)
121. Gaudillière, A., Landim, C.: A Dirichlet principle for non reversible Markov chains and some recurrence theorems. Probab. Theory Relat. Fields **158**, 55–89 (2014)
122. Gaudillière, A., Olivieri, E., Scoppola, E.: Nucleation pattern at low temperature for local Kawasaki dynamics in two dimensions. Markov Process. Relat. Fields **11**, 553–628 (2005)
123. Gaveau, B., Moreau, M.: Metastable relaxation times and absorption probabilities for multidimensional stochastic systems. J. Phys. A **33**, 4837–4850 (2000)
124. Gaveau, B., Schulman, L.S.: Theory of nonequilibrium first-order phase transitions for stochastic dynamics. J. Math. Phys. **39**, 1517–1533 (1998)
125. Georgii, H.-O.: Gibbs Measures and Phase Transitions, 2nd edn. de Gruyter Studies in Mathematics, vol. 9. de Gruyter, Berlin (1988)
126. Gilbarg, D., Trudinger, N.: Elliptic Partial Differential Equations of Second Order, 2nd edn. Grundlehren der Mathematischen Wissenschaften, vol. 224. Springer, Berlin (1983)
127. Glasstone, S., Laidler, K., Eyring, H.: The Theory of Rate Processes. McGraw-Hill, New York (1941)
128. Gois, B., Landim, C.: Zero-temperature limit of the Kawasaki dynamics for the Ising lattice gas in a large two-dimensional torus. Ann. Probab. **43**, 2151–2203 (2015)
129. Grassberger, P., Barkema, G., Nadler, W.: Monte Carlo Approach to Biopolymers and Protein Folding. World Scientific, Singapore (1998)
130. Großkinsky, S., Schütz, G.M.: Discontinuous condensation transition and nonequivalence of ensembles in a zero-range process. J. Stat. Phys. **132**, 77–108 (2008)
131. Großkinsky, S., Schütz, G.M., Spohn, H.: Condensation in the zero range process: stationary and dynamical properties. J. Stat. Phys. **113**, 389–410 (2003)
132. Großkinsky, S., Spohn, H.: Stationary measures and hydrodynamics of zero range processes with several species of particles. Bull. Braz. Math. Soc. (N.S.), 489–507 (2003)
133. Gyöngy, I.: Lattice approximations for stochastic quasi-linear parabolic partial differential equations driven by space-time white noise. I. Potential Anal. **9**, 1–25 (1998)
134. Gyöngy, I., Pardoux, É.: On quasi-linear stochastic partial differential equations. Probab. Theory Relat. Fields **94**, 413–425 (1993)
135. Hairer, M.: A theory of regularity structures. Invent. Math. **198**, 269–504 (2014)
136. Helffer, B., Klein, M., Nier, F.: Quantitative analysis of metastability in reversible diffusion processes via a Witten complex approach. Mat. Contemp. **26**, 41–85 (2004)
137. Helffer, B., Nier, F.: Quantitative analysis of metastability in reversible diffusion processes via a Witten complex approach: the case with boundary. Mém. Soc. Math. Fr. (N.S.) **105**, vi+89 (2006)
138. Helffer, B., Sjöstrand, J.: Multiple wells in the semiclassical limit. I. Commun. Partial Differ. Equ. **9**, 337–408 (1984)
139. Holden, H., Oksendal, B., Uboe, J., Zhang, T.: Stochastic Partial Differential Equations. Probability and Its Applications. Birkhäuser, Boston (1996)
140. Holley, R.A., Kusuoka, S., Stroock, D.W.: Asymptotics of the spectral gap with applications to the theory of simulated annealing. J. Funct. Anal. **83**, 333–347 (1989)
141. Huang, S.: The molecular and mathematical basis of Waddington's epigenetic landscape: A framework for post-Darwinian biology? BioEssays **34**, 149–157 (2011)
142. Itō, K., McKean, H.: Diffusion Processes and Their Sample Paths. Springer, New York (1965)
143. Jacod, J., Shiryaev, A.: Limit Theorems for Stochastic Processes, 2nd edn. Grundlehren der Mathematischen Wissenschaften, vol. 288. Springer, Berlin (2003)
144. Kakutani, S.: Two-dimensional Brownian motion and harmonic functions. Proc. Imp. Acad. (Tokyo) **20**, 706–714 (1944)
145. Kakutani, S.: Markov process and the Dirichlet problem. Proc. Jpn. Acad. **21**, 227–233 (1949), 1945
146. Kallenberg, O.: Foundations of Modern Probability, 2nd edn. Probability and Its Applications (New York). Springer, New York (2002)

147. Kallianpur, G., Xiong, J.: Large deviations for a class of stochastic partial differential equations. Ann. Probab. **24**, 320–345 (1996)
148. Karatzas, I., Shreve, S.: Brownian Motion and Stochastic Calculus. Graduate Texts in Mathematics. Springer, New York (1988)
149. Kauffman, S.: The Origins of Order. Oxford University Press, Oxford (1993)
150. Kemeny, J., Snell, J., Knapp, A.: Denumerable Markov Chains, 2nd edn. Springer, New York (1976)
151. Kemeny, J.G., Snell, J.L.: Finite Markov Chains. The University Series in Undergraduate Mathematics. D. Van Nostrand Co., Inc., Princeton-Toronto-London-New York (1960)
152. Kifer, Y.: Random Perturbations of Dynamical Systems. Progress in Probability and Statistics, vol. 16. Birkhäuser, Boston (1988)
153. Knessl, C., Matkowsky, B.J., Schuss, Z., Tier, C.: An asymptotic theory of large deviations for Markov jump processes. SIAM J. Appl. Math. **45**, 1006–1028 (1985)
154. Kolokoltsov, V.: Semiclassical Analysis for Diffusions and Stochastic Processes. Lecture Notes in Mathematics, vol. 1724. Springer, Berlin (2000)
155. Kotecký, R., Olivieri, E.: Droplet dynamics for asymmetric Ising model. J. Stat. Phys. **70**, 1121–1148 (1993)
156. Kotecký, R., Olivieri, E.: Shapes of growing droplets—a model of escape from a metastable phase. J. Stat. Phys. **75**, 409–506 (1994)
157. Kramers, H.A.: Brownian motion in a field of force and the diffusion model of chemical reactions. Physica **7**, 284–304 (1940)
158. Külske, C.: Metastates in disordered mean-field models: random field and Hopfield models. J. Stat. Phys. **88**, 1257–1293 (1997)
159. Landim, C.: Metastability for a non-reversible dynamics: The evolution of the condensate in totally asymmetric zero range processes. Commun. Math. Phys. **330**, 1–32 (2014)
160. Lawler, G.F.: Intersections of Random Walks. Probability and Its Applications. Birkhäuser, Boston (1991)
161. Lawler, G.F., Schramm, O., Werner, W.: Conformal invariance of planar loop-erased random walks and uniform spanning trees. Ann. Probab. **32**, 939–995 (2004)
162. Lebowitz, J.L., Penrose, O.: Rigorous treatment of the van der Waals-Maxwell theory of the liquid-vapor transition. J. Math. Phys. **7**, 98–113 (1966)
163. Levin, D., Peres, Y., Wilmer, E.: Markov Chains and Mixing Times. American Mathematical Society, Providence (2009)
164. Levin, D.A., Luczak, M.J., Peres, Y.: Glauber dynamics for the mean-field Ising model: cut-off, critical power law, and metastability. Probab. Theory Relat. Fields **146**, 223–265 (2010)
165. Levit, S., Smilansky, U.: A theorem on infinite products of eigenvalues of Sturm-Liouville type operators. Proc. Am. Math. Soc. **65**, 299–302 (1977)
166. Liggett, T.: Interacting Particle Systems. Grundlehren der Mathematischen Wissenschaften, vol. 276. Springer, New York (1985)
167. Lubetzky, E., Martinelli, F., Sly, A., Toninelli, F.L.: Quasi-polynomial mixing of the 2D stochastic Ising model with "plus" boundary up to criticality. J. Eur. Math. Soc. (JEMS) **15**, 339–386 (2013)
168. Lubetzky, E., Sly, A.: Critical Ising on the square lattice mixes in polynomial time. Commun. Math. Phys. **313**, 815–836 (2012)
169. Maier, R.S., Stein, D.L.: Limiting exit location distributions in the stochastic exit problem. SIAM J. Appl. Math. **57**, 752–790 (1997)
170. Maier, R.S., Stein, D.L.: Droplet nucleation and domain wall motion in a bounded interval. Phys. Rev. Lett. **87**, 270601 (2001)
171. Manzo, F., Nardi, F.R., Olivieri, E., Scoppola, E.: On the essential features of metastability: tunnelling time and critical configurations. J. Stat. Phys. **115**, 591–642 (2004)
172. Manzo, F., Olivieri, E.: Relaxation patterns for competing metastable states: a nucleation and growth model. Markov Process. Relat. Fields **4**, 549–570 (1998)
173. Manzo, F., Olivieri, E.: Dynamical Blume-Capel model: competing metastable states at infinite volume. J. Stat. Phys. **104**, 1029–1090 (2001)

174. Martinelli, F., Olivieri, E., Scoppola, E.: Small random perturbations of finite- and infinite-dimensional dynamical systems: unpredictability of exit times. J. Stat. Phys. **55**, 477–504 (1989)
175. Martinelli, F., Olivieri, E., Scoppola, E.: Metastability and exponential approach to equilibrium for low-temperature stochastic Ising models. J. Stat. Phys. **61**, 1105–1119 (1990)
176. Martinelli, F., Sbano, L., Scoppola, E.: Small random perturbation of dynamical systems: recursive multiscale analysis. Stoch. Stoch. Rep. **49**, 253–272 (1994)
177. Martinelli, F., Scoppola, E.: Small random perturbations of dynamical systems: exponential loss of memory of the initial condition. Commun. Math. Phys. **120**, 25–69 (1988)
178. Martinelli, F., Toninelli, F.L.: On the mixing time of the 2D stochastic Ising model with "plus" boundary conditions at low temperature. Commun. Math. Phys. **296**, 175–213 (2010)
179. Mathieu, P.: Spectra, exit times and long time asymptotics in the zero-white-noise limit. Stoch. Stoch. Rep. **55**, 1–20 (1995)
180. Mathieu, P., Picco, P.: Metastability and convergence to equilibrium for the random field Curie-Weiss model. J. Stat. Phys. **91**, 679–732 (1998)
181. Matkowsky, B.J., Schuss, Z.: On the lifetime of a metastable state at low noise. Phys. Lett. A **95**, 213–215 (1983)
182. Matkowsky, B.J., Schuss, Z., Tier, C.: Uniform expansion of the transition rate in Kramers' problem. J. Stat. Phys. **35**, 443–456 (1984)
183. Menz, G., Schlichting, A.: Poincaré and logarithmic Sobolev inequalities by decomposition of the energy landscape. Ann. Probab. **42**, 1809–1884 (2014)
184. Metzner, P., Schütte, C., Vanden-Eijnden, E.: Transition path theory for Markov jump processes. Multiscale Model. Simul. **7**, 1192–1219 (2008)
185. Miclo, L.: Comportement de spectres d'opérateurs de Schrödinger à basse température. Bull. Sci. Math. **119**, 529–553 (1995)
186. Mogul'skiĭ, A.A.: Large deviations for the trajectories of multidimensional random walks. Teor. Verojatn. Primen. **21**, 309–323 (1976)
187. Mourrat, J.-C., Valesin, D.: Phase transition of the contact process on random regular graphs. arXiv:1405.0865 (2014)
188. Nardi, F.R., Olivieri, E.: Low temperature stochastic dynamics for an Ising model with alternating field. Markov Process. Relat. Fields **2**, 117–166 (1996)
189. Nardi, F.R., Olivieri, E., Scoppola, E.: Anisotropy effects in nucleation for conservative dynamics. J. Stat. Phys. **119**, 539–595 (2005)
190. Nardi, F.R., Spitoni, C.: Sharp asymptotics for stochastic dynamics with parallel updating rule. J. Stat. Phys. **146**, 701–718 (2012)
191. Neukirch, R.: Metastability in the Zero-range Process. Diploma thesis, Bonn University (2011)
192. Neves, E.: A discrete variational problem related to Ising droplets at low temperatures. J. Stat. Phys. **80** (1995)
193. Neves, E.J., Schonmann, R.H.: Critical droplets and metastability for a Glauber dynamics at very low temperatures. Commun. Math. Phys. **137**, 209–230 (1991)
194. Newman, M., Barkema, G.: Monte Carlo Methods in Statistical Physics. Oxford University Press, Oxford (1999)
195. Norris, J.: Markov Chains. Cambridge Series in Statistical and Probabilistic Mathematics, vol. 2. Cambridge University Press, Cambridge (1998)
196. Olivieri, E., Scoppola, E.: Markov chains with exponentially small transition probabilities: first exit problem from a general domain. I. The reversible case. J. Stat. Phys. **79**, 613–647 (1995)
197. Olivieri, E., Scoppola, E.: Markov chains with exponentially small transition probabilities: first exit problem from a general domain. II. The general case. J. Stat. Phys. **84**, 987–1041 (1996)
198. Olivieri, E., Vares, M.E.: Large Deviations and Metastability. Encyclopedia of Mathematics and Its Applications, vol. 100. Cambridge University Press, Cambridge (2005)

199. Penrose, O., Lebowitz, J.: Rigorous treatment of metastable states in the van der Waals-Maxwell theory. J. Stat. Phys. **3**, 211–236 (1971)
200. Penrose, O., Lebowitz, J.: Towards a rigorous molecular theory of metastability. In: Fluctuation Phenomena, 2nd edn. North-Holland, Amsterdam (1987)
201. Pollak, E., Talkner, P.: Reaction rate theory: What it was, where is it today, and where is it going? Chaos **15**, 026116 (2005)
202. Presutti, E.: Scaling Limits in Statistical Mechanics and Microstructures in Continuum Mechanics. Theoretical and Mathematical Physics. Springer, Berlin (2009)
203. Prévôt, C., Röckner, M.: A Concise Course on Stochastic Partial Differential Equations. Lecture Notes in Mathematics, vol. 1905. Springer, Berlin (2007)
204. Reed, M., Simon, B.: Methods of Modern Mathematical Physics. I. Functional Analysis, 2nd edn. Academic Press [Harcourt Brace Jovanovich, Publishers], New York (1980)
205. Révész, P.: Random Walk in Random and Nonrandom Environments. World Scientific, Teaneck (1990)
206. Rikvold, P., Tomita, H., Miyashita, S., Sides, S.: Metastable lifetimes in a kinetic Ising model: dependence on field and system size. Phys. Rev. E **49**, 5080–5090 (1994)
207. Rogers, L., Williams, D.: Diffusions, Markov Processes, and Martingales. Vol. 2. Wiley Series in Probability and Mathematical Statistics: Probability and Mathematical Statistics. Wiley, New York (1987)
208. Rogers, L., Williams, D.: Diffusions, Markov Processes, and Martingales. Vol. 1, 2nd edn. Wiley Series in Probability and Mathematical Statistics: Probability and Mathematical Statistics. Wiley, New York (1994)
209. Ruelle, D.: Statistical Mechanics: Rigorous Results. Benjamin, New York-Amsterdam (1969)
210. Schlichting, A.: The Eyring-Kramers Formula for Poincaré and Logarithmic Sobolev Inequalities. Ph.D. thesis, Leipzig University (2012)
211. Schonmann, R.H.: The pattern of escape from metastability of a stochastic Ising model. Commun. Math. Phys. **147**, 231–240 (1992)
212. Schonmann, R.H.: Slow droplet-driven relaxation of stochastic Ising models in the vicinity of the phase coexistence region. Commun. Math. Phys. **161**, 1–49 (1994)
213. Schonmann, R.H.: Theorems and conjectures on the droplet-driven relaxation of stochastic Ising models. In: Probability and Phase Transition (Cambridge, 1993). NATO Adv. Sci. Inst. Ser. C Math. Phys. Sci., vol. 420, pp. 265–301. Kluwer Acad., Dordrecht (1994)
214. Schonmann, R.H.: Metastability and the Ising model. In: Proceedings of the International Congress of Mathematicians (Berlin, 1998), vol. III, pp. 173–181 (1998)
215. Schonmann, R.H., Shlosman, S.B.: Wulff droplets and the metastable relaxation of kinetic Ising models. Commun. Math. Phys. **194**, 389–462 (1998)
216. Schütte, C., Huisinga, W., Meyn, S.: Metastability of diffusion processes. In: IUTAM Symposium on Nonlinear Stochastic Dynamics. Solid Mech. Appl., vol. 110, pp. 71–81. Kluwer Acad., Dordrecht (2003)
217. Schütte, C., Sarich, M.: Metastability and Markov State Models in Molecular Dynamics. Courant Lecture Notes in Mathematics, vol. 24. Courant Institute of Mathematical Sciences/American Mathematical Society, Providence/New York (2013)
218. Sewell, G.: Quantum Theory of Collective Phenomena. Oxford University Press, Oxford (1986)
219. Slowik, M.: Contributions to the Potential Theoretic Approach to Metastability with Applications to the Random Field Curie-Weiss-Potts Model. Ph.D. thesis, Technische Universität Berlin (2012)
220. Slowik, M.: A note on variational representations of capacities for reversible and non-reversible Markov chains. Unpublished, Technische Universität Berlin (2012)
221. Sowers, R.B.: Large deviations for a reaction-diffusion equation with non-Gaussian perturbations. Ann. Probab. **20**, 504–537 (1992)
222. Stroock, D.: An Introduction to Markov Processes. Graduate Texts in Mathematics, vol. 230. Springer, Berlin (2005)

223. Stroock, D., Varadhan, S.: Multidimensional Diffusion Processes. Grundlehren der Mathematischen Wissenschaften, vol. 233. Springer, Berlin (1979)
224. Sugiura, M.: Metastable behaviors of diffusion processes with small parameter. J. Math. Soc. Jpn. **47**, 755–788 (1995)
225. Sugiura, M.: Exponential asymptotics in the small parameter exit problem. Nagoya Math. J. **144**, 137–154 (1996)
226. Sznitman, A.-S.: Brownian Motion, Obstacles and Random Media. Springer Monographs in Mathematics. Springer, Berlin (1998)
227. van den Berg, M.: Exit and return of a simple random walk. Potential Anal. **23**, 45–53 (2005)
228. van Kampen, N.: Stochastic Processes in Physics and Chemistry. North-Holland, Amsterdam (1981)
229. van 't Hoff, J.: Études de Dynamiques Chimiques. F. Muller, Amsterdam (1884)
230. Vanden-Eijnden, E., Westdickenberg, M.G.: Rare events in stochastic partial differential equations on large spatial domains. J. Stat. Phys. **131**, 1023–1038 (2008)
231. Vanheuverzwijn, P.: Metastable states in the infinite Ising model. J. Math. Phys. **20**, 2665–2670 (1979)
232. Varadhan, S.R.S.: Large Deviations and Applications. CBMS-NSF Regional Conference Series in Applied Mathematics, vol. 46. Society for Industrial and Applied Mathematics (SIAM), Philadelphia (1984)
233. Ventcel', A.D.: The asymptotic behavior of the first eigenvalue of a second order differential operator with a small parameter multiplying the highest derivatives. Teor. Verojatn. Primen. **20**, 610–613 (1975)
234. Ventcel', A.D.: Formulas for eigenfunctions and eigenmeasures that are connected with a Markov process. Teor. Verojatn. Primen. **18**, 3–29 (1973)
235. Walsh, J.B.: An introduction to stochastic partial differential equations. In: École d'été de probabilités de Saint-Flour, XIV—1984. Lecture Notes in Math., vol. 1180, pp. 265–439. Springer, Berlin (1986)
236. Weidenmüller, H.A., Zhang, J.S.: Stationary diffusion over a multidimensional potential barrier: a generalization of Kramers' formula. J. Stat. Phys. **34**, 191–201 (1984)
237. Wolfrum, M.: A sequence of order relations: encoding heteroclinic connections in scalar parabolic PDE. J. Differ. Equ. **183**, 56–78 (2002)
238. Wreszinski, W.F., Salinas, S.R.A.: The mean field Ising model in a random external magnetic field. J. Stat. Phys. **41**, 299–313 (1985)

Index

A. Bovier, F. den Hollander, *Metastability*,
Grundlehren der mathematischen Wissenschaften 351,
DOI 10.1007/978-3-319-24777-9

GPSR Compliance
The European Union's (EU) General Product Safety Regulation (GPSR) is a set of rules that requires consumer products to be safe and our obligations to ensure this.

If you have any concerns about our products, you can contact us on

ProductSafety@springernature.com

In case Publisher is established outside the EU, the EU authorized representative is:

Springer Nature Customer Service Center GmbH
Europaplatz 3
69115 Heidelberg, Germany

www.ingramcontent.com/pod-product-compliance
Ingram Content Group UK Ltd.
Pitfield, Milton Keynes, MK11 3LW, UK
UKHW021919270726
14059UKWH00002B/93

* 9 7 8 3 3 1 9 2 4 7 7 5 5 *